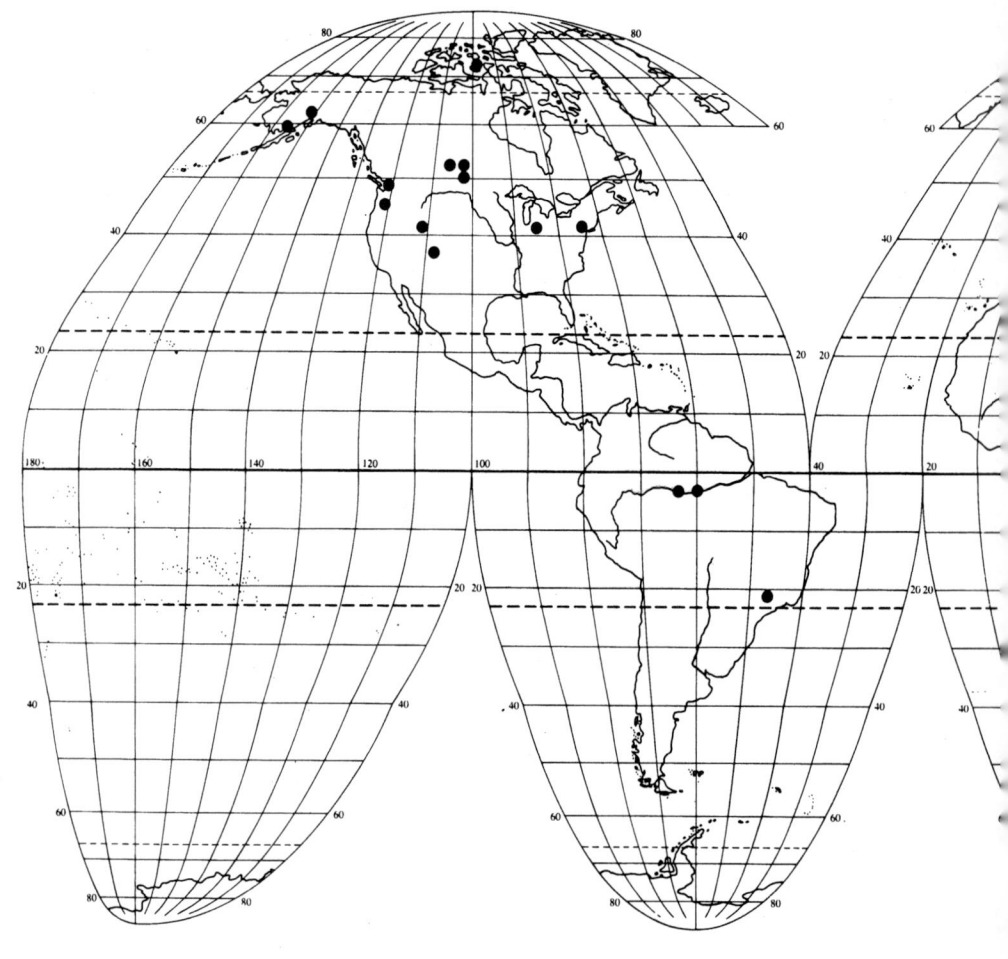

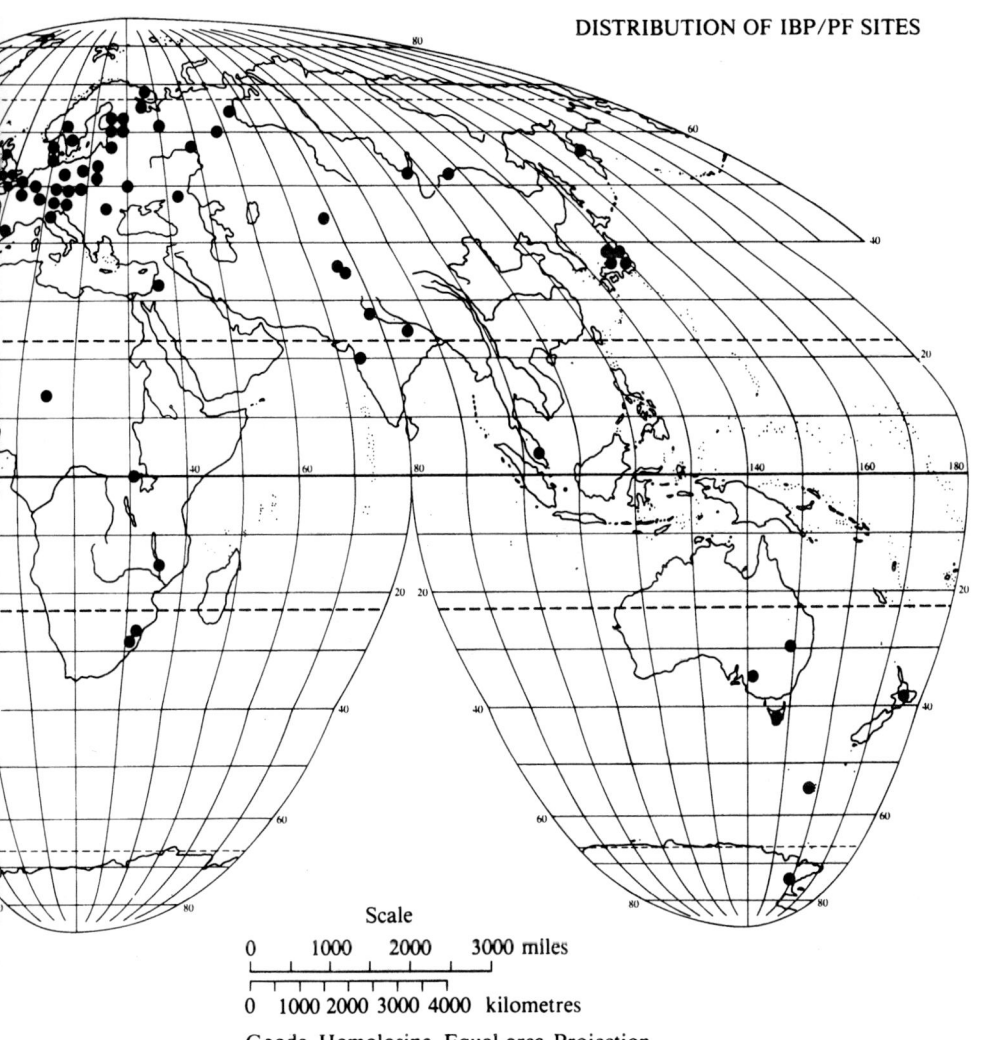

DISTRIBUTION OF IBP/PF SITES

Scale

0   1000   2000   3000 miles

0   1000 2000 3000 4000 kilometres

Goode Homolosine Equal-area Projection

Prepared by Henry M. Leppard

The functioning of freshwater ecosystems

THE INTERNATIONAL BIOLOGICAL PROGRAMME

The International Biological Programme was established by the International Council of Scientific Unions in 1964 as a counterpart of the International Geophysical Year. The subject of the IBP was defined as 'The Biological Basis of Productivity and Human Welfare', and the reason for its establishment was recognition that the rapidly increasing human population called for a better understanding of the environment as a basis for the rational management of natural resources. This could be achieved only on the basis of scientific knowledge, which in many fields of biology and in many parts of the world was felt to be inadequate. At the same time it was recognized that human activities were creating rapid and comprehensive changes in the environment. Thus, in terms of human welfare, the reason for the IBP lay in its promotion of basic knowledge relevant to the needs of man.

The IBP provided the first occasion on which biologists throughout the world were challenged to work together for a common cause. It involved an integrated and concerted examination of a wide range of problems. The Programme was coordinated through a series of seven sections representing the major subject areas of research. Four of these sections were concerned with the study of biological productivity on land, in freshwater, and in the seas, together with the processes of photosynthesis and nitrogen fixation. Three sections were concerned with adaptability of human populations, conservation of ecosystems and the use of biological resources.

After a decade of work, the Programme terminated in June 1974 and this series of volumes brings together, in the form of syntheses, the results of national and international activities.

INTERNATIONAL BIOLOGICAL PROGRAMME 22

The functioning of freshwater ecosystems

EDITED by

E.D. Le Cren
Director, Freshwater Biological
Association, Ambleside, Cumbria

AND

R.H. Lowe-McConnell
Formerly Overseas Research Service

CAMBRIDGE UNIVERSITY PRESS

CAMBRIDGE
LONDON NEW YORK NEW ROCHELLE
MELBOURNE SYDNEY

CAMBRIDGE UNIVERSITY PRESS
Cambridge, New York, Melbourne, Madrid, Cape Town, Singapore, São Paulo, Delhi

Cambridge University Press
The Edinburgh Building, Cambridge CB2 8RU, UK

Published in the United States of America by Cambridge University Press, New York

www.cambridge.org
Information on this title: www.cambridge.org/9780521105583

First published 1980
This digitally printed version 2009

A catalogue record for this publication is available from the British Library

Library of Congress Cataloguing in Publication data
Main entry under title:
The functioning of freshwater ecosystems.
(International Biological Programme; 22)
Text in English with Contents pages in English,
French, Russian, and Spanish.
Bibliography: p.
Includes index.
1. Freshwater productivity. 2. Fresh-water
ecology. I. Le Cren, E.D. II. Lowe-McConnell, R.H.
III. Series.
QH541.5.F7F86 574.5′2632 79-50504

ISBN 978-0-521-22507-6 hardback
ISBN 978-0-521-10558-3 paperback

Contents

Table des matières

Содержание

Contenido

Contributors

Adams, M.S., Department of Botany, University of Wisconsin, Madison, Wisconsin 53706, USA

Backiel, T., Institute of Inland Fisheries, Żabieniec near Warsaw, 05-500 Piaseczno, Poland

Bindloss, M.E., Knott End, Windyhall Road, Windermere, Cumbria, England

Blažka, P., Hydrobiological Laboratory, ČSAV, Vltavská 17, 15105, Prague 5, Czechoslovakia

Bretschko, G., Biologische Station, 3293 Lunz am See, Austria

Brylinsky, M., RR # 3, Canning, Nova Scotia, Canada BOP 1HO

Cummins, K.W., Kellogg Biological Station, Michigan State University, Hickory Corners, Michigan 49060, USA

Duncan, A., Department of Zoology; Royal Holloway College (University of London), Englefield Green, Surrey TW20 9TY, England

Gak, D.Z., Water Problems Institute, Academy of Sciences, Moscow, USSR

Ganf, G.G., Botany Department, University of Adelaide, South Australia 5001.

Gerloff, G.C., Department of Botany, University of Wisconsin, Madison, Wisconsin, USA

Golterman, H. L., Station Biologique de la Tour du Valet, Le Sambuc, 13200 Arles, France

Hammer, U.T., Department of Biology, University of Saskatchewan, Saskatoon, Canada

Hillbricht-Ilkowska, A., Institute of Ecology of Polish Academy of Sciences, Department of Hydrobiology, Twardowska 9, Warsaw 45, Poland

Javornický, P., Hydrobiological Laboratory, ČSAV, Vltavska 17, 15105 Prague, 5, Czechoslovakia

Kajak, Z., Institute of Ecology, Polish Academy of Sciences, Department of Hydrobiology, Dziekanów Leśny K/Warsaw, 05-150 Lomianki, Poland

Kitchell, J.F., Laboratory of Limnology, University of Wisconsin, Madison, Wisconsin 53706, USA

Koonce, J.F., Department of Biology, Cape Western Reserve University, Cleveland, Ohio 44106, USA

Kouwe, F.A., Water Board 'De Dommel', 5281 JV BOXTEL, The Netherlands

Larsson, P., Zoological Museum, Sarsgaten, Oslo, Norway

Le Cren, E.D., Freshwater Biological Association, The Ferry House, Ambleside, Cumbria LA22 OLP, England

Lévêque, C., 163 bis rue de Vaugirard, 75015, Paris, France

Lowe-McConnell, R.H., Streatwick, Streat via Hassocks, Sussex, England

McCracken, M., Department of Biological Sciences, Texas Christian University, Fort Worth, Texas 76129, USA

Marker, A.F.H., Freshwater Biological Association, River Laboratory, East Stoke, Wareham, Dorset BH20 6BB, England

Morgan, N.C., Greenlees, Lasswade, Midlothian, Scotland

Moss, B., Department of Environmental Science, University of East Anglia, Norwich, Norfolk, England

Nauwerck, A., Länsstyrelsen, S-95180 Luleå, Sweden

Park, R.A., Department of Geology & Freshwater Institute, Rensselaer Polytechnic Institute, Troy, N.Y., USA

Pieczyńska, E., Laboratory of Hydrobiology, Institute of Zoology, Warsaw University, Nowy Swiat 67, Warsaw, Poland

Pyrina, I.L., Institute of Inland Water, Academy of Sciences USSR, Borok, Yaroslavl, 152742 Nekouz, USSR

Rzóska, J., IBP Central Office, now 6 Blakesley Avenue, London W5, England

Saunders, G.W., Division of Biomedical & Environmental Research, US Atomic Energy Commission, Washington DC 20545, USA

Schiemer, F., II Zoology Department, University of Vienna, K. Luegerring 1, Vienna, Austria

Steel, J.A.P., Metropolitan Water Board, Queen Elizabeth II Reservoir, Molesey Road, West Molesey, Surrey, England

Straškraba, M.S., Hydrobiological Laboratory, ČSAV, Vltavská 17, 15105 Prague 5, Czechoslovakia

Straškrabová, V., Hydrobiological Laboratory, ČSAV, Vltavská 17, 15105 Prague 5, Czechoslovakia

Taub, F.B., College of Fisheries, University of Washington, Seattle, Washington 98195, USA

Thorpe, J.E., DAFS Freshwater Fisheries Laboratory, Faskally, Pitlochry, Perthshire, Scotland

Tilzer, M.M., Limnologisches Institut der Universität Freiburg 1, BR., 775 Konstanz–Egg, Meinaustraße 212 W. Germany

Tonolli, L., Istituto Italiano di Idrobiologia, 28048 Verbania-Pallanza (Novara), Italy

Walters, C.J., Institute of Animal Resource Ecology, University of British Columbia, Vancouver BC, V6T 1W5, Canada

Westlake, D.F., Freshwater Biological Association, River Laboratory, East Stoke, Wareham, Dorset BH20 6BB, England

Wetzel, R.G., Kellogg Biological Station, Michigan State University, Hickory Corners, Michigan 49060, USA

Winberg, G.G., Zoological Institute, USSR Academy of Sciences, University Embankment, Leningrad B 164, USSR

Preface

The first two chapters of this volume – by Professor Tonolli and Dr Rzóska – between them provide a good introduction to the work of the IBP/PF Section that is summarized in the rest of the book. This preface will therefore concentrate on some comments and explanations essential for those who wish to understand fully what follows.

The final discussion meetings for the preparation of this volume took place in 1974 and drafts of some of the chapters had been completed by then. Why then has the publication taken so long? In part this has been the result of the nature of some of the chapters which required a great deal of collation, editing and cross-checking by the Chapter editors. Some chapters thus took a long time to complete and had to be the spare-time work of people busy on other new projects. Largely, however, the delay has been the result of the first editor taking on a new and demanding professional post and thus having all too little time and energy for editing. He must therefore apologize sincerely to readers who have had to wait, and especially to colleagues in the IBP/PF who completed their share of the work on time. Fortunately the second editor has been able to pick up the large number of loose ends and unpolished MSS and draw them together into a published volume.

In all scientific research, and perhaps especially in ecology, the processes of collecting field data, conducting experiments and analysing the data are only the early stages of completing the whole project. Preliminary analyses and results need checking and polishing, their statistical variabilities estimating, and then their meanings interpreted in relation to other results. No more is this so than in the interpretation of data collected over a wide range of habitats in international cooperative ventures. Some of such difficulties that arose in the IBP/PF are discussed by Tonolli in Chapter 1. The synthesis of the results of such broad projects is a continuing process, in some ways an iteration of successive analyses and syntheses each incorporating some new or better-analysed data or different ways of looking at the results. It must be remembered that not all the results on which this volume is based have yet been published. Ideally any attempt to make a synthesis should await the primary publication of all the component investigations, but this ideal is an impossible one and a compromise is essential. This must result in a volume that is more a collection of interpretations using different approaches, or based on data in different stages of analysis, than a synthesis valid at the moment it is published. It is important for the reader to realize this.

A particular example arises in Chapter 9 by Brylinsky. This chapter is based on an analysis commissioned by the IBP/PF Executive to provide a preliminary statistical analysis of the results from a wide range of lakes. This analysis was carried out in 1972, two or more years before the final 'data reports' were written, during which time significant revisions were made to some of these data. Thus, some of the conclusions reached in this chapter may no longer be valid and this may be the reason for disagreements between this chapter and some of the others.

The form of the volume.   Early on, the IBP/PF Executive Committee agreed that one large synthesis volume should be produced, rather than a series of volumes each dealing with a different aspect of freshwater production-ecology. The volume thus consists of a series of groups of chapters with different approaches.

Chapters 1 and 2 provide an introductory background of the ideas behind IBP/PF, its planning, organization and its history. Livia Tonolli was the Convener of the IBP/PF Committee during its later years and had been in touch with the centre of its activities from the start. Rzóska was its Scientific Coordinator from 1965 onwards and more than any other person was responsible for its central planning and execution; he must therefore be given credit for much of its achievement as a cooperative venture.

The next five chapters deal with the 'meat' of the scientific findings of IBP and synthesize the results in the respective fields of physics, chemistry, primary production, secondary production, and decomposition and microbial production. Straškraba's Chapter 3 adopts a theoretical, mathematical and modelling approach to the physical environmental features of lakes though it draws on a large number of data obtained during the IBP. It provides a basis for the prediction of much of the ecological character of lakes from relatively simple physical variables.

Golterman and Kouwe have analysed the basic chemical data provided by the data reports and provided a brief general account of the chemical environment of lakes with special reference to nutrient pathways and chemical limitations to productivity.

Chapter 5 on primary production has been synthesized and edited by Westlake, but is the work of a large number of co-authors. In the PF programme more attention was paid to primary production and its measurement, than to any other aspect of lake ecology. There were thus a large number of new data to be analysed, interpreted and synthesized. The estimation of primary production and the interpretation of the experimental data obtained are not easy. So this chapter is not just a straight presentation of results, but a series of comparisons and discussions of different aspects of primary production, the processes involved and the factors that may control them.

Chapter 6 on secondary production is also a team effort; the work of synthesis and editing being organized and led by Morgan. The estimation of secondary production is difficult and the consumers in most water bodies are organized in a complex food-web so the analyses are open to different interpretations. The results from some situations where sampling was difficult had to be treated with reservation. The main sections of this chapter deal with zooplankton, zoobenthos and fish.

Chapter 7 deals with the important but difficult and under-researched area of the roles of organic matter, detritus feeders and decomposers.

Chapter 8 adopts a different approach and reviews some aspects of the trophic relationships in the freshwater ecosystem that are not discussed fully in the other chapters and some experiments. This chapter is concerned with physiological and ecological efficiencies and the dynamics of trophic relationships.

The two chapters that follow adopt two different mathematical approaches to the data gained in the IBP/PF projects. As already mentioned, Brylinsky's Chapter 9 reports on a statistical analysis of the preliminary data from several lakes. It is based on correlation analyses of estimations of primary and secondary production in relation to various physical and chemical variables and to each other. Chapter 10 by Walters *et al.* discusses the modelling of lake ecosystems and presents the results of some of the attempts at such modelling carried out during the IBP/PF.

The last chapter, no. 11, is again different from the others. At the start of the IBP, the idea of lakes as trophic-dynamic systems and the measurement of freshwater biological production had advanced further in Russia than in any other country. The Russian PF programme was thus quicker off the mark in its systematic planning than most others. Its results have been extensively published in Russian, but, as this language is not easily read in the West, these publications are not as widely known as they should be. Winberg has therefore prepared an English summary of the results of the Soviet IBP/PF studies for publication in this volume.

The chapters of this book thus present a series of different approaches to the main results of the IBP/PF programme and the study of the functioning of freshwater ecosystems. Many readers will have particular interests and will wish to go straight to the relevant chapter or chapters. They should find that most chapters can be read and comprehended on their own. We have attempted to include cross-references where important information is presented in a different chapter. The chapters have however been arranged in the order that we believe to be the most logical for those who wish to read the whole book and get an overview of all aspects of the IBP/PF.

The IBP/PF in the context of other research. In writing their sections or chapters authors were asked to base their accounts as far as possible on IBP results and to present the main findings of these, but to do so in the context of the large amount of non-IBP research, both past and present. Thus this volume is not just a summary of IBP/PF, but draws strongly on other research as well, though it looks at such research from the IBP viewpoint and primarily as the context in which the results from IBP research must be interpreted.

Readers should find in this volume a fairly comprehensive account of the state of knowledge about production ecology in fresh waters in the 1970s and a great deal that is relevant to limnology and freshwater biology in general, but some areas of these wider disciplines are covered only superficially or from a limited viewpoint. As the chapters were not all completed at the same time, they also vary in the extent to which they are up to date.

The list of references at the end of this volume is a long one and should serve as an excellent introduction to the literature in freshwater biology, especially if supplemented by the references given in one of the larger textbooks or treatises in the subject (e.g. Hutchinson 1957, 1967, 1975).

The management of freshwater ecosystems. The subject of the IBP was defined as: 'the biological basis of productivity and human welfare'; how may the IBP/PF results contribute to human welfare?

The production of biological material in fresh waters usually terminates in the form of fish, many of which can be harvested as human food or provide recreation through angling. Fish are the result of primary production and secondary production and are thus a product of the system and processes discussed in this book. Parts of Chapters 6 and 8 specifically deal with fish production and its relationships to the rest of the freshwater ecosystem.

In general it is the wish of the fisherman and fish farmer to *increase* biological productivity, but in recent years there has been much concern over the *overproduction* of algae in many lakes – the so-called problem of 'eutrophication'. Problems caused by excessive crops of phytoplankton are clearly related to productivity and primary production and there is a great deal of information in this book about such topics. The physical features of lakes relevant to their productivity are discussed in Chapter 3 and the models presented there should help a planner to predict the probable productivity of any new man-made lake. The nutrient cycles in lakes and their role in algal production are discussed in Chapters 4 and 5.

This volume is not a practical guide to the ecological management of fresh waters; its approach is essentially the scientific interpretation of basic and strategic research. But such research provides the understanding fundamental to the control and manipulation of biological production in fresh waters, so its findings are essential to the wise management of inland waters for human welfare.

The Data Reports. As described above and in Chapters 1 and 2, many of the basic data collected by the PF teams working on each lake or river were summarized in a specified format as a 'data report'. These data reports are listed in Appendix I, with a minimum of information about each site. Analyses of some of these data are used in this book, other analyses have been published by the authors working on the sites. The original data reports are now stored in the library of the Freshwater Biological Association at The Ferry House, Ambleside, Cumbria, England, where they may be consulted. The IBP has given an undertaking to the authors of the data reports that the actual data will not be published without their permission and access for consultation will be granted on this understanding.

Throughout this volume there are references to those data reports made, for example, in the form: 'DR 96'; the numbers refer to the list in Appendix I.

The IBP/PF Publications. Appendix II lists all the Handbooks and other publications that were published directly by the PF section. Their provenance is discussed in Chapters 1 and 2.

The List of References. All the citations to literature made throughout the volume have been gathered into one reference list on pages 517 to 570. This list includes many of the hundreds of papers arising directly from IBP/PF research on the sites listed in the Data Reports. There are also many citations to the general limnological literature. The authors and editors have tried to make this list full enough to support all the statements made in the text, but have not attempted to make it comprehensive or a 'bibliography' in the true sense. Nor will all the papers arising from the PF programme have been listed.

Conventions. Several of the conventions followed in this volume are those laid down by the Publications Committee of SCIBP. We have attempted to follow the Système International in the matter of quantities, units and symbols and outside the specifications of SI, we have used the guide given by SCIBP booklet '*Quantities, Units and Symbols. Recommendations for use in IBP synthesis*' published in 1974. We commend this guide to other authors and editors publishing in this field. (The special symbols used in Chapter 3 are listed in table 3.1.)

Epilogue. In this Preface we have attempted to provide a guide to this volume for the prospective reader; to tell him a little of what this book is about and also, so as to avoid his disappointment, to warn him a little of what it is not. Perhaps a few final words from the Editors, as the first readers of the whole book, may be appropriate.

One of us has given a brief review elsewhere of his personal assessment of the successes, limitations and failures of the IBP/PF programme (Le Cren, 1976). It may be a surprise to some that biologists were attempting to measure aquatic production and model aquatic biological systems over fifty years ago. The ideas behind the IBP were not new ones, but by 1960 they were ripe for further development with the aid of new techniques for field work, analysis and modelling. As Tonolli points out in Chapter 1, the development and standardization of PF methodology took place simultaneously with its use in the field. Some data produced thus lacked comparability and rigour. The time, effort and resources needed for the analysis, interpretation and publication of the results after the end of the period of data collection were also underestimated. With hindsight, it is possible now to see that those planning the PF programme were too optimistic over the possibilities of international coordination without the stimulus and control of central funding, and too optimistic of solving the extremely complex problems posed by the freshwater ecosystem without a much larger and more experimental programme. The ideas that were to be explored and tested were too simplistic and too imprecisely defined to be testable hypotheses.

It is always easy to look back and criticize after the event. We believe that this volume, and also the many papers derived from the PF programme, will be a tribute to the energy and enthusiasm of those that started IBP/PF and overcame the scepticism of many of their colleagues in the national and international scientific Establishments. As a result of the IBP, the science of limnology will never be the same. Above all, a great many young and enthusiastic researchers have become limnologists, had experience of international cooperation and made lasting friendships with one another.

E.D. LE CREN
R.H. LOWE-McCONNELL

March 1979

Acknowledgements

A volume like this involves the work of a great many people to all of whom we are grateful.

Professor Livia Tonolli's wise guidance and warm generosity did much towards bringing the PF programme through its final stages to this synthesis. This volume is the result of hard work put in by authors and especially by the chapter coordinators. The IBP Scientific Director, Dr E.B. Worthington, and Gina Douglas and Sue Darell-Brown of the IBP Publications Office have done much to help and keep us at our task. We are grateful to Peter Hargreaves for his careful checking of the references. The FBA Library Staff and our colleagues in the Freshwater Biological Association and at the British Museum (Natural History) have helped us with many detailed problems. Cambridge University Press staff, especially Peter Silver, gave great assistance over many editorial matters and with finding places where the Editors' attention had lapsed.

We are grateful to our very many friends who took part with us in the IBP/PF programme; we hope that we will still be friends after they have read this volume. Finally, we are deeply indebted to the Scientific Coordinator of the IBP/PF, Julian Rzóska; if the PF and this volume belong to any one person, it is to him.

E.D. LE CREN
R.H. LOWE-McCONNELL

1. Introduction

L. TONOLLI

This book is the result of years of a corporate effort by many scientists all over the world to advance the knowledge of production limnology. In my capacity as the last convener of the IBP/PF section I have been asked to introduce the volume and contribute some general remarks on the tasks undertaken, fulfilled or not achieved.

I had already written a different version of this introduction in June 1974 shortly after a 'final meeting of the editors of the chapters, held at my Institute at Pallanza at the end of April of that year. But considerable difficulties arose in collecting, finalizing and editing the great volume of contributions. Five years have passed since that meeting, and only now are we able to present the fruits of the labour of many eminent scientists to the scientific community. In view of this a revision of my introduction seemed advisable.

The origin, history and development of the enterprise is concisely treated in Chapter 2 of this book. The chronicle of organizational efforts therein relates the rise from an idea to the enormous enterprise into which the Freshwater Productivity section (PF) of the International Biological Programme (IBP) has grown. The beginnings were uncertain, the response of limnologists around the world was cautious and not enthusiastic, though favourable. But gradually the contacts with willing collaborators grew and some coherence was achieved on ways and means of elucidating the complex process of production in inland waters of the earth.

In 1968 G.G. Winberg took over the convenership from previous distinguished predecessors and, with his great experience in this field of limnology, appealed to participants to develop a closer communication between workers in similar fields, and he advocated the creation of working groups of international composition. This way of international exchange of opinions and experience became an established method of IBP/PF work.

The biological production of lakes has attracted limnologists for many years. Terms such as 'eutrophy', 'oligotrophy' and others appeared fifty years ago as general denominations of the status of lakes. Such concepts became better defined and were practically investigated. An important attempt was Lindeman's trophic–dynamic hypothesis forty years ago. He tried to analyse the 'energy flow' in the biological community of Cedar Bog Lake in Minnesota (USA). He assessed the production of the three main 'trophic levels' by measuring the sum of organic matter during the growing season and produced a theoretical scheme of this process, a pioneer 'model' for its time.

There were then increasing attempts in some parts of the world to probe further into the process of freshwater production, notably in Russia especially under the direction of G.G. Winberg. Many of these efforts were stimulated by the concept of 'ecosystem' introduced by the English botanist Tansley in 1935. The concept was coined for vegetation patches with their abiotic and biotic components. The concept of 'ecosystem' implies the close interrelations of a recognizable unit of nature and was well suited for production processes in lakes. It spread widely over the world and is used for a variety of habitats.

The IBP/PF section took on the task of attacking the problem of productivity in inland waters by concerted widespread research. Chapter 2 of this book explains that this arose from the general aims of the International Biological Programme. The experiences and results of previous and slender work in this field had to be reviewed and possible guidelines had to be sought. On scrutiny it became obvious that a great effort had to be made to prepare for this extended work.

Preparations had to include consideration of the best methods of assessment in the various trophic levels existing in a lake. Such critical assembly did not exist internationally at the beginning of IBP/PF. Between 1965 and 1969 small groups of limnologists met in Italy, England, the Netherlands, Czechoslovakia and Soviet Russia to discuss and elaborate methods which were at that time available in limnological science or had to be newly proposed. Handbooks of IBP methods were prepared and published by IBP in paperback through Blackwell Scientific Publications, Oxford. These are listed in Appendix II and the publication dates show how long it took to prepare them. Some came late onto the scene and could not influence work already in progress on research sites. During the working meetings some review papers were read and these are also listed in the publications. These small but intense group meetings were to become characteristic of IBP in general and, being held in various parts of the world, they contributed greatly to international scientific relations. Twenty-four handbooks of methods used in the various sections of IBP were on sale and it is gratifying to know that the five manuals of the PF section were 'bestsellers' and some had to appear in revised editions; this process continues even now.

The initiative for these meetings came from the governing body, the Sectional Committee of PF, while the organization was done by the Central Office of the section in London. Indispensable help for these and subsequent meetings came from local organizers in the countries concerned.

The structure and work of the governing body, meeting usually every year, and the permanent secretariat are described in Chapter 2. The secretariat carried out the instructions of the governing body, but increasingly had to take decisions and initiative of its own to foster the accelerating pace of work, and it maintained a vast correspondence with research workers and institutions of many countries. The successive chairmen (conveners) of the section also had to maintain a considerable burden of office. Let it be said at once that the PF Scientific Coordinator's office consisted of only two people.

Difficulties arose when ideas were put into practical application. The manuals of methods were delayed and their advocated methods were not followed, especially in countries with advanced and well-endowed institutions. If final results were to be achieved, all scientific measurements had to be comparable. Units of assessment and symbols used in equations differed in the countries participating and made conversion into comparable results laborious. Even the metric and decimal system was not adhered to everywhere. This became evident in 1970 during a preliminary results symposium in Poland. A small group of specialists under G.G. Winberg as editor elaborated and published in 1971 a booklet containing *Symbols, Units and Conversion Factors in Studies of Fresh Water Productivity*. This attempted to assemble generally acceptable values for physical and chemical properties of lakes and ways of calculating biological productivity and calorimetry. In the event it was to some extent superseded by the general booklet *Quantities, Units and Symbols. Recommendations for use in IBP Synthesis* which was published in 1975 by the Special Committee for IBP (SCIBP) of the International Council of Scientific Unions (and given as an Appendix to IBP Synthesis 1, *The Evolution of IBP*, edited by E.B. Worthington). This latter guide follows more closely the International System of Units (SI) which has been used as far as possible in this volume.

Another improvement was the introduction of Data Reports, a tabular form of recording results of research sites. The usual process in science of extracting results from sometimes lengthy published papers was unsuitable for the quick gathering of the fruits of our extensive work. A small party of limnologists from several countries composed the form of these Data Reports in 1971. Critics may say that these two improvements in communication came too late and should have been introduced from the beginning. But only by experience could we learn how to achieve this kind of international cooperation. Nobody had tried to do it before IBP.

Scientific difficulties appeared in the practical work. Even in arctic lakes with a limited community the interrelations between trophic levels were difficult to elucidate completely; primary production is strongly influenced by respiration making exact measurement uncertain; animals of the secondary trophic level often have mixed food and change their feeding habits. Further, defining an 'ecosystem' as an object of investigation depends on the recognition of all or most factors which influence the characteristics of a water body. In his contribution in this volume (Chapter 9) M. Brylinsky says: 'It is unfortunate that no detailed data on drainage basin characteristics were available for the sites considered in the analysis. Alterations within the drainage basin of a lake or reservoir is one of the more common ways by which man indirectly affects these systems. In this respect, it would have been of considerable interest to attempt to evaluate how factors such as the degree of urbanisation, industrialization, and other forms of land use affect lake biology.' Indeed the character of a lake is determined by many factors, of which the geological substratum and the climate are very decisive. In the USA the term and concept of 'biome' was consequently used to embrace the total number of factors of a landscape surrounding a lake. This approach entails a multidisciplinary attack and teamwork, which is not easily achieved and is costly.

These are some of the difficulties encountered by the PF programme. However, the experience gained by looking back will be useful for the further advance of limnology. This is already evident in a number of activities arising from the IBP/PF impetus, for example the importance of detritus (humic acids) has been recognized, and a deeper probing into zooplankton problems has taken place.

The results and appraisal of our far-flung efforts have to be discussed. In April 1969 250 projects were listed from 42 countries. By the end of our operational phase only half of these projects remained viable and produced the requested results. But never before could nearly a hundred lakes in all climatic zones be compared and analysed in a comprehensive way. Thus some outlines of correlations of abiotic and biotic characteristics have been drawn, into which further efforts may be fitted or amended. As usual in science, the results will take time to influence limnology in general, but there is no doubt about the final impact.

Other results have been immediate. The close cooperation between scientists of many countries in small working meetings was an outstanding feature of our work, and indeed of the whole IBP. A whole generation of younger scientists has been drawn into our work, advanced in their experience and entered into the general field of limnology.

The stimulus provided by IBP/PF is visible inside some of the participating countries. National syntheses have appeared in some countries and possibly more will still follow. Japan has produced a series of books recording results of work in that country, with a handsome volume synthesizing the work on the productivity of a variety of lakes and one river. From Canada comes a concise volume with a chapter on their two lake research sites. The Royal Society in London has published a review of the British contribution to IBP with a survey of PF work. A series of books has presented the results of Russian efforts including contributions on the productivity of their selected lakes and reservoirs. A concise summary of Russian work is included in this volume (Chapter 11) to alleviate the language difficulty of their work published in Russian. Surveys of national syntheses are available also from Poland and Czechoslovakia. In addition many hundreds of papers have appeared in many parts of the world: in Poland alone about 300 papers.

The impetus of IBP/PF has also influenced other organizations. The intergovernmental programme of 'Man and the Biosphere' (MAB) sponsored by UNESCO is in some respects a continuation of IBP.

It must be admitted that we have not succeeded completely. The ultimate aim of producing a general theory of productivity and of biological cycling of matter in inland water bodies advocated by G.G. Winberg has not yet been achieved. But the efforts of the PF section of IBP have pinpointed and narrowed the gaps in our knowledge.

The contents of this book show chapters on all important ingredients of the structure and functioning of lakes, from the physical features, the chemical basis, the contributions of primary producers, to the complex role of animal consumers and the cycling of their metabolic products in the water of lakes. Two chapters apply 'systems analysis' and 'models' to some of the results. Lakes of different latitudes and altitudes have been investigated thus to find correlations between the major factors acting in a lake and its biology. It is interesting to note that of the ninety-three sites examined in PF work, listed in Appendix I, very few were concerned with running water. The problems of biological production in running waters differ from those of standing water bodies because the water medium is under continuous change in time and space.

To present this volume to the scientific scene in general is a privilege which I greatly appreciate. It will be criticized or appreciated and that is the perquisite of scientific advance. Some scientists dislike 'directed' research and believe in unbridled freedom of work and methods; comments have been heard on 'bio-politicians' forcing themselves upon the scientific scene. This point of view is recognized yet the demands for an organized approach to understand the workings of nature will probably increase in the future.

Finally I wish to pay tribute to the many hundred scientists who have made our work possible. First it is to the 'front line' workers in the field who supplied their results from ninety-three sites in different parts of the world. Secondly I appreciate the unselfish work of the chapter editors and those collating this volume: Words of thanks are also due to the central governing body of the International Committee and the services rendered by the PF office in London, and especially to the PF Scientific Coordinator Dr Julian Rzóska to whose devoted activity much of the success of PF must be attributed.

2. History and development of the freshwater production section of IBP

J. RZÓSKA

The origin of IBP as an idea (1959) followed by a closer definition of aims (1960), organization (1962) and development (1964–74) is described in volume 1 of the Synthesis Series, published in 1975 (*The Evolution of IBP*, edited by E.B. Worthington, Cambridge University Press). With the motto 'The biological basis of human welfare' the programme was to be a recognition of the importance of the biological sciences for man's present situation. Sponsored by the International Council of Scientific Unions (ICSU), a Special Committee for IBP (SCIBP) organized the work in seven sections: Productivity of Terrestrial (PT), Freshwater (PF) and Marine (PM) Communities, Production Processes (PP), Conservation of Terrestrial Communities (CT), Human Adaptability (HA), and Use and Management of Biological Resources (UM).

2.1. Early stages of PF section

In August 1962 a working party sat during a congress of the International Association for Limnology in the USA; W. Rodhe (Sweden) chaired, with A. Hasler (USA), H. Sioli (BRD), E.B. Worthington (UK) and H.B.N. Hynes (Canada) as members. In January 1963 a provisional programme was devised with the aim: 'Study of basic parameters of production and metabolism at all trophic levels in standing and running inland waters'. Soon the concept of 'energy flow' was added. The sites of research were to span all climatic conditions, two or three sites in arctic climates, seven or eight in temperate regions, three or four in the tropics. The programme was to have a preparatory phase (1963–5), followed by a research phase (1965–70) with 'minimal' and 'developed' projects. The research was to be financed by a central fund of about $5\frac{1}{2}$ million dollars. This programme was circulated in January 1964 to 254 hydrobiologists and institutions in fifty-four countries. There were 105 replies from thirty-seven countries. An 'Analysis of comments on the Provisional Programme' was made by J. Rzoska, presented to the First General Assembly (IGA) of IBP in Paris (July 1964) by A. Hasler and printed in *IBP News* No. 2, 1964, pp. 46–52. Some of the comments are of interest to the history of scientific enterprise. Of the 105 replies fifty-three came from Europe, three from America, four from Asia, one from the Middle East, twelve from Africa, four from Australia. Only twenty responses were negative,

7

dismissing the plan; these came mostly from institutions with established traditions of research. The other replies were cautious but positive. Attitudes expressed depended on the priority of needs of particular countries. Fish production studies were emphasized in tropical countries, while the need for fundamental research into biological productivity dominated the replies from scientifically developed countries. Phases of work should be regionally adjusted; comparability of results was to be achieved internationally; methods of research in 'developing' countries should be introduced, but only coordinated in advanced countries. Opinion in the USA was strongly in favour of 'new' research people and finance. An international secretariat was to be established for gathering and exchange of information. The first General Assembly accepted in principle the proposals for the work of the PF section; a preparatory phase of 2–3 years was to be followed by an operational period of 5 years. The structure of the governing body of the PF section was established.

2.2. Organization of the PF section

At the head of each section was to be a convener, supported by a sectional committee. W. Rodhe, the first chairman, resigned because he regarded the response to his programme as insufficient. After a short stop-gap by E.B. Worthington, V. Tonolli (Italy) became convener in September 1964 but had to resign in 1966 because of illness (he died in 1967). He was followed by A. Hasler (USA) 1966–8, G.G. Winberg (USSR) 1968–70, and finally Livia Tonolli 1970–4. The sectional committee was composed in succession of S.V. Ganapati (India), S. Mori (Japan), S.D. Gerking (USA), R. Vibert (France), all acting until 1970. Shorter terms of office were occupied by W.H.L. Allsopp (Ghana), J.W.G. Lund (UK), H. Sioli (BRD), F. Evens (Belgium), K. Patalas (Poland), J. Hrbáček (Czechoslovakia), H. Golterman (Netherlands), E.D. Le Cren (UK), Z. Kajak (Poland), A. Bonetto (Argentina), K. Mann (Canada). Some other scientists acted as corresponding members and consultants. The Sectional Committee sat each year until 1970, when for economy and speedier decisions it was replaced by an Executive Committee composed of the Convener Livia Tonolli, H. Golterman, K. Mann, and E.D. Le Cren, with J. Rzoska (Scientific Coordinator from 1965 onwards) as secretary.

2.3. Establishment of international cooperation

Gradually in most adhering countries IBP committees were formed with sections and representatives for particular fields with whom contacts and a flow of information could be established. A list of PF centres and their projects was published in 1969 in *IBP News* No. 17. It contained 232 projects from forty-two countries, later enlarged but totally modified in USA.

We will note here at once the large deviation from W. Rodhe's initial programme. Blue-print ideas changed into more acceptable local concepts. A number of countries with slender resources of scientific manpower and finance rushed in prestige projects, which did not mature. Finance from central funds was not available and national projects had to be funded from national resources.

Noteworthy were bilateral projects between countries: Federal Germany/Brazil, France/Republic of Chad, Japan/Malaya, Britain/Uganda. All these were successful.

2.4. Methods of production research

Very few guidelines existed for conducting research on the productivity of fresh waters at the outset of our work. The collation of methods was therefore decided at an early stage of the PF section. Working groups held sessions for this purpose: on primary production at Pallanza, Italy, in April–May 1965, fish production at Reading, England, September 1966, the chemical environment at Amsterdam, Netherlands, October 1966, secondary production at Liblice, Czechoslovakia, April 1967, aquatic microbiology at Leningrad, USSR, in 1969. The publications resulting from these meetings are listed in Appendix II.

It is obvious from the dates of publication that the time set aside for preparing methods was greatly exceeded. Except in Russia, there was not much experience of methods and consensus on the best approaches was hammered out in discussions. A great effort was demanded from the chief editors of the volumes apart from their normal scientific duties.

2.5. Personal international contacts

Most of the flow of information was by correspondence; over 10 000 letters were issued from the Scientific Coordinator's office and a similar number were received, but personal contacts were also necessary. A lively exchange of ideas was already taking place in the working groups, much more effectively than at congresses or symposia. The Scientific Coordinator undertook a number of journeys to stimulate work in some centres: in Poland 1965, 1968; to the French PF members in 1967; Japan and Thailand 1967; he visited the Dutch team in 1968, 1970; Loch Leven in Scotland in 1968; Southern Africa in 1969. Attempts were also made to stimulate the cooperation of hydrobiologists in areas where it was regarded as necessary. Such meetings were held for Latin America at Santa Fé, Argentina, in March 1969, in Africa at Makerere University, Uganda, in May 1968, in Southeast Asia, at Kuala Lumpur and Malacca, Malaya, in May 1969. These meetings brought together scientists from various countries of each area; all produced scientific contributions and reports, as can be seen from the list of publications (Appendix II). But the hope of further close cooperation between the scientists of the three great inter-tropical areas was not fulfilled.

2.6. Operational phase

The effectiveness of international cooperation was soon to be tested. A symposium on interim results was held in Poland in May 1970, at the invitation of Polish scientific bodies with the help of UNESCO. Two hundred participants from twenty-six countries gathered; reports from eighty-seven research sites were presented. Two volumes of Proceedings with sixty-one papers were published in Poland.

Two years later, in September 1972, a further large meeting in Reading, England, was to clarify the final progress. Participants numbered 195 from thirty-five countries and were to prepare in discussions for the final synthesis of results in the form of Data Reports, according to an easily comparable scheme arrived at by a working group in Amsterdam in October 1971.

The Reading meeting clearly demonstrated the formidable difficulties of reaching a 'synthesis' because of doubts on the comparability of results, expressed by strongly individualistic opinions, and doubts on the assessment of the flow between 'trophic levels'. However, a general statement on scientific policy for the final stages of the PF work was adopted, and the outline of the synthesis volume and chapter editors were agreed.

From then onward only small working groups were asked to cope with the incoming Data Reports. This material, available by the beginning of 1973, was distributed to editorial groups and also submitted to 'systems analysis' and 'modelling'. Zoobenthos was discussed in two sessions at Stirling, Scotland, and again at Lunz, Austria; primary production was treated at the Freshwater Biological Association station, Windermere, England; fish production at a session at Reading; zooplankton was discussed in Stockholm. The contents of these last meetings showed that even at this advanced stage particular components of life rather than their integration were treated. This was partly rectified by the systems analysis and modelling presented in this book (Chapters 9 and 10), and during a meeting at Lunz (Austria) on the interrelations of trophic levels in shallow lakes.

One important event must be mentioned. As our Russian colleagues could not attend the Reading Symposium and subsequent meetings, members of the International Committee went to Leningrad in May 1973 to receive the summarizing reports of the Russian PF work. An overall review of PF work in the Soviet Union is given here in Chapter 11.

There was the usual final pressure to continue some investigations, as in a zooplankton group which still works at the improvement of results (see Bottrell *et al.*, 1976). But with the end of the ICSU-sponsored IBP organization fixed for September 1974, the final formulation of results could not be delayed,

and the chief chapter editors met in April 1974 at the Italian Institute of
Hydrobiology at Pallanza, where under the patronage of the Convener, Livia
Tonolli, the contents of this 'synthesis' volume were agreed upon and
responsibilities for manuscripts were allocated.

2.7. Close down of IBP

By the end of September 1974 the Central Office of IBP in Marylebone Road
London closed down. The PF section issued the last information bulletin,
giving a brief account of the work done. The small Executive Committee of PF
was to be disbanded with the publication of our synthesis of results. J. Rzoska,
the Scientific Coordinator left IBP/PF. The publication of all the work of IBP
was transferred to a small skeleton staff of two people located at the Linnean
Society of London.

Two activities not connected with production biology have to be mentioned
in which the PF section was involved. Conservation and protection of inland
waters was raised at the congress of the International Association for
Limnology (SIL) in Vienna 1959. E.B. Worthington proposed the issue,
curiously neglected in hydrobiology in spite of the rapid deterioration in many
countries. H. Luther (Helsinki, Finland) started collecting data on important
aquatic sites worthy of care and protection. In 1965 the PF office in London
began to help in this work. A preliminary list of such waters appeared as an
IBP report in 1969; further intensive work followed the wide circulation of this
issue and in 1971 a book *Project Aqua* appeared as Handbook No. 21 of the
IBP series of handbooks. After the book was published a considerable number
of new entries were received and a new enlarged edition was planned. But by
that time funds had run out, the PF office was closed and the co-author,
J. Rzóska, was incapacitated by a stroke. Adequate support from the two co-
sponsoring bodies, the International Union for Conservation of Nature
(IUCN) and SIL was lacking and the proposed revised issue did not appear.
The existing 1971 issue has been used successfully in a number of cases where
threats to inland waters became a public issue.

The second venture, a collaboration with UNESCO/IHD, the International
Hydrological Decade, was fruitful and resulted in four important publications
dealing mainly with the serious menace of explosive populations of water
weeds (see list of publications, Appendix II). Inter-sectional liaison within IBP
was also fruitful: macrophytes, wetlands, nitrogen fixation, humic acids and
detritus were all subjects of collaboration with the PF section and all
connected with the functioning of inland waters.

The above is the factual account of the origin, organization and work of the
PF section recorded by the Scientific Coordinator when in office from 1965 to
the end. A more complete diary of all events in chronological order has been
deposited in the archives of IBP. An appraisal of the results, the shortcomings

and achievements of the work of the PF section of IBP is contained in the Introduction by Livia Tonolli, the last head of the section.

Here, in our economically minded times, the cost of the IBP operation should be mentioned. An estimate of how much money was spent by the national participants can only be roughly estimated: it is contained in the first volume of the synthesis series (*The evolution of IBP*, edited by E.B. Worthington). In that volume F. Boulière, the last President of IBP, asserted positively that the effort was worth while: national expenditure had remained in the countries of origin, hundreds of young scientists have emerged from this work and have advanced their various fields of knowledge.

The PF section spent from 1964 to the end of 1974 a total of 206000 dollars on the central organization in London. Of this 59% went on maintaining the coordinator, secretarial help, costs of office, postage and facilities. The remaining 41% was spent on international meetings of the governing body and help for participants to attend working meetings and symposia. Compared with the large sums spent on research in the various countries the expenses of the central offices of IBP accounted for only 1%.

The financial support of UNESCO in Paris, the Commonwealth Foundation, the Canadian and USA centres of IBP is gratefully acknowledged. Professor Livia Tonolli extended her generous hospitality on several occasions.

3. The effects of physical variables on freshwater production: analyses based on models*

M. STRAŠKRABA

3.1. Introduction

The quantitative interpretation of biotic interrelations in freshwater ecosystems, as treated in subsequent chapters, depends on an understanding of the changes in the physical matrix in which they occur.

Most of this chapter is therefore devoted to quantifying the major trends of physical variables on the globe, their reflection in the physical and chemical conditions in water bodies and evaluation of the effects of both on primary productivity. Lakes and reservoirs are of particular concern and simple theoretical models are used. The 'simplicity' of the models is relative to models including more variables, but particularly to more rigorous models of physical hydrology. Interrelations between physical, chemical and biological variables are stressed and trends recognized, rather than details given. Empirical observations are related to the underlying theory as much as possible. Nevertheless, the theoretical approach may seem too superficial to a physically oriented hydrologist and the treatment too mathematical to most freshwater biologists.

A successful application of the theoretical model approach in such a global analysis depends on overcoming its present shortcomings. The most severe limitation of the analytical ecosystem models is connected with their present main objectives – detailed analysis of selected systems for which parameters have been measured directly or derived from similar situations. When conditions change, organisms rapidly become adapted or are replaced by other species. The parameters measured for one system can be of little use for another. One possible way to reflect such changes is to develop analytical models with parameters empirically related to major environmental variables (Straškraba, 1976a). The model used here is still incomplete, and more weight is put on obtaining understanding for the theoretical approach by the reader and the dynamic understanding of the processes involved, rather than on obtaining absolute values.

The analysis of the IBP/PF data cannot be considered completed by this

---

* See table 3.1 for definitions of symbols used in this chapter, as these are somewhat different from those used elsewhere in this volume.

Table 3.1 *List of symbols used in Chapter 3*[a]

Symbol	Definition
a	atmospheric transmission coefficient (dimensionless)
a'	parameter expressing the asymptotic value of the logistic equation Eq. 3.66
$(a*/r*)^2$	earth to sun separation
$A_O; A_{O,S}$	annual mean temperature; the same for surface temperature (°C)
$A_{O,E}$	annual mean E_{GC} (J cm^{-2} d^{-1})
$A_1; A_{1,S}$	semiamplitude of the annual temperature variations; the same for surface temperature (°C)
$A_{1,E}$	semiamplitude of the annual variations of E_{GC} (J cm^{-2} d^{-1})
A_E	surface area of the lower epilimnion boundary (m^2)
A_H	surface area of the lower boundary of hypolimnion (= bottom area of the hypolimnion) (m^2)
A_L	surface area of the lake (m^2)
A_{SE}	surface area of the bottom (lake walls) in the epilimnion (m^2)
A_T	drainage area (m^2)
$A_z; A_{z'}$	area of the lake at depth z; the same for depth z' (m^2)
b	biomass of phytoplankton (mg m^{-3})
$\bar{b}$	average biomass of phytoplankton (mg m^{-3})
$\bar{b}_{max}$	maximum biomass of phytoplankton (mg m^{-3})
b'	parameter in the logistic equation Eq. 3.66
b''	parameter in Eq. 3.35
$b*$	coefficient for the effect of elevation on air mass number
c'	parameter in the logistic equation Eq. 3.66
C	cloudiness (fraction)
chl. a	chlorophyll a concentration (mg m^{-3})
d	day of year
D	depth of the frictional resistance (m)
Dv_T	condensation of water on the surface of a territory (mm a^{-1})
$D_z; D_{zH}; D_{z\,mix}$	vertical eddy diffusion coefficient; the same for hypolimnion and the thermocline region, respectively (m^2 d^{-1})
E	solar radiation (J cm^{-2} d^{-1})
$E'; E'_O; E'_C$	PAR, photosynthetically active radiation; PAR subsurface (J cm^{-2}d^{-1}); PAR underwater, no ice
$E'_{O\,max}$	subsurface PAR at midday (J cm^{-2} min^{-1})
E_{opt}	radiation where maximum P is obtained (J cm^{-2} min^{-1})
$E_D; E_{DC}$	direct solar radiation; the same for cloudiness C (J cm^{-2} d^{-1})
$E_G; E_{GC}$	global radiation reaching the ground; the same for cloudiness C (J cm^{-2} d^{-1})
E_G^*	global radiation reaching the top of atmosphere (J cm^{-2} d^{-1})
$E_S; E_{SC}$	diffuse solar radiation; the same for cloudiness C (J cm^{-2} d^{-1})
En	entrainment of water from the hypolimnion into epilimnion (m^2 d^{-1})
Ev_T	evaporation from a territory (evapotranspiration) (mm a^{-1})
$F_E; F_H$	orthophosphate phosphorus concentration in the epilimnion; same for the hypolimnion (mg m^{-3} P)
g_E	sedimentation rate of algae (m d^{-1})
h	solar hour angle (radians)
I_K	light sensitivity parameter of algae (J cm^{-2} s^{-1})
$L; L'$	length of the lake, and maximum length, respectively (km)
L''	effective length of the lake axis (= $L'W/2$) (km)
L'''	fetch size (km)
L_P	load (g m^{-2} a^{-1})
$m; m_s$	air mass number; same for elevation s (dimensionless)
n'	actual duration of sunshine (hours)

(contd.)

Symbol	Definition
N	number of cases, also North
N'	theoretical maximally possible duration of sunshine (hours)
$N*$	combined morphometric parameter (dimensionless)
p	parameter in Eqs. 3.40–3.41
$p_0; p_s$	atmospheric pressure at sea level; the same for the elevation s (millibars)
$P; P()$	photosynthetic capacity; the same in dependence on variables specified in brackets (mg O_2 (mg chl. a)$^{-1}$ d^{-1})
$P_{max}()$	gross photosynthetic capacity at light saturation in dependence on the variables specified in brackets (mg O_2 (mg chl. a)$^{-1}$ d^{-1})
$P(T)$ gross	temperature dependent gross photosynthetic capacity (mg O_2 (mg chl. a)$^{-1}$ h^{-1})
PP	net phytoplankton production (g C m^{-2} a^{-1})
$P_E; P_H$	phytoplankton in epilimnion, and in hypolimnion, respectively
$Pv_T; Pv_L$	precipitation on a territory, and on a lake, respectively (mm a^{-1})
Q	flow (m^3 s^{-1})
$Q_i; Q_{iT}$	horizontal water inflow to a water body; the same to a territory (m^3 s^{-1})
Q_{oL}	outflow of a lake (m^3 s^{-1})
Q_{oT}	specific outflow from a territory (mm a^{-1})
Q_{PH}	inflow of phosphorus from the hypolimnion (mg m^{-3} d^{-1})
Q_S	annual outflow (m a^{-1})
r	correlation coefficient (dimensionless)
r_E	remineralization of phosphorus by algae
r_P	respiration of algae
R	theoretical water retention time (a^{-1}) = Q/V if Q in m^3 a^{-1}
s	elevation (m.a.s.l.)
S	concentration of the substrate in the lake (μg l^{-1} = mg m^{-3}), also South
$\bar{S}$	average concentration of the substrate (mg m^{-3})
S_i	concentration of the substrate in the inflow (mg m^{-3})
S_{eq}	equilibrium concentration of substrate in the lake (mg m^{-3})
SD	standard deviation of the regression estimate
St	physical stability according to Schmidt (g cm^{-1})
S_E	rate of exchange of a substance with sediments in the epilimnion (mg m^{-2})
S_P	concentration of the substance in precipitation (mg m^{-3})
t	time (days)
T	temperature (°C)
$T(t)$	temperature at day t of year beginning January 1 (°C)
T_A	annual average air temperature
$T_B; T_D$	bottom temperature; annual average bottom temperature (°C)
T_{is}	temperature of the lake when isothermal (°C)
TP	total phosphorus concentration (mg m^{-3} P)
$T_S(\phi', t)$	surface temperature at day t of the year beginning January 1 for the given corrected latitude ϕ' (°C)
T_S max; T_S min	maximum and minimum surface temperature, respectively (°C)
u	wind speed (m s^{-1})
$V; V_E$	volume of the lake, and volume of epilimnion, respectively (m^3)
$V_z; V_{z'}$	volume of the lake below depth z or z' (m^3)
W	maximum width of the lake (km)
W_T	water budget for a territory (mm a^{-1})
z	depth (m)
$\bar{z}$	average depth (m)
z'	normalized depth = z/z_{max} (fraction)
$z*$	zenith angle of the sun (degrees)
z_g	gravity centre of the lake during holomictic conditions (m)
z_1	lower boundary of the metalimnion (m)

(contd.)

Symbol	Definition
z_{max}	maximum depth (m)
z_{mix}	mixing depth = thermocline depth (m)
z_u	upper metalimnion boundary (m)
Z	zooplankton
α	uptake of phosphorus by algae
β^*	ratio of settling velocity (fraction)
β	elevation of the sun (degrees)
$\gamma\Delta$	phase angle for daylength (degrees)
γ_1	phase angle for temperature (degrees)
$\gamma_{1,s}$	phase angle for surface temperature (degrees)
δ	solar declination (degrees)
Δ	daylength-photoperiod (hours)
ε	vertical extinction coefficient (m^{-1} ln units)
ε_q	ε due to dissolved and mineral dispersed matter (m^{-1} ln units)
ε_s	vertical extinction coefficient due to phytoplankton (m^{-1} ln units)
$\mathscr{E}$	ecliptic longitude of the earth in its orbit
θ_a	Birgean annual heat budget (kcal $m^{-2}\,a^{-1}$)
λ	angular velocity of the rotation of the earth
v	shearing stress of the wind
σ	coefficient of sedimentation (dimensionless)
ϕ	geographical latitude (negative for S) (degrees)
ϕ'	geographical latitude corrected for the location of the meteorological equator at $3.4°\,N = \lvert\phi - 3.4\rvert$ (degrees)
ρ	density

[a] These symbols are not entirely consistent with those used elsewhere in this volume.

book. It is hoped that a more detailed model-based analysis of IBP/PF data will be obtained when the basic data are all published. Higher resolution of the model is expected in the near future by stepwise confrontation of its outputs and relevant observations.

U.T. Hammer of the University of Saskatchewan is the co-author of the qualitative part of section 3.2.1 on solar radiation.

3.2. Geographical framework for aquatic ecosystems

When we start to observe our water bodies from space (Henson, Lind & Potash 1975) we need a more quantitative insight into the global meteorological forces determining their presence or absence, shape and mixing, chemical composition and biological productivity. The effect of extraterrestrial forces is both direct, like heating and air mass movement, and indirect via changes in the terrestrial ecosystems.

Biomes, major terrestrial ecosystem units covering the globe, are characterized by climatic and soil conditions interacting with the vegetational cover and animal inhabitants. Life-giving water has also contributed to the plasticity of relief by denuding and transporting materials, limiting vegetation and affecting climatic conditions. Freshwater ecosystems can be considered parts of biomes. The interactions are reciprocal, but we will consider here only the one-sided inputs from space and land to fresh waters.

The external variables driving freshwater systems can be arbitrarily classified as follows:

(*a*) Solar radiation inputs as a direct or indirect cause of geographical variations and seasonal periodicity of all other variables.

(*b*) Air temperature, wind, precipitation and runoff inputs determined by global atmospheric circulation as driven by differential heating of the globe by the sun. Modifications due to the distribution of oceans and continents, their relief and vegetational cover, produce a diversity of physical conditions which cannot be covered by simple global models, but which will be decisive for local differences.

(*c*) Input of dissolved and suspended minerals. The concentration of the main anions and cations is due to dissolution, as well as dilution by rain water and concentration by evaporation. Both geological and meteorological forces are operating. The activity of terrestrial plants modifies the dissolution. For limiting macronutrients, the uptake and release by the complex of soil and terrestrial vegetation is important. Modifications of the terrestrial habitats by human activity and by air and water pollution from industry cause more and more disturbance of the natural situation.

(*d*) Allochthonous organic input. The rich terrestrial vegetation of the humid tropics evidently brings more organic matter (both fresh and refractory) into aquatic systems than does an arctic tundra. More subtle differences in the allochthonous input due to variations of land productivity have to be considered for local conditions.

3.2.1. Solar radiation (with U.T. Hammer)

The following treatment is based particularly on List (1951), Sellers (1965), Robinson (1966) and Monteith (1973).

3.2.1.1. Global extraterrestrial irradiance

Irradiance is the amount of energy received per unit time (radiant flux) per unit area (radiant flux density). Global irradiance, essentially solar, reaches the earth's atmosphere in short-wave radiation in the range of about 300–3000 nm. Moonlight, starlight and cosmic rays contribute only minute amounts of energy.

At the mean sun–earth distance solar radiation is delivered onto a unit area (perpendicular to the incident rays) on top of the earth's atmosphere at a rate expressed by the solar constant (136.1 mW cm^{-2} s^{-1} = 1.95 g cal cm^{-2} min^{-1} = 8.16 J cm^{-2} min^{-1}, where 1 thermochemical g calorie = 4.184 J). Actually, the estimates of the solar constant vary between

1.94 and $2.00 \, \text{g cal cm}^{-2} \, \text{min}^{-1}$, but the most recent value given in Monteith (1973) is used here.

As a result of the inclination of the earth's axis to its orbital movement and the elliptical revolution of the earth around the sun, the amount of solar energy reaching instantaneously a unit of horizontal surface at the top of the atmosphere varies with the elevation of the sun, β, ($\beta = 90 - z^*$), which can be calculated from the latitude of the point on earth (ϕ), day of year (d), and solar hour angle (h) derived from the hour of day (List, 1951; Sellers, 1965; Robinson, 1966; McCullough, 1968; McCullough & Porter, 1971):

$$dE^*_G/dt = 8.16(a^*/r^*)^2 \cos z^* (\text{J cm}^{-2} \, \text{min}^{-1}) \qquad (3.1)$$

where

$$\cos z^* = \cos \phi \cos \delta + \sin \phi \sin \delta \sin h \qquad (3.2)$$
$$\delta = \text{solar declination} = \arcsin(0.3978 \sin \mathscr{E}) \qquad (3.3)$$
$$\mathscr{E} = \text{ecliptic longitude of the earth in its orbit}$$
$$= 2\pi(d - 80)/365 + 0.0335 \sin(2\pi d/365) - \sin(2\pi \, 80/365) \qquad (3.4)$$
$$(a^*/r^*)^2 = 1 + 0.0335 \cos(2\pi d/365) \qquad (3.5)$$

When a daily sum is to be obtained from instantaneous values, the daily

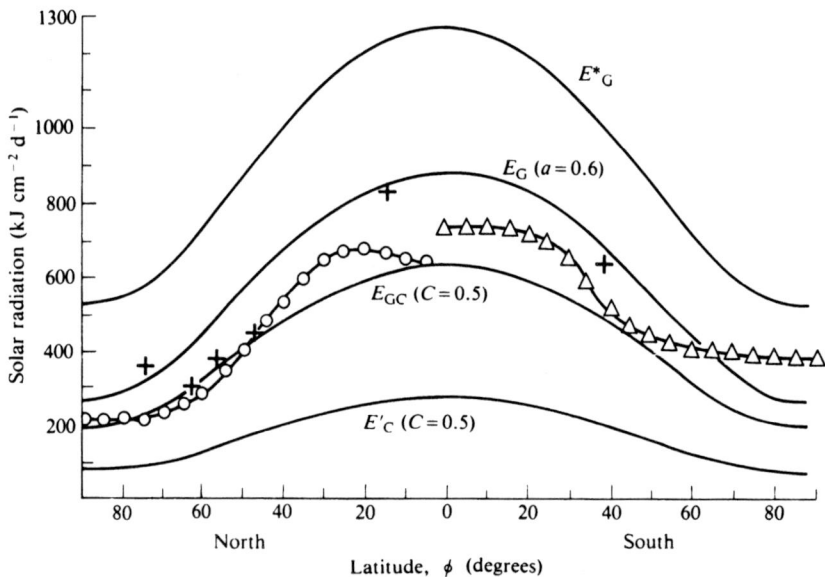

Fig. 3.1. Annual integrals of global radiation reaching the top of the atmosphere (E^*_G) compared with the calculated and observed integrals of global radiation reaching the ground (E_G; with 50% cloud cover, E_{GC}), and of photosynthetically active radiation under water (no ice) (E'_C). Triangles – observations of global radiation reaching the ground by Bridgman (1969). Circles – ditto Houghton (1954). Crosses – values of annual integrals for a few IBP sites.

integral has to be calculated:

$$E_G^*(\text{J cm}^{-2}\,\text{d}^{-1}) = 2 \int_0^{\pi/2} 8.16(a^*/r^*)^2 \cos z^* \, dz^* \qquad (3.6)$$

The values computed by means of equations 3.6 and 3.2–3.5 are tabulated by List (1951) for selected dates and latitudes. Our model computes the values for any date and latitude, using a numerical daily integration with $\Delta z^* = \pi/6$. An annual integral is also obtained as shown in fig. 3.1 upper curve.

3.2.1.2. Global radiation reaching the ground

Global radiation reaching the top of the atmosphere, E_G^*, is accurately specified, being independent of the qualities of the atmosphere. The atmosphere modifies radiation before it reaches the earth and therefore a superficial review of the factors involved and a simplified quantification for the purposes of modelling geographical differences in radiation will be of interest here.

Solar radiation reaches the earth's surface direct, E_D, and diffused by sky and clouds, E_S. In cloudless conditions, the magnitude of E_D and E_S depends on solar elevation and attenuation by gases and aerosols, whereas additional attenuation depends on clouds.

3.2.1.2a. Cloudless conditions

Direct radiation

(A) Effect of the elevation of the sun
The optical path through the atmosphere is greater at low angles than at high angles of the sun further reducing irradiance. The thickness of the atmosphere traversed by the solar beam is expressed as an air mass number m, which is equal to 1 at sea level when the sun is vertical. For sea level, variations of m can be closely approximated by:

$$m \approx \sec z^* (= \operatorname{cosec} \beta) \text{ when } z^* < 80° \qquad (3.7)$$

The thickness of the atmosphere decreases as a function of altitude, s, due to reduced atmospheric pressure:

$$m_s \approx b^* \sec z^* \qquad (3.8)$$

where $b^* = p_0/p_s$

with p_s = atmospheric pressure at altitude s (mb)
p_0 = normal sea level pressure = 1013.23 mb.

The atmospheric pressure at any altitude s can be obtained from:

$$\log p_s = \log 1013.23 - s/18\,400 \qquad (3.9)$$

(B) Attenuation by gases and aerosols

Solar radiation is reduced by molecular scattering by gases (e.g. oxygen, nitrogen, ozone) when the air is clean and dry. Water vapour absorbs and scatters it further. Lambert–Beer's law can be used to calculate the amount of irradiance at a given concentration of the relevant atmospheric constituents. Because of altitude reduction of the vertical extinction coefficients and uneven concentrations, an integration over atmospheric layers would have been necessary (Lettau & Lettau, 1972).

In its most simplified form all the attenuation factors of the atmosphere are combined into an 'atmospheric transmission coefficient' a. For direct radiation the quantity reaching the ground with a cloudless sky is expressed as:

$$E_D = E_G^* \, a^m \tag{3.10}$$

and substituting equation 3.7 we have

$$E_D = E_G^* \, a^{\sec z^*} \tag{3.11}$$

As indicated, the atmospheric transmission coefficient will vary with the amount of gases, water vapour and aerosols for different geographical locations, during the year and also in different years. Long-term changes due to climatic changes and increasing pollution of the atmosphere have to be expected. Regular geographical and annual variations of a were obtained by Burlatskaya & Samoilenko (1962) for the North Pacific Ocean between $0°–50°$ N. Values of about 0.65–0.70 with a low annual change were observed for latitudes below $20°$ N, both the mean value and the annual amplitude increasing towards the north. However, the values obtained for conditions over oceans are not representative for continental conditions with a more complicated pattern of atmospheric circulation (Budyko, 1974). Water vapour being the main scatterer, the annual and geographical variations of evaporation as discussed below will be the main reason for variations in a. Highest values can be expected in the dry arctics (Budyko, 1974 reports $a = 0.9$) and the wetter regions of the temperate and tropical zones will have lower values. Over the sea and large lakes, high evaporation losses reduce irradiance (Hutchinson, 1957).

With increasing altitude the coefficient a also increases systematically, due to reduced integral concentrations of the upper layers. For the related factor, atmospheric turbidity, such observations are summarized in Robinson (1966). A nearly linear increase at a rate of 4×10^{-5} per m results for alpine localities.

Diffuse sky radiation can be computed by utilizing extra-terrestrial radiation and making appropriate allowances for water vapour absorption (7%) and ozone absorption (2%). Half of this diffuse radiation on average is scattered downwards and the resulting approximation (List, 1951) will be:

$$E_S = (0.91 \, E_G^* - E_D)/2 \tag{3.12}$$

This average value will give an overestimate of diffuse radiation at high water vapour and other aerosol content, but for arctic conditions about twice as much diffuse radiation will reach the ground due to the high albedo of snow. Aerosols tend to scatter more solar radiation forwards than backwards so that the loss of direct radiation by aerosols is compensated for by an increase in diffuse radiation.

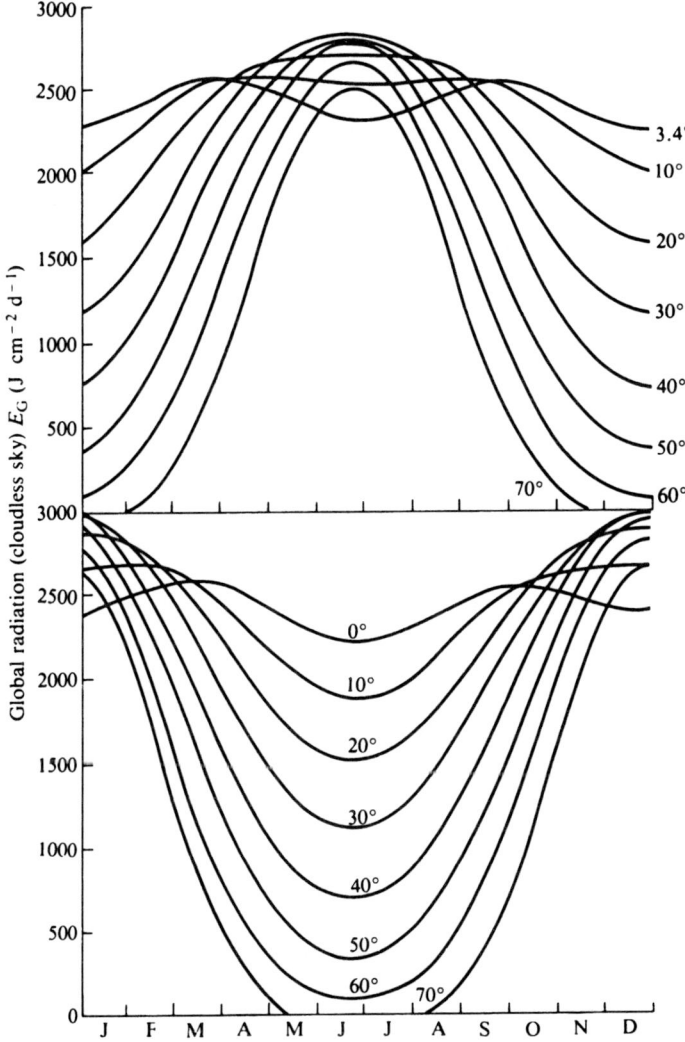

Fig. 3.2. Annual variations of daily integrals of global radiation reaching the ground on clear days calculated from equations 3.6, 3.2–3.5 and 3.11–3.13. The atmospheric transmission factor $a = 0.6$. For northern hemisphere (upper panel) the regular bimodal variations at the meteorological equator ($\phi = 3.4$) are given. Lower panel – southern hemisphere.

Adding diffuse to the direct solar radiation the global radiation reaching the earth is obtained:

$$E_G = E_D + E_S \tag{3.13}$$

Fig. 3.2 shows a plot of the annual variations of daily integrals of E_G during clear days at sea level, for $10°$ intervals of latitude. Values were calculated by means of equations 3.6, 3.2–3.5 and 3.11–3.13, assuming $a = 0.6$. The figure is not symmetrical to the equator, which is reflected in several features of the aquatic environment. The symmetry of the annual distribution is around 3.4 degrees north ($= 3° 24' N$), and this latitude is called the meteorological equator (Linacre, 1969). The southern hemisphere gets less radiation than the northern one. The highest daily values are reached between $30°–40°$.

3.2.1.2b Effect of clouds

Cloudiness would reduce the radiation reaching the ground still further. This factor introduces the largest variation into ground-level total incident radiation. Clouds absorb light and reflect it upwards. Attenuation by water droplets or ice crystals in clouds varies markedly depending on the cloud cover, cloud types and solar elevation. Average levels have been calculated for various types of clouds but vary considerably. Highest levels of incident solar radiation are associated with cirrus and cumulus clouds (65–85%) and the lowest with stratus and nimbus clouds and fog (15–25%). A partial cover of white clouds in an otherwise clear sky always increases the diffuse component E_S but the direct component E_D remains constant where the sun is not obscured. Then the irradiance may be large and even exceed the solar constant in the tropics (Monteith, 1972). When the sky is clear and the sun at high elevation, 85% of the radiation is direct, while the diffuse component increases to 40% of the total as the solar elevation decreases to $10°$.

Quantitatively the long-term value for the effect of clouds is approximated from the average cloudiness ($0 \leqslant C \leqslant 1$). Because measurements of cloudiness are not available for many places, a simpler measure of C [$C \approx 1 - (n'/N')$] is often substituted, with $n' =$ actual duration of sunshine (from heliograph records) and $N' =$ theoretical sunshine duration, obtained from meteorological tables or from the approximation equation 3.18. The use of the value n'/N' over C is preferable, taking partial account of the optical density of different types of cloud (Robinson, 1966).

Global radiation at a cloud cover C is due to direct radiation coming from the uncovered part of sky and diffuse radiation from the uncovered and covered parts of sky:

$$E_{GC} = E_{DC} + E_{SC} \tag{3.14}$$

$$E_{DC} = E_D(n'/N') \tag{3.15}$$

$$E_{SC} = E_S(n'/N') + (E_D + E_S) \cdot f(n'/N') \tag{3.16}$$

Whereas the global radiation from the uncovered portion of sky is sufficiently well approximated by equation 3.15 and the first term of equation 3.16, the approximation of diffuse radiation from the overcast portion of sky is more difficult. Highest E_S is observed at $C = 0.7$ and it drops to both lower and higher C. Robinson (1966) considers that the estimate will be within 5% for

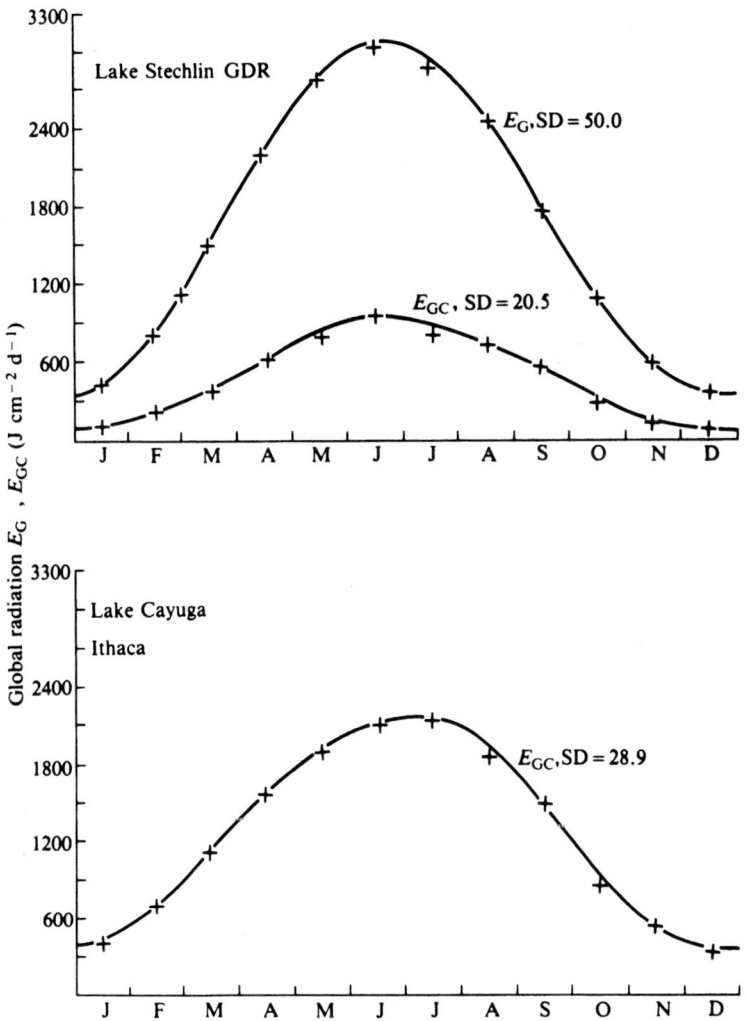

Fig. 3.3. Comparison of the values predicted with the solar radiation model and direct observations for Lake Stechlin (GDR) and Lake Cayuga, New York (USA). Upper panel – Lake Stechlin monthly average values for cloudless days and for average cloudiness ($a = 0.75$). Crosses are direct observations of several-year averages by Heitmann *et al.* (1969). Lower panel – L. Cayuga at Ithaca, several-year averages by Henson *et al.* (1961). The calculated radiation values were obtained with the observed cloudiness used as an input to the model. SD is the standard deviation of the regression estimate of predicted versus observed values.

individual values and more for monthly averages, when a non-linear $f(n'/N')$ is used. The mean of values for two widely separated localities tabulated by Robinson results in a third order polynomial:

$$f(n'/N') = 0.24 + 0.34(n'/N') - (n'/N')^2 + 0.42(n'/N')^3 \qquad (3.17)$$
$$(r^2 = 0.997 \text{ for } N = 12)$$

which is used below for non-arctic localities. The high albedo of snow in the arctics and at higher altitudes results in values of diffuse radiation about twice as high as those given by equation 3.17.

The result of comparison between values predicted by the model used here and direct observations is given in fig. 3.3 for Lake Stechlin (Heitmann, Richter & Schumann, 1969). The long-term mean values for cloudless days compare very well when $a = 0.75$ is used, as demonstrated by the low standard error of the regression estimate of predicted versus observed values (SD). The deviations can be attributed to annual changes of a. When computations were run at the standard value of 0.6, the differences were much larger. For cloudiness the predicted values differed little from the observed ones when the measured annual variations of C were fitted to a harmonic curve and included in the computation.

By means of rockets and satellites we are getting an increasingly detailed picture of the distribution of clouds over the globe for different periods. For the southern hemisphere a relatively recent summary (Bridgman, 1969) indicates an almost continuously half-covered sky at latitudes up to 30 °S. Towards 50 ° – 60 °S both mean cloudiness and annual variation increase with summer maxima and winter minima. Towards the Southern Pole both means and annual amplitudes decrease, and at the South Pole the winter minima reach about $C = 0.3$. As shown in fig. 3.1, the actual distribution of the annual integrals of global total radiation are distorted in comparison with the regular values for cloudless conditions. Similar estimates for the northern hemisphere are of an older date (Houghton, 1954) and yield much lower cloudiness for comparable latitudes, in accordance with the higher proportion of land masses.

3.2.1.2c. Screening by surroundings

Another important aspect limiting the amount of radiation impinging on a water body is the amount of screening produced by the surrounding terrestrial environment. Mountains such as those surrounding the Lunzer Untersee (Sauberer, 1953) and Vorderer Finstertaler See (Pechlaner *et al.*, 1972a, b) or extensive tree cover immediately adjacent to lakes and rivers produce a profound reduction of global radiation particularly on marginal waters and at low sun angles. Other counteracting effects can be caused by increasing albedo of the surrounding slopes, particularly when snow-covered.

3.2.1.3. Spectral energy distribution

For plants, only visible light or 'photosynthetically active radiation' (PAR) extending over the waveband 400–700 nm is of use (Rabinowitch, 1951). Perhaps some of the ultraviolet spectrum should be included in PAR since Halldal (1964, 1966) cites photosynthesis taking place in this range. The utilization of the different wavelengths within PAR is uneven and varies from species to species depending on their photosynthetic response spectra. We should be more concerned about the spectral distribution of the incident solar radiation rather than simple totals. However, this will only complicate both the measurements and estimates, because all the parameters of radiation absorption and scattering are in different ways wavelength dependent. Also, the biological interpretation of such data is not possible with the present inadequate knowledge of photosynthetic response spectra.

The fraction of E'/E_G for the daily integrals varies according to the many factors listed above. It is different for the direct and diffuse component (Robinson, 1966). Šesták, Čatský & Jarvis (1971) give for PAR limited to 386–740 nm a range between 0.38–0.62 in middle Europe depending on meteorological conditions. Extending the geographical range a wider variation will be obtained, particularly because more visible light is reflected at low sun elevations. This will be still more important underwater.

Recently mean estimates for different regions and periods were published (Tooming & Nijlisk, 1967; Rutkovskaya, 1972) and the different underlying factors were begun to be quantified (McCullough & Porter, 1971). However it is difficult until now to present a concise quantification and it is agreed to utilize, for the purposes of IBP, an average estimate of $E'/E_G = 0.46$ which the corresponding authorities consider a good possible average, particularly for temperate regions. The two significant places only presume higher accuracy, and the use of the ratio 1/2 would have the same justification.

3.2.1.4. Photosynthetically active radiation at the IBP sites

We have shown that an approximation of global incident radiation at any spot on the globe is possible. However, because a great many ill-defined variables enter the model, the particular radiation levels computed can only be approximate. They represent mean trends rather than individual years. Independent stochastic variations of individual variables will be superimposed.

For this reason direct measurements of global radiation were obtained for most IBP sites. Fig. 3.4 compares results for selected sites at various latitudes and altitudes with the model output for cloudless sky ($C = 0$) and half-covered sky ($C = 0.5$). The atmospheric transmission factor $a = 0.6$ is used throughout except for arctic and high mountain localities. As evident for most localities the value of $C = 0.5$ will not be far from reality, but considerably higher values are

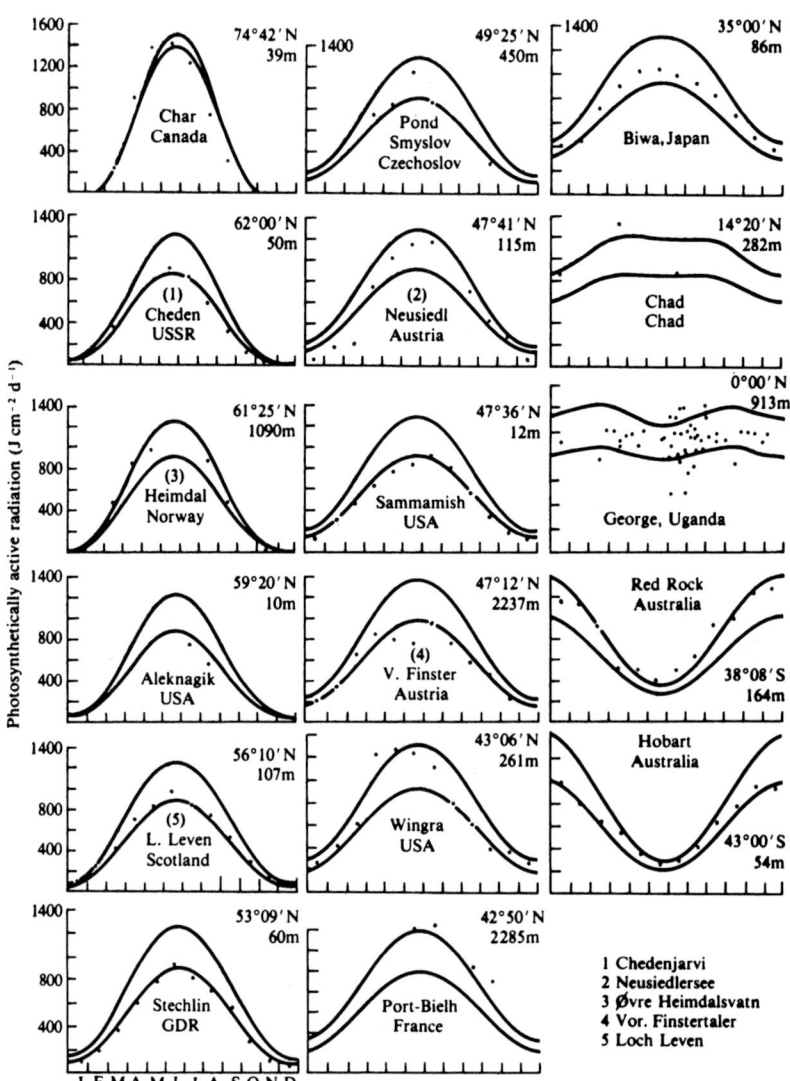

Fig. 3.4. Values of PAR for several IBP sites (dots) compared with model outputs for $C = 0$ and $C = 0.5$. Drawn from data reports and additional information for Sammamish from the Atmospheric Science Department, University of Washington, Seattle, for Red Rock from Walker (1973), for Hobart from Commonwealth Bureau of Meteorology. All data assembled by U.T. Hammer.

shown for the arctic lake Char, Neusiedlersee and Port-Bielh. For the arctic locality Char this is due to inadequacy of the model to account for high albedo of the arctic snow. Also when provision was made for the dry and clean arctic air by using $a = 0.9$, and for the increased diffuse radiation due to high albedo by doubling the $f(n'/N')$, the values were grossly underestimated. The mountainous Port-Bielh was also underestimated by the model, even when the altitudinal correction was applied.

As a most drastic example of differences between latitudinally similar sites, Lago do Castanho, Brazil, at about $4°$ S ($1255-1675$ J cm^{-2}d^{-1}) and Lake George, Uganda, $0°$ ($1778-2092$ J cm^{-2}d^{-1}) have been quoted (Black, 1956). The values for the Brazilian lake are about $1/3-1/5$ lower than at Lake George. However, this is not surprising when we note the differences in altitude, as well as the high cloud cover and air humidity of tropical South America as opposed to the less cloudy dry region of Uganda (Viner & Smith, 1973; Schmidt, 1973). Assuming corresponding values of $C = 0.5$ and $s = 50$ m, and $C = 0.3$ and $s = 900$ m, to Castanho and George, our model agrees with observations, predicting about 30% less radiation for the former lake without considering differences in a due to humid air which will further decrease radiation.

A major difference seems to exist between two high-altitude IBP lakes: Port-Bielh at $43°$ N and 2300 m and Vorderer Finstertaler at $47°$ N and 2200 m. The difference in altitude results in only about 3% summer reduction for the higher locality, but according to Pechlaner *et al.* (1972b) 41% of the theoretically possible sunshine is shaded by the mountains surrounding Vorderer Finstertaler. In addition exceptionally clean and dry air must exist at Port-Bielh as seen in fig. 3.4.

3.2.1.5. Daylength variations

For the life of organisms not only the daily integrals are of concern but also the daily distribution of radiation. Fig. 3.5 is a rough approximation of the annual variations in daylength at latitudes up to 60 degrees. In contrast with the stochastic character of incident radiation reaching the ground, the values are fixed, and a more detailed calculation can be obtained from meteorological equations. Mean daylength is almost constant everywhere, approximately 12 hours. Whereas this is the actual daylength throughout the year in the tropics, the amplitude of daylength variations increases toward the poles. At $\phi > 67°$ days and nights longer than 24 hours start to occur (polar days and nights). The values in fig. 3.5 were calculated from the empirical formula for latitudes between $60°$ N and $60°$ S:

$$\Delta = 12 + (9.14 \times 10^{-2} + 4.03 \times 10^{-2}\phi + 1.05 \times 10^{-3}\phi^2 - 3.39 \times 10^{-5}\phi^3$$
$$+ 6.06 \times 10^{-7}\phi^4)\sin(t + \gamma\Delta) \tag{3.18}$$

where $\gamma\Delta = 280$ for northern hemisphere
$= 100$ for southern hemisphere

Daylength obtained from equation 3.18 at latitudes over $60°$ is less precise and

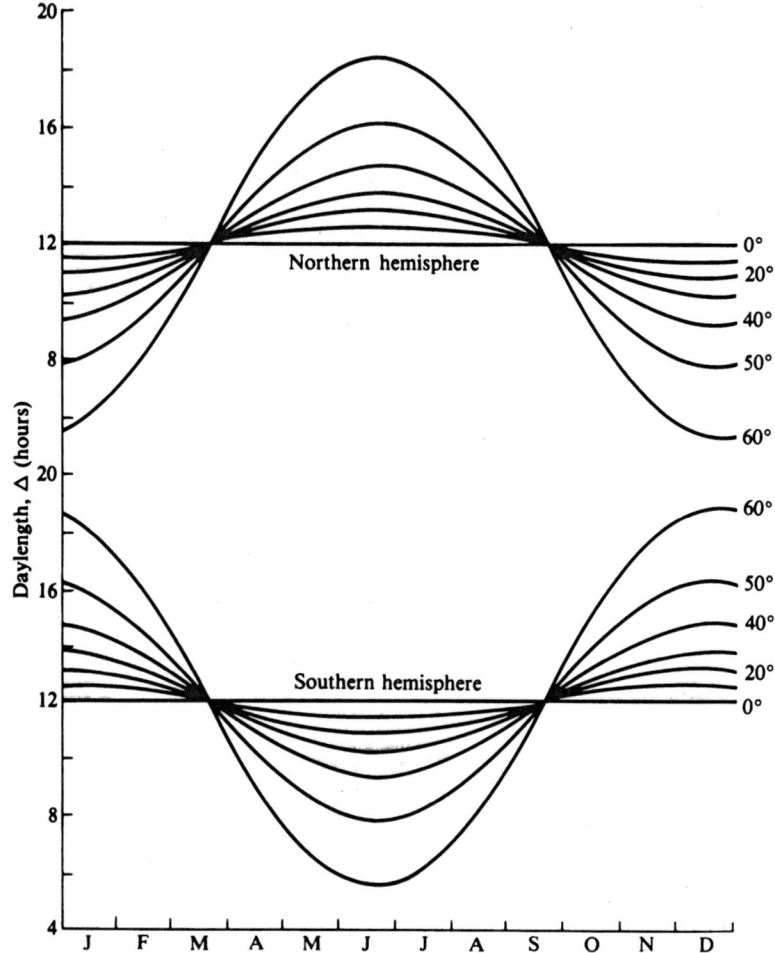

Fig. 3.5. Daylength variations over the globe approximated by equation 3.18.

additional conditions for $\Delta = 0$ and $\Delta = 24$ have to be included for polar nights and polar days.

The difference between the longest and shortest day, the daylength range as defined in Chapter 9 for characterizing light available for photosynthesis can be easily computed from equation 3.18. An approximately exponential increase of the daylength range with latitude results (fig. 3.6), as obtained by Brylinski (Chapter 9).

3.2.2. Global circulation in the atmosphere and hydrological budgets

The continuous intensive heating at the equator results in the expansion and rising of air. The rising air is distributed toward the poles and a substitution

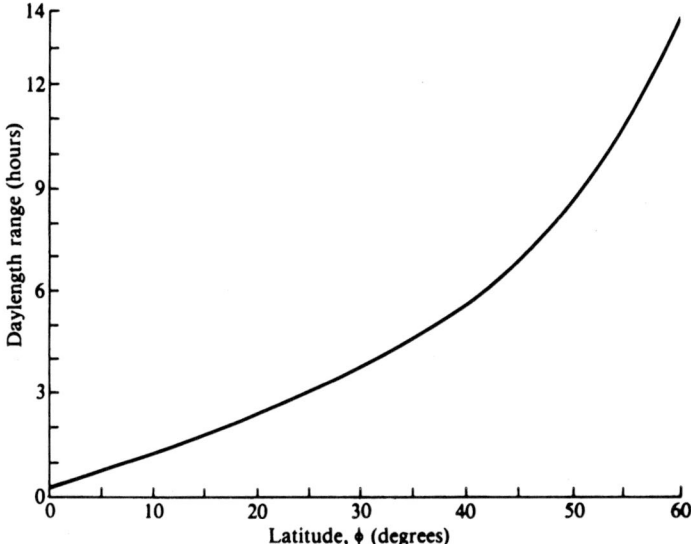

Fig. 3.6. Difference between the shortest and longest day as a function of latitude, calculated from equation 3.18.

from adjacent areas occurs at ground-level. Trade winds replacing the raised air move toward the equator. The raised air falls again at about 30° and spreads in both directions. The general circulation pattern is shown in the upper part of fig. 3.7. Considerable seasonal differences occur.

During its warming the tropical air rises and expands, due to decreased atmospheric pressure at higher altitudes. Due to losses of energy from expansion it cools down at a rate called the adiabatic lapse rate, which is theoretically calculated as 6 °C per km for dry air and 10 °C per km for wet air (MacIntosh & Thom, 1969; MacArthur, 1972). The actually observed mean air temperature drops range between 6–8 °C per km (Nakamura, 1967; Linacre, 1969; Lewis, 1973). Because cold air is capable of holding less water vapour, the rising air above the tropics, and more locally in mountains, loses water as precipitation. The observed rates of precipitation increase with altitude, noted in Chebotarev (1960), are locally different due to the origin (water vapour saturation) of the air masses, but amount to about 300 mm per km. In fig. 3.7 we see that the cold air bereft of water is transported to latitudes 20°–30°. During descent it is heated again and is absorbing rather than releasing water. At the high radiation inputs water evaporates rapidly. Evapotranspiration exceeds precipitation and negative budgets occur, resulting in arid zones with negative runoff, as evident from fig. 3.7. At higher latitudes of 50°–60° the rising air masses dominate again, precipitation being correspondingly increased and evapotranspiration decreased. Total runoff remains positive, dropping steeply toward the arctic circle.

The character of global atmospheric circulation and resulting hydrological

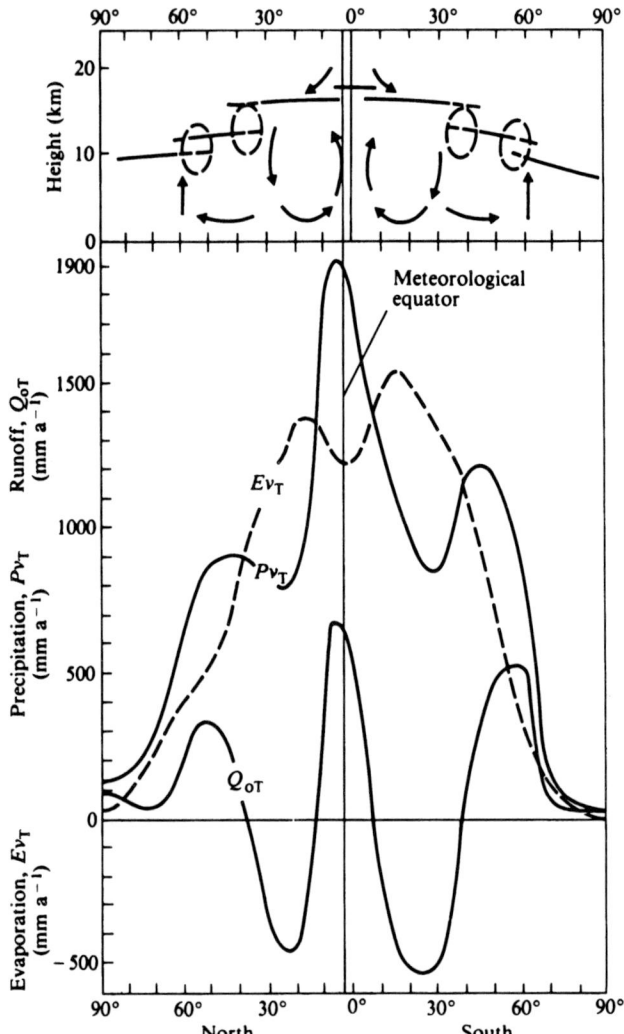

Fig. 3.7. World pattern of atmospheric circulation (upper panel) and the corresponding average annual mean evaporation, Ev_T, precipitation, Pv_T, and runoff, Q_{oT} (lower panel). Location of the meteorological equator is indicated by a thin line. Modified from Sellers (1965) and MacArthur (1972).

budgets outlined in fig. 3.7 are reflected in several ways in freshwater sites:

(1) In accordance with the runoff values freshwater sites are extremely rare at latitudes $20°-30°$. The construction of reservoirs in these latitudes is an ecologically ill-considered activity leading to wastage of water and to other difficulties (for example the Aswan Reservoir: Collier, Cox, Johnson & Miller, 1973).

(2) Mixing conditions in lakes are affected by the uneven latitudinal distribution of wind speed (see for example maps by Lauscher, 1951) and, more importantly, by the geostrophic Coriolis force. This can be explained as on a rotating planet a body moving on the equator has a much higher velocity than it would have at higher latitudes, where it would travel a much shorter distance in a comparable time interval. This force causes any moving object to deflect right (clockwise) in the northern hemisphere and left (anticlockwise) in the southern hemisphere. When an object is moving north in the northern hemisphere it attains a higher speed and deflects right. As a result, the shearing stress of wind, v, will be a function of latitude in addition to that of wind speed:

$$v = \lambda \sin \phi \cdot f(u) \qquad (3.19)$$

The same wind produces half the shearing stress at latitude 30° than at the equator (sin 30° = 0.5). The Coriolis force also causes differences of stratification conditions in lakes.

(3) Local water budgets are driven by the global pattern indicated, but modified according to local meteorological and geomorphological forces as well as by the ratio of lake area to catchment area (Sokolov, 1966).

For any area the water budget, W_T, can be represented by the difference between the positive terms precipitation, Pv_T, condensation of water on the surface, Dv_T, and horizontal water inflow, Q_{iT}, – and negative ones – evaporation, Ev_T, and horizontal outflow, Q_{oT}.

$$W_T = Pv_T + Dv_T + Q_{iT} - Ev_T - Q_{oT} \qquad (3.20)$$

For land areas the term Ev_T represents evapotranspiration, suggesting the importance of plants against the solely physical process – evaporation s. str. from the free surfaces. The term Dv_T is negligible, maximum values amounting to 1 mm per night, except for tropical rain forests.

The hydrological budget of a water body is similar to equation 3.20 except that surface and ground water in- and outflows have to be separated because of the different importance for thermal, chemical and biological phenomena.

3.2.3. Mechanisms controlling world water composition

Major trends of the basic mineral composition of large rivers and lakes, as recognized by Gibbs (1970; see also Feth, 1971), are reproduced in fig. 3.8. With a decreasing precipitation/evaporation ratio the total dissolved salt content increases (no exact quantification being given in Gibbs, but see Langbein & Dawdy, 1964). This increase is accompanied by systematic shifts from 80–100% NaCl (precipitation dominance series) towards 10–50% of $CaHCO_3$ (rock dominance series) and back again to high NaCl percentages at very high salinities (evaporation dominance series). Three controlling processes are recognized: (1) equilibrium dissolution from underlying rocks, (2) dilution by

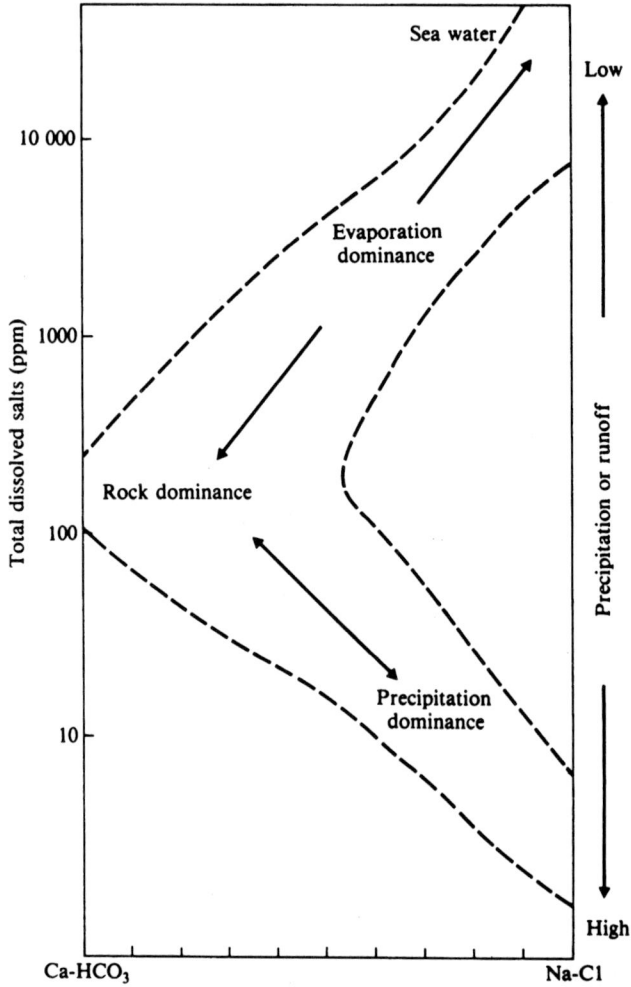

Fig. 3.8. Schematic representation of the major trends in the mineral composition of large rivers and lakes. Modified from Gibbs (1970).

precipitation, and (3) fractional crystallization during evaporation. In arctic localities freeze-out mechanisms are also important.

The recent development of thermodynamic equilibrium models (Sillen, 1961; Kramer, 1964; American Chemical Society, 1967; Stumm & Morgan, 1970) covers the first process. The approach revolutionized the understanding of inorganic water chemistry by explaining quantitatively the mutual relationships of major ions. However, it does not account for the more dynamic changes due to the activity of organisms (Broecker, 1971). Also, the equilibrium concept assuming infinite contact time between water and rocks corresponds more closely to conditions in subsurface waters, where White, Hem & Waring (1963) demonstrated the close relations between the chemical composition of major ions and their genesis on extensive USA material. In surface waters, this is just part of the problem because of the dilution–concentration effects of precipitation and evaporation.

From a simple balance consideration we can hypothesize that the concept of 'mean river water composition' (Clarke, 1924; Conway, 1942; Rodhe, 1949; Livingstone, 1963) is applicable to hydrologically balanced regions, coinciding with the rock dominance series of Gibbs. According to this hypothesis regions with evaporation greatly exceeding precipitation will be the most aberrant (corrected latitude $10° - 40°$). And this is exactly where the inapplicability of the 'mean river water composition' was stressed recently: in Australia (Williams & Wan, 1972). Major geographical differences in surface water chemistry which have been stressed by limnological investigations are well explained on this hydrological basis (Baranov, 1962 – USSR; Talling & Talling, 1965 – east and central Africa; Schnitnikoff, 1973 – Eurasia; Schindler, Welch, Kalff, Brunskill & Kritsch, 1974 – Arctic). Also the latitudinal ranges of Baranov (1962) and Voronkov (1970) agree well with the above hypothesis.

It should not be forgotten that the mean river concept was intended for large rivers. Here the local differences in geological formations are integrated and approach the mean composition of the earth's crust. For smaller watersheds a much larger diversity is to be expected. In addition to latitudinal trends in salinity ranges, increased chloride contents and their systematic drop with distance from the sea have been given by Jackson (1905), Drischel (1940), Brooks & Deevey (1966), Straškraba *et al.* (1969).

The concentration of suspended solids also shows consistent drops with increasing annual precipitation (Langbein & Dawdy, 1964). In the USA the mean concentration drops four orders of magnitude when precipitation increases from zero to 230 mm per year. An important consequence is a decrease in dissolved and suspended load with increasing precipitation or runoff. The ratio of suspended to dissolved solids varies with runoff from less than 1 : 10 to more than 6 : 10. The irregularity of the precipitation distribution greatly increases load as shown in different empirical and analytical observations, and as is well known from high silting in semi-arid regions.

3.2.4. Organic matter input from terrestrial ecosystems

Terrestrial IBP studies have resulted in an empirical model of world primary productivity (Lieth, 1972, 1975). The available productivity data as well as driving variables were summarized in maps (Lieth, 1964) computerized with recent additions by Lieth & Box (1972).

The organic matter production by terrestrial vegetation is closely correlated with the temperature and precipitation conditions (fig. 3.9, equations 3.21 and

M. Straškraba

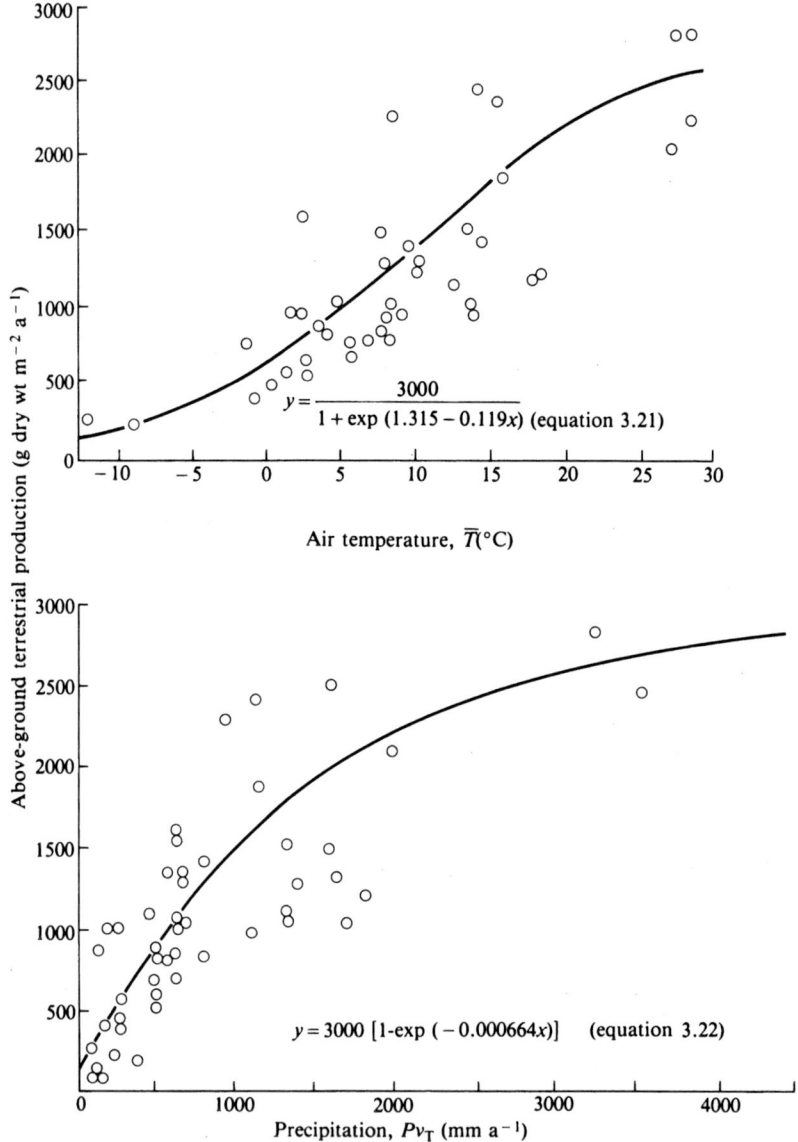

Fig. 3.9. Annual net production of terrestrial vegetation correlated with the annual mean air temperature (upper panel) and annual mean precipitation (lower panel). After Lieth (1972).

3.22). We have approximated the corresponding precipitation curve in fig. 3.7 by a polynomial with the resulting equation 3.23:

$$Pv_T(\text{mm a}^{-1}) = 1952 - 7.8545\,\phi' - 8.4319\,\phi'^2 + 0.482\,\phi'^3$$
$$- 1.0386 \times 10^{-2}\,\phi'^4 + 9.7297 \times 10^{-5}\,\phi'^5$$
$$- 3.3371 \times 10^{-7}\,\phi'^6 \qquad\qquad (3.23)$$

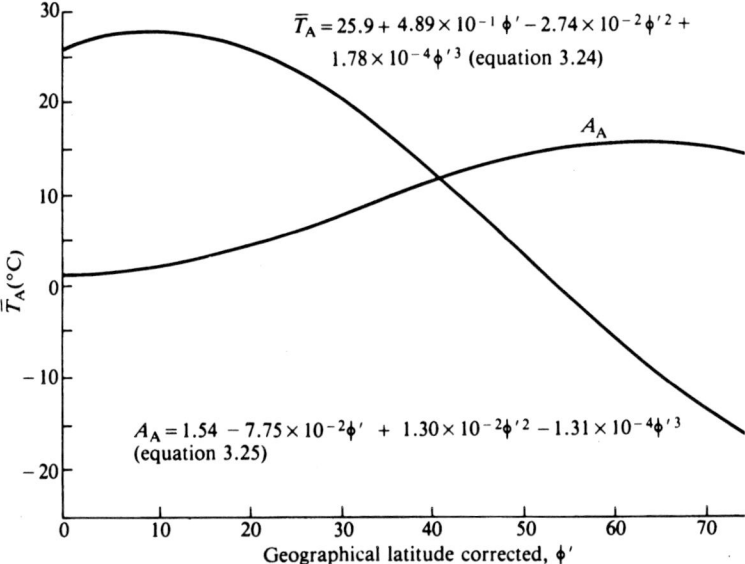

$$\bar{T}_A = 25.9 + 4.89 \times 10^{-1}\phi' - 2.74 \times 10^{-2}\phi'^2 + 1.78 \times 10^{-4}\phi'^3 \quad \text{(equation 3.24)}$$

A_A

$$A_A = 1.54 - 7.75 \times 10^{-2}\phi' + 1.30 \times 10^{-2}\phi'^2 - 1.31 \times 10^{-4}\phi'^3 \quad \text{(equation 3.25)}$$

Fig. 3.10. Annual average air temperature, $\bar{T}_A$, and semiamplitude of the annual temperature variations, A_A, are approximated by polynomials according to data from Linacre (1969), supplemented for arctic conditions from a few additional sources.

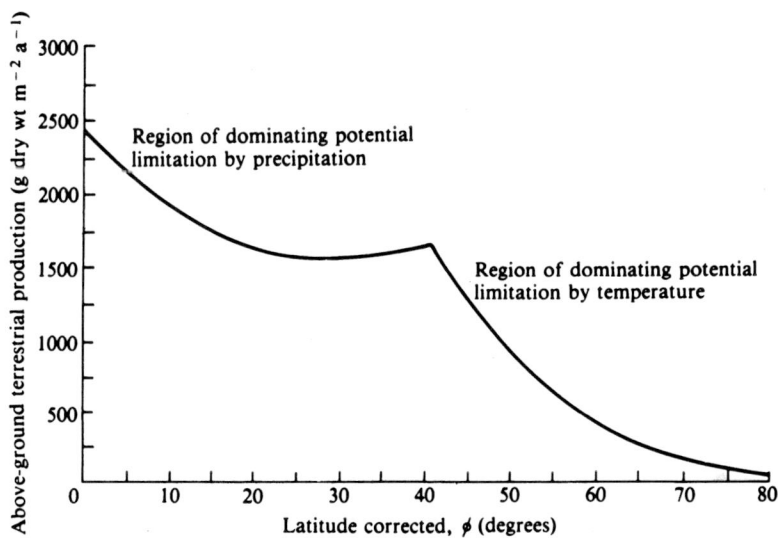

Fig. 3.11. Geographical distribution of annual net production of terrestrial vegetation obtained from equations 3.24 and 3.25 using approximated temperature and precipitation pattern.

Similarly, the average annual mean air temperature was approximated as a function of the corrected latitude (fig. 3.10, equations 3.24 and 3.25). Latitudinal differences of the average production of terrestrial vegetation were then calculated using equations 3.21 and 3.22 and selecting the lower of the two values in agreement with Lieth (fig. 3.11). Two regions can be distinguished as to the factors potentially limiting terrestrial production: precipitation at latitudes below 40° and temperature at higher ones. The highest annual production, that of the wet tropics, amounts to about 3 kg dry matter per m², whereas the lowest, from cold and dry arctic sites produces only about one hundredth of this value.

The fraction of terrestrial primary production reaching water bodies is not known and depends on the transport to lakes and decomposition, as well as on the drainage area/lake area ratio during transport. It is highly modified by man's activity due to plant cultivation and erosion (Likens & Borman, 1974). As discussed in Farnworth & Golley (1974) the greater terrestrial production in the tropics can be accompanied by increased destruction and cancel the differences between temperate and tropical areas. This was actually observed in the study by Malaisse *et al.* (1975). The energetic contribution of the organic matter from land does not seem to be important for most large lakes, but its share increases with decreasing size (Rau, 1976) and increased flushing of the water body. At the extreme, small flowing waters receive most of their energy from terrestrial sources (Fisher & Likens, 1972). However, the refractory organic matter which will be more directly related to the terrestrial net production has potentially important effects on altering nutrient availability for phytoplankton and light penetration into water. This is clearly manifested in the 'black waters' of Amazonia (Sioli, 1964).

3.3. Physical conditions in geographically different lakes and theoretical productivity comparisons

The quantification of basic meteorological and terrestrial input variables in the preceding section is a basis for proceeding towards a quantitative description of physical conditions in water bodies as they vary in geographical respects. This analysis is restricted to lakes of similar morphometry because of the large morphometric and nutrient load differences, to be analysed in section 3.4.

3.3.1. Temperature conditions in the world lakes

The relatively successful use of the heat budget approach by engineers for calculating the exchange at the air–water interface indicates close relationships between the radiation inputs and water temperatures for well-mixed waters, particularly rivers. For stratified waters the second part of the problem, quantification of the redistribution of heat within the water body, is far less known and complicated by other effects. We will not attempt to use this approach for understanding geographical differences in lake temperatures, although we anticipate considerable progress in this direction in the near future. Nevertheless, when studying lake temperatures empirically, we can expect that the fascinating simplicity of the sinusoidal pattern of solar radiation inputs shown in figs. 3.2–3.4 will be retained to some degree for lake temperatures. An excellent figure by Talling (1969) for a few African lakes suggests that nature respects our anticipation.

Surface temperatures from about fifty IBP and other lakes distributed from 26° S to 74° N were examined by a statistical programme of periodic regressions based on Bliss (1958). Selected examples of approximation are shown in fig. 3.12. Least-squares fitting of the equation 3.26 is obtained:

$$T(t) = A_0 + A_1 \sin(t + \gamma_1) \pm SD \qquad (3.26)$$

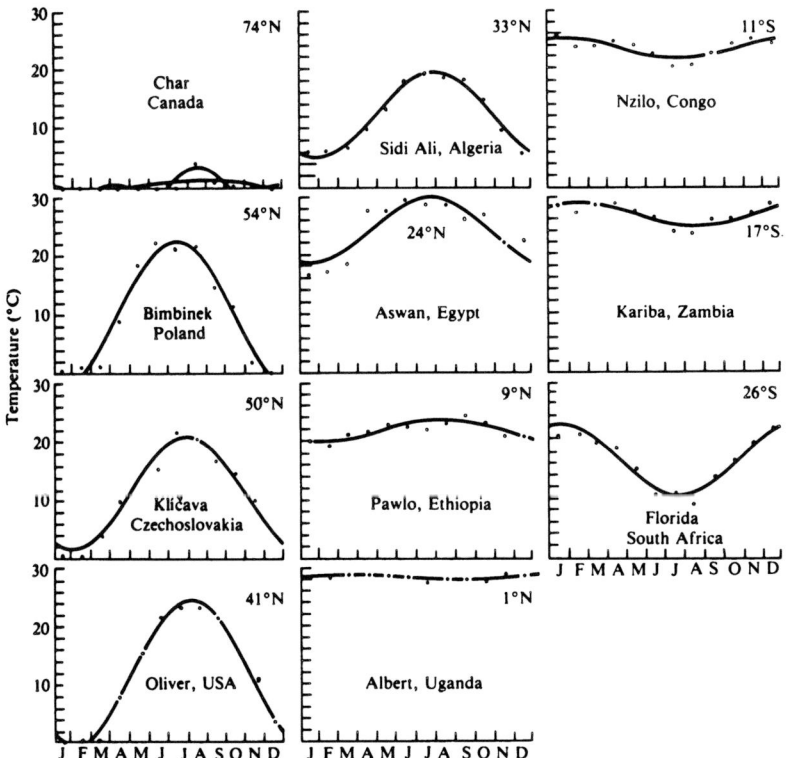

Fig. 3.12. Approximation of the annual temperature variations of selected IBP sites by the periodic regression equation 3.26. Latitudes of the localities are given. For the arctic lake Char a more adequate approximation with a periodic curve including the half-year harmonic oscillation is also given.

M. Straškraba

The results to be published in detail elsewhere suggest that for surface temperature of temperate and subtropical lakes between 85–95% of total variance is explained by the simple annual sinusoidal wave. In tropical lakes with a very low annual wave irregular variations are more pronounced than the annual cycle. A low percentage of total variance due to the annual wave is also observed at higher latitudes with surface freezing for prolonged periods (Krasnoye, Char). Southern hemisphere lakes differ by the phase shift, γ_1.

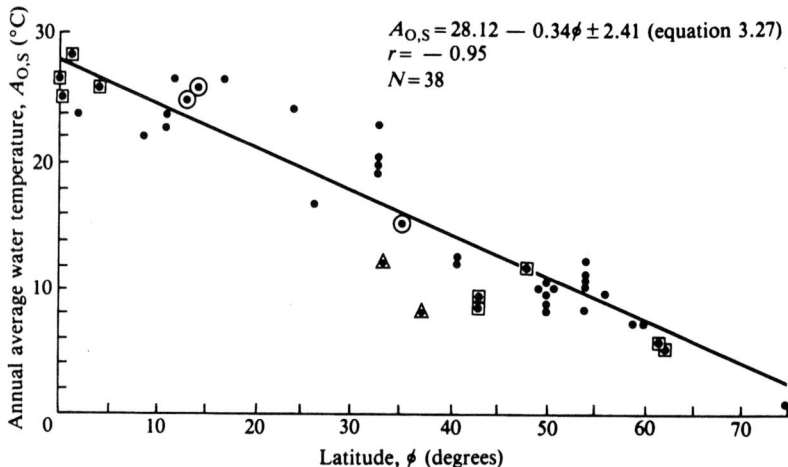

Fig. 3.13. Mean annual surface temperature ($A_{0,s}$) of several IBP lakes plotted against the corrected latitude and the linear approximation for medium size lakes at altitudes less than 2000 m. Squares – large lakes, circles – shallow lakes, triangles – lakes above 2000 m.

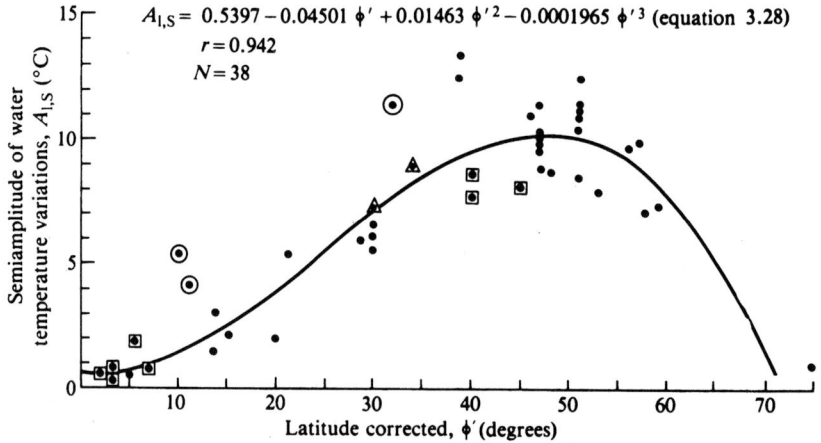

Fig. 3.14. Semiamplitude of the annual surface temperature variations ($A_{1,s}$) of the same lakes as in fig. 3.13 and the polynomial approximation for medium size lakes at altitudes less than 2000 m.

Maxima are reached about half a year earlier, but mean values and amplitudes are otherwise similar to those in the northern hemisphere.

The mean annual surface temperature $A_{0,s}$ and the semiamplitude $A_{1,s}$ are plotted against latitude in figs. 3.13 and 3.14. The latitudes are corrected for the meteorological equator ($\phi' = \phi - 3.4$) (see section 3.2.1.2a). Preliminarily three major morphometric groups were distinguished, medium depth and size lakes, very shallow lakes and very large lakes. In addition, lakes situated above 2000 m were noted. In spite of variability, a clear nearly linear drop of mean temperatures ($A_{0,s}$) and non-linear non-monotonous rise and drop of amplitudes ($A_{1,s}$) with increasing latitude is observed. The regression equations 3.27 and 3.28 fitted for medium depth lakes are substituted for equation 3.26 to describe the mean geographical trends of annual temperature variations at the surface of medium size lakes of the world:

$$T_s(\phi',t) = 28.1 - 0.34\,\phi' + (0.54 - 0.045\,\phi' + 0.0146\,\phi'^2$$
$$- 1.97 \cdot 10^{-4}\,\phi'^3)\sin(t + \gamma_{1,s}) \qquad (3.29)$$

where $\gamma_{1,s} = 240$ for $\phi > 0$ (northern hemisphere, maximum in mid-August. In reality $\gamma_{1,s}$ varies from 213–249 for the lakes with $\phi > 10$ included.)

$= 60$ for $\phi < 0$ (southern hemisphere, maximum in mid-February. In reality $\gamma_{1,s}$ varies from 46–69, but only five lakes were covered.)

Shallow lakes show temperatures higher, and large deep lakes lower, than the computed line would seem to indicate. Effects of continentality, altitude and some more local effects as quantified by Linacre (1969) for world air temperatures are suggested in figs. 3.13 and 3.14 but can be analysed only when more data become available.

The stochastic variations of temperatures, as expressed by the standard deviation, also show consistent geographical trends. The trend for the absolute value seems to be rather similar to this shown in fig. 3.14 for the semi-amplitude, that is it increases from low values of less than 0.2 near the meteorological equator up to about 1.0 at latitudes $40°-60°$. The high arctic locality (Char Lake) again shows a low standard deviation. However, for the density-related physical phenomena (stratification, mixing) the short-term temperature changes are more important relative to the annual cycle. When expressing standard deviation as a fraction of the annual temperature amplitude ($SD/2\,A_{1,s}$) the value is up to 0.7 for lakes with $\phi' < 20°$ and drops approximately hyperbolically with latitude, except for the arctic lake Char which has a high value.

Fig. 3.15 demonstrates the geographical trends of surface temperature variations of the medium size lakes at low elevations calculated from equation 3.29 for selected latitudes. The southern hemisphere picture would be an exact

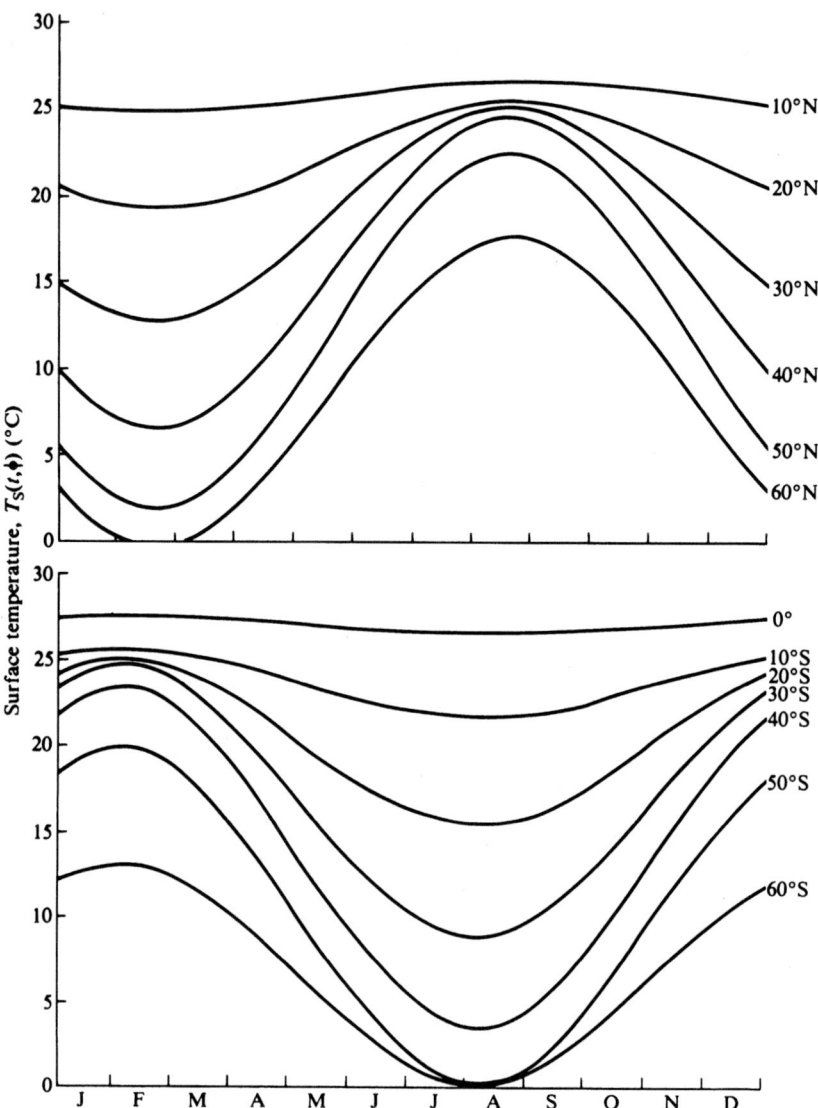

Fig. 3.15. Annual surface temperature variations for selected latitudes ϕ calculated from equation 3.29.

mirror image were it not for the asymmetric position of the meteorological equator and slight shift of maxima and minima. At the meteorological equator the model is discontinuous or rather bivalent – as shown above the regular annual variations are here not important.

To indicate to what extent the model (equation 3.29) represents the actual lake temperatures, a comparison is made in fig. 3.16 for a few lakes with data obtained after the approximation was calculated.

For *bottom temperatures* of lakes two authors have independently quantified clear-cut geographical trends (Löffler, 1968, Lewis, 1973), both for tropical lakes and their altitudinal variations. However, as pointed out by Lewis, altitudinal and latitudinal effects are interconvertible. When we base the altitudinal correction on lake bottom temperature approximately 4–5 °C per km as obtained by Lewis and Löffler, respectively, results. Both treatments, by Löffler (1968) and Lewis (1973), irrespective of quantitative differences in the altitudinal effect extrapolate to 28.1 °C at $s = 0$ which is exactly the same value as for $A_{0, s}$ in equation 3.29. At latitudes around 40° the bottom temperatures drop to the maximum density temperature and do not change further north or south, except at very high latitudes (Char Lake, 74° N,

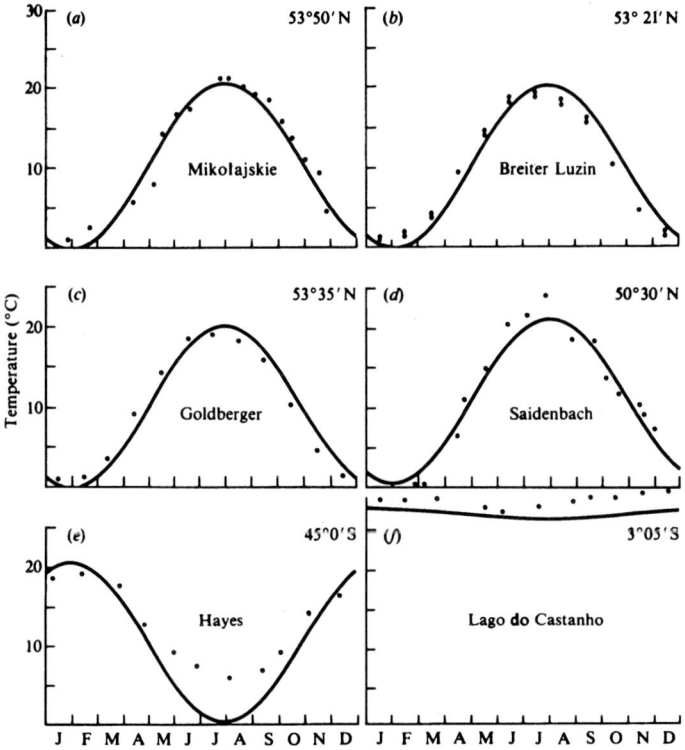

Fig. 3.16. Comparison of temperatures obtained from equation 3.29 and measured in localities not included in model computations. The localities do not exactly correspond to the size of the category for which equation 3.29 was derived. (*a*) Data by Rybak (1972); (*b*, *c*) data by Karbaum (1966), monthly averages for 3 and 5 years (the lower points in (*b*) are for a depth of 10 m); (*d*) unpublished data by E. Höhne, Technical University, Dresden; (*e*) unpublished observations by C. Burns, New Zealand; (*f*) data by Schmidt (1973).

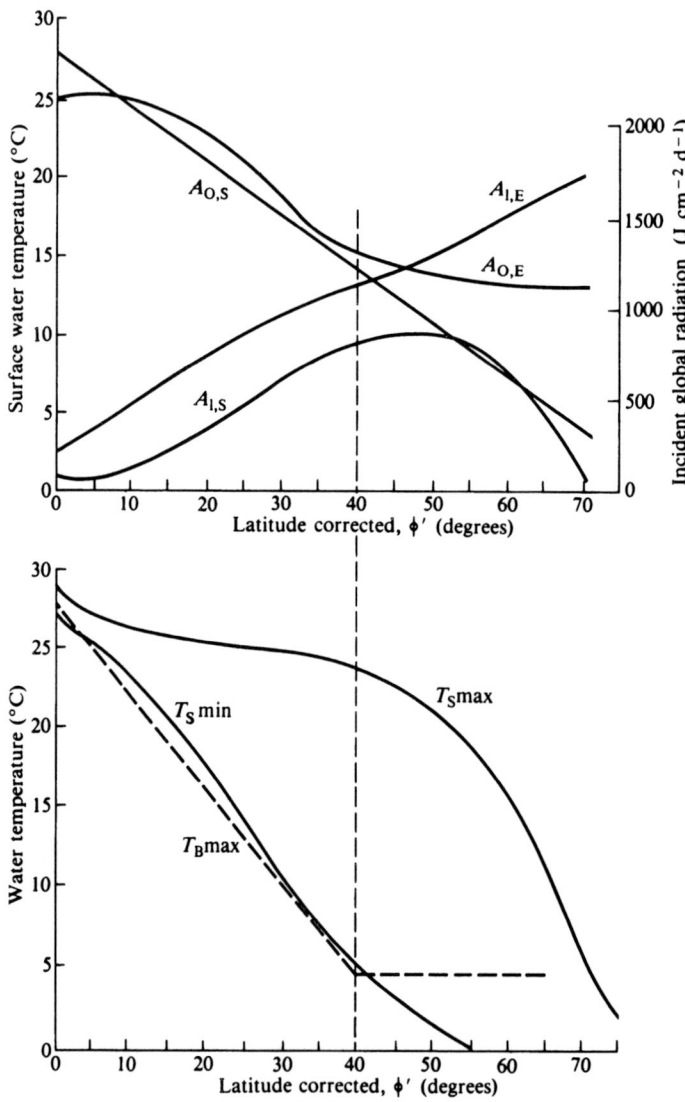

Fig. 3.17. Comparison of latitudinal trends of water temperatures and solar radiation. Upper panel – annual average surface water temperature $A_{o,s}$ from fig. 3.13, and semi-amplitude of annual temperature variations $A_{1,s}$ from fig. 3.14 compared to similar trends for the near-ground global radiation as computed from data obtained by Bridgman (1969) for the southern hemisphere ($A_{o,E}$ and $A_{1,E}$). Lower panel – maximum and minimum surface temperature T_s max and T_s min obtained from equation 3.29, and maximum bottom temperatures calculated from equation 3.30. Dashed line indicates latitude at which trends for water temperature start to deflect from those for radiation.

bottom temperatures for most of the year 1–2 °C). The annual variations in medium-sized lakes at 30 m are not large (maximally 4 °C in middle latitudes) and the maxima are shifted systematically in time in comparison with the surface. A somewhat more complicated expression than equation 3.26, including higher harmonic oscillations would be necessary for an adequate description of the annual course of bottom temperatures. For the present purposes, however, the annual variations will be neglected and bottom temperatures (at 30 m) assumed to be constant over the year. The bottom temperatures are supposed to vary geographically according to:

$$\begin{aligned} \bar{T}_B &= 28.1 - 0.6\,\phi' \qquad && \phi' \leq 40^\circ \\ &= 4 \qquad && \phi' > 40^\circ \end{aligned} \tag{3.30}$$

In fig. 3.17, the observed trends of water temperatures are compared with trends in incident radiation. As evident particularly for semiamplitudes ($A_{1,s}$ and $A_{1,E}$), from the equator to about 40° the trends are parallel, deflecting considerably at higher latitudes. The reason for the deflection of water temperatures at about 40° is the drop of bottom temperatures to maximum density (4 °C), when inverse stratification enables the water-surface to freeze. The annual average radiation does not drop linearly with latitude, and this seems to be also indicated for water temperatures in fig. 3.13.

3.3.2. Geographical differences of mixing and physical stability

A problem often discussed in comparative limnological studies is whether tropical lakes with higher temperatures but lower temperature gradients have greater physical stability than temperate ones with lower temperatures and higher gradients (Ruttner, 1931).

Based on the above approximation of world trends of lake temperatures we will attempt to simulate a world pattern of the degree of mixing and physical stability for an idealized medium-size lake (area 100 ha, maximum depth 30 m, conical shape). The trends will be more important than absolute values for several reasons: (1) temperatures were only roughly approximated, (2) stochastic variations of temperatures were neglected, (3) the present theories of mixing and stability have considerable gaps.

Three measures will be considered (a) the coefficient of the vertical eddy diffusion, D_z, as a measure of the exchange of water and substances between the epilimnion and hypolimnion, (b) mixing depth z_{mix} identical with the depth of thermocline, i.e. a depth where $d^2 T/dz^2 = 0$, as an important parameter for depth distribution of organisms and light available to the phytoplankton population, (c) stability of Schmidt, St, as the degree of displacement of the gravity centre of the stratified lake from holomictic conditions. To avoid confusion with the now popular concept of ecological stability we are adding the adjective *physical* here. For a recent limnological explanation of the concepts used see Hutchinson (1957) or Cole (1975).

Theoretical deviations will be related to the period of maximum surface temperature only. For other periods, empirical assumptions will be adopted. Limitations of the computation methods used, however serious they might appear, cannot be discussed here and reference is made to special papers. The typology of lake mixing, so popular in the early periods of productivity studies, is intentionally avoided here.

3.3.2.1. Turbulent diffusion and mixing depth

The method of calculating the *eddy diffusion coefficients* for the present purpose of gross geographical comparisons was selected from several ones now in use in addition to the classical McEven–Hutchinson method (Hutchinson, 1957; Wright, 1961; Khomskis, 1969; Lerman & Stiller, 1969; Bella, 1970; Lerman, 1971; Khomskis & Filatova, 1972; Blanton, 1973; Tzur, 1973). I have selected the simplest computations based on sinusoidal temperature approximations not only because of the availability of the corresponding curves but, as Lerman & Stiller (1969) compared, also for low sensitivity to small temperature changes and applicability to all depths. Theoretically the uniformity of D_z with depth is far from justified (Darbyshire & Colclough, 1972; Sundaram & Rehm, 1973) and the consequences of this assumption were paraphrased by Idso & Cole (1973).

For the period of maximum temperatures the vertical eddy diffusion coefficient is calculated from equation 3.31 in fig. 3.18 according to Lerman & Stiller (1969). By this method the authors obtained the highest values of any method. As evident indirectly from reasoning by Khomskis & Filatova (1972) this is at least partly due to assuming an unrealistic rectangular shape of the lake basin. As a matter of simplification a stepwise switch from surface to bottom temperature is assumed to occur at a particular depth, z_{mix}. In fact the changes are stepwise and their description by the error function (Goodling & Arnold, 1972), rather than by a simple step function, would be more adequate. The depth z_{mix} is calculated here indirectly from Ekman's theory of the effect of geostrophic winds (equations 3.19 and 3.32) for the depth of frictional resistance, D. This depth is related to z_{mix} by a constant factor discussed in section 3.4.2. The use of a constant factor for different conditions of the heat and wind energy input does not seem to be justified, but more detailed data are not available. The values of A_z for the 1-km^2 conical lake were obtained from equation 3.41 below, and temperature-dependent densities from an approximating polynomial.

The resulting mutually dependent equations 3.31 and 3.32 in fig. 3.18 were solved by an iterative computer program, starting from an arbitrarily selected value $z_{mix} = 5$ m. At low latitudes the convergence of the method was poor, but progressively higher accuracy was obtained at increasing latitudes.

Results of simulations for the idealized lake are evident from fig. 3.18, upper panel. The mixing depth increases rapidly at $\phi' < 20°$ and particularly below 5°, down to the bottom of the simulated lake (30 m). This is due to much increased wind stress in the denominator of equation 3.32, which is not compensated for by the decreased D_z connected with lower temperature

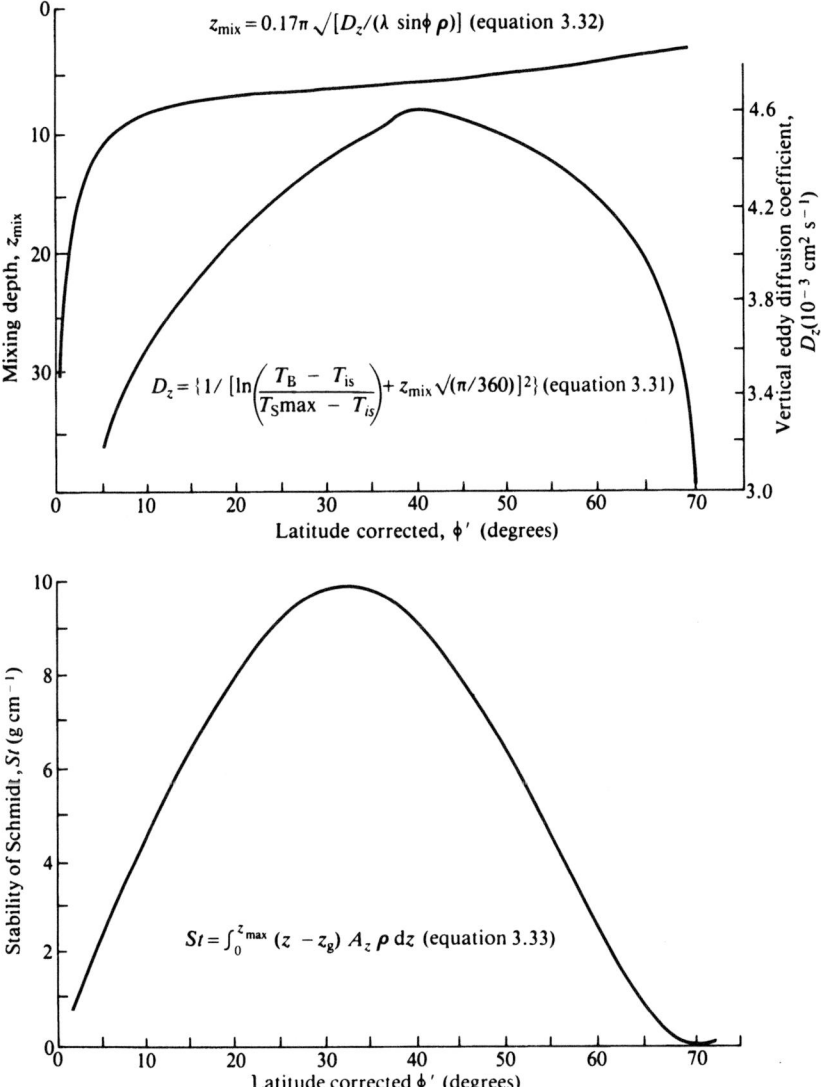

Fig. 3.18. Calculated latitudinal variations of mixing depth, z_{mix}, coefficient of vertical turbulent diffusion, D_z, and physical stability, St, as computed from equations 3.31–3.33.

gradients at higher tropical temperatures. Above $\phi' = 25°$ there is an almost linear increase in summer mixing depth with latitude. The eddy diffusion coefficient increases from the equator to latitudes of $40°$ and then drops again, but the absolute values range only from 3.9×10^{-3} to 4.6×10^{-3} cm^2 s^{-1}.

Simulations under such a number of simplifying assumptions, anyone of which can seriously bias the values obtained, can be doubted as to their applicability. Nevertheless, the geographical changes of z_{mix} obtained seems to be in agreement with direct observations of much higher mixing depths in the tropics (Lewis, 1973). For D_z the comparison is much more difficult because of the lack of adequate data. For values from the temperate region see section 3.4.2.

3.3.2.2. Physical stability

Using Schmidt's original formula (see Hutchinson, 1957) equation 3.33, we have computed the corresponding values of St for our idealized medium-size lake during the period of maximum temperatures. The integral was computed from 0.5 m intervals by means of the Simpson numerical integration method.

The gravity centre of the lake during holomictic conditions, z_g, is a function of the lake basin shape only, but the gravity centre of the lake for the period of maximum stratification also depends on temperature-induced density differences. St is a measure of work needed to shift the lake from maximum stratification to the holomictic state, and is measured in dyne-centimetre per square centimetre of the lake surface (reduces to g cm^{-1}). Idso (1973) replaced Schmidt's formula with a physically and computationally more correct one, but the integral values are identical with equation 3.33.

The lower panel of fig. 3.18 suggests that under similar morphometric conditions the tropical and northern lakes would have identical physical stability. A distinct maximum is obtained at about 30 degrees of corrected latitude. Hence the combined action of the higher temperatures but lower gradients in the tropics is compensated for by the differences in mixing depth. Comparing the upper and lower panels of fig. 3.18 suggests a close relation of D_z and St as shown for a few lakes by Khomskis & Filatova (1972), Darbyshire & Colclough (1972), Blanton (1973) and others.

The depth-dependence of St on lake size is well known from the classical investigations by Ruttner (1931) in Java. Smaller lakes have much greater stability. However, the relative latitudinal dependence would not be much changed if we used sizes and lake shapes other than those selected. More detailed simulation has to await corresponding temperature data.

It is very difficult to compare our figures directly with published empirical data, as no stability computations have been made during IBP investigations. However, an order of magnitude comparison will be made with the only lake of the corresponding size category for which data have been published.

Ruttner (1931) gives for Ranu Lamongan ($A_L = 44$ ha, $z_{max} = 28$ m, $\phi' = 11°$) maximum stability equivalent to 4 g cm^{-1}. This is in good agreement with our figure of 4.7 g cm^{-1} for the corresponding latitude.

3.3.3. Theoretical geographical productivity comparisons

Geographical differences of freshwater production have been of interest to limnologists since the time of Ruttner's Java studies and Worthington's studies of the African Rift Valley lakes. Speculative opinions as to the importance of different causative factors were expressed, and a few attempts to use theoretical simulation models remained at a preliminary level (Patten, 1966; Cole, 1967). Patten showed that with increasing latitude seasonal cycles in plankton become more pronounced as a result of the increasing amplitudes of the solar radiation and temperature inputs, and not as a result of the simplification of food webs postulated by the ecological stability theory of MacArthur (1955).

Empirical quantification of the geographical differences has been made possible by the extensive IBP studies, but opinions differ strongly as to whether the radiation input (Brylinski & Mann, 1973, and Brylinsky Chapter 9) or nutrient load (Schindler & Fee, 1974, 1975) is the dominant factor determining productivity. Recently Idso & Foster (1975) used three different photosynthetic models to calculate the effect of radiation on daily and annual theoretical integral photosynthetic capacity for selected dates and latitudes.

Let us use a theoretical phytoplankton model developed previously (Steel, 1978; Straškraba, 1976a) to determine how far geographical differences of energy-related resources affect integral ($=$ areal, column) photosynthetic capacity of phytoplankton in a theoretical lake of the same size situated at different latitudes.

The equations of the photosynthetic capacity model used for simulating geographical differences driven by physical variables are as follows:

$$P(T)\,\text{gross} = 1.2/\varepsilon \times P_{max}(T) \times \Delta \times f(E) \tag{3.34a}$$

$$
\begin{aligned}
f(E) = {}& 1.333 \tan^{-1}[E'_{O\,max}/(2I_K)] - I_K/E'_{O\,max} \\
& \times \ln[1 + (E'_{O\,max}/(2I_K))^2] - \{1.333 \tan^{-1}[E'_{O\,max}\exp(-\varepsilon z_{mix}) \\
& /(2I_K)] - I_K/(E'_{O\,max}\exp(-\varepsilon z_{mix})) \times \ln[1 + (E'_{O\,max}\exp(-\varepsilon z_{mix}) \\
& /(2I_K))^2]\}
\end{aligned}
\tag{3.34b}
$$

$$P_{max}(T) = 3.1\exp(0.09\,T) \tag{3.34c}$$

$$I_K = 0.8\,P_{max}(T)/b'' \tag{3.35}$$

$$b'' = 75.4$$

The model used is given in equations 3.34–3.35. In its present form it is intended for phytoplankton of a constant composition and photosynthetic response spectra uniformly distributed over depth. The photosynthesis–light curve and analytical daily integral assuming sinusoidal pattern of incident radiation and exponential drop of E'_O with depth is from Steel (1978). The exponential increase of maximum photosynthesis with temperature is recalculated from Steel's oxygen-based determinations to chlorophyll assuming its constant percentage. This shape of temperature dependence is in agreement with recent summarizations of the temperature effect on algal growth by Eppley (1972), Goldman & Carpenter (1974) and Canale & Vogel (1974).

The most important modification is in treating the parameter I_K as an *analytical* function of temperature. The function is derived for the phytoplankton photosynthetic model by Vollenweider–Steel in a way similar to equation 38 of Straškraba (1976a) and given here as equation 3.35. The confusing notion of P_{max} is used here in a more specified version $P_{max}(T, S, b)$, i.e. maximum photosynthetic capacity (at optimum illumination) at the given temperature, T, given limiting nutrient status, S, and given biomass, b.

Now this model, however inadequate, is used to evaluate the simultaneous effect of the primary geographically-dependent driving variables $E'_{O\,max}$, Δ, T and z_{mix} as they are described by equations 3.1–3.17, 3.18, 3.29 and 3.31–3.32 respectively. The radiation is computed for half-covered sky. In addition underwater radiation is computed according to quantification similar to that treated in Chapter 10, including the effect of snow and ice. The mixing depth varies seasonally assuming tentatively a jump to and from holomictic conditions at the surface-bottom temperature difference predescribed for each latitude. The extinction coefficient of water ε is assumed constant for all latitudes. It may be argued on the basis of sections above that on average ε will vary geographically and seasonally, due to hydrologically-based variations of organic content and suspended matter. However, clear and humic lakes do exist both in the tropics and high in the north.

Fig. 3.19 shows computations of the driving variables as well as daily integral gross photosynthetic capacity $P(T)$ (i.e. assuming that S is unlimiting and b low) for selected latitudes. The lake is again our idealized lake ($A_L =$ 100 ha, conical shape, $z_{max} = 30$ m), situated at a low altitude. An extinction coefficient of 0.4 equivalent to a Secchi disc reading of about 5 m as considered a reasonable average for IBP lakes is used here. Of the driving variables the least reliable is the simulation of the annual changes of z_{mix} which jump about in comparison with the smoother transition in nature. In the autumn the thermocline descends rapidly because of advection due to cooling. During summer, too, the thermocline does not remain constant but varies, usually descending towards the fall.

The sinusoidal character of the energy inputs is distorted at higher latitudes by the formation of ice and snow. Photosynthetic capacity retains the shape of

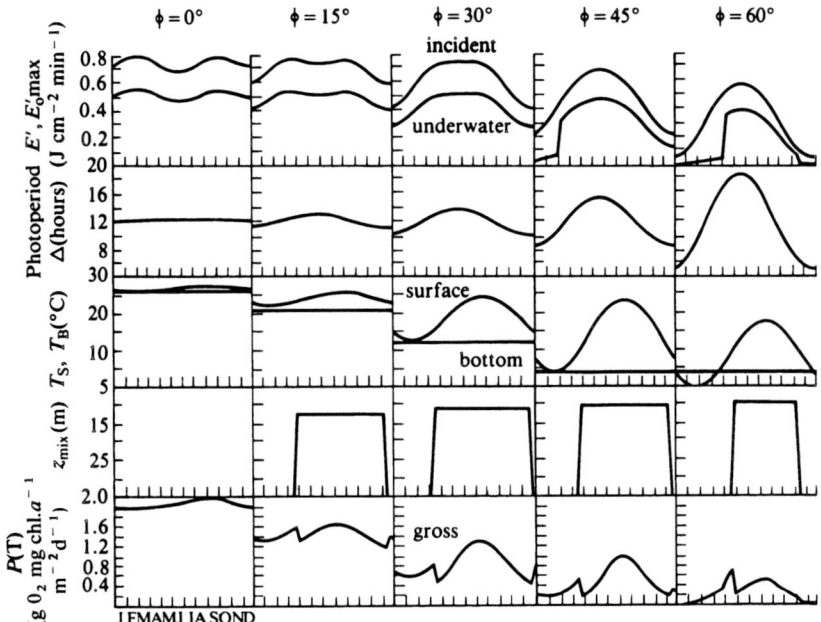

Fig. 3.19. Annual variations of daily maximum photosynthetically active incident solar radiation reaching water surface E'_{max}, and underwater $E'_{o\,max}$, surface and bottom temperature, T_S and T_B, mixing depth z_{mix}, and gross areal photosynthetic capacity $P(T)$ computed for selected latitudes, for conditions specified in the text.

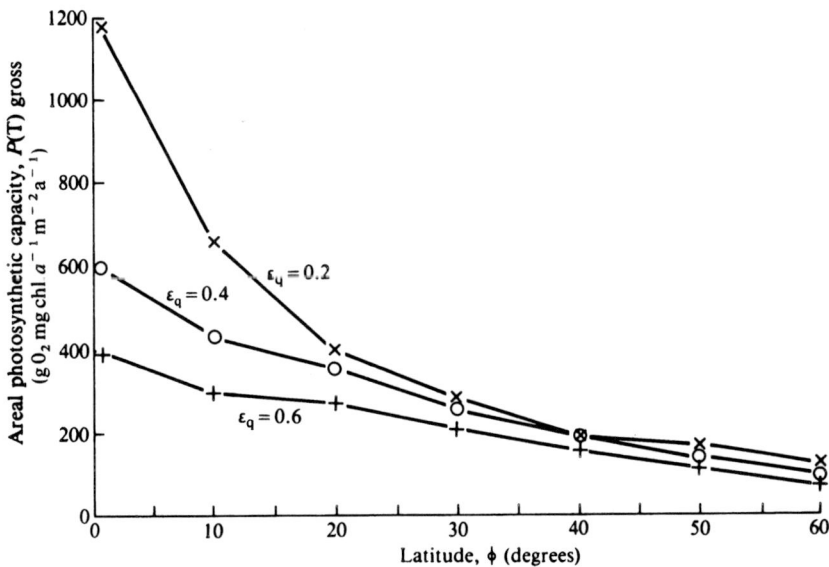

Fig. 3.20. Annual integrals of areal photosynthetic capacity $P(T)$ at different latitudes calculated for different values of the abiotic vertical extinction coefficient ε_q.

the energy inputs, except for being modified by the differences in mixing depth. We can recognize how P per growing season will be related to latitude. High values are retained at the equator and 10° for a whole year (= growing season). At 20° the maxima are higher, but because they are only maintained for three months the per-season value is far lower. It can also be recognized that the annual maximum photosynthetic capacity remains higher at low latitudes despite the fact that daylength range decreases towards the equator.

The same kind of computation was also performed to obtain the annual integral values of photosynthetic capacity, this time for different values of ε_q (fig. 3.20). The decreased transparency (increased ε_q) reflects a more profound effect on $P(T)$ at low than at high latitudes. Whereas the upper curve corresponding to fairly transparent lakes shows a very steep drop of $P(T)$ with latitude, this is not so marked for higher extinction coefficients. Evidently due to only one additional variable, the water transparency, a high scatter of values for a given latitude is produced. If cloudiness is varied, as actually observed in different localities, the scatter will increase.

3.4. Model series of freshwater lakes of varying depth and nutrient load

Until several years ago fishery problems were the only driving force and source of money for the application of theoretical productivity studies in fresh water. To feed the world still seemed to be the main objective at the beginning of the IBP period. Then the loading of the environment by the products of man's activity increased, and water-concerned management problems became centred around getting sufficient amounts of clean water for man's direct use, and preserving the surroundings for his survival. Eutrophication became a word used almost as often by laymen as by limnologists, but this at least gave new impetus to studies concerning the variables most suspected of producing local differences between lakes: depth and nutrients. It became urgent to understand the operation of these variables quantitatively, both on an empirical and on a theoretical basis.

The approach used is similar to that used above, i.e. breaking down the problem into understandable bits: physical characteristics of lakes of different depth and size followed by a theoretical–empirical treatment of the effect of these variables on the processes of phytoplankton production. The inter-related effects of nutrients inevitably have to be included.

Quantification of lake morphometry is exemplified by the detailed treatment of the ratio mean depth to maximum depth, which is shown to be a parameter useful for characterizing lake basin shapes. Morphometric determination of physical variables like mixing, turbulent diffusion, temperatures and heat budgets is then treated in later sections.

A number of empirical studies, particularly those primarily concerned with fisheries, have shown that depth, size or morphometry in general, are important site variables, determining productivity in several ways: Deevey (1940), Rounsefell (1946), Rawson (1952, 1953a, 1955, 1961), Carlander (1955), Nyggard (1955), Northcote & Larkin (1956), Ohle (1956), Hayes (1957), Larkin & Northcote (1958), Hayes & Anthony (1964), Ryder (1965), Sakamoto (1966), Findenegg (1967), Jenkins & Morais (1971), Henderson, Ryder & Kudhongania (1973), Kerekes (1975). Selected relationships will be dealt with in section 3.5.1.

Several attempts to analyse the depth–productivity relationship on a theoretical basis have been made: Gorham (1958), Murphy (1962), Larkin (1964), Sakamoto (1966), Vollenweider (1969a), Ahlgren (1970), Straškraba (1972), Steel (1975), Lorenzen & Mitchell (1973). Although the body of theory accumulates, all the interacting phenomena are not sufficiently understood and a more profound analysis is needed (see section 3.5.2).

The situation is similar concerning the effect of nutrients on production, as well as the effect of depth on nutrient loads and trophic degree. Recent empirical observations followed the classical limnological trophic classification approach of Naumann and Thienemann, while struggling with mixed-up, insufficiently specified concepts. Two lines of quantifying the nutrient–productivity relationships will be followed here:

(1) The nutrient budget concept, based on the idea that lakes change between the three qualitatively vaguely defined trophic categories at constant critical concentrations of limiting macronutrients, P and N.

(2) The nutrient–phytoplankton relationship concept.

3.4.1. Mathematical formulations of lake basin morphometry

Hutchinson (1957) listed 76 types of lake origin, most of which are hydrologically distinct, i.e. have characteristic drainage area/lake area relationships and theoretical replacement times as well as basin morphometry. On one hand, a glance at a map of regions like the Baltic lowlands or the Canadian Shield shows how imaginative nature is in shaping lakes. Geomorphological forces seem to operate, on the other hand, in unifying lake morphometry to some degree (Livingstone, 1954; Mason, 1967). In connection with productivity studies it became of interest to limnologists to look for average lake shapes and mutual interrelations between different hydrological and morphometric parameters. Of the important hydrological characteristics of lakes we have selected the mean and maximum depth as an example of the degree of interrelations.

Based on data assembled by Hutchinson (1957), Neumann (1959) analysed the $\bar{z} : z_{max}$ relationship for 107 lakes with a maximum depth of up to 614 m from all over the world. He found a mean ratio of 0.467 and concluded that the average lake shape is that of an elliptical sinusoid. Most of the lakes varied between 0.33 and 0.67, i.e. they have shapes approximated by elliptical cones ($\bar{z}/z_{max} = 0.33$), elliptical paraboloids (0.5) and ellipsoids (0.67). The adjective elliptic relates to the elongated shape of lake surface. Variations of the $\bar{z}/z_{max}$ ratio found by different authors for particular groups of lakes are summarized in table 3.2 and fig. 3.21. In addition to published correlations we have correlated morphometric parameters of the 231 Japanese lakes tabulated by Horie (1962) as well as the 39 New Zealand lakes by Irwin (1972). Evidently different mean ratios do exist for different locations in spite of large variability, ranging from the near-conical Baltic lowland lakes studied by Ventz (1973a) to the near-paraboloidal lakes of Japan. The ratio (0.455) derived from data on morphometric control of heat budgets of world lakes by Gorham (1964), is almost identical to the mean world value of Neumann. This is not surprising because almost all the same lakes were included in both treatments. The mean value is also similar for 39 alpine and highland mid-European lakes (0.46) (Ventz, 1973a). Comparisons by Gorham (1958) of Scottish lochs suggest a difference in lake shape in different geological formations as well as differences related to the overall size of the water body. The relation of basin shape to size is evident from Hayes (1957) who correlated morphometric

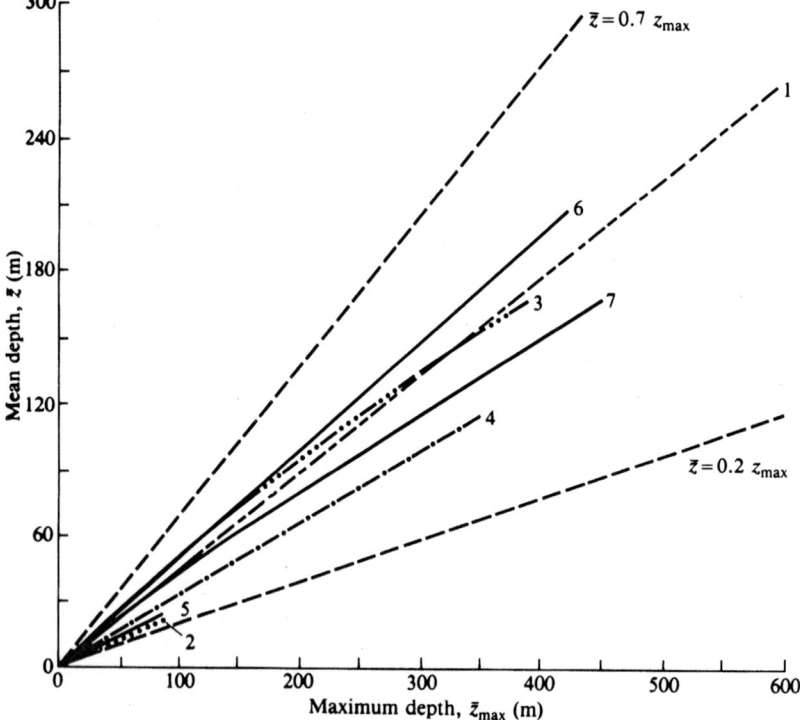

Fig. 3.21. Relationship of mean depth to maximum depth for different sets of localities. The numbers correspond to curve numbers in table 3.2.

Table 3.2 *The relationships of mean depth to maximum depth in lakes of different geographical regions*

Authors	Region	Regression equation $\bar{z} =$	Maximum lake depth included (m)	Number of lakes	Curve in fig. 3.21
Neumann (1959)	World	$0.467\ z_{max}$	614	107	1
Gorham (1958)	Scotland rock basins	$0.46\ z_{max}^{0.95}$	350	262	4
Gorham (1958)	Scotland drift basins	$0.75\ z_{max}^{0.80}$	80	137	5
Gorham (1964)[a]	World	$0.455\ z_{max}$	1741	71	—
Ventz (1973a)	N. Europe Baltic lowland	$0.65\ z_{max}^{0.82}$	83	85	2
Ventz (1973a)	Middle Europe 'Mittelgeb.-Alpen'	$0.617\ z_{max}^{0.95}$	370	39	3
Horie (1962)[a]	Japan	$0.420\ z_{max}^{1.03}$	425	231	6
Irwin (1972)[a]	New Zealand	$0.588\ z_{max}^{0.93}$	444	39	7

[a]Calculations done by author.

material for over 500 lakes in the United States, unfortunately only for z_{max} and A_L. Also the parabolic relationships of Ventz (1973a) suggest that lakes with lower maximum depths tend to have more U-shaped basins (higher $\bar{z}/z_{max}$ ratio) than do deeper lakes, and the same is evident for data from Gorham (1958). In fig. 3.22 this is represented by slightly convex relationships. The reverse is observed for Japanese lakes of rather variable origin as well as for the Canadian Shield lakes more uniform in origin studied by Koshinsky (1970). The lakes are reported to be progressively more U-shaped from small to larger lakes. Sorokin (1968) found systematic morphometric differences between groups of small lakes in north-western USSR.

From the example of $\bar{z}/z_{max}$ it becomes evident how far the morphometric parameters of lakes are interrelated. However, the relationships are of a statistical character only. Due to a combination of incompletely understood geographical, geological and local forces, any lake can differ considerably from the statistical relations derived from local groups of lakes. It is simply not possible to transfer experience between regions.

To understand the effect of morphometry on productivity, two approaches to a mathematical formulation of lake basin morphometry seem possible: (1) selection of an idealized 'average' series of lakes of gradually increasing size and depth, (2) a sufficiently generalized mathematical description of lake basin shapes which allows representation of any shape desired by changing as few parameters as possible.

As to the *idealized average lake series*, we can select the average basin shape as represented by the elliptic extreme half-sinusoid, whose surface area will be an ellipse with the semi-axes = L'/W ratio $3:2$. Its area A_L and maximum length L' will be given by:

$$A_L = \pi(L'/3)(L'/2) \tag{3.36}$$

$$L' = \sqrt{6A_L/\pi} \tag{3.37}$$

The maximum and mean depth is also related to L', and from data treated by Gorham (1958) we have obtained, respectively, the following approximations:

$$z_{max} = 26.6\sqrt{L'} \tag{3.38}$$

$$\bar{z} = 12.1\sqrt{L'} \tag{3.39}$$

It is interesting to note the great similarity between equation 3.38 and an approximation of the 'maximum mixing depth' in the alpine meromictic lakes which were treated by Berger (1955). This supports the statement by Mason (1967) that geomorphological forces would tend to keep this average lake shape. Nevertheless, it is clear that although the 'average lake series' is based not only on statistical treatment, but also on actual operation of external forces, it does not exist in reality. Very few lakes will approach particular points in the series. For instance, for the Japanese lakes mentioned above we were unable to find any statistical relationship between A_L and $\bar{z}$, nor between A_L and L'/W.

A generalized mathematical representation of lake basin shapes suitable for our purpose is that of Younge (in Hrbáček, 1966). He showed in a comparative geometrical treatment, that the ratio $\bar{z}/z_{max}$ can be used as a general parameter of lake basin shapes. By using this parameter the volumes, $V_{z'}$ and areas, $A_{z'}$ at particular depths for any regular lake shape can be calculated from equations 3.40 and 3.41 respectively:

$$V_{z'} = [6z' - 3(1-p)z'^2 - 2pz'^3]/(3+p) \tag{3.40}$$

$$A_{z'} = pz'^2 + (1-p)z' \tag{3.41}$$

where $z' = z/z_{max}$, the normalized depth
and $p = 6(\bar{z}/z_{max}) - 3$

The parameter p equals -1 for cones, $p=0$ for paraboloids, $1 \geqslant p \geqslant 0$ for hyperboloids, $p=1$ for the extreme ellipsoid, but any other shape is given by the $\bar{z}/z_{max}$ ratio.

3.4.2. Turbulent diffusion and mixing depth

The difficulties of measuring *vertical turbulent diffusion coefficients* were discussed in section 3.3.2.1, different methods resulting in different absolute values. Nevertheless, using both temperature and chemical data Mortimer (1942) suggested a close relationship of the coefficients of vertical eddy diffusion in the hypolimnion, D_{zH}, of eleven temperate lakes to their size (area, maximum depth and mean depth). Recently Blanton (1973) suggested a simple method of determining 'entrainment of heat from hypolimnion to epilimnion', *En*, from the increase in volume of epilimnion during the period of stabilized summer lake temperatures. He found a very close power curve relationship of *En* to $\bar{z}$ for fifteen northern temperate lakes ranging from $3.2 \leqslant \bar{z} \leqslant 740$. The numerical values of *En* are not exactly identical to the vertical exchange coefficient in the region of z_{mix}, $D_{z\,mix}$. In our opinion this is because *En* is actually an integral value whereas $D_{z\,mix}$ is a point value. Snodgrass & O'Melia (1975) supposed that Blanton's *En* refers to a region just above thermocline. The authors estimated $D_{z\,mix}$ from literature data for fourteen temperate lakes and Lake Tiberias, correlated them successfully with $\bar{z}(r = 0.924)$, and compared the results to previously mentioned relations.

Fig. 3.22 (equations 3.42–3.44) compares the relations of D_{zH}, $D_{z\,mix}$ and *En* to the mean depth of lakes. As seen, the empirical observations suggested a continuous increase of both thermocline and hypolimnic vertical diffusion coefficients with the depth of the lake. However, Khomskis & Filatova (1972) showed for about twenty lakes very close non-linear relationships of D_z at the

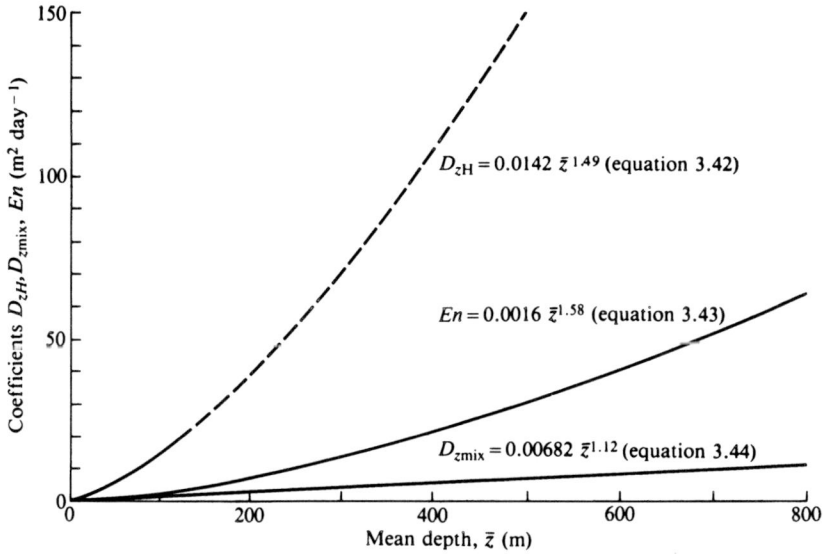

Fig. 3.22. Relations of the coefficients of vertical turbulent diffusion to mean depth of lakes. D_{zH} – coefficient of vertical turbulent diffusion in the hypolimnia of lakes as obtained by Mortimer (1942). The size of lakes included extended only as far as full line indicates. *En* – entrainment of heat from hypolimnion into epilimnion by Blanton (1973). $D_{z\,mix}$ – coefficient of vertical turbulent diffusion in the mixing depth region (metalimnion) of lakes as obtained by Snodgrass & O'Melia (1975).

lower boundary of the metalimnion to a combined morphometric parameter N^* to be discussed below.

Mixing depth, z_{mix}, is considered here to be the depth at which the thermocline defined as $d^2 T/dz^2 = 0$ is located in July–August. Due to continuous tilting of thermoclines (from which En of Blanton is calculated) we should have compared only values from a fairly restricted period, say late July to early August, but this was not possible for the present material. In table 3.3 different empirical relations of z_{mix} to lake size were derived for local groups of northern temperate lakes. Z_{mix} seems to be related to lake size as expressed by the maximum length of lake (L'), to the effective length of lake axis ($L'' = 0.5$ (maximum length + maximum width)) or fetch size (L''', for computations see Smith & Sinclair, 1972). Original units are given in table 3.3 as well as the recomputation to L' based on an idealized mean lake area shape. All the corresponding approximations are shown in fig. 3.23, irrespective of differences between the three measures of L used. Each curve is drawn only for the observed extent of L. We are neglecting the different meanings of L (some judgement of this effect being possible on the basis of table 3.3), errors inherent to regression approximations and possible inclusion of values from slightly different periods of year. Nevertheless, there is remarkable similarity in the curves for both close and remote areas having similar climatic conditions (equations 3.45, 3.46, 3.47, see table 3.3, fig. 3.23). Conditions are also fairly similar in Japanese lakes slightly further south (equations 3.48, 3.49, see table 3.3, fig. 3.23). In geographical regions with higher wind speeds and at higher elevations the mixing depth tends to be higher as seen from equation

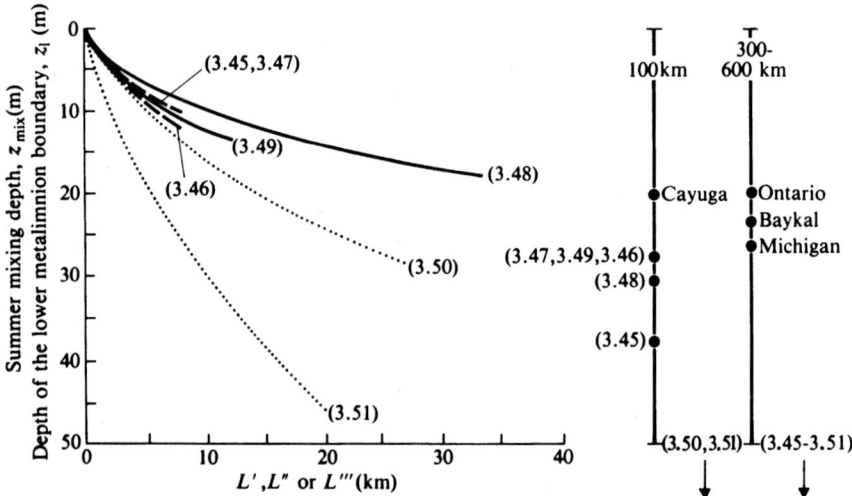

Fig. 3.23. Summary of the empirical relations of mixing depth and depth of the lower metalimnion boundary to the length of a given lake. The equations are specified in table 3.3.

Table 3.3 *Empirical relations of summer mixing depth, z_{mix}, and depth of the lower metalimnion boundary, z_1, to the maximum length of lake (L), effective length (L''), or fetch size (L'''), respectively, in temperate lakes. Values of L in km, z_{mix} in m*

Author or data source	Region	Equation number	Regression $z_{mix} =$	r	Extrapolation $L = 100$ km	Recomputation for L'(a)(d)	Max L (km)	Number of lakes
Patalas (1960a, 1961)	Poland Baltic lowland	3.45	$4.55\,L''^{0.455}(a)$	0.94	37.0	$3.79\,L'^{0.455}$	6.5	53
Arai (1964)	Japan	3.49	$6\,L''^{0.33}(e)$		27.4	$5.46\,L'^{0.33}$	12	32
D. Schlindler (personal communication)	Canadian Shield lakes	3.46	$4.98\,L'^{0.38}(a)$	0.85	28.7	$4.98\,L'^{0.38}$	7	67
Yoshimura (1936)	Japan (c)	3.48	$3.56\,L'^{0.46}(a)(b)$	0.69	29.6	$3.56\,L'^{0.46}$	36	19
Ventz (1973b)	E. Germany Baltic lowland	3.47	$4.72\,L'^{0.39}$		28.4	$4.04\,L'^{0.39}$	7.5	30
I. Smith (personal communication)	Scottish highlands	3.50	$4.66\,L'^{0.55}$	0.64	59.2	$4.66\,L'^{0.55}$	28	59
Nydegger (1957) (f)	Switzerland, subalpine	3.51	$z_1 = 8.31\,L'^{0.57}(b)$	0.99	115	$8.31\,L'^{0.57}$	20	11

(a) Own recalculations.
(b) L' recalculated from surface areas assuming elliptic surface with axis ratio 3 : 2.
(c) Only lakes from areas with mean wind speed between 2–3 m s^{-1}.
(d) Assuming elliptic surface with axis ratio 3 : 2, of which $L''' = 0.75\,L'$ (when wind is blowing along the major axis) and $L'' = 0.67\,L$.
(e) The author defined fetch as square root of surface area.
(f) See also Ambühl (1967).

3.50 in fig. 3.23. The values for subalpine lakes by Nydegger (1957) give the depth of the lower metalimnion boundary (equation 3.51).

However, if we extrapolate the empirical formulae to lakes with $L \approx 100$ km and to the largest lakes with $L \approx 300 - 600$ km, all give results far higher than those actually observed (Cayuga – Henson, Bradshaw & Chandler, 1961, $L' = 61$ km, $z_{mix} = 20$ m; Baykal – Khomskis & Filatova, 1972, $L' = 636$ km, $z_{mix} = 24$ m; Michigan – Van Oosten, 1960, $L' = 500$ km, $z_{mix} = 25$ m; Ontario – Lee, 1972, $L' = 300$ km, $z_{mix} = 20$ m). As is evident, z_{mix} tends to increase with lake size only up to about $L' = 25$ km, and then remains constant at about 20–25 m up to the largest temperate lakes. However, in temperate seas the so-called seasonal thermocline lies deeper.

Khomskis & Filatova (1972) suggested an empirical formula for the upper and lower metalimnion boundaries, z_u and z_l, respectively, which uses a combined morphometric parameter N^* (equations 3.52, 3.53 in fig. 3.24). As can be seen, the formula accounts excellently for the summer mixing depths in a variety of temperate lakes ranging from small lakes on the Karelian isthmus

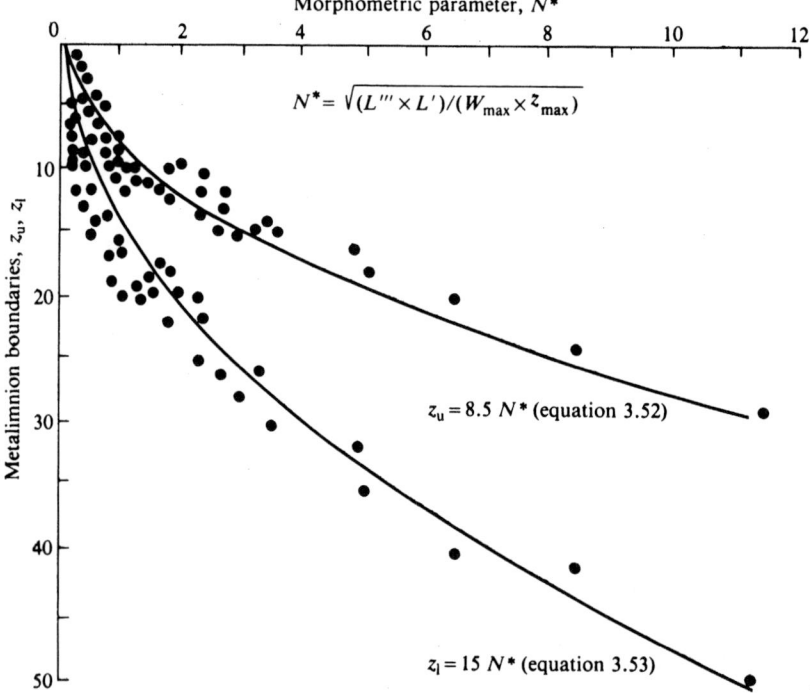

Fig. 3.24. Relation of the upper metalimnion boundary, z_u, and lower metalimnion boundary, z_l, to the morphometry of the lake as expressed by a combined morphometric parameter, N^*. After Khomskis & Filatova (1972).

up to lake Baykal.† The figure also demonstrates the metalimnion thickness increasing with increasing N^*. The parameter N^* would indicate that the depth of mixing is related not only to the direct effect of wind, as would seem from the above relations to L' and L''', but rather to the basin morphometry relative to L'. Deeper mixing would occur in lakes narrower and shallower in relation to their length. As given by equation 3.38 larger lakes are on average relatively much shallower and the observed correlations with L' (equations 3.45–3.51) would be only indirect. This would be a possible explanation of the much deeper thermoclines in mountain lakes which are usually longer and shallower. In addition, high flushing seems to increase thermocline depths as indicated by Ambühl (1967) for several subalpine lakes.

The empirical observations on turbulent diffusion and mixing depth summarized here imply that the effect of wind is not likely to be the only determining force, as this would lead to relationships with fetch only. Fetch is decisive, but it operates with other factors.

Theoretically, recent research indicates the operation of two forces: mechanical energy due to shearing stress by wind and heat energy (Kraus & Turner, 1967; Mortimer, 1974). Sundaram, Rehm, Rudinger & Merritt, (1970) and Sundaram & Rehm (1973) have shown the importance of the non-linear interaction between the wind-generated turbulence and the stable density gradients in water bodies for the formation of thermoclines. When considering the turbulent diffusion coefficient to be dependent on a gradient Richardson number, they obtained for Lake Cayuga an excellent agreement between predicted and observed seasonal changes of thermocline depth and temperature distribution. Other current theories are based on differential absorption of solar radiation at various depths (e.g. Dake & Harlemann, 1969), but computations based on this principle depend on entering empirical values of turbulent diffusion coefficients from 'similar' water bodies. Orlob & Selna (1968) obtained successful prediction for a few reservoirs and their procedure is becoming part of the large ecosystem lake models (Chen & Orlob, 1972). This approach is also currently used in engineering practice for reservoir temperature prediction.

The suitability of Sundaram & Rehm's (1973) theory for explaining the morphometric and geographical differences of turbulent diffusion and mixing depth has to be verified. For biological purposes the explicit inclusion of the mechanisms of differential radiation absorption seems to be wanted from the point of view of the effect of dense phytoplankton crops on absorption and its consequence for increasing physical stability, resulting in increased photosynthetic capacity and higher chlorophyll (Foster & Idso, 1975).

† Dr Filatova, Leningrad, kindly gave me a copy of her figure including the names of lakes, which shows two tropical lakes, Victoria and Nyasa, as the lakes with largest N^*, but fitting the relations. This seems to be in contradiction with our previous statements about deeper tropical thermoclines, but see e.g. Talling (1969, fig. 10) or Beadle (1974, figs. 6.8, 6.10 and 6.12).

3.4.3. Temperatures and heat budgets of morphometrically different lakes

Both surface and deep water temperatures of lakes of different sizes in similar climatic regions are different, as revealed by several empirical observations (Yoshimura, 1936; Findenegg, 1953; Nydegger, 1957; Gorham, 1958; Patalas, 1960a; Bogoslovskii, 1960; Arai, 1964; Straškraba, 1972). We do not consider the question in detail here, since the effect of such differences on phytoplankton production is low, as shown below. The generally higher temperatures of small shallow lakes and lower temperatures of large deep lakes were noted in section 3.3.1.

Heat budgets have been recognized to reflect the morphometric differences sensitively (Gorham, 1964). For seventy-one temperate lakes the regressions of annual heat budgets (kcal m^{-2}a^{-1}) with mean depth, area and volume were as follows:

$$\theta_a = -2.52 + 18.4 \log \bar{z} \quad \text{(m)} \qquad r = 0.77 \qquad (3.54)$$

$$= -28.3 + 14.8 \log \sqrt{A_L} \quad \text{(m}^2\text{)} \qquad r = 0.74 \qquad (3.55)$$

$$= -29.7 + 18.8 \log \sqrt[3]{V} \quad \text{(m}^3\text{)} \qquad r = 0.82 \qquad (3.56)$$

Arai (1964) obtained for Japanese lakes a relationship between morphometrically determined wind fetch and θ_a. The morphometric factors considered included both depth and lake basin shape. Other investigations include lakes in Colorado (Reed, 1970), Canada (Schindler, 1971a; Kerekes, 1974), USSR (Kirillova & Smirnova, 1972) and Australia (Timms, 1975).

From the preceding treatment of turbulent diffusion it would seem interesting to relate heat budget to the combined morphometric parameter N^* as well as to distinguish between the morphometric and solar radiation effects.

3.4.4. Nutrient concentrations and lake depth

One of the explanations of hyperbolic drops of production with depth (Vollenweider, 1969a) is based on decreasing nutrient concentrations with increasing depth of lake. Such a relationship was observed empirically by Sakamoto (1966) – see fig. 3.25.

Let us first consider an *unstratified* lake for which Vollenweider (1969a) developed a substance budget. Vollenweider assumed that the concentration is a steady state function of supply and losses due to outflow and sedimentation, proportional to the concentration. He verified the steady state ($t \to \infty$) solution of his differential equations for the total and orthophosphate phosphorus budget for eight subalpine lakes. It appeared that the retention of phosphorus by sedimentation is evidently a function of the load and outflow values, but by a factor of $\beta^* = 0.4$ higher than calculated. For calcium carbonate and mineral nitrogen there was no agreement, but data were scarce and less accurate.

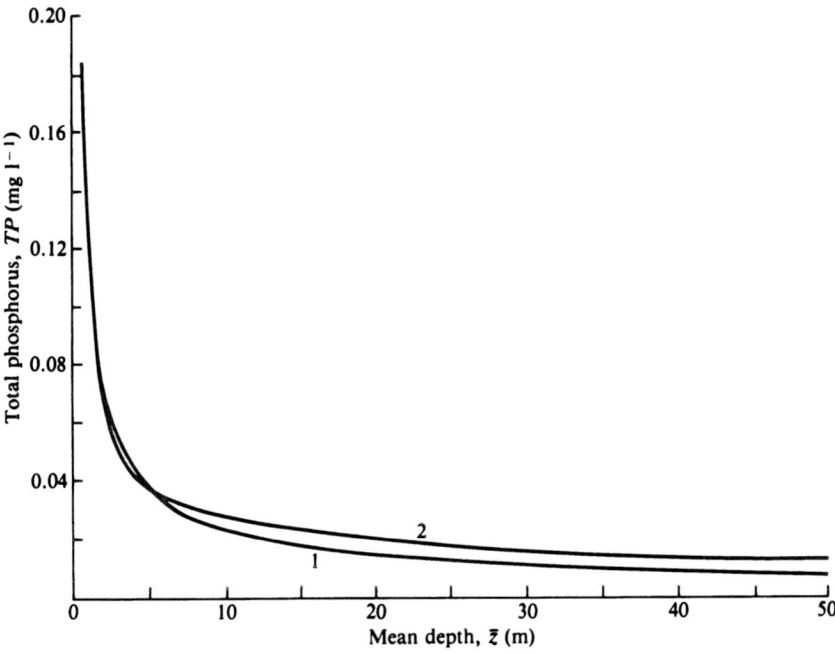

Fig. 3.25. Comparison of the relation of total phosphorus in lakes of different depth as observed by Sakamoto (1966) in Japanese lakes (equation 3.60), with the values computed according to the nutrient budget model by Vollenweider (1969a) (equation 3.57): $\beta^* = 0.4$, $\sigma = 0.1$, $R = Q_S/\bar{z}$, $Q_S = 5$, $L_p = $ constant. Curve 1 – Sakamoto, observed. Curve 2 – Vollenweider, calculated.

On these assumptions, Vollenweider obtained a relationship of the mean concentrations, $\bar{S}$, with load, L_p, and morphometric factors:

$$\bar{S} = L_p[1 - \beta^*(1 - R)]/[\bar{z}(\sigma + R)] \tag{3.57}$$

Equation 3.57, rearranged to give $\bar{S} \times L_p = f(\bar{z})$, is solved in fig. 3.25 for a theoretical retention time varying with mean depth of the lake, and other parameters as specified. The curve drops hyperbolically with depth, and the striking agreement with the tail of the empirical curve by Sakamoto is shown. However, the beginning of the calculated curve is very different. Unfortunately the assumption about sedimentation being proportional to substance concentration proved unrealistic (Snodgrass & O'Melia, 1975) and further refinement is necessary.

Recently Vollenweider's model is being gradually extended to cover the more complex conditions in *stratified* lakes (O'Melia, 1972; Imboden, 1973, 1974; Krambeck, 1974; Snodgrass & O'Melia, 1975; Straškraba, 1976b). When Vollenweider's assumption that sedimentation losses occur throughout the whole lake volume is substituted by a more realistic representation of losses to the lake bottom, the depth dependence of concentrations vanishes

(Imboden, 1973). The changes of particulate phosphorus in the epilimnion, F_E, of a stratified lake during stagnation period are represented by:

$$\frac{dF_E}{dt} = \frac{1}{V_E}[L_p + Q_{PH} - (Q_i + En)F_E] + S_E\frac{A_{SE}}{V_E} - \frac{g_E A_E F_E}{V_E} + \alpha F_E - r_E F_E \quad (3.58)$$

All rates are per day. A similar equation holds true for the hypolimnion, but there is no loading and direct inflow of water, sedimentation represents an input from the epilimnion and output to sediments, and there is no algal uptake. Simulation of the hypolimnic oxygen conditions affecting phosphorus solubility was attempted by Imboden. Concentrations of phosphorus rising in lakes with increasing lake depth resulted from the model, contrary to direct observations. The epilimnion depth and parameters g_E, α and r_E as well as the value of Q_{PH} and En were held constant and not dependent on size.

To override the discrepancy between observed concentrations decreasing with depth contrasting with the model output, Snodgrass & O'Melia (1975) stressed the dependence of En on depth of the lake via vertical turbulent diffusion (section 3.4.2). They accounted for flocculation of phosphorus particles increasing with depth. However, the most important process of

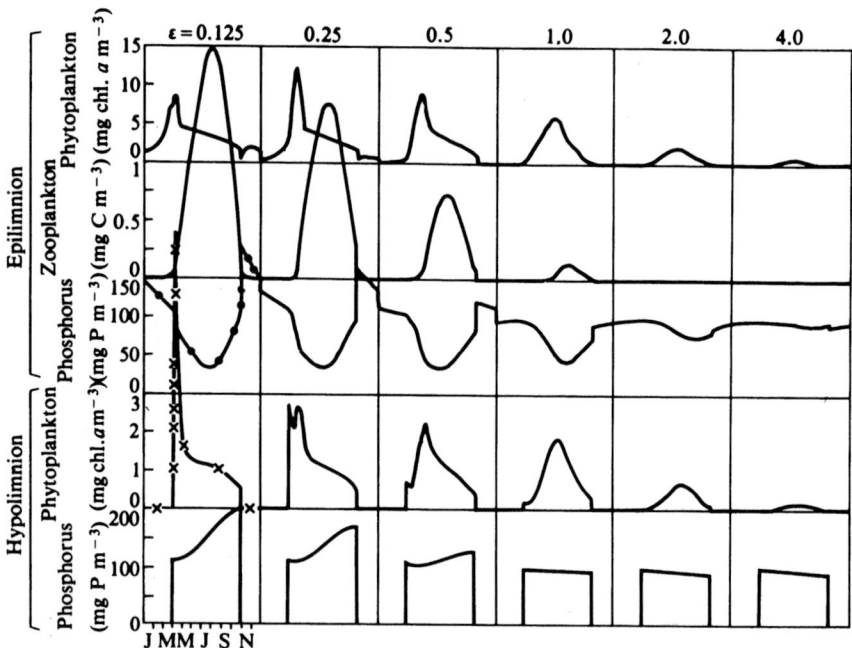

Fig. 3.26. Output of the theoretical two-layer model of phosphate–phosphorus, phytoplankton and zooplankton for a lake of maximum length 3 km, situated in mid-European climatic conditions. For cases 1–6 representing one year simulation each, the coefficient of vertical extinction ε was varied as given at the top of the figure. After Dvořáková (1976).

particle formation will be that of incorporation into algal cells and hence depth–production dependencies have to be taken into consideration.

Dvořáková (1976) and Straškraba (1976b) stressed the importance of phosphorus uptake by organisms in simulations with a theoretical two-layer model, coupling equation 3.58 with the phytoplankton–zooplankton model. Phytoplankton growth rates are limited by temperature, light and nutrients, losses are due to sedimentation, excretion, respiration and grazing by zooplankton. In one run of the model for a physically identical lake receiving the same phosphorus load, the vertical extinction coefficient ε was varied, as representing different amounts of refractory organic matter ('humic' substances) or different silting. All the assumptions of the model have not yet been sufficiently verified, nevertheless the preliminary results in fig. 3.26 stress the importance of differences in phytoplankton activity, caused by differences in light availability, for nutrient budgets. Phosphorus concentrations are rather different in spite of identical phosphorus load.

Recently Vollenweider (1976) extended the critical loading concept to incorporate quantitatively differences in retention times which he recognized theoretically in earlier papers. This is a topic which we have not covered here, although we have recognized its prime importance for reservoirs (Straškraba, 1972, 1973, 1976b).

3.4.5. Effect of depth and nutrients on phytoplankton production

Sakamoto (1966) was the first to distinguish empirically the most relevant trends of primary production and chlorophyll concentration in relation to both mean depth and nutrient concentrations. But more than that, he also carried out an advanced theoretical analysis of the mechanisms underlying the relations obtained.

As to depth effects, the most widely accepted recent empirical analysis is that by Vollenweider (1968, 1969a). It is summarized in the well-known three-dimensional plots of lake trophic degree versus phosphorus load and mean depth, suggesting that shallow lakes will reach meso- and eutrophy at considerably lower loads than deep lakes. Consequently, shallow lakes were called 'morphometrically eutrophic' in contrast with 'morphometrically oligotrophic' deep lakes. For a time much of the current eutrophication practice was based on Vollenweider's plots, and many investigations have added new lakes in agreement with the concept (e.g. Gächter & Furrer, 1972; Ahl, 1973; Welch, 1974). Kerekes (1974, 1975) demonstrated morphometric and light penetration differences of primary production in five lakes in an oligotrophic region of Newfoundland, Canada.

A few of the most appealing statistical relations of productivity phenomena with depth found for local groups of lakes are summarized in fig. 3.27. As

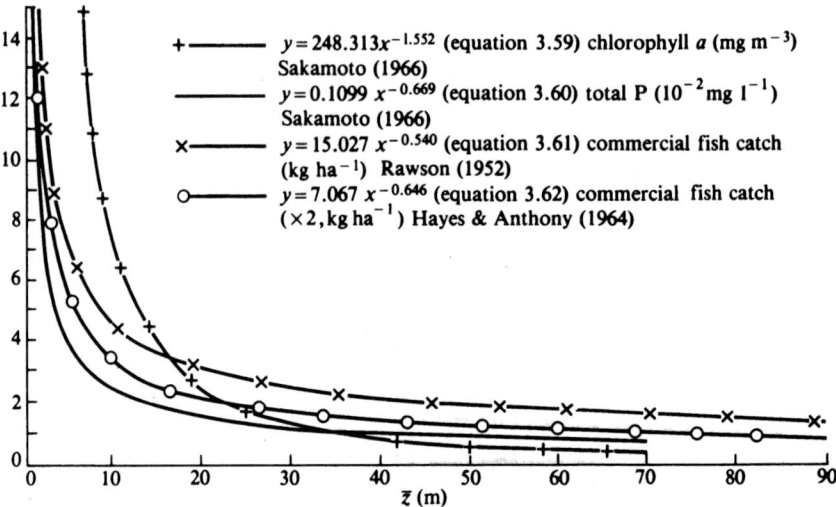

Fig. 3.27. Some empirical relations of productivity variables to mean depth of lakes, according to the authors indicated.

discussed in Chapter 8, the relations concerning commercial fish catch use 'per effort' data and hence are affected by the greater difficulty in catching fish from diluted populations in deeper lakes. Nevertheless, dilution is exactly the phenomenon to be taken into consideration for the phytoplankton–zooplankton–fish relationship.

The curve for the chlorophyll to depth relationship expressed by equation 3.59 is only a power curve approximation, remaining good for greater depths but not for very shallow localities. As suggested by Sakamoto the curve rises steeply from zero until the total photosynthetic profile is realized and only then approaches equation 3.59.

The nutrient–phytoplankton relationships are treated theoretically in papers based on the kinetics of algal growth, mathematically described by the microbial continuous culture theory of Monod: Dugdale (1967), Droop (1968), Eppley, Rogers & McCarthy (1969), Uhlmann, Benndorf & Albert (1971), Müller (1972), Caperon & Meyer (1972), Paasche (1973). The theory is based on single-species populations under experimental conditions and warnings have been expressed about transferring the approach to multispecies populations (Williams, 1973). The Monod growth limitation approach is included in the mathematical simulation ecosystem models (Di Toro, O'Connor & Thomann, 1971; Chen & Orlob, 1972; Park et al., 1974; Middlebrooks, Falkenborg & Maloney, 1974, Chapter 10). Most of the ecosystem models are intended for engineering application for eutrophication prediction in single water bodies, rather than aiming at theoretical understanding. More theoretically orientated models were those by O'Brien (1974) and Lehman, Botkin & Likens (1975). Although the kinetics of growth under nutrient limitation seems to be known, the dynamics of the multispecies systems and competition phenomena under varying natural conditions are less well-known.

Because the nutrient–productivity relationship has been treated empirically only very recently, we will concentrate on this topic in the next section. The theoretical analysis will be distinguished in two parts, dealing first with the question as to how the shape of the chlorophyll to lake depth relation is produced, and then how the spreading of the chlorophyll–phosphorus values originates.

3.4.5.1. Empirical analysis of the nutrient–productivity relationship

Limitation of primary productivity by macronutrients, microelements and promotive organic substances has been demonstrated under specific natural conditions. Although limnologists are aware of the need for a multi-nutrient limitation approach, both urgent environmental problems and strategic reasons for tackling the complex theoretical problems push towards concentrating on macronutrients. However, shifting to macronutrients does not diminish our problems, as shown by the carbon–phosphorus controversy which is just over, due mainly to the thorough data collection and analysis by Schindler and his associates (Schindler & Fee, 1975). It seems that in

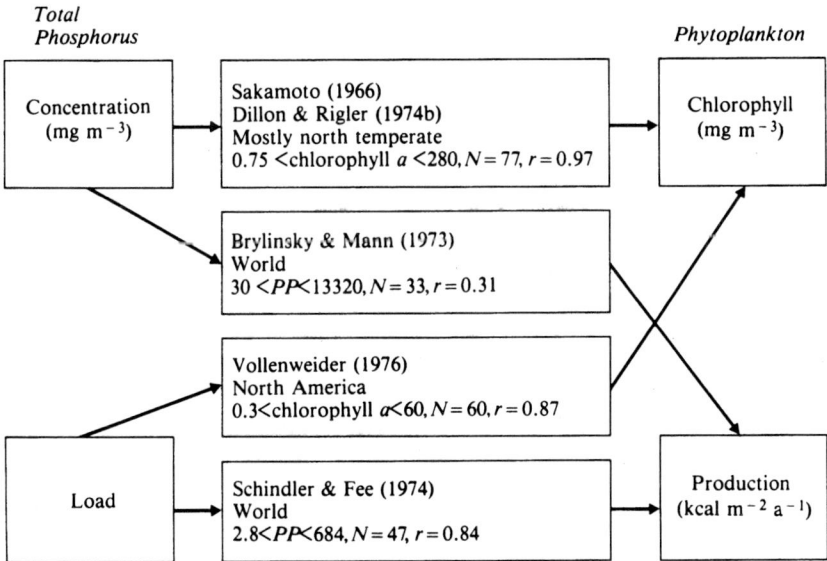

Fig. 3.28. The different empirical phytoplankton to phosphorus relations.

broadening our interest to cover tropical conditions we are coming to the early stages of a new nitrogen–phosphorus controversy (EUTROSYM '76, 1976).

The empirical relations between total phosphorus and phytoplankton can be based on their mass or on rates, for which the usual measures are phosphorus load, phosphorus concentration, chlorophyll *a* and primary production. As shown in fig. 3.28, all four possible straightforward phosphorus → phytoplankton relations have recently been investigated. The correlation coefficients, number of cases, range of measurement of variables, as well as the geographical ranges given in fig. 3.28, are very different for the four relationships. For these reasons it would be impossible to infer simply from the correlation coefficients that anyone of these relationships is superior to the others. Nevertheless, fig. 3.28 indicates that Dillon & Rigler's relationship with a geographically restricted range is more suitable for a more detailed analysis directed towards understanding the mechanisms involved. Because of the coverage of a broad geographical spectrum, based on the IBP data, and the superiority for management purposes of relationships based directly on loads, as well as the claim by Schindler & Fee (1974) and Schindler (1978) that the rate parameter relationship of load to production should be theoretically superior, we will look first at this relationship.

Results by Schindler (1978) are replotted here in fig. 3.29. Two versions of the same data are plotted: part (a) is the original log–log plot including the standard errors (SD); part (b) is a linear plot of the same curves. Although the log–log plot may seem promising (actually it shows that no straight line is obtained with this transformation, as well recognized by Schindler & Fee, 1974), the linear plot of the same data clearly demonstrates that nothing more than a feeble suggestion of a positive relationship between primary production and the phosphorus load is obtained. The highest *PP* value corresponds to a very low *TP* load, whereas at highest loads medium *PP* were found. The relatively high correlation coefficient (when leaving out the statistical problem of the selected regression model not corresponding to data) suggests that the relationship is a real phenomenon in spite of the wide geographical coverage and in other respects rather broad spectrum of conditions.

Perhaps when geographically more restricted sets of data are used, the confidence intervals will be narrower. But at present it would be of little use to predict for a given load primary production covering more than half of the observed total span.

Dillon & Rigler's (1974b) revival of Sakamoto's (1966) relationship is particularly interesting from the heuristic point of view – the concept being originally combined with theoretical analysis. The pragmatic point is the applicability of the empirical equation derived by Sakamoto (1966) for thirty-three Japanese lakes and ponds (averages for May, June and September) to another set of forty-six temperate, mainly North American, lakes (with one Middle American and one arctic one), data mostly as spring maximum values

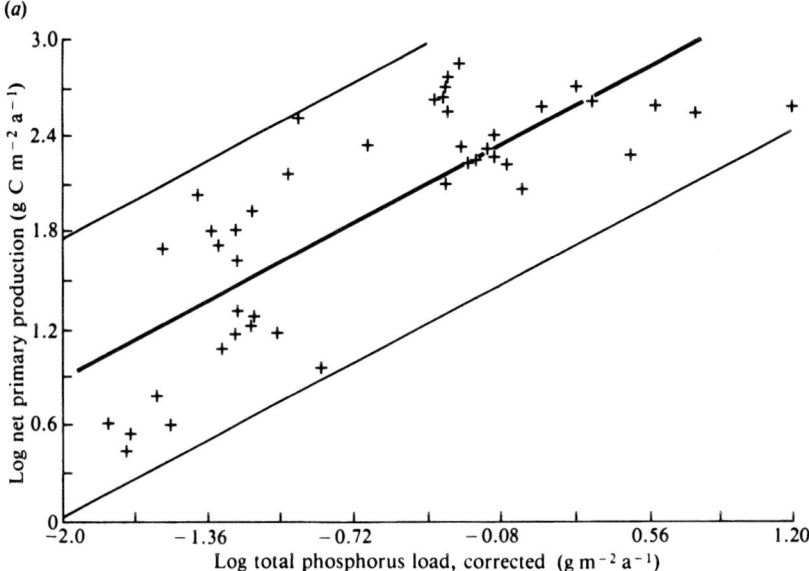

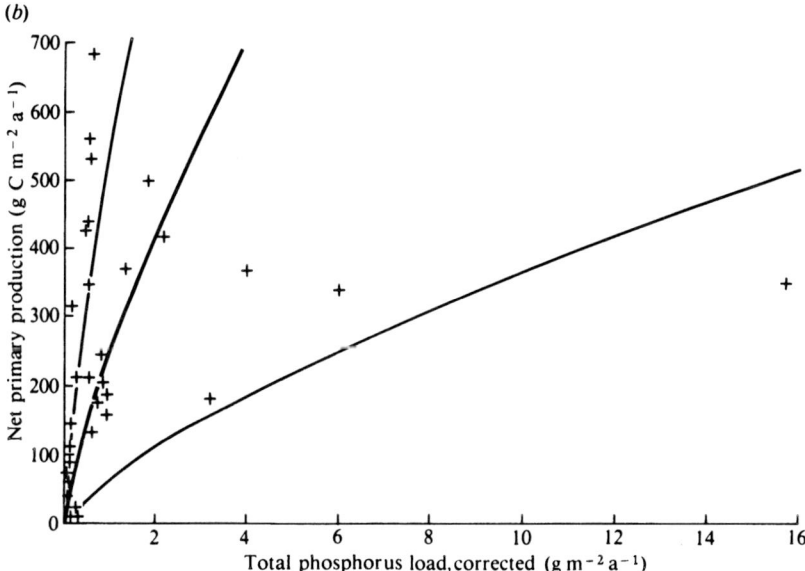

Fig. 3.29. Relationship of the annual phytoplankton production (g C m^{-2} a^{-1}) of the IBP lakes to the annual total phosphorus load corrected for retention times according to Schindler & Fee (1974). (*a*) Log–log scale from Schindler (1978); the lines enclose all data. (*b*) Arithmetic scale, with the two boundary lines corresponding to the 95% confidence limits as derived for the log–log regression. (Based on data kindly supplied by D. Schindler.)

for *TP* and summer averages of chlorophyll *a* (Dillon & Rigler, 1974b), as well
as for seventeen additional lakes from the same continent (Dillon & Rigler,
1975). Our data for Czechoslovakian reservoirs, summer averages, are also in
fair agreement. Not to be unfair to Schindler & Fee's relationship, we have also
to show a linear plot (fig. 3.30) because log–log plots always give a good
impression. Only data tabulated by Dillon & Rigler are included, with the
value for the Swedish not very deep lake Norrviken corrected to be consistent
with the other measurements. In the log–log plot there is no curvature similar
to that in fig. 3.29. The spread of chlorophyll *a* values increases with *TP* and as
a result of the power curve regression model the confidence limits become
progressively larger. An extrapolation to high values of *TP* would give
misleading results. A superficial theoretical consideration would suggest that
the power curve $y = ax^{b''}$ with the accelerating, upwards curvature $(b'' > 1)$ can
be at best a section of a higher order curve. It is well known that some upper
limit of chlorophyll *a* exists, although the theoretical estimates of its exact
value are continually updated (Steele & Menzel, 1961; Steemann Nielsen,
1962b; Talling, Wood, Prosser & Baxter, 1973).

When we correlate partial sets of data with different ranges of *TP*, the slope
b'' reverses from highly positive through zero to negative values (Straškraba,
1976b). As a result we concluded that the power curve model is inadequate and
some kind of asymptotic curve would correspond more accurately.

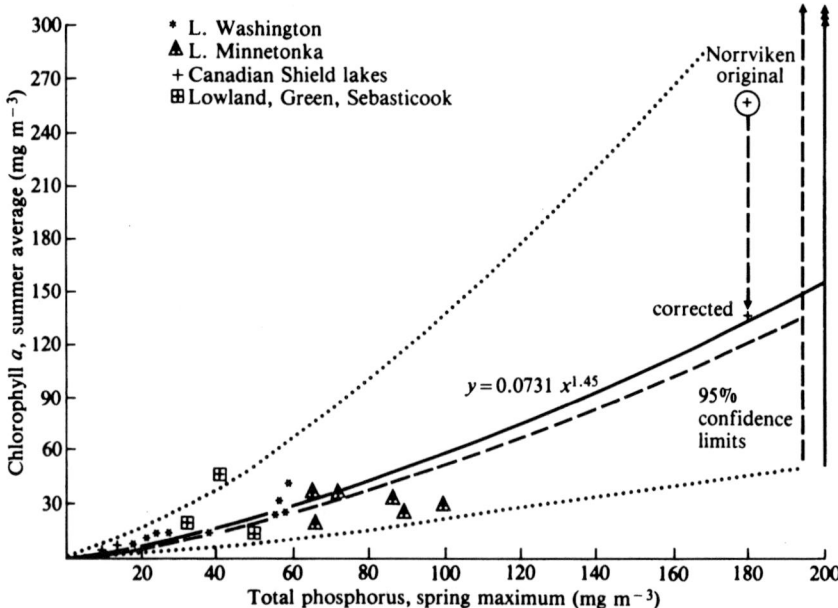

Fig. 3.30. Relationship of the chlorophyll *a* concentration in the epilimnion to total phosphorus
concentration according to Dillon and Rigler (1974b). Linear scales.

Considering the upwards curvature of the lower portion of the curve by Sakamoto and Dillon & Rigler as realistic, because this portion is well covered by the original data for Canadian lakes by Dillon, we have selected one of the logistic type models as adequate:

$$y = a'/[1 + b' \exp(-c'x)] \qquad (3.66)$$

Its use would be suggested both biologically and cybernetically by the similarity to the phytoplankton growth curve and the curve for transition characteristics of regulation circuits, respectively. Non-linear least-squares fitting of the Dillon & Rigler data to equation 3.66 is shown in fig. 3.31(*a*), with a non-statistical estimate of lines corresponding approximately to confidence limits of the equation. This shows that the total phosphorus does not extend very high, and the asymptotic character is not evident. In fig. 3.31(*b*) the logistic curve with parameters derived in the upper panel is confronted with data for Czechoslovakian reservoirs and well-fertilized Czech carp ponds. In fig. 3.31(*c*) parameters of the model (equation 3.66) are calculated independently for data from phosphorus-limited lakes of Florida by Brezonik & Shannon (1971). Clear asymptotes are seen both for Czech ponds and Florida lakes.

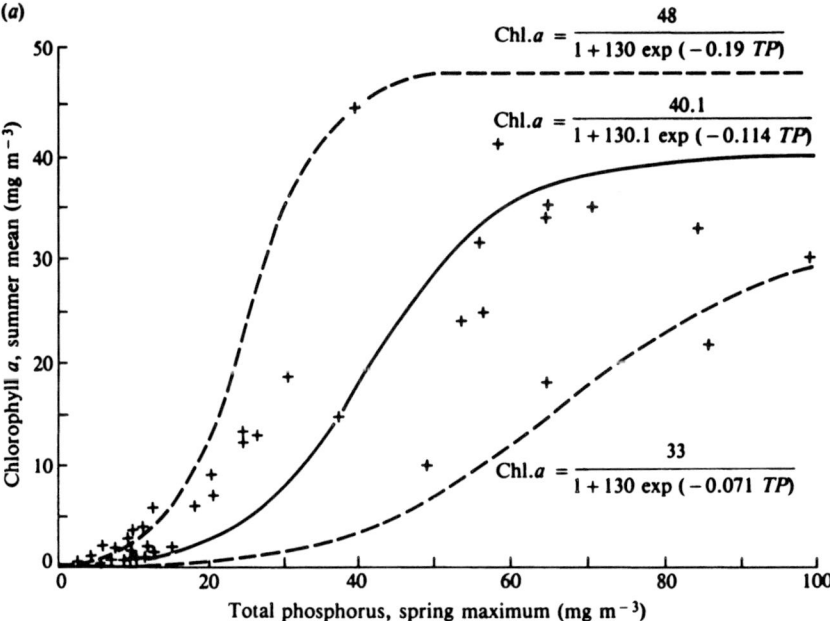

Fig. 3.31. Data on the chlorophyll *a*–total phosphorus relationship fitted to the logistic type model. (*a*) Data tabulated in Dillon & Rigler (1974b). (*b*) Curve from (*a*) compared with data from Czechoslovak reservoirs (+) and carp ponds (O). (*c*) The logistic model fitted independently to data from phosphorus-limited Florida lakes by Brezonik & Shannon (1971).

(b)

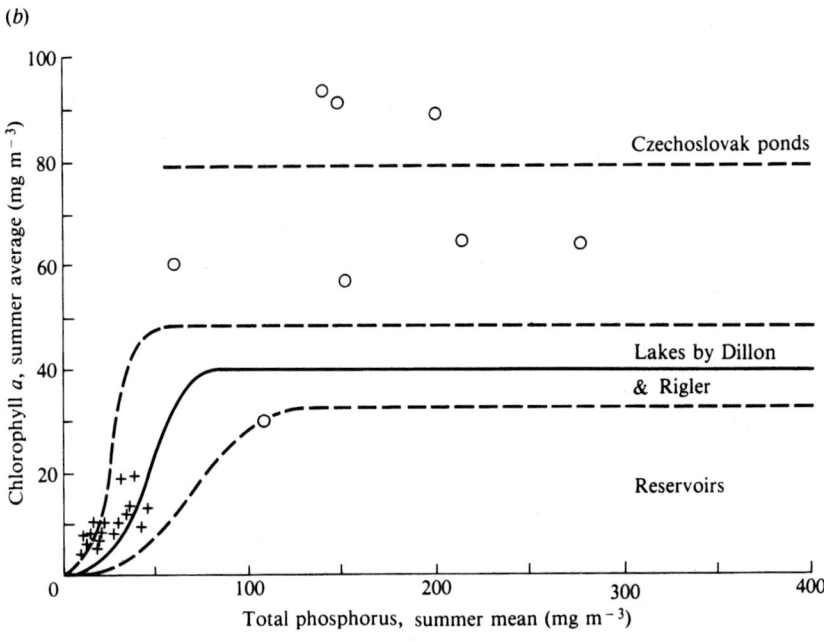

(c)

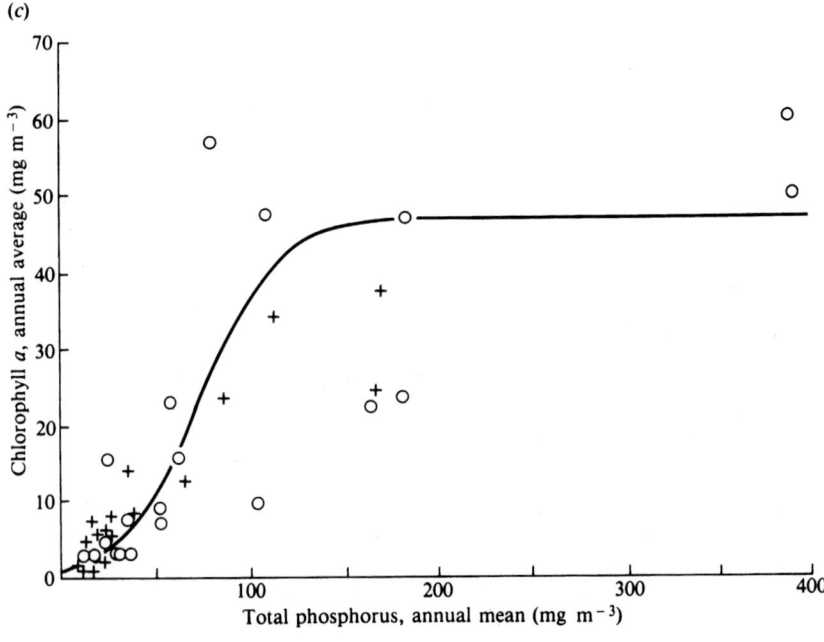

3.4.5.2. Theoretical analysis of depth–productivity relations

This section is aimed at a theoretical understanding of the empirically observed shape of the chlorophyll–mean depth curve shown in the previous section, or, more simply, at answering the question; how does the relation of chlorophyll concentration to the mean depth of lake originate?

We will show first the theoretical derivation of the effect of mixing depth and of the abiotic extinction coefficient of water by Steel (1975, 1978). The model is as in equations 3.34a and 3.34b, the value of I_K being considered constant. Respiration is proportional only to photosynthesis. The extinction coefficient ε actually covering extinction of pure water, solutes, non-living particles and plankton is separated into two components, ε_q due to non-living matter and ε_s due to phytoplankton. Steel introduced the term 'extinction depth' for the product of mixing depth and ε_q, $z_{mix}\varepsilon_q$. The equivalence of mixing depth and ε_q for primary production of homogeneous phytoplankton populations is stressed by this term. The maximum algal biomass is obtained from the differential equation of which equation 3.34 is the solution. The extreme of a function is obtained when its derivative, that is the rate of change, equals zero. For the extreme to be a maximum, the non-zero biomass is a sufficient additional condition. This can be reached only when the gains equal losses, here when the integral (over depth and day) gross photosynthetic capacity equals daily column respiration. When ε is split into

$$\varepsilon = \varepsilon_q + \varepsilon_s b \tag{3.67}$$

the maximum biomass for which the identity of photosynthesis and respiration holds true is calculated:

$$b_{max} = \frac{P_{max}(T)\Delta f(E)(1/24\,r_p) - \varepsilon_q z_{mix}}{\varepsilon_s z_{mix}} \tag{3.68}$$

For details see Steel (1978).

Results of calculations of equation 3.68 for values of E_0', Δ and T corresponding to mid-European summer conditions are given in fig. 3.32. The values of I_K and ε_s considered reasonable average values were given. Solid lines show a linear drop, with increasing depth, of biomass per unit surface, and a steep hyperbolic drop for algal concentrations per unit volume.

For biomass per unit volume the hyperbolic curve strongly resembles this computed from field measurements in fig. 3.27 by the power curve regression. As shown above, this is inadequate for very shallow depths. At these depths density-dependent physiological responses as discussed in Chapter 5 possibly take place. Steel used the approximation

$$P_{max}(b, T) = P_{max}(T)\exp[-2.3 \times 10^{-3}(chl.a)] \tag{3.69}$$

with the results shown in fig. 3.32 by dashed lines. Here the identity of the curve

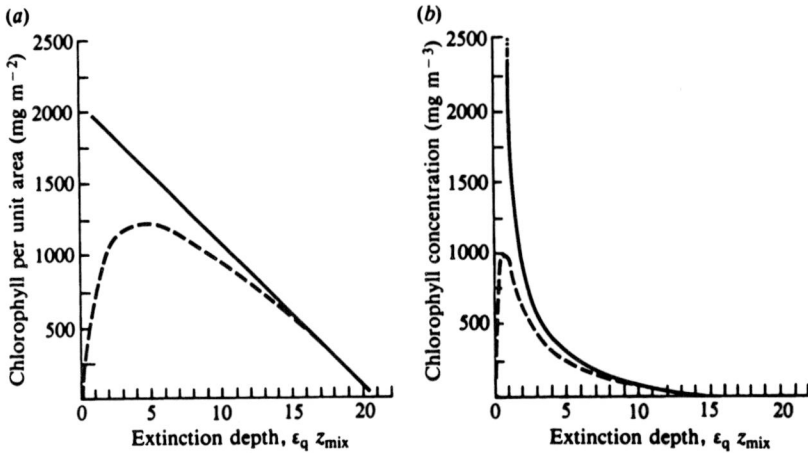

Fig. 3.32. Relation of algal biomass per square metre of the lake surface (*a*) and per cubic metre of the mixing volume (*b*) to the extinction depth. Solid lines – unrestrained response. Dashed line – taking into consideration density-dependent physiological response. From Steel (1978).

shape observed by Sakamoto is very good, the absolute values of the peak being rather different. When $\varepsilon_q = 1$, the *x*-axis scale in fig. 3.32 will represent directly mixing depth. We can easily see that when ε_q decreases, increased chlorophyll values will be obtained for the same depth and the hyperbolic drop will be less steep. The potential for producing large algal concentrations is much greater in shallow water. In deep, clear lakes considerable further increase in mean depth is required before any great reduction of concentration occurs.

The theoretical model response by Steel reached considerable agreement of the shape with observations when only the direct light effect via $\varepsilon_q z_{mix}$ was included. However, we have shown above that mixing depth, surface temperatures and phosphorus concentrations drop systematically with the increasing mean depth or length of the lake. Straškraba (1972) used a simpler model, but included a Monod function for the effect of nutrient concentration, for relating the integral photosynthetic capacity divided by mixing depth (= average mixing depth photosynthesis) directly to lake size. Both mixing-depth effects and such effects combined with other accompanying changes were examined. Empirical relations (equations 3.39, 3.45 and 3.60) were used for $\bar{z} = f(L')$, $z_{mix} = f(L')$ and $TP = f(\bar{z})$ respectively, as well as a preliminary approximation of the surface temperature–mean depth relation.

Fig. 3.33 shows the two responses comparable to the unrestrained response of Steel, i.e. valid only for $\varepsilon_q z_{mix}$ higher than about 1–2. Comparison of the two curves shows that the inclusion of the additional variables named does not seem to change the shape much. Also, the responses are rather similar irrespective of the difference in models used.

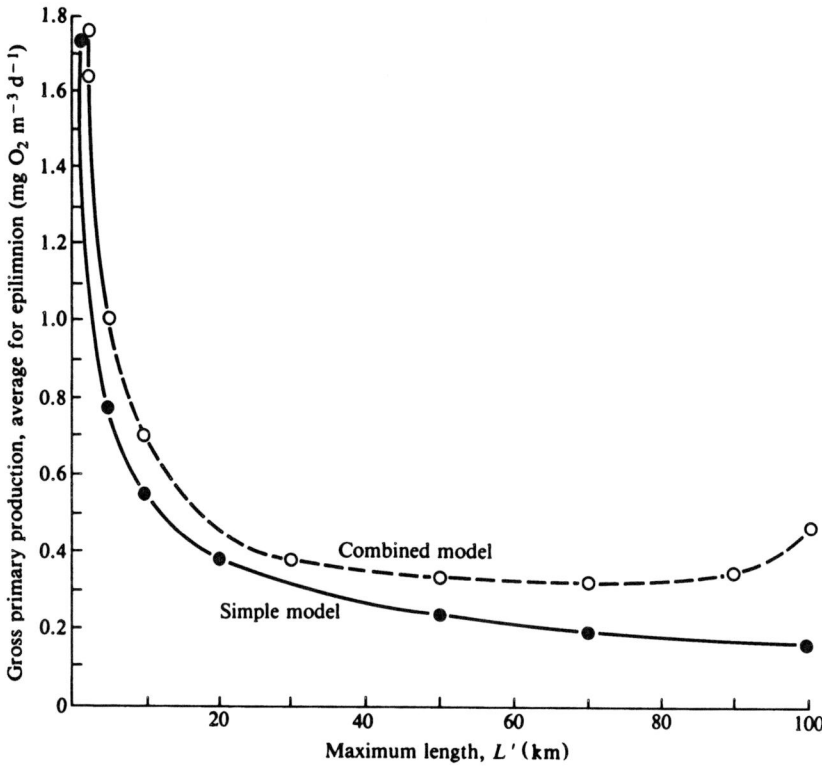

Fig. 3.33. Average mixing depth photosynthesis (per unit mixing depth volume) plotted against maximum length of the lake. Full line – simple model including only light effects induced by changes of z_{mix}. Dashed line – combined model including changes of temperature and phosphorus concentrations. From Straškraba (1972).

Quantitative differences of the restrained model response by Steel and chlorophyll concentrations by Sakamoto are due partly to different seasons being included – the season of maximum temperature, radiation and day-length in the calculation and autumnal conditions during observations. Maximum chlorophyll *a* concentrations by Sakamoto given as averages for May, June and September correspond to about 500 mg m^{-3}, which is much closer to values by Steel. Essential for both the understanding of the restricted response as well as for correct numerical values is the function $P_{max} = f(b)$. If a hyperbolic relation such as that observed by Javornický (Chapter 5) is used instead of equation 3.69, lower b_{max} values will be obtained. Also, higher values of ε_s (Straškraba, 1976a, table 3 gives values between 0.01 and 0.04) will reduce the results by up to one quarter.

The major omission of equation 3.34 is in neglecting losses to the population other than through respiration. We may easily extend equation 3.68 for inclusion of other losses, but we have mainly followed the more satisfactory

integration with an annual dynamics model to account for the other effects
preliminarily.

In accordance with the above theoretical reasoning we can suggest that the
shape of the chlorophyll–depth relationship is due to light-limitation of
phytoplankton photosynthesis, increasing with the lake size via changes of
mixing depth. For this reason the optical qualities of water are rather
important in modifying the response obtained, and the same is true for the
radiation inputs relative to I_K. Much more attention has to be paid to changes
of z_{mix} within years, during the season and among localities.

3.4.5.3. Theoretical analysis of the chlorophyll–nutrient relation

Here we would like to investigate theoretically how the scatter of the empirical
observations of the chlorophyll–phosphorus relationship is produced, and
what is the origin of the different asymptotic levels.

Two features of the logistic model approximation (equation 3.66) are of
interest: (1) the asymptotic character of the upper portion of the curve, (2) the
accelerated type of reaction in the lower portion of the curve.

As to the first, the previous section results in the idea of the light-limited
maximum biomass being different for each mixing depth and abiotic
extinction coefficient combination. With the approximation of $P_{max} = f(b)$, see
equation 3.69, the absolute maximum in a well-mixed column is reached at
$\varepsilon_q z_{mix} \approx 1$, which might be in a pond of 1 m depth and $\varepsilon_q = 1$, but also in a clear
lake with $\varepsilon_q = 0.2$ and $z_{mix} = 5$. When the amount of nutrient is insufficient to
cover the needs of the light-limited photosynthesis, the photosynthetic activity
and resulting standing crop decrease. Correspondingly, for each $\varepsilon_q z_{mix}$, the
nutrient-limited lower portion of the curve can be distinguished from the light-
limited upper portion. In agreement with this hypothesis the asymptotic level
for shallow water carp ponds is higher than for deep reservoirs (fig. 3.31(b)). In
fig. 3.31(c) shallow or clear-water lakes seem to be situated above the
computed average curve, whereas deep or coloured lakes are below average.
As we have shown, shallow water localities will also usually have higher
nutrient concentrations (Straškraba, 1976b).

We have no quantitative explanation of the stimulatory type reaction in the
lower portion of the curve. Individual algae react according to the non-
stimulatory resource utilization function of Monod, whereas the system
reaction seems to be of a stimulatory type. We suspect that the reaction is not
due to phytoplankton growth kinetics of individual species, but to the
dynamics of the plankton system. Most probably a restructuring of the system
takes place, species with increased resource utilization appearing at higher
nutrient levels. Other system mechanisms may cooperate.

The phenomenon called accelerated eutrophication represents an accel-
eration with time of the rate of biomass increase and is explained as due to

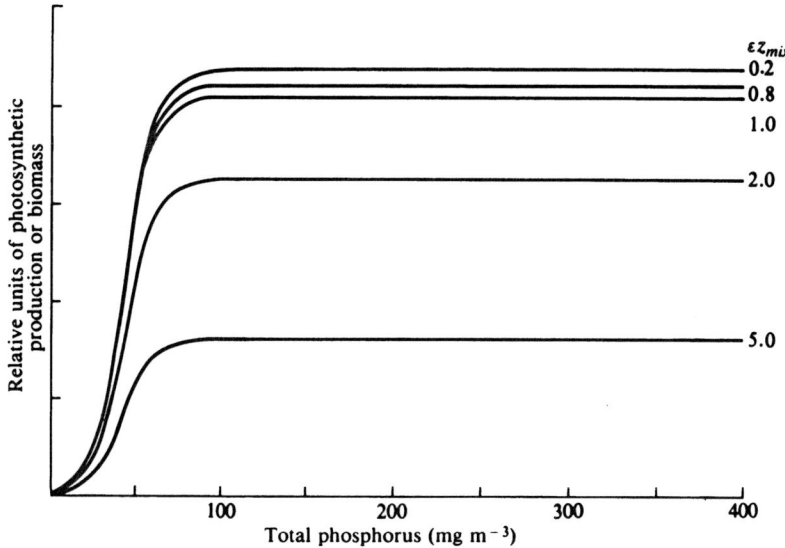

Fig. 3.34. Combined effect of phosphorus concentration, mixing depth and light extinction on chlorophyll concentration obtained theoretically.

increased liberation of phosphates from sediments when the system switches to hypolimnic anoxia. The phosphorus scale is linear in fig. 3.31, and no accelerated liberation is taking place. Nevertheless, an accelerated reaction is seen.

When we combine the empirical logistic curve with the asymptotes calculated as in the previous section, we obtain fig. 3.34. This explains how we imagine the scatter of the chlorophyll–phosphorus relation being produced. Evidently, as will be discussed below, other mechanisms will cooperate to extend the picture.

3.5. Discussion

The quantification of the effect of physical variables on phytoplankton production considered previously (sections 3.3–3.4) may seem unimportant because of the non-existence, or at least extreme rarity, of such schematic average situations for which they were obtained. But more important than the numerical results obtained is the attempt to show how following the operating processes helps to take steps towards understanding what happens in specific situations – real water bodies.

3.5.1. Lake productivity projected into the multispace of complex axes

The reasons for the difficulties encountered when attempting to recognize empirically some trends in aquatic productivity have been discussed sufficiently in connection with the IBP data analysis by Brylinsky & Mann (1973), Richardson (1975), Horne, Newbold & Tilzer (1975). More arguments can be raised suggesting that global empirical observations can result in a clear quantification of complicated trends only under conditions which are almost impossible to satisfy. Namely, the selection of localities representative in respect of most major variables, very complete and methodically comparable

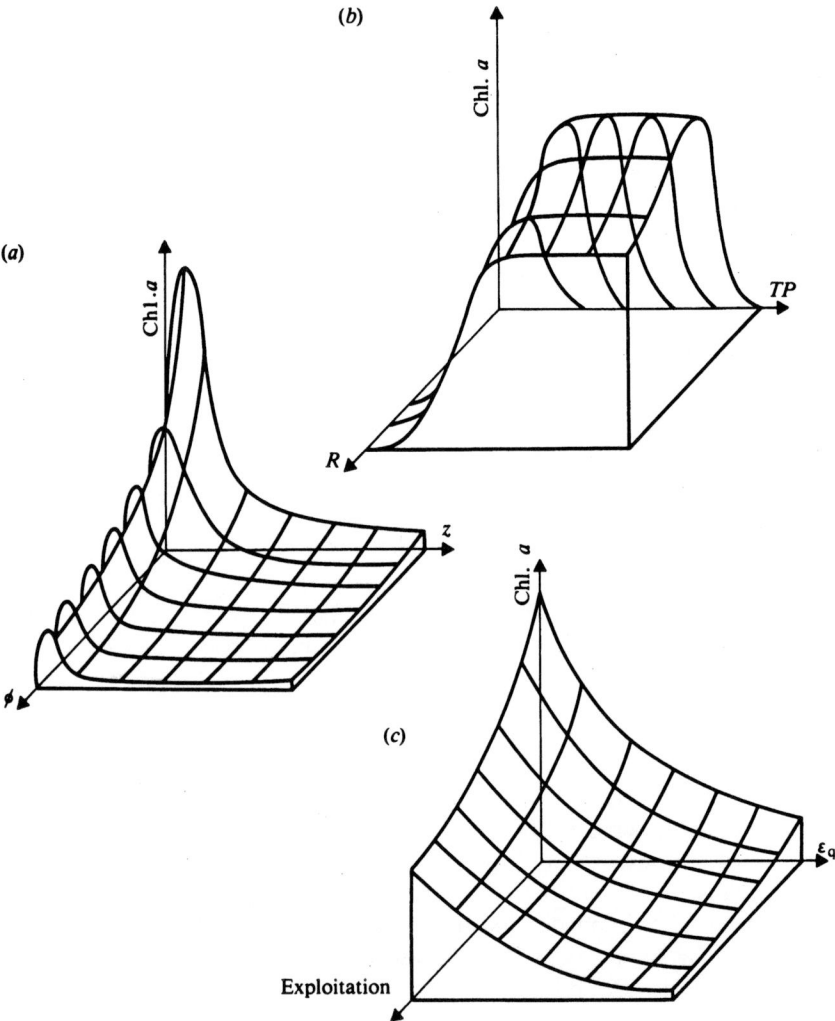

Fig. 3.35. Hypothesis of lake productivity as the realization of a point in a multispace of complex axes, (a) Three-dimensional projection of chlorophyll *a* concentration (chl. *a*) against morphometry (z) and latitude (ϕ), (b) Inserted subspace with dependent variables limiting nutrient concentration (TP) and theoretical water retention time (R), (c) Inserted subspace with dependent variables abiotic extinction coefficient (ε_q) and biotic control by predation pressure (exploitation).

observations at all selected localities, results expressed adequately for the purpose, and statistical treatment enabling non-linear multiple analysis of interrelations. The reason is that several complex major axes intersect: namely *geographical, morphometric, hydrologic, nutrient status* and *biological balance of populations.*

Theoretically we can view the productivity of any water body as a realization of a point within a multispace characterized by at least the five major complex axes named. Fig. 3.35 is an imperfect attempt to visualize the hypothesis of the determination of primary production of a water body within this multidimensional space. This is achieved by inserting subspaces into the three-dimensional space. The figure is a mere schematization. Although some of the curves inserted will resemble those derived theoretically in previous sections, others are pure imagination. The figure is not intended to represent actual relations but to demonstrate how all these (and probably many more) axes determine productivity *contemporaneously.* Not shown is a feature we have repeatedly recognized above – that the axes are not independent but highly interrelated. Water bodies in different geographical situations will be hydrologically rather different, as will their nutrient status and morphometry, and man's activity will be associated with the geography. As shown below (section 3.5.2) the complex axes named here are only apparent axes and understanding is obtained only when analysing directly the individual resources as they are modified by geography, morphometry, etc.

The approach in fig. 3.35 is also purely deterministic, leaving no room for the effect of unconsidered variables. We have shown how variations of a single additional variable like penetration of solar energy into deeper layers due to solutes and mineral particles can distort the regular picture. Actually there are many more such variables, without mentioning stochastic variations due to local weather variations, minor differences in space and time and others, which can result in considerable differences in productivity.

Looking now at the complex figure obtained from a point of view of a statistically ensured experimental sampling design, it becomes evident to anyone at all familiar with the analysis of variance how difficult it would be to uncover the scheme by a global analysis. Greater success is to be expected from more restricted, but clearly specified, analyses.

3.5.2. Hypothesis of multiple resource kinetics

The view of the above multispace of complex axes is only a superficial projection of our world into the world of phytoplankton. The axes of Hutchinson's multispace niche of phytoplankton are different. These are represented by those qualities of the environment directly perceived by organisms, like temperature, radiation reaching the chloroplasts, etc. These qualities, resources of recent ecology (Watt, 1973), can be classified into three categories: energy, matter and diversity. Diversity we understand in the broadest sense of the biotic interrelations among individual organisms and species. Energy acts in water bodies mainly in three forms: light, temperature and mechanical energy. Light available to phytoplankton is more important than light incident to water surface. In addition to incident radiation the light available is determined also by light penetration characterized by the extinction coefficient of water ε, and how deeply the algae are mixed (z_{mix}). Not only the total amount of photosynthetically available radiation is important, but also how this is distributed during the day. Immediately we see that for one form of energy not only are the other forms co-acting but also other resources. The mixing depth is determined not only by radiation penetration but also by mechanical energy, and ε is affected by matter entering water bodies.

The present ideas about the quantitative effects of individual resources are based on observations on growth kinetics of individual algae. Responses were measured in experimental conditions when only one or a few variables were varied systematically, others remaining seemingly constant (but unfortunately mostly rather unspecified). The resulting models are directed towards *empirically* describing responses of *individual* algae to *single* resources.

To demonstrate the shortcomings of the ecological applications of the present kinetic models, one aspect of the photosynthetic models can be mentioned in somewhat more detail. In recent literature and IBP studies much care was devoted to what was called adaptation of phytoplankton to light conditions (Steemann Nielsen & Hansen, 1961 according to Steemann Nielsen, 1975; Kalff, 1969; Kalff & Welch, 1974; Platt, Denman & Jassby, 1975; Tilzer & Schwarz, 1976). Daily, seasonal and latitudinal changes of I_K were demonstrated. While preparing this chapter I have assembled data for correlating I_K with both incident light and available light. Correlations were significant for most sites analysed individually, but not at all for several sites combined. However, it appears that the idea of 'light adaptation' is largely an artifact due to inadequate photosynthetic models. The heart of the problem was recognized by Talling (1957a), Steele (1962) and revived by Takahashi, Fujii & Parsons (1973), but is continually neglected. If the absolute value of P_{max} is under otherwise similar conditions a function of temperature ($P_{max}(T)$), and the initial slope of the photosynthesis–light curve remains constant, then I_K represents an analytical function of temperature (fig. 3.36), as explained in Straškraba (1976a, fig. 11). However, if this is true, then any other variable changing P_{max} (nutrients, biomass, population age, etc.) will also effect I_K systematically. If so, no wonder we are unable to recognize any empirical relations for different localities, whereas individual localities with a narrower range of variables show some trends.

We can raise similar criticisms concerning 'temperature adaptation' and the Monod substrate kinetics concept, transferred to algae from the study of heterotrophic microorganisms. The basic difference is that for algae the

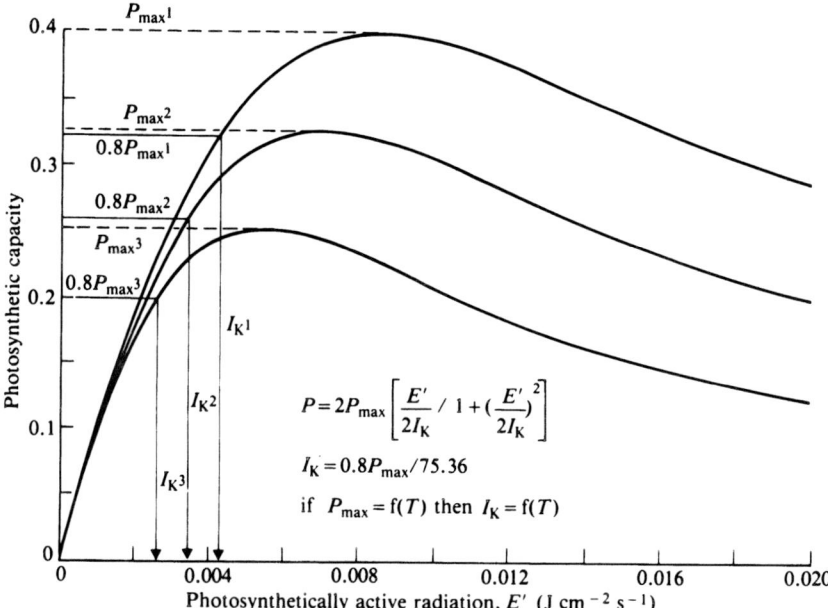

Fig. 3.36. Dependence of photosynthesis on light and temperature. The light sensitivity parameter I_K is an analytical function of temperature.

substrate, e.g. phosphate, is *not* a source of energy. Energy is gathered independently, from light.

In addition to energy and matter as resource categories usually considered more attention should be paid to diversity. One aspect generally not considered in physiological experimental studies, but recently shown to modify the growth rates under near-natural conditions rather deeply, is the density-dependent physiological response (Steel, 1978; Javornický Chapter 5, section 5.4.2.6). Theoretical computations by Javornický suggest that this response is due in most instances to nutrient limitation rather than to self-shading, but the actual mechanism remains unspecified. The response is probably connected with the lowered 'specific substrate concentration' of Benndorf & Stelzer (1973).

For ecological purposes we need models with attributes opposite to those mentioned above for the present kinetic models: *causal* descriptions comparing responses of *different* algae to *combinations* of resources.

To advance towards this goal we would like to formulate the following hypothesis of multiple resource kinetics:

(1) Functional responses of photosynthesis to various resources (E, S, T, b) are mutually interrelated.

(2) Parameters of the present functional response–kinetic models are analytical functions of the effect of other resources.

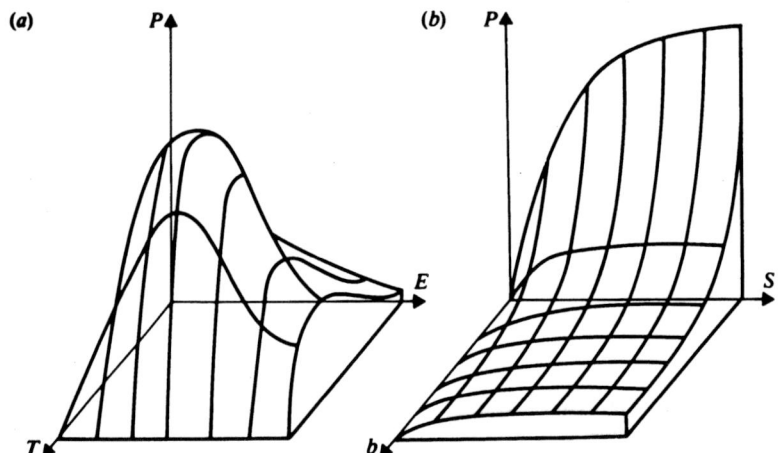

Fig. 3.37. Preliminary quantification of the hypothesis of multiple resource kinetics. (a) Three-dimensional representation of the effect of light (E) and temperature (T) on production (P). (b) Subspace with axes limiting nutrient concentration (S) and biomass (b).

A preliminary quantification of this hypothesis as obtained from a detailed treatment of the functional responses of algae (Straškraba, 1976a) is given in fig. 3.37. The figure is intended to demonstrate the trends of the interrelations rather than specific shapes, which have to be investigated in detail in the immediate future. The following specifications are obvious:

(a) The basic factor is the light limitation of photosynthesis at low light intensities. Photosynthesis is practically independent of other resources at low light, but highly sensitive when light increases. The light sensitivity parameter I_K is an analytical function of temperature, biomass and nutrients.

(b) The response intensity to other resources increases from low to optimum temperature, but decreases thereafter. Optimum temperature is a function of light intensity.

(c) The response intensity to other resources decreases from low to high biomasses, but increases from low to high nutrients.

If the hypothesis of the multiple resource kinetics is verified and more exactly quantified, we will have to re-evaluate limnological data about P_{max}. $P_{(Eopt, T, b, S)}$ has to be substituted for P_{max}, and coefficients of the empirical relations to resources purified from the effect of other resources. Statistical relations of P_{max} are based on observations of $P_{(Eopt, T, b, S)}$ or for $P_{max}(T)$ on $P_{(Eopt, T, b, S)}$ where the effect of $\bar{b}$, $\bar{S}$ or T remains obscure. The values are surely much lower than would be the optimum obtained for b low and S unlimiting, or $T = T_{opt}$.

Instead of treating parameters of the present models as functions of other resources, a model of comparative resource kinetics should be formulated, based on parameters different from the present ones and more suitable for direct comparison and understanding of response surfaces of different species of algae. This can also shift our interests towards quantifications of the real adaptation of algae to light (by changing chlorophyll content), to temperature and to substrates.

3.5.3. Functional dynamics of plankton systems

The kinetic approach which we have been mostly using above considers only one aspect of the dynamics of phytoplankton population changes, namely gains of population due to gross or net phytoplankton growth. As shown in fig. 3.38, this is only one side of the functional dynamics, the one which mostly affected ideas of plankton ecology. Recent investigations suggest that the same attention should be paid to losses due to both internal mechanisms of phytoplankton respiration, excretion and natural mortality, and systems mechanisms of cell sinking, grazing and outflow. In comparison with the straightforward nutrient–productivity relations represented in fig. 3.28 the actual mechanism is rather different.

Nutrients in connection with other resources directly affect only the specific gross growth rates of phytoplankton. Simultaneously nutrients are released by phytoplankton, incorporated by zooplankton and excreted by it, as well as entering the pelagic region from inflows, sediments and the hypolimnion, and leaving it via the outflow and to sediments. But phytoplankton primary

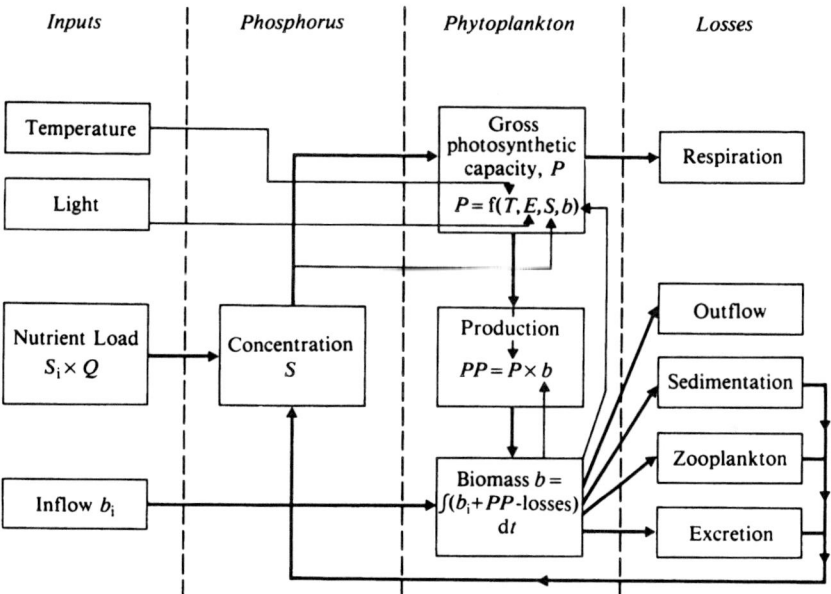

Fig. 3.38. Balance equation for phytoplankton and nutrients.

production is a product of specific rates and biomass present. Biomass itself is a result of integration over time of the changes resulting from the balance between gains and losses. At any instant of time, conditions in previous periods determine productivity more than do the instantaneous conditions, as is seen from primary production being most strongly correlated with phytoplankton biomass. This is a feature usually not considered by limnologists, but having profound consequences for understanding many phenomena observed.

In the light of the dynamic approach the explanation by Schindler & Fee (1975) that phytoplankton production as a rate parameter should be related more strongly to nutrient load than to the static nutrient concentration is far from unequivocal. A simple rate parameter would be specific photosynthetic rate per unit biomass, but primary production is a rather composite rate. Moreover, the specific rate is directly affected by concentration and not by load. Thomann (1977) suggests that a direct relation between nutrient load and phytoplankton biomass can be expected only under fairly constant yield coefficients. In spite of not being able to agree with the straightforward argument of Schindler & Fee we can produce dynamic system arguments that load can be rather important in the long run. Nutrient concentration is rather a consequence of production as well as its cause. This was shown by means of a dynamic model (fig. 3.26).

Another phenomenon we can understand more clearly on a dynamic basis is the photosynthetic efficiency of phytoplankton. A strong direct relationship of efficiency to biomass has been observed (Chapter 9 and Horne *et al.*, 1975). This can be explained by the model: production is a multiple of biomass and photosynthetic capacity. Biomass varies within broad limits and the decrease of photosynthetic capacity does not negate the biomass increase. Hence, production is always related to biomass and higher total production means also higher efficiency of solar radiation utilization. All factors leading to high biomass and high photosynthetic activity will tend to increase efficiency. To separate the light effects from the others entails studying integral specific photosynthetic rates. In addition to incident light and light attenuation studied by Horne *et al.* (1975) the effect of the mixing depth should be considered.

Quantitative understanding of the effect of losses on production has been obtained both on an empirical and a theoretical basis (e.g. Bella, 1970; O'Brien, 1974). Field experiments by Hrbáček, Dvořáková, Kořínek & Procházková (1961) have suggested that the effect of selective fish feeding and indirect effects alter zooplankton pressure on phytoplankton and the community metabolism is changed. Gliwicz (1975, 1976) suggested that the effect of grazing on phytoplankton composition and production is more clearly recognizable under tropical conditions where other variables are more constant.

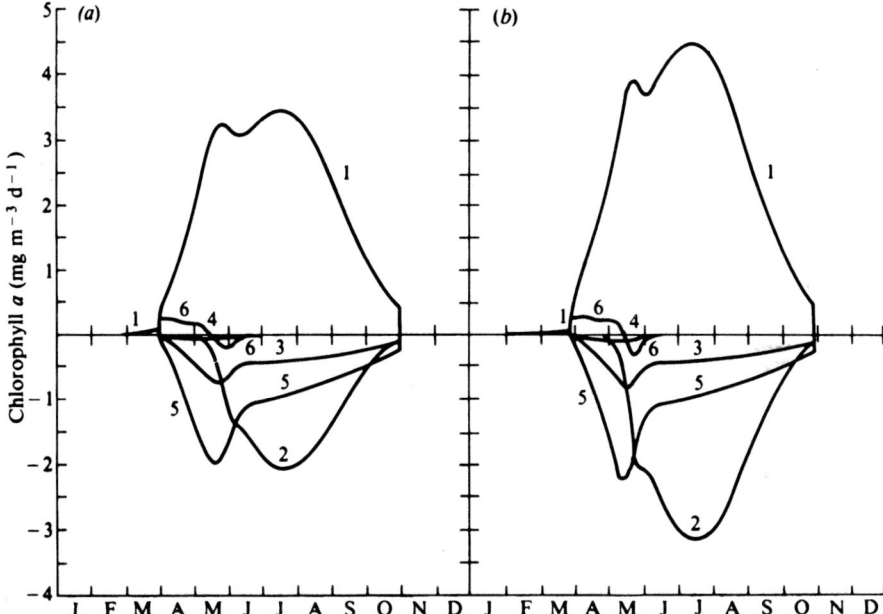

Fig. 3.39. Partition of gains and losses in a dynamic two-layer plankton system model. Two years in the same lake are given, assuming different phosphorus limitation: (*a*) strong limitation ($K_s = 10$), (*b*) low limitation ($K_s = 1.25$). 1, gross population growth; 2, grazing by herbivorous zooplankton; 3, respiration; 4, turbulent exchange with hypolimnion; 5, sedimentation losses to hypolimnion; 6, the sum of processes 1–5, the resulting growth of phytoplankton populations. After Dvořáková (1976).

A dynamic combination of the simultaneous effects of a number of mutually interrelated variables on phytoplankton population changes is possible by means of a dynamic model. Fig. 3.39 shows computed population rates for gross production, respiration, grazing, sedimentation and turbulent exchange for two situations in the same lake. The changes were computed by means of a dynamic model. The enormous difference between the gross rates and the resulting rate of population change is seen in both situations, the relative proportion of the various losses being rather different for specific conditions and during different periods.

3.6. Epilogue

By concentrating on theoretical aspects of the mechanisms of processes determining aquatic ecosystem reactions, rather than on apparent correlations, we are starting to understand the functional dynamics of our systems. Although the coefficients of the model are no more constant when the hypothesis of multiple resource kinetics is utilized, the structure of the model remains constant. We understand the structure of the system as represented by different species present (Uhlmanr, Weisse & Gnauck, 1974). Functional and structural properties are interrelated, the functional response surfaces depend on properties of the species present. On the other hand, the species present can be viewed as resulting from competitive mechanisms of resource utilization, or maximization of population growth within the changing multidimensional space of resources. Different species have different competitive advantages. Structural dynamics and diversity were unfortunately largely neglected during the IBP/PF studies. Although many specific instances have been studied qualitatively, the factors governing species composition remain to be quantified. What we need to advance the theoretical analysis is a *comparative* multiple resource kinetic hypothesis.

The section on effects of depth and nutrients on production (section 3.4.5) is based in part on the ideas of J.A.P. Steel, Metropolitan Water Board, London. I. Smith, Nature Conservancy, Edinburgh and D. Schindler, Fisheries Research Board of Canada, Winnipeg, kindly submitted data for the section on turbulent diffusion and mixing depth. D. Schindler also kindly sent original data concerning the primary production to phosphorus load relationship.

4 Chemical budgets and nutrient pathways

H.L. GOLTERMAN & F.A. KOUWE

4.1. Introduction

In ecosystems, the primary production of organic matter provides two of the most essential items for all other biological and chemical processes. It supplies both the energy source for all endothermic reactions and also the power (electrons) for all reducing reactions, which encompasses most of the fundamental processes which one finds in ecosystems.

Solar energy is converted to the chemical energy of organic material by photosynthesis. The subsequent dispersion of this energy through the ecosystem and the associated recycling of primary nutrients are the two important mechanisms by which the components, both living and non-living, of any ecosystem may be related to one another. Thus, to understand the main pathways of these processes gives great insight into the functioning of a given ecosystem.

These processes are in fact almost overwhelmingly complex, but a systematic approach to their rationalisation is, firstly, to try to grasp how the photosynthetic fixation of energy can lead to a particular level of total organic material or of primary production. For this an appreciation is necessary of the chemical environment surrounding the organisms which are directly concerned.

In this chapter the two-fold function of the chemical environment will be discussed. Firstly attention will be given to the chemical composition of the water – the environment of all aquatic organisms – secondly the chemical environment will be considered as that component through which recycling of nutrients takes place.

4.1.1. Nature of the chemical environment

Besides the physical necessities of light and temperature, and the fifteen to twenty chemical elements so far known to be essential nutrients for the growth of organisms, the chemical environment also appears to impart an apparently non-nutritional 'selective mechanism', which can favour one assemblage of organisms rather than another. In freshwater ecosystems the major dissolved ions of calcium, magnesium, sodium, potassium, (bi)carbonate, sulphate and chloride – with the pH – are known to delineate these selective mechanisms, but other elements may further define the environment. The particular pattern in which the total ionic composition is distributed amongst the elements concerned (see table 4.1) very often has an influence over and above the sum of the component elements.

Unfortunately in IBP studies, no data were collected specifically for the study of this selective mechanism, but doubtless a search through the list of species from the IBP lakes would reveal information on the subject. The selective mechanism of the general ionic environment for algae has been reviewed by Lund (1965), and for animals in African waters by Beadle (1974). The complexity of the problem of the selective mechanisms has been discussed by Golterman (1975a), from whom the following extract is reproduced.

Planktonic desmids are for example normally found only in calcium-poor waters, although the distinction between calciphobe and calciphile should not be taken too rigidly. Moss (1972) attempted to establish factors affecting the distribution of algae in hard and soft waters by studying the mineral requirements of these algae. Some oligotrophic (calcium-poor) desmids were found to be unaffected by levels of calcium as high as those which actually occur in eutrophic lakes, while calcium-poor media could be used for in-vitro growth both of eutrophic and oligotrophic species. Lakes with a different calcium content will also have a different pH and bicarbonate content.

The types and quantities of lake organisms present are due to a complex of factors and not simply to the concentration of one ion. Thus Provasoli, McLaughlin & Pinter (1954) found that the diatom *Fragilaria capucina* Desmaz grew better when the calcium/magnesium ratio was greater than one and the ratio of monovalent to divalent ions (M/D) was low, whilst some *Synura* species grew better when M/D was high and concentration of total solids was low. However, *Synura* also includes species with the opposite behaviour.

Pearsall (1923) suggested that diatoms are abundant only in calcareous waters having an M/D ratio below 1.5. Nevertheless, it is now known that in African waters, where the calcium content is low, diatoms bloom in waters which have a high M/D ratio. Thus the situation is probably much more complex than was originally supposed, and greater consideration should be given to other factors such as erosion. For example, most of the silicate supply derives from erosion and erosion will in itself affect and be correlated with the M/D ratio. M/D ratios in natural waters are therefore not independent of the silicate concentrations. Furthermore, since calcium will erode from marine limestone, there will often be more phosphate in hard than in soft waters.

The presence of chelating substances may greatly affect certain ionic ratios. Miller & Fogg (1957) demonstrated an antagonistic effect of high concentrations of calcium ions on magnesium uptake by *Monodus*. The addition of chelating agents such as EDTA – but also of naturally occurring chelators, such as glycine – had an enhancing effect on magnesium accumulation and uptake, this being due probably to their preferential chelation of the calcium ions. Because chelation processes occur in most natural waters, only a very cautious attempt should be made to relate data obtained from culture studies to actual field conditions. Growing algae, e.g. blue-greens, may excrete organic chelating compounds into a culture solution which originally contained only inorganic compounds (Fogg & Westlake, 1955).

An example illustrating the relative importance of potassium or sodium comes from work on Russian fish-ponds (Braginskii, 1961). Application of potassium phosphate caused a change in the dominance of a population from blue-green algae towards desmids. The addition of sodium phosphate caused colonies of the blue-green algae *Microcystis* and *Aphanizomenon* to disintegrate but this did not occur when potassium phosphate was added.

The hydrogen ion is always present in fresh waters, but it contributes significantly to the ionic balance only in waters where the pH is four or less and which have a low salinity. This acidity has a strong selective influence on the flora and fauna, especially because of the low concentration of other ions. Hydroxide ions may occur in strongly alkaline waters but only together with much larger concentrations of bicarbonate or carbonate. It will therefore seldom contribute significantly to the ionic balance, but during periods of intensive photosynthesis it may increase significantly, for example to $0.1 \, mmol \, l^{-1}$ in Lake George (Uganda) (pH 10), and in Lake Aranguadi (Ethiopia) (pH 10.3) (Prosser *et al.*, 1968), and also in fish-ponds as recorded in Bohemia (Fott, 1972); it may then become a selective mechanism for community composition.

The rate of algal growth, and in many cases the final yield as well (see p. 119), is controlled by the so-called 'minor' constituents, especially phosphate, nitrogen compounds and silicate. Such constituents are dealt with separately below. The terms major and minor refer to the quantities found and not to their biological significance. A summary of this quantitative classification is given in table 4.1, in which most of the other constituents, such as trace elements, organic compounds and gases are also included. In this table the percentages of the major elements are calculated on the basis of a hypothetical 'mean' global freshwater composition (Rodhe, 1949). Unlike sea water, the composition of fresh water is not 'standard', but is always changing, both in time and place. These changes can be seasonal, such as in regularly stratifying and destratifying lakes but can also follow a slower progressive trend. This can be induced by human activities, for example for the St Lawrence lakes. Beeton (1969) showed that in Lake Erie and Lake Ontario calcium, sodium plus potassium, chloride and sulphate increased considerably between 1900 and 1960 (table 4.2). Sulphate and chloride concentrations increased even in Lake Huron and Lake Michigan, while only in Lake Superior did the concentrations remain the same. It seems unlikely that the lack of uniformity between the lakes in these developments could be due to natural differences in their basins, even though the lakes were formed at somewhat different times. The changes during the period involved were so fast that they can hardly be

Table 4.1. *Chemical constituents of fresh water*

Major elements (mole %)				Minor elements	Trace elements and organic compounds	Gases
Ca^{2+}	63.5	(34)[a]	HCO_3^- 73.9 (82)	N: as NH_3, NO_3^-	Fe, Co, Cu, Mo, Mn, B, Zn,	O_2
Mg^{2+}	17.4	(33)	SO_4^{2-} 16.0 (8)	P: as inorganic	vitamins, humic compounds,	N_2
Na^+	15.7	(26)	Cl^- 10.1 (10)	and organic P	inhibitory and growing	CO_2
K^+	3.4	(7)		Si: as SiO_2, $HSiO_3^-$	substances (?), metabolites etc.	

[a] In brackets: values if L. Tanganyika data are included.

Table 4.2 *Chemical characteristics of the North American Great Lakes*

Year	Ca^{2+} (mmol l^{-1})				$Na^+ + K^+$ (mg l^{-1})				SO_4^{2-} (mmol l^{-1})				Cl^- (mmol l^{-1})			
	1900	1920	1940	1960	1900	1920	1940	1960	1900	1920	1940	1960	1900	1920	1940	1960
L. Superior	0.65	0.65	0.65	0.65	2.5	2.5	2.5	2.5	0.22	0.22	0.22	0.22	0.06	0.06	0.06	0.06
L. Michigan	1.78	1.78	1.78	1.78	—	—	—	—	0.61	0.80	0.98	1.19	0.09	0.12	0.12	0.19
L. Huron	1.27	1.27	1.27	1.27	3.8	3.8	3.8	3.8	0.43	0.59	0.74	0.88	0.14	0.14	0.14	0.18
L. Erie	1.57	1.61	1.87	1.94	6.6	6.6	7.8	10.9	0.78	0.94	1.25	1.53	0.19	0.27	0.44	0.65
L. Ontario	1.57	1.66	1.88	1.97	6.3	6.9	8.8	11.6	0.98	1.09	1.35	1.80	0.21	0.27	0.43	0.65

After Beeton (1969).

explained as being natural processes. Differences due to the application of newer analytical techniques can, however, not be excluded; it seems worth while to compare the present results with results obtained by applying the old methods to the present lake water. Although the old concept that fresh waters tend to develop towards 'standard' fresh water is nowadays abandoned, the concept is still in use for example for global erosion calculations and is often the basis for preparing algal cultures. However, the deviations from 'standard' water are probably more important than the 'standard' water itself.

The 'standard' composition of fresh water mentioned in Rodhe (1949) originated from Clarke's (1911) collection of composition of fresh waters. Clarke calculated the average composition for each of the continents as well as for the whole world. However, at the time of Clarke's writing the majority of available data were derived from lakes and rivers on the continents of Europe, North America and South America (the latter not being known very well). Apparently no data from Australian fresh waters were then available. For Africa the River Nile, and for Asia a few rivers in India and Java were used. By assuming a resemblance in character between South American waters and African waters and using the figures for the Nile, Clarke calculated an average for Africa. For Asia he had to make an even rougher estimate.

Nowadays more data from all the continents are available (see Meybeck, 1976), and the present analytical methods give greater reliability to the data, so that if one were to recalculate a new global freshwater composition, one might very likely arrive at quite another conclusion from that of Clarke. Furthermore, in Clarke's calculations one of the world's largest freshwater bodies, Lake Tanganyika, containing roughly 20% of the world's fresh water, is lacking. This water contains 8 mmol l^{-1} of cations of which Mg = 45%, Na = 35%, K = 11% and Ca = 9%. Bicarbonate (plus carbonate) represents 89% and chloride 10% of the total of 7.8 mmol l^{-1} of anions (Talling & Talling, 1965). With these percentages for the additional Lake Tanganyika water a new global freshwater composition can be calculated by taking four volumes of Clarke's water to one volume of water from Lake Tanganyika. Bicarbonate

Table 4.3. *Percentage of Ca^{2+}, Mg^{2+}, Na^+ and K^+ in some sodium-dominated lakes*

	Cations (mmol l^{-1})	Ca^{2+}	Mg^{2+}	Na^+	K^+
		(%)			
Lake Chilwa	24	2.7	2.0	93.7	1.6
Echo Lake	24.4	18.2	32.1	46.8	2.9
Pasqua Lake	24.1	19.6	29.3	48.2	2.9
Buffalo Pound	9.2	16.7	27.3	51.6	4.4
Katepwa Lake	23.8	14.0	33.4	49.1	3.5

Table 4.4 *The major ionic composition (in mmol l^{-1}) of waters in widely distributed lakes (mostly from IBP studies)*

No.	Lake	Country	Ca^{2+}	Mg^{2+}	Na^+	K^+	Cations	$HCO_3^- + CO_3^{2-}$	Cl^-	SO_4^{2-}	Anions	pH	Conductivity ($\mu S\,cm^{-1}$ at 25°C)	Period
1	Aleknagik	Canada	0.25	0.79	0.03	0.01	1.08	0.20				7.1	38	ions: summer 1962, 1963; pH: 1962–1970 (VI–IX); conductivity: 1965–1972 (VI–IX)
2*	Aranguadi	Ethiopia	0.67	<0.6	67.0	8.1	76.37	51.4	22.0	0.7	74.1	10.3	8635[a]	1963: IV, V 1965: VI
3*	Bete Mengest	Ethiopia	0.34	3.94	23.9	1.31	29.49	26.8	5.67	0.4	32.9	9.2	3207[a]	1963: IV, V 1965: VI
4*	Bishoftu	Ethiopia	0.37	5.8	16.0	1.5	23.67	20.0	4.0	0.35	24.4	9.2	2487[a]	1963: IV, V 1965: VI
5*	Bolsena	Italy	1.06	1.21	1.92	1.11	5.30	4.07	0.81	0.42	5.30	8.6	564	IX'67–IV'70
6	Bolshoy Kharbey	USSR	0.15	0.07		0.20[j]	0.42	0.19	0.09	0.14	0.52	6.9	52[a]	1968–1969
7*	Bracciano	Italy	1.09	0.92	2.37	1.05	5.43	3.23	1.34	0.66	5.23	8.5	567	IX'67–IV'70
8*	Buffalo Pound	Canada	1.53	2.50	4.73	0.40	9.16	4.07	1.30	4.40	9.80	8.8	1116[a]	1967
9	Chad	Chad	1.06	0.99	1.55	0.39	3.99	3.90						
10	Chedenjarvi	USSR	0.22	0.14				0.47	0.04	0.08	0.59	7.8		1969: I–IX
11	Chester Morse	USA	0.08	0.03	0.06	< 0.005	0.17	0.17	0.02	0.01	0.2	6.8	28	summer 1971
12	Chilwa	Malawi	0.65	0.47	22.45	0.39	23.96	11.96	10.07	0.73	22.76	8.8	2000	IX'70–IX'71
13	Coragulac	Australia	0.40	19.70	128	7.20	155.30	52.60	115.2			9.9	10698[b]	8 July 1970
14	Corangamite	Australia	3.40	90	609	5.30	707.7	18	677	8.2[c]	703.2	8.9	39535[b]	9 July 1970
15	Dusia	USSR	1.68	1.15	0.21	0.06	3.10	2.75	0.16	0.19	3.10	8.4	328[a]	summer 1971
16	Echo	Canada	4.44	7.83	11.43	0.73	24.43	6.91	4.23	17.55	28.69	8.7	3308[a]	1965–1967
17	Findley	USA	0.06	0.03	0.03	< 0.003	0.12	0.16	0.02	0.01	0.2		21	summer 1971
18	Galstas	USSR	1.72	1.22	0.19	0.06	3.19	2.90	0.14	0.15	3.19	8.4	335[c]	summer 1971
19	George	Uganda	0.85[c]	0.62	0.87	0.15[c]	2.49	1.75[c]	0.23	0.34	2.32	8.5–10.2	210[c]	IV 1967–IV 1968
20	Himmasjön	Norway	0.22	0.14	0.26	0.02	0.64	0.04	0.24	0.35	0.63	6.1	69	1971
21	Katepwa	Canada	3.33	7.93	11.68	0.83	23.77	3.74	3.25	14.80	21.80	8.6	2838[a]	1966
22	Kiev Res.	USSR	2.12	0.71		0.35[j]	3.18	2.41	0.32	0.45	3.18	7.7	355[a]	1967–1968
23*	Kilotes	Ethiopia	0.7	<0.6	70.5	4.5	76.3	63.4	13.6	0.4	77.4	9.6	8104[a]	1963: IV, V 1965: VI
24	Krivoye	USSR	0.33	0.19		0.56[j]	1.08	0.49	0.46	0.13	1.08		132[c]	1968–1969
25	Krugloye	USSR	0.13	0.14		0.21[j]	0.48	0.22	0.17	0.09	0.48	6.5	57[c]	1968–1969
26	Loch Leven	Scotland	1.26	0.61	0.37	0.05	2.29	1.18[c]	1.08[c]	0.52[c]	2.78	6.7–9.6	246	1968–1971
27	Locomotive Spring	USA	5.35	4.08	21.00	0.69	31.12	3.5[c]	27.07	1.81	32.38	7.6	5685[c]	III'70–XII'71
28	Neusiedlersee	Austria	1.36[f]	7.98[c]	9.22[c]	0.59[c]	19.15[c]	9.7[c]	3.72[c]	7.25[c]	20.67[c]	8.6	1606[c]	1968–1971
29	Obelija	USSR	2.24	1.20	0.19	0.06	3.69	3.20	0.22	0.27	3.69	8.4	395[c]	summer 1971
30	Øvre Heimdals-													

31	Pasqua	Canada	[illegible]	[illegible]	[illegible]	[illegible]	[illegible]	[illegible]	[illegible]	[illegible]	[illegible]	[illegible]	[illegible]	[illegible]
32[d]	Pawlo	Ethiopia	0.55	4.65	5.5	1.05	11.75	10.20	0.9	< 0.1	11.20	9.2	1186[a]	1963: IV, V 1965: VI
33	Pink Lake	Australia	2.55	657	6816	26.5	7512.05	11.75	7696	—	—	8.4	270 349[b]	1970: 1–V + 7–VII
34	de Port-Bielh	France	0.19	0.01	0.02	0.01	0.23	0.21	< 0.001	—	—	8.0	28	1968: IV–X
35	Red Rock Tarn	Australia	0.65	8.10	348	22.40	379.15	169	132	—	—	9.7	32 035[b]	1970: 1–V + 8–VII
36	Rybinsk	USSR	1.47	0.48		0.16[f]	2.11	1.80	0.04	0.35	2.19	7.5	235[a]	
37	Saidenbach Res.	DDR	0.95	0.75	0.28	0.10	2.08	0.5	0.5	1.11	2.11	6.6	268[a]	1970–1971
38	Sammamish	USA	0.38	0.25	0.19	< 0.026	0.85	0.68	0.07	0.14	0.90	8.1	102	summer 1971
39	Slavantas	USSR	2.04	1.14	0.18	0.06	3.42	3.08	0.14	0.20	3.42	8.4	362[a]	summer 1971
40	Tjeukemeer	Netherlands	2.63	0.97	2.72	0.25	6.57	1.82	3.43	0.98	6.23	8.5	526[a]	1970–1972
41	Trummen	Sweden	0.77	0.28	0.51	0.1	1.66	0.38	0.66	0.62	1.66	7.3	192	1969–1972
42	Vechten	Netherlands	2.38	0.70	0.48	0.04	3.60	2.69	0.47	0.29	3.45	8.1	387[a]	1968
43[e]	Vico	Italy	1.19	0.94	1.37	0.64	4.14	2.52	0.43	0.84	3.79	8.4	400	IX'67–IV'70
44	Washington	USA	0.33	0.29	0.20	< 0.026	0.84	0.59	0.08	0.15	0.80	—	104	summer 1971
45	Yunoko	Japan	0.80	0.06	—	—	—	0.39	0.19	0.65	1.04	7.1	—	VII 1968–VI 1969

[a] Calculated conductivity [b] Conductivity has been calculated and is larger than the measured conductivity. [c] Approximate value.
[d] Data taken from Prosser, Wood & Baxter (1968). [e] Data taken from Gerletti (1974). [f] Values are the sum of Na^+ and K^+.

(plus carbonate) is still the dominating anion in this new 'standard' composition although its percentage is changed from 74% to 82% (table 4.1). Calcium and magnesium are present in roughly equal amounts, but are also changed considerably and are now 34% and 33% respectively, from about 64% and 17%. Like magnesium, sodium and potassium are also much increased. Unlike the 'standard' fresh water, in which calcium is the dominating cation, and (bi)carbonate the dominating anion, sodium and magnesium may dominate in many African waters (e.g. Lake Tanganyika), while in some other waters sodium may be the dominating cation. A very clear example of this is the sodium dominance over calcium as well as over magnesium in Lake Chilwa in Malawi. This lake contains about $24\,\mathrm{mmol\,l^{-1}}$ cations of which sodium is 94%. Less pronounced than in Lake Chilwa is the sodium dominance in four lakes of the upper Qu'Apelle River system in Canada, Echo Lake, Pasqua Lake, Buffalo Pound and Katepwa Lake (see table 4.3). Bayly & Williams (1972) and Williams & Hang Fong Wan (1972) found that also in many waters in Queensland (Australia) sodium dominates the cations and chloride the anions both for dilute and saline waters. Some incidental data from some IBP saline lakes, Coragulac, Pink Lake, Red Rock Tarn and Corangamite in Victoria (Australia) show the same feature as mentioned above (see table 4.4). Many such lakes were also found in Tasmania (Australia) by Buckney & Tyler (1973). No explanation is given by these authors for the frequently high sodium chloride content, of which the normal sources are ocean spray, natural or artificial dissolution of marine evaporites or, in delta and other coastal areas, seawater penetration. From the ratio $(Ca^{2+} + Mg^{2+})/(Na^+ + K^+)$ (fig. 4.3) it is apparent that in half of the lakes the divalent cations calcium and magnesium dominate over the monovalent cations sodium and potassium, although in a few lakes the amount of divalent cations is balanced by monovalent cations. Only in saline lakes (high conductivity) do sodium and potassium predominate. In most European lakes in table 4.4 the ratio $(Ca^{2+} + Mg^{2+})/(Na^+ + K^+)$ is larger than 1 and divalent cations and bicarbonate dominate. In most of the lakes outside Europe this ratio is less than 1. In intermediate types of water the chloride is roughly equal to the bicarbonate content.

The electrical conductivity of any water is roughly proportional to the concentration of its dissolved major elements. It varies strongly between water bodies ranging in value between 10 and over $100\,000\,\mu\mathrm{S\,cm^{-1}}$ ($\mu\mathrm{S} = \mu$ Siemens = $\mu\mathrm{m\,hos\,cm^{-1}}$ ($20°\,\mathrm{C}$)). An approximate estimate of the quantity of dissolved ionic matter in $\mathrm{mg\,l^{-1}}$ in a water sample may be made by multiplying the specific conductivity by an empirical factor varying from 0.55 to 0.9 depending on the nature of the dissolved salts. A similar estimate for $\mathrm{mmol\,l^{-1}}$ may be made by multiplying the conductance by 0.01.

Talling & Talling (1965) gave examples of many waters in Africa with exceptionally low and high mineral concentrations. They divided the lakes in

Africa into three classes according to their conductivity value. Class I had a conductivity $< 600\,\mu S\,cm^{-1}$, Class II $600-6000\,\mu S\,cm^{-1}$ and Class III $> 6000\,\mu S\,cm^{-1}$. Some of the lowest conductivities (Class I) are found either in the Zaire Basin, or in lakes whose inflow drains through swampy regions. These lakes are frequently dark-coloured and contain a high concentration of organic compounds (compare with the black waters of Amazonia, see review by Golterman, 1975b). Class II includes large waters such as Lake Turkana (Rudolf), one of the most saline in this group, which occupies a closed basin which formerly contained a larger lake with an outflow to the Nile. The salinity in these lakes is due largely to sodium, chloride and bicarbonate. Class III includes lakes in closed basins with no outflows; under these conditions salts accumulate. The highest conductivity which Talling & Talling measured was $160\,000\,\mu S\,cm^{-1}$ with alkalinities up to 1.5 to 2.0 mmol l^{-1}.

Because many waters are calcium bicarbonate dominated if one considers them as a whole a strong correlation exists between conductivity and calcium, bicarbonate and carbonate content. IBP results for major ions are summarised in table 4.4 and figs. 4.1 and 4.2. From these figures it can be seen that both calcium (fig. 4.1a) and bicarbonate (fig. 4.2a) concentrations are around

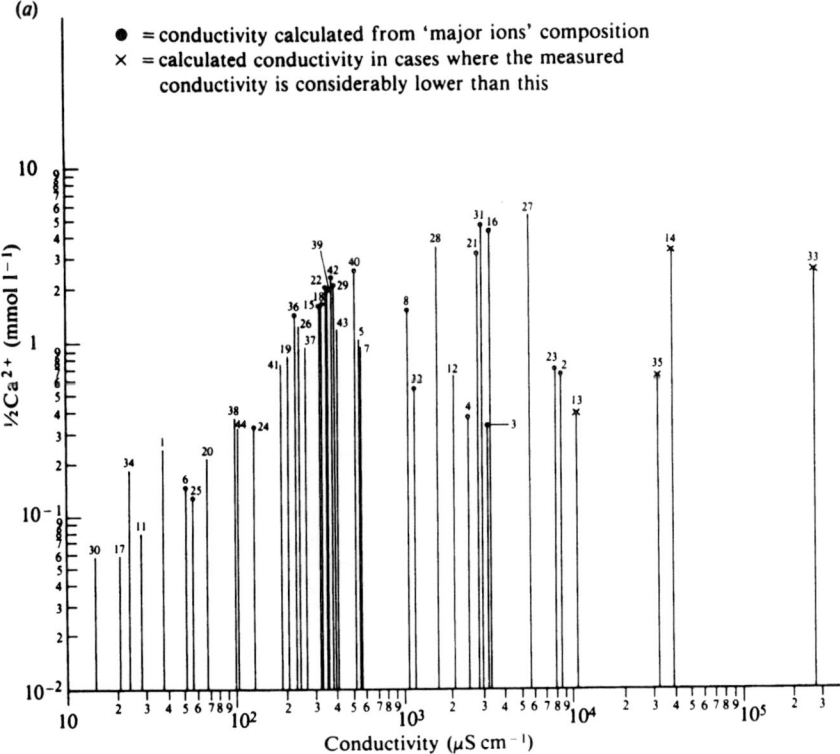

(a)

● = conductivity calculated from 'major ions' composition
× = calculated conductivity in cases where the measured conductivity is considerably lower than this

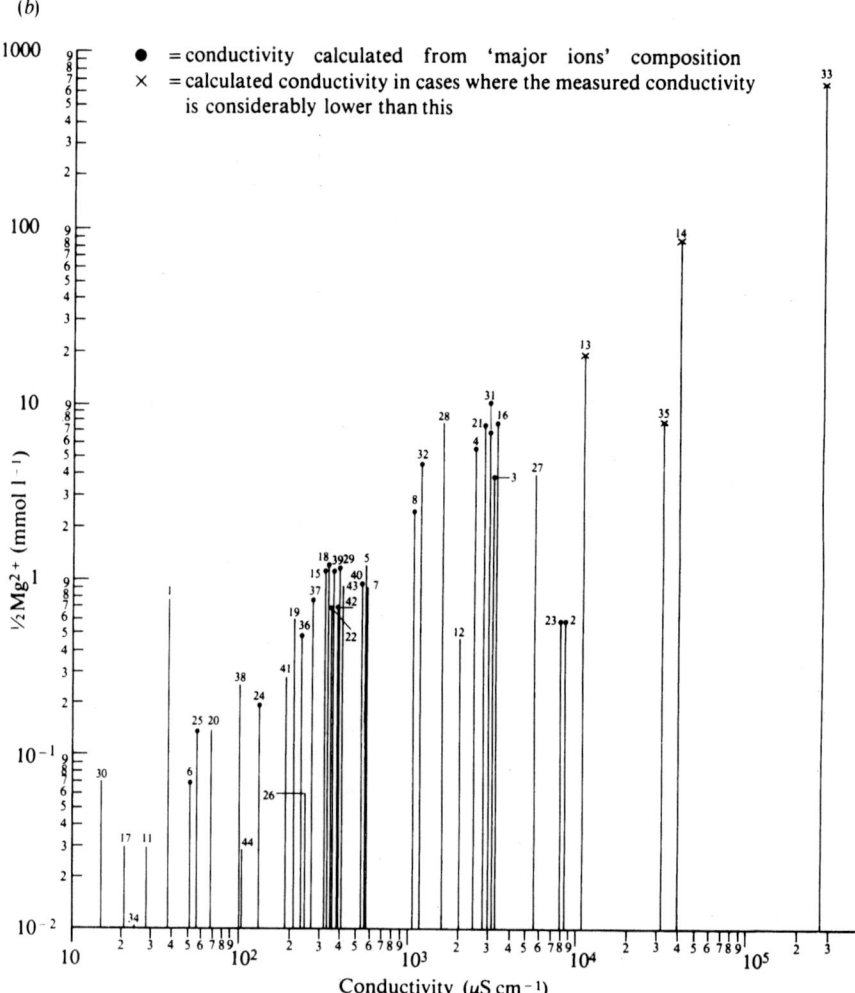

Fig. 4.1. The concentration of the major cations (*a*) Ca^{2+}, (*b*) Mg^{2+}, (*c*) Na$^+$ and (*d*) K$^+$ (mmol l^{-1}) in worldwide lakes arranged in order of increasing electrical conductivity (μS) (log–log scale). Numbers indicate lake names as listed in table 4.4.

(c)

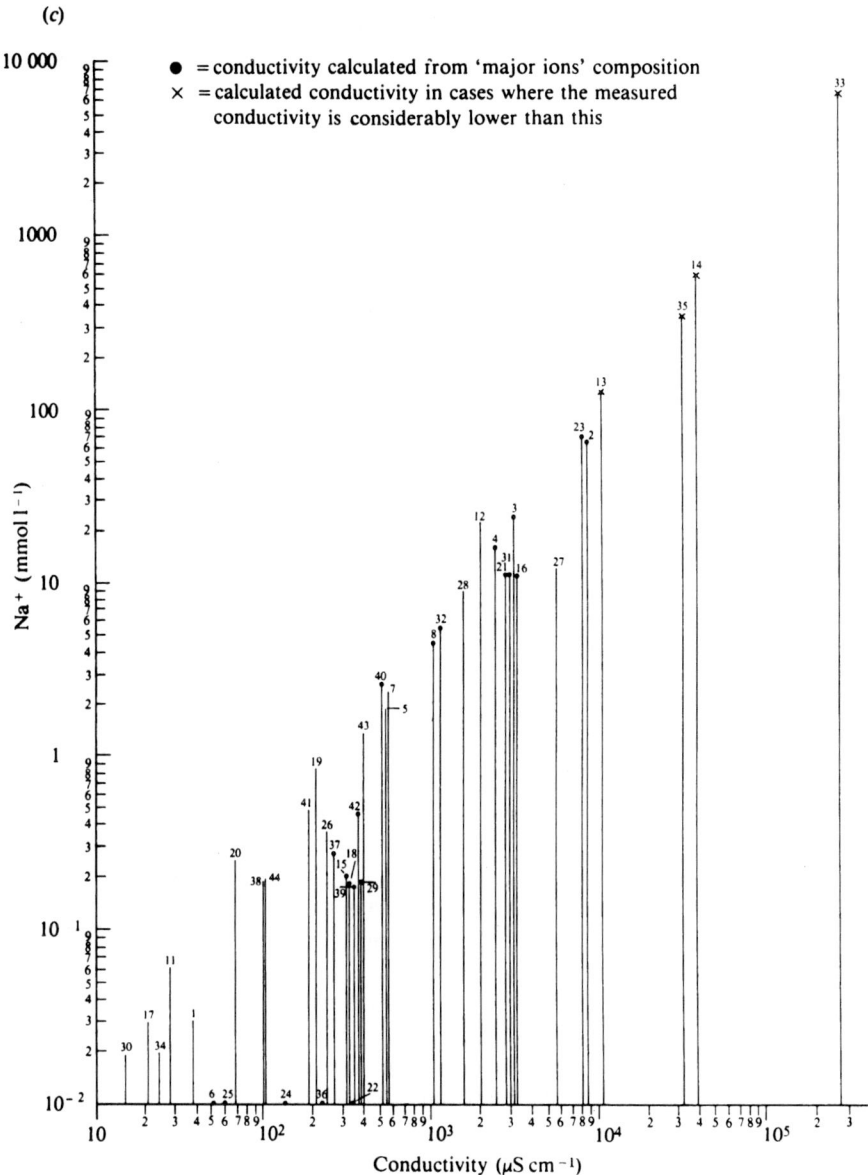

(d)

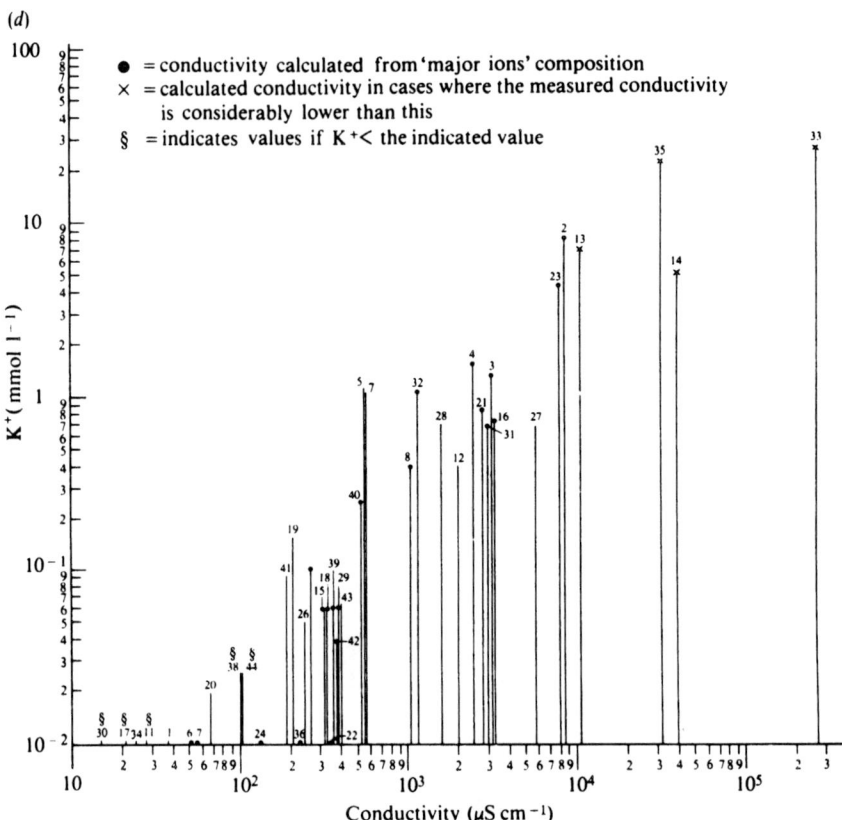

0.2 mmol l^{-1} at the lower conductivity values and increase up to conductivity values between 100–500 μS cm^{-1}, above which calcium (generally) decreases and bicarbonate (and carbonate) still increase. In fig. 4.1c a pronounced continuous increase of sodium with increasing conductivity is apparent, demonstrating that in saline waters sodium replaces calcium in the relationship to conductivity.

Unlike the situation in African lakes as described by Talling & Talling (1965) (and for only one of which chemical data are represented in the IBP collection), magnesium does not follow the same pattern as calcium, and increases with increasing conductivity (fig. 4.1b). Potassium (fig. 4.1d) is apparently always present in noticeably smaller amounts than sodium and the difference between the two tends to become larger as conductivity rises, although potassium follows the same pattern as sodium with increasing conductivity. Among the anions (fig. 4.2) the sum of carbonate and bicarbonate dominates, and these two ions account for the correlation between alkalinity and conductivity (see fig. 4.2*a*). The proportion of bicarbonate and carbonate in the total ions is not very regular, with the majority of lakes falling in a 0.20–0.45 range. This proportion does not vary much with increasing salinity. Talling & Talling (1965) found a range of 0.6–0.8 in African lakes.

From fig. 4.4 it can be seen that the pH generally rises with increasing carbonate plus bicarbonate alkalinity. Lakes with pH > 10 are not very common and most of the lakes lie within the pH range 7.5–9.0. Values between

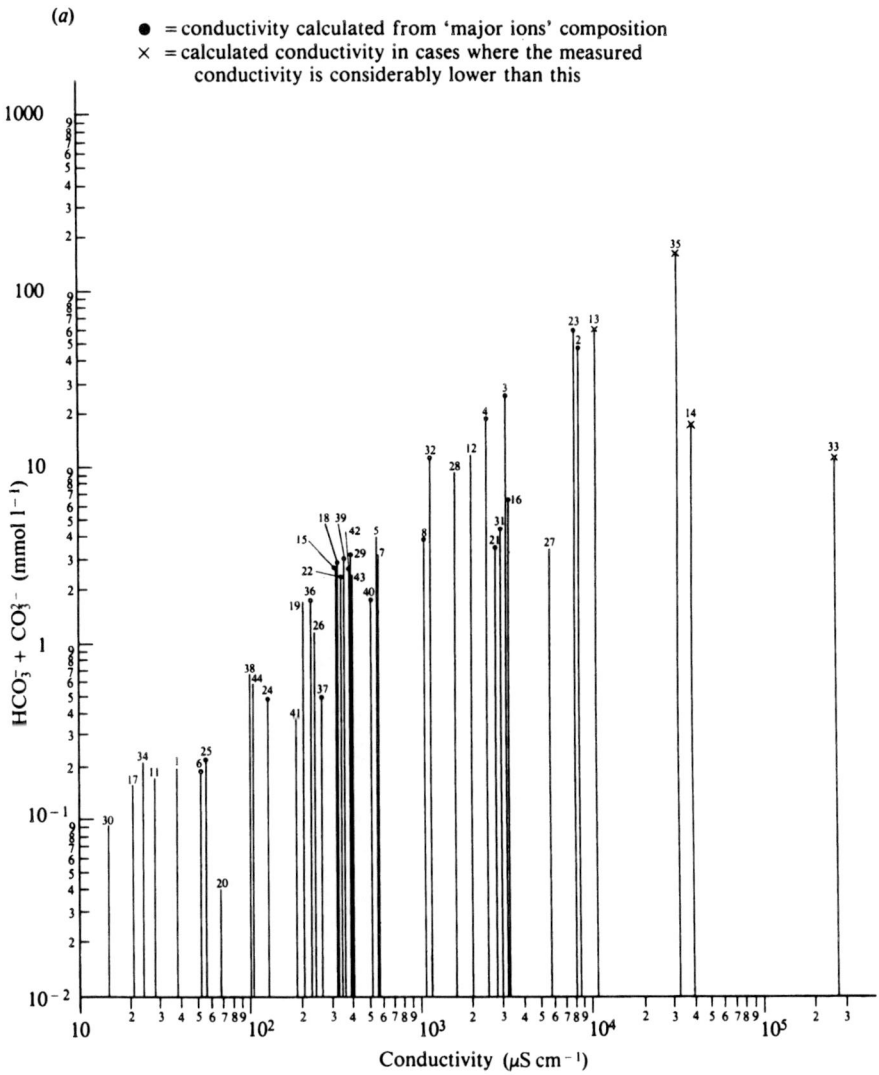

(*a*)

● = conductivity calculated from 'major ions' composition
× = calculated conductivity in cases where the measured
     conductivity is considerably lower than this

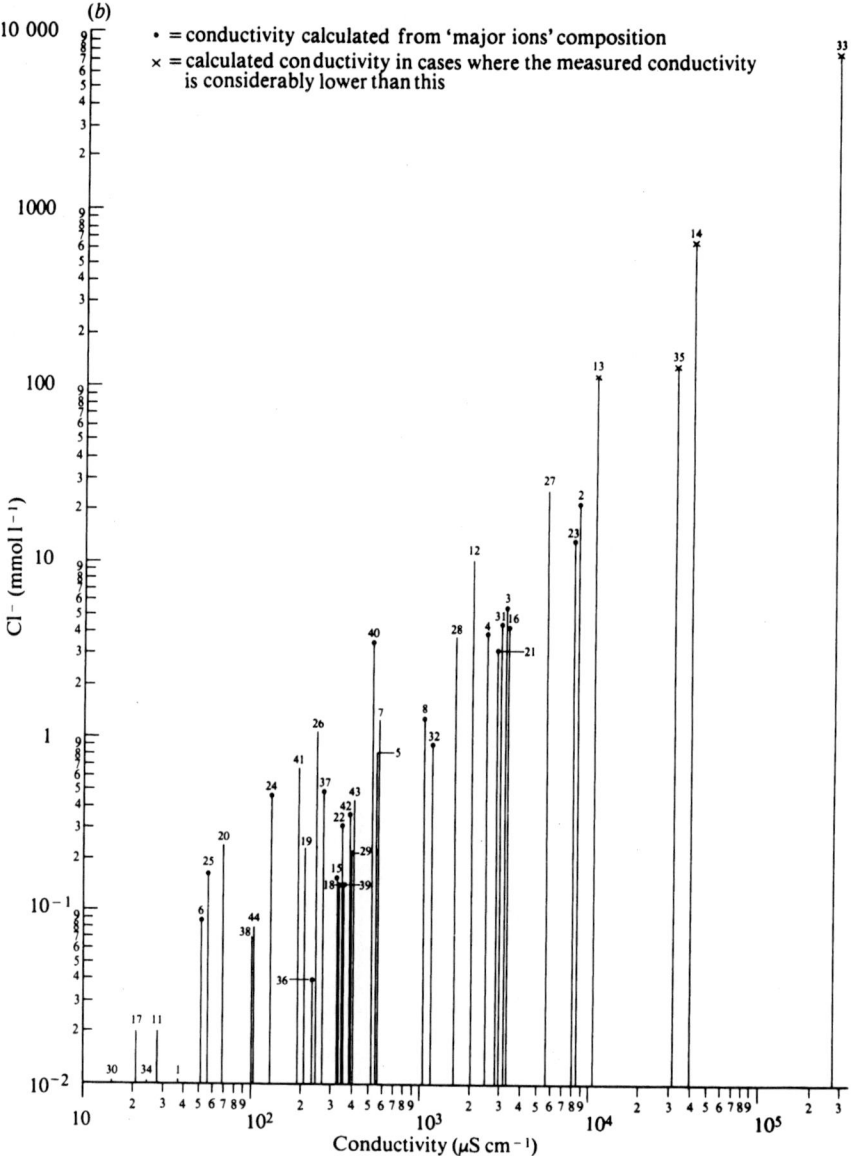

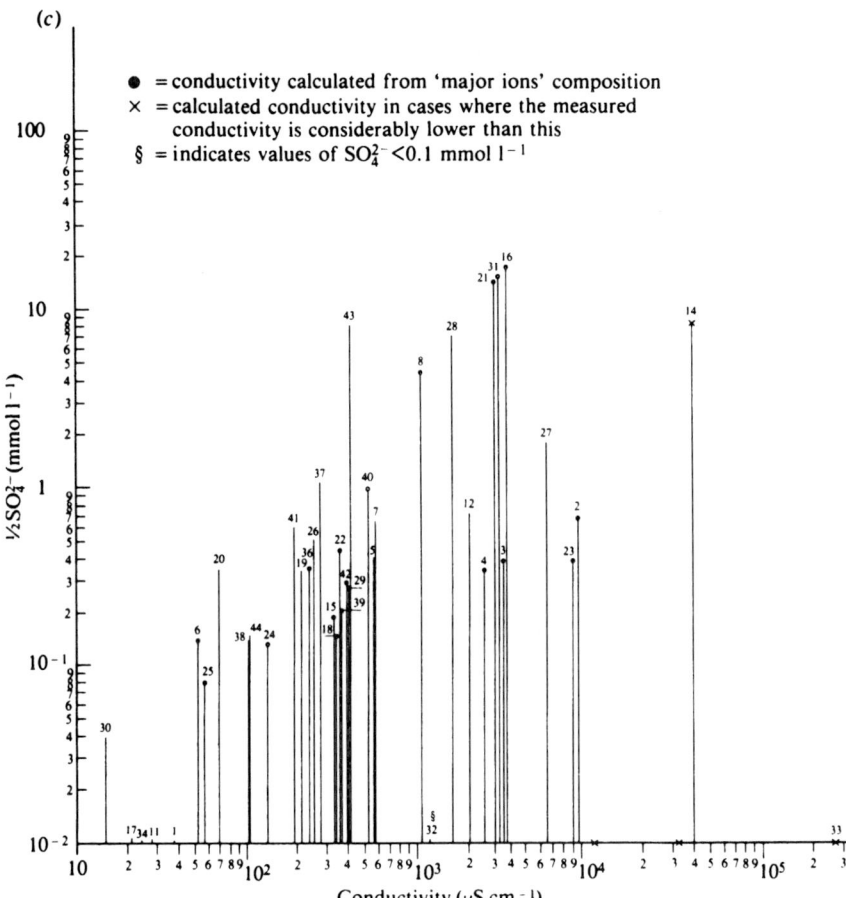

Fig. 4.2. The concentration of the major anions (in mmol l^{-1}) (a) $HCO_3^- + CO_3$, (b) Cl^-, (c) SO_4^{2-}, arranged as in fig. 4.1. Numbers indicate lake names as listed in table 4.4.

four and seven are not uncommon in the weakly buffered lake and river waters that have a conductivity less than 50 μS cm^{-1} (Berg, 1961). In such waters the bicarbonate alkalinity may be undetectable, and hydrogen ions will contribute significantly to the total ionic concentration when pH = 4 or less (Berg, 1961, 1962). In a survey of lakes in northern Sweden, A. Nauwerck (personal communication) found a pH value as low as 2.8 in one case. Chloride increases with increasing conductivity (figs. 4.2b, c) and especially in Coragulac Lake, Pink Lake and Lake Corangamite. Chloride exceeds the other anions in a way parallel to that in which sodium dominates the cations in the same lakes. Although these data were collected incidently they are enough to show that these waters can be considered good examples of NaHCO$_3$–NaCl lakes.

Sulphate follows quite an irregular pattern compared to the other anions in relation to conductivity (see fig. 4.2c) and as a proportion of the total ionic concentration, varies within a range of 0.001–0.1 for the majority of the lakes. In some lakes this fraction exceeds 0.2 e.g. Echo Lake, Buffalo Pound, Katepwa Lake, Pasqua Lake, Lake Trummen (data reports for 1971 and 1972).

The only major ions that were intensively studied in IBP projects were calcium and bicarbonate. They are generally thought to be essential to measure for the understanding of primary production, chiefly because they are important in controlling and buffering the pH of the water.

4.1.2. Origin of and interactions between constituents

Many interactions between major and minor elements occur due to solubility products, adsorption processes and chelating mechanisms. Important interactions are those between calcium, carbonate and phosphate.

Calcium may enter a lake from two sources: erosion of primary minerals and dissolution of sedimentary marine limestone. The first process will bring

(a)

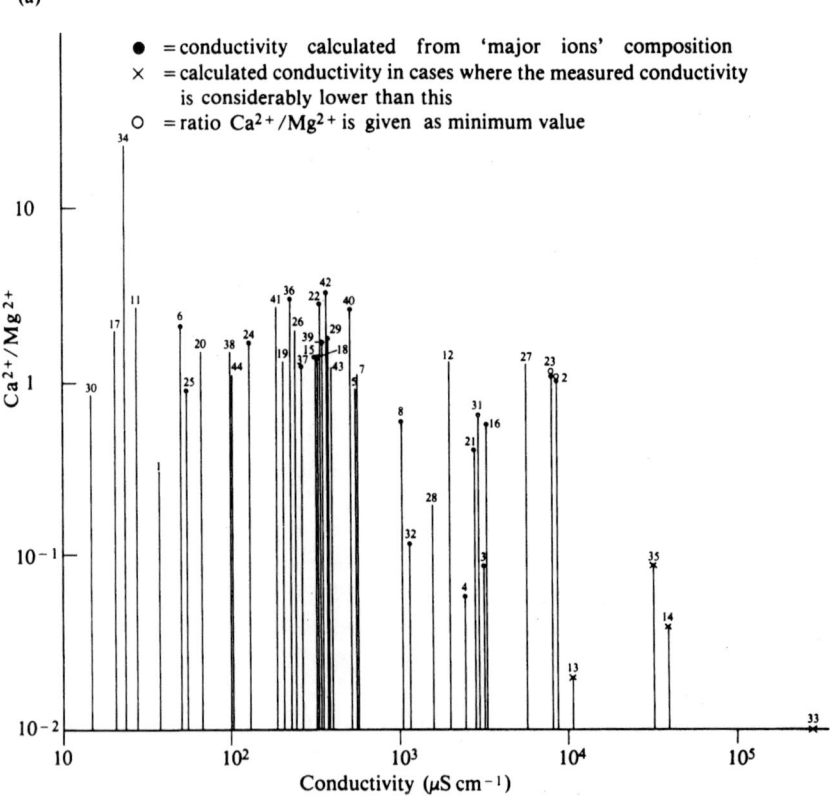

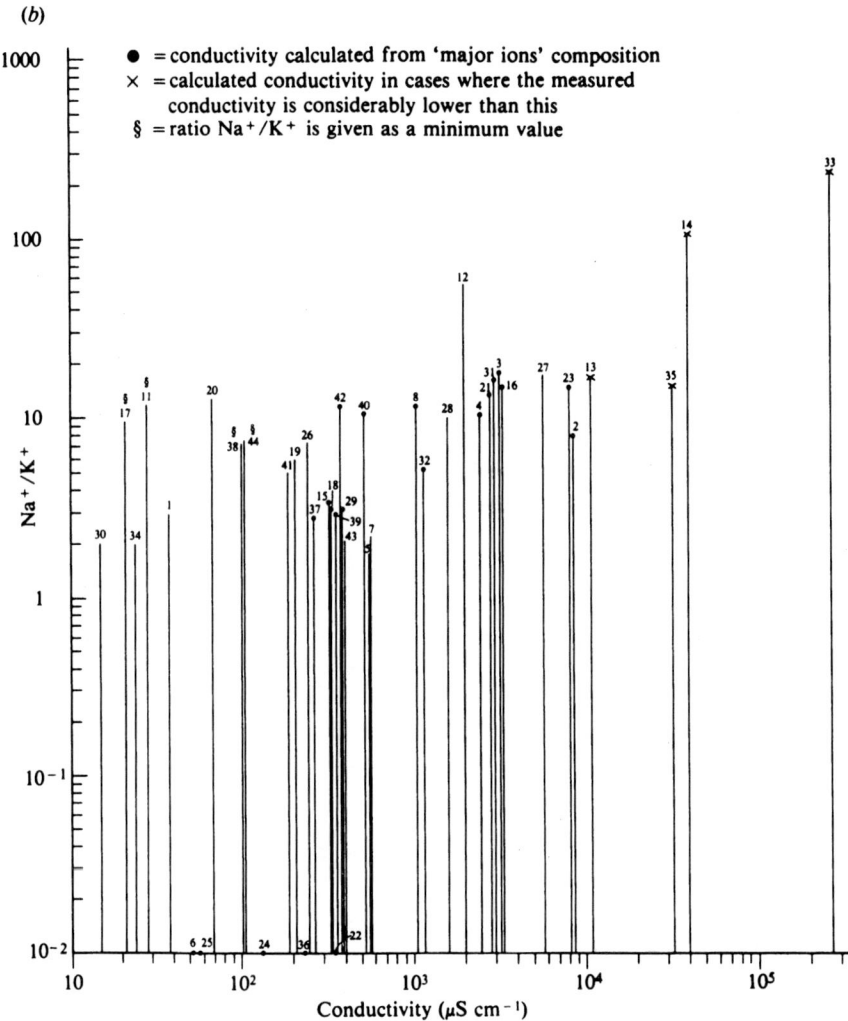

Fig. 4.3. Ratios of concentrations of (*a*) Ca^{2+}/Mg^{2+}, (*b*) Na$^+$/K$^+$ and (*c*) (Ca^{2+} + Mg^{2+})/ (Na$^+$ + K$^+$), arranged as in figs. 4.1 & 4.2. Numbers indicate lake names as listed in table 4.4.

(c)

low concentrations into solution because of the low erosion rate of processes concerned, such as:

$$2NaSi_3AlO_8 \cdot CaSi_2Al_2O_8 + 6CO_2 + 17H_2O \rightarrow$$
$$\text{anorthite}$$

$$3Al_2Si_2O_5(OH)_4 + 4H_4SiO_4 + 2Na^+ + 2Ca^{2+} + 6HCO_3^-$$
$$\text{kaolinite}$$

while the dissolution of limestone may deliver up to 2 mmol l^{-1} to lakes which are in equilibrium with air. For inflowing rivers, however, in which time has been insufficient to effect complete equilibrium with air much higher concentrations may be found, e.g. in the Rhine up to 6 mmol l^{-1}, 5.3 in Locomotive Spring, 4.7 in Pasqua Lake, 4.4 in Echo Lake. This quantity is

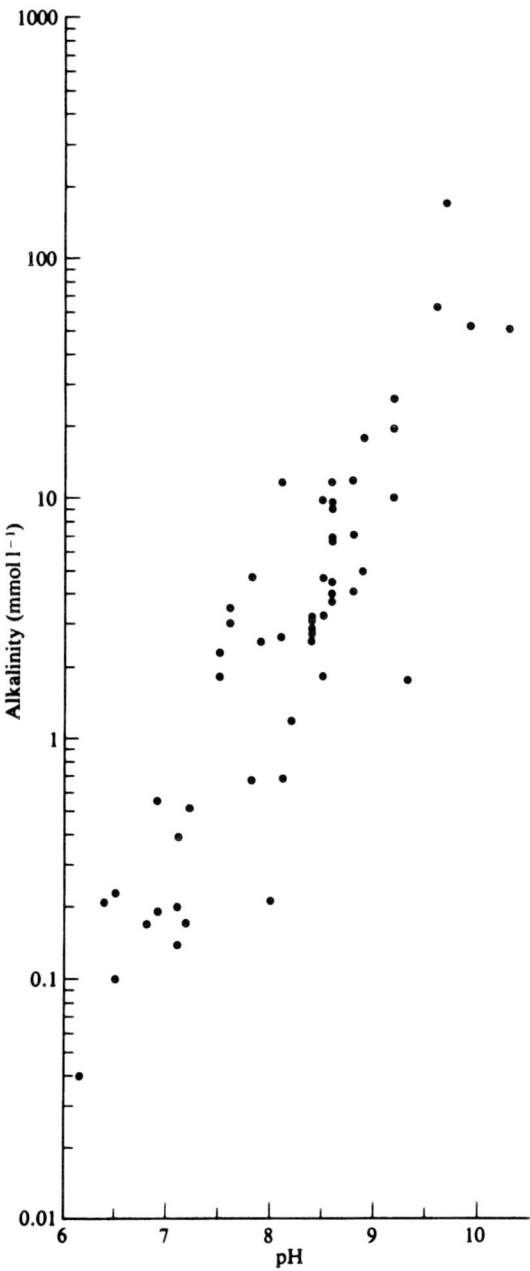

Fig. 4.4. The relationship between pH and alkalinity (log–linear scale). Values are taken from table 4.4, though in some cases data for each year separately were taken instead of mean values as listed in table 4.4.

kept in solution due to an excess of free carbon dioxide, which in the case of the Rhine seems to originate from pollution. Production of free carbon dioxide from organic waste has seldom been considered to be due to this source, although its existence can be demonstrated by the concomitant lack of dissolved oxygen in spite of oxygen diffusion from the air. Normally, however, the excess of free carbon dioxide comes from soils through which rainfall destined for rivers has percolated, and from (volcanic) gases. Dissolved gases will fully equilibrate with the air only after the river water has subsequently experienced a period of relatively sustained stable conditions in a lake. The excess carbon dioxide will escape and the excess calcium carbonate will precipitate. During photosynthesis more carbon dioxide is taken up and a further shift from bicarbonate to carbonate occurs, causing more calcium carbonate to precipitate. There are numerous published examples of this, for instance Wetzel, Rich, Miller & Allen (1972) estimated a precipitation of calcium carbonate of $196 \, \mathrm{g \, m^{-2} a^{-1}}$ at the central depression of Lawrence Lake. In calcium-rich waters, therefore, the calcium concentration is often controlled by the carbonate concentration. However, a supersaturation is often found, i.e. the measured product of $[Ca^{2+}][CO_3^{2-}]$ is higher than 10^{-8}, sometimes considerably so. Fig. 4.5 has some examples of calcium carbonate supersaturation in lakes showing the range through which this occurs throughout the annual productive cycle. Supersaturation can thus reach considerably high values, up to 1360 times the solubility product of calcium carbonate (taken as 1×10^{-8} at 20 °C) in the case of Red Rock Tarn (period May–July 1970.) Some lakes e.g. Chilwa, Pasqua, Echo, Sammamish, Lago Maggiore and the four Australian saline lakes: Coragulac, Pink Lake, Red Rock Tarn and Corangamite, are supersaturated throughout the whole year or growing season. However, the annual extremes for these lakes are not as separated as they are in the other lakes which are apparently supersaturated only during certain parts of the year. For instance in Loch Leven the difference between the extremes is a factor of about one hundred i.e. the range being 0.12×10^{-8} to 12×10^{-8}. An especially interesting case in this respect is Lake Trummen in Sweden. During the years 1970 and 1971 a restoration programme was carried out by sludge dredging. Before restoration the $[Ca^{2+}][CO_3^{2-}]$ product ranged from 6×10^{-11} to 4×10^{-8}. During and after restoration a pronounced change was brought about in that no supersaturation occurred from 1970 onwards (highest summer value in 1972: 7.5×10^{-10}) and in association with this pH decreased from 9.7 to 8.4. It might be suggested that this lower pH and supersaturation was caused by decreased algal growth.

Golterman (1973a) found the supersaturation to increase with increasing pH values in two Dutch lakes. Values from Tjeukemeer an eutrophic, organic-rich lake, and oligotrophic Lake Vechten fitted roughly on the same curve of pH against supersaturation. Collating this with hard-water IBP lakes it is seen

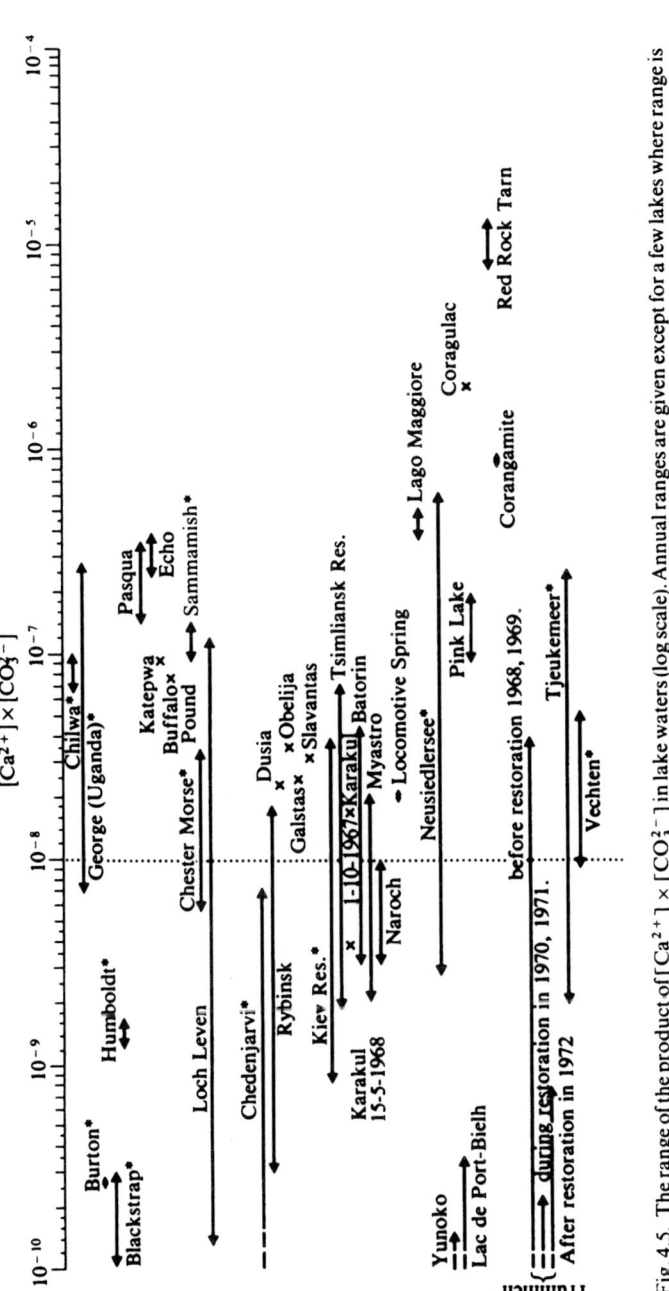

Fig. 4.5. The range of the product of $[Ca^{2+}] \times [CO_3^{2-}]$ in lake waters (log scale). Annual ranges are given except for a few lakes where range is for growing season (these are marked with an asterisk). Solubility products: 1×10^{-8} at 20 °C. The $[Ca^{2+}] \times [CO_3^{2-}]$ product was less than 10^{-10} for several lakes: these were Aleknagik, Øvre Heimdalsvatn, Krugloye, Krivoye, Bolshoy Kharbey, Hakojärvi (Finland), Lipno Res. (Czechoslovakia), Vorderer Finstertaler See (Austria) and Pääjärvi (Finland).

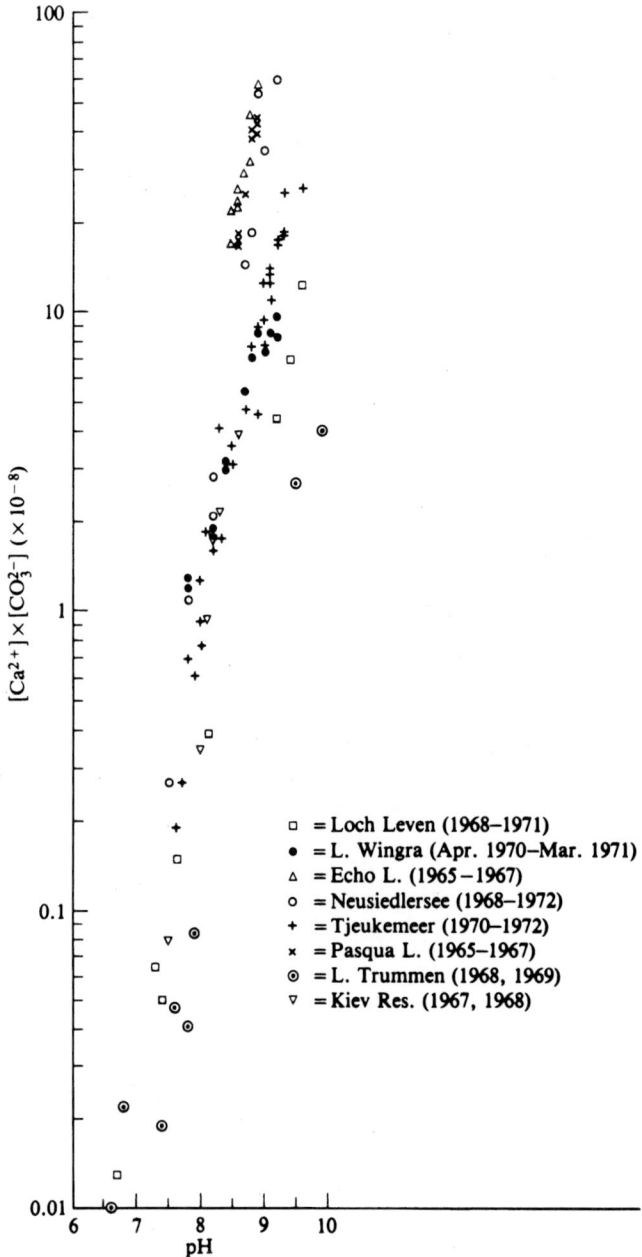

Fig. 4.6. The relation between the product of $[Ca^{2+}] \times [CO_3^{2-}]$ and the pH in several lakes (log–linear scale).

that Lakes Echo, Pasqua, Wingra, Trummen, Neusiedlersee, Kiev Reservoir and Loch Leven all follow the same pattern (fig. 4.6). Unfortunately not enough data from lakes were available to compare the above phenomena more extensively. In these calculations the influence of total ionic strength is not included. Morton & Lee (1968) calculate these activity coefficients, which are 0.74 for Ca^{2+} and 0.96 for HCO_3^- and CO_3^{2-} respectively, if $\mu = 0.0045$. In eutrophic and calcium-rich lakes the calcium concentration may influence the phosphate concentration through the solubility product of hydroxyapatite, which is:

$$[Ca^{2+}]^{10} \times [PO_4^{3-}]^6 \times [OH^-]^2 \simeq 10^{-105} \qquad (4.1)$$

In this formula (PO_4^{3-}) stands for the real trivalent *o*-phosphate concentration and is:

$$[PO_4^{3-}] = \tfrac{1}{f}[HPO_4^{2-} + H_2PO_4^- + PO_4^{3-}] \qquad (4.2a)$$

where

$$f = 1 + 2.8 \times 10^{12}[H^+] + 1.4 \times 10^{19}[H^+]^2 \qquad (4.2b)$$

so that at pH 8: $f = 3 \times 10^4$, while at pH 10: $f = 3 \times 10^2$. From this formula it can be calculated that an increase of pH (e.g., due to photosynthesis) may cause the calcium concentration to decrease (due to the solubility product of calcium carbonate) but does not however increase the *o*-phosphate concentration, because of the greater influence of pH on the *f*-factor. Secondly if calcium carbonate is precipitating, phosphate may be adsorbed as well, causing thus a further reduction of the phosphate concentration (Golterman, 1973b; Otsuki & Wetzel, 1972). Active photosynthesis, encouraged by high phosphate concentrations, will thus lead to a mechanism which may remove phosphate from solution. Nevertheless, the way in which calcium carbonate can exist at supersaturation may be considered to retard this phosphate self-regulatory mechanism. Viner (1975a) showed that in Lake George (Uganda) calcium had a great influence on the phosphate concentration, especially at high pH values and suggested that hydroxyapatite co-precipitation was involved.

Calcium and iron, which both form insoluble phosphates, enter lakes in particulate form as well as in dissolved form. Very little is known about the quantities of these suspended loads. This particulate matter, especially when containing iron, may be active in sedimenting phosphate, and thus have a large influence on the availability of this nutrient, besides the effects of calcium and iron which are present in the dissolved state in the lake.

4.2. The minor elements

4.2.1. Sources of the minor elements

Erosion is the basic mechanism by which elements find their way into natural waters. On average, rock contains 35 phosphorus atoms and 2 nitrogen atoms per 10^4 silicon atoms, so that one may expect that, assuming complete dissolution of the rock by weathering, the rough average (weight) composition of the elements in fresh waters would be $10^4 : 35 : 1$ for Si : P : N. In reality part of the rock silicate goes to form soil so that the proportion of phosphorus and nitrogen relative to silicon delivered to the drainage will be higher than those ratios. Nitrogen will further be greatly increased by the development of a terrestrial ecosystem within the catchment area with the consequent high likelihood of some degree of microbial nitrogen fixation there. Although some ammonium may be adsorbed onto clay, oxidation to nitrate will however prevail and as nitrate is not adsorbed it will largely be washed out, especially

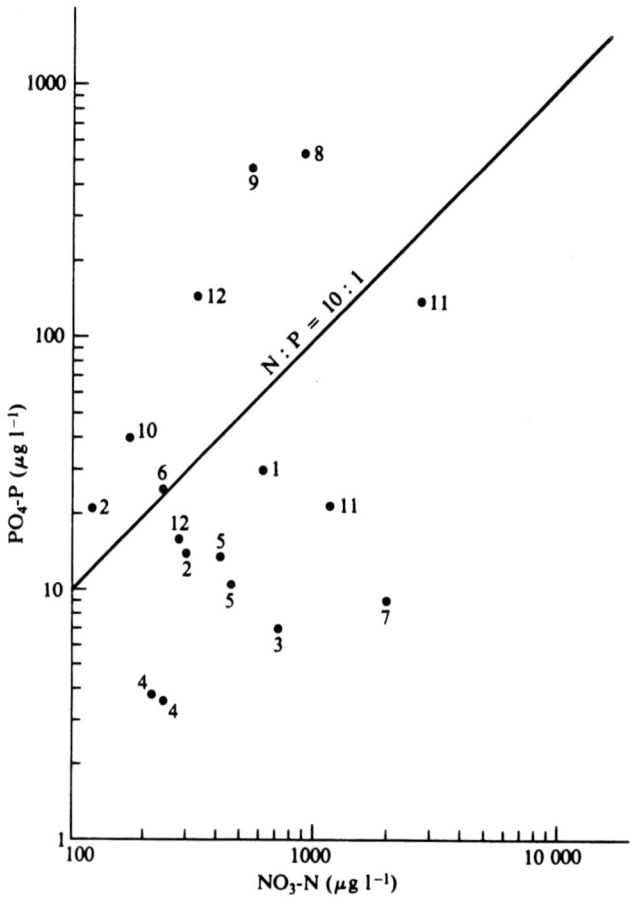

Fig. 4.7. The maximum winter concentration of PO_4-P compared with maximum winter concentration of NO_3-N in several lakes (log–log scale). 1, L. Wingra, 2, L. Aleknagik; 3, L. Maggiore; 4, Vorderer Finstertaler See; 5, L. Sammamish; 6, L. Krasnoye; 7, L. Chedenjarvi; 8, L. Burton; 9, L. Blackstrap; 10, L. Yunoko; 11, Tjeukemeer; 12, L. Vechten.

with heavy showers as often occur in the tropics. Because phosphate may be more extensively adsorbed onto clay it will solubilise more slowly than will nitrogen, so that nitrogen is the more quickly mobilised.

Indeed Lund (1970) found evidence that in the nutrient-poor lakes (in the English Lake District) the N : P ratio was considerably higher than ten showing the importance of phosphate as a key element for algal growth, for which a ratio of N : P between ten and sixteen is needed, and showing the greater rate of leaching of nitrogen. Man can modify this natural relationship and for example in lakes receiving sewage both the nitrogen and phosphate concentration will increase and the ratio N : P in the lake will approach ten, i.e. roughly that of the incoming (domestic) sewage. Although there is no simple relationship between winter concentrations and algal growth, these concentrations seem to have some predictive value. In fig. 4.7 the maximum winter concentrations of PO_4-P and NO_3-N are compared for several lakes (unfortunately not enough useful data could be derived from the data reports for a very comprehensive comparison). In just a few lakes NO_3-N and PO_4-P winter concentrations were found to have a ratio of about ten so that probably neither nutrient would be a limiting factor for algal growth in the succeeding summer. However, for the majority of the lakes listed here, the N : P is above ten suggesting phosphate might become limiting before nitrogen would. Nonforested tropical areas provide an exception to this generalisation. In the tropics, due to differences in soil formation and soil erosion by rainfall, lack of nitrogen enrichment by nitrogen fixation and larger losses in the soils by denitrification, tropical lakes in non-forested regions such as found in E. Africa (and even reservoirs in Spain) will tend to have relatively low N : P ratios, and consequently algal growth may be limited by nitrogen (see Talling & Talling, 1965; Talling, 1966a; Moss & Moss, 1969 for African waters; Sioli, 1975 for the Amazon region).

The maximum winter values could indicate the algal growth to be expected in the succeeding summer, but unfortunately not enough data on algal yield in terms of chlorophyll *a* are available from the IBP lakes to check whether there is a relationship between PO_4-P, NO_3-N and algal growth in these lakes. This relationship is discussed later in section 4.2.3.

In spite of its importance a theoretical quantification ('prediction') of the amount of phosphate entering a lake (and its biological consequences) is still a dubious exercise. Perhaps this is more true for a 'scientific' approach than for the practical water manager. The estimation of the amount of phosphate derived from the population in a watershed is relatively easy. The amount of phosphate in human faeces and urine may amount to 2.0 g of phosphorus per capita per day entering sewage. If detergents containing phosphate are included a further 2-3 g of phosphorus will be added (W. Europe, USA mainly). Furthermore data are needed on the percentage of the population having sewers; these figures can be obtained from statistical sources. More difficult to obtain are data on phosphorus-release by septic tanks. Hetling & Sykes (1973) made such a calculation carefully for Lake Canadarago.

The amount of phosphate eroded naturally is more difficult to estimate. Two approaches are possible for this purpose:

(a) An estimate can be made using the erosion in tonnes per km^2 per year, for which figures are available for several watersheds, and the average phosphate content of the rock.

(b) By comparing with a 'standard' element which is less influenced by other sources.

The first method has been applied for lakes in an undisturbed area, by Rigler (1975) who showed that forested igneous watersheds export an average of PO_4-P of 4.7 mg m^{-2} a^{-1}, with a range of 0.7 to 8.8 mg m^{-2} a^{-1}. Rigler, citing Kirchner (1975), said that this variation can be explained as a function of drainage density, the formula being:

$$P \text{ export} = 1.32 + 5.54\, D$$

where P export = the PO_4-P exported from a watershed (mg m^{-2} a^{-1})
        D = drainage density (km^{-1}; river length per km^{-2} catchment area

Besides the difficulty of the interpretation of the dimensions of this formula, it should be noted that such a formula cannot be used in agricultural areas where soil erosion takes place. (See, for example, Dillon & Kirchner, 1975; McColl, White & Waugh, 1975). The second method is probably the most accurate one, although not many data are available. Viner (1975b) measured silicate and phosphate in several rivers in unpolluted areas in Uganda and Golterman (1973b, 1975b) suggested that the ratio, SiO_2-Si : PO_4-P, which Viner found to be around 110 in rivers, could be used to predict phosphate concentrations (sum of dissolved and particulate) from the silicate content. In special cases often involving secondary processes (e.g. erosion from phosphate-rich deposits) this trend may, however, be modified. With this ratio (110) Golterman (1975b) calculated the naturally occurring phosphate in the Rhine to be around 15% of the total phosphate, (while the other sources were human waste 60–80% and agriculture 1.5 to 15%). Comparing the ratio of SiO_2-Si/PO_4-P = 110 with that in rocks (which is 285) it may be suggested that the solubilisation rate of silica is two to three times lower than that of phosphate. The advantage of using silicate as a 'reference' element is that to some extent it takes soil erosion into account.

The situation for nitrogen is much more complex, because nitrogen metabolism in the soil cannot yet be described quantitatively due to the existence of processes such as nitrogen fixation, denitrification and nitrogen accumulation in refractory material (such as humus). Apparently many more

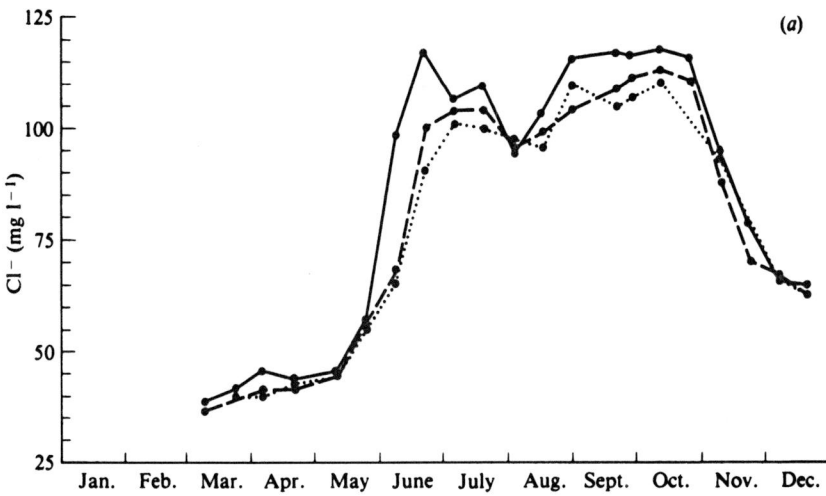

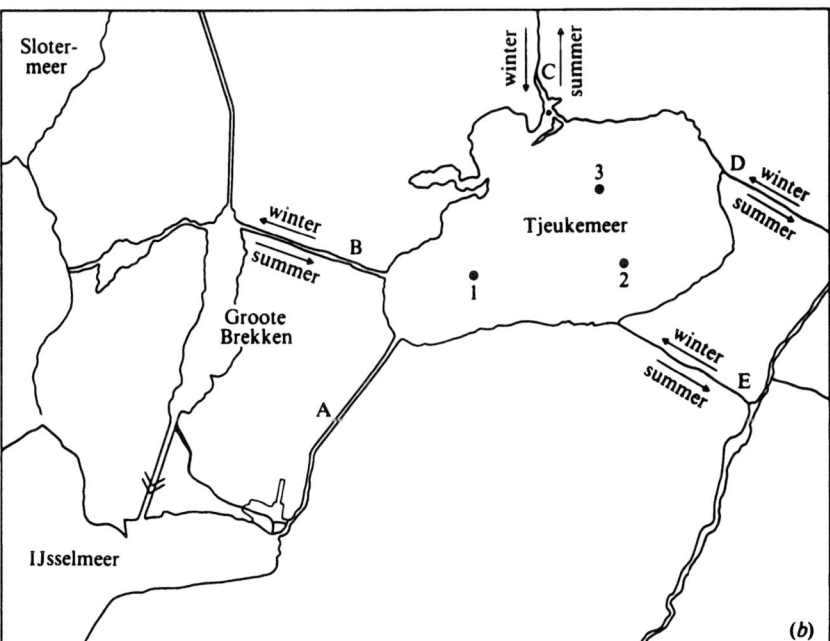

Fig. 4.8. (*a*) Concentration of Cl⁻ at sampling points 1(● —— ●), 2(● — ●), 3(● --- ●) in Tjeukemeer, 1971. Water coming from the west (sampling point 1 western-most) may contain up to 200 mg l⁻¹, that from the east about 40 mg l⁻¹. (*b*) Geographical position of Tjeukemeer (DR 21) showing sampling stations and indicating winter and summer water movements.

data on nitrogen, phosphate and silicate contents of rivers are needed to provide for a satisfactory calculation for nitrogen using this method.

Another standard element that could be used is the chloride concentration, because chloride is a conservative element and can be easily measured accurately (error $= 0.2\%$). As differences in concentration often exist between the lake water and the inflowing river water, renewal times and thus nutrient concentrations can easily be derived by proportionality to the chloride. This type of calculation is easier for lakes situated in the lower reaches of rivers where chloride concentration tends to be higher. For example the water balance from the Dutch IBP site of Tjeukemeer can be calculated from the rate of change in chloride concentration with time (see fig. 4.8). The summer input from the IJsselmeer which receives much chloride from the Rhine, may contain up to 200 mg l^{-1} of chloride, while the winter input from the surrounding polders contains about 40 mg l^{-1}. Horizontal mixing takes place gradually upon entry of the inputs into Tjeukemeer, and can be followed by comparing concentrations at the different sampling points.

4.2.2. Relationships between nutrient loading and concentration

If the annual supply of a chemical element per surface area (mostly in g m^{-2}a^{-1}), the so-called loading of a lake, is known the potential total concentration can be calculated from sedimentation and outflow. If no sedimentation takes place (e.g. sulphate or chloride) the relation between concentration and loading in a vertically and horizontally mixed lake is:

$$-\frac{d[m_o]}{dt} = \frac{Q}{V}([m_o]-[m_i]) = \frac{Q}{V}[m_o] - \frac{L}{\bar{z}} \qquad (4.3)$$

in which $[m_o]$ = concentration in lake (g m^{-3})
            $[m_i]$ = concentration in inflow (g m^{-3})
            Q = through flow (m^3 d^{-1})
            V = lake volume (m^3)
            $\bar{z}$ = mean lake depth
            L = loading (g m^{-2} d^{-1})

This equation is similar to that of Biffi (1963):

$$-\frac{dM}{dt} = \frac{Q}{V} \cdot M_o - J \qquad (4.4)$$

in which J = daily supply (kg d^{-1})
            M_o = total amount in lake (kg)
It must be noted that $L \times S = J$.

No simple relation however exists between concentration and loading, if or when chemical and biological 'sedimentation' also play a role. If sedimentation does occur, as it may for PO_4^{3-} or Ca^{2+}, two alternative equations can again be suggested for the loading calculation assuming either that sedimentation is linearly related to the total daily input (Piontelli & Tonolli, 1964) or to the concentration (Vollenweider, 1964, 1969a).

Piontelli & Tonolli:

$$\frac{dM_o}{dt} = (1 - r)J - Q_d[m_o] \qquad (4.5)$$

Vollenweider:

$$\frac{dM_o}{dt} = (1 - s)\frac{J}{V} - \gamma[m_o] \qquad (4.6)$$

where r = fraction of supply sedimenting
               s = fraction of supply in outflow
               γ = proportion of concentration sedimenting
               Q_d = water outflow ($m^3 \, d^{-1}$)
               V = lake volume (m^3)

Golterman (1975a, pp. 360, 361) has shown that if the concentration in the lake does not show large annual fluctuations (is in near equilibrium conditions), the two equations are normally identical.

However, these equations have the shortcoming that no chemical mechanism can be suggested by which a constant proportion of the concentration of any solute can sediment. If a solution of $Ca(HCO_3)_2$ supersaturated with carbon dioxide enters a lake, calcium carbonate will precipitate until the saturation concentration of calcium carbonate is reached (but see page 104). In calcium-rich lakes the phosphate concentration may be controlled by the solubility product of hydroxyapatite, so that if it is present in excess phosphate will precipitate. During the growing season most phosphate will be taken up by algae, and after mineralisation the refractory components will sediment. Due to carbon dioxide removal and increasing pH during photosynthesis, the calcium carbonate precipitate may adsorb phosphate or form hydroxyapatite. Winter and summer situations are therefore apparently different, and in a parallel manner the English Lake District (calcium-poor water) is different from the Swiss lakes (calcium-rich waters). Indeed Stewart & Markello (1974) showed different correlation lines (chl.*a* against P-loading) for six lakes in western New York and lakes from the English Lake District. These authors showed furthermore the importance of short-lasting spikes due to severe thunder-storms. These short-lasting inputs are easily overlooked in monitoring programmes, causing disturbances in a possible correlation. They may have a lesser biological consequence than the same amount arriving over a long term, because the system may not be able to consume these short loadings biologically. Nitrogen will not participate in an inorganic chemical precipitation; sedimented nitrogen will be in biological materials. Silicate does not precipitate chemically either. The silicate sediment consists of diatom frustules and will therefore accumulate maximally only during short periods of the year.

In spite of these theoretical objections several attempts have been made to relate the phosphate concentration (e.g., the spring concentration) with the phosphate input. Vollenweider (1969a) used the formula:

$$[P] = \frac{L}{\bar{z}(\sigma + \rho)} \quad \text{or} \tag{4.7a}$$

$$[P] = \frac{L}{\sigma\bar{z} + q_s} \tag{4.7b}$$

where $[P]$ = phosphate concentration in lake (g m^{-3})
L = phosphate loading (g m^{-2} a^{-1})
$\bar{z}$ = mean depth
σ = sedimentation rate coefficient (fraction of phosphorus content sedimented yearly; t^{-1})
ρ = replenishment coefficient (t^{-1})
q_s = flow throughrate expressed in height of water column (m a^{-1})

the two formulae are identical because $\bar{z}\rho = q_s$.
Dillon & Rigler (1974a) tested this formula, but because σ cannot be measured directly they calculated the retention coefficient (R) instead.

$$R = 1 - \frac{\rho[P]}{L/\bar{z}} = 1 - \frac{\rho}{\rho + \sigma} \quad \text{or} \quad \sigma = \frac{R\rho}{1 - R} \tag{4.8}$$

They determined R experimentally as

$$R_{exp} = 1 - \frac{q_o[P_o]}{\Sigma q_i[P_i]} \tag{4.9}$$

where q = annual water flow (m^3 a^{-1})
$[P]$ = phosphate concentration (g m^{-3})

and the indices o and i stand for out and in.
Equation 4.7 can thus be written as

$$[P] = \frac{L(1 - R_{exp})}{\bar{z}\rho} \tag{4.10}$$

Dillon (1975) showed the importance of flushing rate for the phosphate concentration resulting from the loading and introduced the parameter $L(1 - R)/\rho$ to express the effects of ρ and R on $[P]$. Dillon & Rigler (1974a) found an excellent agreement between the measured and predicted $[P]$ values, if the results of two of their thirteen Ontario lakes were omitted and also if four oligotrophic Swiss alpine lakes were included. However, Dillon & Rigler (1974a) made a tautology when demonstrating the validity for practical use of equation (4.7a), although they said that they had avoided this danger. Because σ, the sedimentation coefficient cannot be measured directly, they derived σ from R, the retention coefficient, which they defined as the proportion of the total phosphate entering the lake which is retained, or:

$$R = 1 - \frac{\rho[P]}{L/\bar{z}} \tag{4.11a}$$

which can be shown to be identical to

$$R = 1 - \frac{\rho}{\rho + \sigma} \tag{4.11b}$$

In order to avoid a tautology, they intended to calculate R_{exp} from independent parameters, i.e. from equation (4.9). Unfortunately this tautology remains since this formula is identical to the formula (4.11a), because this may be written as

$$R = 1 - \frac{\rho[P]V}{LV/\bar{z}} = 1 - \frac{\rho[P]V}{LS} = 1 - \frac{\text{output}}{\text{input}} = 1 - \frac{q_o[P_o]}{\Sigma q_i[P_i]} \tag{4.12}$$

Real independent means would have been to measure σ directly from the phosphate in the sediment (which is in principle impossible) or to calculate from solubility products, formation of refractory phosphate etc. Attempts to do this only got as far as assuming the amount sedimenting to be a fraction of the input or of the concentration. This latter is the assumption on which Vollenweider calculated the formula Dillon & Rigler used (their formula 1), but the σ thus defined is different from that which Dillon & Rigler used in their formula (2). In this formula σ is defined as

$$\sigma = \frac{P_{in} - P_{out}}{P_w} \tag{4.13}$$

which is numerically different from Vollenweider's definition

$$\frac{dM_s}{dt} = \sigma M_w \tag{4.14}$$

It is perhaps worthwhile to note that Dillon & Rigler's formula (1) does not *need* to be proved true by experimental means, because under the four assumptions made it *is* a mathematical truth, based on an input–output balance. The results of Dillon & Rigler's measurements show that the assumptions made are partly true, or compensate each other in the given situation. Furthermore it must be noted that the predicted Ontario values all fall in the range between 6 and 9 mg m^{-3} of PO_4-P, in which range two of the four Swiss lakes also fall, so that they effectively measured only one point on the curve showing the full relationship between loading and concentration. In nutrient-rich lakes the relationship breaks down because one of the conditions of the formula (equation 4.7b) is not fulfilled, i.e. the lake is not in a steady state. If a lake is not in a steady state – and so many eutrophic lakes are not – the concentration can be calculated only from the (annual) mean balance formula:

$$P_{in} = P_{out} + P_{sed} + \Delta[P]_{lake}$$

All the above cited formulae relating the phosphate concentration in the lake, $[P]_{lake}$, with the loading, (L), assume that $\Delta[P]_{lake}$ is zero, which is not the case in the eutrophic Swiss lakes. For example, in Lake Constance, Wagner (1976) found that the concentration as well as the loading roughly double in 10 years, so the amount of phosphate in the lake should not be neglected in the above-mentioned formulae. From a statistical point of view it seems likely, however, that a quite different curve fitting for the Ontario values would be just as possible (see fig. 4.9). This implies that before these models can be further used criteria have to be found to evaluate goodness of fit.

For Swiss lakes Vollenweider (1969a) found that the few data available did not accord with the expected results and that sedimentation was greater than the predicted amount. He obtained better results if he assumed that sedimentation depends on both concentration and annual net loading:

$$\frac{dM_s^*}{dt} = \sigma[m_w]\bar{Z} + \tau'L \tag{4.15}$$

where M_s^* = nutrient (e.g. P) per m^2 sediment
             $[m_w]$ = concentration of a nutrient in lake water (g m^{-3})
             τ' = fraction of L sedimenting

Golterman (1975a) argued that the results obtained were caused by a 'petitio principi' and that τ' should be obtained from the lake properties instead of being derived from the loading after correction for the outflow. In this respect it should be noted that using Dillon & Rigler's data, it can be shown that R has a negative linear correlation with σ, so that phosphate retention in lakes seems to be a function of the number of flushings per year. Larsen & Mercier (1976) showed that P retention was negatively related to water flow either expressed as q or as ρ_w. They stressed the important point of not only considering the phosphate loading but also the phosphate concentration in the inlet water.

The present authors feel that although these models may be useful for the practical water manager, they have no use for the precise scientific understanding of phosphate dynamics. The practical water manager will be restricted in using these models because of the apparently large confidence limits (although

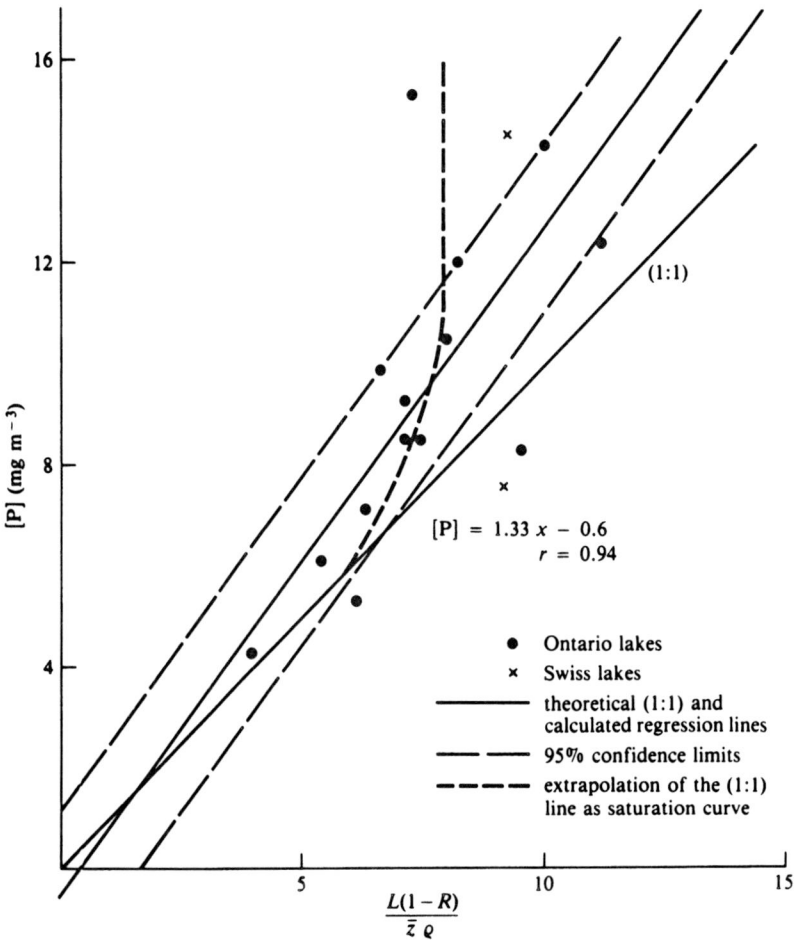

Fig. 4.9. Predicted and measured phosphate concentrations in fifteen Ontario lakes and two Swiss lakes (Dillon & Rigler, 1975). Indicated are the lines showing the theoretical ratio (1 : 1), the best-fitting linear relationship, with its 95% confidence limits, and the best-fitting saturation curve (extrapolation of the 1 : 1 line).

they are never calculated), and because of the lack of knowledge of the boundary limits in which these models may be applied. One of the more 'scientific' phosphate models is that of O'Melia (1972), which takes vertical transport into account. It is summarised by Golterman (1975a) as follows:

$$V_e \frac{d[P_t]_e}{dt} = W - Q[P_t]_e + k_z A_e \frac{d[PO_4\text{-}P]}{dz} - \sigma V_e[PP]_e \qquad (4.16)$$

where $W = $ input
$V_e = $ volume of epilimnion (1)

$Q[P_t]_e$ = rate of removal in lake discharge (from epilimnion)
$\sigma V_e[PP]_e$ = rate of sedimentation of part-P to the hypolimnion
σ = sedimentation coefficient (a^{-1})
$[PP]_e$ = concentration of part-P in epilimnion $(\mu g\, l^{-1})$
k_z = vertical mixing coefficient $(m^2\, a^{-1})$
A_e = horizontal area of thermocline (m^2)
$d[PO_4\text{-}P]/dz$ = gradient of phosphate across the thermocline

The term $k_z A_e (d[PO_4\text{-}P])/dz$ describes the input of phosphate to the epilimnion by diffusion from the hypolimnion. This vertical flux can exceed input of phosphate from land runoff during the summer period of stagnation. O'Melia calculated that if $k_z = 0.05$ cm^2 s^{-1} in a thermocline 5 m thick with a gradient of 20 $\mu g\, l^{-1}$ per 5 m, then the vertical flux is:

$$(0.05\,\text{cm}^2\,\text{s}^{-1})(20 \times 10^{-6}\,\text{g}\,\text{l}^{-1})\left(0.2\,\text{m}\frac{10^3\,\text{l}}{\text{m}^3}\frac{\text{m}^2}{10^4\,\text{cm}^2}\right)(3.16 \times 10^7\,\text{s}\,\text{a}^{-1})$$

$$= 0.6\,\text{g}\,\text{m}^{-2}\,\text{a}^{-1} \text{ of } PO_4\text{-P}$$

This is equal to the estimated rate of addition of phosphate from the land to Lake Lucerne. It seems likely that k_z is really much greater; eddy diffusion has been ignored, while gradients may be much higher than 20 $\mu g\, l^{-1}$ per 5 m. Direct solution of equation 4.16 is not possible as $[P_t]_e$, $[PP]_e$ and $d[PO_4\text{-}P]/dz$ are interrelated, vary with time, and are not measurable directly.

Assuming steady-state conditions, $V_e(d[P_t])/dt = 0$, and that input, W, equals lake discharge, $Q[P_t]$, equation 4.16 becomes:

$$k_z A_e \frac{d[PO_4\text{-}P]}{dz} = \sigma V_e[PP]_e \quad \text{or} \quad \frac{[P_t]_e}{[P_t]_h} = \frac{k_z}{Z_e Z_t} \qquad (4.17)$$

as $[PO_4\text{-}P]_h = [P_t]_h$ and $[PP]_z = [P_t]_e$. With $\sigma = 0.02\,\text{d}^{-1}$ and using measurements obtained by Gächter (1968) O'Melia found that $[P_t]_e/[P_t]_h = 0.3$, which seems a reasonable value and suggests that equation 4.16 is a fairly good description of the real situation.

If total phosphate behaves conservatively, then:

$$\bar{t}_{H_2O} = \bar{t}_{P_t}$$

where $$\bar{t}_{H_2O} = V/Q.$$

$$\bar{t}_{P_t} = \frac{P_t}{dP_t/dt}$$

if phosphate sediments, $\bar{t}_{P_t}$ will be less than $\bar{t}_{H_2O}$. If the phosphate concentration increases, the reverse will be true. When $\bar{t}_{P_t}/\bar{t}_{H_2O}$ is greater or less than 1, $[P_t]_e/[P_t]_h$ should be greater or less than 1, which is consistent with equation 4.17. O'Melia suggested that $[P_t]_e/[P_t]_h$ and therefore $\bar{t}_{P_t}/\bar{t}_{H_2O}$ are both functions of k_z/σ.

Slow sedimentation (small σ) and rapid vertical mixing (large k_z) will result in accumulation of phosphate in the water and the reverse situation will result in loss of phosphate from the water towards the sediments.

Sedimentation rate depends not only on Stokes's law but also on water current, calcium concentration, and mineralisation rate in the epilimnion. Vertical mixing depends on eddy diffusion and other local currents, so that the results of calculations for the Swiss lakes should be applied only very cautiously, if at all, to other situations. The (theoretical) predictive model was later extended and published by Snodgrass & O'Melia (1975) for lakes with oxic hypolimnetic waters. These models do not allow for the possibility that water from the hypolimnion may escape upwards round the edges of an oscillating thermocline.

A more pragmatic way of calculating a nutrient budget is to make daily or weekly measurements of the inputs and outputs. Clearly this is only possible in cases where there are few inputs and outputs.

4.2.3. Relationships between nutrient loading and concentration with algal growth

Loading and concentrations are important factors in controlling primary production and primary productivity.* Much confusion has been caused by comparing *loading* with *yield* or *rate* of primary production on different time scales. Apparently several different cases can be distinguished. Phytoplankton 'primary production' is often measured as a rate of photosynthesis, but if a long study period is available it is much preferable to calculate, for purposes of comparison with other lakes, the *total* annual production or the so-called yield. Thus the rate of production times time is yield, or:

$$\frac{d\,C_{org}}{dt} \times t = \text{yield} \tag{4.18}$$

which may often be linearly related to the *total load* of a given limiting nutrient. Over a longer period there will be a compensation for those days in which the inflowing nutrients could not be utilised to the maximum by days in which metabolism is insufficiently high to utilise even the unused nutrients of previous days. Thus annual production is theoretically more related to the annual load of nutrients, while instantaneous growth rate is more related to their transitory concentration (see also p. 209) or to fluctuations in the underwater light conditions on which photosynthesis depends. Instantaneous growth rate (i.e. the change in biomass, $dB/dt = B'$, controlled by transitory

---

* Following the IBP definitions production is expressed as an amount of substance or energy. Productivity is a non-quantitative term used to describe in a general way the summation of all those processes which are concerned in the production of biological material (SCIBP, 1974).

concentration*) and yield (production over a prolonged period, thus $t \times dB/dt$) must thus be clearly distinguished. As yield is apparently related to concentration multiplied by time, it is related to loading (mean concentration $\times t$ is a function of loading). Instantaneous algal growth rates must however be compared with the concentration of that nutrient that is simultaneously present in limiting amounts.

Instantaneous integral photoassimilation per surface area (i.e. $m^2 : \Sigma A$) may be described as (Talling, 1970):

$$\Sigma A = A_{max} \frac{\ln I_0 - \ln(0.5 I_k)}{\varepsilon_{diss} + \varepsilon_{chl}} = A_{max} z' \qquad (4.19)$$

where A_{max} = maximal photoassimilation in the photosynthesis–depth profile (per unit volume, e.g. m^3)

ε_{diss} = light attenuation coefficient due to the water itself + the dissolved compounds (ln m^{-1})

ε_{chl} = light attenuation coefficient due to algal pigment and other particles (ln m^{-1})

I_0 = irradiance at surface layer (e.g.: $J\ m^{-2} s^{-1}$)

I_k = irradiance at onset of light saturation

z' = depth, where $I = 0.5\ I_k$

This formula is not directly concerned with the relationship between algal growth rate and nutrient concentration. It may be assumed however, that A_{max} is related to the concentration of the growth rate limiting nutrient in the same way as is the growth rate constant, μ, for algal cultures. The empirical curve of Monod has been found to be a reasonable approximation to the relation between this growth rate constant and the concentration of the growth rate limiting nutrient:

$$\mu = \mu_{max} \frac{[Nutr]}{K_m + [Nutr]} \qquad (4.20)$$

where

μ = instantaneous (or daily) relative growth rate (t^{-1})

μ_{max} = maximal value for μ (reached when $[Nutr] \gg K_m$)

$[Nutr]$ = concentration of nutrient limiting the growth rate ($g\ m^{-3}$)

K_m = a constant (the nutrient concentration where $\mu = 0.5\ \mu_{max}$)

Thus, assuming that A_{max} is related to the limiting nutrient concentration in the same way as is μ, A_{max} is linearly related to the lower nutrient concentration, as long as $\varepsilon_{chl} < \varepsilon_{diss}$. With increasing nutrient concentrations

---

* It might be argued that 'physiological history' controls growth rate as well. However in nature these effects are less likely to occur than in cultures, probably because algae in nature grow more constantly under nutritional stress conditions.

Monod's equation predicts that saturation will be reached, the influence of ε_{chl} becoming even greater. Either another nutrient becomes limiting, or light saturation will occur.

Of course a strict linear correlation between A_{max} and [Nutr] will not always be found, the natural situation normally being more complex than the simple culture conditions, for which Monod's equation was developed. Sudden dark days and unusual turbidity may easily destroy simple patterns; on such days the A_{max} may not be reached. The relations should, therefore, be studied over long periods in one lake, so as to encompass the ideal circumstances under which the maximal A_{max} can occur. In a series of shallow lakes a good correlation between $[A_{max}]_{max}$ and ΣA and phosphate could be found (unpublished results of a meeting on shallow IBP lakes, Lunz 1972) (see table 4.5), the relations being

$$[A_{max}]_{max} = 11 \times \text{Tot-P(column 1} \times 11 \approx \text{column 2)}$$
$$\Sigma A = 100 \times \text{Tot-P(column 1} \times \frac{100}{1000} \approx \text{column 4 or 5)} \qquad (4.21)$$

where $[A_{max}]_{max}$ = maximal value of A_{max} (mg m^{-3} d^{-1})
Tot-P = total phosphorus in unfiltered water (in mg m^{-3})

Sudden high values of A_{max} over short periods (some days up to a fortnight) are a major concern for the practical water manager, but they can apparently be moderated by decreasing the phosphate input.

Contrarily, the relationship between the average rate of production and the concentration of the limiting growth factor over long periods will be less apparent. Over longer periods growth rate can no longer simply be converted into yield. This characteristic of primary production is of importance in food chain studies. In these the best food transfer efficiency can probably be reached by a low average rate of production sustained over a long period. Two

Table 4.5 *The maximal values of A_{max} and phosphate concentration in several shallow lakes*

Column	1 Total P (mg m^{-3})	2 A_{max} (mg O$_2$ m^{-3} h^{-1})	3 A_{max} per mg chlorophyll a	4 ΣA (g O$_2$ m^{-2} d^{-1})	
				mean	high
Neusiedlersee	25	300	12–18	0.5	1.9
Loch Leven	70	1000	4–15	8	21
Lake Chad	130	900	33–37		20
Tjeukemeer	160	1500	15–20	8	25
Lake George (Ug.)	220	3000	23	12	16
Kilotes	2500	4000–9500	20–25		

Data from IBP symposium in Lunz 1972 (unpublished).

approaches are possible for studying the relationship between production and the growth limiting factor, e.g. phosphate, for many lakes:

(1) Comparison of the mean or maximal daily rate of primary production with the mean Tot-P concentration, or with the PO_4-P concentration in winter.
(2) Comparison of the yield of primary production over the growing season with the loading of a critical nutrient. When comparing lakes at different latitudes, the length of the growing season should be taken into account.

For the first approach Lund (1970) compared the maximum concentration of PO_4-P in winter with the annual algal population maximum in summer expressed in terms of chlorophyll *a* for a number of British lakes and rivers. Because Lund had no data on production he used chlorophyll *a* data, correctly assuming that these two parameters are strongly related. Apparently the chlorophyll concentration is an integrated property in which the daily fluctuations are balanced out (see Chapter 9, p. 429). Between the chlorophyll *a* concentration and phosphate concentration there seemed to be a linear relationship up to a PO_4-P concentration of about 50 $\mu g\,l^{-1}$ (PO_4-P) above which there was no marked increase of the chlorophyll *a*.

Vollenweider (1968) compared differences between winter and summer values of alkalinity with corresponding winter values of total phosphorus from forty-six Swiss lakes. Alkalinity was used as a measure of productivity because no other criterion was available. Vollenweider's results show a remarkable resemblance to these of Lund. Above 50 $\mu g\,l^{-1}$ of Tot-P no marked increase in seasonal change of alkalinity was noticeable and a significant correlation with phosphorus was demonstrated below this concentration. In his report Vollenweider was dealing with the indirect relationship between algal growth and nitrogen and phosphorus on the one hand and the acceptability of a particular level of algal growth on the other hand. This distinction is clearly dependent upon nutrient loading (and depth) of a lake. However no possible direct correlation was investigated by him. Although a similar general relationship was found between primary production and phosphorus concentration the situation in the Swiss lakes studied by Vollenweider is different from the British ones studied by Lund. The winter phosphorus input in the British lakes is mainly washed in by the winter rains draining the soil, whereas in the Swiss lakes the water is mainly derived from melting glaciers, and the phosphate input is derived mainly from the human population (and agriculture?), thus is more constant. Furthermore the water in the Swiss alpine lakes is hard, so that during summer, calcium carbonate formed during photosynthesis may adsorb phosphate, the phosphate thus being withdrawn from algal growth. Although the overall picture of phosphate limiting algal growth is the same in the two situations, unfortunately this cannot be shown quantitatively, because no comparison is possible between the two criteria for productivity (chlorophyll *a* concentration and alkalinity difference).

The first who used the second approach was Thomas (1956/57), who showed that in five Swiss lakes the nutrient concentration as well as the trophic status increased with increasing loading. Thomas pointed out the importance of the lake depth.

Vollenweider (1968) strengthened the 'loading concept' and stressed the importance of lake depth. His semi-quantitative approach in which the vertical distance from the line of dangerous loading is a measure for the degree of eutrophication has recently been too widely used without consideration of the limitations as outlined by himself.

For the second approach Schindler & Fee (1974) found a linear relationship between phosphorus surface loading (it is not clear whether Tot-P or PO_4-P was meant) and annual primary production for several lakes, in which phosphorus was known to be the growth controlling nutrient. However, it must be noted that the confidence limits are again broad (see p. 116). This is understandable, two processes are now included:

P-loading → P-concentration and **P-concentration → algal growth.**

Therefore the relation is more important for practical water management than for scientific understanding. The large differences (e.g. twenty-fold in production at a loading of $\log[P] = -1$) must still be explained scientifically.

In this respect it is interesting to compare data from Sakamoto (1966), who was one of the first to compare chlorophyll data with nutrient concentrations and lake depth, with those of Bachmann & Jones (1974), Jones & Bachmann (1976) and those of Dillon & Rigler (1974b). Bachmann & Jones compared the chlorophyll concentrations with a 'potential' phosphorus concentration, i.e. the total phosphate input divided by the lake volume (which, by the way, is equal to the surface loading per m^2 divided by the mean depth: input/volume = input/$S \times \bar{z} = L/\bar{z}$). Dillon & Rigler compared the summer average chlorophyll concentration with the measured total phosphate concentration at spring overturn. In Bachmann & Jones's calculated potential phosphate concentration, the precipitation is thus not taken into account; in Sakamoto's measured data the effect of precipitation is automatically included. From their tables or formula the following comparison can be made:

PO_4-P conc. ($mg\ m^{-3}$)	chlorophyll *a* concentration ($mg\ m^{-3}$) calculated from		
	Sakamoto	Bachmann & Jones	Dillon & Rigler
500	1000	200	600
100	100	40	58
10	2.5	2	2
5	0.6	1	0.75

It can be seen that the agreement in the nutrient-poor lakes is excellent, but that in the eutrophic lakes Sakamoto's data are much higher, which can partly be explained by precipitation of phosphate, which lowers Bachmann & Jones's actual concentrations, while Dillon & Rigler's data fall somewhere in between. More difficult to explain however are the slopes (and the mutual differences) of the regression lines, which indicate strong deviations from the linear regressions which should be found assuming a constant chlorophyll to phosphate ratio. But even in Bachmann & Jones's and Dillon & Rigler's figures there is a strong tendency for larger values of chlorophyll over phosphate concentration for the higher phosphate concentrations, the fitted regression lines being

$$\log Y = 1.21 \log X + 2.93 \text{ (Bachmann \& Jones) and} \qquad (4.22a)$$
$$\log Y = 1.45 \log X - 1.14 \text{ (Dillon \& Rigler)} \qquad (4.22b)$$

where $Y =$ chlorophyll concentration (mg m^{-3})
            $X =$ potential phosphate concentration (Bachmann & Jones; PO$_4$-P, μg l^{-1}); measured phosphate concentration (Dillon & Rigler; PO$_4$-P, mg m^{-3})

It is especially difficult to understand how the chlorophyll a concentration can be related with [P]$^{1.45} \approx$ [P] × [P]$^{0.5}$. In the first instance one would expect the phosphate to be more efficiently converted into algal matter (measured as chl. a) at lower loadings (because of chemical precipitation at higher loadings), while apparently the opposite seems to be true. This may be explained by assuming a lower availability at lower concentrations, or a more efficient recycling at higher phosphate concentrations, because the bacteria which mineralise algal organic matter are limited by their substrate concentration (see Saunders, Chapter 7). This would mean that recycling is more efficient in eutrophic lakes. The IBP results, although not quantitatively interpretable in this sense, point strongly in this direction. Dillon & Rigler have already pointed out the wide confidence limits, for example, for the 100 mg m^{-3} PO$_4$-P concentration one may expect chlorophyll concentrations between 20.6 and 162 mg m^{-3}, which means values between 'good' or 'acceptable' water quality and very 'bad' indeed. Even with 50% confidence limits the range of the expected chlorophyll concentration is between 40.8 and 81.9 mg m^{-3}. This is illustrated even by the nineteen Ontario lakes, which have a mean PO$_4$-P value of 8.7 mg m^{-3} only, with a range of 4.1–15.3 mg m^{-3}.

  If this variability is now combined with the variability of the phosphate-loading against phosphate concentration, the uncertainty for the water manager becomes unacceptably large, because the errors may add up, leaving uncertainties of an order of magnitude, especially in the problem area, the eutrophic situations. It is quite clear that before these models will be acceptable for water management there must first be a much greater understanding of the background mechanisms, which ultimately must lead to conceptual models. Some promising results have been obtained in oligotrophic situations. Thus Scavia & Chapra (1977) found good agreement between the simple chlorophyll a relation with P-loading, equations 4.7 and

4.22b, and a more complex ecological model for L. Ontario. At higher phosphate concentrations no agreement was found. No criteria, however, for 'agreement' have been defined in their paper, while several questions about the relevant technical details remain unanswered. One of the improvements that could be made in the P-loading models is the distinction between apatite and non-apatite phosphate although this is clearly chemically more difficult than is often thought. Burns, Williams, Jaquet *et al.* (1975) demonstrated that apatite phosphate forms 44% of the phosphate loading of Lake Erie. If we assume that the apatite phosphate is not available for algal growth, this observation is important for the evaluation of the relative effects of the diversion of phosphate loadings from sewage water.

Brylinsky (see Chapter 9) did not find a good correlation between annual production and nutrient concentrations for the IBP lakes. The lack of correlation is presumably caused primarily by comparing annual production with concentration instead of with loading. In general such a correlation will be poor for lakes with high loading and high water renewal rates. A lack of correlation between annual loading of a growth-limiting nutrient and algal yield could be caused by the fact that in winter the nutrient will be partly washed out, without having been used by algae, or be sedimented before growth sets in. Because of sedimentation it is very difficult to estimate the input of the actual amount of the nutrients that remain available in the water column. However, the outflowing amount can be measured and should be subtracted from the inflow. Outflow of phosphate in summer will be mainly in the form of cellular phosphate, i.e. in a form which has passed through biological processes.

In several lakes – and certainly before the heavy impact of man on lakes due to the installation of sewers, as mentioned above for the English Lake District – the amount of phosphate available during the summer is probably mainly the amount washed in during the winter. This may mean that the winter rains transport more phosphate into these lakes than do the summer rains. Both periods now include human inputs. The rate of production during the growing season was originally mainly maintained by recycling of the nutrients; at present the human input continues a supply during summer. Thomas (1969, 1973) pointed out the important fact that the original winter input was divided over the whole lake, while the input due to sewers goes mainly straight into the productive euphotic zone. Possible losses will consist only of sinking detritus and outflow. Phosphate thus lost into the hypolimnion and by outflow will be replenished by the external input, which will be taken up immediately by algae. However, in summer another source of phosphate has to be considered because the lake will be internally loaded by nutrients previously trapped in the sediments, the winter sedimentation as mentioned above. Mainly in shallow lakes this source will make a considerable contribution to the amount of nutrients actually metabolised.

When comparing the total production (either annual or seasonal) of different lakes with the annual nutrient loading it is often overlooked that the length of the growing season has a major influence. Thus the higher production in the tropical lakes shown in Brylinsky's Chapter 9 is mainly explained by the longer duration of the growing season. This means that biological production (e.g. in food chains) may indeed be higher and that a better usage is made of the available nutrients. But if one correlates the algal growth with phosphate loading for the study of eutrophication the results are partly misleading.

4.3. Nutrient cycles

During the working groups in both the IBP meetings at Kazimierz Dolny and Reading the discussions always drifted back to phosphorus, although the chairman often pointed out that other elements are equally important in understanding algal growth and that such elements are also involved in other processes of scientific interest. Nevertheless the working groups felt that phosphate is the key element in the process of eutrophication and concluded in particular that the rapid increase of the phosphate loading has caused the present eutrophication problem; although in a minority of lakes nitrogen may hold this position, the working groups persisted in their attempts to understand phosphate metabolism as the group felt that phosphate is the only element which can be controlled to the point at which it becomes limiting to algal growth. This is so for two reasons. In the first place nitrogen enters the lakes from more diffuse sources than phosphate and is thus more difficult to control, and secondly, eutrophication is frequently accompanied by blue-green algal growths which often are able to sustain their nitrogen demands by fixation of atmospheric nitrogen.

4.3.1. The phosphate cycle

Processes in the phosphate cycle may be summarised by a so-called phosphate flow diagram (see fig. 4.10) in which recycling through the biological compartments is depicted together with the processes of inflow, outflow and sedimentation. In the central part of this diagram the recycling of phosphate between the dissolved and particulate compartments is shown. In most lakes phosphate is not present in excess with regard to the demands of photosynthesis so that the uptake is a fairly rapid process. Even if phosphate is present above growth-rate limiting concentrations the uptake rate may still continue to be rapid to the extent that the phosphate may be stored in the algal cells as polyphosphate.

No adequate mathematical model for uptake kinetics has yet been devised, but a pragmatic approach is to calculate the phosphate uptake from the data

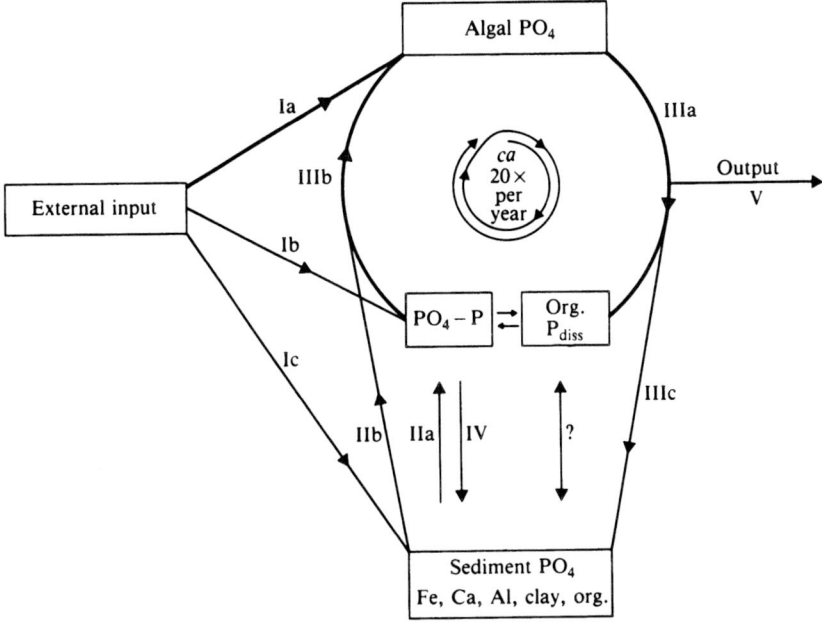

Fig. 4.10. Phosphate flow diagram visualising the recycling of phosphate through the biological compartments. The indicated turnover time per year may vary between five and forty times per year. (After Golterman, 1975a.)

on primary production assuming a more or less constant ratio between cellular carbon and phosphorus. Most IBP data for this ratio fall in the range between fifty and hundred, with sixty as a likely mean value if there is a well-defined seasonal production phase. Healey (1975) found a value of around sixty for phosphate-limited phytoplankton.

If the demand is estimated in this way and if no mineralisation were to occur, the pool of PO_4-P would rapidly be exhausted and growth would cease. However, mineralisation does occur and, through processes Ia and IIa as shown in fig. 4.10 the primary producer part of the system receives the necessary PO_4-P.

In table 4.6 an estimation is made of the duration for which dissolved PO_4-P could theoretically support photosynthesis in the various IBP lakes, i.e. the time in which all the PO_4-P in a lake could be used by the phytoplankton for photosynthesis (disregarding mineralisation). For this the PO_4-P consumption was calculated from the productive season mean carbon fixation data by assuming a C : P ratio of 60 : 1 and taking the depth of the epilimnion for estimating the PO_4-P which could be available per unit surface area (if no stratification for any particular lake occurred, the mean depth was taken). It will be seen from the results that a great diversity of periods for phosphorus-supported photosynthesis can occur between the lakes listed. In most lakes

Table 4.6 *Estimation of the time that the dissolved PO_4-P could theoretically support photosynthesis, disregarding mineralisation*

Lake	Country	Method	Growing season	Period	Calculated consumption of P (10^{-3} g m^{-2} d^{-1})	Available amount of PO_4-P (10^{-3} g m^{-2})	t (days)	
Aleknagik	Canada	^{14}C	June–Sept.	1962–70	28.8	15	0.53[b]	
Biwa	Japan	—	—	Aug. 1971–June 1973	18.8–37.2	<40	<2.1[c]	
Blackstrap	Canada	^{14}C	—	1968:June–Sept.	3.66 (h^{-1})	1045	285 h[b]	
Bolshoy Kharbey	USSR	O$_2$	July–Aug.	1968–9	3.6	41.4	11.5[b]	
Buffalo Pound	Canada	^{14}C	—	1967:June–Aug.	29.3	930	31[d]	
Burton	Canada	^{14}C	—	1968:June–Sept.	1.54 (h^{-1})	1410	915 h[d]	
Chad	Chad	O$_2$	Jan.–Dec.	June 1968–Mar. 1970	28.3	46.8	1.7[b]	
Chedenjarvi	USSR	O$_2$	June–Aug.	1971	7.8–30.3	≤129.2	≤16.6[b]	
Chester Morse	USA	^{14}C	—	1971:May–Dec.	6.3	32	5[d]	
Chilwa	Malawi	O$_2$	Jan.–Dec.	1970–1	33.8–83	2700	32–83[b]	
Echo	Canada	O$_2$	—	1965:June–Sept.	36.9	6533	177[d]	
George	Uganda	O$_2$	Jan.–Dec.	—	75	<12.5	<0.17[b]	
Hakojärvi	Finland	^{14}C	May–Sept.	1969	0.7	88.2	125[b]	
Humboldt	Canada	^{14}C	—	1968:June–Sept.	9.95 (h^{-1})	1050	105 h[d]	
Katepwa	Canada	^{14}C	—	1966:June–Sept.	51.7	840–3080	16.4–58.8[d]	
Kiev Res.	USSR	O$_2$	May–Oct.	1967, 1968	16.1	74	4.6[b]	
Krasnoye	USSR	O$_2$	May–Oct.	1970	10.8	114	10.5[b]	
Loch Leven	Scotland	O$_2$	—	1968–71	44.9	45	1[c]	
Locomotive Spring	USA	O$_2$	—	1971	7.5	—		
Neusiedlersee	Austria	^{14}C + ^{14}C	—	1968–70	3.3	2.2[a]	0.7[c]	
Pasqua	Canada	O$_2$ + ^{14}C	—	1965–7:May–Sept.	53.9	5480	102[d]	
de Port-Bielh	France	^{14}C	July–Nov.	1966–8	2	22.5	11.2[b]	
Obelija	USSR	—	—	1971:Summer	7.5	3.4	0.5[d]	
Rybinsk	USSR	^{14}C	May–Oct.	1955–70	7.3	12–60	1.6–8.3[b]	
Sammamish	USA	^{14}C	—	1970,1971	9.9	61.6	6.2[c]	
Tjeukemeer	Netherlands	^{14}C + O$_2$	(Mar.–Apr.)–Oct.	1970–3	37.1	24.6	0.7[b]	
Trummen	Sweden	^{14}C	Apr.–Oct.	1969	28.8	186	6.5[b]	before restoration of the lake
			Apr.–Sept.	1972	22.7	11.4	0.5[b]	after restoration of the lake
Vechten	Netherlands	^{14}C	(Mar.–May)–Oct.	1972–4	8.2	31	3.8[b]	
V. Finstertalersee	Austria	^{14}C	July–Nov.	1968–70	2	6.7	3.4[b]	
Wingra	USA	^{14}C	Apr.–Nov.	1970	28.8	15	0.53[b]	
Yunoko	Japan	—	—	May 1968–July 1969	6.1	67.5	11[d]	

$$t = \frac{\text{actual available } PO_4-P}{PO_4-P \text{ consumption}}\left(\frac{\text{g m}^{-2}}{\text{g m}^{-2}\ \text{day}^{-1}}\right)$$

[a] 1968 data were used for 1969 and 1970.

[a] t calculated for growing season. [b] t calculated for the whole year. [c] t calculated over the period indicated in column 'Period'. [d] in other cases t has been calculated over the period indicated in column 'Period'.

Table 4.7. *The number of days that PO_4-P can support photosynthesis in Tjeukemeer and Lake Vechten*

Month	Jan.	Feb.	Mar.	Apr.	May	June	July	Aug.	Sept.	Oct.	Nov.	Dec.
Tjeukemeer:												
1970	—	—	—	6.6	1.7	0.07	0.2	0.4	0.5	2.0	18	> 64
1971	102	84	3	1.5	0.3	0.07	—	—	0.1	0.08	—	4
1972	36	2.8	0.6	0.2	0.1	0.10	0.4	0.6	0.2	0.6	—	—
Vechten:												
1972	—	—	34	6	7.5	3.8	2	3.4	3.7	10.7	—	—
1973	29	21	9.5	5	0.2	4	—	3	3	2.7	—	—
1974	—	19	5.5	8.6	10	4.6	4.9	1.4	1.8	6.5	7.5	—

$$t = \frac{\text{actual available } PO_4-P}{PO_4-P \text{ consumption}} \left(\frac{\text{g m}^{-2}}{\text{g m}^{-2} \text{ day}^{-1}} \right)$$

this period is larger than 1 day and in some lakes it is very high e.g. Hakojärvi, Pasqua and Echo Lakes. The majority (fourteen) of the lakes fall within a range of 1–20 days and in seven lakes PO_4-P can support photosynthesis for more than 20 days. In only five lakes is this period less than 1 day; the lowest value obtained was found for Lake George (Uganda) i.e. < 0.17 days indicating that the available PO_4-P can support photosynthesis for only a very short period. Note that in Lake Trummen the available PO_4-P decreased by 94% after the lake had been restored. This resulted in concomitant decrease of the phosphorus-supporting period from 6.5 days to 0.5 days. The two Dutch lakes listed in table 4.7 are Tjeukemeer, a shallow, eutrophic, organic-rich polder lake, and Lake Vechten a deep, stratifying, oligotrophic sand-pit. The figures for the supporting period as calculated from individual measurements during the growing season show a pronounced difference between the lakes (April–October). In Lake Vechten periods smaller than 1 day are rare (in fact known just once, in May 1973) whereas in Tjeukemeer values less than one day are common during summer.

Recycling of phosphate within the epilimnion has been demonstrated by Bloesch (1974) and Bloesch, Stadelmann & Bührer (1977). They showed, by measuring phytoplankton production and sedimentation, that in Lake Lucerne (Horwer Bay) and in Rotsee 35–75% of the nitrogen and 55–85% of the phosphorus needed for the measured primary production could be re-generated in the epilimnion. These values parallel those of Golterman (1976), who estimated that 80% ± 20% of the organic matter derived from primary production is oxidised in the epilimnion. Internal recycling in the water column was also shown by Barica (1974) in shallow pothole lakes in south-west Manitoba.

Supply of PO_4-P comes from external input, recycling within the plankton in the water column and, in shallow lakes, from the sediments. Input has been measured in only a few IBP studies (Schindler & Fee, 1974). Schindler & Fee found a good correlation between phosphorus loading and rate of annual primary production. The influence of the length of the growing season was not included in their paper. It might be suggested that such a linear relationship between phosphorus loading and rate of production means that phosphorus recycling as a phosphorus source is of less importance than loading or that recycling is a function of algal concentration and thus indirectly of phosphorus loading. It does seem likely that mineralisation of dead algal material is dependent on the concentration of this dead material, because bacterial growth in lakes is controlled by the substrate concentration. (See Saunders, Chapter 7.)

It is difficult to estimate the mineralisation of algal material. If photo-synthetic production has been measured by the oxygen exchange method then the 'dark' bottles of such experiments which would have recorded community oxygen uptake, can give some useful information. It should be made clear that the word mineralisation as used here means any oxidising process which reverses the reductive one of photosynthesis, and as such includes the respiratory activity of living algae as well as of bacteria. Allochthonous organic material will also be broken down and its oxygen equivalent recorded within the oxygen decrease, but the products of such breakdown cannot legitimately be termed part of the recycling. The technique should be used cautiously and with due consideration for the peculiarities of the conditions in the lake concerned.

Also extrapolations of the dark bottle oxygen uptake from a conventional 4 hours' experiment to a 24-hour period can lead to large errors, and further inaccuracies can be introduced in an extrapolation of the results for the small amount of material in a bottle, located in only one zone of a water mass to obtain a figure for the whole lake with its numerous limnologically different zones which may possess varying amounts of organic material.

Two examples amongst IBP lakes for which this technique was used are Tjeukemeer and Lake George (Golterman, 1971; Ganf, 1974c). Mineralisation was measured directly by estimating the decrease of particulate carbon and nitrogen in darkness. Golterman found the O_2 uptake in Lake George to be about 30%, and in Tjeukemeer 50–200% larger than the decrease of particulate organic content (expressed as O_2 equivalents). This suggests that in the figure for O_2 uptake, oxidation of dissolved organic matter was included to some extent. In Tjeukemeer this organic matter probably comes from allochthonous sources, but in Lake George it could be mainly from algal excretion products and dead algae. In both cases it means that the estimation of the algal recycling suggested by the figures for O_2 uptake is too high. By measuring the decrease of cellular carbon and nitrogen recycling, turnover times ranging from 5 to 10 days were calculated, which is probably a factor of two too high, due to the presence of (allochthonous) detritus. The detrital carbon, the amount of which in some lakes may be equal to the cellular carbon, has of course a much longer turnover time than the cellular carbon, but in the measurements it cannot be distinguished from the cellular carbon. To be able to differentiate living from non-living material would be of great interest. It would also be useful if detrital carbon could be distinguished from fresh non-living carbon by confining the word detritus to refractory (dead) organic carbon. A second way to calculate the mean expected survival time of algae is to compare photosynthetic rate with the biomass of the standing stock of the algae. In this way we calculate an 'apparent' population life expectancy or an 'expected' mean duration for the whole population, which is quite different from the life expectancy for a single cell. The apparent population life expectancy – subsequently called 'apparent longevity' – is useful for calculating nutrient recycling rates, not for the population dynamics of algal populations.

In most lakes no important changes will occur between the mean annual

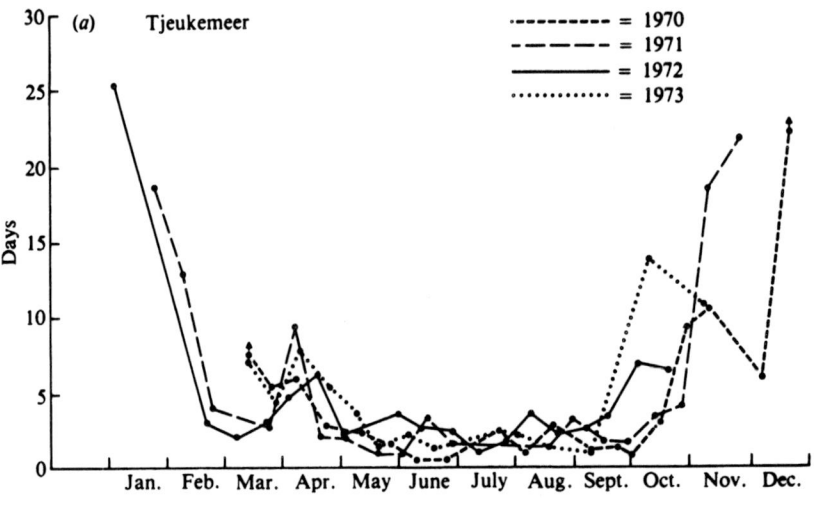

(a) Tjeukemeer

─ ─ ─ ─ ─ = 1970
── ── ── = 1971
───────── = 1972
·········· = 1973

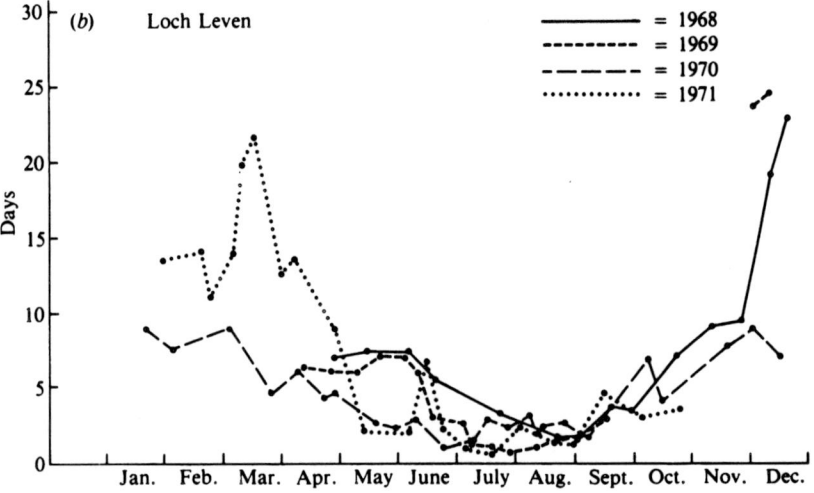

(b) Loch Leven

───────── = 1968
─ ─ ─ ─ ─ = 1969
── ── ── = 1970
·········· = 1971

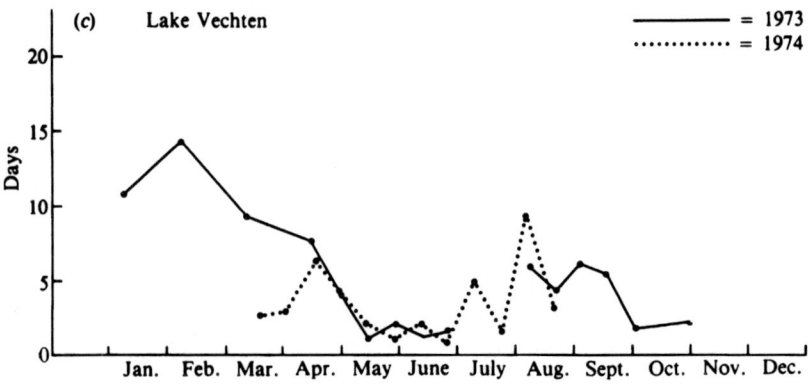

(c) Lake Vechten

───────── = 1973
·········· = 1974

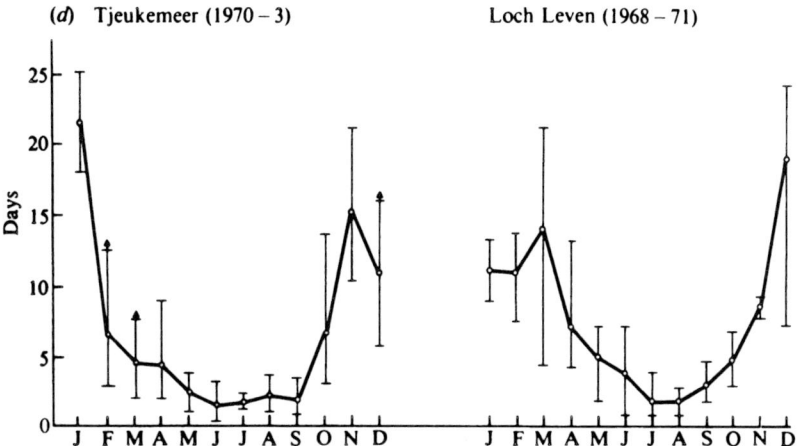

Fig. 4.11. The annual variation of the apparent longevity of algal cells in (*a*) Tjeukemeer (1970–3), (*b*) Loch Leven (1968–71), (*c*) Lake Vechten (1973, 1974), (*d*) average values for Tjeukemeer and Loch Leven.

algal concentrations as calculated between two successive winters, although the variation around this mean value between samples taken during the year will of course be large. The amounts of algae in most IBP studies have not been assessed directly; an indirect estimation can be made by assuming that chlorophyll *a* is 1.5% of dry weight and carbon is 50% of organic dry weight. Cellular carbon is thus estimated as thirty-five times the chlorophyll content. In fig. 4.11 the course of the 'apparent longevity' of algal cells throughout the year can be seen for Tjeukemeer, Lake Vechten and Loch Leven. In Tjeukemeer and Loch Leven an increase in this value is noticeable in early autumn: going up from low summer values (0.5–4 days) to winter values as high as 25 days for Tjeukemeer. During the second part of winter the apparent mean algal longevity decreases in Tjeukemeer, but in early spring rises again to values as high as 9 days. This increase parallels the spring bloom of diatoms in the lake. In Loch Leven the high value, up to 22 days, in the period January– May (1970, 1971) are probably also due to diatoms in this period as these were the dominant algae. Throughout the second half of spring and the whole of summer an average value of about 2 days was reached in Tjeukemeer.

In Loch Leven this level was on the average reached one month later, whereas in 1968 the summer value of algal longevity was only arrived at in August. Such differences between Loch Leven and Tjeukemeer might be due to differences in climate, caused by a different geographical position. These values indicate that the rate of mineralisation of algal cells varies considerably throughout the year. In summer mineralisation takes place at a rate equal to the mean biomass per 2 days, i.e., a high rate compared to winter rates, so that in summer phosphorus for instance, should be recycled many times more than in winter. (The reciprocals of this apparent longevity are equivalent to the fractions of the biomass mineralised per day during the year.) In Tjeukemeer, Lake Vechten and Loch Leven they show that the rate of mineralisation in summer can be about ten times higher than in winter. In table 4.8 the calculated apparent algal longevities are listed for several lakes. It can be seen that in most lakes presented here this value ranges from 1 to 5 days. These theoretically obtained recycling times are in agreement with the experimentally obtained values if it is accepted that the latter are probably too high by a factor of two. It is also of interest to calculate the mean algal longevity using the maximal value of the oxygen production (A_{max}) per mg of chlorophyll a per hour, which is estimated to be 25 mg of O_2, or 9.4 mg of C, per mg of chlorophyll a per hour (see Chapter 5). This means that the production is 9.4 mg of carbon per 35 mg of cellular carbon per hour or $(9.4 \times 12)/35 = 3.2$ mg of carbon per mg of cellular carbon per day, if a mean daylength

Table 4.8 *Daily photosynthesis, algal standing stock and* (t) *apparent mean duration of life of algal cells in different lakes* (t *is derived by dividing the standing stock by the daily photosynthesis*)

Lake	Daily photosynthesis (g C m^{-2} day^{-1})	Chlorophyll a (g m^{-2}) 10^{-3}	Standing stock g C (= 35 chl a) (g C m^{-2})	t (days)
Tjeukemeer 1970	3.23	98	3.43	1.1
1971	1.88	114	3.99	2.1
1972	1.98	162	5.67	2.9
1973	1.82	150	5.25	2.9
Aleknagik 1962–70				
1967 excl. (June–Sept.)	0.15	40	1.40	9.3
Trummen 1972	1.36	152	5.32	3.9
Sammamish 1970	0.71	48	1.68	2.4
1971	0.47	58	2.03	4.3
Chester 1971				
Morse (May–Dec.)	0.38	14	0.49	1.3
Yunoko				
(May 1968–July 1969)	0.37	143	5.0	13.5
Loch Leven 1968	2.18	332	11.6	5.3
1969	3.79	384	13.4	3.5
1970	2.85	256	8.9	3.1
1971	1.95	332	11.6	5.9
Chad				
June 1968–Mar. 1970	1.70	66	2.3	1.4
Red Rock Tarn				
Oct. 1969–July 1970	7.27	194	6.79	0.9
Corangamite				
Nov. 1969–July 1970	2.46	184	6.44	2.6
Coragulac				
Oct. 1969–July 1970	1.23	101	3.54	2.8
Pink Lake				
Oct. 1969–July 1970	0.06	17	0.6	10

throughout a year is taken as 12 hours. Respiration may be estimated by assuming an O_2 uptake of 1 mg of O_2 per mg of chlorophyll a, or 1 mg of O_2 per 0.38 mg of carbon, and thus a carbon loss of:

$$\frac{0.38 \times 24}{35} \times \frac{\text{depth of respiration}}{\text{depth of euphotic zone}}.$$

For a depth ratio of two, a value of 0.52 mg of carbon per mg of cellular carbon per day is found. This means that the net photosynthesis or daily mean specific growth rate at the depth of A_{max} can be calculated to be about $3.2 - 0.52 \simeq 2.7$ mg carbon per mg cellular carbon per day.

If the integral photosynthesis, ΣA is approximated by:

$$\Sigma A = A_{max} \cdot z' = \bar{A} \cdot z_{eu} \qquad\qquad \text{(see (4.17))}$$

where $z' =$ depth of I_k
$z_{eu} =$ depth of euphotic zone
$\bar{A} =$ mean A in water column,

and as z' often extends to near 15–20% of I_0, and z_{eu} to 1% of the I_0, it follows that $z_{eu} \simeq 2.6z'$. Therefore the column mean specific growth rate for algal cells will be about 1 mg of carbon per mg of cellular carbon. Thus under optimal conditions (where $A_{max} = 25$ mg of O_2 per mg of chlorophyll a per hour and respiration is 4% of A_{max} as above) the mean annual value for the apparent algal life duration is 1 day. It also means that the daily mineralisation over the whole year must be roughly the annual mean biomass per day. As A_{max} per mg of chlorophyll a in temperate waters may normally be lower than 25 mg O_2 per hour, the apparent longevity of the algae will be several days and mineralisation will have to take place at the rate of the annual mean biomass per several days.

The short apparent algal longevity during the summer and the highly efficient and rapid process of phosphate mineralisation form a major contribution to the maintenance of photosynthesis and thus gross primary production during the summer season. During mineralisation of dead algal material only a small part (*ca.* 5%) of the liberated phosphate may be converted into refractory phosphate, but even this phosphate is not necessarily lost for algal growth. These compounds will sediment to the bottom, where mineralisation will proceed further and, as the C–O–P bond is thermodynamically weak, over a prolonged period most of the organic phosphates will disappear. This does not necessarily lead to a release of free orthophosphate because many sediments may contain high concentrations of calcium or iron in their interstitial water and the phosphate may then remain in the sediment as calcium or iron phosphate.

In shallow lakes most of these compounds can be sources of phosphates for algae. In lakes which stratify, during the period in which a hypolimnion is well established, calcium phosphate may dissolve due to a higher acidity induced by increased carbon dioxide concentration in the adjacent bottom water, while iron phosphate may dissolve due to this lower acidity and reducing conditions. The intensity of the transfer process (IIa) in fig. 4.10 is controlled by water movements induced by wind.

The hypothesis concerning the availability of these sediment phosphates has often been discussed in various contexts since the early work of Mortimer and Einsele, but has never actually been proved correct. It is often stated that the phosphate released from the sediments will diffuse upwards by molecular and eddy diffusion, while movements of animals may enhance this movement. Although molecular diffusion is thought to be an exceedingly slow process it should not be overlooked altogether. In a stratifying sand-pit (depth, 11 m; thermocline, 5–7 m; PO_4-P above sediment, 0.5 mg l^{-1}) it was found (unpublished results) that, by using Megard's formula for molecular diffusion, it was still possible to obtain a phosphate loading of 1 g $m^{-2} a^{-1}$.

The upward movement of phosphate may be counteracted by a ferric hydroxide precipitate formed from the oxidation of ferrous iron diffusing upward from anoxic into oxygenated water, the phosphate being adsorbed onto this precipitate.

4.3.2. Phosphate recycling through sediments

Contrary to the phosphate recycling in the water column (the so-called 'internal' or 'metabolic' P-cycle), the phosphate recycling through the sediments (the so-called 'geochemical' or 'external' P-cycle) is a relatively slow process. Processes in the metabolic cycle are now fairly well understood, or at least well defined, although quantitative knowledge is lacking. The situation is more obscure in the external P-cycle, where the compounds involved are not yet well defined and the reaction rates *in situ* are completely unknown. In the IBP programme no studies were planned specifically to investigate the role of sediments, but in the course of the work this aspect became more and more important. If there is any area of research where IBP has a significant impact on future studies it is certainly in the field of research on sediments.

Although no papers were published before the first draft of this chapter (1975) a few published since then have involved IBP lakes, e.g. the papers of Viner (1975c, d). Other studies – not from IBP lakes, but certainly related, are those of Andersen (1974), Lean *et al.* (1975), Kouwe & Golterman (1976) and Golterman (1977a), all showing the importance of internal loading from the sediments. These papers definitely prove the fact that sediments may release phosphate in a manner contrary to what might be thought if considering them to be only a sink. They surely do act as a sink as long as external loading takes place (reaction I in fig. 4.10), but it should be clear from this figure that as soon as the external loading is cut off, a shift in equilibria renders reaction II relatively more important. This process may delay lake restoration after phosphate diversion. It is also clear from fig. 4.10 that reaction II will be less important the deeper the lake, and dependent on pH and oxygen concentration. Nevertheless Kamp Nielsen (1977) showed that even in Lake Esrom (20 m) this reaction adds a considerable amount of phosphate to the pool of phosphate in the epilimnion.

The recently increasing flow of papers on sediments is exemplified by a special UNESCO/SIL symposium on sediments (Golterman, 1977b). In this symposium twenty-one papers were published on phosphate exchange dynamics. Several papers were devoted to deciphering experimentally physico-chemical factors which influence the nutrient release, and appeared to substantiate what was originally proposed by the classical work of Einsele (1936, 1937, 1938) and Mortimer (1941, 1942) except that the detailed mechanics may differ from the original proposals. For instance the role of iron in controlling the phosphate dynamics was demonstrated to be much less general than often thought; in hard waters calcium carbonate may have an even more important central function. Also the reduction of the iron–phosphate complex was questioned, although the idea that it cannot be reduced has not yet been proved true. A more important aspect is that the different extents to which those factors such as forms of iron, oxygen concentration, pH, calcium carbonate, etc., operate in different limnological environments cannot readily be predicted quantitatively without prior investigation of the particular water body.

In the preface of the UNESCO/SIL symposium proceedings Golterman, Viner & Lee (1977) defined a series of new bearings arising for the future and several special problem areas. This symposium volume also contrasts the great gap between practical models for the water manager (see for example Jørgensen, Kamp Nielsen & Jacobson, 1975; Jørgensen, 1977) and models for the purpose of scientific understanding; a scientific or synthesis model should start from processes and measured rates and predict phosphate kinetics in different limnological environments. Such a model (or submodel, as it deals with only a portion of the phosphate model for a complete lake) has been published by Kamp Nielsen (1977), which predicts the seasonal variations in exchangeable sedimentary phosphate and in interstitial phosphate. One of the most serious problems is that such submodels can only very seldom be tested in another lake. Furthermore the weakest point is that, in attempting to use such submodels for the prediction of what might happen in a lake as a whole, it is assumed that the total phosphate released from the sediments arrives in the epilimnion, which apparently does not happen. More submodels are thus necessary, and the work involved makes it in practice impossible to test such models simultaneously in several lakes. The models are however powerful detectors of still unknown significant processes.

4.3.3. The nitrogen cycle

Two distinct differences can be found between the nitrogen cycle (fig. 4.12) and phosphate cycle, both of which make the calculation of a nutrient budget for nitrogen more complicated. In the first place nitrogen may enter and leave the cycle as gaseous N_2 by nitrogen fixation and denitrification respectively. In the second place amino acids provide an important part of the energy available to bacteria and also for zooplankton which assimilate algae. This energy will be derived from oxidation of these compounds. As the protein content of algae is often 50–60% of the organic weight and only a small part of the digested protein will be used as such for the newly formed bacterial or zooplankton biomass a considerable quantity of ammonia will be produced.

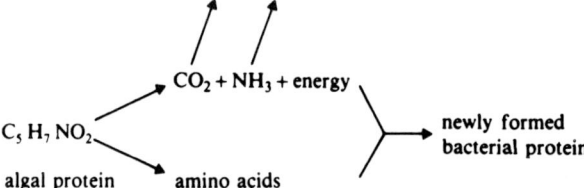

Assuming that 10–30% of the digested protein will be converted into newly formed proteins, the remaining 70–90% will be mineralised to ammonia (Golterman, 1976).

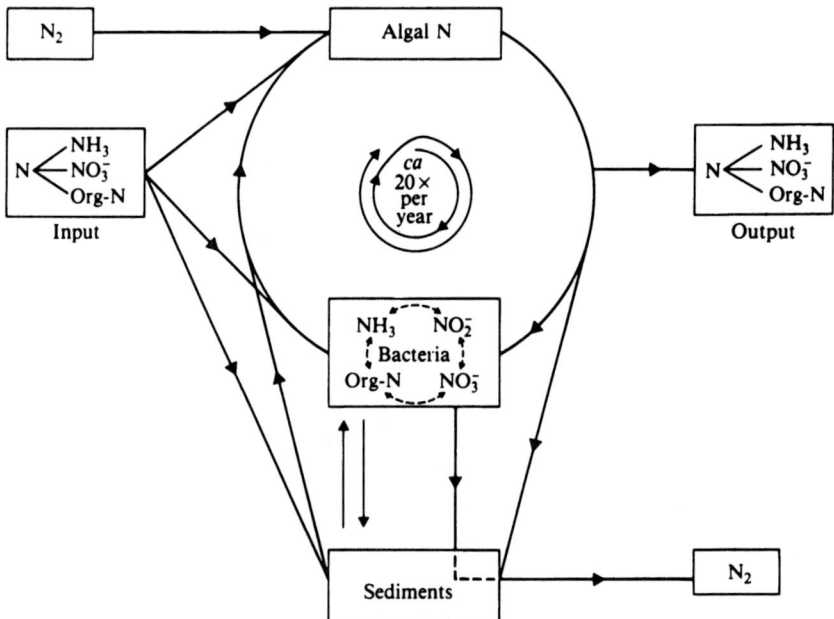

Fig. 4.12. Schematic representation of the nitrogen cycle. (After Golterman, 1975a.)

No predictions can yet be made about the relative percentages of algal cells digested by bacteria and zooplankton. Zooplankton feeding is discussed in Chapter 6 of this book. It is relevant to note here that a few studies have revealed that zooplankton respiration contributes only a small part of the community respiration (see, e.g., Ganf, 1974c for Lake George, and Chapter 5). Although daily uptake of algae by zooplankton may be equivalent to the greater part of primary production during short periods (e.g. summer blooms of Chlorococcales) it is probable that spring blooms of diatoms and blooms of blue-green algae are not eaten often enough to produce a zooplankton-derived mineralisation, although in the Lake George example quoted above the blue-greens were to some extent digested. Studies on this part of the recycled primary production have been discussed by Golterman (1972).

Nitrogen fixation has been measured in Tjeukemeer and Lake George (Uganda) although no long series of measurements have been made. In the former lake nitrogen fixation contributed a small fraction of the total nitrogen budget although in Lake George it was considerable (approximately 30%, see Viner & Horne, 1971). Some doubt may be expressed on the validity of the extrapolation from the experimental period to a whole year. It does seem unlikely that nitrogen fixation can form a significant contribution in general to the nitrogen cycle.

A process working against nitrogen fixation is the so-called denitrification. This is the process whereby

$$\text{organic matter} + NO_3^- \rightarrow N_2 + CO_2.$$

It can only take place under oxygen-poor conditions, for example, in partly deoxygenated hypolimnia or sediments. It seems likely that these occur only in eutrophic lakes so that denitrification may be restricted to such lakes. It seems that nitrogen fixation may similarly be restricted to the eutrophic lakes. To date the only technique available to estimate denitrification *in situ* is deduction from total nitrogen budgets. Since the pertinent data are not available this cannot be applied to the IBP studies.

During the IBP studies several reports were presented concerning a decrease of nitrogen content of shallow eutrophic lakes, such as Loch Leven, Tjeukemeer and fish-ponds. It was noticed that the algal organic nitrogen content was considerably lower than the sum of ammonia and nitrate in the previous winters. Whether denitrification can occur under the conditions in the waters is open to doubt, although Hrbáček (1972) strongly suggested that it was possible. It has certainly also been shown in many polluted rivers. Jannasch & Pritchard (1972) mentioned that bacterial populations growing on particulate matter may denitrify by establishing a micro-anaerobic zone around these particles, but such a process has not yet been demonstrated in a natural environment. It seems possible that losses due to NO_2^- formation during the reduction of NO_3^- by algae and subsequent reactions with amino nitrogen are also possible. Another possibility is denitrification in the sediments. Andersen (1974, 1977) found that denitrification contributed significantly to the degradation of organic matter in the sediment but only if the concentration of nitrate was artificially considerably increased. Tirén (1977) found denitrification in experimental bell-jars *in situ,* but warned that although the data may indicate the existence of the process to be possible, it says little about what actually happens in the lake as a whole. Brezonik & Lee (1968) reported a denitrification loss of 11% of the annual nitrogen input in Lake Mendota. It must be mentioned however, that in such balance studies all the errors may aggregate in the calculation of this process. Keeney (1973) and Isirimah, Keeney & Dettmann (1976) have reviewed usefully the processes in the nitrogen cycle.

The authors wish to thank cordially all colleagues who provided them with reports (unpublished or not) and reprints with data related to this chapter. Specially are mentioned all the IBP Data Reports, which are listed in Appendix I. The authors acknowledge the sometimes controversial advice from their colleagues, Dr A.B. Viner, Dr D. Westlake and Dr J. Talling. The help of Dr L. Lyklema with some of the mathematical problems, of Mrs G. Würtz for the references, of Mr J. Landstra for the general preparation of the manuscript and of Mrs L. Williboordse for the typing is respectfully acknowledged.

This chapter, submitted in 1975, was partly updated in 1977.

5. Primary production

Coordinator: D.F. WESTLAKE

5.1. Introduction (Marker & Westlake)

Photosynthesis by green plants adds chemical energy and organic matter to ecosystems, much of which becomes available to consuming organisms. The general theme of this chapter, in accordance with the general conclusions of Chapter 3, is that the rate of production of organic matter is governed in the first place by the rate of input of light and the capacity of the organisms present to use this energy for gross photosynthesis. True gross photosynthesis (A_t) is the rate of transformation of radiant energy to chemical energy, accompanied by equivalent conversions of carbon dioxide and water into organic carbon and oxygen. Various concurrent respiratory activities (R) consume oxygen and convert some of this organic matter back to carbon dioxide, releasing energy. The remainder is the net photosynthesis $(N = A_t - R)$.

If ample nutrients are present, and there are only respiratory losses, the rates of net photosynthesis and of accumulation of biomass $(w = N)$ can be high, and the photosynthesis per unit surface area increases with biomass (B). The penetration of light into the water is, however, dependent on the biomass. As this increases, more and more of the population in a mixed water column becomes self-shaded and the respiration per unit surface area increases until net photosynthesis becomes zero and growth stops. If nutrient supplies are low, respiration and losses such as excretion and mortality may increase, thus reducing net photosynthesis and biomass accumulation. Such effects may be more sensitive to nutrient deficiencies than gross photosynthesis. If grazing and bacterial decomposition are high, so that nutrients are recycled rapidly, it may be possible for net photosynthesis to be rapid even though nutrient concentrations and plant biomass are low.

The increase in oxygen concentration $(A_t - R_{com}$, where R_{com} is the community respiration) in an illuminated enclosure cannot measure true gross photosynthesis because of oxygen consumption by the plants (R) and other, heterotrophic, organisms $(R_h: R_{com} = R + R_h)$. Traditionally the oxygen decrease in a darkened enclosure $(_d R_{com})$ has also been measured, which is the sum of dark respiration by plants $(_d R_d)$ and the consumers $(_d R_h)$, and a correction made $[A_t = (A_t - R_{com}) + {_d R_{com}}]$. This assumes that total plant respiration and the respiration of consumers are the same in the light and dark enclosures (i.e. $R = {_d R_d}$, $R_h = {_d R_h}$). However, it is now clear that in many

Contributors: D.F. Westlake, M.S. Adams, M.E. Bindloss, G.G. Ganf, G.C. Gerloff, U.T. Hammer, P. Javornický, J.F. Koonce, A.F.H. Marker, M.D. McCracken, B. Moss, A. Nauwerck, I.L. Pyrina, J.A.P. Steel, M. Tilzer & C.J. Walters.

plants at least one metabolic pathway taken by the early products of photosynthesis allows a considerable additional conversion of organic matter back to carbon dioxide with losses and transfers of chemical energy and consumption of oxygen (photorespiration R_p; $R = R_p + R_d$; Goldsworthy, 1970; Zelitch, 1971; see also section 5.5). Photorespiration varies with carbon and oxygen as well as with species and irradiance, and does not occur in the dark. Furthermore the physiology and anatomy of different species affects the recycling of carbon dioxide produced by photorespiration, presenting very real problems for the measurement of both photorespiration and photosynthesis.

Hence the dark respiration correction generally gives serious underestimates of gross photosynthesis, and routine techniques for measuring and correcting for photorespiration are not available. However, if the dark respiration components remain unchanged in the light it can be shown that the sum of the oxygen changes in light and dark enclosures gives the true gross photosynthesis minus photorespiration:

$$A_t - [R_p + R_d + R_h] + [_d R_d + _d R_h] = A_t - R_p$$

This may be regarded as the rate at which organic carbon becomes available for growth and dark respiration, or the rate at which it becomes assimilated into the plant, and will therefore be called photoassimilation (A) throughout this chapter, emphasising that it is not the true rate of gross photosynthesis.

There is some evidence that the dark respiration and the non-plant respiration in the dark enclosure may also not be the same as in the light, because they may vary with the current and previous exposure to light ($R_d \neq {_d}R_d$; $R_h \neq {_d}R_h$; see section 5.5). This would introduce errors into the measurement of photoassimilation.

The increase in oxygen concentration in an illuminated enclosure ($A_t - R_{com}$) may not measure true net photosynthesis [$N = A_t - (R_p + R_d) = A - R_d$], because of the oxygen consumption by heterotrophic organisms such as bacteria and zooplankton or zoobenthos (R_h). It is equivalent to the overall net rate of gain of autochthonous (internal) organic matter to the whole biological biocoenosis but does not measure any allochthonous (external) gains (U). In experiments with macrophytes non-plant respiration is usually negligible and net photosynthesis is the best value to use as a criterion of photosynthetic activity. For experiments with phytoplankton or periphyton the non-plant respiration is usually large and photoassimilation is preferred.

The increase in radiocarbon fixed in the plant and excreted material can theoretically be anything between true gross photosynthesis and net photosynthesis (of the period of exposure). This depends on the extent of recycling of respiratory carbon dioxide and the degree to which this becomes labelled, and hence on the plant's anatomy and biochemistry, the environmental conditions and the duration of the exposure. For radiocarbon fixation to equal gross photosynthesis all respiratory carbon must be lost and must be unlabelled; for it to equal net photosynthesis all respiratory carbon (from both dark and photorespiration) must be refixed or lost labelled. If radiocarbon fixation is to measure photoassimilation, unlabelled carbon must be lost equivalent to dark respiration and the remaining respiratory carbon must be refixed, or lost labelled as carbon dioxide. For phytoplankton it is generally considered that radiocarbon fixation rates are close to photoassimilation. On the assumption that lower surface to volume ratios and internal air spaces favour recycling, it has been anticipated that fixation may be closer to net photosynthesis for macrophytes.

Since dark respiration continues overnight, the net production over twenty-four hours ($N_{24} = A_{24} - R_{d, 24}$), or multiples of that period, is considerably less than the net photosynthesis found during the daylight periods. Twenty-four hour net production is more relevant for studies of growth and food supply, but because of the difficulty in measuring algal respiration little good information is available. Even for macrophytes photosynthetic estimates of net production have rarely been used and most data are from biomass changes over long time intervals (greater than a week). In situations where the decomposition of allochthonous material is a significant component of community respiration the difference between photoassimilation and dark community respiration ($A_{24} - R_{d, com, 24}$), which is the value most easily found in practice, tells us little about autochthonous productivity and may even be negative.

The increase in biomass (w) of photosynthetic organisms may not equal the net production because of losses (L_{tot}) by processes such as excretion, mortality, grazing, sedimentation and wash-out ($A - R_d = w + L_{tot}$). For natural algal populations the losses are normally so large and difficult to measure that this is rarely a practical method. For natural macrophyte populations the losses are often less, and easier to measure, so the increase in biomass is used to measure net production. However, the losses are often neglected, which usually introduces a significant error.

Since the development of the radiocarbon technique many comparisons have been made with phytoplankton photosynthesis measured by oxygen or total carbon method; but interpretation is difficult because editorial policies frequently lead to inadequate descriptions of procedures, and there is little consistency about the assumptions and corrections that should be involved. These include biological factors such as excretion, internal recycling of carbon dioxide and assumed photosynthetic quotients, and analytical corrections such as self-absorption and isotope discrimination (Fogg, 1974). Steemann Nielsen (1965), Javornický (DR 93*) and Jewson (1977b) have found good average agreement provided the barium carbonate calibration correction is

---

* See list of data reports (DR) given in Appendix I.

taken into account. On the other hand the results of Ross & Kalff (1975) and Tilzer *et al.* (DR 38) showed that the oxygen method sometimes indicates a carbon uptake over twice the radiocarbon results and that this factor was affected by irradiance. Tilzer *et al.* thought excretion was the main source of the discrepancy but the effects of light suggest that photorespiration and recycling are also involved. Ganf & Horne (1975), who corrected for excretion, also found some effect of light. Their average factor relating the two methods was about 1 but the range was from 0.53 to 2.43. Recent laboratory studies on a macrophyte (J. Titus, personal communication) found that radiocarbon fixation was about 1.2 times net photosynthesis, but the relationship depended on experimental conditions.

In general it is therefore difficult to make accurate comparisons between different sets of data, since different methods, or even a single method used in various circumstances, may measure different criteria of productivity. On the other hand sampling errors also prevent significance being given to small differences. Limited comparisons between results obtained by a single method are best, but wider comparisons are often necessary if any general conclusions about different ecosystems and the roles of different organisms and different environmental factors are to be reached.

To facilitate such comparisons, and back-calculation to the original data, the authors agreed on a number of general assumptions, standards and conversion factors to be used whenever disparate data are assembled for comparisons. Brylinsky adopts a similar approach in Chapter 9, taking the fixation of carbon-14 during the daylight period as standard and converting oxygen results, by a factor of 0.32, to an approximate carbon equivalent. When such approximations and assumptions are made the results of the comparisons should be treated with caution. Conversions made in this chapter, indicated by an asterisk, are based on:

(1) Photosynthesis and respiration results expressed in terms of oxygen exchange per unit area, volume or chlorophyll *a*.

(2) Results obtained using carbon-14 with phytoplankton are multiplied by three to obtain the oxygen output of photoassimilation, which includes an addition of approximately 10% and a photosynthetic quotient (PQ) of 1.0 (very similar to Brylinsky's assumptions).

(3) Results obtained using carbon-14 with higher plants are multiplied by 2.5 to obtain the net photosynthetic oxygen output, which includes a subtraction of approximately 10% and a PQ of 1.0.

(4) Photosynthetically active radiation (PAR) at zero depth (I'_0) is approximately 0.4 of the total incident irradiance (I_0), which includes 0.46 to convert total to PAR, and a 10% reflection loss.

Many further investigations are needed into the relations between results obtained by different methods and into better methods of measuring the ecologically important rates. It will be unfortunate if the differences between researchers become matters of irreconcilable principle rather than phenomena that can help to resolve the complexities of field photosynthesis.

The relationships between the components of a system carrying out and utilising primary production are summarised in fig. 5.1 (see also section 5.6). This figure shows a typical phytoplankton biocoenosis and the emphasis will differ in other types of community (see DR 30 for a river system). Since phytoplankton, periphyton and macrophytes have very different relationships with their environment, their behaviour and contributions to the total primary

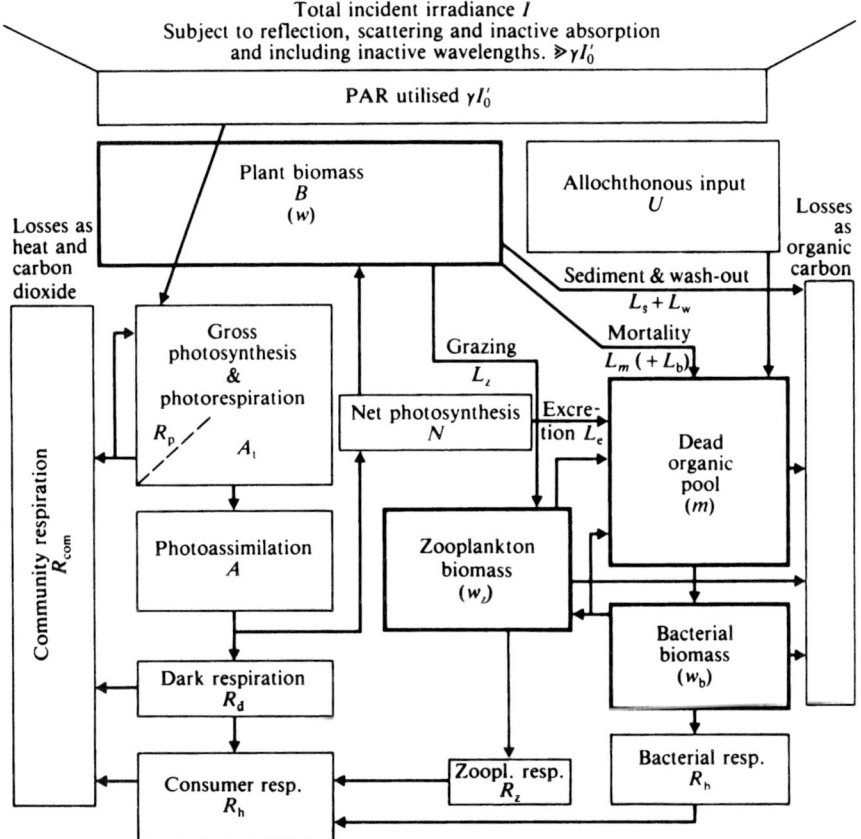

Fig. 5.1. The flow of energy and organic material in a planktonic ecosystem. The areas of rate boxes (thin outlines) are approximately in the proportions that might be found in some real situations. Boxes with thick outlines are accumulators; the rates of accumulation are indicated in brackets.

There are two practical problems of scale. The efficiency constant $\gamma(A_t/I_0)$ is normally much smaller than 0.1, so the total incident radiation is much greater than the proportion actually used in photosynthesis. Rates that are probably less than $0.1\,A_t$ also cannot be conveniently shown and appear without boxes.

The allochthonous input in this example is probably relatively high.

production are treated separately throughout this chapter. However, they often interact and the factors affecting their relative importance in the ecosystem are discussed. For the purposes of IBP synthesis the emergent macrophytes are treated in the Wetlands volume (Květ, Szczepański & Westlake, in preparation) and only submerged macrophytes are included in this volume. Production by the marginal emergent macrophytes is regarded here as an allochthonous input to the main body of open water.

5.2. Biomass

5.2.1. Introduction (Moss)

The species composition of the biomass depends on many factors. Certain common properties of groups of species (not necessarily taxonomic) shed light on their association with particular environments – e.g. the motility and circadian rhythms of algae living on unstable mud surfaces (Round & Eaton, 1966), the low growth rates and relatively high carbon dioxide requirements of oligotrophic desmids (Moss, 1973a), and the association of araphidinean plankton diatoms with nutrient enriched conditions (Stockner, 1971). More specific information on why a given species occurs in particular places rarely exists.

Amount of living matter can be expressed on an areal basis, or as a concentration, and its species composition and overall quantity may be expressed as dry ash-free organic weight, carbon or chlorophyll a, or in many other ways. In IBP investigations of algal populations cell volumes and chlorophyll a measurements have been most widely used as measures of biomass, but dry or ash-free dry weight have generally been used for macrophytes. Despite the acknowledged variability of chlorophyll a content per unit cell carbon or ash-free dry weight under different conditions in the laboratory few comparisons have been made in natural populations. Simple calculation of the chlorophyll (uncorrected for degradation products) to organic carbon ratio gives the superficial appearance of a similar variability to that in cultures, e.g. for Abbot's Pond phytoplankton a range of 2 to 43 mg chl. $a g^{-1}$ (g total C) was found (Moss, DR 6). Regression of corrected chlorophyll a on organic matter, however, permits calculation of the detritus component and may show the ratio of chlorophyll a to cell organic carbon to be much less variable than has previously been supposed. The ratio for Abbot's Pond was 21.3–24.4 mg chl. a (g algal C)$^{-1}$. Similar constancy of the ratio of chlorophyll a to carotenoids was also found (Moss, 1970a).

The total chlorophyll to dry weight ratio in whole macrophytes of normal ash content is generally between 5 and 15 mg chl. (g dry wt)$^{-1}$ (reviewed in Westlake, 1975b) with highest values found in young shoots. Assuming 0.15 of the dry weight is ash, 0.47 of the organic weight is carbon and 0.75 of the total chlorophyll is in the *a* molecule, this would be approximately $10-25$ mg chl. $a\ (\mathrm{g\,C})^{-1}$, considerably less than for many algae.

5.2.2. Phytoplankton biomass (Moss)

Biomass is here considered on the basis of (*a*) maximum attained, (*b*) horizontal variation, (*c*) vertical variation. The minimum biomass may be important as the level from which exponential growth starts (Lund, 1954) but sufficiently accurate measurements of such low concentrations are difficult and values are few.

(*a*) Maximum biomass. Some interest attaches to the maximum attained in view of possible self-shading effects on photosynthesis (cf. sections 5.3.4 and

Table 5.1 *Maximum phytoplankton biomass*

Lake	Mean depth (m)	b_{max} (mg m^{-3})	B_{max} (mg m^{-2})	B_{max} (mg m^{-2} euphotic)	Reference
		Chlorophyll *a*			
L. Kilotes (Ethiopia)	2.6	412	—	194	Talling *et al.* (1973)
L. Aranguadi (Ethiopia)	18.5	2170		325	Talling *et al.* (1973)
L. George (Uganda)	2.25	440	1084	350	Ganf (1972, 1974d)
Jezárko Pond (Czechoslovakia)	0.7	177[a]	124[a]	—	Fott *et al.* (1974)
Velký Pálenec Pond (Czechoslovakia)	1.4	135[a]	190[a]	—	Fott *et al.* (1974)
Velký Bezděkovsky Pond (Czechoslovakia)	1.0	1800[a]	1800[a]	—	Fott (pers. comm.)
R. Thames (UK)	2.3	199	458	—	DR 80
R. Kennet (UK)	1.0	34	34	~34	Kowalczewski & Lack (1971)
L. Batorin (USSR)	3	74	222	—	DR 68
L. Yunoko (Japan)	12 max	34	379	197	DR 8
L. Trummen (Sweden)	1.8	173	398	351	DR 28
L. Chilwa freshwater phase	2	1400	—	—	DR 5
(Malawi) drying phase	<2	436	—	—	
Abbot's Pond (UK)	<2	300	460	—	Hickman (1973)
Loch Leven (UK) at b_{max}	3.9	217	846	260	DR 3
at B_{max}		170	663	456	
Lago do Castanho (Brazil)	1–12	100	400	—	DR 41
L. Biwa (Japan)	41.2	26	484	255	DR 36
Corangamite L. (Australia)	2.9	226	—	—	DR 14
Red Rock Tarn (Australia)	1.4	1050	—	—	DR 14
L. Werowrap (Australia)	1.4	810	—	—	Walker (1973)

[a] Uncorrected for phaeopigments, chlorophyll *a* probably $10-15\%$ less.

5.9.2). The largest values reported for IBP projects, plus some others are shown in table 5.1.

Values above 400 mg m^{-3} are rare, generally involve blue-green algae and perhaps reflect the concentration of these algae by flotation and windrowing as much as association of the group with nutrient-rich conditions.

Data expressed as milligrams per square metre reveal values up to just over 1000, but high values for the euphotic zone are generally in the range 150–450.

(*b*) Horizontal variation. Horizontal variation in phytoplankton biomass may be expected in large lakes and will reflect hydrographic and possibly even latitudinal conditions. Lake George (Uganda) frequently has the greatest biomass (324 mg m^{-3} chlorophyll *a* on December 9, 1969) in the centre, with concentric zones of decreasing biomass surrounding it (down to 18 mg m^{-3} on December 9, 1969) (Burgis *et al.*, 1973). This situation is attributed to a circular current movement (Ganf, 1974d). Bays where exchange of water with the main lake is slow may also maintain different biomasses, frequently as a result of local drainage of nutrient-rich rivers, and whole inshore areas may be richer than offshore areas for the same reason (e.g. Holland, 1968). In Lake Kariba enrichment of inshore water in the wet season by decomposing dung dropped on lakeside flats by grazing mammals in the dry season leads to localised algal blooms (A.J. McLachlan, personal communication). In Lake Chad (DR 4) a horizontal variation of 8–20 mg m^{-2} was found in December 1970, and of 10–48 mg m^{-2} in 1971. A homogeneous horizontal distribution is often assumed, but this is clearly unwise.

(*c*) Vertical changes in biomass. Whenever thermal stratification occurs there is some concomitant stratification of phytoplankton. At its simplest this consists of a uniform suspension of cells in the epilimnion, perhaps some accumulation of sedimenting cells at the thermocline, and a transitory sinking population in the hypolimnion. From the available examples, however, the vertical pattern is usually much more complex, perhaps because of chemical gradients forming around the thermocline and in the epilimnion during still periods. There is a marked stratification of different species and also of chlorophyll *a*, though the often rather sharply defined species pattern is much blurred by the use of overall biomass parameters. The stratification may be stable over quite long periods (Moss, 1969b), but may also be diurnally altered by vertical movements of photosynthetic flagellates (Happey & Moss, 1967; Tilzer, 1973).

Interpretation of the reasons for differential depth distributions of species often demands knowledge of changes over a period of time. Species may accumulate at the surface or just below it because they are phototactic, or towards the bottom because of sinking or because their particular nutrient requirements are met there. Populations whose growth has ceased in the epilimnion may take several weeks to disappear entirely from the water column and during this period may form a distinct layer in the hypolimnion (e.g. *Fragilaria crotonensis* Kitton in Gull Lake, Michigan, Moss, 1973b). Other populations (e.g. *Oscillatoria agardhii* var. *isothrix* Skuja; Saunders, 1972d) may begin growth in the deoxygenated hypolimnion and perhaps later move upwards. Photosynthetic bacteria may persist only in regions of the water column where hydrogen sulphide or other necessary reduced hydrogen donors are present. Aggregations in mid-column may result from a need for intermediately deoxygenated conditions around the thermocline or from the presumed abundance of bacterially produced vitamins in this region, where organic detritus tends to collect. *Cryptomonas* spp. are frequently clustered in this region.

When thermal stratification is insufficient to suppress turbulent mixing water movements can impose vertical patterns (e.g. Lake George, Ganf, 1974b).

5.2.3. Macrophytes (Westlake)

In comparison with algae a considerable proportion of the biomass of macrophytes is relatively non-photosynthetic, such as stems, and storage and supporting tissues. The underground parts are usually sampled and included in the biomass, but are only a large part of the whole in rosette plants such as *Eleocharis, Littorella* and *Isoetes* (Kansanen *et al.*, 1974).

5.2.3.1. Maximum and minimum biomass

The maximum biomass found at different temperate zone sites has a special significance because it may be close to the annual cumulative net production, losses before the maximum often being less than 0.2 of the maximum. In terms of ash-free dry weight their maximum biomass does not exceed $700 \mathrm{~g~m^{-2}}$ as an average for particular sites (table 5.2). Values of over $1000 \mathrm{~g}$ dry wt $\mathrm{m^{-2}}$ can be found in local accumulations or in stands of species such as *Chara* (Forsberg, 1960) which may have over $0.5 \mathrm{~g~g^{-1}}$ of calcareous deposits present in the dry biomass samples. In terms of total chlorophyll the biomasses summarised in the table could correspond to $200-5000 \mathrm{~mg~m^{-2}}$, or rather less in terms of chlorophyll a (assuming $4-8$ mg chl. (g dry wt)$^{-1}$; Westlake, 1975b). The only direct determinations are $100-3000 \mathrm{~mg~chl.}a\mathrm{~m^{-2}}$ (Odum, 1957b; Ikusima, 1965, 1966; Maier, 1973; Kansanen *et al.*, 1974). This shows a tendency to higher values than in natural algal communities (cf. table 5.1).

The minimum biomass is rarely studied, but will be important in communities where the plants must grow some way towards the surface to reach the euphotic zone, or where the first species to establish a canopy is enabled to dominate. Some species show large seasonal differences in biomass, but others change little (see section 5.6).

Table 5.2. *Some recent determinations of the seasonal maximum biomass of submerged or floating macrophytes*

Water body	Dominant species	Extent of zone in depth (m)	Biomass (g org. wt m^{-2}) Macrophyte zone	Whole area	Reference
Pääjärvi (Finland)	Isoetes lacustris	0–1.5	27	2.4	Kansanen et al. (1974)
Jezárko Pond (Czechoslovakia)	Batrachium aquatile	0.75	46[a]	5[a]	DR 10
R. Thames (England)	Nuphar lutea	0.5–2	180	7	Lack (1973)
Marion L. (Canada)	Nuphar polysepela	0–4	110	25	Mann et al. (1972) Davies (1970)
Ø. Heimdalsvatn, (Norway)	Isoetes lacustris	0–7	110[a]	40[a]	Brettum (1971)
Mikolajskie L. (Poland)	Elodea canadensis	—	210[a]	13[a]	Kajak et al. (1972)
L. Wingra (Wisconsin, USA)	Myriophyllum spicatum	0.1–2.5	220	68	Adams & McCracken (1974)
Lawrence L. (Michigan, USA)	Scirpus subterminalis	1–6	300	130	Rich et al. (1971)
L. Naroch (Byelorussia, USSR)	Nitella spp.	100–260 0–10	320	95	Winberg et al. (1972)
Bere Stream (England)	Ranunculus calcareus	0–1	100–260		Westlake et al. (1972)
L. Biwa, Shoizu Bay (Japan) 1962–5	Hydrilla verticillata Elodea nuttallii	0.5–6	360–880	8	Ikusima (1975)
L. Biwa, S. Basin	Egeria densa	1.5–2	160	9	Tanimizu & Miura (1976)

[a] Typical ash contents of ~ 20% and calorific contents of ~ 19 kJ (g org. wt)$^{-1}$ assumed.

5.2.3.2. Horizontal and vertical distribution

Within the community the biomass in individual samples may vary from zero to over 1500 g ash-free wt m^{-2}, for most macrophytes show very contagious distribution patterns. On a larger scale the rooted macrophytes are controlled by permanently differentiated features of the environment such as depth, exposure and sediments, which are more evenly distributed, often changing regularly in one particular direction. This is very different from the phytoplankton. The highest biomasses are found on rich sediments in fairly shallow, sheltered, clear water, though the maximum is usually not in the shallowest water because of wave action and level changes (e.g. Brettum, 1971; Rich *et al.*, 1971; Adams & McCracken, 1974; Spence, 1976). In a river with relatively stable flow Dawson (1976) has found an inverse regression relating biomass and depth between 0.3 and 3 m. Once the depth becomes so great that the light reaching the bottom is insufficient for growth, or the plants cannot grow up towards sufficient light, macrophytes are absent. Some Charophytes and mosses can sometimes maintain stands down to ~ 35 m (Spence, 1976) but angiosperms rarely approach 10 m. Free-floating macrophytes such as *Lemna* and *Eichhornia* are distributed by wind, wave and current action and often accumulate on lee shores or at obstacles (Mitchell, 1969).

Apart from the limitations imposed by factors such as the decrease of light with depth and the increase in exposure to water movement towards the surface, macrophytes have considerable control over the vertical distribution of biomass within the plant stand, achieved by their morphological features which give buoyancy, attachment to the bottom and resilience to water movement. There are three main types of distribution:

(1) Concentrated at the water surface, either free-floating or attached, e.g. *Lemna*, *Nuphar* (Ikusima, 1970).

(2) Concentrated fairly near the water surface e.g. *Potamogeton pectinatus, P. crispus, Myriophyllum* spp. (Westlake, 1964; Ikusima, 1965; Owens, Learner & Maris, 1967; Rich *et al.*, 1971; DR 12).

(3) Concentrated in a sward at the bottom, e.g. *Vallisneria, Nitella* (Odum, 1957b; Ikusima, 1965; Tanimizu & Miura, 1976; DR 12).

Stratiotes aloides and *Lemna trisulca* have a limited ability to adjust their vertical distribution seasonally by means of buoyancy changes (Sculthorpe, 1967).

In addition to these gradients of total shoot weight the proportion of leaves to stem increases towards the top of the stand (Ikusima, 1965; Brettum, 1971; Adams *et al.*, 1974).

5.2.4. Periphyton (Moss)

One group of periphyton communities is firmly attached to substrata of various kinds, forms dense felts and harbours many old and dead cells as well

D.F. Westlake

as live ones. A second group consists of massed filaments or lumps of gelatinous material. A third group includes communities whose members are not firmly attached, nor entangled together, but which move freely over a substratum, typically sediment.

Table 5.3 *Biomass of periphyton communities*

	Maximum values recorded (mg chlorophyll a m^{-2})	Reference
(i) *Free-living communities*		
Pennate diatom communities	9.4, 22.9, 124.8,	(a)
living on sediment, *Euglena*	82.1, 2.6, 3.4, 66.5,	
populations and various	3.8, 15.0, 32.0,	
Oscillatoria populations	79.8, 51.3	
Diatoms plus desmids:		
Mlungusi dam, Malawi	9.3	Moss (1970b)
Diatoms + blue-green algae:		
L. Chilwa, Malawi	126	Moss & Moss (1969)
Diatoms: Shearwater, UK	14	Hickman & Round (1970)
Diatoms + blue-green algae:		
Priddy Pool, UK	21	Hickman (1971b)
Diatoms: Abbot's Pond, UK	92	Hickman (1971b)
(ii) *Attached communities*		
Various communities on rocks,	800, 260, 1200, 381,	(a)
higher plants, sand grains	350, 1085, 303, 224,	
and sea ice	2350, 186, 12	
Largely green algal community,		
Equisetum: Priddy Pool, UK	78	Hickman (1971a)
Diatoms attached to sand		Hickman & Round
grains: Shearwater, UK	170	(1970)
Diatoms attached to sand		Bailey-Watts
grains: Loch Leven, UK	800	(1973, 1974)
Diatoms on gravel:		
Bere Stream, UK	250–300	Marker (1976a)
Diatoms and other algae on rocks:		
Deep Creek, Idaho, USA	180	DR 30
Diatoms on macrophytes:		
Mikołajskie Lake, Poland	33–200	DR 38
(iii) *Massed, gelatinous and mat-forming communities*		
Various filamentous green, and	269, 189, 380, 299,	(a)
blue-green algal communities	518, 590, 1960	
Spirogyra: Mlungusi dam, Malawi	308	Moss (1970b)

(a) Summarised from tables in Moss (1968). Other data conform to the analytical standards used to compile the tables in Moss (1968)

These functional considerations partly explain the differences in biomass attained by various periphyton communities. Free-living communities are subject to erosion by water movements, whereas the firmly attached and aggregated groups are able to accumulate biomass because their growth form gives resistance to disruption. Table 5.3 shows available data for biomass of periphyton communities.

The horizontal and vertical distribution of periphyton is obviously very dependent on the distribution of adequately illuminated substrata. In comparison with both macrophytes and phytoplankton, periphyton populations are confined to a relatively thin layer, within which very high concentrations of biomass may be obtained.

5.2.5. Discussion (Moss)

As it usually physically overlies any periphyton, or submerged macrophyte communities, phytoplankton receives incident light first and it thus has a competitive advantage. In fertile lakes with high phytoplankton biomass the periphyton or macrophyte communities are unlikely to cover much of the lake area. Conversely where shading by phytoplankton is diminished the proportionate importance of the periphyton and macrophytes is likely to be greater (see section 5.8). In Abbot's Pond, UK, decrease of the epipelic algal flora occurred as phytoplankton biomass increased (Moss, 1969b) and transparent oligotrophic lakes such as the Mlungusi dam, Malawi (Moss, 1970b) and the Vorderer Finstertaler See, Austria (Pechlaner *et al.*, 1972b; Pfeifer, 1974) have very large periphyton biomasses.

The average biomass per square metre of free-living algae on sediments increases with nutrient richness of the water (Moss, 1969a), as does that of phytoplankton. This is because these communities are subject to net loss of cells by burial and sinking once population growth ceases; thus biomass reflects current net production which is determined by nutrient availability. In attached and aggregated communities and, judging by the substantial biomass of higher plants in soft water and bog lakes, in macrophyte communities also, no such correlation exists since biomass can accumulate, even if slowly, by virtue of the mechanical stability of these communities; this may give an erroneous impression of high production rate.

5.3. Irradiance in the water column (Tilzer)

In closed terrestrial plant stands most light energy is absorbed by photosynthetic tissues. Contrary to this, in aquatic environments the medium itself absorbs considerable fractions of the incoming light. On an areal basis, therefore, the utilisation of energy for photosynthesis in aquatic ecosystems is frequently much lower than in terrestrial ones (e.g. Tilzer *et al.*, 1975). It also follows that aquatic primary production is often light-limited and consideration of the light climate is of major importance.

5.3.1. The light climate in water (Tilzer)

5.3.1.1. Surface effects

(a) Open water.   Surface losses are caused by reflection from the surface itself and by back-scattering from suspended particles (Sauberer & Ruttner, 1941). Reflection of direct sunlight depends on the sun angle and increases markedly in the early morning and evening. In such situations wave action has a strong influence on reflection (Angström, 1925; Kirillova, 1970). Some 6–8% of diffuse light is reflected, which is little affected by sun angle (Hutchinson, 1957). When the sun angle is between 55° and 5° total reflection losses (direct plus diffuse) are between 5 and 25% of the total incident irradiance (I_0) (Anderson, 1952; Kirillova, 1970).

In most IBP studies total surface losses were assumed to be 10% (Vollenweider, 1974), but an average of 13% was measured for the Neusiedlersee (DR 50). Although the importance of the sun angle for reflection is minor at low latitudes and in temperate zones, at high latitudes considerable losses are probable (Thomasson, 1956; Winberg, 1973) and should not be neglected when considering regional variations of primary productivity.

(b) Frozen lakes.   The occurrence of a winter cover may have profound influence on the annual radiant energy inputs and thus the seasonal pattern of photosynthesis (pp. 170, 176). The few direct measurements available (Mokievskii, 1961; Pechlaner, 1966; Goldman *et al.*, 1967; Holmgren, 1968; Capblancq, 1972; Sherstyankin, 1975) have revealed drastic reductions of subsurface light to less than 1% of the summer values. Underwater irradiance is reduced by enhanced reflection and by absorption. While clear ice layers behave similarly to liquid water, thick and turbid ice reflects 35–40% (Ambach & Mocker, 1959) and snow 70–80% (Pechlaner, 1966). Reflection is reduced if the layers are thin; to approximately 50% of the maximum when 4 cm thick.

Light transmission through the winter cover caused only insignificant spectral effects in the visible range (Ambach & Habicht, 1962; Thomas, 1963). Attenuation coefficients (log base e) range from values similar to unfrozen water to 1.6 m^{-1} in clear ice, from 10 to 20 m^{-1}, depending on the density, in dry snow, and increase to 35 m^{-1} in old packed snow (Ambach & Habicht, 1962).

Occurrence and duration of winter cover depend on latitude and altitude as well as on morphometry and mixing type of the lake. Deep holomictic lakes in the temperate region rarely or never freeze (Lake Constance and Lake Tahoe, respectively) while shallow or meromictic lakes in similar climates normally have a winter cover (e.g. Carinthian lakes, Austria). The amount of winter precipitation is of great importance for the structure of the winter cover and, because of the high attenuation of light by snow, for the subaquatic light climate. In dry climates highly transparent clear ice covers occur (e.g. Char Lake, Canadian Arctic, Ivano-Arakhley Lakes, Central Asia, USSR).

5.3.1.2. Light underwater

The incident irradiance is attenuated underwater, depending on water colour and turbidity. In pure water and very clear oligotrophic lakes attenuation is least in the blue region of the spectrum (cf. Smith, Tyler & Goldman, 1973). With increasing trophic status or load of dissolved and particulate materials (especially when these are organic) the least attenuated irradiance may be green, yellow or even red. Ultra-violet and infra-red irradiance is rapidly attenuated. Conventionally the radiation most active in photosynthesis (PAR) is taken to be between 400 and 700 nm, approximately the same as the visible spectrum, although there is often some photosynthetic response to irradiance between 320 and 400 nm (Halldal & Taube, 1972).

At depth z the subsurface irradiance, I_0' is reduced to:

$$I_z = I_0' \cdot e^{-\varepsilon z} \tag{5.1}$$

This equation is strictly valid only for a specified wavelength. The vertical attenuation coefficient (ε) varies with wavelength, and at any depth the value obtained from equation 5.1 is not equal to the average of the attenuation coefficients for every wave-band. With increasing depth the light most attenuated becomes less relative to the light which has the lowest attenuation coefficient (ε_{min}) and the overall effective value ($\bar{\varepsilon}$) decreases, i.e. $\bar{\varepsilon}$ is itself a function of depth. In many situations, particularly if the attenuation close to the surface is excluded, the effective coefficient varies little with depth and has a value between 1.1 and 2 times the minimum coefficient (Vollenweider, 1960; Schindler, 1971b; Talling, 1971; Ganf, 1974a, 1975). Variation in $\bar{\varepsilon}$ with depth means that calculations of integral column photosynthesis ($\bar{A}$), as a function of the photosynthetically available irradiance (see also section 5.9.2) will be inaccurate if either ε_{min}, or $\bar{\varepsilon}$ obtained from readings at one particular depth, are used to model PAR. Talling (1957b, 1971) has found that observed values of column photosynthesis in a variety of waters can be reconciled with calculated values if $\bar{\varepsilon}$ is calculated as $\sigma\varepsilon_{min}$ where σ is 1.33, although this factor is expected to vary with water colour. Recent studies have found a range of only 1.15 to 1.33 for σ (Steel, 1978; Ganf, 1975; Jewson, 1975; Bindloss, 1976). This empirical factor might also include other modelling errors.

The total attenuation coefficient in natural waters can be attributed to attenuation by pure water, by dissolved colour, and by suspended particles (James & Birge, 1938; Hutchinson, 1957). Since particles both absorb and scatter light, their colour *and* size distribution are important. When considering primary productivity it is more convenient to distinguish non-plant contributions (ε_q) and plant contributions (ε_b) to total attenuation (ε_{tot}) as

$$\varepsilon_{tot} = \varepsilon_q + \varepsilon_b \qquad (5.2)$$

Since light absorption by plants is their sole source of photosynthetic energy, the ratio $\varepsilon_b/\varepsilon_q$ is important (Tilzer *et al.*, 1975).

5.3.2. Effects of non-living material (Tilzer)

Such material either originates from the watershed (Vorderer Finstertaler See, Tilzer, 1972, Bretschko, 1975; Lake Tahoe, Goldman, 1974; Tilzer *et al.*, 1976, Lake Chad; DR 4; Lemoalle, 1973), or is stirred up from the lake bottom by wind action in some shallow lakes (Ganf, 1974a; Dokulil, 1975; Jewson, 1977a). Dissolved colour is high in waters draining from peaty watersheds (Jewson, 1977a) or sometimes in polluted rivers (Westlake, 1966a). Suspended particles often cause rapid attenuation in river waters (Westlake, 1966a). Light attenuation by suspended particles may be constant throughout the spectrum, or selectivity may increase in certain regions, but attenuation by dissolved material is usually highest for blue light resulting in brown water.

5.3.3. Effects of zooplankton and bacterioplankton (Tilzer)

Little direct evidence is available on the optical effects of zoo- and bacterio-plankton. Transmission through zooplankton is high in general because of large body sizes and insignificant pigmentation. Dense populations with carotenoid accumulation may be exceptions. Scattering by bacterioplankton is considerable because bacterial cells are usually very small. Dense layers of pigmented bacteria are likely to have marked effects, as suggested by results of Ruttner & Sauberer (1938) in Lunzer Obersee (Austria).

5.3.4. Effects of phytoplankton   (Tilzer & Bindloss)

Inverse correlations between phytoplankton density and underwater light penetration have been described by Talling (1960, 1971), Bindloss (1974) and Ganf (1974a). This 'self-shading effect' imposes an upper limit to photo-synthetic productivity (Talling *et al.*, 1973; see also section 5.9.2).

   In most densely populated waters the minimum attenuation coefficient (ε_{min}) is in the orange-red spectral region while the phytoplankton has most marked effects on light in the blue spectral region (Talling, 1960, 1970; Ganf, 1974a). If ε_{min} is plotted against biomass concentration (as $mg\,chl.\,a\,m^{-3}$) regression lines can help to separate algal from non-algal effects (Pyrina *et al.*, 1972; Ganf, 1974a; Bindloss, 1974). The slope of the regression line indicates the mean increment in ε_{min} associated with unit increase in chlorophyll concentration b (ε_s Talling, 1960; in ln units m^{-1} (mg chl. $a)^{-1}m^{-3}$, i.e. ln units (mg chl. $a)^{-1}m^{-2}$):

$$\varepsilon_s = \Delta\varepsilon_{min}/\Delta b \tag{5.3}$$

where Δ means corresponding changes in value. If ε_q is negligible,

$$\varepsilon_s = \varepsilon_{min}/b \tag{5.4}$$

The intercept of the regression line on the y-axis is a measure of the average background level of non-algal attenuation (ε_q). Since the proportion of total attenuation due to ε_q increases with decreasing biomass concentration, self-shading effects are most clearly analysed where ε_q is low and constant and relatively large variations in biomass concentration occur.

Field measurements have yielded values of specific attenuation coefficients (ε_s) varying between 0.006 and 0.020 m^{-1} (Talling, 1960, 1971; Steel, 1972; Bindloss, 1974; Ganf, 1974a; Berman, 1976; Jewson, 1977a). There is some evidence that lower values of ε_s are often associated with large or small cell sizes. Light attenuation by phytoplankton is caused by both pigment absorption and scattering and differences in their relative importance may change the relation with size. Low ε_s values, e.g. those found for *Peridinium* populations in L. Kinneret (Israel), may be due to both large cell sizes and low pigment concentrations within the cells (Berman, DR 88). In Abbot's Pond (England) the influence of biomass was obscured by large and fluctuating amounts of non-algal material and by seasonal variations in phytoplankton species composition (Moss, 1970a).

5.3.4.1. Phytoplankton biomass and euphotic depth

At light levels of less than 1% of surface irradiance, net photosynthesis is conventionally assumed to be insignificant. The depth of the 1% light level is therefore assumed to be the lower limit of the euphotic zone (euphotic depth z_{eu}). This depth can be determined from the minimum attenuation coefficient by:

$$z_{eu} = \beta/\varepsilon_{min} \tag{5.5}$$

where β is a constant. Talling (1965, 1971; Talling *et al.*, 1973) has found $\beta = 3.7 \pm 15\%$ and subsequent determinations have also found values between 3.2 and 4.2 (Ganf, 1974a; Bindloss, 1976; Jewson, 1977a). Since $z_{eu} = \ln 0.01/\varepsilon_{eu}$ (from equation 5.1, where ε_{eu} is an overall effective attenuation coefficient appropriate to the depth of 1% of subsurface PAR), it follows that $z_{eu} = 4.6/1.25\, \varepsilon_{min}$.

When phytoplankton is distributed uniformly with depth, the amount of biomass in the euphotic zone per unit area (B_{eu}) is given by

$$B_{eu} = bz_{eu} \tag{5.6}$$

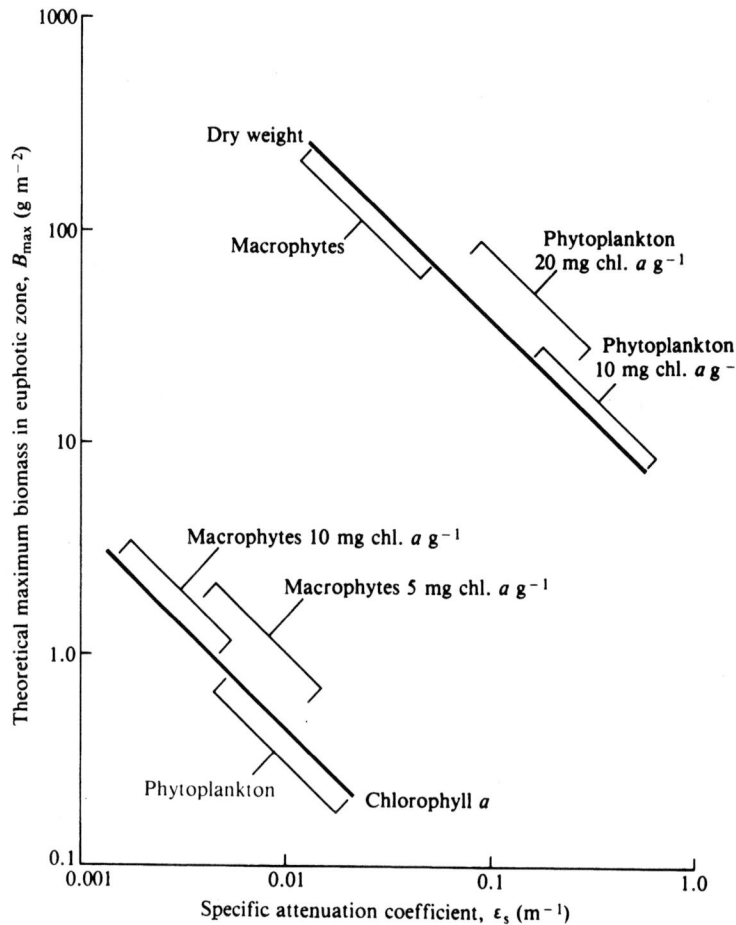

Fig. 5.2. Maximum possible chlorophyll cover or biomass (B_{max}) in the euphotic zone for natural phytoplankton and macrophyte populations at different values of the specific attenuation coefficient (ε_s).

(Calculated from $3.7/\varepsilon_s$ as that biomass which allows 1% of the surface irradiance to be transmitted, see p. 159. It is assumed that non-plant attenuation is negligible and that the full ranges of observed values of ε_s are compatible with the full ranges of chlorophyll contents used.)

The upper line is based on observed values of ε_s for macrophytes expressed in dry weight units, and comparable values for phytoplankton calculated from the observations in the lower line and a typical range of 10–20 mg chl. a (g dry wt)$^{-1}$ in phytoplankton.

The lower line is based on observed values of ε_s for phytoplankton expressed in chlorophyll units and comparable values for macrophytes calculated from the observations in the upper line and a typical range of 5–10 mg chl. a (g dry wt)$^{-1}$ in macrophytes.

For references see text.

Theoretically ε_s controls the upper limit of the size of the euphotic biomass
(B_{max}). If algae were the sole source of light attenuation in the water column
($\varepsilon_q = 0$) $\varepsilon_s = \varepsilon_{min}/b$ and, if the population is uniformly distributed vertically,

$$B_{max} = 3.7/\varepsilon_s \tag{5.7}$$

(from equations 5.5 and 5.6; cf. fig. 5.2). The maximum possible euphotic
biomass per unit area is independent of biomass concentration per unit
volume (Talling, 1970; Talling *et al.*, 1973). In general, biomass concentrations
have greater fluctuations than biomass per unit area, but the latter is
considerably reduced in cases where non-plant attenuation is high (e.g. the
turbid lake Chad, DR 4; Tilzer *et al.*, 1975). B_{max} will be approached as non-
plant attenuation becomes negligible relative to attenuation by phytoplankton.

5.3.4.2. The ratio of euphotic depth to mixing depth and phytoplankton stratification

Vertical phytoplankton stratification is influenced by reproduction rate (light
and nutrient dependent), passive sinking, resuspension and active migrations
(all influenced by water turbulence). High turbulence in the epilimnion favours
homogeneous distribution while in the less turbulent metalimnetic and
hypolimnetic layers stratification is possible. Two general situations can be
distinguished:

(a) $z_m < z_{eu}$: the whole euphotic phytoplankton is circulating. This occurs in
most lakes during isothermal mixing, in polymictic lakes, and in lakes with
light attenuation sufficiently high to restrict the euphotic zone to the
epilimnion. Although uniform vertical distributions of the phytoplankton
might be expected, in most cases slight inhomogeneities are observed. Near the
surface, sedimentation may reduce the biomass (Hutchinson, 1967), or
accumulations of buoyant blue-green algae may be found (Fogg & Walsby,
1971). Decreasing sinking velocities due to increases in the density and
viscosity of the water may cause higher concentrations of biomass towards the
bottom of the epilimnion. Differences in sinking speeds between species may
result in vertical changes in species composition.

(b) $z_m \leq z_{eu}$: the euphotic zone extends into the metalimnion and sometimes
into the hypolimnion. Water layers containing phytoplankton can thus be
separated from each other, and taxonomic and/or physiological differen-
tiations can develop as consequences of the different environmental conditions, especially with regard to light. This occurs in wind-protected,
frequently meromictic, lakes with shallow epilimnia, and in transparent
oligotrophic lakes. In many highly transparent lakes (e.g. Baykal, USSR;
Tahoe, Crater Lake, USA) the epilimnetic phytoplankton is extremely dilute
because of nutrient depletion and light inhibition, and the maximum biomass

occurs well below the thermocline. In arctic and high-altitude lakes the low thermal stability allows only weak stratification of immobile species, but there are deep populations mainly consisting of flagellates. Negative phototaxis to escape from light inhibition (Rodhe, 1962; Nauwerck, 1966, 1968; Holmgren, 1968; Tilzer, 1972, 1973) as well as better nutrient availability near the lake bottom (Pechlaner, 1967, 1971) have been proposed as explanations. In lakes where the euphotic zone extends into the hypolimnion, vertical changes in light adaptation (Tilzer & Schwarz, 1976) may allow a positive net photosynthesis balance below the 1% light level. Such populations are subject to considerable biomass losses by sinking into the aphotic zone.

Under ice phytoplankton biomass is concentrated near the surface in the highest irradiance available and where the release of nutrients from the water, as it freezes, presumably increases their concentrations.

5.3.4.3. Diurnal migrations

Diurnal alterations in phytoplankton stratifications are only possible if their migratory velocities exceed the rate of vertical displacement of water by turbulence. Two mechanisms have been described:

(a) Migrations by changing buoyancy. Many blue-green algae change their densities by gas-vacuole formation or resorption, which leads to vertical migrations. During high irradiance the gas-vacuoles disappear and the cells sink. By gas-bubble formation under low light conditions the buoyancy of the cells increases and the algae ascend (Talling, 1957c; Fogg & Walsby, 1971; Reynolds & Walsby, 1975). The form of resistance of small filamentous species (e.g. Oscillatoria) is too great to allow a distinct migratory pattern, and large colonies of e.g. Aphanizomenon, Anabaena, and Microcystis migrate most.

In Lake George, Uganda, Ganf (1974b) has observed migrations of blue-green algae in connection with diurnal changes in temperature stratification. Homothermy during evening, night and early morning causes the phytoplankton to be homogeneously distributed, while migrations are possible during the day when the water is stratified. Colonies then concentrate in the depth of the lake and near the surface only aggregates persist, which are apparently light-damaged. Deep-living cells, exposed to low irradiances, form gas-vacuoles, which cause them to rise. In general, sinking movements dominate during periods of high irradiance. Since the intensities of the water movements decrease with depth, migratory effects dominate in deeper water, while nearer the surface turbulence effects are more important. In non-turbulent cylinders Ganf measured vertical migratory movements up to 3.9 m during a four-hour experimental period.

(b) Active migrations by flagellates. Soeder (1967) interpreted flagellate migrations as phototactic, being negative in high and positive in low irradiances. In general, the algae descend as irradiance increases (morning) and ascend as it decreases. It can be assumed that alterations in the phototactic reactions are related to light adaptations (Tilzer, 1973). The maximum velocities observed in Vorderer Finstertaler See were 1 m h^{-1} and were almost independent of cell size. This type of migration can be observed under conditions of low turbulence, either under ice (L. Erken; Nauwerck, 1963), or during stable, thermal, stratification (L. Kinneret; Berman & Rodhe, 1971), or in the hypolimnion (Tilzer, 1973).

Both mechanisms of phytoplankton migration lead to surprisingly similar patterns. Inhibitory irradiances which may be destructive are avoided and low light (morning and evening) is more effectively utilized. The movements also enable algae to take up nutrients more efficiently by steepening concentration gradients around the cells (Tilzer, 1973). Moreover, more concentrated nutrients in aphotic water can be taken up, stored, and assimilated later under euphotic conditions (Fogg & Walsby, 1971).

5.3.5. Effects of macrophytes (Westlake)

The spatial distribution of macrophytes (see section 5.2.3.2) means that their effects on the underwater light climate are localised and their seasonal variation means their effects change with time. Emergent macrophytes and floating macrophytes shade the whole water column, which may then be nearly entirely aphotic and species with maximal biomass near the surface severely limit the irradiance in deeper water (Westlake, 1966a; Ikusima, 1970; Straškraba & Pieczyńska, 1970; Dykyjová, 1973). Sward species on the other hand live in a light climate governed by non-living material, phytoplankton and other macrophytes. Macrophytes provide substrata for periphyton, which may be close to the surface (Wetzel & Allen, 1972; Adams & McCracken, 1974), and they often allow light to penetrate between clumps to phytoplankton or benthic algae.

Their specific attenuation coefficients (ε_s) are relatively easily measured, since determinations of the non-macrophyte attenuation coefficient (ε_q) can be made directly in open water (ε_q here includes attenuation by both non-living and phytoplankton components). Field measurements give different values from laboratory measurements probably because the structure of the canopy is an important factor affecting light transmission. Unfortunately few values have been reported and these have been expressed in terms of fresh or dry weights or leaf area, not chlorophyll a (Westlake, 1964; Owens *et al.*, 1967; Ikusima, 1970). The morphology of the stems and leaves affects the attenuation coefficients, which are lower for plants with filiform or dissected leaves than for plants with broader leaves (Ikusima, 1970). Values comparable with phytoplankton can be calculated if chlorophyll a is assumed to be between 5 and 10 mg chl. a (g dry wt)$^{-1}$ (see section 5.2.1). They are generally appreciably lower for macrophytes than for phytoplankton (see fig. 5.2). The spectral distribution of radiant energy changes rapidly to green within a weed bed (Westlake, 1964).

Stands of many growth forms of macrophytes may be able to adjust their canopy structure to minimise the effects of self-shading and attenuation by water (Westlake, 1966a; Adams *et al.*, 1974). New growth is directed towards the surface, leaves that are in the aphotic zone are shed, and there may be some alignment of leaves and stem that aids efficient utilisation of irradiance. Thus most of their biomass is within the euphotic zone of the stand and there is little biomass which can only respire. The extreme development is seen in floating-leaved plants which can grow in turbid water, provided they have enough material in reserve to reach the surface at the start of the growing season. Some sward rosette species have their maximum biomass concentration near the bottom. These have most chlorophyll in their upper parts and the deeper tissues are relatively inactive (Ikusima, 1966).

5.3.6. Periphyton (Marker)

Periphyton, which usually occurs in quite thin layers on stone, boulders, sand and mud, is directly affected by the attenuation of light passing through the water and phytoplankton above. Within the periphyton high concentrations of algae and accumulated silt and dead algae, from both the periphyton and phytoplankton, all lead to a very rapid attenuation of light; sometimes 95–99% (Marker, 1976c). Certain epipelic communities exhibit a diurnal migration through the sediment (Round & Eaton, 1966) so that the amount of light they receive will be changing continuously as they migrate through the sediment.

The high densities of epiphytic algae, which have been found on numerous occasions (Edwards & Owens, 1965; Szczepańska, 1968; Straškraba & Pieczyńska, 1970; Hickman, 1971a; DR 6), must lead to considerable self-shading of the periphyton–macrophyte community at lower depths.

5.3.7. Actual and optical depth (Tilzer)

The aquatic producers are distributed within various gradients among which temperature, water movements, irradiance, concentrations of nutrients, and dissolved oxygen are the most important. Light, which decreases exponentially with depth, is of particular importance for the vertical zonation of the primary producers. Since plants represent the primary food source for most consumers, the latter are also, at least partly, controlled by irradiance.

The 'optical depth', defined by the subaquatic irradiance relative to the surface irradiance, can frequently contribute more to the understanding of the vertical zonation of lakes than the morphometric depth. Highly turbid or coloured, but morphometrically shallow lakes can be optically deep with distinct euphotic and aphotic layers. In clear oligotrophic lakes, on the contrary, the vertical differentiation of environmental conditions can be less pronounced, even if they are significantly deeper morphometrically.

Consequently, shallow but optically deep lakes have frequently been considered as 'compressed versions' of morphometrically deep lakes (e.g. Brylinsky & Mann, 1973). Although this holds for the light climate considered alone, it is an oversimplification when dealing with all factors controlling plant growth. The nutrient supply in general is higher in shallow lakes because turbulent exchange processes are faster and the surface area of the benthic zone is larger relative to the lake volume. Morphometrically shallow lakes frequently have a higher biomass per unit volume as well as per unit area, which in turn decreases the transparency of the water and thus the optical depth.

Sustained thermal stratification is usually absent which restricts the development of vertical differentiation of phytoplankton, even though there are steep gradients in irradiance. Benthic primary producers are often important in shallow lakes. They are more affected by the optical depth than is the phytoplankton because turbulent mixing cannot affect their vertical distribution directly.

5.4. Photosynthesis

5.4.1. Introduction (Westlake)

This section discusses the photosynthetic behaviour of the three main groups of primary producers; in particular the factors affecting their photosynthetic capacity, the vertical distribution of photosynthesis in the water column, the integral photosynthesis under unit surface area and diurnal and seasonal variations. The submerged macrophytes and the periphyton are contrasted with the better known and understood phytoplankton.

Most studies of phytoplankton photosynthesis in the field have been made by placing samples of the population in light and dark bottles, suspended at different depths in the water body, and measuring the uptake of radiocarbon or the exchange of oxygen. Similar laboratory incubations have also been made. Modifications of these techniques have been used for periphyton and macrophytes, but work on macrophytes is concentrated on laboratory experiments with shoots. A few interpretations of oxygen changes in the free water have been made to provide data on field photosynthesis.

5.4.2. Phytoplankton (Tilzer with Pyrina & Westlake)

5.4.2.1. Photosynthetic capacity

Photosynthetic capacity (P_{max}) is defined here as the maximum specific rate of photoassimilation ('gross' photosynthesis) reached at high irradiances, ex-

Table 5.4 *Annual means of specific photosynthetic rates and corresponding generation times at optimum depths; annual mean photosynthetic capacities at temperature in situ and at 10°C*

Lake	Annual mean depth integral of specific photoassimilation, P_y^a (mg O$_2$ (mg chl. a)$^{-1}$ h^{-1})	Biomass concentration, at depth P ($b_{z,opt}$) (mg chl. a m^{-3})	Annual mean generation time, at depth of P_{opt} ($D_{opt,y}$)[b] (h)	Annual mean photosynthetic capacity, $P_{max,y}$ at: in situ temp. (mg O$_2$ (mg chl. a)$^{-1}$ h^{-1})	10°C[c]
Chad, Chad (DR 4)	20.5	18.1	4.4	33.04[d]	10.1
Sammamish, USA (DR 11)	4.1[e]	6.0	22.2	14.5[e]	11.8[e]
Loch Leven, UK (DR 3)	2.9	70.0	31.1	5.9	5.5
Aleknagik, Alaska (DR 60)	1.1[e]	1.3	78.9	3.2[e]	3.1[e]
Vorderer Finstertaler See, Austria (DR 1)	1.8[e]	1.3	50.8	4.2[e]	7.0[e]

[a] From $\Sigma_n^1 [(\int_z^0 A_z)/B]/n$, where n is the number of observations in the year.
[b] From $D_{opt,y} = [\Sigma_n^1 (\ln 2/G)]/n$, where $G = \int [P_{opt,24} - 0.15 (P_{max_h})24]$ and $f = 0.01$, which converts oxygen exchanges into equivalent chlorophyll gains or losses; assuming that chlorophyll a is 15 mg (g org. wt)$^{-1}$ and respiration is 0.15 of P_{max}.
[c] Assuming a Q_{10} of 2.1.
[d] Probably an overestimate due to erroneous chlorophyll measurements (Lemoalle, 1973).
[e] Standard conversions were needed (see section 5.1).

pressed per biomass unit of chlorophyll *a*, or sometimes dry or wet weight. All values in this section have been converted to mg oxygen (mg chlorophyll *a*)$^{-1}$ h^{-1} but these conversions suffer from considerable uncertainties. These units can be further converted into the form of specific growth rates, (*G* mg g^{-1} h^{-1}) by assuming equivalents between oxygen, chlorophyll and dry organic weight, and making an appropriate deduction for respiration. P_z is the rate of photoassimilation per unit volume at depth $z(A_z)$ divided by the biomass per unit volume (b_z), and the specific integral rate for the column ($\bar{P}$) is obtained by dividing the column photoassimilation per square metre ($\bar{A}$) by the total biomass under a square metre (*B*).

The IBP data reveal photosynthetic capacities on particular days ranging between 0.5 and 38 mg O_2 (mg chl. *a*)$^{-1}$ h^{-1} (DR 1, tables 5.4 and 5.19), which are equivalent to specific growth rates corresponding to generation times (*D*) between 180 and 3 hours in the optimum layer, assuming that net photosynthesis is 85% of photoassimilation. Some of the higher values may be biased by underestimation of the chlorophyll. Table 5.4 includes some annual mean values. Temperature influences photosynthesis less than other processes, since the light reaction is independent of temperature. However, changes of temperature affect photosynthetic capacity since it is often influenced by the enzymatic processes in photosynthesis (Jørgensen & Steeman Nielsen, 1965), depending on which reaction is limiting. Assuming an exponential response which does not change within the temperature range considered, photosynthetic capacities should then vary with temperature according to

$$_{T_1}P_{max} = {}_{T_2}P_{max} \exp \cdot \pi(T_1 - T_2) \tag{5.8}$$

where T_1 and T_2 are temperatures and π is the temperature coefficient (ln Q_1) per °C (cf. Q_{10} which is the factorial increase over 10 °C). In 24-h measurements Pyrina *et al.* (1975) found Q_{10} values of about 1.5 ($\pi = 0.04$). Experiments by Talling (1957b, 1961) and Jewson (1976) revealed values from 2.1–2.3 ($\pi = 0.07$–0.08) and Jewson found that seasonal changes in photosynthetic capacity were largely governed by temperature. The highest photosynthetic capacities have been found in the tropics (table 5.4; cf. Talling, 1965; Talling *et al.*, 1973). Near the freezing point, temperature responses seem to be enhanced (Goldman *et al.*, 1963; Albrecht, 1966). If the capacities measured are converted to one temperature (10 °C), using equation 5.8 and a Q_{10} of 2.1 it appears that temperature plays an important role in climatic differences (table 5.4).

Apart from temperature, the photosynthetic capacity can be markedly affected by nutrient depletion, including carbon dioxide in extreme conditions, and by toxins, including inhibitory extracellular substances released by the algae or other organisms (Wright, 1960; Bindloss *et al.*, 1972). However such effects are not always found (Talling, 1965). See also section 5.4.2.6.

5.4.2.2. Vertical profiles

Plots of specific photoassimilation (P_z) versus depth (fig. 5.3a) reveal the responses of phytoplankton to various gradients, among which irradiance and spectral composition changes are the most important.

(a) Light limitation. At low irradiances the specific photoassimilation rates should follow the exponential decrease of irradiance with depth, thus

$$P_z = P_{lim} \cdot e^{-z} \, (z > z_i)$$ (5.4)

where z_i is the depth below which light becomes the main limiting factor (Megard, 1972) and where the specific rate of photosynthesis is P_{lim} (fig. 5.3a). Vollenweider (1960) found agreement between total photosynthetically active radiation and photosynthesis while Rodhe (1965, 1972) has shown that

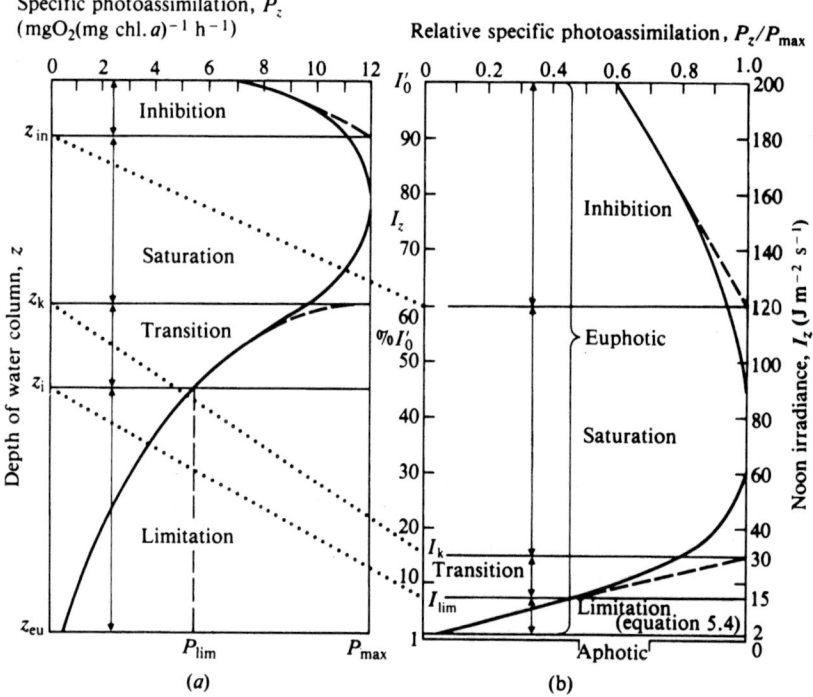

Fig. 5.3. Generalised diagrams of specific photoassimilation (P_z) in phytoplankton populations: (a) a typical depth profile for a homogeneous population; (b) a typical specific photoassimilation v. irradiance relationship for phytoplankton.

Based on values found in IBP studies, showing the effects of light limitation, light saturation and light inhibition. (The absolute values in any individual situation will differ from these.) Note the wide range of irradiances which are inhibitory or saturating and the increased importance of the light-limited region resulting from the exponential attenuation of irradiance underwater. For further explanations see text.

frequently certain spectral components are more important, often the most penetrating component.

(*b*) Light saturation. At higher irradiances, approaching optimum levels, other factors become limiting. In oligotrophic lakes a saturation plateau with low photosynthetic capacity is frequently formed, where nutrients limit photosynthesis. In eutrophic lakes a sharp maximum at much higher photosynthetic capacities is formed at the depth of optimum light (Findenegg, 1964). The parameter locating the onset of light saturation (I_k) is defined from plots of specific photosynthetic rates against irradiance (fig. 5.3b), as the irradiance at the intersections of extrapolations of the initial linear region and the final light-saturated level (Talling, 1957b). From IBP data it seems that this is about 15% of surface irradiance around noon, or 30 J m^{-2} s^{-1}, expressed as photosynthetically active radiation (PAR). However, the variation of values measured is great. Conversions of Pyrina's (1967a) 24 h values are in good agreement.

There is a transition region between full saturation and the depth of full limitation (z_i). The latter frequently corresponds to $0.5\,I_k$.

(*c*) Light inhibition. Excessive irradiances are inhibitory. From IBP data it appears that levels $> 60\%\ I_0'$ around noon, or 120 J m^{-2} s^{-1}, are inhibitory (fig. 5.3b). In clear lakes inhibition sometimes reaches great depths and has a major influence on integral photosynthesis, while in densely populated lakes light attenuation only allows light inhibition in a shallow surface layer. High total irradiances lead to photoautoxidation of pigments (Yentsch & Lee, 1966). The inhibitory effect of relatively high levels of ultra-violet radiation is not entirely proven (Findenegg, 1966) and evidence exists that visible radiation not utilised in photosynthesis can be inhibitory (Steemann Nielsen, 1962a). Nutrient depletion (Lund, 1965) and adaption to low light (Steemann Nielsen & Jørgensen, 1968; Tilzer & Schwarz, 1976) enhance the sensitivity of phytoplankton to high irradiances. Part of the 'inhibition' observed is due to the lack of natural circulation of plankton in the exposure flasks (Rodhe, 1958b; Talling, 1961), which has been confirmed directly by measuring photosynthesis in circulating phytoplankton (DR 21; Jewson & Wood, 1975).

5.4.2.3. Photosynthesis and biomass

In lakes with unstratified phytoplankton $(z_m > z_{eu})$ the profile of photo-assimilation per unit volume (A_z) follows photoassimilation per unit biomass (P_z) and shows the idealised form (cf. fig. 5.3a) (Rodhe, 1965). If considerable vertical changes in biomass occur (e.g. $z_m < z_{eu}$) the photosynthetic profile becomes irregular (e.g. the metalimnic maxima of *Oscillatoria rubescens*, Findenegg, 1964; deep maxima of flagellates in high altitude and latitude lakes, Pechlaner, 1967; Tilzer, 1972). Some effects of biomass changes on profiles are shown in fig. 5.4 (cf. also Talling, 1965, 1966b).

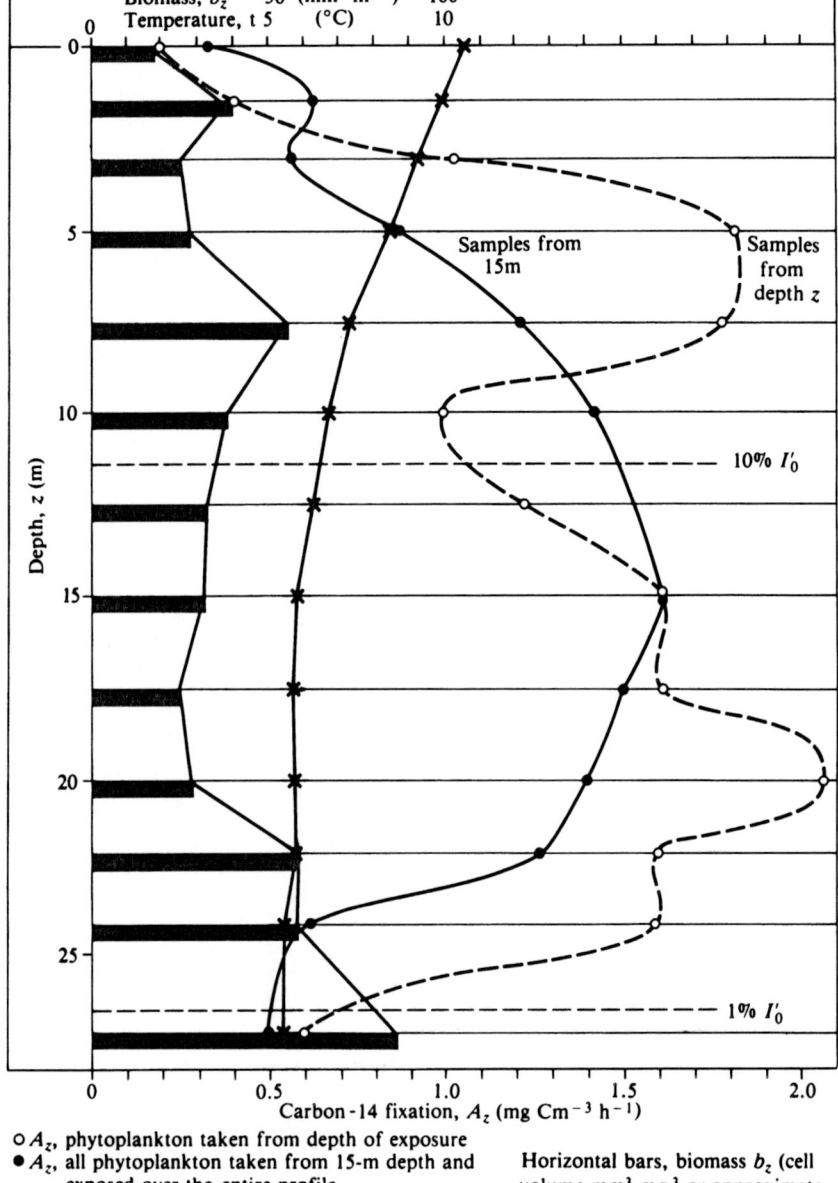

Fig. 5.4. Effects of the vertical distribution of phytoplankton biomass on depth profiles of photosynthesis (A_z) in Vorderer Finstertaler See, Austria. (Data on carbon-14 fixation from DR 1 and Tilzer & Schwarz, 1976.)

5.4.2.4. Species composition

Different photosynthetic behaviour of different species can be due to (1) cell size (2) nutrient requirements (3) light and temperature requirements (4) age and history of populations.

Indirect evidence has been found by Findenegg (1971) and Tilzer (1972) by comparing the photosynthetic rates of populations dominated by one species. Direct data on differences in photosynthetic activity between species have been provided by autoradiography (Watt, 1971; Stull *et al.*, 1973). Inverse size relationships exist as well as differences between algal groups. The relationship of cell size to photosynthetic activity is also apparent from comparisons of net- and nannoplankton. These indicate that nannoplankton contributes more to total primary productivity than their relative biomass indicates (Gliwicz, 1967; Malone, 1971; table 5.5). Decreasing trophy is linked with increasing proportions of nannoplankton (Pavoni, 1963; Gliwicz, 1967; Hillbricht-Ilkowska & Spodniewska, 1969; Hillbricht-Ilkowska *et al.*, 1972) leading to pure nannoplanktonic assemblages in ultra-oligotrophic lakes at high altitudes and latitudes (Nauwerck, 1966). Nannoplankton is more effectively grazed by filter-feeding zooplankton and therefore plays a major role in the transfer of energy to the secondary productivity level (Monakov, 1972).

Findenegg (1943) explained seasonal succession of phytoplankton by the four possible combinations of light and temperature requirements of the species involved. Although this may be an oversimplification, strong indications for such differences exist. Pyrina (1967a, 1974) and Kalff & Welch (1974) have found that the light responses of mixed populations altered significantly with species composition. Pyrina noted that when blue-greens

Table 5.5 *Relative contributions of nannoplankton to total biomass and photoassimilation in Mikołajskie Lake, Poland*

Nannoplankton as per cent of total biomass (% B)	Contribution of nannoplankton to integral photoassimilation (% A̅)	Ratio of specific activities (P_{nanno}/P_{net})
59	86	1.46
53	79	1.49
39	63	1.62
25	39	1.56
24	57	2.38
17	47	2.76
15	84	5.60
7	23	3.29
5	13	2.60

After Kowalczewski *et al.* (DR 104).

and diatoms were dominant, optimum light levels were significantly lower than in populations rich in green algae. Accessory pigments (phycocyanin in blue-greens, chlorophyll c and fucoxanthin in diatoms) may aid in utilising the central portions of the visible spectrum, which penetrates to greater depths in many lakes.

5.4.2.5. Light and shade adaptations

Changes in the photosynthetic behaviour due to different photic histories are due to two possible mechanisms: alteration in pigment content which influences the slopes of light limitation, and changes in photosynthetic enzymes altering the maximum photosynthetic rates (Jørgensen & Steemann Nielsen, 1965). Both can act alone or together to change the efficiency of utilisation of light energy by shifting the onset of light saturation, (I_k; Vollenweider, 1970). Furthermore, the response to high irradiances is altered (Steemann Nielsen & Jørgensen, 1968; Tilzer & Schwarz, 1976). Decreasing surface irradiance does not necessarily lead to lower column photosynthesis ($\bar{A}$) by phytoplankton adapted to low light if the decrease in surface inhibition has greater effects on column photosynthesis than the decrease of photo-synthesis caused by lower irradiances in deeper water. This is particularly true in very clear lakes, where the zone of inhibition may form a considerable portion of the total lake volume.

Dramatic changes in light adaptation are found in ice-covered lakes, which allow the utilisation of extremely low light levels. In the arctic Char Lake the chlorophyll concentration per unit biomass falls when light starts to penetrate the water in the early spring (Kalff et al., 1972) and the saturating irradiance was much higher in the summer than in the spring and autumn (Kalff & Welch, 1974). Similar chlorophyll changes were seen in Rybinsk reservoir (Elizarova, 1974) and in Vorderer Finstertaler See (Tilzer & Schwarz, 1976). In stratified euphotic layers ($z_m < z_{eu}$) vertical changes in adaptation are found because algae have different photic histories (Steemann Nielsen & Hansen, 1959; Tilzer & Schwarz, 1976).

5.4.2.6. Density-dependent effects (Javornický)

There is some conflict of evidence over the effects of increasing population density in phytoplankton ecology. For example, Pyrina (1967b) and Schindler and Holmgren (1971) believe that photosynthesis in their reservoirs and lakes increases linearly with phytoplankton biomass (cell volume). Similarly Talling (1965) and Welch (1969) reported linear relationships between photosynthesis and chlorophyll concentrations in the water. On the other hand, numerous investigators have found that specific photosynthesis (i.e. per unit biomass) decreased with increasing biomass, measured either as cell volume or as chlorophyll (e.g. McQuate, 1956; Rodhe *et al.*, 1958; Wright, 1960; Mikheeva, 1970; Bindloss *et al.*, 1972; Ganf, 1972; Javornický & Komárková, 1973). Sometimes such effects have been regarded as artifacts arising from the enclosure of the populations in bottles for measurement of photosynthesis.

It is probable that all three phenomena occur and merge into each other or become distinct depending on conditions. It is unlikely that photosynthesis can increase indefinitely with chlorophyll concentration, or that specific photosynthesis can remain constant over the whole range from the lowest to highest biomass. Javornický (1979) has used numerous observations throughout many growing seasons (i.e. between March and October over 7–10 years)

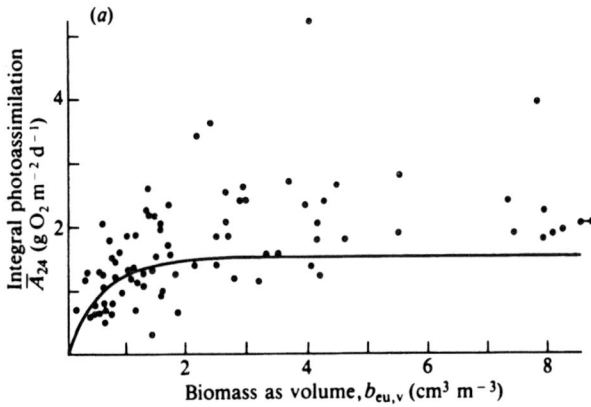

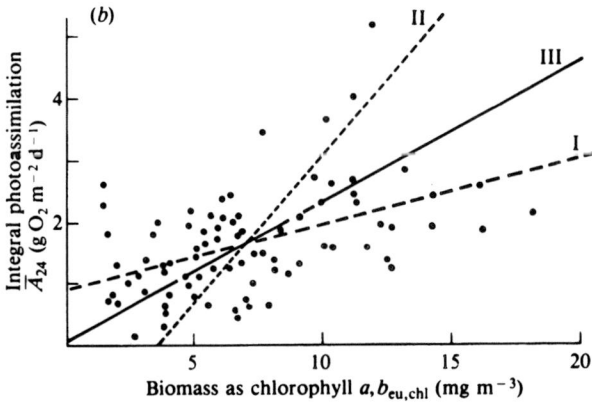

Fig. 5.5. Daily integral photoassimilation in Klíčava Reservoir against mean biomass concentration in euphotic zone. (*a*) Biomass as cell volume, mean square regression: $A_{24} = 1/(0.625 + 0.176/b_{eu,v})$: $r = 0.89$, probability $\ll 0.001$. (*b*) Biomass as chlorophyll *a*, geometric mean regression: $A_{24} = 0.055 + 0.221\, b_{eu,\,chl}$: $r = 0.47$, probability < 0.001. Regressions: I, $\bar{A}_{24}$ on $b_{eu,\,chl}$; II, $b_{eu,\,chl}$ on $\bar{A}_{24}$; III, geometric mean.

in Slapy and Klíčava reservoirs to elucidate these problems. Firstly the best fit
for the correlation between daily integral photoassimilation $\bar{A}_{24}$) and
biomass, in terms of volume, was curvilinear, starting to level out at quite low
cell volumes (fig. 5.5a). When using chlorophyll a concentrations as the
measure of biomass the best fit was linear (fig. 5.5b). If a similar curvilinear
relation were fitted, this would only start to level out at high chlorophyll
concentrations. It follows that the daily mean specific photoassimilation,
expressed in terms of algal volume ($_{v}\bar{P}_{24}$), declined rapidly as cell volume
increased but the decrease was much less marked when specific photoassimilation, in terms of chlorophyll ($_{chl}\bar{P}_{24}$), was plotted against chlorophyll

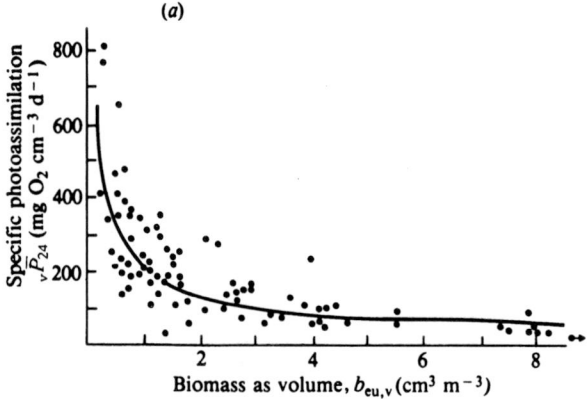

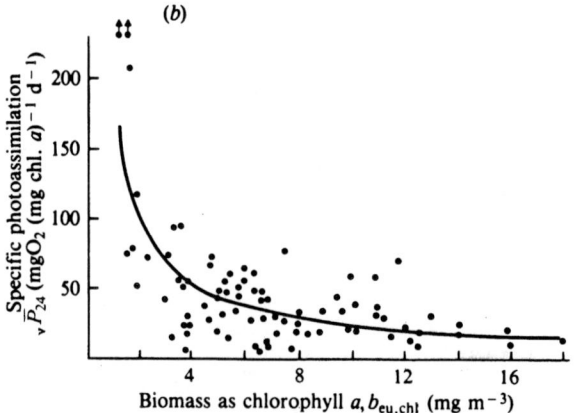

Fig. 5.6. Specific column mean photoassimilation in Klíčava Reservoir in relation to mean
biomass concentration in euphotic zone. (a) Biomass as cell volume, mean square regression:
$\bar{P}_{24} = 210(b_{eu,\,chl})^{-0.63}$: $r = -0.84$, probability $\ll 0.001$. (b) Biomass as chlorophyll a, mean
square regression: $_{v}\bar{P}_{24} = 12.3 + 180/b_{eu,\,chl}$: $r = -0.76$, probability $\ll 0.001$.

concentration (figs. 5.6a and b). Similar relationships were found at the optimal depth. The chlorophyll content of the algae, compared with their biomass concentration (as volume), showed a negative parabolic correlation much resembling that between specific photoassimilation and biomass concentration (fig. 5.7). Similar findings have been made by Mikheeva (1970). This means that technical artifacts cannot be responsible for the decrease in specific photoassimilation (the chlorophyll was extracted immediately after collection). Specific photoassimilation in terms of volume increased with increasing chlorophyll content in the biomass.

Wright (1960), Steele & Baird (1962) and Javornický (1979) therefore suggest that the deterioration of certain factors in the environment of planktonic algae leads to a decrease of their chlorophyll a content. Chlorophyll synthesis probably lags behind cell growth. This leads to the decline of specific photoassimilation, in terms of volume or carbon, in dense and growing populations, while in terms of chlorophyll it remains more or less constant as long as chlorophyll continues to fall with the limiting environment factor.

These experiments and observations seem to indicate two main factors: self-shading and nutrient deficiency. Javornický, Straškraba & Dvořaková (Javornický, 1979) have developed an analytical model which shows that nutrient limitation was probably the most important factor in Slapy and Kličava reservoirs, as well as in some other water bodies. The lower the concentration of the limiting nutrient, the lower the biomass at which specific photoassimilation started to decline. In the water bodies studied, self-shading seemed to play a more important role only in shallow fish-ponds, i.e. turbid and nutritionally rich waters.

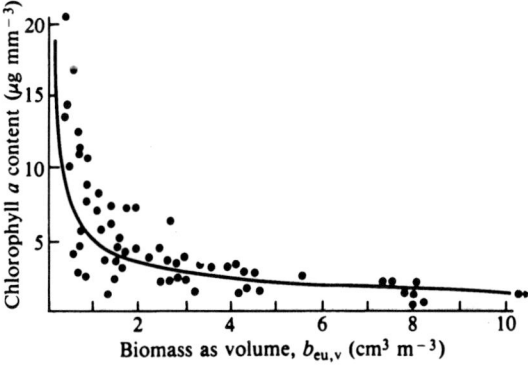

Fig. 5.7. Chlorophyll a content of phytoplankton (μg mm^{-3}) against mean biomass concentration (cm^3 m^{-3}) in euphotic zone of Kličava Reservoir; mean square regression: $b_{eu, chl}/b_{eu, v} = 5.17$ $(b_{eu, v})^{-0.64}$: $r = -0.80$, probability $\ll 0.001$.

5.4.2.7. Diurnal photosynthesis

Due to great variation in incident irradiance the shapes of graphical plots of column photosynthesis ($\bar{A}$) are very variable during the course of a day (Vollenweider & Nauwerck, 1961). Diurnal phytoplankton migrations (sections 5.2.2 and 5.3.4.3) normally decrease photosynthesis close to the surface around noon, in addition to inhibitory effects, and at the same time lead to enhanced photosynthesis in deeper water (Tilzer, 1973; Ganf, 1974c). In shallow high-latitude lakes light inhibition can lead to inverse relations between integral photosynthesis and irradiance (Goldman *et al.*, 1963). Photosynthesis in the afternoon is often lower than in the morning at similar irradiances, which can be due to nutrient depletion (Ohle, 1958; Ganf, 1975), accumulation of inhibiting substances, increasing respiration (Doty & Oguri, 1957; Lorenzen, 1963, also suggested for Lake George, Ganf, 1974c), bleaching of pigments (Goldman *et al.*, 1963; Yentsch & Scagel, 1958) and grazing by diurnally migrating zooplankton (McAllister, 1963). There is little evidence of light adaptation (changes in I_k) over 24 hours, but there is some evidence for internal clocks which change some nutrient uptake constants diurnally, thus indirectly affecting rates of photosynthesis (Stross, Chisholm & Downing, 1973; Stross & Pemrick, 1974). However such asymmetric photosynthesis–time curves are not always found.

Analytic depth–time integrations to obtain daily integrals of photo-assimilation ($\bar{A}_{24}$) attempt to account for some of these factors, but the variability of possible diurnal patterns make them very difficult (cf. Talling, 1957b; Vollenweider, 1965 and section 5.9). Comparisons of daily integrals, calculated from short time exposures with values observed with 24-hour experiments are scarcely possible. Total respiration does not necessarily vary proportionally with photosynthetic rates, inhibition is not always proportional to incident irradiance (Tilzer, 1973) and furthermore inhibition as an artifact increases with the duration of the exposure (Vollenweider & Nauwerck, 1961). Values of daily integrals are discussed in section 5.10.

5.4.2.8. The seasonal pattern of photosynthesis

Variation in solar light inputs seems to be the major reason for seasonal changes of daily integral photoassimilation ($\bar{A}_{24}$; fig. 5.8; Jewson, 1976). Besides irradiance changes, the day length has a profound influence on the 24-hour net production, the final result of fixation and losses, and hence on the accumulation of biomass. Other factors of importance are temperature, turbulent mixing, nutrient availability, total biomass, species composition and biomass stratification.

The range of variation in daily integral photoassimilation thus depends on the seasonal changes of environmental conditions. In the equatorial Lake George, Uganda, photoassimilation does not vary significantly throughout

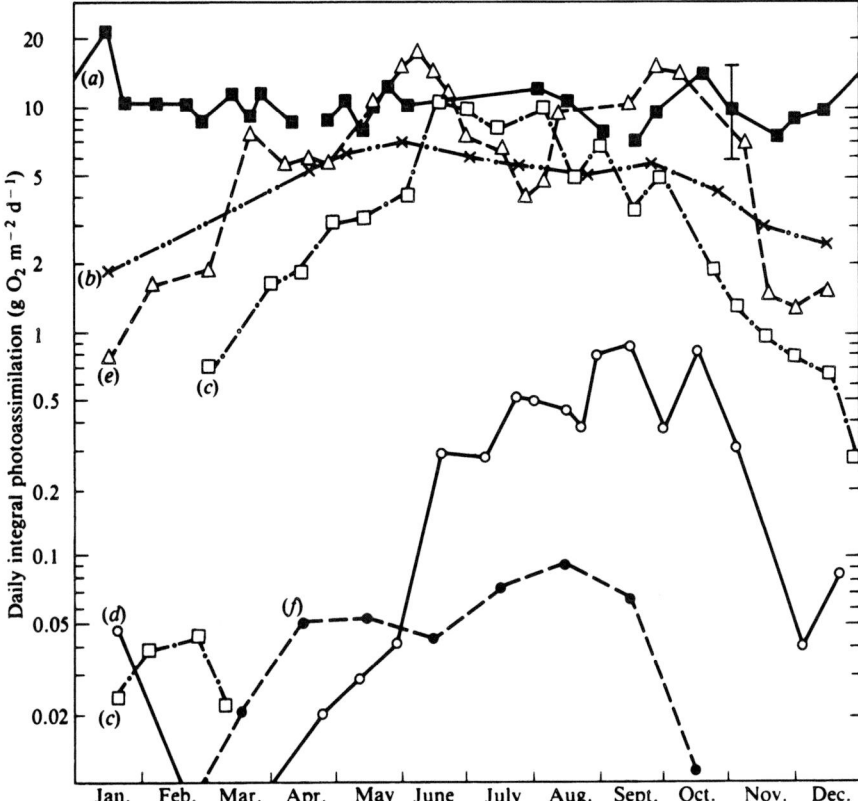

Fig. 5.8. Seasonal fluctuations of daily integral photoassimilation by phytoplankton ($\bar{A}_{24}$) in six lakes at different latitudes and altitudes. (*a*) L. George, Uganda 1967–8 913 m 0° (DR 77). Data for days of observation at central lake site converted to whole lake (noon rates for whole lake approximately 0.87 central site). Points calculated by author on the basis of the total daily irradiance and the irradiance of the period of exposure. Typical 95% confidence limits on October point based on variation between twenty sites.

(*b*) Bol, L. Chad, Chad 1969 282 m 13° N (DR 4). Data for days of observation at one site. Successive 3- or 4-h exposures summed.

(*c*) L. Wingra, Wisc., USA 1970 262 m 43° N (DR 13) (values converted from carbon-14 fixation by standard factor of three). Points calculated by authors from exposures around noon for days of observation at single site. No corrections for saturation, inhibition not observed. Data as corrected in supplement to Data Report.

(*d*) V. Finstertaler See, Austria 2237 m 47 N (DR 1). Points calculated by authors from exposures around noon for days of observation, converted to daily values on the basis of total daily irradiance and proved relationship between around-noon and daily photoassimilation, (cf. Tilzer, 1972, 1973).

(*e*) L. Leven, Scotland 1970 107 m (DR 3; Bindloss, 1974). Points calculated by authors on the basis of mean total daily irradiance for 10-day periods, irradiance of the periods of exposure around noon at a single station, and the onset of saturation (I_k; cf. Talling, 1965).

(*f*) Char L., Canada 1970 30 m 74° N (Kalff & Welch, 1974) (values converted from carbon-14 fixation by standard factor of three, see section 5.1). Monthly means calculated by authors from data on the distribution of irradiance with depth, on the monthly mean irradiances in 4-h periods and on the seasonal changes in the relations between photoassimilation and irradiance.

the year (Ganf, 1975; fig. 5.8). The lack of distinct seasons and the high metabolic activity causes diurnal patterns comparable to the seasonal patterns of temperate lakes (Ganf & Horne, 1975). At slightly higher latitudes dry and rainy seasons cause alterations in temperature, insolation and often lake levels (Chad, DR 4). In this lake turbidity is generally high, which explains the relatively low level of photoassimilation (fig. 5.8). The mixing events in temperate lakes cause alternating nutrient-depleted and nutrient-enriched conditions.

Transparent winter covers have only a slight effect on photosynthesis, as in the Ivano–Arakhley lakes (Central Asia, USSR) where 25–35% of the annual primary production occurs under ice-covered conditions (G.G. Winberg, personal communication). Under thick and highly light-absorbing snow and ice covers, however, winter photoassimilation approaches zero. Net rates may be negative and lead to decreasing biomass.

The annual fluctuations of the daily integral photoassimilation normally follow those of biomass, but are frequently greater (table 5.19). In the alpine Vorderer Finstertaler See the annual range of biomass (minimum : maximum) was only 1 : 5, while the range of daily integral photoassimilation was from 1 : 270 (Tilzer, 1972; fig. 5.8). Loss rates were directly proportional to photoassimilation, leading to limited biomass increments in summer and relatively small biomass decreases in winter. Arctic lakes (e.g. Char L., Kalff & Welch, 1974; fig. 5.8) also show wide seasonal fluctuations and low daily integral photosynthesis, since both low temperatures and nutrients are unfavourable. In some densely populated lakes (e.g. Loch Leven) specific integral photoassimilation rates ($\bar{A}_{24}/B$) fall with increasing biomass due to self-shading as well as biomass-dependent decreases of P_{max}, which reduces the variations of daily integral photoassimilation (Bindloss et al., 1972; DR 3; fig. 5.8).

In many temperate lakes daily integral photosynthesis is limited by the daily totals of irradiance in winter and by nutrients for much of the summer. The low irradiance in winter may be due to daylength alone or also to ice and snow cover. The changes from winter to summer conditions frequently lead to the highest photosynthetic rates of the year which level off as nutrients are exhausted (Pechlaner, 1970). The 'spring bloom' of phytoplankton populations is due to the favourable combination of nutrient and light availability. Sometimes, however, if nutrients are being recycled only within the pelagic zone (Windermere, Talling, 1966b; Vorderer Finstertaler See, Tilzer, 1972), photoassimilation remains high in the summer, but does not lead to further biomass increments. More than one peak in daily integral photoassimilation may occur during the summer, due to changing environmental conditions (e.g. water transparency, DR 1), or to phytoplankton succession (Lake Sammamish, DR 11), or to density-dependent changes in photosynthetic activities (Loch Leven, DR 3).

5.4.3. Macrophytes (Westlake)

5.4.3.1. Photosynthetic capacity

The data for macrophytes are almost entirely from laboratory experiments and are expressed per unit dry weight or occasionally leaf area. The photosynthesis measured by enclosing macrophytes is very much affected by duration of exposure, diffusion from intercellular air spaces, and water movement (Wetzel, 1974), and many results are probably too low. Most determinations of specific net photosynthetic capacity ($N_{s, max}$) are between 6 and 40 mg O_2 (g dry wt)$^{-1}$h^{-1} and about 10 may be regarded as typical (Hammann, 1957; table 5.6). To compare their capacities with phytoplankton requires assumptions about respiration and chlorophyll content. If a chlorophyll a content of 5 mg (g dry wt)$^{-1}$ and an average dark respiration rate of 0.1 of P_{max} are taken as typical (cf. section 5.2.3 & 5.5), this would suggest a typical rate of photoassimilation of about 2 mg O_2 (mg chl. a)$^{-1}$h^{-1}, compared with about 15 for phytoplankton (Talling, 1975; cf. section 5.4.2). Direct measurements by Hammann (1957), Ikusima (1966) and Spence & Chrystal (1970b) have found 0.2–4 mg O_2 (mg chl. a)$^{-1}$h^{-1}.

The temperature coefficients reported for macrophyte photosynthesis show considerable variation ($Q_{10} = 1.1–3.1$; $\pi = 0.01–0.1$; Blackman & Smith, 1911; Steemann Nielsen, 1947; Westlake, 1967; Stanley & Naylor, 1972; Tanimizu & Miura, 1976), probably depending on the region of the total temperature response examined, adaptation and which stage of carbon uptake and fixation is limiting for the particular species and experimental conditions.

Water velocity and turbulence are important factors (Steemann Nielsen,

Table 5.6 *Net photosynthetic capacities of submerged macrophytes* ($N_{s, max}$)

Species	Capacity (mg O_2(g dry wt)$^{-1}$h^{-1})	Reference
Egeria densa	0.6–7.6	Tanimizu & Miura (1976)
Elodea canadensis	3.1	Wetzel (1964)
Hydrilla verticillata	35	Ikusima (1965)
Myriophyllum spicatum	4.4–16	McGahee & Davis (1971) Stanley & Naylor (1972) Adams *et al.* (1974)
Nitella spp.	21–60[a]	Kumano (DR 76) Stross (DR 12)
Potamogeton obtusifolius	26	Spence & Chrystal (1970b)
P. polygonifolius	37	Spence & Chrystal (1970b)
Ranunculus calcareus	13	Westlake (1967)

[a] Standard conversion, 2.5 × carbon-14 rate (see section 5.1).
Experimental conditions have not been included in this table because they have often been inadequately or incompletely defined. Most of these results are for 5–20 cm shoot tips in stirred natural waters, at summer temperatures and irradiances above I_k.

1947), affecting the specific rates of photosynthesis by macrophytes much more than phytoplankton. The macrophytes are larger and fixed in relation to the water, so boundary layers offer a resistance to the uptake of carbon dioxide sources; phytoplankton cells move with the water and are so small that hydrodynamic boundary layer concepts are inapplicable (Munk & Riley, 1952; Gavis 1976). There is some evidence (Westlake, 1966c; 1967; *Ranunculus calcareus*) that the relation approximates to $N_s = (k\sqrt{v}) + N_0$ where v is the linear velocity of the water and N_0 is the specific rate of net photosynthesis at zero velocity. The response of photosynthesis by *Myriophyllum spicatum* to flow (DR 12) has a similar shape up to the flow-saturating velocity. This appears to be somewhere between 6 and 10 mm s^{-1} depending on the carbon dioxide demand.

5.4.3.2. Vertical photosynthetic profiles

Theoretically the relations between irradiance, photosynthesis, biomass and non-plant attenuation are basically similar for macrophytes and phytoplankton and similar profiles should be found (Westlake, 1966a; Ikusima, 1967). In macrophyte communities there are vertical changes in total biomass concentration, leaf weight, photosynthetic capacity and micro-environmental conditions which are as important as the light profile (Carr, 1969; Ikusima, 1970; Adams, Titus & McCracken, 1974; fig. 5.9). The stratification of factors such as temperature, nutrients, oxygen and water velocity within a plant stand is also influenced by the plants themselves. The photosynthetic capacity of sections of shoots decreases in the deeper parts of the stand, which is apparently due more to morphological differences than directly to age. The maximum photosynthesis ($N_{z, \max}$) is therefore often near the surface of the stand, even when the maximum biomass concentration is further towards the base (Ikusima, 1966; Adams *et al.*, 1974; fig. 5.9).

The onset of light-saturation (I_k) is typically at about 40–80 J m^{-2} s^{-1} (PAR; $\sim$ 10–20 000 lux surface light) but there are wide variations between species and plants adapted to light or shade conditions. The approach to complete saturation is often gradual and slight increases may be observed up to 160 J m^{-2} s^{-1} (Steemann Nielsen, 1947; Ikusima, 1966, 1967; Westlake, 1967; Spence & Chrystal, 1970b; Tanimizu & Miura, 1976; DR 12; DR 76). In dense stands most of the biomass is light-limited (fig. 5.9).

Few recent studies have found pronounced inhibition at the surface and maximum specific photosynthesis has been observed at the surface with irradiances above 150 J m^{-2} s^{-1} (Adams *et al.*, 1974; McCracken *et al.*, 1975; Tanimizu & Miura, 1976). Adams *et al.* found evidence of the adaptation of upper parts of the *Myriophyllum spicatum* shoots to utilise high irradiances. Such adaptations are unlikely in a vertically mixed phytoplankton. However some inhibition (about 15%) has been found in populations of *Utricularia* sp.

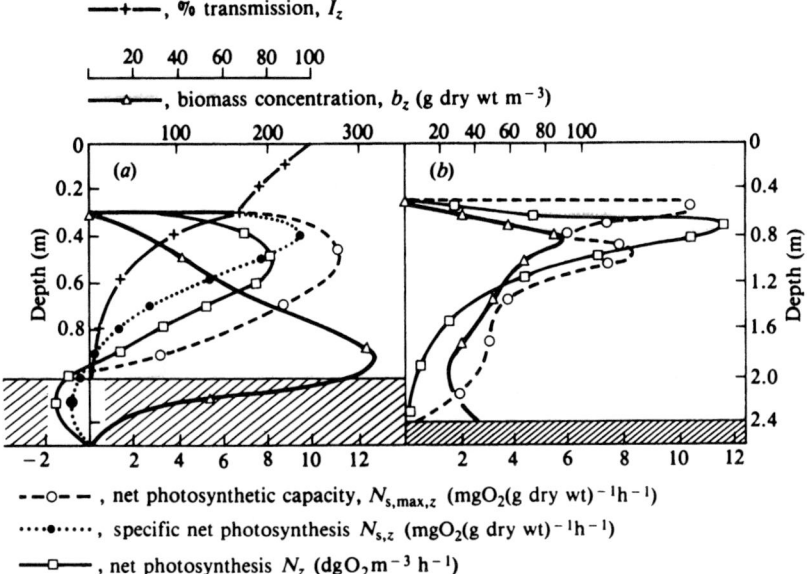

—+—, % transmission, I_z

—△—, biomass concentration, b_z (g dry wt m^{-3})

Fig. 5.9. Diagrams of vertical profiles in stands of submerged macrophytes.

(a) Sward species. *Vallisneria denseserrulata*, July. Note how the profile of specific photosynthesis is mainly affected by irradiance and photosynthetic capacity, and the small influence of the biomass profile on the profile of net photosynthesis.

Based on smoothed data from Ikusima (1966).

(b) Species with subsurface maximum. *Myriophyllum spicatum*, May. Note the general similarity of the photosynthetic profiles to those of the sward species despite the very different biomass profile.

Based on smoothed carbon-14 and biomass data from Adams & McCracken (1974) and Adams et al. (1974). Carbon converted to oxygen by standard factor of 2.75.

and *Nitella* sp., which had low saturating irradiances (DR 76). Ikusima (1970) reported inhibition for submerged leaves of *Nuphar japonicum* exposed to irradiances above 5000 lux (~ 20 J m^{-2} s^{-1}), but not for the floating leaves. This suggests that auto-oxidative processes causing inhibition can be prevented by an improved supply of carbon dioxide.

5.4.3.3. Differences between species and forms

Good evidence of specific differences can only be obtained by comparisons of rates of photosynthesis under the same conditions, using material taken at the same time of year, or in the same phase of their growth cycle, from similar sites. Few investigations have aimed at this. It has often been suggested, but never convincingly proved, that finely divided leaves are more efficient than broad leaves (Whitwer, 1955; Sculthorpe, 1967).

Differences between light-species and shade-species (and between analogous forms of one species) are better known (Gessner, 1938; Spence &

Table 5.7 *Example of sun- and shade-species and adaptations from Scottish fresh waters*

Type and depth range as mean depth ± SD (cm)	Grown in	Specific leaf area (cm² mg⁻¹)	Chlorophyll content (mg g⁻¹)	Relative specific metabolic rates[a] (Sun *P. polygonifolius* as 1) Net photosynthesis N_s				Dark resp. C_d per unit dry wt	Ratio $\dfrac{N_s \text{ at } 16J}{N_s \text{ at } 82J}$	Onset of saturation I_k (J m⁻² s⁻¹)	Comp. point $P=C$ (J m⁻² s⁻¹)
				per unit dry wt[b]		per unit chlorophyll (16 J m⁻² s⁻¹)y	per unit leaf area (82 J m⁻² s⁻¹)y				
				(16 J m⁻² s⁻¹) ~3 kluxf	(82 J m⁻² s⁻¹) ~16 kluxf						
Sun-species *Potamogeton polygonifolius* 0–40	Sun	0.5	14	1.0	1.0	1.0	1.0	1.0	0.24	>80	1.6
	Shade	1.4	17	1.8	1.7	1.4	0.7	1.8	0.25	~80	1.2
Shade-species *Potamogeton obtusifolius* 60–200	Sun	1.9	24	3.2	1.5	1.8	0.4	0.9	0.50	30	0.6
	Shade	2.1	27	2.2	0.7	1.1	0.2	0.02	0.77	15	0.05

After Spence & Crystal (1970a, b).

[a] Original data in μl O_2 (cm leaf area)⁻² h⁻¹.

[b] All plant weights are oven-dry weights of leaf only.

[c] All irradiances are for photosynthetically available radiation.

Chrystal, 1970a, b). Shade adaptations include broader, thinner leaves, a higher chlorophyll concentration, low respiration and a low value for the onset of light-saturation (table 5.7). Exposed to low or decreasing irradiance, shade species and forms tend to maintain their net photosynthesis better and cease net production at much lower irradiances. The shape of the light curve and the respiration levels are more important than the absolute levels of photosynthesis, and it is not yet entirely clear how these responses affect interspecific competition. Probably sun-species, with generally higher absolute rates of net photosynthesis, grow faster in well-illuminated habitats but cannot achieve higher growth rates than shade-species in dim habitats (cf. N_s for *P. polygonifolius* and *P. obtusifolius* at low and high irradiances in columns 5 and 6 of table 5.7). Other things being equal, the more adaptable species should dominate a particular habitat. Within a stand of a single species, parts or leaves at different depths may be adapted to their prevailing light climate (Ikusima, 1966; Adams *et al.*, 1974).

5.4.3.4. Diurnal and annual photosynthesis

Diurnal patterns, with an afternoon or mid-day depression of photosynthesis, have been reported (reviewed in Goulder, 1970; Hough, 1974; McCracken *et al.*, 1975), but many are probably experimental or environmental in origin. The results of Edwards & Owens (1962) obtained from oxygen changes in a natural river, showed no obvious depression but Goulder found afternoon depression in stagnant conditions where local depletion of total carbon dioxide and raised pH were likely. Hough & Wetzel (1972) proposed that an increase in photo-respiration could be involved when photosynthetic capacity is depressed.

Changes in integral photosynthesis ($\bar{A}$) during a day or from day to day, are mainly the result of changes in incident irradiance (Edwards & Owens, 1962; Ikusima, 1966; Westlake, 1966a). Over longer periods, or between different sites, biomass changes, transparency and photosynthetic capacity changes are also important. Capacity of temperate macrophytes changes in response to seasonal environmental changes and also with the inherent growth pattern of the plants. Typically capacity is low in the winter and high in the spring, though late summer or autumn maxima or two maxima have been observed (Wetzel, 1964; Ikusima, 1966; Westlake, 1967; Carr, 1969; Spence & Chrystal, 1970b; Kaul & Vass, 1972; Adams & McCracken, 1974; Tanimizu & Miura, 1976). These patterns broadly correspond with the annual cycle of growth and biomass accumulation (Forsberg, 1960; Wetzel, 1964; Westlake, 1965; Rich *et al.*, 1971; Westlake *et al.*, 1972; cf. section 5.6.3).

The values determined in these studies show a very wide range, and photosynthetic integrals tend to be higher than biomass accumulation rates. In fertile waters, during the period of rapid growth, average rates of net photosynthesis are usually between 2 and $10\,g\,org.\,wt\,m^{-2}\,d^{-1}$ (3–

$15 g O_2 m^{-2} d^{-1}$), in terms of the area occupied by macrophytes (Westlake, 1975b). For the temperate zone annual integrals are usually obtained from the seasonal maximum biomass (see section 5.2.3), sometimes with a correction for losses, and on this basis range from below 20 to nearly $1000 g org. wt m^{-2} a^{-1}$ ($30-1500 g O_2 m^{-2} a^{-1}$). Tanimizu & Miura (1976) found over $1300 g O_2 m^{-2} a^{-1}$ by integrating photosynthetic rates, but this considerably exceeded the equivalent of the seasonal maximum biomass.

Even in warmer waters seasonal changes in productivity can be found (Odum, 1957a; Ambasht, 1971; Unni, 1976). Biomass and turnover are higher and production can probably exceed $2000 g m^{-2} a^{-1}$.

5.4.4. Periphyton (Marker)

5.4.4.1. Methodological difficulties

Considerable effort has been concentrated on basic methodology because of the variable nature of the substrata, the irregular distribution of the periphyton and the importance of water movement. Quite complex enclosures for measuring photosynthetic rates *in situ* have been designed (Thomas & O'Connell, 1966; Hansmann, Lane & Hall, 1971; Berrie, 1972a; Bombówna, 1972; Marker, 1976b). In rivers the movement of water will affect metabolic rates (Whitford & Schumacher, 1961, 1964). In lakes water movements also affect photosynthetic rates (Wetzel, 1964; DR 13, 14). The limnologist still cannot measure current velocity and turbulence in the boundary layers, let alone reproduce them accurately in small enclosures.

The upstream–downstream method of Odum may avoid these problems in rivers but is rarely suitable because macrophytes frequently dominate the system (Edwards & Owens, 1965). Some data have been obtained when macrophytes appeared to be absent (Odum, 1957b; Stockner, 1968; Flemer, 1970; Kelly, Hornberger & Cosby, 1974) but it is not always clear if they were completely absent. Frequently the algal periphyton is not sufficiently well developed to produce detectable oxygen changes. In most studies adequate replication has not been possible.

5.4.4.2. Photosynthetic capacity

The light-saturated specific rate (P_{max}; $mg O_2 (mg chl. a)^{-1} h^{-1}$) is a useful physiological parameter. It is not readily determined on periphyton (however see Tominaga & Ichimura, 1966), and even integral specific rates for the whole population ($\bar{P}$) at high irradiances are not always available. The values in table 5.8 are based on estimates of biomass and photoassimilation per unit area of the entire algal mat. An inverse relationship between rates of integral

Table 5.8 Rates of integral specific photoassimilation by periphyton

Site and value	Population or substrate	Photoassimilation		Comments	Reference
		per unit dry wt mg O_2(gm dry wt)$^{-1}$h^{-1}	per unit chlorophyll mg O_2(mg chl.a)$^{-1}$h^{-1}		
Lake Wingra (P_{opt})	Oedogonium sp. Cladophora sp.	3 27 1-6			(DR 12)[a]
Tasek Bera (P_{max})	Batrachospermum sp. 2 Batrachospermum sp. 3	20-25 40 50	16-20 23 29	net, but $C_{com} < 5\%$	(DR 76)[b]
Bere Stream (P_m)[c]	epilithon		0.7 2.3 1.3 6.0 2.6-6.7	November–February March May June–August	(DR 40)
Logan River (P)	concrete rocks		0.07-2.0		McConnell & Sigler (1959)
Mikołajskie Lake (P)	glass slides		0.07 2.0		Szczepańska (1968)[d]
River Raba (P_m)	epilithon		0.64 1.80		Bombówna (1972)
Kiev Reservoir (P_m)	filamentous algae	33		June–July	Gak et al. (1972)

[a] Carbon-14 uptake data converted to oxygen using a standard factor of three (see section 5.1).
[b] Species 2, 1.27 mg chl.a(g dry wt)$^{-1}$; Species 3, 1.78 mg chl.a(g dry wt)$^{-1}$.
[c] P_m Specific photoassimilation of active community exposed around mid-day.
[d] Original data in terms of carbon dioxide.

specific photoassimilation and biomass (chlorophyll a) has been observed (Marker, 1976b) and at least part of this effect will be due to self-shading. Specific rates obtained from the fixation of radiocarbon by *Oedogonium* in Lake Wingra fell from 2.1 mg C (g dry wt)$^{-1}$ h^{-1} at the surface to 0.4 at 100 cm (DR 12, 13). Phinney & McIntyre (1965) showed how closely irradiance and temperature affected photosynthesis between 8–20 °C and between 5000–20 000 lux ($\sim$ 25–100 J m^{-2} s^{-1} PAR).

Within some populations the penetration of light must be very low indeed. Tominago & Ichimura (1966), studying the primary production of a montane river, found that surface layers of periphyton have photosynthetic capacities (P_{max}) of about 10 mg O$_2$ (mg chl. a)$^{-1}$ h^{-1} at 20 000 lux ($\sim$ 85 J m^{-2} s^{-1} PAR). Not only did algae taken from the lower layers have lower capacities (3 mg O$_2$ (mg chl. a)$^{-1}$ h^{-1}) but surface algae taken at different seasons also had different capacities (from 10–12 in July to 2 in January). The higher rates observed approach phytoplankton rates (cf. table 5.19).

5.4.4.3. Integral photosynthesis

Most measurements of rates of photoassimilation by periphyton (table 5.9) have been made on the entire algal community and are equivalent to the photosynthesis of the entire water column of phytoplankton ($\bar{A}$). Periphyton usually occurs as a thin layer and even biomasses as high as 200 mg chl. a m^{-2} would be less than 5 mm thick. With this degree of compression, diffusion gradients become increasingly important. In addition interpretation of the data is complicated by the nature of the substratum. Biomass and production may be very different when expressed in terms of substrate surface instead of littoral surface (DR 91).

Diurnal changes in photosynthesis broadly follow changes in solar irradiance (Thomas & O'Connell, 1966; Odum, 1957a; Stockner, 1968; McCracken *et al.*, 1975; Marker, 1976b). In the well-buffered hard waters of southern England and Wisconsin post-noon suppression of oxygen output was not observed. As so little is known of diurnal changes in photosynthetic rate of periphyton, most calculations of day rates from hourly rates have been based on principles derived from phytoplankton data. Undoubtedly epipelic populations, which migrate diurnally through sediments and are at the surface during the morning, would exhibit a marked photosynthetic diurnal asymmetry. Broadly, high daily rates of photosynthesis are associated with dense populations such as algal mats (Odum, 1957a; Stockner, 1968), epilithon (Marker 1976b) or epiphyton (Odum, 1957a; DR 6). The lower hourly rates associated with epipelon appear to be due to low population density, (Hickman & Round, 1970; DR 6; DR 82). Hickman & Round (1970) found that in the sediments of Shearwater, production by epipsammon was greater than by epipelon.

Table 5.9 *Rates of photoassimilation by periphyton communities* (A)

Site	Population	Photoassimilation per hour (mg O_2 m^{-2}h^{-1})	Photoassimilation per day (g O_2 m^{-2}d^{-1})	Comments	Reference
Shearwater	epipsammon	60 & 146 / 300–600		1967 and 1968 ann. mean / max. rate	Hickman & Round (1970)[a]
	epipelon	1.6 & 5.2 / 12–24		1967 and 1968 ann. mean / max. rate	Hickman & Round (1970)[a]
Priddy Pool	epiphyton	107 & 236 / 300–600		1967 and 1968 ann. mean / max. rate	Hickman (1971a)[a] (DR 6)
	epipelon	5.6 & 6.5		1967 and 1968 ann. mean	Hickman (1971b)[a] (DR 6)
Borax Lake	periphyton		0–17.3		Wetzel (1964)[a]
A pond	*Chara*		20.3–47.6		Wetzel (1964)[a]
Pääjärvi	epipelon	3–12			(DR 82)[a]
Lawrence Lake	*Scirpus* epiphyton *Najas* & *Chara* epiphyton		0.59	mean	Allen (1971a)[a]
Marion Lake	epipelon		5.42	mean	Hargrave (1969)
Truckee River	*Cladophora*, *Oscillatoria*		0.04–1.5	0.5 m	Thomas & O'Connell (1966) (DR 66)
Kiev Reservoir			4.7–16.5		Marker (1976b)
Bere Stream	epilithon	20–160 / 110–690 / 100–400	0.56–3.35	November–February / March–May / June–August	
Kurakhov Reservoir			0.73 / 0.87	unheated / heated	Pidgaiko et al. (1972)
River Thames	mixed population		about 12 / 3–6	max. / August	Berrie (1972a)
River Raba	epilithon	240–1600	0.10–0.80	substrate surface	Bombowna (1972)
Mikolajskie Lake	epiphyton on *Phragmites communis*		0.09–0.90	littoral surface	Spodniewska et al. (1975) & DR 91
	epiphyton on sub. macrophytes		0.09–0.90	substrate surface	Kowalczewski (1975)
			0.08–4.90	littoral surface	
Taltowisko Lake	epiphyton on *Phragmites communis*		0.10–1.00	substrate surface	(DR 91)
			0.80–1.10	littoral surface	

[a]Carbon-14 uptake data converted to oxygen using a standard factor of three (see section 5.1)

5.4.5. The exploitation of light (Westlake)

When light-limited the more light a plant can usefully absorb by chlorophyll the faster it can grow. Water, non-plant material in the water, other species and its own structural materials compete in absorption of light. If the irradiance is more than the plant's metabolism can assimilate, some of the light will be absorbed by chlorophyll, but not utilised to produce organic materials. If the irradiance is less than sufficient to supply the demands of respiration, its absorption achieves no net gain of organic matter. Advantageous adaptations enable the plant to absorb light efficiently, to have little of its biomass exposed to excessively high, or inadequately low irradiances, to compete well with other plants and other absorbing materials, and to have a high proportion of chlorophyll.

Often these conflict and some particular compromise occurs. For example floating-leaved plants are exposed to high irradiances and require a relatively high proportion of structural materials to maintain a surface cover against wind and water movements. However, the effect of this high surface concentration is to absorb most of the light, preventing competition from other plants, and yet avoiding excessive shaded biomass. Hence, although they are relatively inefficient and poorly productive, they can dominate in some circumstances.

Phytoplankton organisms usually have a high chlorophyll content, a minimum of structural material and relatively high specific attenuation coefficients facilitating competition with other absorbing materials. They are at the mercy of water currents and are rarely able to control their vertical distribution, which usually means a considerable proportion of the biomass is poorly illuminated.

Periphyton is restricted by the distribution of substrata, but is able to exploit surfaces in favourable positions. The biomass concentration is very high, but within the community there is often high light attenuation by other materials and chlorophyll cover is not high. Algae that can migrate can achieve a large change in irradiance within a short distance.

Submerged macrophytes have some ability to maintain their biomass in high irradiances and appear to be able to tolerate high irradiances without inhibition, either by physiological features or stand structures which allow lower irradiances at leaf surfaces. Although the biomass is high, their low attenuation coefficients allow good penetration of light into the stand and little biomass is poorly illuminated. Since they can remain at a particular depth, light and shade adaptations are more common and can be more exploited than by the phytoplankton.

As has been discussed earlier (section 5.3.4.1), phytoplankton can theoretically have a maximum chlorophyll cover of $\sim 180-620$ mg chl. a m^{-2} in the euphotic zone. Any chlorophyll in excess of this in the mixed column will represent non-productive and respiring biomass and any increase in chlorophyll concentration will reduce the depth of the euphotic zone, the biomass in the euphotic zone remaining unchanged. The optimum leaf area index for macrophytes (DR 12) is a similar concept.

The optimum situation for phytoplankton is a high chlorophyll cover in a euphotic zone similar to the mixed depth. The actual value of this chlorophyll cover (B_{opt}) depends on the specific attenuation coefficient of the alga and the water, and the lower these are, the higher the maximum chlorophyll cover can be (see fig. 5.2 and section 5.9.2). The algal coefficient is affected by factors such as cell size, cell aggregation and cell structure, and the coefficient based on dry weight will be affected by the chlorophyll content (and vice versa). Large algae, other things being equal, tend to have lower coefficients, which may be attributed to the fact that a higher proportion of light paths penetrate to a given depth without encountering plant material. This trend is more marked when algae and macrophytes are compared, and the latter have much lower coefficients (fig. 5.2). Stand structure is probably a contributory factor. Hence the maximum chlorophyll cover and biomass in the euphotic zone can be much higher for macrophytes than phytoplankton despite their lower chlorophyll content. Values from the upper end of the range are about 200 g dry wt m^{-2} or 2 g chl. a m^{-2} for macrophytes and about 50 g dry wt m^{-2} or 500 mg chl. a m^{-2} for phytoplankton (fig. 5.2). These compare well with high values found in the field, though higher values can be found in the whole column, and they correlate well with the striking difference in magnitude of the biomasses of these two types of plant. This ability of macrophytes to maintain a high biomass in the euphotic zone enables them to resist adverse conditions, damage and grazing and gives greater stability to macrophyte communities in the growing season.

5.5. Oxygen uptake

5.5.1. Introduction (Ganf)

Rates of photosynthesis are frequently measured using either the carbon-14 or oxygen light and dark bottle method. The advantages of the carbon-14 method are well known but the disadvantage is that no reliable estimates of respiration are possible. The oxygen method allows an estimate of oxygen consumption but the data are difficult to interpret. The problems encountered are:

(1) The assumption that respiration rates are the same in light and darkness is being increasingly challenged by the data on photorespiration, and the influence of the previous light history on rates of dark respiration (Golterman, 1971; Padan *et al.*, 1971; Lex *et al.*, 1972; Ganf, 1974c). This prevents the calculation of true gross photosynthesis by the conventional method.

(2) Point (1) is complicated by algal cells that are freely circulating through a

strong vertical light gradient, especially where the euphotic depth is less than the mixed depth, and by the varying ratio of daylight to dark hours.

(3) The oxygen consumption in the dark enclosure includes plant respiration, animal respiration, and the bacterial oxidation of both dissolved and particulate organic material (as well as other, non-respiratory oxidations, Abeliovich & Shilo, 1972; Serruya & Serruya, 1972). The rate of oxygen consumption is therefore a measure of *community* respiration. Thus neither plant respiration (R) nor net photosynthesis (N) can be measured and a further obstacle is placed in the way of calculating true gross (A_t) or net photosynthesis (N) from the relation: $A_t - R = N$. This problem is normally less important for measurements on macrophytes.

The problems resolve into a need firstly for reliable techniques for measuring respiration rates in the field, and secondly to relate respiration rates to a function of solar irradiance and vertical circulation.

Values of algal respiration given in IBP data reports, and most of those given in the ecological literature, refer to community respiration. These measurements have shown that large variations of community respiration occur with regard to both time and depth, suggesting that the commonly held assumption that respiration is a constant fraction of gross photosynthesis is dubious.

5.5.2. Phytoplankton (Ganf)

5.5.2.1. Community respiration

Volume (or area) based rates (R_{com}) are dependent upon the concentration of material present, and are of limited use for inter-lake comparisons. Specific rates based upon unit plant biomass (C_{com}) (e.g. dry weight, chlorophyll a) are more useful. The most commonly used is that based upon chlorophyll a estimations ($mg\,O_2\,m^{-3}\,h^{-1} \div mg\,chl.\,a\,m^{-3} = mg\,O_2\,(mg\,chl.\,a)^{-1}\,h^{-1}$).

The laboratory work of Ryther & Guillard (1962) on axenic cultures suggested a specific respiration rate of $\simeq 1\,mg\,O_2\,(mg\,chl.\,a)^{-1}\,h^{-1}$ for algae. A selection of field results for specific rates of community respiration are given in table 5.10. Results from both Jezárko Pond and Tjeukemeer generally fell within the range 0.1–4.5 but rates as high as $10\,mg\,O_2\,(mg\,chl.\,a)^{-1}\,h^{-1}$ were measured during periods when high bacterial activity was suspected.

Seasonal variations of community respiration were observed in Loch Leven, Tjeukemeer and Queen Mary Reservoir. Maximum values normally occurred during the summer, and low values during the cold winter months, thereby suggesting some temperature response. However, there was often a marked lag between temperature and respiration changes (fig. 5.10) and the correlation between temperature and respiration was poor, which indicates that factors other than temperature were acting. For Lake George, Uganda, Ganf (1974c)

Table 5.10 *'Specific' dark respiration rates of plankton communities measured in the field*

Lake	Respiration $C_{com, d}$ (mg O_2(mg chl. a)$^{-1}$ h^{-1})	Relative to max. photo-assimilation r	Reference
Aranguadi, Ethiopia	~ 1	0.05–0.09	Talling *et al.* (1973)
Kilotes, Ethiopia	~ 1	0.06–0.3	Talling *et al.* (1973)
George, Uganda	0.2–4.5	0.01–0.22	Ganf (1974c)
Jezárko Pond, Czech.[a]	0.3–1.8(7.6)		Fott (DR 10)
Loch Leven, Scotland	0.1–4.0	0.07–0.5	Bindloss (1974)
Lough Neagh, Ireland	0.06–2.9	0.02–0.12	Gibson (1975) Jewson (1976)
Queen Mary Res., England	0.5–3.5	0.04–0.10	Steel (1978)
Tjeukemeer, Holland	~ 1–3(+)	0.05–0.10	Golterman *et al.* (DR 21)
Windermere, England	~ 1	0.02–0.10	Talling (1957a)

[a] Chlorophyll uncorrected for phaeopigments.
() Bacterial respiration known to be very high.

came to a similar conclusion. Jewson (1977a), however, found a good correlation in L. Neagh with a Q_{10} of 2.6 ($\pi = 0.096$).

For Loch Leven Bindloss (DR 3) found no correlation between species composition, cell size or nutrient concentration and respiration rates. There was evidence, however, in both Loch Leven and Tjeukemeer (DR 3, 21) that high bacterial numbers and zooplankton may have contributed towards the

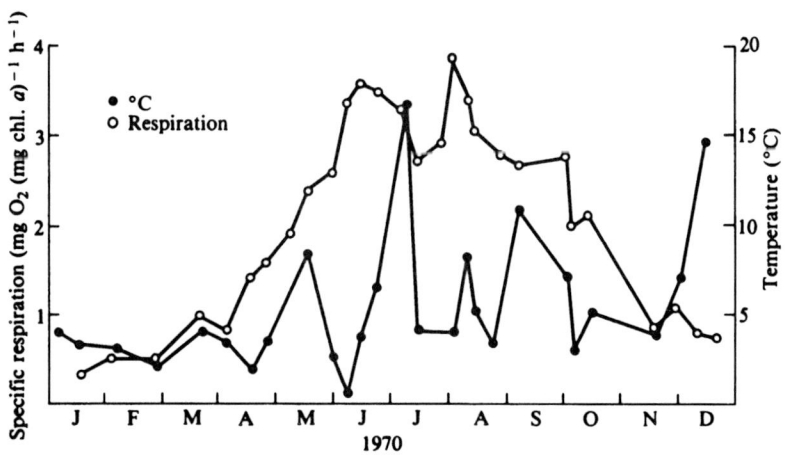

Fig. 5.10. Seasonal variation of 'specific' community respiration and water temperature in Loch Leven, Scotland (Bindloss, DR 3).

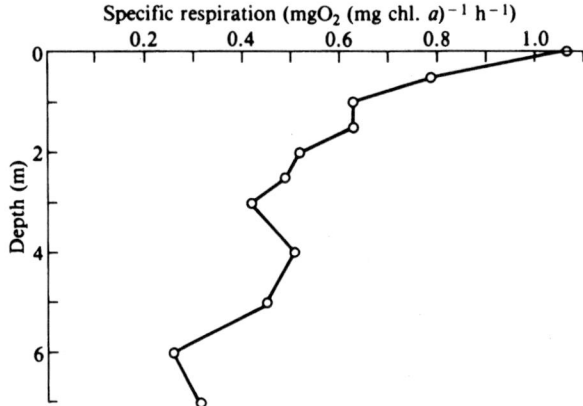

Fig. 5.11. Depth profile of 'specific' community respiration Antrim Bay, Lough Neagh. Material collected 11.00 hours 15 August 1973, incubated at 15 °C for 48 hours. Samples dominated by *Oscillatoria agardhii* and *O. redekei* (Gibson, 1975).

seasonal variation. Steel has suggested that part of the seasonal variation of respiration in Queen Mary Reservoir could be dependent upon cell size (DR 81). Such a response is well known for animal respiration (Zeuthen, 1953).

Gibson (1975) working with natural populations of *Oscillatoria agardhii* and *O. redekei* from Lough Neagh, has demonstrated a depth dependence of community respiration (fig. 5.11). Ganf (1974c) found a similar depth dependence of community respiration which was interpreted as a function of the photosynthetic history of the phytoplankton. A depth dependence of re-

Table 5.11 *Depth distribution of respiratory electron transport activity (ETS act.)*

Phytoplankton + bacteria Lake Washington Station 1 (Packard, DR 45)		
Irradiance ($\% I_0$)	Depth (m)	ETS act. ($mg\,O_2\,m^{-3}h^{-1}$)
100	0	5.560
50	1.5	4.820
25	3.0	4.280
10	4.5	3.395
1	9.0	3.165
< 1	15.0	0.892
	20.0	0.640
	25.0	0.475
	30.0	0.407
	40.0	0.299
	50.0	0.275

spiration was also reported by Packard *et al.* (1971, DR 45) for Lake Washington (table 5.11).

5.5.2.2. Fractionation of community respiration

The fractionation of community respiration was attempted by Machenko *et al.* (DR 66) for Kiev Reservoir. Biomass estimates of the fauna, phytoplankton and bacteria were obtained and oxygen consumption rates for individual plankters were either measured or taken from the literature. Bacteria were assumed to respire at a rate of 0.1×10^{-9} mg O_2 cell^{-1} d^{-1}. These calculations showed that 50% of community respiration was bacterial, 30% algal and 20% faunal.

Lewis (1974) assumed that size was the only variable affecting the rates of respiration of autotrophs and heterotrophs and thus calculated their relative contributions to community respiration from volume measurements of the populations present. This suggested that phytoplankton accounted for 0.8 of the community respiration and protozoa, large zooplankters and bacteria accounted for 0.12, 0.05 and 0.03 respectively. It seems likely that this method underestimates bacterial respiration.

Ganf and Blažka (1974) found that in equatorial Lake George 5–15% of community respiration was due to zooplankton. In L. Neagh removal of large zooplankters actually increased the oxygen demand and the total zooplankton demand was usually probably less than 10% (Gibson, 1975; Jewson, 1976).

The use of enzyme analysis to measure community respiration is a comparatively new technique in ecological research (e.g. electron-transport activity, Klingenberg, 1968; Packard, 1971). A number of authors have suggested the use of succinate dehydrogenase assays to estimate zooplankton respiration (Curl & Sandberg, 1961; Pearre, 1964; Packard & Taylor, 1968). Such enzymatic techniques may be a useful area for further research when attempting to fractionate community respiration.

Specific chemical inhibitors and antibiotics have also been used to fractionate community respiration. By the use of antibiotics Hargrave (1969) showed that $\sim 30\%$ of the total oxygen uptake by the benthic flora and fauna in Marion Lake was due to bacterial respiration. Golterman (1971, DR 21) estimated the rate of dark respiration in the light by the use of 3, (3,4-dichlorophenyl)-1, -dimethyl urea (DCMU) which inhibits algal photosystem II, thus blocking the photosynthetic production of reduced nicotinamide adenine dinucleotide phosphate (NADPH); it does not, however, block cyclic photophosphorylation which may continue to be a source of adenosine triphosphate (ATP). Since DCMU completely inhibits photorespiration (Hoch *et al.*, 1963; Lex *et al.*, 1972) the ecological significance of these results is debatable. Lex *et al.* (1972) found that potassium cyanide (10^{-4} M) did not inhibit photorespiration but stopped a possible dark respiration. They also suggest that hydroxyethane sulphonate (8.8×10^{-3} M) will inhibit photorespiration, providing that it enters the cell. To stabilise bacterial numbers the use of chloroamphenicol may be useful (Mahler & Cordes, 1966).

A further method for the fractionation of community respiration was proposed by Golterman (1971). The assumption is made that the normal respiratory substrate in phytoplankton is carbohydrate or fat (Gibbs, 1962). Thus a decrease of particulate and dissolved organic nitrogen is a measure of bacterial activity. Therefore, by determining the total oxygen uptake and decreases of both dissolved and particulate organic carbon and nitrogen, the partition of community respiration into algal and bacterial respiration is possible. From such experiments in both Tjeukemeer and Lake George (Golterman, 1971; Ganf, 1974c) a fraction as high as 50% of total respiration was thought to be due to bacterial activity. Such experiments are complicated, however, by the loss of particulate material due to non-oxidative causes.

Parsons and Strickland (1962) suggested that the uptake of carbon-14 labelled organic compounds could be used to estimate heterotrophic bacterial activity. Subsequently various workers have used and modified this method (e.g. Wright & Hobbie, 1966; Hobbie & Crawford, 1969; Allen, 1971a; Wetzel & Allen, 1972; Overbeck, 1972). Although two uptake mechanisms have been distinguished it does not appear as though either is specific for bacteria or phytoplankton (Allen, 1971a; Miller, Cheng & Coleman, 1971; Tarant & Colman, 1972). Thus such uptake rates are not always a measure of heterotrophic bacterial activity.

5.5.2.3. Laboratory experiments

Photosynthesis in C_3-plants has a three-carbon metabolic cycle which is linked to the photorespiration pathway. C_4-plants have special histological features and a four-carbon pathway giving efficient uptake of carbon dioxide, recycling of respiratory carbon dioxide, and no apparent photorespiration. In unicellular and colonial algae these histological characteristics are not possible and the biochemical characteristics of C_4-plants have not been found. No ecological data are available on photorespiration but laboratory work suggests that its influence on measurements of photosynthesis by all kinds of C_3-plants is considerable (Brown & Webster, 1953; Hoch *et al.*, 1963; Brown & Tregunna, 1967; Bunt & Heeb, 1971; Lex *et al.*, 1972; Goldsworthy, 1970; Jackson & Volk, 1970; Šesták *et al.*, 1971; Hough, 1974). Lex *et al.* (1972) have shown for a blue-green alga that photorespiration may be twenty times the dark respiration rate, and may approach the rate of photoassimilation under optimal conditions. These optimal conditions (high light intensity, low partial pressure of carbon dioxide, and high partial pressure of oxygen) frequently occur in the euphotic zones of eutrophic lakes during periods of high phytoplankton densities (e.g. Bindloss, 1974; Ganf & Horne, 1975). A rather lower value of 3-4 times the dark respiration rate has been suggested by Šesták *et al.* (1971) as a rate of photorespiration in large terrestrial C_3-plants. Also many aquatic plants may excrete a three-carbon compound, glycolate, rather than oxidise it (see Hough & Wetzel, 1972; Fogg, 1975).

Other laboratory experiments have shown that dark respiration is enhanced by nutrient additions to starved cells (e.g. Stewart & Alexander, 1971). Ried (1970) found that dark respiration varied according to the age of the cell. Padan *et al.* (1971) demonstrated that dark respiration in *Plectonema boryanum* was influenced by the previous light history of the cells. Stewart & Pearson (1970), Antia & Cheng (1970) and Dokulil (1971) have all obtained evidence to suggest that dark respiration in blue-green algae is markedly reduced under conditions of stress.

5.5.2.4. General comments

Models of phytoplankton production that incorporate respiration rates usually express respiration as a fraction (r) of the maximum rate of 'gross' photosynthesis or photoassimilation (RP_{max}^{-1}; Talling, 1957b, 1971; Vollenweider, 1970; Steel, 1978). A selection of relative respiration rates are given in table 5.10.

It should be noted that these values were all obtained from community respiration determinations in dark enclosures, and are probably an overestimate of algal respiration. A characteristic of phytoplankton models is their extreme sensitivity to changes of the relative respiration rate (Talling, 1971; Fott, 1972; Ganf & Viner, 1973). Considerable biological insight may be gained from the use of such models in the prediction of the upper limits of photosynthetic production and algal densities (Steel, section 5.9.2).

Since community respiration rates are frequently the only estimate of respiration available some consideration should be given to their ecological interpretation. In a system where the allochthonous input of organic material is negligible the major substrate for bacterial respiration will come from the phytoplankton. If community respiration, as measured by experimental enclosures, is assumed to be primarily due to the phytoplankton–bacterial complex, then the difference between community respiration (R_{com}) and true gross photosynthesis (A_t), which is the value actually measured in a light enclosure, could be used as a measure of the food available (F) to the grazing organisms in such a system. However such differences are often negative (Riznyk & Phinney, 1972; Ganf, 1972, 1974c; Bindloss, 1974), and if the algal crop continues to expand under these conditions, the conclusion must be that a substantial proportion of the energy required for bacterial and animal respiration is derived from an allochthonous input of organic material.

One of the aims of this section was to illustrate how tenuous are our production estimates because of the lack of information on respiration. Similar conclusions have been reached by workers in IBP/PP (e.g. Šesták *et al.*, 1971; Talling, 1975).

5.5.3. Macrophytes (Westlake)

Since macrophytes can usually be separated from most other associated organisms their respiration rates are known, from laboratory experiments, better than those of algae, although there may be an appreciable residual error. As in the case of photosynthesis, specific rates are usually determined as oxygen uptake based on the dry weight or leaf area. In the field the problems of separating the components of community respiration remain and the best estimates of macrophyte respiratory losses are probably obtained from the product of laboratory values for the specific respiration and the biomass (e.g. Edwards & Owens, 1962). Studies made with carbon-14 give no information on respiration and hence no valid values for 24-h net photosynthesis.

5.5.3.1. Specific respiration rates

Some rates of dark respiration (C_d) in terms of dry weight are given in table 5.12 and about 1 mg (g dry wt)$^{-1}$ h^{-1} may be regarded as typical. Values are usually lower at lower oxygen concentrations and temperatures and in unstirred conditions (Owens & Maris, 1964; Westlake, 1967; McDonnell, 1971; Tanimizu & Miura, 1976). However Hough (1974) did not find an effect of oxygen on the dark respiration of *Najas flexilis*.

There are a few results indicating seasonal changes (Ikusima, 1965; McDonnell, 1971; Dawson, 1973; Tanimizu & Miura, 1976) with low values found in the winter, especially in perennating organs. Temperature seems to be the main factor, but there may also be internal changes.

The relative respiration $(r$; dark respiration, C_d/photosynthetic capacity, $P_{max})$ is typically about 0.1–0.15, but very large variations occur, and values of between 0.001 and 0.4 have been found for apparently healthy material (Ikusima, 1965, 1970; Westlake, 1967; Spence & Chrystal, 1970b; Tanimizu & Miura, 1976; DR 76). Shade-species have relatively low ratios.

5.5.3.2. Respiratory losses

Ikusima (1965) studied vertical profiles of photosynthesis and respiration by measuring the oxygen exchanges of submerged plant material taken from different depths in the stand and exposed in enclosures at the same depth. For much of the daylight period on a fine day the column respiration $(\bar{R})$ was 0.3 to 0.5 of the photoassimilation $(\bar{A})$. Over twenty-four hours, and adding the respiration of underground organs, one stand respired 0.8 of the total

Table 5.12 Specific dark respiration (C_d) of macrophyte leaves and shoots

Species	Temperature (°C)	Oxygen conc. (mg l⁻¹)	Dark respiration (mg O_2 (g dry wt)⁻¹ h⁻¹)	Reference
Potamogeton spp. (sun and shade forms)	20	~9	0.02–1.5	Spence & Chrystal (1970b)
Utricularia sp. and *Nitella* sp.	30	~7	0.7–2.2	(DR 76)
Myriophyllum spicatum	25	~8	1.5–3.6	McGahee & Davis (1971)
Vallisneria denserrulata and other species (roots and stems)	~25	~8	0.4–3.7 (0.6–0.9)	Ikusima (1965, 1966, 1970)
Ranunculus calcareus	15 / 15	1–2 / ~10	0.5–1.0 / 1.5–2.2	Westlake (1967)
Ranunculus calcareus and other species	10 / 20	1–2 / ~10	0.3–0.8 / 1.2–2.7	Owens & Maris (1964)
Egeria densa	5–30	~12–7	0.3–2.1	Tanimizu & Miura (1976)

photoassimilation and another, which was young and probably using stored material, respired 1.9 of the total photoassimilation. In his 1966 paper he used measurements of the irradiance, irradiance–oxygen curves and biomass distributions to calculate profiles. The values of the ratio $\bar{R}_{24}/\bar{A}_{24}$ ranged between 0.35 for fine days to 0.9 for rainy days, and between 0.5 and 0.7 as monthly averages. Studies of a floating-leaved community (Ikusima, 1970) showed a value of 0.43 for a clear June day. The actual rates of loss were in the range $2-16 \text{ g O}_2 \text{ m}^{-2} \text{ d}^{-1}$.

Tanimizu and Miura (1976) studied the seasonal variations in photo-synthesis and respiration of a stand of *Egeria densa*. On fine days during the spring the ratio $\bar{R}_{24}/\bar{A}_{24}$ was about 0.27 to 0.33 but it rose to over 0.8 and sometimes over 1.0 in the late summer and autumn. On cloudy days the ratio was 0.34 to 0.39 in the spring and exceeded 1 throughout the summer and autumn. The monthly average was about 0.3 in the spring and around 1 (little or no growth possible) in the late summer and autumn. The annual average was 0.67 and actual rates of loss varied from 1.4 to $24 \text{ g O}_2 \text{ m}^{-2} \text{ d}^{-1}$, with an annual average of 7.8.

Edwards & Owens (1962) made direct measurements of community dark respiration in a river dominated by macrophytes using the upstream–downstream method. After subtracting the benthic respiration (as determined from the uptake of oxygen by mud samples in the laboratory) the residual rate of respiration was about $9 \text{ g O}_2 \text{ m}^{-2} \text{ d}^{-1}$ in the summer. Plant respiration calculated from biomass and specific respiration was about $6 \text{ g O}_2 \text{ m}^{-2} \text{ d}^{-1}$. The ratio $\bar{R}_{24}/\bar{A}_{24}$, calculated from these rates and the daily rates of photoassimilation during the summer, was about 0.4–0.6.

Similar studies by Odum (1957b) for a subtropical spring community of periphyton on *Sagittaria lorata* gave an annual average, $\bar{R}_y/\bar{A}_y$, of 0.57.

The model developed by Westlake (1966a) showed that the ratio $\bar{R}_{24}/\bar{A}_{24}$ could be expected to show very wide fluctuations with the value of relative respiration rate (r), stand structure, transmission and incident irradiation, from less than 0.1 for low r on clear days in clear water to over 5 for high r on dull days in fairly turbid water. For a typical stand the respiratory loss would be about $7 \text{ g O}_2 \text{ m}^{-2} \text{ d}^{-1}$.

5.5.3.3. Photorespiration

Although it seems probable that most submerged plants are C_3-plants, there are many anomalies arising because of morphological, and perhaps physico-chemical, features of submerged macrophytes (Hough, 1974). He has assem-bled evidence which makes it clear that several species of submerged macrophyte show a C_3-metabolism and histology, and some apparent photorespiration. However the measured loss of carbon dioxide to the medium from carbon-14 labelled plants may be less in the light than in the dark, (Carr, 1969; Hough, 1974), which is probably mainly due to a high level of refixation, facilitated by the internal lacunae and the high external resistance to the diffusion of carbon dioxide through water. The refixation makes quantitative evaluation of photorespiration very difficult. The low carbon dioxide compensation points found (Hammann, 1957; Stanley & Naylor, 1972; Hough 1974) are also atypical of C_3-plants, and are better evidence of a relatively low level of photorespiration. Black (1973) however found high compensation points for three *Potamogeton* species. Compensation points expressed in terms of free carbon dioxide concentrations (for comparison with aerial photosynthesis) may not be relevant to the aquatic situation where plants may be capable of responding to bicarbonate when the free carbon dioxide is negligible. There may be a decrease in net photosynthetic oxygen output at higher oxygen concentrations (Westlake, 1967; Hough, 1974) but this could be accounted for quantitatively by effects of oxygen on dark (or community) respiration rather than on photorespiration.

5.5.4. Periphyton (Marker)

The measurement of the uptake of oxygen by periphyton and associated organisms presents similar problems to those of phytoplankton, but certain points need extra emphasis. There can be few benthic communities which do not contain large proportions of secondary producers. Indeed in some sediments algae may be a very minor component. Consequently the oxygen uptake measured in a field respirometer (R_{com}) will need to be interpreted with very considerable caution. It is not known how far measured photo-assimilation departs from true gross photosynthesis because little or nothing is known about the significance of photorespiration in periphyton. In these communities a considerable part of the uptake of oxygen may be due to the respiration of allochthonous material. In lakes the allochthonous input from outside may be relatively small, but the settling out and decomposition of phytoplankton within the periphyton will still make it difficult to interpret the significance of respiration values. In streams there may be a very large input of organic detritus from outside. Within the stream, the movement of decaying material from one part of a system to another will also affect local respiration values. Such movements are much more complex than either the mixing of phytoplankton in a lake or its sedimentation to the bottom, both of which have been the subject of much controversy. It is important to realise just how far the community uptake rates shown in table 5.13 may diverge from algal respiration rates. The size of the relative respiration (r) is some indication of this (cf. r in table 5.10 with table 5.13). Because the non-algal respiration may be so large, there is normally little point in expressing it as a specific rate (per unit chlorophyll). Hargrave (1969) attempted to partition the respiration of a benthic community. Using antibiotics he found 30–45% was bacterial

Table 5.13 Dark oxygen uptake rates of periphyton communities

Site	Population	Oxygen uptake R_{com} ($mgO_2 m^{-2} h^{-1}$)	Relative to max photo-assimilation r	Reference
Laboratory streams	epilithon	67–138[a]	1.0–2.6	McIntire et al. (1964)
Laboratory streams	epilithon 6.5 → 16.6 °C	41 → 132		Phinney & McIntire (1965)
	17.5 → 9.4 °C	105 → 63		
Kirakhov Reservoir	glass slides heated water	22[a]	1.38	Pidgaiko et al. (1972)
	unheated water	42[a]	0.87	
River Thames		42–208[a]		Berrie (1972a)
Truckee River	Cladophora and Oscillatoria	92–546[a]		Thomas & O'Connell (1966)
River Raba	epilithon	60–460	3.3–4.0	Bombówna (1972)
Marion Lake whole community	epipelon	5–65[a]	~ 1[a]	Hargrave (1969)

[a] Average over 24 h.

throughout the year, and in June algal respiration was calculated (by difference) to be 20–25% of the total community respiration (R_{com}).

5.6. Biomass changes and net photosynthesis

5.6.1. Introduction (Westlake)

As outlined in the introduction to this chapter changes in biomass are the result of an intricate balance between gross photosynthesis, respiration and a variety of losses:

$$B[P_t - C] - [L_e + L_b + L_z + L_s + L_w + L_m] = w \qquad (5.9)$$

where: B is the biomass of photosynthetic organisms; P_t is the specific rate of true gross photosynthesis (per unit biomass); C is the specific rate of respiration of photosynthetic organisms (per unit biomass); $B(P_t - C) = N$, the rate of net photosynthesis; L_e is the rate of loss as excreted material; L_b is the rate of loss to consumers such as bacteria and fungi; L_z is the rate of loss to larger consumers such as zooplankton and zoobenthos; L_s is the rate of loss by sedimentation; L_w is the rate of loss by wash-out; L_m is the rate of loss by other unspecified mortality (for phytoplankton $L_e + L_m$ is practically indistinguishable from L_b; for macrophytes there may be, for example, recognisable mortality of individual shoots or leaves); w is the rate of change of biomass of photosynthetic organisms.

Note that the growth rate (w) has a direct feed-back to biomass (B) as well as indirect feed-back effects on specific net photosynthesis ($P_t - C$), and may also affect losses through changes in biomass. Some of the losses will change with changes in definition of the community considered; for example algae reaching the sediment have left the phytoplankton but may remain alive in the lake as a whole.

From the point of view of the botanist the rate of change of plant biomass is the main interest but the ecologist is more interested in the rate at which new organic material (energy) is entering the system:

$$N - (L_s + L_w) = w + L_m + L_z + L_b + L_e \qquad (5.10)$$

Over a long period, when the biomass starts from and returns to a low value, or when the biomass fluctuates about a constant average, the positive and negative values of w will cancel out and, in the absence of sedimentation and wash-out losses, the integrated net production of the cycle (N_c) equals the sum of the mortality losses:

$$N_c - \Sigma_c(L_e + L_b + L_z + L_m) \qquad (5.11)$$

The only values in these equations which are normally well known are B and w. An equation describing the rate of change of all organic matter in the system has a better chance of being quantified:

$$w + m + w_h = [A - {_d}R_{com}] + U - [{_t}L_s + {_t}L_w] \qquad (5.12)$$

where: m is the rate of change of dead organic matter; w_h is the rate of change of the biomass of heterotrophic organisms; A is the rate of photoassimilation; ${_d}R_{com}$ is the rate of community respiration in the dark; $(A - {_d}R_{com})$ may be assumed to be equivalent to the rate of change of oxygen in a light enclosure; U is the rate of input of allochthonous organic matter (usually of unknown origin); ${_t}L_s$ is the rate of sedimentation of all organic matter; ${_t}L_w$ is the rate of wash-out of all organic matter.

Changes in biomass may not only affect the rate of photosynthesis but may also vary the concentration of food, alter the shelter and substrata available, change the rate of supply of nutrients and modify other chemical and physical features of the environment.

5.6.2. Phytoplankton (Nauwerck)

5.6.2.1. Patterns of biomass change

The classical picture of biomass changes in the phytoplankton of a lake shows a spring maximum, usually of diatoms and chrysomonads and a second maximum, usually of green or blue-green algae in the summer or of diatoms in the late summer, both apparently primarily controlled by nutrient limitations and sedimentation (or buoyancy) once the winter limitations of light and temperature are overcome (Hutchinson, 1967). This generally remains true of temperate lakes that are not particularly oligotrophic or eutrophic, but many different patterns have been found as a greater variety of lakes has been studied, especially during the International Biological Programme. As lakes become more oligotrophic the maximum biomass tends to decrease (max : min usually $< 30 : 1$, cf. table 5.14) and the seasonal course of biomass becomes smoother. Little is known about such lakes in the tropics, but seasonal changes are probably slight. In colder water there is usually one summer maximum, though an early spring maximum may be found under the ice in arctic–alpine lakes. Species diversity becomes high in temperate oligotrophic lakes, but tends to be lower in arctic–alpine and tropical oligotrophic lakes. The importance of grazing tends to increase towards oligotrophy, but decreases towards the extremes of very cold or very oligotrophic waters. Sedimentation is often very important, especially under the ice cover of arctic–alpine lakes.

As lakes become more eutrophic the maximum biomass tends to increase (max : min usually $> 30 : 1$, cf. table 5.14) and the seasonal course of biomass becomes more irregular with frequent fluctuations. In the tropics the range of fluctuations is not large (max : min $< 5 : 1$, cf. table 5.14), with no particular season for the maximum, though the dominant species may be influenced by wet and dry seasons. In arctic–alpine lakes the maximum occurs in summer but in temperate eutrophic lakes there are often a number of maxima occurring irregularly throughout the growing season, dominated by different species. Nevertheless overall species diversity is generally low except in arctic–alpine eutrophic lakes. Environmental factors, particularly wind and temperature, influencing turbidity and sedimentation have relatively large effects, but grazing is occasionally important. Very dense biomasses produce self-shading and other auto-inhibiting effects, leading to decay and sedimentation of the population.

Some examples supporting these conclusions may be found in Rodhe *et al.* (1958), Talling (1965, 1966a, b), Findenegg (1965), Hutchinson (1967), Gliwicz (1969a, b), Schindler and Holmgren (1971), Andronikova *et al.* (1972), Bindloss *et al.* (1972), Capblancq and Laville (1972), Lévêque *et al.* (1972), Pechlaner *et al.* (1972b), Holmgren and Lundgren (1972–74), Andersson *et al.* (1973), Ganf (1974d), Ilmavirta *et al.* (1974) and Welch *et al.* (1975).

5.6.2.2. Mechanics of loss

Whatever the primary causes of biomass fluctuations they must act by changing the rate of growth or the rate of loss. In many situations, as emphasised in the above summary, sedimentation and grazing appear to be the main losses affecting the biomass of the phytoplankton. However, it is not yet certain if this is generally true and other factors of largely unknown quantitive importance are damage and destruction of cells by turbulence, diseases, parasites, or toxic substances released from organisms. In certain cases, wash-out of plankton from a lake may be a major factor for biomass regulation. Sudden changes of physical and chemical environmental conditions (e.g. heating up of the water or lowering of pH) are also likely to cause mass death of certain species, but have not been studied particularly so far. Obviously the hydrology, chemistry and trophic level largely determine the nature of regulatory mechanisms in specific lakes.

(*a*)Wash-out effects. Drainage of water from plankton-rich layers, and wash-out effects may play a dominant or significant role in lakes with a low retention time (e.g. IBP-lake Ø. Heimdalsvatn, Larsson & Tangen, 1975) and in reservoirs (Steel *et al.*, 1972; Javornický & Komárková, 1973; Straškraba & Javornický, 1973). For actively migrating organisms like many dinophyceans, this type of regulation may work selectively.

(*b*) Sedimentation effects. Settling of cells to the bottom is to be expected, particularly for large and heavy species like green algae and diatoms. Such algae are kept suspended by turbulence but will sink quickly during periods of stratification, with limited water mixing. Qualitatively, successive sedimentations of phytoplankton populations have been recorded by many authors. Daily losses of the order of magnitude of the daily production, and even more, have been calculated (Grim, 1950; Thomas, 1950). On the other hand, wind-drifted accumulations of floating blue-green algae decaying in thick layers at the shore of many eutrophic lakes during warm seasons is a well known phenomenon. For example Reynolds (1972, 1973) describes a population of *Anabaena circinalis* that fell from 32–1000 to 4–20 filaments ml^{-1} and populations of *Microcystis aeruginosa* that were reduced by over 60% after these blooms were blown ashore by light winds.

(c) Grazing effects.   During recent years, much attention has been paid to zooplankton grazing. Nevertheless, there is no unanimous opinion concerning the quantitative role of this factor upon the phytoplankton. In many lakes successional development of phytoplankton and zooplankton or depressions in phytoplankton development corresponding with zooplankton maxima have been observed (e.g. Nauwerck, 1963; Jónasson & Kristiansen, 1967; Aasa, 1970; Hillbricht-Ilkowska *et al.*, 1972; Sorokin, 1972; Winberg *et al.*, 1972; Bailey-Watts, 1973; Grönberg, 1973). Statistical relationships are obvious, but the details do not always satisfy grazing effect theories; for example if the zooplankton species present are not able to feed directly on the algae of the preceding phytoplankton peak for anatomical reasons. It is noteworthy that all these observations stem from more or less eutrophic waters of the temperate zone. No examples of food depression by zooplankton have been shown so far from oligotrophic lakes, in spite of the fact that algae are usually a larger proportion of all particles in the seston in oligotrophic than in eutrophic lakes, and have been considered to be the most important food source for zooplankton in the former (Gliwicz, 1969b).

Evidence has been recorded for depression of phytoplankton growth by zooplankton grazing on low concentrations of algae in ponds (J. Fott & V. Kořínek personal communication), and also for changes in food quality due to selective grazing by different zooplankters (Bailey-Watts, 1973; Bailey-Watts & Lund, 1973) which is to be expected from the results of feeding experiments using particles of different size (Rigler, 1972; Gliwicz, 1969a).

Organisms in the bottom fauna and fish are also able to control phytoplankton, particularly in highly eutrophic waters. The impact of, for example *Chironomus anthracinus*, is in the first place on algae settled to the bottom (Jónasson & Kristiansen, 1967), and direct plankton feeding, (recorded for zooplankton, see p. 255 Chapter 6), has not been observed for phytoplankton. The impact of plankton-feeding fish on phytoplankton may be considerable. Silver carp (*Hypophthalmichthys molitrix*) is regarded to be such an efficient plankton feeder that the species has been taken into consideration for managed control of plankton blooms (Opuszynski, 1971; Sobolev, 1971; B. Ahling, personal communication). *Tilapia* spp. have been found to keep a dense population of *Chlamydomonas* in steady-state conditions through several months in an African irrigation pond (Nauwerck, unpublished).

(d) Effects of other biotic factors.   Besides zooplankton, bottom fauna and fish, heterotrophic and phagotrophic species of phytoplankton may act as phytoplankton feeders (Nauwerck, 1963). Parasites are frequently found to be a factor controlling growth in phytoplankton populations (Canter & Lund, 1948, 1966; Lund, 1971b). Growth inhibition due to extracellular products causing auto-antagonism (in aged populations of algae) or hetero-antagonism between mutually exclusive species (e.g *Aphanizonemon* and *Scenedesmus*) have been shown to result in breakdowns of plankton peaks or in sudden alternations between competing species (Lefèvre *et al.*, 1952). Also algae which for some reason (size, shape, taste) are not eaten by grazers may disappear because of a 'natural' death. Starving or an excess of metabolites inhibits growth, and old algae are less resistant against diseases and parasites.

(*e*) Effects of other non-biotic factors. Most dramatic changes in phytoplankton biomass and species composition seem to be generally related to abrupt changes of the non-biotic environment. The spring peak of small diatoms which can be observed in northern eutrophic lakes (Rodhe *et al.*, 1958) disappears before the arrival of grazers. In many cases peaks of blue-green algae or diatoms develop and disappear, apparently without the interaction of zooplankton, but are closely connected to changing weather conditions (e.g. Talling, 1966a). Reasons may include nutrient depletion, departure from favourable temperature limits and damage by turbulence or even by light. Autolysis, bacterial attack and rapid decay of the algae follow.

(*f*) Conclusions. Quantitative data on the relative importance of different factors controlling phytoplankton are scarce. Obviously, one or the other factor may be found dominating at one time or the other, annihilating a standing crop within days or even within hours. However, from data available it seems that algae are far more often destroyed by means other than grazing, and that grazing is most important in eutrophic lakes but seldom works as a growth limiting factor.

5.6.3. Macrophytes (Westlake)

5.6.3.1. Seasonal patterns

In large lakes and rivers the species composition is usually fairly stable but small water bodies, especially ditches or places where an annual emergent vegetation can develop, show large seasonal differences which may follow quite complicated cycles (e.g. Hoogers & van der Weij, 1971; DR 40).

In general the biomass of macrophytes in temperate waters increases in the spring to reach a summer maximum and then decreases during the late summer or autumn to a winter minimum (see examples in table 5.14). Species that die right down show a greater range in biomass than those that are ever-green. Data on the minimum biomass for true annual species are lacking but the range is probably even greater. Where such patterns are well defined, and non-respiratory losses are not large, the maximum seasonal biomass can then be regarded as a first approximation to the cumulative annual net production, which becomes available to detritus consumers after death. This has meant that many workers have concentrated on the seasonal maximum; there are few detailed studies of growth cycles and even fewer of annual cycles. There are some indications that there is a trend towards smaller seasonal differences at lower latitudes (table 5.14; Odum, 1957a; DR 52; DR 76) and possibly also at very high latitudes.

Much depends on the life cycle of individual species, or peculiarities of the environment which may cause other seasonal patterns of net photosynthesis. For example *Ranunculus calcareus*, a plant of spring-fed chalk streams, starts regrowth from rhizomes in the autumn, grows rapidly in the spring to flower in May, or earlier, and reaches a relatively early maximum in June (Dawson, 1976). These streams are relatively less turbid, warmer and less prone to violent floods than other rivers in the winter, while they may be too warm and slow flowing at midsummer for this species.

In some circumstances there may be more than one peak in the seasonal pattern of biomass. In Lake Wingra the *Myriophyllum spicatum* population reached the first maximum in May and the second in August, corresponding to two flowering seasons (Adams & McCracken, 1974; DR 12). In Lawrence Lake the mean biomass of *Scirpus subterminalis* showed two not very pronounced maxima in late winter and late autumn, which were the overall result of a complex interaction of different seasonal cycles at different depths (Rich *et al.*, 1971). In shallow water (2–3 m) there was a pronounced midsummer minimum and an autumn maximum; in deeper water the slight minimum was in winter and there was a broad spring–summer plateau. *Nitella flexilis* in L. George (NY) also showed at least two growth periods, out of phase at different depths (DR 12).

In Lake Beloie *Elodea canadensis* had a midsummer minimum and an autumn maximum, due to a combination of a specific tendency to overwinter as shoots, the long cold season, the decay of much of the overwintering population before much new growth had occurred and damage by swimmers during the summer (Borutskii, 1950). In L. Biwa *Egeria densa* showed a January maximum biomass arising from high loss rates during the growing season (Tanimizu & Miura, 1976).

5.6.3.2. Mechanisms of change

The typical temperate pattern arises from the spring increases in light, temperature and photosynthetic capacity (p. 181) and the often concomitant decrease in physical disturbances. These lead to the accumulation of biomass which itself accelerates the rate of accumulation per unit area. Later self-shading, and possibly internal rhythms, increased grazing, excessive temperatures and increasing stagnation, produce decreasing rates of accumulation. Rooted plants are probably little affected by nutrient changes in the water, but free-floating plants may suffer, like phytoplankton, from nutrient depletion (Goulder & Boatman, 1971). Towards the end of the productive phase, usually after the seasonal maximum biomass is reached, respiration may approach or exceed photosynthesis (Tanimizu & Miura, 1976), the plant becomes un-healthy, and losses by disease, damage and mortality increase, sometimes accelerated by higher grazing pressures and, in the autumn by increasing disturbance (Dawson, 1976). Over the year losses are approximately equal to gains.

Biomass changes observed, even in the accumulation phase, are rarely precisely equal to net production. Firstly some of the fixed carbon may be excreted. In experimental conditions plants in axenic cultures may lose more than half in this way (Wetzel, 1969). Under more natural conditions (where some excretion and uptake by epiphytic organisms may also occur) losses of dissolved organic material of 1–10% have been found (Hough & Wetzel, 1972; Wetzel & Allen, 1972; DR 12).

Parts of the plant, usually leaves which are seriously shaded by younger leaves, may die and be shed (Geus-Kruyt & Segal, 1973; Dawson, 1976). Adams & McCracken (1974) give an example of this, where leaf fall caused a decrease in biomass of *Myriophyllum*, although the photosynthetic capacity of shoot apices was near its maximum. Mechanical damage may break off shoots, particularly around flowering time (DR 12; Dawson, 1976) and biotic damage (apart from grazing) may also occur. For example stems of *Ranunculus* are severed by the case-making activities of caddis larvae; and *Elodea* is easily damaged by swimmers (Borutskii, 1950). Pieczyńska (1972a) found that 2% of the total net production of submerged plants in Mikołajskie Lake reached the shore as detached fragments and the losses, preceding the maximum biomass, from *Ranunculus calcareus* in a river were 16% of the maximum (Dawson, 1976). In extreme cases weed control measures may remove a large part of the biomass (e.g. Westlake, 1968) and if this occurs early in the season regrowth may develop. In some communities herbivores eating live macrophytes are unimportant, but in others herbivorous invertebrates, fish or birds may cause significant losses (Uspienski, 1965; Kajak *et al.*, 1972; Dobrowolski, 1973). Shelter from wave-action and wild-fowl by enclosures increased the autumnal biomass of *Potamogeton filiformis* from 1.2 to 22 g dry wt m^{-2} (Jupp, Spence & Britton, 1974) but this difference includes the effect on growth related to a higher biomass throughout the season.

Unusual patterns of biomass change may arise from specific internal rhythms of growth, flowering and perennation, environmental factors giving unusually high losses, such as storms in lakes or floods in rivers, or biotic factors such as especially high grazing pressures or human interference. Some examples have been mentioned above. In Lake Biwa net production and loss rates of *Egeria densa* were simultaneously high (5–14 and 4–11 g dry wt m^{-2}d^{-1} respectively), so biomass accumulation was no indication of net production and the seasonal maximum biomass, which occurred in midwinter, was only 40% of the annual net production (Tanimizu & Miura, 1976). Unfortunately the sources of these losses are unknown.

5.6.4. Periphyton (Moss)

The biomass of periphyton is known not to remain constant throughout the year in temperate lakes but no data are available for equatorial sites to compare with the relatively constant biomass recorded for equatorial phytoplankton (e.g. DR 4 for Lake Chad, Ganf (1972) for Lake George, Uganda and Talling (1966a) for Lake Victoria). In a tropical montane dam (Moss, 1970b), with a high replacement rate, large changes in biomass of algae living on the mud were related to spates in the wet season.

In temperate regions biomass changes (measured as chlorophyll *a*) have been recorded by Moss & Round (1967), Hickman & Round (1970) and Bailey-Watts (1973, 1974) for diatoms attached to sand grains; by Moss (1969a), Hickman (1971b), Moss & Round (1967), and Hickman & Round (1970) for algal populations living on mud surfaces (epipelon); by Wetzel (1964) for an algal community partly free living and partly attached to gravel, and by Minshall *et al.* (DR 30), Westlake *et al.* (1972), Bombówna (1972) and Marker (1976a) for algae attached to rocks in streams and rivers.

In temperate regions a peak of biomass frequently occurs in spring or early summer, followed by a midsummer decline. However, this pattern is often seen to be much more complex when sampling is carried out more frequently. The 'spring peak' may be initiated in December. Clumping in periphyton communities, and the necessity, not always realised, for many replicate samples to be taken, also makes interpretation of recorded seasonal changes difficult.

Reasons for the usual rapid growth early in the year are obscure but probably lie partly in increasing incident radiation coupled in many cases with availability of nutrients following replenishment by the inflows during winter. Reasons for decrease in biomass are equally unknown. Possibilities include mechanical wash-out (Moss, 1970b), nutrient limitation, shading by phytoplankton (Moss, 1969a) and grazing by invertebrates. In the Bere Stream the summer decrease in epilithic algae seems unrelated to nutrient deficiency (Marker, 1976a), light (DR 40), phytoplankton or flow. Selective grazing of periphyton species by *Ancylus fluviatilis* (Mull) has been shown (Calow, 1973).

5.6.5. Community stability (Moss & Westlake)

Large and rapid changes in the biomass of free-living communities such as phytoplankton, mud-living algae (epipelon) and free-floating macrophytes

Table 5.14 *Stability of communities measured by ratio of maximum to minimum biomass over a period covering at least midsummer and midwinter*

Site	Notes	Biomass max : min	Reference
Phytoplankton			
L. George, Uganda	Tropical	1.5–2.5	Ganf (1974d)
L. Chad, Chad	Tropical	2	Lemoalle (1969)
L. Victoria E. Africa	Tropical	4	Talling (1966a)
Walensee, Austria	Oligotrophic	3	Findenegg (1965)
V. Finstertaler S., Austria	Alpine	5	Pechlaner et al. (1972b)
Port-Bielh, France	Alpine	7	Capblancq & Laville (1972)
L. Balaton, Hungary	Saline	13	(DR 44)
Borax L., Calif., USA	Saline	18	Wetzel (1964)
Pääjärvi, Finland	Oligotrophic	20	Ilmavirta et al. (1974)
Priddy Pool, UK	Mesotrophic	34	Moss (1969a)
Zellersee, Austria	Mesotrophic	32	Findenegg (1965)
L. Sammamish, Wash., USA	Mesotrophic	44	Welch et al. (1975)
Abbot's Pond, UK, 1966, 1967, 1968	Eutrophic	60–82	Moss (1969a) Hickman (1973)
Loch Leven, UK	Eutrophic	83[a]	Bailey-Watts (1974)
Epipelon			
Abbot's Pond, UK, 1966–7		7	Moss (1969a)
Priddy Pool, UK, 1966–7		11	Moss (1969a)
Shearwater, UK		14	Hickman & Round (1970)
Priddy Pool, UK, 1969		26	Hickman (1971b)
Abbot's Pond, UK, 1969		90	Hickman (1971b)
Attached communities			
Loch Leven, UK	epipsammon	4.5	Bailey-Watts (1973, 1974)
Borax L., Calif., USA	epilithon	5.7	Wetzel (1964)
Shearwater, UK	epipsammon	8.5	Hickman & Round (1970)
Deep Creek, USA	epilithon	16	(DR 30)
Bere Stream, UK	epilithon	16	Westlake et al. (1972)
Priddy Pool, UK	epiphyton	18	Hickman (1971a)
Macrophytes			
Pääjärvi, Finland	*Isoetes lacustris*[b]	1.5	Kansanen et al. (1974)
Lawrence L., USA	*Scirpus subterminalis*[b]	1.5	Rich et al. (1971)
	Characeae[b]	2.6	
Doodhadhari L., India	*Hydrilla* sp. and *Ceratophyllum demersum*[b]	2.6	Unni (1976)
Marion L., Canada	*Isoetes occidentalis*[c]	5.3	Davies (1970)
	Potamogeton epihydrus[c]	5.5	
	Nuphar polysepela[c]	8.1	
L. Wingra, USA	*Myriophyllum spicatum*[d]	15[e]	Adams & McCracken (1974)
Ösbyjön, Sweden	*Myriophyllum verticillatum*[d]	20[e]	Forsberg (1960)
Bere Stream, UK	*Ranunculus calcareus*[d]	190	Dawson (1976)

[a] Very variable from year to year, this is maximum. [b] Evergreen. [c] Annual regrowth. [d] Partial annual regrowth
[e] Very approximate values suggested by extrapolation and assumption that winter biomass underground is approximately 5% of seasonal maximum standing crop of green shoots.

(pleustophytes) are characteristic, but attached algae, as is to be expected as a result of protection from mechanical disturbances, appear to be more stable. Within their growing season rooted macrophytes show greater stability, with infrequent and less pronounced changes, but some growth forms have very marked changes over the annual cycle. The ratio of maximum to minimum biomass over a period including midwinter and midsummer is a crude measure of stability and a number of examples are given in table 5.14. There is clearly some tendency for lower values to be found for attached algae and rooted evergreen species of macrophytes, and higher values to be found for the phytoplankton, epipelon and macrophytes of annual regrowth. The apparent instability of this latter group is more a reflection of the inadequacy of the criterion than a true representation of their behaviour, but their very low minimum biomass must make them vulnerable to competition early in the growing season.

5.7. Nutrients and growth

5.7.1. Introduction (Adams, Gerloff, Koonce, Westlake)

The relations between primary productivity and nutrients have generally been studied in terms of mean concentrations of dissolved nutrients and annual productivities, or at most in terms of rates of photoassimilation and current concentrations. It has become apparent that such rather static approaches are inadequate to describe the dynamic and interacting behaviour of aquatic ecosystems, but as yet it is rarely possible to do more than speculate about the probable complications.

The external concentration of a nutrient may limit the specific rate of gross photosynthesis directly though, apart from situations where carbon is deficient, this may not be common. Photorespiration and dark respiration may also be affected directly by nutrient concentrations, which will then influence the specific rates of photoassimilation and net photosynthesis. Nutrient concentrations may affect the degree of excretion, which will further change the relative rate of gain of biomass. Finally effects of minerals on the relative rates of mortality, sedimentation, senility, or disease losses may influence the relative rate of change of biomass. However, the ability of many plants to store nutrients, especially phosphorus, may prevent any swift response to changes in the media, and nitrogen-fixing algae may be independent of mineral nitrogen. Rooted plants may also be independent of concentrations in the open water when there are nutrients available in the sediments.

The relative rate of change of biomass affects the concentration of accumulated biomass and hence the rate of net photosynthesis per unit area, which ultimately integrates to the total net production achieved in a growing season. If the nutrient concentrations are always above limiting levels they will not affect the total production, but once they become limiting the total production is decreased below the potential. In a closed system in a temperate climate, where the nutrients went straight from the medium to the plant and remained there for the rest of the year, their gradual exhaustion would decrease the rate of net photosynthesis until it ceased and the final net production would be fixed by the total mass of the limiting nutrient. In a similar, open, system the annual net production might be fixed by the total nutrient supplied in the year, i.e. the nutrient loading, but if the concentrations attained limit the rate of approach to the potential maximum, that may never be reached before climatic conditions end the growth. In practice some of the nutrients will be released by the breakdown of dead material during the year i.e. nutrient turnover. In an equable climate a steady state of continual net photosynthesis could then be accompanied by a relatively high constant biomass and low nutrient concentration, as appears to happen in Lake George, Uganda (Ganf, 1974d, 1975).

Much discussion of growth limitations is confused because it is rarely possible under natural conditions to say any single factor is limiting. Over twenty-four hours, or over a growth cycle, limiting factors may be constantly changing and different factors will affect different parts of the system (subcellular processes, differentiation, cell division, mortality, species selection etc.) each having different feed-back effects. So, many factors appear to be partially limiting and two apparently contradictory statements may both have some validity (e.g. Schindler & Fee, 1974). Also, because of the exponential nature of growth, quite small differences in photosynthetic rates, especially in the lag phase, can make large differences in final growth.

5.7.2. Phytoplankton (Koonce)

The nutrients required to support algal growth have been documented for many species (Gerloff, 1963; Epstein, 1965) but the role of nutrients in regulating phytoplankton population dynamics is much more controversial. Considerable discussion has been generated around the relative role of phosphorus, carbon and nitrogen in limiting growth of phytoplankton in natural waters (Likens, 1972).

Studies of nutrient limitation of phytoplankton growth have fallen into three general categories (cf. Goldman, 1972): (1) correlations of nutrient concentration and phytoplankton biomass; (2) determinations of internal nutrient concentration and comparisons with critical internal nutrient concentrations, for a variety of nutrients; and (3) utilisation of nutrient enrichment bioassays to determine which nutrients are in fact limiting algal growth. Unfortunately, these studies may provide conflicting evidence for the nutrient limitation of algal growth.

The time course comparison of nutrient concentrations and phytoplankton standing crops is a standard technique employed by most of the IBP projects. To varying degrees such data indicate phytoplankton peaks following nutrient peaks. Furthermore the general pattern indicates that attendant on the phytoplankton peak, there is a marked depletion of one or more nutrients (usually N, P or Si), a situation that then remains for the remainder of the growing season (e.g. Gibson *et al.*, 1971; Serruya & Pollingher, 1971; Lund, 1972; Lund *et al.*, 1975). To identify the nutrient limiting algal growth requires some type of correlative analysis. One approach is to treat algal biomass or production as a function of some index of nutrient loading or of nutrient concentration of the lake prior to algal blooms (e.g. Vollenweider, 1968; Schindler, 1971b; Edmondson, 1972b; Schindler & Fee, 1974). Such studies often indicate that phosphorus can determine the maximum biomass or annual production but this cannot be assumed to be generally true.

Critical internal nutrient concentrations (cf. p. 211) of some nutrients have been established for some algal species in culture (e.g. Gerloff & Skoog, 1957; Gerloff, 1975; table 5.15) but accurate field measurements are often difficult. If this level is not maintained growth decreases or stops. This technique has indicated that nitrogen was limiting the growth of *Microcystis aeruginosa* in some lakes in Wisconsin (Gerloff & Skoog, 1957). In a review of the literature, using slightly different criteria, Healey (1975) found rather higher values for a wide variety of cultured algae (70 mg N (g dry wt)$^{-1}$, 10 mg P g^{-1}).

Theoretically, nutrient enrichment bioassays using growth measurements (Lund *et al.*, 1971, 1975; Goldman, 1972) have the best potential for determining nutrient limitation of phytoplankton growth at a particular time, but many bioassays rely on brief measurements of photosynthesis (e.g. Goldman, 1963). Reporting on the results of nutrient enrichment bioassays on a series of lakes varying in productivity, Powers *et al.* (1972) found carbon-14 uptake responded positively to additions of nitrogen and phosphorus, both alone and in combinations. Frequently, their results indicated that the

Table 5.15 *Critical concentrations of various essential elements in several algal species*

Species	Element (critical concentration in mg g^{-1} dry wt)				
	N	P	Ca	Mg	K
Chlorella pyrenoidosa	—	—	0.0	1.5	4.0
Scenedesmus quadricauda	—	—	0.6	0.5	2.5
Draparnaldia plumosa	23	1.8	0.3	2.0	24.0
Stigeoclonium tenue	—	—	0.3	2.5	19.0
Microcystis aeruginosa	40	1.2	0.4	3.0	5.0
Nostoc muscorum	—	—	0.2	2.5	8.0

After Gerloff (1975).

addition of both nutrients produced a larger carbon-14 uptake response than either alone. A similar response has been observed by Schelske & Stoermer (1972) for combined nitrogen, phosphorus and silicon additions to Lake Michigan water. Lund *et al.* (1975), using growth bioassays of water from Blelham Tarn, a moderately eutrophic lake, found differences at different times of the year and between years. Generally silicon was limiting at the time of the spring maximum and subsequently phosphorus was also necessary. In the summer a further need for chelated iron developed.

Short-term bioassay experiments thus frequently indicate that more than one nutrient is limiting and that combinations of nutrients will yield the greatest response. A variety of enrichment responses is probably to be expected because bioassays are usually applied to communities near a steady state. Yet, for those systems with sufficient data, correlations between summer phytoplankton biomass and winter concentrations have often supported the inference that the total loading of a single nutrient, often phosphorus, sets long-term upper limits on phytoplankton biomass.

5.7.3. Macrophytes (Gerloff with Westlake)

Growth of macrophytes has been related to environmental concentrations of specific elements and critical concentrations have been established, but in total there has been relatively little research on aquatic macrophytes.

Since rooted macrophytes have access to nutrients in both sediment and water (Denny, 1972; Bristow, 1975), and little is known about available concentrations in the sediments, it is difficult to quantify environmental concentrations. It appears probable that phosphorus, nitrogen and possibly potassium may be insufficient for optimum growth in oligotrophic water bodies but that ample supplies are available in most eutrophic water bodies (Westlake, 1975a). Campbell & Spence (1976) used nutrient enrichment bioassay to show that phosphorus was limiting the rate of uptake of carbon-14 by leaves of *Potamogeton* spp. in Scottish loch waters with less than $18 \mu g PO_4 P l^{-1}$ but found that the stimulation was less marked in water containing dense phytoplankton.

Interpretation of nutrient supplies and growth-limiting nutrients by plant analysis requires the critical concentration of each essential element of interest (Gerloff, 1969), i.e. the minimum concentration of an element in a plant, or more often slightly less than the minimum concentration, which permits maximum yield. Critical concentrations have been established for most essential elements in *Elodea occidentalis* and for several elements in *Ceratophyllum demersum* and *Myriophyllum spicatum*, showing that they may differ considerably among species (table 5.16). The critical phosphorus concentration in *Elodea* is 1.4 mg (g dry wt)$^{-1}$ but in *Myriophyllum* it is only 0.75 mg g^{-1}. However, for quick assessments of the importance of nitrogen or

Table 5.16 *Critical concentrations of nitrogen, phosphorus and other essential elements in index segments of* Elodea occidentalis, Ceratophyllum demersum *and* Myriophyllum spicatum

Element	Critical concentration in index segment $mg\,g^{-1}$ dry wt		
	Elodea	*Ceratophyllum*	*Myriophyllum*
N	16	13	7.5
P	1.4	1.0	0.70
S	0.8	—	—
Ca	2.8	2.2	—
Mg	1.0	1.8	—
K	8.0	17	3.5
Fe	60×10^{-3}	—	—
Mn	4.0×10^{-3}	—	—
Zn	8.0×10^{-3}	—	1.8×10^{-3}
Mo	0.15×10^{-3}	—	—
B	1.3×10^{-3}	2.8×10^{-3}	—

After Gerloff (1975).

phosphorus for the nutrition of macrophytes, critical concentrations of $13\ mg\,N\,g^{-1}$ and $1.3\ mg\,P\,g^{-1}$ in whole plants may be assumed (Gerloff and Krombholz, 1966).

In Lake Wingra (Adams & McCracken, 1974) internal concentrations of phosphorus are close to this critical value in the summer, and in some other Wisconsin lakes (Gerloff & Fishbeck, 1973) either nitrogen or phosphorus is limiting or close to limiting. In the nutrient-rich chalk streams of southern England the annual and the growing season throughputs of dissolved nitrogen, phosphorus and potassium are many times the total of these nutrients in the maximum biomass (Casey & Westlake, 1974) and the internal concentrations are far in excess of limiting (Casey & Downing, 1976). Only in small densely weeded waters does macrophyte growth have any appreciable effect on nutrient concentrations, and potassium is most affected. The review of internal concentrations of plant nutrients by Hutchinson (1975) shows that the mean concentrations are always above the critical concentrations in table 5.16 but that the bottom end of the range is always below the critical concentration.

5.7.4. Periphyton (McCracken)

Except for well-studied macroalgae such as *Cladophora*, there is little information on relationships between nutrients and periphyton growth. There is some evidence that phosphorus or nitrogen may be limiting for some periphytic algae communities. In a periphyton dominated by diatoms Stockner & Armstrong (1971) found that internal concentrations of nitrogen were usually below the critical concentrations found by Gerloff & Skoog

(1954, for *Microcystis aeruginosa*), but phosphorus concentrations were usually higher. Internal concentrations of chlorophyll *a* were high relative to phosphorus in the spring but low in the autumn, suggesting autumnal storage and utilisation in the spring.

Recently there have been many reports of *Cladophora* becoming a serious nuisance, which is thought to be related to the effects of cultural eutrophication (Neil & Owen, 1964; Whitton, 1970; Adams & Stone, 1973; Bolas & Lund, 1974). In general high nitrate, and especially high phosphate, seem to be important but the observations have many anomalies. Adams & Stone found that *Cladophora* in Lake Michigan exhibited increased rates of photosynthesis when supplied with nutrient-rich water and accumulated nitrogen and several metals, but not phosphorus. Similarly McCracken & Adams (DR 12) found the most extensive growths of *Oedogonium* in Lake Wingra to be associated with near-by sewer outfalls.

Marker (1976a) found no evidence that nutrients affected the seasonal cycle of epilithon biomass or photosynthesis in a stream where nitrate-nitrogen always exceeded $3 \, mg \, N \, l^{-1}$ and phosphate-phosphorus always exceeded $20 \, \mu g \, P \, l^{-1}$.

There is no evidence in general that periphytic algae will differ significantly from phytoplankton in terms of nutrient requirements, growth responses, or nutrient sources. However it is possible that diffusion limitations caused by the boundary layers enclosing periphyton and nutrient relationships between epiphytic periphyton and their host substrates may prove to be of particular interest. In waters with low nutrients *Cladophora* grows better where flow or wave action are experienced. Fitzgerald (1969) found that epiphytes and their hosts competed for nitrogen and other nutrients. Linskens (1963) and Harlin (1973) have reported the exchange of phosphorus-32 between marine macrophytes and epiphytes and this could mean that epiphytes have access to sediment nutrients.

5.7.5. Carbon supply (Westlake)

Although carbon is nearly half the dry organic weight of plants it has been rather neglected as an essential nutrient. Since carbon dioxide can be supplied from the atmosphere, it is unlikely to limit the maximum biomass ultimately attained, but there is no doubt that in some circumstances carbon supply can limit the rate of photosynthesis, the rate at which the maximum biomass is reached, the annual rate of production and the occurrence of plants (Frantz & Cordone, 1967; Bindloss, 1974; Schindler & Fee, 1973; Shapiro, 1973; Westlake, 1975a; Talling, 1976).

'Carbon supply' in this context is not a simple concept. Firstly, the rates of exchange of carbon dioxide at the air–water or sediment–water interface and of its transfer throughout the water body, in relation to the demand, may govern its local concentrations. Then the effective concentration depends on the ability of the plant to respond to the concentration of free carbon dioxide or of bicarbonate ion, which appears to vary with species. Thus *Myriophyllum spicatum* and *Potamogeton praelongus* and *lucens* can photosynthesise in water containing bicarbonate but little free carbon dioxide, while *Myriophyllum verticillatum* and *Potamogeton polygonifolius* require free carbon dioxide; which corresponds with the distribution of these species (Steemann Nielsen, 1947; Gessner, 1959; Black, 1973). Mosses appear to be unable to use bicarbonate, many benthic algae can. Both types are found in the phytoplankton and blue-green algae in particular are often able to use bicarbonate. The data are reviewed in some detail by Hutchinson (1967, 1975) and discussed by Talling (1976).

Kinetic considerations are also involved (Talling, 1976). The half-saturation constant for bicarbonate uptake is probably higher than for free carbon dioxide, so low concentrations of free carbon dioxide can support rates that require considerably higher concentrations of bicarbonate and, in comparable low concentrations, rates of photosynthesis may be increased much more by an increase in free carbon dioxide than by the same increase in bicarbonate (Steemann Nielsen, 1947). The concentration of free carbon dioxide in bicarbonate solutions is in an equilibrium governed by the pH, becoming very low above pH8 in many waters. A plant with a very low carbon dioxide compensation point may be able to use free carbon dioxide in waters of high pH, which is then replaced by dissociation of bicarbonate. The rate of photosynthesis may then be limited by the rate of dissociation. *Myriophyllum spicatum* is reported to have a low compensation point (Stanley & Naylor, 1972) but it is not clear if the experimental conditions excluded the possibility of bicarbonate uptake and there is plenty of evidence that this species can use bicarbonate.

Rooted plants may obtain carbon dioxide directly from the sediments (Wium-Anderson, 1971; DR 59).

5.8. The relative importance of different producing communities (Moss)

Over a century of observations on phytoplankton, and the relative ease with which this community can be studied has led to its emphasis as the major primary producer of lakes. In terms of total energy fixed per lake this is probably true of very large lakes such as the St Lawrence Great Lakes and the larger East African Rift Valley lakes, though sufficient data do not exist for this to be certain. Even in moderate sized lakes such as Windermere, L. Mendota and L. Erken phytoplankton is probably the dominant producing community.

Budgetary studies involve much work and hence few comprehensive sets of data exist. Table 5.17 lists those complete sets available at present together with a few with one component missing. Straškraba (1968) has reviewed earlier work on the relative importance of macrophytes and phytoplankton, showing twelve lakes where the phytoplankton production was more than twice the macrophyte production and only two (out of fifteen) where macrophyte production was greater.

With the exception of studies from Eastern Europe and the USSR it has not been routine practice to measure the photosynthesis of all the primary producing communities of a lake under study and in Russian work it is rare for emergent and submerged production to be separated. It follows that in many studies lakes have been chosen by investigators particularly interested in the non-phytoplanktonic communities, and the lakes in question do not represent a random sample. Generalisations on the relative importance of the various communities on a world scale are therefore inadvisable. The data however give ample evidence of large non-planktonic contributions, and the abundance of small, relatively shallow and weedy lakes over much of North America, Europe and Asia in particular suggests that overall the macrophytes and periphyton may be at least as important as phytoplankton in terms of photosynthesis.

Some hypotheses can be posed. Other things being equal, an increase in the depth of a lake is expected to increase the phytoplankton and a decrease in depth is expected to increase the periphyton and macrophytes. Increasing nutrient input might be expected to result in progressive dominance of phytoplankton owing to shading effects (Spence, 1976). This is perhaps shown in the interconnected lake series of Naroch, Myastro and Batorin in the USSR (Winberg et al., 1972) where a trend from mesotrophy to eutrophy leads to an increase from 31% to 94% in the planktonic contribution to photosynthesis. More controlled examples are shown by the experimental ponds of Hall, Cooper & Werner (1970) and Moss (1976), where replicated experimental ponds were variously fertilised. In the most fertile ponds heavy blooms of phytoplankton were associated with a relatively low maximum biomass of *Elodea*, *Cladophora* and *Oedogonium* compared with less well fertilised ponds, where crops of *Chara* were greater and phytoplankton smaller. Such relationships are modelled in section 5.9.4.

5.9. Models

5.9.1. Introduction (Steel & Westlake)

Mathematical modelling is an attempt to express formally the concepts that the modeller has of the systems studied. Such concepts can never, however, be more than simplifications of the real system complexes. There are many classifications of models, which emphasises that they are constructed for many

Table 5.17 *Relative importance of different producing communities in water bodies*

Water body lat. and long.	Reference	Mean depth (m)	Maximum depth (m)	Area (m or m²)	Photosynthesis (see notes 1–4) Macrophyte	Periphyton	Phytoplankton	Notes
(a) Rivers and streams								
Bere stream UK 2°12'W 50°44'N	Westlake et al. (1972) & DR 40		0.3–0.7	240 × 7.5	1.4 organic matter m⁻² a⁻¹ net (90%)	0.17 g organic matter m⁻² a⁻¹ net (10%)	Negligible	Chalk stream; allochthonous particulate input of 40 000 g organic matter m⁻² a⁻¹
Root Spring USA 71°W 42°N	Teal (1957)		0.1–0.2	3	0	Filamentous mats 3.0 MJ m⁻² a⁻¹ gross (100%)	0	Allochthonous particulate input 9.8 MJ m⁻² a⁻¹
Silver Springs USA 82°W 29°N	Odum (1957b)		14	500 × 50	1900 g dry wt m⁻² a⁻¹ gross (30%)	Epiphytes 4500 g dry wt m⁻² a⁻¹ gross (70%)	0	Thermostatic, chemostatic spring
R. Thames UK 0°55'W 51°30'N	Mann et al. (1972)		4.5	40–80 wide	0.18 MJ m⁻² a⁻¹ (2%)	0.66 MJ m⁻² a⁻¹ (8%)	8.0 MJ m⁻² a⁻¹ (90%)	Allochthonous input 1.2 MJ m⁻² a⁻¹
(b) Lakes and reservoirs in order of area								
Abbot's Pond UK 2°40'W 51°28'N	Hickman (1971b, 1973)	2	4	3.7 × 10³	?	Epipelic algae 6.2 g C pond⁻¹ h⁻¹ net	110 g C pond⁻¹ h⁻¹ net	Macrophytes and epiphytes probably not negligible
Priddy Pool UK 2°39'W 51°15'N	Hickman (1971a, b)	0.6	1.2	7.3 × 10³	?	Epipelon 0.28 mg C m⁻² h⁻¹ net Epiphytes 55 mg C m⁻² h⁻¹	0.3 mg C m⁻¹ h⁻¹ net	86% of area colonised by Equisetum so macrophyte photosynthesis

Location	Reference			Area				Remarks
Japan 140°05'E 37°39'N	moto (1975)				net		net	turnover; allochthonous input 32 MJ m^{-2} a^{-1}
Lawrence L. USA 85°22'W 42°27'N	Wetzel et al. (1972)	5.9	13	5×10^4	88 g C m^{-2} a^{-1} net (51%)	Epiphytes 38 g C m^{-2} a^{-1} net (22%) Epipelic algae 210 g C m^{-2} a^{-1} net (1.1%)	43 g C m^{-2} a^{-1} net (26%)	Highly calcareous infertile marl lake; particulate allochthonous input 4.1 g C m^{-2} a^{-1}
Marion L., Canada 122°33'W 49°19'N	Hargrave (1969) Davies (1970)	2.4	7	13×10^4	18 g C m^{-2} a^{-1} net (27%)	Epipelic algae 40 g C m^{-2} a^{-1} gross (61%)	8.0 g C m^{-2} a^{-1} net (12%)	Acid bog lake; stream input 143 g m^{-2} a^{-1}
L. Krugloye USSR 35°E 65°N	Alimov et al. (1972)	12	1.5	43×10^4	0.017 MJ m^{-2} growing season^{-1} net	?	0.15 MJ m^{-2} growing season net	Brown water bog lake
L. Yunoko Japan 135°26'E 36°48'N	Tanaka, Mizutani & Hanya (1975)	7.4	15	35×10^4	3.4 g C m^{-2} a^{-1} net	?	100 g C m^{-2} a^{-1} net	Eutrophic; allochthonous input ~74 g C m^{-2} a^{-1} ~50% particulate
Borax L. USA 122°55'W 39°00'N	Wetzel (1964)	0.7	1.5	43×10^4	1.4 kg C lake^{-1} d^{-1} net (0.7%)	Epipelic and epilithic algae 76 kg C lake^{-1} d^{-1} net (43%)	100 kg C lake^{-1} d^{-1} net (57%)	Alkaline saline lake; estimates allow for fluctuating water levels
L. Krivoye USSR 35°E 65°N	Alimov et al. (1972)	11	30	50×10^4	0.033 MJ m^{-2} net	?	0.56 MJ m^{-2} growing season^{-1} net	Steep-sided rocky lake poor in nutrients
L. Latnjajaure Swedish Lapland 18°49'E 68°12'N	Bodin & Nauwerck (1968)	16.5	43	73×10^4	Net (20%) (aquatic moss)	Bottom diatoms net (15%) Epiphytes net (5%)	Net (60%)	Extremely oligotrophic arctic lake; total net photosynthesis 3500–4000 g C d^{-1}

(contd.)

Table 5.17 (contd.)

Water body lat. and long.	Reference	Mean depth (m)	Maximum depth (m)	Area (m or m²)	Photosynthesis (see notes 1–4)			Notes
					Macrophyte	Periphyton	Phytoplankton	
Rybinsk Reservoir USSR 39°02'E 59°14'N	Sorokin (1972)	5.6	25	4.6×10^6	0.11 MJ m^{-2} in May–Oct. gross (5%)	?	2.1 MJ m^{-2} in May–Oct. gross (95%)	Mesotrophic; allochthonous input of 5.4 MJ m^{-2} in May–Oct., emergents 3% of total prod.
Karakul, USSR 73°30'E 39°00'N	Khusainova et al. (1973) DR 63	1.4	3.8	1.4×10^6	40 MJ m^{-2} a^{-1} net (93%)	?	2.9 MJ m^{-2} a^{-1} net (7%)	Saline, low turnover; emergents 60% of macrophyte prod.
Mikołajskie L., Poland 21°35'E 53°50'N	Kajak et al. (1972)	11	28	5×10^6	1.4 MJ m^{-2} a^{-1} net (12%)	0.78 MJ m^{-2} a^{-1} net (6.5%)	9.4 MJ m^{-2} a^{-1} net (82%)	Eutrophic; emergents 70% of macrophyte prod.
Batorin L. USSR 26°E 54°88'N	Winberg et al. (1972)	3	5.5	6.3×10^6	0.28 MJ m^{-2} in May–Oct. (3.6%)	0.18 MJ m^{-2} in May–Oct. (2.2%)	7.4 MJ m^{-2} in May–Oct. (94%)	Eutrophic; emergents 46% of macrophyte prod.
L. Krasnoye (Punnus-Yarvi) USSR 29°08'E 60°16'N	DR 62	7	13	9×10^6	0.67 MJ m^{-2} a^{-1} gross (12%)	'Microphyto-benthos' 0.57 MJ m^{-2} a^{-1} gross (7.6%)	4.4 MJ m^{-2} a^{-1} gross (80%)	Mesotrophic; allochthonous particulate input 1.4 MJ m^{-2} a^{-1}, includes emergents
L. Kojima, Japan 135°56'E 34°35'N	Hata (1975)	1.6	9	11×10^6	0.1 g C m^{-2} d^{-1} net	?	1.2 g C m^{-2} d^{-1} net	Saline, polluted; allochthonous input 3.8 g C m^{-2} d^{-1} 9–10% particulate; emergents 78% macrophyte prod.
Myastro L. USSR 26°E 54°88'N	Winberg et al. (1972)	5.4	11	13×10^6	0.56 MJ m^{-2} in May–Oct. gross (7.4%)	0.42 MJ m^{-2} in May–Oct. gross (5.5%)	6.6 MJ m^{-2} in May–Oct. gross (87%)	Mesotrophic–eutrophic; emergents 64% of macrophyte prod.

Location	Reference			Area				Notes
Päijärvi Finland 25 08'E 61 04'N	Sarvala (1974)	14	87	13 × 10⁶	0.27 MJ m⁻² a⁻¹ net (18%)	0.066 MJ m⁻² a⁻¹ net (77%)	1.1 MJ m⁻² a⁻¹ net (77%)	Oligotrophic; allochthonous input 1.3 MJ m⁻² a⁻¹; emergents 86% of macrophyte prod.
Kurakhov Reservoir, USSR 37°15'E 48°N	Pidgaiko et al. (1972)	6	9	15 × 10⁶	1.4 MJ m⁻² a⁻¹ gross (9.7%)	'Phytomicrobenthos' 1.8 MJ m⁻² a⁻¹ gross (13%)	11 MJ m⁻² a⁻¹ gross (78%)	Heated reservoir
Naroch L., USSR 26°E 54 88'N	Winberg et al. (1972) DR 5	11	35	80 × 10⁶	2.5 MJ m⁻² in May–Oct. gross (38%) 4.2 × 10¹⁰ MJ lake⁻¹ a⁻¹ net	2.1 MJ m⁻² in May–Oct. gross (32%) ?	2.0 MJ m⁻² in May–Oct. gross (31%) 2.4 × 10¹⁰ MJ lake⁻¹ a⁻¹ net	Mesotrophic; emergents 1% of macrophyte prod.
L. Chilwa Malawi 35°45'E 15°15'S		2 (when full)	7 (when full)	700 × 10⁶ (plus 700 × 10⁶ swamp)				Endorheic soda lake with occasional blue-green algal blooms; emergents most of macrophyte prod.
Kiev Reservoir USSR 30°30'E 50°30'N	Gak et al. (1972)	4	19	9.9 × 10⁸	0.74 MJ m⁻² a⁻¹ gross (6.4%)	'Phytomicrobenthos' 2.9 MJ m⁻² a⁻¹ gross (26%) Filamentous algae 2.8 MJ m⁻² a⁻¹ gross (24%)	51 MJ m⁻² a⁻¹ gross (44%)	Eutrophic; emergents 54% of macrophyte prod.

1. Gross and net designations refer to individual author's opinion.
2. Percentages of total photosynthesis are only calculated where all communities have been examined and do not take into account allochthonous contributions.
3. Values per square metre refer to whole lake area, i.e. allowance has been made for proportions of littoral and open water zones.
4. Data presented have been rounded to two significant figures after calculating percentages.

different purposes, each requiring different simplifications, and different types are needed at different stages in investigations. Modelling is therefore a process of informed, selective condensation.

Often the first steps will correlate a mass of data to synthesise patterns. For example data on lakes or species may lead to discrete classifications enabling predictions to be made about the probability of one species occurring in association with another, or two lakes having similar associations. Or measurements describing the system may be sorted into a series of quantified interrelationships thus enabling predictions, for example of rates of production and the relative importance of individual environmental factors (see Brylinsky, Chapter 9). Such models often guide the isolation of a distinct system, and its analysis into components with defined properties and interrelationships which can be used to construct a model simulating the behaviour of the system. The components or compartments, may vary in size with the scale of the investigation, from the concentration of a metabolite, through the photoassimilation of the phytoplankton to the net primary production of a water body. Ecologically successful constructive models require a considerable degree of biological insight (Ganf, 1975) but modelling itself may help to demonstrate biological weaknesses. Most models of the processes of primary production show very considerable sensitivity to plant respiration which is still much less well quantified than photosynthesis.

Entering real data into such a model leads to predictions about the behaviour of the system which can be compared with observations. If this shows that the model is capable of valid predictions the mathematical relationships may be used to illuminate general properties and previously poorly defined or unrecognised interactions within the system (Vollenweider, 1970). Models can never lead to anything fundamentally new, for all the conclusions which may be drawn from them must have been implicit in the mathematical model, but they help us to increase the depth of our perception.

Throughout this book models are used for various purposes; often to elucidate the nature of the responses of the system to some single or multiple external influences. In the present context the greatest effort will be directed to characterising some of the responses of photoassimilation and respiration and from this infer some possible generalisations about biomass and primary production.

5.9.2. Phytoplankton models (Steel)

The approach is exactly that of Talling (1957b), relating the population's growth to its photosynthesis and light climate. All other sources and sinks are assumed to be negligibly small. It will be further assumed that the phytoplankton is always distributed evenly vertically and horizontally within an optically deep column (see Fee, 1973b for a numerical approach allowing for vertical inhomogeneities). Photosynthesis is reduced to zero at or above the greatest depth of the distribution. This will be conditioned by: $z_{eu} \leqslant z_{mix}$; z_{mix} is the depth of vertical mixing, z_{eu} is defined as the depth at which the photosynthetically active radiation (PAR) has been reduced to 1% of its immediate subsurface intensity. From equations 5.2 and 5.5 it follows that $z_{eu} = 3.7/(\varepsilon_s b + \varepsilon_q)$ or, when $\varepsilon_s b = 0$,

$$z_{eu} \varepsilon_q = 3.7 \tag{5.13}$$

When $z_{mix} \varepsilon_q \geqslant 3.7$ the condition $z_{eu} \leqslant z_{mix}$, is fulfilled. If $z_{max} \varepsilon_q \geqslant 3.7$ then the bottom deposits will not be significantly illuminated.

To eliminate differences due to the variations in irradiance with depth and during the day the model will consider the 24-h column integral photosynthesis, assuming both insolation and physiological response to be symmetrical about local noon. The basic function relating the photoassimilation rate (P) of an alga to the PAR (I) is of the form:

$$P = P_{max} f_1(I) \tag{5.14}$$

This function contains elements of the light and dark reactions of photosynthesis in that it initially increases linearly with increase in irradiance, until an eventual, saturation rate (P_{max}) is obtained:

$$P \propto I; \quad I \rightarrow 0$$
$$P \rightarrow P_{max}; \quad I \rightarrow I_{opt}$$

The photosynthetic rate per unit volume will then be

$$A \text{ mg O}_2 \text{ m}^{-3} \text{h}^{-1} \sim b P_{max} f_1(I) \tag{5.15}$$

Dimensional consideration of equation 5.15 requires that $f_1(I)$ be dimensionless and Talling (1957b) found that this was fulfilled if the irradiance was considered as relative to an irradiance defining the onset of the saturation plateau (I_k). For underwater photosynthesis this function is often of the form:

$$A \text{ mg O}_2 \text{ m}^{-3} \text{h}^{-1} \sim b P_{max} I \cdot I_k^{-1} [1 + (I \cdot I_k^{-1})^2]^{-0.5} \tag{5.16}$$

but requires modification when there is inhibition at higher irradiances or where only P_{opt} can be measured because factors other than light are limiting (Vollenweider, 1965). Since $I_z = I_0' \exp(-\sigma z \varepsilon_{min})$ (see section 5.3) photoassimilation by the algal population in an optically deep mixed column will be:

$$\int_0^\infty A_z d_z \text{ mg O}_2 \text{ m}^{-2} \text{h}^{-1} = b P_{max} f_2(I \cdot I_k^{-1})(\sigma \varepsilon_{min})^{-1} \tag{5.17}$$

This solution is known from Talling (1957b) and Vollenweider (1965). If the duration of daylight hours is Δ, then the daily integral photoassimilation would be approximated by:

$$\bar{A}_{24}\,\mathrm{mg\,O_2\,m^{-2}\,d^{-1}} = \int_0^{\Delta}\int_0^{\infty} A_z d_z\,\Delta b P_{max}\,f_3(I_{0,\,max}\cdot I_k^{-1})(\sigma\varepsilon_{min})^{-1}$$

$$(5.18)$$

It seems that if $I_{0,\,max} \geqslant 15 I_k$ it is often possible to take $f_3(I_{0,\,max}\cdot I_k^{-1}) \sim$ 2.5–2.7; but Fee (1973a) found a wider range of values. Then, since $\sigma \sim 1.3$,

$$\bar{A}_{24}\,\mathrm{mg\,O_2\,m^{-2}\,d^{-1}} \sim 2\Delta\,b\,P_{max}(\varepsilon_q + \varepsilon_s b)^{-1}$$
$$\text{(see section 5.3.1 for } \sigma \text{ and } \varepsilon_{min}) \qquad (5.19)$$

It can now be demonstrated that, with respect to biomass, the maximum photoassimilation occurs when the biomass is a maximum, assuming for the present no population-density-dependent response in P_{max}. Thus from:

$$\bar{A}_{24,\,max}\,\mathrm{mg\,O_2\,m^{-2}\,d^{-1}} \sim 2\,\Delta\,b_{max}P_{max}(\varepsilon_q + \varepsilon_s b_{max})^{-1}$$
$$B = z_{mix}b$$
$$\bar{A}_{24,\,max}\,\mathrm{mg\,O_2\,m^{-2}\,d^{-1}} \sim 2\,\Delta\,B_{Max}P_{max}(z_{mix}\varepsilon_q + \varepsilon_s B_{Max})^{-1} \quad (5.20^*)$$

When respiration is the only significant loss the population for which the net photosynthesis is zero will be at the maximum biomass possible under the current conditions. Let:

$$\bar{N}_{24}\,\mathrm{mg\,O_2\,m^{-2}\,d^{-1}} = \bar{A}_{24} - R_{d,\,24} \qquad (5.21)$$

In order to write this equation it is necessary to obtain the dark respiration rate (R_d) in the same dimensions as A. If it is assumed to be independent of irradiance and time an approximation is very easily achieved:

$$R_{d,\,24}\,\mathrm{mg\,O_2\,m^{-2}\,d^{-1}} = \int_0^{24}\int_0^{z_{mix}} R_z d_z\cdot d_t \sim b z_{mix}\,24\,C_d \qquad (5.22)$$

where C_d is the specific dark respiration rate.
Then:

$$\bar{N}_{24}\,\mathrm{mg\,O_2\,m^{-2}\,d^{-1}} \sim 2\,\Delta\,b\,P_{max}(\varepsilon_q + \varepsilon_s b)^{-1} - b z_{mix}\,24\,C_d$$
$$\sim 2\,\Delta\,B\,P_{max}[(z_{mix}\varepsilon_q + \varepsilon_s B)^{-1} - 12 r\,\Delta^{-1}] \qquad (5.23)$$

where $r = C_d P_{max}^{-1}$. When respiration is the only significant loss the population for which the net photosynthesis is zero will be the maximum biomass possible under the current conditions, therefore:

$$2\Delta\,B_{Max}P_{max}[(z_{mix}\varepsilon_q + \varepsilon_s B_{Max})^{-1} - 12 r\,\Delta^{-1}] \sim 0$$

and because, within the conditions of the model, it is not sensible for

* Note that B_{Max} is the maximum biomass in the mixed column as distinct from the B_{max} of section 5.3 which is the maximum biomass in the euphotic zone.

$2\Delta B_{Max} P_{max}$ to be zero:

$$(z_{mix}\varepsilon_q + \varepsilon_s B_{Max})^{-1} - 12r\Delta^{-1} \sim 0$$

Therefore

$$B_{Max}\ \mathrm{mg\ chl.}\,a\,\mathrm{m}^{-2} \sim \varepsilon_s^{-1}[\Delta(12r)^{-1} - z_{mix}\varepsilon_q] \qquad (5.24)$$

and correspondingly (from equation 5.20):

$$\bar{A}_{24,\,max}\ \mathrm{mg\,O_2\,m}^{-2}\,\mathrm{d}^{-1} \sim 2\Delta P_{max}\varepsilon_s^{-1}[1 - 12r\,z_{mix}\varepsilon_q\Delta^{-1}] \qquad (5.25)$$

Let the dimensionless group $12r\,z_{mix}\varepsilon_q\Delta^{-1} = j$, which is related to respiration in the mixed depth and is clearly of some importance in determining phytoplankton growth. Now, if $\Delta_{max} \sim 16$ and $r_{min} \sim 0.04$ (cf. table 5.12) $(\Delta \cdot r^{-1}) \sim 400$; however, as a high value for $(\Delta \cdot r^{-1})$, 240 may be somewhat more realistic, then with $(z_{max}\varepsilon_q)_{min} \sim 3.7$ (see equation 5.13), $j \sim 0.18$, and:

$$\bar{A}_{24,\,max}\ \mathrm{mg\,O_2\,m}^{-2}\,\mathrm{d}^{-1} \sim 1.6\,\Delta P_{max}\varepsilon_s^{-1} \qquad (5.26)$$

A working group (Lunz), on the shallow lakes of IBP, found values of $\Delta P_{max,\,max}$ to be remarkably close to 280 mg O_2 (mg chl. $a)^{-1}\mathrm{d}^{-1}$, in association with values for ε_s close to 0.015. The apparent similarity of results for $\Delta P_{max,\,max}$ in different lakes seemed to be due to the higher P_{max} of tropical situations being offset by shorter daylengths, whereas in temperate areas the somewhat lower P_{max} values are enhanced by the longer daylengths of summer. These values imply:

$$\bar{A}_{24,\,max} \sim 30\,\mathrm{g\,O_2\,m}^{-2}\,\mathrm{d}^{-1} \quad \text{or} \quad 10\,\mathrm{g\,C\,m}^{-2}\,\mathrm{d}^{-1}$$

i.e. that photoassimilation could potentially reach $30\,\mathrm{g\,O_2\,m}^{-2}\,\mathrm{d}^{-1}$ in favourable conditions. However some higher values have been measured (table 5.19, Talling *et al.*, 1973; Melack & Kilham, 1974).

The expression for B_{Max} (equation 5.24) also suggests that there is an ultimate maximum standing crop. Again choosing $(\Delta \cdot r^{-1}) = 240$ and $(z_{mix}\varepsilon_{q,\,min}) = 3.7$:

$$B_{Max,\,max}\ \mathrm{mg\,chl.}\,a\,\mathrm{m}^{-2} \sim 16.3\ (\varepsilon_{s,\,min})^{-1}$$

Since $\varepsilon_{s,\,min} \sim 0.008$, $B_{Max,\,max} \sim 2040$ mg chl. a m^{-2}

At present it appears that either very small ($< 5\mu m$) or very large algae ($> 40\,\mu m$) may have low values of ε_s, whereas algae intermediate in size are associated with values closer to 0.02 (Bindloss, 1974; see also p. 157) therefore for many algae lesser maxima are all that could be reasonably achieved. All the data in table 5.1 are well below this limit.

Whilst predictions of the upper limits of photoassimilation and biomass are of some interest, it is also pertinent to inquire whether or not they are likely to be achieved in practice. It may be, for instance, that the larger biomasses would only be attainable if the conditions for growth remain stable for some reasonably long periods of time. In such circumstances it may be that low rates of net photosynthesis could not be sustained long enough to produce high biomasses in relatively unstable temperate climates, but could allow considerable biomass to accumulate when maintained under tropical conditions.

Equation 5.23 implies that as the biomass increases from very low levels, toward the maximum, net photosynthesis would be low initially, rise toward a maximum and then reduce. There would therefore be some biomass (B_{opt}) for which the net photosynthesis is maximal. The expression which results from the required manipulations is non-algebraic, but some approximation may be attempted (Steel, 1978).
Then:

$$\bar{N}_{24,\,max}\, mg\, O_2\, m^{-2} d^{-1} \sim 2\Delta P_{max}\varepsilon_s^{-1}\quad (1 - j^{1/2})^2 \qquad (5.27)$$

and

$$B_{opt}\, mg\, chl.\, a\, m^{-2} \sim \Delta(12r\varepsilon_s)^{-1}\quad (j^{1/2} - j) \qquad (5.28)$$

Using the values of the variables as for previous estimates of B_{Max} and $\bar{A}_{24,\,max}$:

$$(\bar{N}_{24,\,max})_{max} \sim 12\, g\, O_2\, m^{-2} d^{-1} \quad or \quad 4\, g\, C\, m^{-2} d^{-1}$$

By suitable manipulation it is possible to suggest:

$$\bar{N}_{24,\,max}(\bar{A}_{24,\,max})^{-1} \sim (1 - j^{1/2})(1 + j^{1/2}) \qquad (5.29)$$

and:

$$B_{opt}(B_{Max})^{-1} \sim j^{1/2}(1 + j^{1/2}) \qquad (5.30)$$

$\bar{N}_{24,\,max}$ may be of some aid in assessing the largest potential daily nutrient demand on the environment.

The 'assimilation efficiency' (Γ) of the phytoplankton may be defined as the ratio between the net photosynthesis and photoassimilation:

$$\Gamma = N \cdot A^{-1} \quad (N\ and\ A\ usually\ in\ g\, O_2\, m^{-2} d^{-1})$$
$$= 1 - R_d A^{-1}$$

By substituting for R_d and A from equations 5.19 and 5.22, and by using the definitions of r and j from p. 222–3, it can be shown that

$$\Gamma \sim 1 - j - 12r\varepsilon_s b\Delta^{-1}$$

Thus $\Gamma_{max} \sim 1 - j$, when $b \sim 0$, but the mean conversion of photoassimilation into net photosynthesis will be reduced by a factor dependent upon the magnitudes and course of the population production. It would seem that high physiological and environmental advantage and relatively small biomasses would be needed for Γ to be consistently close to 0.75–0.80.

Within the general assumptions used here, other system properties, previously explored by Talling (1971), may be inferred. Thus $B^*_{max} \sim 3.7 \cdot \varepsilon_s^{-1}$ which implies $B_{max} \sim 400$ mg chl. $a\,m^{-2}$, for algae with low observed values of ε_s (see section 5.3.4 and fig. 5.2). Talling has also indicated an interesting relationship between the mixed and euphotic depths:

$$z_{mix}(z_{eu})^{-1} \sim z_{mix}\varepsilon_{min}(3.7)^{-1}$$

$$\sim (z_{mix}\varepsilon_q + \varepsilon_s B)(3.7)^{-1} \tag{5.31}$$

Thus:

$$[z_{mix}(z_{eu})^{-1}]_{max} \sim (z_{mix}\varepsilon_q + \varepsilon_s B_{max})(3.7)^{-1}$$

$$\sim \Delta(45r)^{-1} \tag{5.32}$$

Hence the mixed depth should be no greater than $\Delta(45r)^{-1}$ times the euphotic depth if any net photosynthesis is to occur. This is discussed in detail in Vollenweider (1970). In transparent water this limiting depth is much greater than the normal mixed depths, but in less transparent waters it can easily be less than the mixed depth.

Although an initial simplification was that the mixed depth was never less than the euphotic depth, it is clear that this may frequently not be the case. Steel (1978) attempted some simple analysis of the system as $z_{mix}\varepsilon_q$ approaches zero. In these situations the predicted B_{Max}, for example, is still a finite quantity; however if $z_{mix}\varepsilon_q$ approaches zero because z_{mix} approaches zero this implies near infinite phytoplankton concentrations, for: $b_{max} = B_{Max}z^{-1}$. This corresponds in practice to the very high biomasses reported in very shallow, stirred mass cultures (Tamiya, 1957; Šetlik *et al.*, 1970). For ease, let $z_{mix}\varepsilon_q$ remain sufficiently great to ensure an optically deep column, but let this be conferred by large ε_q, so that z_{mix} may be considered as small (say, $z_{mix} < 5$ m). It will be clear that in such situations responses not previously considered would have to be taken account of. Although the response would no doubt be by virtue of a complex of factors, here only observed relationships between P_{max} and population density will be considered (Bindloss *et al.*, 1972; Ganf, 1972; see also section 5.4.2.6). Let the effect on P_{max} be similar to that observed in Lake George, Uganda (Ganf, 1972):

$$P_{max}\,mg\,O_2\,mg\,chl.a\,h^{-1} \sim \hat{P}e^{-kb}$$

$$\sim 25\,e^{-2.3 \times 10^{-3}b} \tag{5.33}$$

In a similar situation, when mixing is throughout the depth:

$$\bar{N}_{24}\,mg\,O_2\,m^{-2}\,h^{-1} \sim [2\Delta B\hat{P}e^{-10^{-3}B}(z\varepsilon_q + \varepsilon_s B)^{-1}] - 24\,C_d B$$

$$\sim 2\Delta B\hat{P}[e^{-10^{-3}B}(z\varepsilon_q + \varepsilon_s B)^{-1} - 12r\Delta^{-1}] \tag{5.34}$$

If $z_{mix}\varepsilon_q$ is taken as ~ 5; $\varepsilon_s \sim 0.018$ and $\Delta = 12$

$$\bar{N}_{24}\,mg\,O_2\,m^{-2}\,h^{-1} \sim 2\Delta B\hat{P}[e^{-10^{-3}B}(5 + 0.018B)^{-1} - r] \tag{5.35}$$

* Note B_{max} is the maximum in the euphotic zone, not in the column (B_{Max}).

Then the maximum self-supporting areal biomass would be:

$$B_{Max} = 180 \text{ mg chl. } a^{-2} \quad \text{when} \quad r = 0.10 \quad (278)$$
$$= 450 \qquad\qquad\qquad\qquad r = 0.05 \quad (833)$$
$$= 550 \qquad\qquad\qquad\qquad r = 0.04 \quad (1111)$$

Also indicated, in parentheses, are estimates derived from the expression for B_{Max} as used earlier (equation 5.24). The population-density-dependent formulation also suggests a defined maximum photoassimilation of about $13 \text{ g O}_2 \text{ m}^{-2} \text{d}^{-1}$. The potential maximum net photosynthesis associated with the lower respiratory rates would be very slightly greater than $5 \text{ g O}_2 \text{ m}^{-2} \text{d}^{-1}$, with $B_{opt} \sim 150-200 \text{ mg chl.} a \text{ m}^{-2}$.

The climatic stability of a tropical situation would seem to increase the chances of populations reaching close to their potential maxima, although at the higher concentrations the net photosynthetic rates would be considerably reduced. In the situations in which the populations exhibit density-dependent effects, the specific net photosynthesis $(N \cdot B^{-1})$ responds much more rapidly to changes in biomass than when there is no such dependency. Thus if any process tends to decrease the standing crop then the specific net photosynthesis will increase and act in opposition to that stress. Where there are density-dependent effects compensation for the stress will be achieved when the population has been reduced by a much smaller amount than would be the case in its absence. Similarly, should the biomass be increased by some process other than net photosynthesis, then the stress on the enhanced biomass will build up much more rapidly with population dependency than in the non-dependent case, hence reducing the apparent effect of the process and tending to return the population very rapidly to self-supporting levels once the process ceases. It would seem, therefore, that population-density-dependent responses, whilst reducing the magnitudes attainable, could well endow the population with a greater stability, in the sense of severely restricting the range of the observable biomasses. It appears furthermore, that climatic stability would be an important element in allowing the manifestation of this type of response.

In the preceding paragraphs the philosophy has been essentially one of maximisation, and so little or no consideration has been given to nutrient limitation, density-dependent physiological responses, grazing and the other energy fluxes contained within the ecosystem. Little would be gained by attempting to explore them fully within the limits of the section. However, if these 'energy-limited' maximum rates are to occur, sufficient nutrients must be available. In 'nutrient-limited' situations the effects of nutrient concentrations may conveniently be treated in terms of depressions of rates relative to the energy-limited maxima (Steel, 1978). Results and concepts derived from one type of extreme situation may be quite inappropriate for the other, and there will be a spectrum of situations, difficult to define, between the extremes.

It is hoped that this selection has been useful in postulating some of the interactions within a photosynthetic system, and by inference therefrom the order of some of the rates and yields that might be attained. They are not offered as requirements of systems, but as possibilities against which systems might be judged.

5.9.3. Macrophyte models (Adams & Westlake)

Modelling of macrophyte systems is less advanced than that of phytoplankton systems because biomass and photosynthetic capacity cannot be assumed to be distributed evenly vertically, which are basic assumptions of most phytoplankton models, and not enough is known about the vertical distributions to derive mathematical formulations.

An early attempt to overcome the problem of biomass distribution used graphical techniques (Westlake, 1966b). The primary data were plots of observed values of light transmission (I_z) and biomass concentration (b) against depth (z), specific net oxygen exchange (N_s) against irradiance and surface irradiance (I'_0) against time. These enabled the net oxygen exchange for unit volume (N_z) to be found for any depth and surface irradiance and a set of depth profiles of photosynthesis corresponding to a range of surface intensities was plotted. The integrals under each curve were measured to plot net oxygen exchange per unit area $(\bar{N})$ against irradiance. This graph could be used with graphs of irradiance against time for a range of days to plot a set of graphs of net oxygen exchange $(\bar{N})$ against time and each of these could be integrated to give values of daily net oxygen exchange $(\bar{N}_{24})$.

Thus it could be demonstrated that net oxygen exchange in this particular situation should always be light-limited and that on very dull days respiration should always exceed photosynthesis. Further developments allowed the transmission at any depth in a weed bed to be calculated from the attenuation coefficients of the water and plant and biomass concentrations. Using the appropriate primary data for a stream, for which the relation between hourly oxygen exchange and irradiance was known from measurements, gave a similar curve and similar absolute values. Some effects of changing some of the variables were demonstrated starting from the primary data for a population of *Potamogeton pectinatus* in a polluted stream. For example, the vertical distribution of the biomass and the shape of the river bed had marked effects on the net oxygen exchange, removal of suspended material from the effluent could nearly double the net oxygen output on an average day, and sun-plants could only survive in this river if their biomass maximum was within 0.2 m of the surface. It was also possible to show that an optimum biomass could be predicted which was dependent (among other factors) on the horizontal distribution of biomass.

Ikusima (1967, 1970) used a similar relation to Talling (1957b) for photosynthesis–irradiance modelling and assumed sine-curve functions for irradiance against time to obtain a relation for the daily specific photo-assimilation per unit area ($\bar{P}_{24}$), when the biomass was evenly distributed, somewhat analogous to equation 5.18. He dealt with the effects of biomass and capacity distributions by dividing the column into finite layers, assuming that the concentration and physiological behaviour was constant within each layer, calculating for each layer separately by substituting the appropriate values and summing the results. This model predicted that daily integral photoassimilation ($\bar{P}_{24}$) was proportional to daylength and inversely pro-portional to the total attenuation coefficient in the community (ε_{min}). Higher rates were predicted in communities with the biomass concentrated near the surface and the maximum rate of photoassimilation expected for *Elodea nuttallii* in Lake Biwa in summer was about 30 g O_2 m^{-2} d^{-1} with about 500 g dry wt m^{-2} of plant.

Owens *et al.* (1969) developed a model primarily to predict the changes in oxygen concentration produced in rivers by the photosynthesis and res-piration of plants. They started from field observations of community photoassimilation and incident irradiance obtained by the upstream–downstream method and derived an empirical relationship. Biomass had relatively little effect on photoassimilation once adequate cover was estab-lished, and exerted its effect by increasing respiration and thus decreasing net photosynthesis. Respiration was measured in the laboratory and expressed as a function of plant biomass and oxygen concentration. High net photo-synthesis was predicted under high irradiance with low biomass. When biomass was high, very low minimum oxygen concentrations were predicted during dull weather.

A much more extensive model covering macrophyte photosynthesis and growth has been developed as part of the Eastern Deciduous Forest Biome Project, US-IBP (Titus *et al.*, 1975), based on field and laboratory physiologi-cal experiments and observations. As does Ikusima's model, it evaluates the physiological processes in depth classes throughout the water column, but daylength and temporal changes of irradiance and temperature within the day are not included except in so far as they contribute to the daily total incident irradiance ($I_{0, 24}$) and the mean temperature of the water column ($\bar{T}$) which are the primary forcing variables. Both act on a maximum photosynthetic capacity ($P_{max, max}$) while only temperature acts on specific respiration by the leaves, stems and roots (C_l, C_s, C_r). Each twenty-four hours is simulated by a twelve-hour daylight run and a twelve-hour night run.

The attenuation of irradiance (I_z) by the water with its phytoplankton and by the macrophytes themselves, and the variation of leaf biomass con-centration ($b_{l, z}$) and physiological activity ($P_z, I_{k, z}$) with depth, temperature and temperature history are incorporated into a depth-integrated value for the rate of input of photosynthetic products by the leaves ($\bar{N}_l$). The balance between this input, excretion, non-leaf respiratory losses and the demands of growth produces a labile carbohydrate pool which is treated as an intermediate compartment that controls plant growth. This model differs from the others, which treat the products of net photosynthesis directly as growth.

Growth rates ($\bar{g}$) are derived by summing sets of equations for the growth of the leaves and stems in each depth class, and for the growth of the roots. These are based on observed maximum growth rates which are subject to certain constraints. In all cases the labile carbohydrate pool must exceed a minimum level for growth to occur and then growth is related hyperbolically to the state of this pool. It is assumed that the plant can optimise the input of photosynthetic products by responses which approach the optimum leaf area in all depth classes, giving rise to the observed changes in the vertical distribution of biomass. Leaf growth proceeds asymptotically to the optimum leaf area, which is the surface area of leaves (per square metre of ground) which results in the maximum input of photosynthetic products. Greater leaf surface results in a reduced input because increased self-shading means there is little effect on photoassimilation, while respiration continues to increase with biomass. If continued growth in shallow depth classes (near the lake surface) reduces the optimum leaf area in deeper depth classes, sloughing losses of leaves occur on lower portions of the stem. Stem and root growth are similarly affected in the model by the carbohydrate pool, and cannot exceed the maximum observed in the field.

Seasonal changes in rate of growth ($\bar{g}$) and of change in biomass ($\bar{w}$) are mediated by seasonal changes in irradiance and temperature and by post-flowering sloughing losses of both stems and leaves.

The overall state of the model is subject to numerous internal feed-backs and self-controlling mechanisms. An increase in the labile carbohydrate pool increases growth, hence leaf, stem and root biomass increase and therefore photosynthesis and respiration of the total plant increase. All these effects will tend to change the carbohydrate pool, and the processes tending to reduce leaf growth and biomass may be brought into operation by the increase in leaf area.

Testing of this model has been carried out by taking parameters from field and laboratory data collected in 1970 and using them to obtain predictions for comparisons with field data for 1972. General agreement was obtained for vertical and seasonal distributions of biomass and photosynthesis. Areas for future research, possible relations between plant form and environment and some points of relevance to weed management were highlighted.

5.9.4. Production pathways in relation to nutrient loading (Walters)

The simple compartment model outlined in fig. 5.12 and table 5.18 can be used

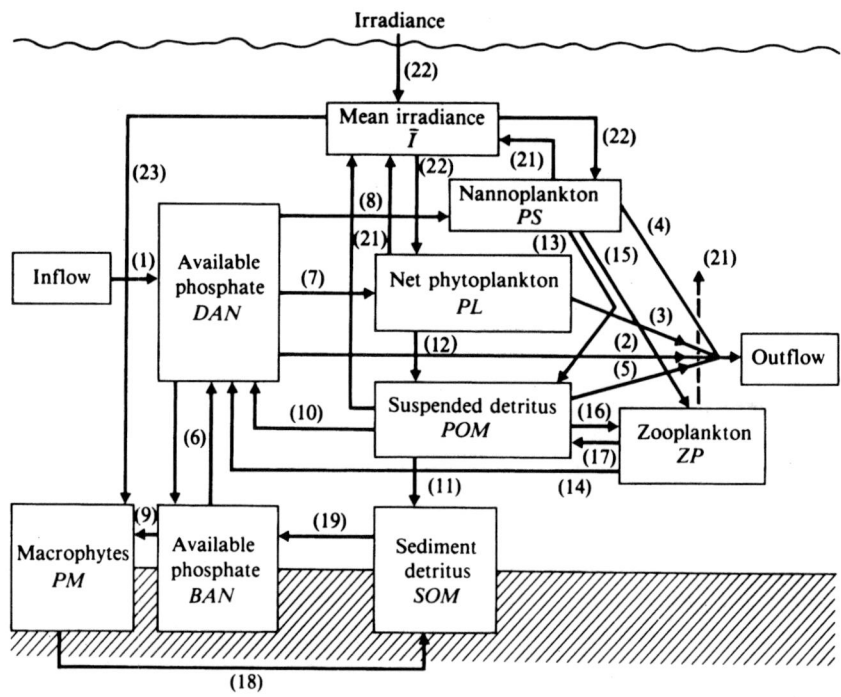

Fig. 5.12. Components and pathways used in a simple model to predict production patterns as a function of water turnover time and phosphate concentration in inflow water. (The symbols are those used in the equations of the model, which are numbered against the corresponding lines and are defined in table 5.18.)

to draw together and explore the overall consequences of some of the production relationships discussed in this book. This model is essentially a simplified version of the general biomass model discussed in Chapter 10, with the following basic assumptions:

(1) a homogeneous, vertically mixed water column with no lateral effects from shallower or deeper water;

(2) biomass transfer rates, converted to phosphorus units, are predictably related to biomass levels when those levels are near the growing season average;

(3) biomass equilibria evaluated without consideration of seasonal environmental change are representative of average conditions across several growing seasons;

(4) plant growth is affected only by light penetration and phosphate concentration; very low free phosphate concentrations are necessary before primary production would decrease significantly;

(5) algae are rapidly mixed (every few hours) through the water column, there is no inhibition, but saturation of photosynthetic rate occurs at high irradiances:

(6) biomass pools not considered explicitly (fig. 5.12) exert constant relative effects (e.g. per cent daily decomposition rate is constant) on the overall biomass levels of the pools that are considered;

(7) the macrophytes are sward species (p. 151), or are controlled early in their growth cycle.

Close inspection of the equations used (table 5.18) will reveal other relatively minor assumptions that are not likely to affect predictions about equilibrium conditions. The model as outlined would definitely not be appropriate for examining seasonal biomass changes; thus only predictions about average conditions (see assumption 3 above) will be considered below. At least five major nutrient pathways are implicit in the model equations (fig. 5.13). All transfer pathways leading to plants are also affected in the model by light penetration, which in turn is assumed to be affected by the plants.

Values for transfer rates in the various pathways were estimated from data scattered throughout this book, though considerable guesswork was necessary. Test simulations showed the general (equilibrium) patterns predicted by the model to be especially sensitive to estimates of the rate of diffusion of phosphate from the sediment to the water and to the slope of the line relating the light attenuation coefficient to algal concentrations: these two parameters are critical in determining the relationship between benthic and planktonic primary production. Equilibrium predictions of the model are relatively insensitive to parameters related to individual biomass pools (e.g. growth rates, metabolic rates), provided that the same ranking of turnover times is maintained between pools (nannoplankton turnover faster than net phytoplankton or zooplankton turnover, macrophyte turnover slowest).

As one would expect intuitively, the model predicts that water exchange rates and phosphate concentrations in the inflow should be major

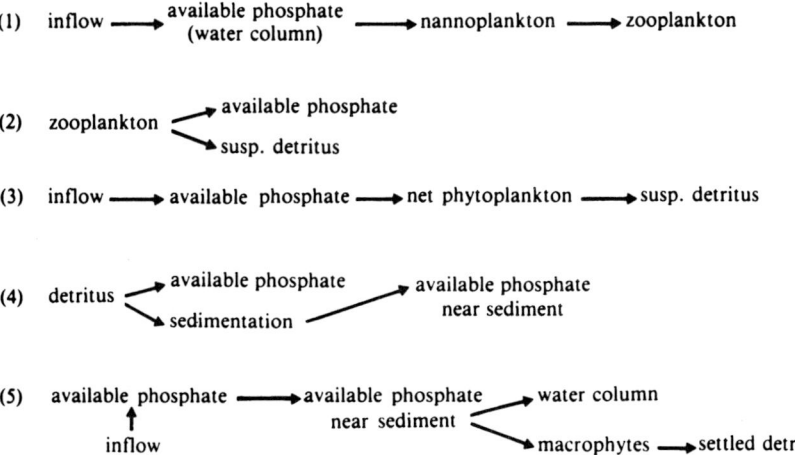

Fig. 5.13. Major pathways of phosphorus in model lake system.

Table 5.18 *Definitions and rate transfer equations for model of production pathways and nutrient loading (numbers in brackets refer to transfer rate equations)*

	Symbol	Definition
Forcing variables		
water exchange rate	γ	held constant over time at selected value
inflow PO_4^{3-} conc.	DIN	held constant over time at selected value
water depth	Z	held constant over time at selected value
State variables		
(21) light attenuation coefficient	ε	$= 1.5 + 0.055\,(PL + PS + POM + ZP)$
(22) mean irradiance in water column	$\bar{I}$	$= (1 - e^{-\varepsilon z})(\varepsilon Z)^{-1}$
(23) mean irradiance at bottom	I_z	$= e^{-\varepsilon z}$
available PO_4^{3-} in column	DAN	$dDAN/dt = (1) + (10) + (14) - (7) - (8) - (6) - (2)$
available PO_4^{3-} near sediment	BAN	$dBAN/dt = (6) + (19) - (9)$
suspended particulate organics	POM	$dPOM/dt = (12) + (13) + (17) + (20) - (10) - (16) - (5) - (11)$
sediment particulate organics	SOM	$dSOM/dt = (11) + (18) - (19)$
nannophytoplankton	PS	$dPS/dt = (8) - (13) - (15) - (4)$
net phytoplankton	PL	$dPL/dt = (7) - (3) - (12)$
macrophytes	PM	$dPM/dt = (9) - (18)$
zooplankton	ZP	$dZP/dt = (15) + (16) - (14) - (17)$
Transfer rates		
(1) PO_4^{3-} import rate $= \gamma \cdot DIN$		(12) net phytoplankton death rate $= 0.07 \cdot PL$
(2) PO_4^{3-} export rate $= \gamma \cdot DAN$		(13) nannoplankton death rate $= 0.015 \cdot PS$
(3) net phytoplankton export rate $= \gamma \cdot PL$		(14) zooplankton respiration and excretion rate $= 0.15 \cdot ZP$
(4) nannoplankton export rate $= \gamma \cdot PS$		(15) zooplankton feeding rate on nannoplankton $= 0.05 \cdot PS \cdot ZP$
(5) detritus export rate $= \gamma \cdot POM$		(16) zooplankton feeding rate on detritus $= 0.05 \cdot POM \cdot ZP$
(6) nutrient vertical diffusion rate $= 0.3 \cdot (DAN\text{-}BAN)$		(17) zooplankton death and defecation rate $= 0.5 \cdot (15) + 0.9 \cdot (16)$
(7) net phytoplankton nutrient uptake rate $= 0.0125 \cdot DAN \cdot PL \cdot \bar{I}/(0.2 + \bar{I})$		$\qquad + 0.01 \cdot ZP$
(8) nannoplankton nutrient uptake rate $= 0.05 \cdot DAN \cdot PS \cdot \bar{I}/(0.15 + \bar{I})$		(18) macrophyte death rate $= 0.005 \cdot PM$
(9) macrophyte nutrient uptake rate $= 0.008 \cdot BAN \cdot PM \cdot I_z/(0.03 + I_z)$		(19) sediment detritus decomposition rate $= 0.1 \cdot SOM$
(10) suspended detritus decomposition rate $= 0.1 \cdot POM$		(20) organic detritus input rate $= 0.0$ for comparisons
(11) detritus sedimentation rate $= 0.02 \cdot POM$		

Predictions are made for the equilibrium condition when all inputs are in balance with their outputs (sum equals zero).

determinants of primary production. To test the consequences of different nutrient input rates, the model was simply solved many times for equilibrium biomass values, using different input rates each time. The results of this exploration are presented in terms of production isopleth diagrams in fig. 5.14. These diagrams can be interpreted in the same way as contour maps, but with production criteria as the 'altitudes' and with water exchange time and phosphate input concentration as the horizontal dimensions. This method of presentation allows quick comparison of many alternative water management strategies.

Several basic predictions can be drawn from the model outputs presented in fig. 5.14 and these predictions are in general agreement with the limnological literature:

(1) (Fig. 5.14a) The depth of macrophyte penetration should be greater in

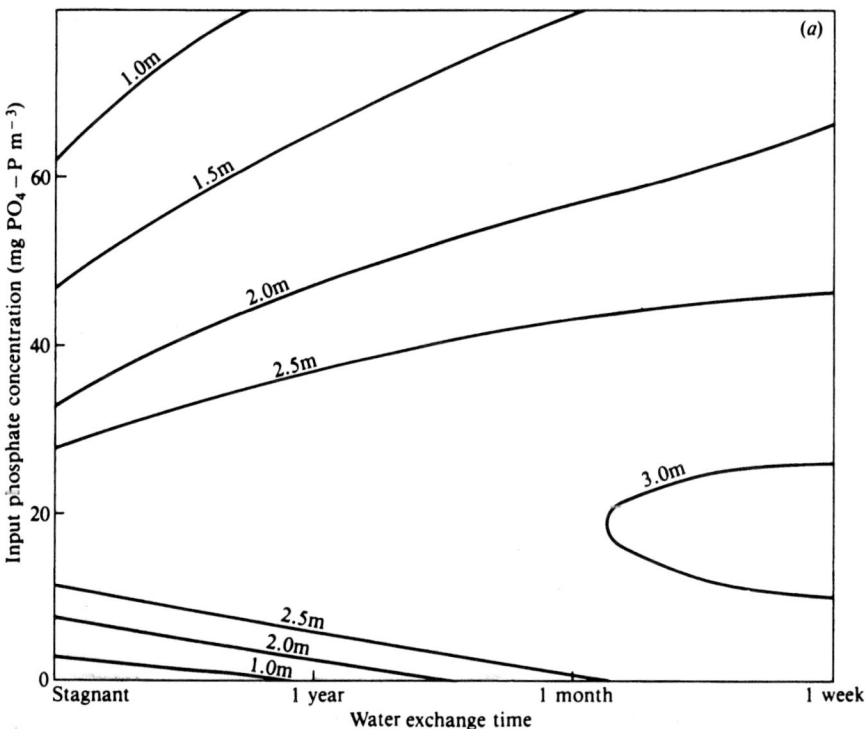

Fig. 5.14. Simulated behaviour of macrophytes, phytoplankton and zooplankton as a function of water turnover time and phosphate concentration in inflowing water. (a) Depth of macrophyte penetration (m). (b) Biomass of net phytoplankton (mg dry wt m^{-3}) (total shows similar relative values). (c) Zooplankton biomass (mg dry wt m^{-3}).
?—indicates that the model becomes unreliable because of instability or the presence of known biological effects not included in the model.

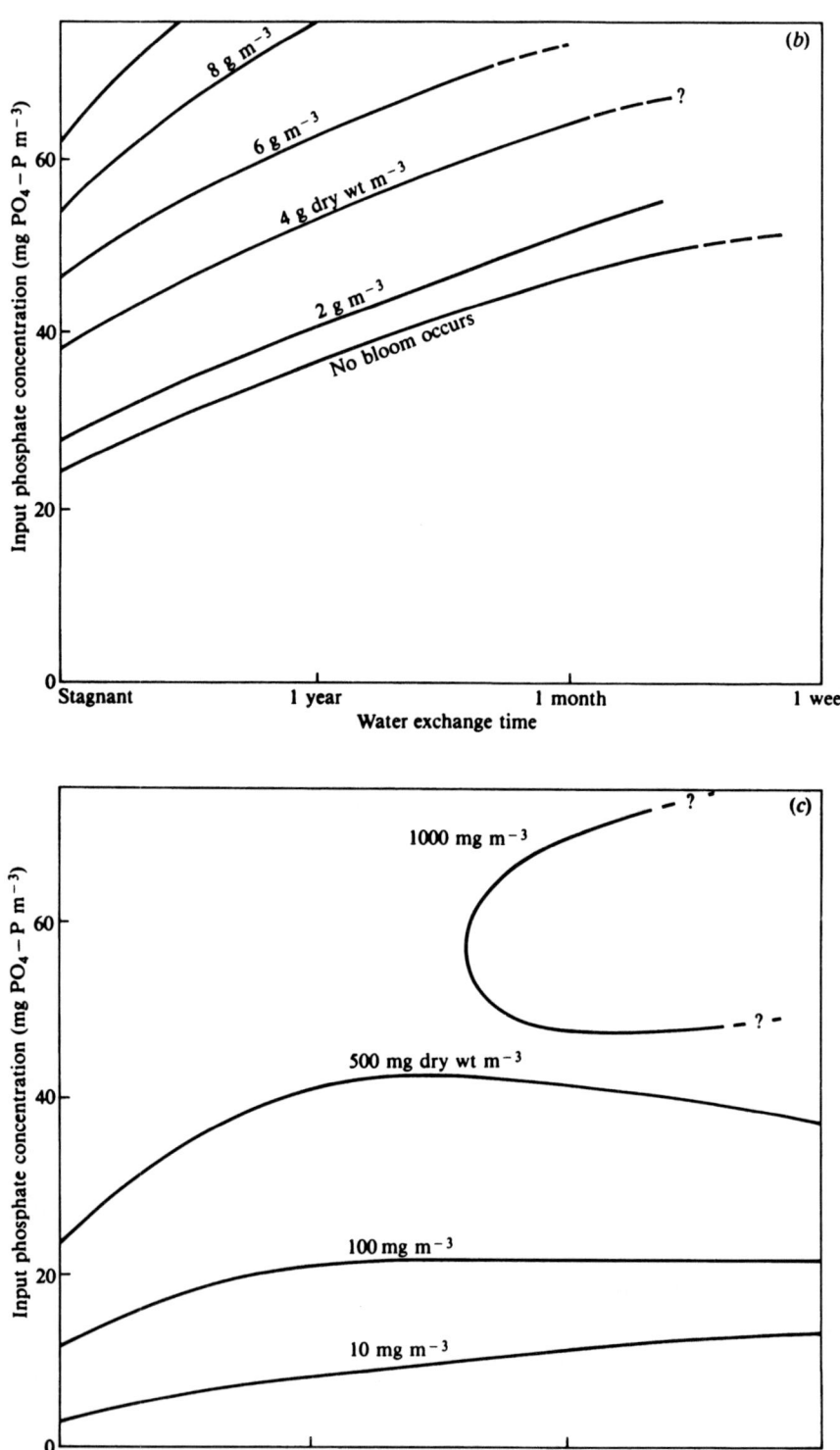

more productive lakes up to a point, but should decrease in lakes having very high nutrient input (due to shading by net phytoplankton blooms); more rapid water exchange should also favour macrophytes (cf. Spence, 1976).

(2) (Fig. 5.14*b*) Net phytoplankton blooms should only occur when phosphate input concentrations exceed some critical threshold; this threshold concentration should be lower in lakes with slow water exchange (high water exchange should tend to flush out the slower growing, large forms).

(3) (Fig. 5.14*b*) Water bodies with short retention times, such as short rivers, should not develop dense algal populations.

(4) (Fig. 5.14*c*) Low phosphate input levels should favour nanno-phytoplankton production expressed as the concomitant zooplankton standing crop.

(5) (Fig. 5.14*c*) At equilibrium, in the absence of strong seasonal disturbances, zooplankton production should not be high in lakes that have high total phosphorus content but turnover slowly. Some of these predictions will be discussed again in Chapter 10 in the context of seasonal environmental change; essentially the same general conclusions arise from more complex models.

Though this model is highly oversimplified in many respects and leaves out many of the subtle mechanisms thought to operate in aquatic systems, it appears able to predict the general productivity changes that are usually associated with eutrophication. Though the parameter estimates used to construct the isopleth diagrams are too crude to be taken very seriously, the general approach appears promising as a way to bring together basic biological information into a framework that could be useful for the aquatic resource manager.

5.10. Geographical variations (Hammer)

Geographical variation in phytoplankton photosynthesis has long been of interest to limnologists. During the decade before IBP most production studies were carried on in the temperate zone, mainly in Europe and North America. Very little research was done on this aspect in tropical and subtropical lakes. A few measurements were also carried out at high latitudes mainly in the Arctic. Most measurements of photosynthesis have been done in lakes at low altitudes.

Thienemann (1931) believed tropical lakes to be productive as well as eutrophic. He cited favourable conditions of a tropical environment including a full year's growing period and continuous warm water temperatures. Ruttner (1952) argued that the shorter tropical day mitigated against higher production as compared to long temperate days.

In this section differences in maximum photosynthetic capacity ($P_{max, max}$), maximum photoassimilation per unit volume ($A_{max, max}$), ranges of daily integral photoassimilation ($\bar{A}_{24}$) and annual production will be related to the latitude, altitude and trophic nature of IBP and other lakes.

5.10.1. P_{max} relationships

Photosynthetic capacity (P_{max}, light saturated rate of photoassimilation per unit biomass) may be used to compare the efficiency of different phytoplankton populations or species, particularly if measured under standard conditions. If measured on populations selected by the environment, under the current environmental conditions, comparisons also indicate the influence of limiting factors other than light.

5.10.1.1. P_{max} and latitude

The maximum photosynthetic capacity varied widely over the whole range of latitude from 1.8 (Hakojärvi, 61° N) to 116.8 (Red Rock Tarn, 38° S) $mg O_2 (mg chl. a)^{-1} h^{-1}$ (table 5.19) and no significant relationship between $P_{max, max}$ and latitude exists statistically. Low or high values are not restricted to particular latitudes. Some high values might be due to incomplete extraction of chlorophyll.

Manning & Juday (1941) found maximum rates varying from 5.8 to 13.7 $mg O_2 (mg chl. a)^{-1} h^{-1}$ for a series of Wisconsin lakes. Talling *et al.* (1973) report maximum values of 33.7 and 18 mg for Lakes Kilotes and Aranguadi respectively, while Talling (1965) found $P_{max, max}$ for Lake Victoria was 25. Schindler & Holmgren (1971) report P_{max} values up to 70* mg $O_2 (mg chl. a)^{-1}$ in the ELA lakes (Experimental Lakes Area, Northwestern Ontario, Canada).

5.10.1.2. Variations of P_{max} and altitude

There is no significant relationship between these parameters. Two of the lower values are attributable to high Vorderer Finstertaler See (8) and 4.7 for Lake Kinneret (Israel) below sea level.

5.10.1.3. Variations of P_{max} and trophic lake type

The oligotrophic lakes have $P_{max, max}$ values below 11 $mg O_2 (mg chl. a)^{-1} h^{-1}$ but eutrophic Lakes Humboldt (Canada) and Kinneret have values within this range and oligotrophic Lake Chester Morse has a value of 18.8. A so-called mesotrophic lake has one of the two highest $P_{max, max}$ values (Sammamish, USA; 77.7). On the other hand all the other eutrophic lakes have $P_{max, max}$ about 20 or higher.

---

* All values marked by an asterisk have been converted from carbon by a standard factor of three.

Talling (1966b) found a P_{max} value of 12.5 for the Windermere North Basin which might be considered oligotrophic. Clear Lake, Canada, had a value of 65.7* mg O_2 (mg chl. $a)^{-1}h^{-1}$ (Schindler & Nighswander, 1970), while the ELA lakes, Canada, had up to 70* (Schindler & Holmgren, 1971). These lakes were considered by the authors to be oligotrophic to mesotrophic. Hepher (1962a) found that the P_{max} value increased from 23.4 to 34.5 when fish-ponds were fertilised. The high P_{max} values in eutrophic African lakes (above) also support the conclusion that P_{max} values for eutrophic lakes generally exceed 20.

5.10.2. A_{max} relationships

The maximum rate of photoassimilation per unit volume per hour (A_{max}), determined from a series of photosynthesis–depth profiles within a lake, is an expression of the response to optimal irradiance and available nutrients at a particular time at a specific depth by the biomass present. The biomass is however an integration of factors affecting growth in the whole column over a longer period and so $A_{max, max}$ may be regarded as a measure of the capacity of a lake to produce and sustain algae. To permit selection of a meaningful $A_{max, max}$ value a sufficient number of profiles must be measured throughout the growing season and the phytoplankton must be evenly distributed in depth. Where rates of photosynthesis are high, short-term experiments will give a fair approximation to the instantaneous maximum. Where they are low, the longer-term experiments needed will lead to serious underestimates of $A_{max, max}$ (Vollenweider & Nauwerck, 1961; Hammer *et al.*, 1973). The maxima in table 5.19 were obtained from Data Reports with exposures (when cited) lasting from 2.7 to 48 h.

5.10.2.1. Variations of A_{max} with latitude

The lowest $A_{max, max}$ values reported (table 5.19) are 2.1, 7.1, 7.28 and 9.4 mg O_2 $m^{-3}h^{-1}$ for Lakes Char, Canada (75° N), Øvre Heimdalsvatn, Norway (61° N), Aleknagik, Alaska (59° N) and Chester Morse, USA (47° N) respectively. Five lakes in the temperate zone lower than 57° have maximum values less than 10 mg O_2 $m^{-3}h^{-1}$. All are deemed oligotrophic by their investigators except for Yunoko, Japan. Nutrient limitations thus prevail.

The highest $A_{max, max}$ value (table 5.19) is 18 616* mg O_2 $m^{-3}h^{-1}$ for Red Rock Tarn, Australia (38 °S). Other very high values are 12 000, 5800 and 2456* in Lakes Chad, Chad (14 °N), George, Uganda (0°) and Trummen, Sweden (57 °N) respectively. Three of the sites are in very high-radiation regions, but Lake Trummen is at a relatively high latitude. Other lakes with $A_{max, max}$ values over 500 are located near 50 °N in Canada, United Kingdom, United States and Czechoslovakia and in subtropical Australia and tropical Brazil.

The correlation coefficient for the relationship of $A_{max, max}$ to latitude is =

Table 5.19 *Primary productivity in water bodies studied during 1 BP[a] (in order of latitude N to S)*

Water body	DR	Years	Latitude	Altitude (m)	Trophic status	Maximum photosynthetic capacity $P_{max.max}$ (mg O_2 (mg chl. a)$^{-1}$ h^{-1})	Maximum rate of photo-assimilation $A_{max.max}$ mg O_2 m^{-3} h^{-1} (exposure, h)	Range of daily integral photo-assimilation $\bar{A}_{24}$ (mg O_2 m^{-2} d^{-1})	Annual net production (except where specified) (g C m^{-2} a^{-1})
Char. Canada[b]	79, Kalff & Welch (1974)	1969–72	74 42'N	30	oligotrophic	2	2.1 (5)	0–210	4.2
Meretta. Canada[b]		1969–72	74 42'	7	mesotrophic?	8	30 (5)	0–2400	11
Krivoye, USSR	70	1968–9	67 30'	7	oligotrophic	—	—	160–330	9.7
Krugloye, USSR	71	1968–9	67 30'	7	oligotrophic	—	—	18–130	3.6
Bolshoy Kharbey. USSR	72	1968–9	67 30'	200	oligotrophic	—	42 (48)	160–520	13
Krasnoye, USSR	62	1970–1	65 00'	40	eutrophic	—	68 (24)	170–3700	106
Chedenjärvi, USSR	58 & 64	1961–71	62 00'	151	oligotrophic?	—		–5100	107
Heimdalsvatn, Norway[b]	2	1971–2	61 25'	1090	oligotrophic	—	71 (10)	3.7–190	6.9
Hakojärvi, Finland[b]	39	1969	61 15'	146	dystrophic	1.2–1.8	4.8–18	30–230	5.1
Aleknagik, USA[b]	60	1962–70	59 20'	10	oligotrophic	9.1	7.3 (4)	220–620	13
Rybinsk, USSR[b]	69	1971–2	58 00'	102	mesotrophic	33	160 (16)	200–2800	—
Trummen, Sweden[b,c]	28	1969	56 52'	161	eutrophic	—	2700 (8)	370–11 000	260
		1972			mesotrophic	10	1300 (1.5)	1500–7300	180
Leven, Scotland[c]	3 Bindloss (1974)	1968–71	56 10'	107	eutrophic	20	1030 (3)	400–21 000	340–620[d]
Neagh, Ireland[e]	Jewson (1976)	1971–2	54 36'	15	eutrophic	14	1980 (3)	25–16 000	80[f]
Stechlinsee, DDR[b]	43	1970–1	53 09'	60	oligotrophic	—	17 (8.5)	72–2900	100–140
Burton, Canada[b]	16	1968–9	52 17'	530	mesotrophic	—	180 (4)	800–4200	—
Humboldt, Canada[b]	16	1968–9 1972	52 09'	531	eutrophic	8.5	1300 (3)	1000–40 000	680
Blackstrap, Canada[b]	16	1968–9	51 50'	535	eutrophic	—	340 (4)	20–19 000	—
Abbot's Pond, England	6	1968–9	51 28'	102	eutrophic	8.5	930 (3)	—	—
River Thames, England	80	1967–8	51 20'	ca 45	eutrophic	17	1000 (15)	20–15 000	460

Location		Year	Position		Trophic				
River Kennet, England	80	1967–8	51°20'	ca 45	eutrophic	7.9	100 (13)	110–1400	−28[f]
Echo, Canada[b]	15	1965–7	50°48'	479	eutrophic	—	720 (6)	1200–15 000	210–520
Pasqua, Canada[b]	15	1965–7	50°48'	479	eutrophic	—	1420 (12)	1700–27 000	450–460
Buffalo Pound[b], Canada	15	1967	50°40'	509	eutrophic	—	710 (7)	1100–10 000	210
Katepwa, Canada[b]	15	1966	50°40'	478	eutrophic	—	500 (6)	1200–16 000	560
Kiev Res., USSR	66	1967–70	50°30'	103	?	—	—	2600–10 000	—
Jezárko Pond, Czechoslovakia	10	1968	49°25'	460	eutrophic	21	1600 (6)	20–7000	100
Marion, Canada	Davies (1970)	1966	49°19'	305	oligotrophic	—	—	<21–36	120
Neusiedlersee, Austria[c]	50	1968–70	47°50'	115	mesotrophic	21	260	30–2200	41
Sammamish, USA[b]	11	1970, 71	47°36'	12	mesoeutrophic	78	110 (3.4)	120–3800	170–260
Chester Morse, USA[b]	11	1971	47°20'	473	oligotrophic	19	9.4 (4)	90–4200	47
Vorderer Finster-taler See, Austria[b,c]	1	1968–70	47°12'	2237	oligotrophic	8	120 (4)	3–840	23–31
Maggiore, Italy[b]	74	1970–3	45°57'	194	mesoeutrophic	19	115	700–5400	365
George, USA[b]	13	1970	43°31'	97	oligotrophic	—	93 (4)	75–680	16–28
Wingra, USA[b]	13	1970–1	43°05'	261	eutrophic	—	820 (4)	9–11 000	880
Port Bielh, France[c]	32	1967–8, 1970	42°53'	2285	oligotrophic	4.5–6.5	15 (4)	0.3–725	25
Yunoko, Japan	8	1968–9	36°43'	1478	eutrophic	39	98 (6)	400–3000	110
Biwa, Shiozu Bay, Japan	36, 37 & 46	1971–3	35°00'	85	oligotrophic	11	110 (10)	18–1170	60–90
Kinneret, Israel[b]	88	1970–3	32°53'	−209	eutrophic	4.7	760 (12)	2400–11 000	—
Ramgarh, India	57	1967–8	25°40'	95	eutrophic	—	—	59–630	56[f]
Siddhanath, India	9	1971	20°15'	23	eutrophic	—	—	380–16 000	130; 1500 (gross)
Chad, Chad[c]	4	1968–3	14°20'	282	eutrophic	33–38	900–12 000 (3)	2300–7200	—
George, Uganda[c]	77, 78	1967–8	0°00'	914	eutrophic	25	5800	10 000–15 000	290; 400 (gross)
Lago do Castanho, Brazil[c]	41	1967–8	3°00'S	30	eutrophic	47	600 (12)	−5700	—
Chilwa, Malawi	5	1970–1	15°30'	623	eutrophic	—	—	5400–13 000	730–1300

(contd.)

Table 5.19 (contd.)

Water body	DR	Years	Latitude	Altitude (m)	Trophic status	Maximum photosynthetic capacity $P_{max,max}$ (mg O_2 (mg chl. a)$^{-1}$ h^{-1})	Maximum rate of photo-assimilation $A_{max,max}$ mg O_2 m^{-3} h^{-1} (exposure, h)	Range of daily integral photo-assimilation $\bar{A}_{24}$ (mg O_2 m^{-2} d^{-1})	Annual net production (except where specified) (g C m^{-2} a^{-1})
Pink, Australia[b]	14	1969–70	38°02'	115	oligotrophic	11	57 (5)	14–55	24
Corangamite, Australia[b]	14	1969–70	38°05'	116	eutrophic	24	1100 (7.1)	560–10 000	760
Coragulac, Australia[b]	14	1969–70	38°05'	105	eutrophic	39	970 (5.6)	380–8800	350
Red Rock, Australia[b]	14	1969–70	38°05'	164	eutrophic	120	19 000 (2.7)	180–53 000	2200
Leake, Tas., Australia[b,c]	51	1970–1	42° app.	571	oligotrophic	—	—	21–180	9
Tooms, Tas., Australia[b,c]	51	1970–1	42° app.	464	oligotrophic	—	—	9–90	9
Sorell, Tas., Australia[b,c]	51	1970–1	42°10'	820	mesotrophic	—	—	62–290	17
Crescent, Tas., Australia[b,c]	51	1970–1	42°10'	820	eutrophic	—	—	100–750	45

[a] Much of the data presented had to be extracted from data report graphs and tables and suitable conversions made to equivalent values; rounded to two significant figures.
[b] Production determined by the carbon-14 method – values multiplied by three to obtain O_2 production (authors' committee decision).
[c] Additional data from investigators.
[d] Respiration 5% of light-saturated photosynthesis.
[e] Received too late for inclusion in regression.
[f] Clearly a community net value.

$-0.386(n = 36, t = 2.37)$, which is significant at the 95% level, but the scatter of points is very great so the relationship is probably not real.

Goldman, Mason & Hobbie (1967) report maximum A_{max} values of 220 and 510* for Algal and Skua Lakes (78 °S) in Antarctica. Rodhe (1958a, b) implies low A_{max} values for lakes of Swedish Lapland while Lake Erken had a photosynthetic rate as high as 112*. Talling (1966b) illustrates a value of 46 for the North Basin of Windermere, UK, while Vollenweider (1960) shows a maximum A_{max} of 15* for Lago Maggiore, Italy. High values have been reported by Talling *et al.* (1973) for Lakes Aranguadi and Kilotes in Ethiopia of 30 and 10 $g\,O_2\,m^{-3}\,h^{-1}$ respectively. The former is the highest value reported to date. Vollenweider's fig. 3 (1960) for Lake Mariut, Egypt, illustrates an A_{max} of 14* $g\,O_2\,m^{-3}\,h^{-1}$, a value very similar to that for Red Rock Tarn.

The literature and IBP results indicate that high latitude lakes, i.e. over 59°, have low $A_{max, max}$ values with few exceptions. Lakes with high $A_{max, max}$ values span the temperate and tropical regions. However, the highest rates measured were in lakes within 40° of the equator.

5.10.2.2. Variations of A_{max} with altitude

For the most part lakes investigated for IBP/PF are located at relatively low altitude, i.e. 200 m or less. However, some lakes, located either in mid-continent in North America, Europe and Africa, or in central Tasmania, lay between 200 and 1000 m elevation. Only four lakes were investigated at still higher altitudes.

The four high IBP lakes had relatively low $A_{max, max}$ values, between 7.1 for Øvre Heimdalsvatn to 117 $mg\,O_2\,m^{-3}\,h^{-1}$ for Vorderer Finstertaler See. Although the former lake has the lowest altitude it is located at the highest latitude. Lac de Port-Bielh (France), at the same altitude as Vorderer Finstertaler See but at a lower latitude, has a lower $A_{max, max}$ (15). Lake Yunoko has a lower altitude and latitude than Vorderer Finstertaler See but also a lower A_{max}. Fabris & Hammer (1975) found $A_{max, max}$ values of 280*, 130*, 68* and 123* $mg\,O_2\,m^{-3}\,d^{-1}$ in four small mountain lakes between 1615 and 2318 m in Banff National Park, Canada.

At lower altitudes the variations in $A_{max, max}$ cover a broad range of values. A correlation coefficient relating $A_{max, max}$ values for lakes at all altitudes was not significant.

Although high altitude lakes usually have low A_{max} values, Lakes Aranguadi and Kilotes at 2000 m (Talling *et al.*, 1973) have some of the highest values on record.

5.10.2.3. Variations of A_{max} with trophic lake type

Lakes designated as oligotrophic by their investigators or by extrapolation from the data make up 40% of the total (table 5.19). They have $A_{max, max}$ values which range from 2.1 to 117 mg O_2 m^{-3}h^{-1}. Dystrophic Hakojärvi also fits this range. However, lakes considered eutrophic by their investigators also have $A_{max, max}$ values in this range, i.e. Lake Krasnoye, Lake Yunoko.

The eutrophic lakes, omitting Krasnoye and Yunoko, have $A_{max, max}$ values ranging from 338 to 18 616 mg O_2 m^{-3}h^{-1}. Although the lakes (table 5.19) were more often eutrophic below 60° there is no indication that the lakes chosen represent the area from which they were selected. For example, Loch Leven does not typify Scottish lochs. These choices tend to bias general conclusions.

Sakamoto (1966) found that A_{max} could be increased from 1 to 10 to 1000 mg O_2 m^{-3}h^{-1} by increasing the phosphorus level in the water from 0.005 to 0.01 to 0.1 mg^{-1}. Similarly Hepher's (1962a) fertilised fish-ponds had A_{max} values 4–5 times as great as unfertilised fish-ponds. The A_{max} values for Lake Trummen were reduced to one-third by decreasing the nutrient level.

5.10.3. Ranges of daily integral photoassimilation

The daily ranges $(\bar{A}_{24}$; mg O_2 m^{-2}d$^{-1})$ in table 5.19 are statistically unrelated to latitude or altitude.

The maximum daily integral photoassimilation for a majority of the oligotrophic lakes is less than ca. 800 mg O_2 m^{-2}d^{-1}. However, six lakes considered oligotrophic have maxima between 2500 and 6000.

Eutrophic waters have very broad daily photosynthetic ranges with many minima well within the oligotrophic range. Most of them (84%) have maxima exceeding 5000 and extending to 53 000 (Red Rock Tarn). The exceptions are the River Kennet (1430), Lake Ramgarh (631), Lake Krasnoye (3740) and Lake Yunoko (3009).

Rodhe (1958a) reports Lapland lakes producing less than 300* mg O_2 m^{-2}d^{-1}. Schindler and Holmgren (1971) cite daily values ranging from 150* to 7700* for lakes on the Canadian Precambrian Shield and 540*– 3310* for lakes in the Experimental Lakes Area. Most of the maxima are less than 3000*. However, these maxima of Schindler and Holmgren for the unpolluted Pre-Cambrian Shield lakes of Canada, which vary from oligotropic to mesotrophic, are generally much higher than those in the IBP/PF results.

The highest maximum for daily integral photoassimilation reported is 57 g O_2 m^{-2}d^{-1} for Lake Aranguadi, Ethiopia (Talling et al., 1973), using diurnal oxygen changes. This lake and Red Rock Tarn are both soda lakes with abundant blue-green algae. Many other eutrophic lakes have high daily maxima: Lake Mariut, 32* (Aleem & Samaan, 1969); fertilised Israeli fishponds 40* (Hepher, 1962a); Lake Sylvan, 15* (Wetzel, 1966a, b); Clear Lake, 7.3* (Goldman & Wetzel, 1963); Lake Erken, 6.6* g O_2 m^{-2}d^{-1} (Rodhe, 1958a).

5.10.4. Annual net production relationships

Table 5.19 contains data given as net production by the authors ($\bar{N}_a$). Since there has been little general agreement on how to measure net production this may be based on unaltered carbon-14 values (see section 5.1) or daytime net oxygen values (both of which will not take account of overnight respiration), community net values (subtracting non-plant respiration) or various fairly arbitrary adjustments of the primary results. All values have been converted to carbon using Winberg's recommended factors (Winberg, 1971a).

5.10.4.1. Variations in annual net production with latitude and altitude.

The annual net production of phytoplankton was usually less than $15\,\mathrm{g\,C\,m^{-2}\,a^{-1}}$ above $59°$ latitude. Eight of the ten lakes were below 13 and averaged $8.27\,\mathrm{g\,cm^{-2}\,a^{-1}}$. Between $40°$ and $59°$ six of twenty-two lakes and one river have net production below 35. At still lower latitudes only a few annual net production values are available and only one of nine is below 35 (Pink Lake, an aberrant saline lake).

Approximately half the net production values greater than 100 $\mathrm{mg\,C\,m^{-2}\,a^{-1}}$ were distributed between 100 and 500 while the other half were above 500, up to 2200 for Red Rock Tarn ($38°$S). In addition to Red Rock Tarn, only Lake Chilwa ($16°$S) produced over $1000\,\mathrm{g\,C\,m^{-2}\,a^{-1}}$ net.

The annual net production of forty-seven lakes and latitude were not correlated significantly. In addition to correlating all values for the two parameters, oligotrophic lakes and eutrophic lakes and lakes where the carbon-14 method was used were dealt with separately, and values were averaged in $10°$ groups. No significant relations were found.

Brylinsky (Chapter 9) found a significant relationship between ^{14}C-fixation and latitude using forty-three lakes. Although different criteria of productivity were used there is no obvious reason to suspect a systematic difference between the two criteria. However the two sets of data had only twenty-eight lakes in common, for a variety of reasons. Brylinsky used a number of non-IBP lakes, there were differences in the availability of data to the authors partly because the chapters were written at different times, sometimes different revisions were used and it was often necessary to decide if the data could be used to give an annual value of the chosen criterion, which was a matter for individual judgement. Both sets of data had relatively few values at low and high latitudes. In very general terms the majority of lakes at high latitudes are likely to have relatively low productivities because mean temperatures, irradiance and nutrient supply are likely to be low, but there may be some lakes where a good nutrient supply enables the phytoplankon to exploit the long days in the growing season. At low latitudes, high mean temperatures and irradiance will allow sustained production where nutrient supply is good, but there will often be lakes with a poor nutrient supply, or at high altitudes, or with high non-plant colour or turbidity, which may have very low production. In between are a large number of temperate lakes with a wide range in productivity with altitude and trophic status. In these circumstances selection of relatively few lakes at extreme latitudes may have a large influence on regression analyses.

Previous research on lake production at high latitudes tends to support the conclusion that low production is the rule. Hobbie (1964) found that Lake Schrader (Alaska) produced $6.56-7.54$ g C m^{-2} a^{-1}, much of it under ice. Kalff (1967) found still lower rates in Alaskan ponds of $0.38-0.85$.

Annual production in temperate lakes has run from 104 g C m^{-2} a^{-1} in Lake Erken (Rodhe, 1958a) to 571 in Sylvan Lake (Wetzel, 1966b). Walker (1973) reports 435 in Lake Werowrap, Australia. In the subtropics Hepher (1962a) reported up to 780 but the fish-ponds were filled with water for part of the year only. Talling (1965) estimated gross production in Lake Victoria, East Africa (0° lat.) as 950 while temperate Windermere (North Basin), (54° 20′N) produced (net) a meagre 20.

Few lakes were studied at high altitudes. Annual net production values were available for only three of these, and they varied from 6.9 g C m^{-2} a^{-1} for Øvre Heimdalsvatn to $23-31$ for Vorderer Finstertaler See and 34 in Lac de Port-Bielh. The latter two are located within 4° of each other. They also have much lower production than other lakes at the same latitude. Øvre Heimdalsvatn (1090 m) has an annual production not much different from other lakes at low altitude ($7-200$ m) but at the same high latitude.

Little other work has been carried out at high altitudes regarding annual production. Fabris and Hammer (1975) found that the net production (carbon-14) for the ice-free period of four Banff National Park lakes in Canada varied through 48.5, 2.5, 6.9 and 5.5 g C m^{-2} a^{-1} as the altitude increased from 1615 m to 2316 m.

Lakes at high latitudes and high altitudes tend to have low annual net production. At low latitudes and altitudes the annual production rate varies tremendously from rates as low as those of high latitudes and altitudes to rates up to two orders of magnitude higher.

5.10.4.2. Variations in annual net production and trophic lake type

The set of lakes in table 5.19 fits Rodhe's (1969) scheme of separating lakes on the basis of production into trophic types remarkably well. According to his classification annual production of oligotrophic lakes is up to 25 g C m^{-2} a^{-1}. Eleven oligotrophic lakes fit this category, and Sorell (Australia) should probably be listed in this category as well, on the basis of low maximum daily production. Rodhe suggested that categories probably overlap to a certain extent, so that George (USA) at 28 and Lac de Port-Bielh at $34 \, g \, C \, m^{-2} a^{-1}$ properly fit this category. Lakes Biwa (90), Chedenjarvi (106), Chester Morse (47) and Stechlin (100–140) exceed the oligotrophic limits considerably. All of these lakes are deep and clear so that the limiting depth should always exceed the mixed depth (6.9). Stechlin is warmed by cooling water from a power plant so that high temperatures may promote recycling at a higher rate similar to tropical lakes.

Only one of four lakes classified as mesoeutrophic (Sammamish–170–260) exceeds Rodhe's 25–75 $m \, C \, m^{-2} a^{-1}$ range for this lake type. It is also a relatively deep lake.

Eutrophic lakes exceed $75 \, g \, C \, m^{-2}$ annual production in Rodhe's scheme. Omitting Crescent Lake, 89% of the lakes so characterised (table 5.19) fit Rodhe's eutrophic category. The River Kennet (-28) and Ramgarh Lake (56) fail this test. Presumably high community respiration reduces the 'net' production below real values for the phytoplankton.

Annual production of oligotrophic lakes reported in the literature varies from $1 \, g \, C \, m^{-2} a^{-1}$ for Alaskan ponds (Kalff, 1967) to about 7 for Lake Schrader, Alaska (Hobbie, 1964) to 12 for Lake Ogac, Baffin Is., Canada (McLaren, 1969) to 250 in Clear Lake, Ont., Canada (Schindler & Nighswander, 1970). These lakes fit Rodhe's limits except for Clear Lake. Like the oligotrophic IBP lakes above, this lake fits the general oligotrophic picture (Schindler & Nighswander, 1970), and is deep and clear.

The highest annual production reported is that for Mariut, Egypt, with a mean of $7.308 \, g \, C \, m^{-2} d^{-1}$ or $2667 \, g \, C \, m^{-2} a^{-1}$ (Aleem & Samaan, 1969). This is somewhat higher than Red Rock Tarn.

5.10.5. Conclusions

Maximum photosynthetic capacity ($P_{max, max}$) varies widely latitudinally and between oligotrophic and eutrophic lakes. The tendency is for mesotrophic and eutrophic lakes to exceed maxima of $20 \, mg \, O_2 \, (mg \, chl. \, a)^{-1} h^{-1}$ while oligotrophic lakes have lower rates. Exceptions occur in both directions. Although it has been thought that oligotrophic lake phytoplankton were more efficient photosynthesisers (per mg chlorophyll a) than those of eutrophic lakes this hypothesis does not appear to be borne out.

Maximum daily ranges of integral photoassimilation are unrelated to latitude or altitude statistically. Suitable ranges may be as high as $3 \, g \, O_2 \, m^{-2} d^{-1}$ for oligotrophic lakes and minimal maxima of $5 \, g \, O_2 \, m^{-1} d^{-1}$ for eutrophic lakes. Perhaps mean daily rates for the growing season (Rodhe, 1969) may be a better index.

Maximum photoassimilation ($A_{max, max}$) varies considerably latitudinally. Low rates are typical of high latitudes and high altitudes but also occur at low latitudes and altitudes. The highest rates were encompassed by a zone within $40°$ of the equator. Oligotrophic lakes have maximal A_{max} values up to *ca* 100 mg O_2 m^{-3} h^{-1} while eutrophic lake rates exceed 300 mg O_2 m^{-3} h^{-1} during the year.

No significant correlation was found between annual net production and latitude or altitude. Oligotrophic lakes tended to produce less than 25 g C m^{-2} a^{-1} while eutrophic lakes produced more than 75 g C m^{-2} a^{-1}, following Rodhe's (1969) scheme. Oligotrophic lakes tend to predominate at latitudes higher than $59°$ and altitudes above 1000 m but are not excluded from low latitudes or low altitudes.

Rather than latitude and altitude, a combined parameter expressing the climate of the locale would be more useful. Such factors as growing season, day-length, water temperature and light penetration could be included in this parameter. Nutrient concentration is essential in characterising water type. Lakes typifying these parameters could then be more precisely compared.

The parameters $A_{max, max}$, $P_{max, max}$, maximum daily areal rates and annual net production were all significantly interrelated. This indicates that they all tend to measure variations of a similar related parameter. All of these parameters are useful in arriving at relations with trophic lake type and perhaps also climatic type.

However, many measurements are required to arrive at maximal or annual rates. A_{max} is probably the best estimator of lake photosynthetic capability from only a few samples if periods of optimal photosynthesis can be selected on the basis of easily measurable parameters such as Secchi disk readings. Since $A_{max, max}$ is highly correlated with $P_{max, max}$, it may then be concluded, since P_{max} is related to chlorophyll *a* concentration, that maximum chlorophyll *a* concentration is probably the simplest predictive parameter.

6. Secondary production

Coordinator: N.C. MORGAN

6.1. Introduction (Morgan)

6.1.1. Taxonomic diversity

Under the heading of secondary production this chapter covers studies on the zooplankton, zoobenthos and fish. Within themselves these are very varied and highly organised groups covering a great diversity of life modes. Zooplankton, zoobenthos and fish are very distinct taxonomic as well as functional groups. The zooplankton are adapted to live a relatively passive existence distributed throughout the water mass feeding on small food particles and are generally themselves small organisms. They are poorly developed in running waters except where these flow out of a lake. The zoobenthos consists of larger invertebrates adapted to exist in a wide range of conditions at the bottom of both static and running water bodies on soft organic to hard rock substrates. Their mode of life varies greatly from burrowers in sediments and borers in plants to forms which make cases on firm substrates or are free living over the surface of the bottom or plants. They are generally more active than zooplankton and capable of directional movement and grow up to several centimetres long. Fish on the other hand are vertebrates and are functionally and socially more highly organised than the zooplankton and zoobenthos. They are widespread in both standing and running waters with adaptations to surface, mid-water and bottom dwelling. Many species may grow over a metre in length and some are capable of long migrations over hundreds of kilometres.

6.1.2. Trophic diversity

In reality one is dealing here with tertiary as well as secondary producers but because of the complex and variable network of relationships between species within the groups it is impractical to separate them into rigid trophic levels. Whereas some carnivores are always carnivorous in their feeding habits, because of their methods of food capture many species which have traditionally been considered as carnivores, herbivores or detritivores have been found to be facultative herbivores or carnivores. Thus a species which feeds on plant material in one lake may feed principally on animal material in another lake, or vary in its food intake from plant to animal material at different places

Contributors: N.C. Morgan, T. Backiel, G. Bretschko, A. Duncan, A. Hillbricht-Ilkowska, Z. Kajak, J.F. Kitchell, P. Larsson, C. Lévêque, A. Nauwerck, F. Schiemer & J.E. Thorpe.

within the one lake or between different times of year. Many examples of such problems have been revealed by the zooplankton, zoobenthos and fish feeding work.

The principle is shown in table 6.1. Adult cyclopoids have generally been considered as carnivores but some IBP studies show that small cyclopoids can be almost entirely herbivorous (e.g. *Thermocyclops hyalinus* in Lake George, Uganda). *Cyclops scutifer* is a facultative herbivore in Lakes Krivoye and Krugloye (Alimov & Winberg, 1972) and in Stugsjön and Hymenjaure (Persson, 1973). *Cyclops abyssorum tatricus* is mainly carnivorous in Vorderer Finstertaler See (P. Gollmann, personal communication) but is mainly herbivorous (*Cyclotella*) in the nearby Gossenköllesee (Eppacher, 1968). At Loch Leven *Cyclops strenuus abyssorum* has occasionally been found with the gut full of diatoms (A.E. Bailey-Watts, personal communication).

Among the larger calanoids Monakov (1968) found that *Heterocope saliens* consumed bacteria, algae and tree pollen when zooplankton was scarce but switched to zooplankton when this was available. In Lake Krasnoye, *H. appendiculata* is a predacious species whereas in Stugsjön it is a facultative herbivore.

Thus the predatory effect of some species which have traditionally been considered to be predators may have on occasion been overestimated.

On the other hand 'herbivorous' filter feeders such as *Daphnia magna*, *D. hyalina* and *D. pulex* may feed on fragments of detritus derived from benthic animals in London reservoirs (A. Duncan & D.E. Hurley, personal communication) where they normally feed on planktonic detritus.

Pronounced cannibalism is recorded from lakes where one species of *Cyclops* is the predominant zooplankter as in Loch Leven and Latnjajaure where *C. strenuus abyssorum* and *C. scutifer* adults, respectively, feed primarily on their own young and may also feed on each other at certain times of year. Such behaviour complicates the determination of predator/prey relationships and trophic efficiencies.

Amongst the zoobenthos the food of the 'carnivorous' Tanypodinae larvae is chiefly small chironomids, other invertebrates and algae; mostly large cells. Algae seem to be a more important and complete food for young instars but most species require animal food to complete the larval cycle. On the other hand *Psilotanypus imicola* cannot complete its larval development on animal food alone. Many 'herbivorous' species of the genera *Cricotopus*, *Psectrocladius*, *Endochironomus* and even *Chironomus* are facultative predators taking weak or damaged animal food to supplement their diet.

The diversity of feeding behaviour in fish is better known. Major changes during their life history occur in most species. The well-known piscivore, *Esox lucius*, feeds on zooplankton as a larva. In Lake George the 'herbivorous' *Tilapia nilotica* feeds as a fry on various floating and attached particles including detritus, rotifers and copepods and later becomes almost exclusively a phytoplankton feeder (Moriarty *et al.*, 1973). Up to being several centimetres long, silver carp, *Hypophthalmichthys molitrix*, feed on zooplankton after which they change to phytoplankton, changing over to bottom detritus when there is not enough phytoplankton. Grass carp, *Ctenopharyngodon idella*, which feed primarily on macrophytes, eat mainly animals when fry (Opuszynski, 1969). There are also major seasonal changes as in the chub, *Leuciscus cephalus*, which in the Vistula River takes mostly plant material, including large algae, in the summer and in the cold seasons becomes piscivorous (Backiel, 1971).

The size of secondary producers varies so greatly from newly hatched larvae to adults that it is perhaps not surprising that the diet may vary so much throughout the life history. Some species may remain in one trophic level throughout their lives although it is clear that this is not so for most. However with our limited knowledge of feeding, for many species we must at present treat all the secondary producers together, although in future it should be possible to consider herbivorous, omnivorous and carnivorous stages separately.

6.1.3. Location in the ecosystem

Organisms may change their position in the ecosystem during their life history. Thus the first two instars of *Chaoborus* are planktonic, whereas the third and fourth instars become increasingly benthic, only swimming into the water mass at night. *Mysis relicta* also makes diurnal movements between the sediments and open water and D. Lasenby (personal communication) in Stony Lake, Ontario, has recorded it feeding on *Daphnia*. *Hydra* is sometimes found in large numbers in the water mass (B. Davies, personal communication) but it is not known whether it feeds whilst being dispersed in this way. Several lamellibranchs, such as *Dreissena*, have planktonic larvae which periodically form an important component of the zooplankton. Chironomid larvae may also be abundant in the water mass, usually during the first instar but also when migrating in other instars. Such movements lead to problems both in determining their numbers and the proportion of their food taken from different sections of the ecosystem. They may also bring about interchange of nutrients by feeding and defecating in different regions. A classical example is sockeye salmon (*Oncorhynchus nerka*) which spawns in lakes, feeds and grows there for one or two years, then migrates to the ocean and later comes back to the lakes to spawn and to die bringing in organic matter produced in the other ecosystem (Ricker & Foerster, 1948).

6.1.4. Methodology

The accuracy of quantitative sampling of secondary producers is less advanced than for primary producers. This is related to the mobility of the animals and their ability to redistribute themselves in response to stimuli in an active manner. The zooplankton is probably the most uniformly distributed component of the secondary producers and, relatively, is easier to study than the benthos and fish. The bottom habitat may vary considerably from one square metre to the next and some substrates such as stones or beds of large macrophytes present serious sampling difficulties. On the other hand fish are highly mobile, except in the larval stages, and many species form shoals. It is therefore difficult to relate net catches to area. The basis for production measurements is the estimation of numbers, biomass and growth rates and the accuracy of the production calculation is dependent on the accuracy of these measurements. The choice of sampling technique and design of sampling procedure and consequent treatment of the material are extremely important in determining the accuracy of results. It is essential that the order of accuracy is determined if the production estimates are to be compared meaningfully with elsewhere. The IBP has focussed on this problem but by no means solved it. For many sites no confidence limits have been given for the data, which makes the comparison of geographical regions difficult. The IBP did, however, make the first concerted effort to measure biomass and production in different areas of the world, thus enabling comparisons to be made and consistent differences to be revealed. The IBP has brought about important advances in methodology and an awareness of the problems through both the IBP handbooks on assessment of fish production (Ricker, 1968) and secondary production (Edmondson & Winberg, 1971) and other related books on methodology such as Elliott (1971) and Winberg (1971b). The strategy of sampling has been studied in great detail in some IBP studies (Charles et al., 1974).

In general the accuracy of sampling population size declines from zooplankton to zoobenthos to fish. The size range of fish, say 1 mg to 20 kg ($1 : 20 \times 10^6$) is such that although various methods of sampling have been tried no single one can apply to the entire fish stock. On the other hand growth rates are most readily measured in certain fish from the temperate zones and the longer-lived larger invertebrates, and are difficult to measure in the smaller zooplankton and benthos. Even in well-mixed lakes the density of the zooplankton, zoobenthos and fish populations may vary horizontally and there may be movement of species from the deeper water to the littoral seasonally or at different life stages. Vertical distribution of organisms both in the water column and the mud is also important, e.g. where *Cyclops* is important it is necessary to sample carefully the bottom of the water column as the bulk of the population may be concentrated here. This is particularly important in oligotrophic lakes. Diurnal vertical movements may also influence population estimates. Also young and adults of some migratory fish live most of their life stages in different habitats.

As well as the problem of obtaining an adequate set of samples the zoobenthos also presents the difficult problem of separating the animals, particularly first and second instars, from the substrate material collected with the sample. In soft organic substrate tubificids and chironomids may penetrate down to 25 cm and very large quantities of organic material are collected compared with the volume of animals. Various techniques of subsampling and flotation are used to remove the animals but flotation methods are not satisfactory for all groups and straightforward sorting of the material by eye may have to be adopted if all components of the fauna are being measured. Whole lake studies of benthos therefore require big teams of people if accurate measurements are to be made and this has limited the number of lakes where whole lake studies have been attempted.

6.1.5. General

Thus the secondary producers consist of invertebrate and vertebrate components of the freshwater fauna and portray a complex of trophic relationships which may change during the life cycle of a species or from one site to another. It would therefore be surprising if a few general principles could be applied to the group as a whole. Thus although there has been an accumulation of a large amount of information on secondary production during IBP which has increased our understanding of certain freshwater systems we must be wary of drawing too broad-reaching conclusions until information is available for a wider range of water bodies. Nevertheless the data presented form a useful basis for formulating hypotheses for testing in other situations.

Zooplankton, zoobenthos and fish are dealt with in sequence in the following sections. The contents relate primarily to production studies and the factors affecting production. Wherever possible the text is illustrated by data from IBP studies but in order to be more comprehensive it has been necessary also to draw on literature from outside IBP. There has inevitably been selectivity with emphasis laid on certain groups such as Chironomidae, Salmonidae and carp. Because of their wide distribution and relative ease of collection the zooplankton has been most comprehensively covered.

In terms of ecosystems running waters have been very sparsely covered by secondary production studies although the few that have been covered are dealt with comprehensively.

Throughout the chapter measurements of biomass are as dry weight (dry wt) for zoobenthos and as wet weight (wet wt) for the zooplankton and fish.

6.2. Zooplankton (Nauwerck, Duncan, Hillbricht-Ilkowska, Larsson)

6.2.1. Introduction

This section is primarily concerned with variations in zooplankton biomass and composition as expressions of production under various conditions.

Absolute values of debatable significance will not be stressed in detail, but an attempt is made to interpret findings and to explain, with their help, the production mechanisms in freshwater zooplankton.

Besides IBP data reports and earlier IBP publications, unpublished material has been made available for use in this chapter by different authors. Only a minor part of this material can be presented here.

Unpublished contributions came from R.S. Anderson, G. Andersson, A.E. Bailey-Watts, H.H. Bottrell, Z. Brandl, R.H. Britton, M.J. Burgis, J. Darlington, M. Doohan, J.A. Eie, Z.M. Gliwicz, P. Gollmann, B. Grönberg, A. Herzig, J. Hrbáček, M.B. Ivanova, W. Lampert, T. Narita, R. Pechlaner, G. Persson, F. Rigler, M. Straškraba, M. Tevlin and T. Węgleńska.

The working base of this chapter is the forty-seven (out of a total of eighty-six) IBP/PF reports which include information on zooplankton. They have been supplemented with data from seven projects for which Data Reports were not provided, and from two projects not included in IBP but done at the same time and according to similar principles.

6.2.2. The lakes and aspects studied

Several IBP reports deal with more than one site. Altogether, there are data from ninety water bodies, most of them (sixty-one) in Europe and in European parts of the Soviet Union, nine in North America, seven in South America, six in Asia, five in Africa and two in Australia (Tasmania). Generally speaking, the majority of the lakes can be classified as mesotrophic–eutrophic. Oligotrophic lakes from northern cold regions or from high mountains are also well represented. Highly productive water bodies, tropical lakes and humic brown water lakes are less well represented.

Most projects provided taxonomic data, usually only for dominant species. Numbers per cubic metre or square metre for groups (copepods, cladocerans, rotifers) or important species are available from some sixty water bodies, but comparable curves of annual or seasonal growth are only available for little more than ten. Only four reports do more than mention the importance of protozoans.

Many reports give examples of seasonal variation in total numbers or single species. Seasonal variation of total biomass is available from fifteen water bodies. Estimation of total zooplankton production, annual or for the growth period, the proper core of IBP, has been done for thirty-one water bodies; nineteen in the Soviet Union, six in Poland and six in other regions. Brylinsky (Chapter 9) has treated a number of these lakes statistically.

Numerous reports touch on questions of population dynamics and feeding. Division into predators and non-predators is done in calculations of production in the Soviet and Polish material, while a few other reports briefly discuss the problem of such a concept. Several projects present long series of zooplankton counts and deal with long term fluctuations and successions. Some take up questions concerning predation, spatial distribution and sampling methods or other methods. In a few cases measurements of respiration and assimilation have been carried out. Very little attention has been paid to zooplankton in running waters, to the role of parasites, to competition, and to interrelationships between plankton and benthos or plankton and littoral communities.

6.2.3. Main faunal components

6.2.3.1. Taxonomic composition

The groups of zooplankton organisms considered are almost exclusively crustaceans and rotifers. It is obvious that crustaceans are usually responsible for the major part of zooplankton production. Rotifers sometimes surpass them in relative and absolute importance. *Chaoborus* and *Dreissena* larvae are occasionally reported to reach quantitatively comparable importance.

Nevertheless, the available data demonstrate that the biomass and production of protozoans at times equals that of other groups. The number of taxa of Protozoa in one lake is similar to that of the rotifers, with usually 5–10 predominant species.

Apart from the protozoans slightly more than 200 taxa of zooplankters are mentioned as dominant or prominent secondary producers in the range of waters investigated. Even if the number of possible finds is restricted by the geographical restriction of the material, this number is low. Where more complete species lists are given, they usually contain between 50 and 100 taxa per lake (the most extreme biotopes excluded), and well-studied European or Soviet lakes may well have more than 100 taxa, the majority of them commonly distributed in many water bodies. The figure of 200 important taxa is reduced further if taxonomic confusion is taken into account, and if closely related forms are considered as species *sensu lato* (e.g. the *Bosmina coregoni* group). A fair statement is, that around 100 taxa of zooplankters, mainly crustaceans, are responsible for the major part of zooplankton production in fresh waters. The most widespread important species throughout the world seems to be *Bosmina longirostris*.

Though the number of species in a lake is to a certain degree a function of the intensity of investigation and of the investigators taxonomic competency, there is little doubt that the well-studied Northern European–Baltic lake type is characterised by a high species diversity. This is true not only for the total number of producers found during the year in one lake but also for the number of important producers present together at one time. The zooplankton community in this type of lake is usually composed of the following dominants: 1–3 planktonic cyclopoids, (among them *Mesocyclops leuckarti* and usually also a *Cyclops* of the *strenuus* group), 1 diaptomid, (*Eudiaptomus gracilis* or *E. graciloides*, the former more frequently in more eutrophic waters, the latter in more oligotrophic ones), accompanied by *Eurytemora* (in warmer and more eutrophic waters) or *Heterocope* (in colder and oligotrophic waters) and 3–10 cladocerans. To these is added a number of rotifers which in individual numbers frequently, and in biomass and production occasionally, can surpass the crustaceans, particularly in late winter and spring. The same composition of zooplankton is found also in many Siberian lakes.

Similar population structures, that is one or two cyclopoids, one or a few calanoids, several cladocerans and a number of rotifers, characterise the zooplankton community of temperate lakes in North and South America (Pennak, 1957; Patalas, 1971), but substitute species, and in some cases different genera are found. The most frequent cyclopoid in North American plankton seems to be *Diacyclops bicuspidatus thomasi* whose European form only exceptionally appears in the pelagic zone of lakes (e.g. Lake of Constance and Vltava reservoirs). It seems to play the role of *Mesocyclops leuckarti* in Eurasian waters. An obvious difference compared with Old World lakes is that several *Diaptomus* species are commonly found together as important parts of the same zooplankton community. The cladocerans show similar composition as in northern Eurasia, but *Daphnia cristata* and *D. cucullata* are replaced by *D. retrocurva* and *D. galeata mendotae*, and *Diaphanosoma brachyurum* by *D. leuchtenbergianum*. Among the rotifers are some endemic American species like *Keratella taurocephala*, *K. crassa* and *Kellicottia bostoniensis*. The latter has recently been spread to Europe (Arnemo *et al.*, 1968).

From the temperate zone of South America *Mesocyclops longisetus* and *Acanthocyclops robustus* are reported as planktonic cyclopoids. *Notodiaptomus incompositus* is a common calanoid and *Boeckella gracilipes* is reported as a dominant. The northern *Bosmina coregoni* is replaced by *Eubosmina hagmanni*.

To the north of the northern temperate zone (no information from extreme southern waters is available) a general reduction in the number of species sets in. In particular the variety of cladocerans diminishes, and among the remaining zooplankters are several circumpolar species, such as *Holopedium gibberum*, *Kellicottia longispina*, *Keratella hiemalis*, *Cyclops scutifer* and *Limnocalanus macrurus*. In the most arctic lakes of the IBP material, more than 95% of the zooplankton production is made up of two (Latnjajaure: *Cyclops scutifer* and *Bosmina coregoni obtusirostris*) or one species (Char Lake: *Limnocalanus macrurus*).

Moving from the temperate zone towards the equator, first the number of quantitatively important cladocerans increases, and among the rotifers typical warm-water genera become more common (*Brachionus*, *Trichocerca*, *Tetramastix*, *Hexarthra*). Similar zooplankton communities, however, can be found frequently in shallow, eutrophic and temporary well-heated bodies of water elsewhere, e.g. many 'summer-warm' ponds in Central Europe, or lakes and reservoirs in the southern Soviet Union.

It seems that within natural temperature limits, the number of possible secondary producers increases with temperature, but abrupt succession of species in temperate (winter-cold–summer-warm) waters is replaced by permanently high diversity in tropical ones, while low diversity rules in arctic waters and shows little change during the short cool summers. This basic picture, however, is strongly modified as factors like predators, food availability and quality, turbidity or extreme chemical conditions affect the life conditions of the plankton. Extreme chemical conditions are commonly found in arid regions bordering south of the northern temperate zone. It is conspicuous that geographically isolated, chemically extreme (conductivity, pH), and often also low production lakes, are characterised by specific calanoids as the main producers in the zooplankton.

As only three tropical lakes (all African) are included, no generalisations are possible concerning tropical waters. Burgis (1971, 1973), Rzóska, Brook & Prowse (1955) and Talling & Rzóska (1967) summarise some earlier work, mainly from tropical Africa: 'In Africa the Cyclopoidea are mainly cosmopolitan except for the genus *Thermocyclops*. Calanoidea are zoogeographically different and the Cladocera are similar to circumtropical species except for the *Daphnia* species.' Typical for the African lakes investigated is the dominance of one or two small planktonic cyclopoids plus a number of cladocerans with *Moina* sp., *Ceriodaphnia cornuta*, *Diaphanosoma excisum* and *Daphnia lumholtzi* as typical representatives. In addition in Lake Chad the calanoids *Tropodiaptomus incognitus* and in Lake Chilwa *T. kraepelini* are important. The quantitative role of rotifers seems to be subordinate.

6.2.3.2. Feeding types

A general division into predacious and non-predacious species has been carried out for the Soviet and Polish lakes. Spatial and temporal changes in food availability and preferences for different types of food are to be expected in omnivorous species which feed selectively. Also mechanical feeders may occupy different positions in the food chain depending on the food available. A mechanical filter feeder grazes mainly on nannoplanktonic algae in oligotrophic waters, but in eutrophic waters becomes a detrito- and bacteriovore simply because of the natural dominance of bacteria and detritus among the particles available. Similarly, oligotrophic waters usually offer relatively much more animal food of suitable size to raptors than do eutrophic waters, which generally provide a higher share of 'net' planktonic algae. Thus, basically omnivorous *Cyclops* species tend towards plant feeding in eutrophic lakes.

The type of food taken by a given species is largely dependent on three

Table 6.1 Zooplankton food according to different authors (G, gut analysis;

		Size of available food particles (μm)	
		optimum	maximum
Ciliata			
Dreissena veligers			
Rotifers with malleate and malleoramate mastax	*Kellicottia longispina*		4–6
	Keratella cochlearis	1	4–12
	Keratella quadrata	2	12
	Notholca spp.		12
	Conochilus unicornis	1	3–12
	Epiphanes senta		15
	Brachionus angularis		5–15
Small non-predatory cladocerans	*Bosminopsis deitersi*		10
	Bosmina coregoni		15
	Bosmina longirostris	2–5	18
	Chydorus sphaericus	2–5	5–18
	Ceriodaphnia cornuta		18
	Ceriodaphnia quadrangula		20
	Daphnia longiremis		20
	Daphnia longispina	3–12	22
	Daphnia cucullata	3–12	22
	Diaphanosoma brachyurum	5	25
	Daphnia dubia		28
	Daphnia galeata mendotae		35
	Moina rectirostris		40
Larger non-predatory cladocerans	*Daphnia pulex*	18	40
	Daphnia obtusa	18	40
	Daphnia schødleri		45
	Daphnia magna		70
	Daphnia rosea		80
Calanoid and cyclopoid nauplii			
Small raptorial rotifers with virgate mastax	*Polyarthra vulgaris*		30
	Trichocerca spp.		> 30
	Gastropus stylifer		30
	Chromogaster spp.		30
	Synchaeta pectinata		
Non-predatory calanoid copepods	*Diaptomus oregonensis*	7	
	Diaptomus gatunensis	5–10	25
	Eudiaptomus graciloides	10 and 30	20–50
	Eudiaptomus gracilis		20–50
	Limnocalanus macrurus	7 and 30–40	50
	Acanthodiaptomus denticornis		
	Heterocope appendiculata		
	Hemidiaptomus amblyodon		
	Tropodiaptomus incognitus		
	Thermodiaptomus galebi		
Predatory calanoid copepods	*Eurytemora velox*		200
	Heterocope saliens		
	Diaptomus shoshone		
	Diaptomus arcticus		
	Diaptomus nevadensis		
Large raptorial rotifers	*Asplanchna priodonta*		300
	Asplanchna brightwelli		
	Asplanchna herricki		
Non-predatory cyclopoid copepods	*Thermocyclops hyalinus*		
	Thermocyclops neglectus		
	Eucyclops agilis		
	Eucyclops macruroides and *macrurus*		
	Mesocyclops oithonoides		
	Mesocyclops leuckarti		
	Acanthocyclops viridis		
	Cyclops abyssorum tatricus		
	Cyclops strenuus s. lato		
Predatory cyclopoid copepods	*Cyclops scutifer*		
	Thermocyclops incisus		
	Cyclops bicuspidatus thomasi		
	Cyclops vernalis		
	Cyclops vicinus		
	Cyclops gigas		
	Macrocyclops fuscus		
	Macrocyclops albidus		
	Polyphemus pediculus		400
	Leptodora kindtii	700	
	Bythotrephes longimanus	700	
Diptera larvae	*Chaoborus flavicans*		

Right-margin bracket categories (top to bottom):
TYPICAL SEDIMENTORS — FINE PARTICLE FEEDERS — TYPICAL FILTER FEEDERS — SUSPENSION FEEDERS — LARGER PARTICLE FEEDERS — RAPTORS — TYPICAL HUNTERS

Left-margin gradient labels (arrows):
Increase of food selectivity · Increase of animal food importance · Intensification of hunting behaviour · Increase of non-living algae importance · Increase of the most convenient size of food particles · Increase of range of food particles size · Increase of live algae importance · Decrease of detritus and bacterial food importance

Food	
bacteria, detritus, μ-algae: *Cryptomonas, Chrysochromulina, Mallomonas*	E, O
bacteria, detritus, μ-algae: *Quadrigula, Cryptomonas*	E, O
bacteria, detritus, μ-algae: *Chlamydomonas, Phacus*	E, O
bacteria, detritus, μ-algae: *Cryptomonas*	O
bacteria, detritus, μ-algae: flagellates	E, O
bacteria, detritus, μ-algae	O
bacteria, detritus, μ-algae: *Chlorella, Phacus, Synura*	E, O
bacteria, detritus, small nannoplanktonic algae	E
bacteria, detritus, nannopl./small net algae in fragments: greens/	G, E
bacteria, detritus, nannopl./small net algae: *Eudorina*, greens, blue-greens	G, E
bacteria, detritus, small nannopl./large phytopl. colonies when feeding as raptor/	G, E, O
bacteria, detritus, nannopl., smaller net algae: blue-greens	G, E
bacteria, detritus, small nannopl. including small blue-greens	G, E
	E
bacteria, detritus, nannopl.	G, E
bacteria, detritus, nannopl., net algae in fragments: diatoms, greens, blue-greens	G, E
bacteria, detritus, nannopl., net algae in small fragments	G, E
	E
	E
bacteria, detritus, nannopl., smaller net algae, fragments of animal food	G, E
	E
	E
	E
	E
? probably nanno- and net algae, small rotifers, protozoans, small cladocerans	
nannoplankton	E, O
net algae: diatoms, desmids, filamentous blue-greens	E, O
net algae: diatoms, peridineans	E
net algae	E, O
nanno- and net algae	E, O
nannopl.: *Cyclotella*	E
nannopl.; small desmids and diatoms	E
nannopl., small net algae, bacteria, detritus	G, E
nannopl., small net algae, bacteria, detritus	G, E
nanno- and net smaller algae: flagellates, greens, diatoms, blue-greens, rotifers, bacteria	G, E
nanno- and net smaller algae	E
nanno- and net algae	E
nanno- and net algae, rotifers, small cladocerans and copepods/bacteria, detritus/	E
net algae: smaller blue-greens	G
net algae: blue-greens	G
nanno- and net algae; rotifers, cladocerans, copepods/bacteria, detritus/	E
nanno- and net algae, rotifers, cladocerans, copepods/bacteria, detritus/	E
cyclopoid and calanoid copepods, rotifers	E
cyclopoid and calanoid copepods, rotifers	E
cyclopoid and calanoid copepods, rotifers	E
net algae: *Ceratium, Peridinium*, diatoms; rotifers, small cladocerans	G, E, O
net algae, rotifers: up to *Brachionus*	G, E
net algae, rotifers, small cladocerans	G, E
net algae: blue-greens	O
net algae: blue-greens; cladocerans	G
net algae: diatoms, greens, blue-greens/cladocerans, copepods, small benthic fauna/	G
net algae: diatoms. greens/blue-greens, copepods, benthic small fauna/	G
net algae: peridineans, greens; cladocerans	G, E
net algae, cyclop. and calan. nauplii, rotifers/adult copepods, cladocerans/	G, E, O
cyclopoid copepods, benthic fauna/cladocerans, rotifers, net algae/	G, E
rotifers, benthic small fauna, detritus, algae	O
calanoid copepods/cyclopoid copepods, cladocerans, rotifers, net algae/	G, E, O
rotifers, protozoans, small cladocerans	E
cladocerans, cyclopoid copepods, rotifers	G
nauplii of calanoid and cyclopoid copepods/rotifers/	E
calanoid and cyclopoid copepods	E
protozoans	E
nauplii and copepodites of copepods	O
cladocerans, cyclopoid copepods/cladocerans, rotifers, net algae/	G
cyclopoid and calanoid cooepods, cladocerans, rotifers, protozoans, small benthic fauna	G, E
cladocerans, calanoid and cyclopoid copepods, rotifers, protozoans	E
cladocerans, calanoid and cyclopoid copepods, rotifers, protozoans	E
cladocerans, calanoid and cyclopoid copepods, rotifers, protozoans	E
cladocerans, calanoid and cyclopoid copepods	E, O

factors: food uptake mechanisms, behaviour of food and feeder, and abundance of different food types in the environment. While the first two factors are essentially specific to a species, they may change from stage to stage of individual development (e.g. in copepods), but also from environment to environment: e.g. *Chydorus sphaericus* acts as a 'scraper' when feeding on epiphyton, larger algae etc. (Fryer, 1968), but as a filter feeder in a limnetic environment lacking larger forms of phytoplankton (Gliwicz, 1969a, b).

Table 6.1 lists a large number of species and their main food based on the observations of different authors. In general, an increase of food selectivity and of the importance of animal food takes place along with an increase of optimal food particle size and also with changing feeding mechanism and feeding behaviour from sedimentation to filter feeding and to raptorial feeding. It is obvious that the predatory effect of some species which have traditionally been considered to be predators may have been overestimated.

6.2.4. Standing crop

6.2.4.1. Numbers

The main difficulty in comparing data presented by different IBP teams is the variation in sampling frequency and method of calculation of biomass and production.

In most projects weekly or bi-weekly sampling was carried out during the growing season with lake averages based on some 5–20 sampling stations. Where standard deviation has been calculated, most values given are between 20 and 30% at a confidence level of 95%. Extreme variable horizontal distribution patterns are not recorded, but in one case there are figures which indicate that this can be considerable (fig. 6.1: *Bosmina*, Latnjajaure).

An obvious problem is that of sampling close to the bottom. In particular *Cyclops* species are often concentrated in the immediate vicinity of the sediments. For this reason, in Finstertaler See calculations of production of *Cyclops abyssorum tatricus* were impossible. J.P.E.C. Darlington (personal communication) has compared vertical haul samples and benthos samples from Lake George, Uganda. She found that not only the semi-planktonic *Chaoborus* but also *Cyclops* and even *Daphnia* can be more or less concentrated in the bottom 10 cm of water, a part of the water column which is not taken by normal plankton sampling procedures. The overall ratios of mean numbers per litre in the bottom 10 cm : mean numbers per litre in the water column found by her were 1.51 for *Daphnia* and 9.65 for *Thermocyclops*.

Hillbricht-Ilkowska & Węgleńska (1970) found that values of standing stock and production of crustaceans and rotifers obtained at a sampling frequency greater than six days differed essentially from values based on daily sampling. In extreme cases the values differed 2–3 times. Further they stated,

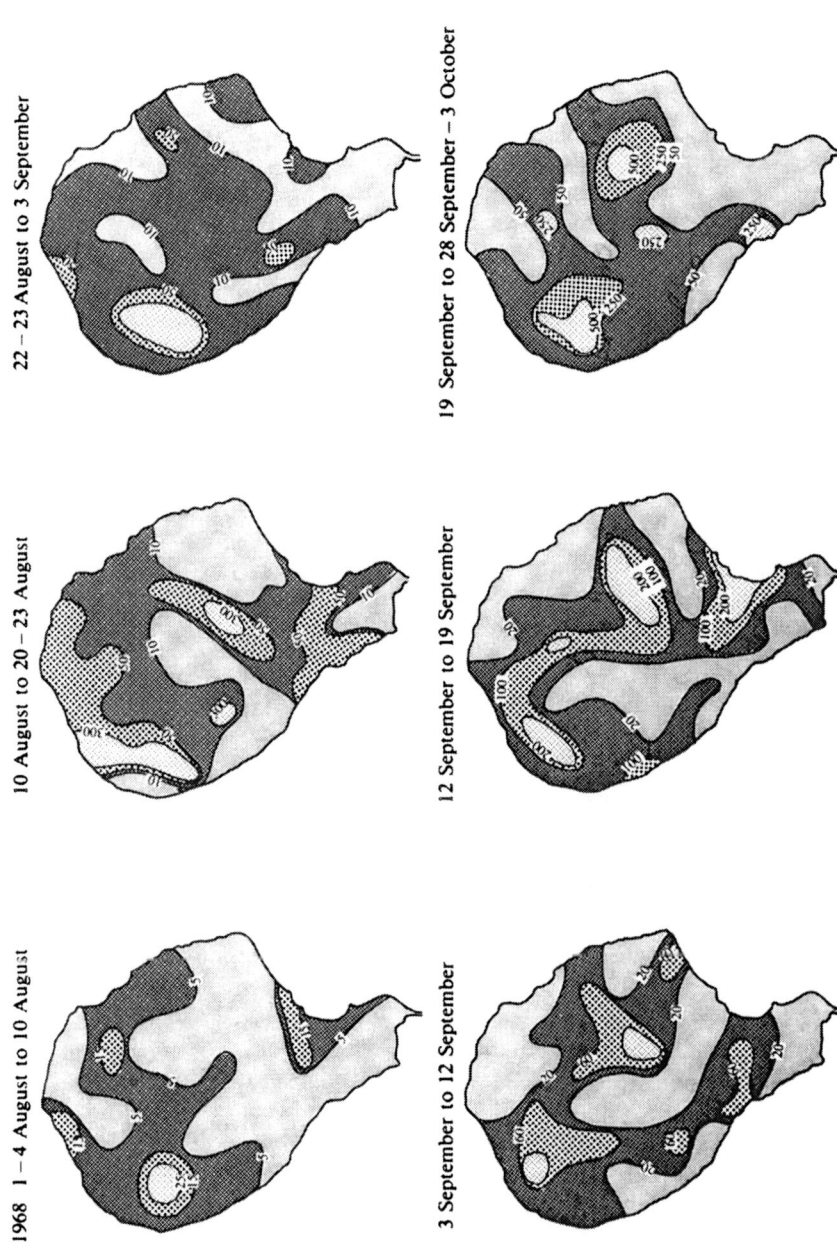

Fig. 6.1. Relative distributions of *Bosmina coregoni obtusirostris* during the growing season in Lake Latnjajaure. Original from Nauwerck. Trap catches individuals per day.

that in short-time investigations or during periods characterised by sharp directional changes of a zooplankton population, sampling frequencies have to be greater than in cases where the whole seasonal period of occurrence of the species or community can be examined.

Because of the heterogeneity of research interests in different IBP projects, basic data were handled and presented in most variable ways, and although the above conditions were fairly satisfactorily applied for some fifty water bodies, only very crude or rather isolated detailed comparisons of numbers are possible. The best comparisons can be obtained by examining maximum values.

Among rotifers the highest numbers recorded are of the order of magnitude $10^4 \, l^{-1}$. Highest values are recorded from eutrophied, temperate waters, usually due to mass developments of *Keratella cochlearis*. The lowest values, on the other hand, are not from the most oligotrophic lakes, where rotifers sometimes may dominate, but from shallow, warm and rather turbid lakes, for instance the African ones.

Protozoans, though mainly smaller than rotifers, occur in similar maximum numbers. Surprisingly enough, similar numbers are recorded from the rather eutrophic Volga reservoirs and from the arctic Char Lake. It is likely that a broader study of particularly the smaller forms among protozoans will produce higher densities.

Values for cladocerans reach orders of magnitude of $10^3 \, a^{-1}$, but generally are rather below than above $10^2 \, l^{-1}$. The highest values are from 'summer-warm', eutrophic lakes or reservoirs in Europe.

The copepods, including nauplii, are in approximately the same range as the cladocerans, but maximum values are lower, probably owing to a lower tendency to swarm (Hobbie *et al.*, 1972; J. Vijverberg, personal communication). The highest abundances are achieved by relatively small cyclopoids in shallow, warm, turbulent waters, where they can exceed cladocerans in numbers. The calanoids, being mainly larger than the cyclopoids, rarely exceed $10^2 \, l^{-1}$. Top values for calanoids are generally one order of magnitude below those of cyclopoids, even when they are dominant.

Larger cladocerans (e.g. *Leptodora, Bythotrophes*) and calanoids (e.g. *Limnocalanus, Heterocope*) reach a maximum of a few individuals per litre, but usually stay at much lower levels. As a rule the maximum (and consequently also average) abundance reached by a zooplankton species is inversely proportional to the size of the animal.

In principle densities can fluctuate between zero and the potential maximum value in one and the same water body. This is more apparent the more the environmental factors fluctuate in the biotope in question. Arctic tarns which freeze to the bottom or ponds which dry out must start every season from zero, from resting stages or by immigration. Species which start and finish their cycles with resting eggs, periodically disappear even in permanent water bodies. However, the more balanced the environmental conditions, the more stable becomes the zooplankton standing crop and production. An ideal steady-state situation would be realised in a permanently turbid lake with unchanged temperature and continuous and uniform nutrient exchange. Among the lakes studied, Lake George, Uganda, comes closest to this situation. For instance, the *Thermocyclops* numbers of this lake only fluctuate during the year by 1 : 2. Changes in zooplankton abundance in the lake are related to periods of rainfall as is the more pronounced case in Lake Chilwa.

On the other hand, stability is also found in zooplankton populations of arctic–alpine lakes. Differences between the lowest and highest numbers of *Cyclops scutifer* in Latnjajaure (six years' observations) are 1 : 5, in Øvre Heimdalsvatn for the same species (over two years) 1 : 10, and for *Limnocalanus macrurus* in Char Lake (two years) also 1 : 10. This is in comparison with fluctuations of 1 : 100 to 1 : 1000 in perennial species of temperate lakes.

6.2.4.2. Biomass

Computing numbers to biomass necessarily introduces particular sources of error. Different methods of calculation have been used by different authors and sometimes have to be used when different groups of animals are concerned. For small zooplankters such as rotifers or nauplii calculations of specific volumes often are the only possible way of estimating biomass (Grönberg, 1973). For larger species dry weight measurements are easier to carry out and give reliable results. The relation between dry weight and fresh weight (or volume) may not be strictly linear in all species (P. Larsson, personal communication). However, the conversion factor of 1 : 10 as recommended by Winberg (1971) seems generally acceptable.

In many cases quite good length–weight relationships have been found in crustaceans which allow simple translations (Smyly, 1973; P. Larsson, personal communication). Formulae used for calculation of copepod volumes (Klekowski & Shushkına, 1966; Grönberg, 1973) also give reasonable comparability. Chislenko (1968), has worked out nomograms for volume calculations which have been used to estimate rotifer wet weights in Polish lakes among others. In most of the IBP studies conversion factors have been adopted from elsewhere. So far as direct measurements were made they reveal variations which can be significant in crustaceans. The weight of *Limnocalanus macrurus* of Char Lake, appears to be 4–5 times that of the same species from Ekoln. As shown from Neusiedlersee, even adult *Diaptomus* weights can differ at any one time or throughout the year by 1 : 2. In rotifers, on the other hand, discrepancies seem to originate from the approximations made by different authors. Fresh weight (as μg) for *Keratella cochlearis* for instance differs between 0.05 (Ekoln), 0.122 (Balaton), 0.2 (Zelenetzkoye/Akulkino) and 0.64

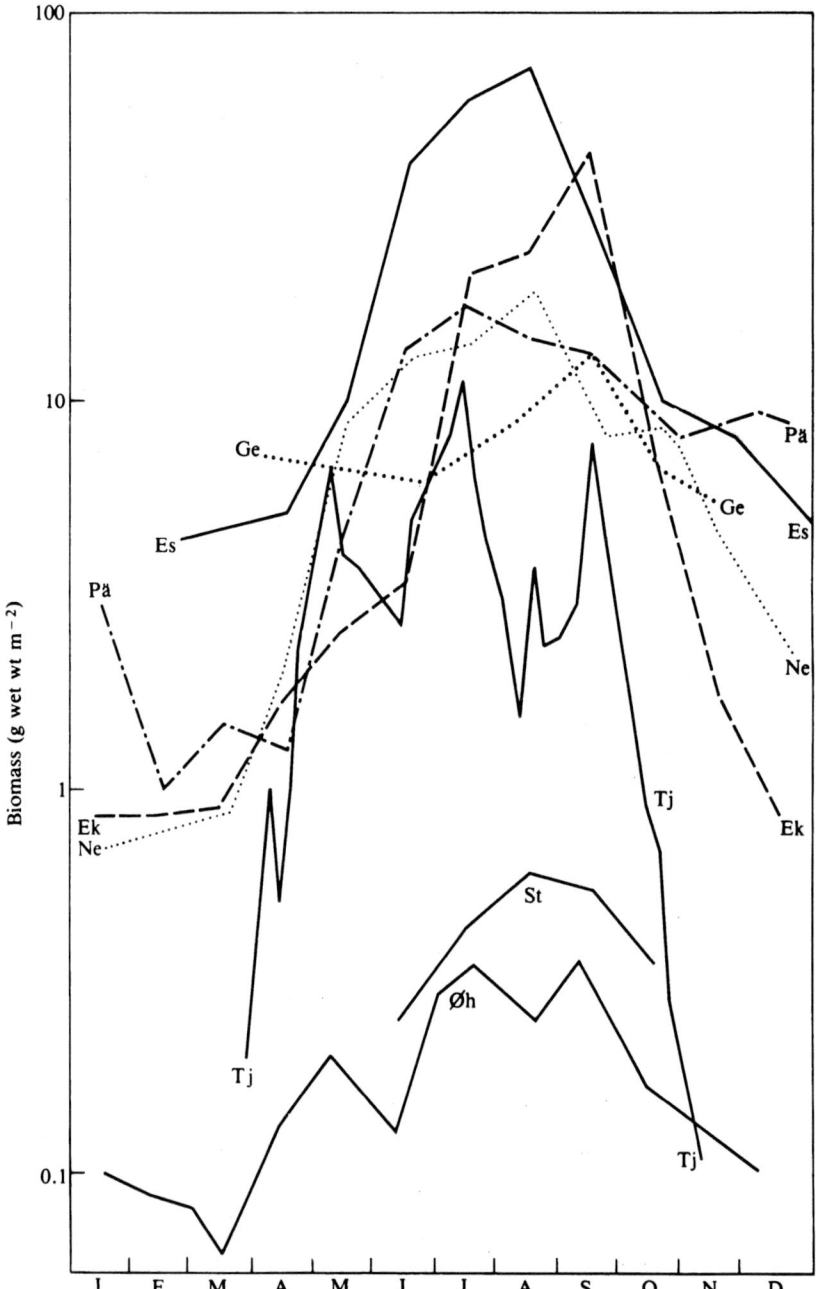

Fig. 6.2. Seasonal development of zooplankton biomass in some IBP lakes. (Es, Esrom 1970; Ek, Ekoln 1968; Pä, Päajärvi 1972; Ne, Neusiedlersee 1971; Ge, Lake George, Uganda 1969/1970; Tj, Tjeukemeer 1968; St, Stugsjön 1971; Øh, Øvre Heimdalsvatn 1970.)

(Belgian fish-ponds). *Asplanchna priodonta* fresh weights vary between 6 (Biwa) and 100 (Pääjärvi). Even if size variation is large in *Asplanchna*, it seems unlikely that the differences reflect real ranges of weight variation.

Unfortunately, the material available does not allow more detailed comparisons of average or maximum biomass per cubic metre but only allows comparisons per square metre. Since within one water body biomass per unit area shows clear dependence upon depth (Brandl, 1973), it is likely that depth effects overrule, within certain limits, the effect of the trophic state of the water on its zooplankton biomass per surface area.

Total biomass figures, taken for what they are, scatter over 4–5 orders of magnitude in different water bodies (fig. 6.2). The lowest winter values in cold, oligotrophic lakes are below 0.01 g wet m^{-2} (Øvre Heimdalsvatn, Latnjajaure) and the highest summer values, in 'summer-warm' Baltic lakes, exceed 100 g m^{-2} (e.g. several Polish lakes). Winter–summer amplitude can stretch over 4 orders of magnitude in a single water body (Tjeukemeer) but normally keeps within 2–3. It might be expected that eutrophic ponds would have a higher zooplankton biomass, if not per square metre at least per cubic metre of water volume but this is not the case in the material presented. The reason may be predation by fish.

According to data provided by M.B. Ivanova (personal communication) biomass per cubic metre generally decreases in Soviet lakes from south to north. Rotifers are obviously more important in northern than in temperate lakes and in certain lakes biomass values are higher in winter than in summer.

It is striking that maximum biomass in Lake George, Uganda, and in other tropical waters, remains considerably lower than maximum values in temperate lakes. In the temperate zone the start of biomass development in spring as well as the times of maxima can shift within a couple of months in either direction, both within and between lakes, even if the lakes are situated within the same geographical region. The primary reasons are local or year-to-year differences in climate (ice break-up etc.), which demonstrates the delicacy of the definition of 'growing season' and the problems of comparing results from different lakes studied during different years.

The pattern of normal zooplankton development in natural water bodies is not followed in reservoirs, where drainage periods, volumes drained, and depth of drawdown essentially control its development, as shown for example in London reservoirs studied by Steel, Dunčan & Andrew (1972) and in Czechoslovakian reservoirs by Straškraba & Javornický (1973) and Brandl (1973). Also Tjeukemeer cannot be considered to be a natural water body.

It would be interesting to compare the number of zooplankton peaks occurring in different water bodies during a comparable period of time. Because of different sampling frequencies and other statistical limitations in different projects, this comparison has to be subjective. However, repeated outbursts and breakdowns of zooplankton populations seem to be most typical for eutrophic lakes of the temperate zone.

Summarising biomass observations one can state that eutrophic and warm
waters have high average standing crops and cold and oligotrophic ones have
low ones. At comparable trophic levels, deep lakes have larger standing crops
than shallow ones. Standing crop fluctuations are more pronounced in
temperate lakes than in stenothermic ones, and the more pronounced the
fluctuations in environmental factors then the more pronounced are the
zooplankton population fluctuations. Under natural conditions, fluctuations
are larger in eutrophic and shallow waters than in oligotrophic and deep ones.
The retention time of the water mass governs the size of standing crop: the
more frequent the renewal of water the smaller the standing crop.

6.2.4.3. Long-term variations

Several projects have two years' investigations, and some have longer periods.
In practically all cases development patterns, zooplankton maxima and
averages are rather different in different years even if some general principles
seem to be fairly stable.

The more constant characteristics are the composition of dominant species
and the successions of different species or groups whereas the more variable
characters are the relative proportions of dominant species and the absolute
levels and time of their maxima in different years. No significant trends or
definite changes in zooplankton communities occur except in cases of lake
fertilisation, regulation or other human interference.

Long series of zooplankton counts exist from many lakes in the Soviet
Union (Smirnov, 1973). Lake Dalnee, for instance, shows fluctuations of the
mean seasonal biomass of dominant species of 1 : 4.7 during a 32-year period
(fig. 6.3). The proportion of *Cyclops scutifer* of the whole zooplankton varies
between 47.2% and 93.4%, that of *Neutrodiaptomus angustilobus* between 3.3%
and 39.9% and that of *Daphnia longispina* between 3.3% and 27% in different
years. In Lake Baykal, during a 24-year period (1947–71), extremes in annual
average biomass are 1 : 8.1. In Rybinsk Reservoir during a 14-year period
(1956–70) values differ by 1 : 5.9. From outside the Soviet Union Lake
Aleknagik varies from 1 : 3.6 during 10 years (1961–71), Latnjajaure from 1 : 5
during 6 years (1964–9), and Balaton from 1 : 6.7 in 11 different years between
1936 and 1966. The differences, may be larger if only growing season averages
are compared rather than annual averages. This is shown from Czechoslovakian reservoirs (Kličava 1961–70, Slapy 1959–69), where average biomass for
April–October can differ in extreme cases from 1 : 20, but annual averages
only differ 1 : 2.1. It must be kept in mind, however, that reservoirs are not
quite comparable with natural lakes.

Even if certain rhythms become visible in long series of observations,
extreme differences often appear between neighbouring years. The extreme
values of the Lake Baykal series for instance are only 3 years apart, but

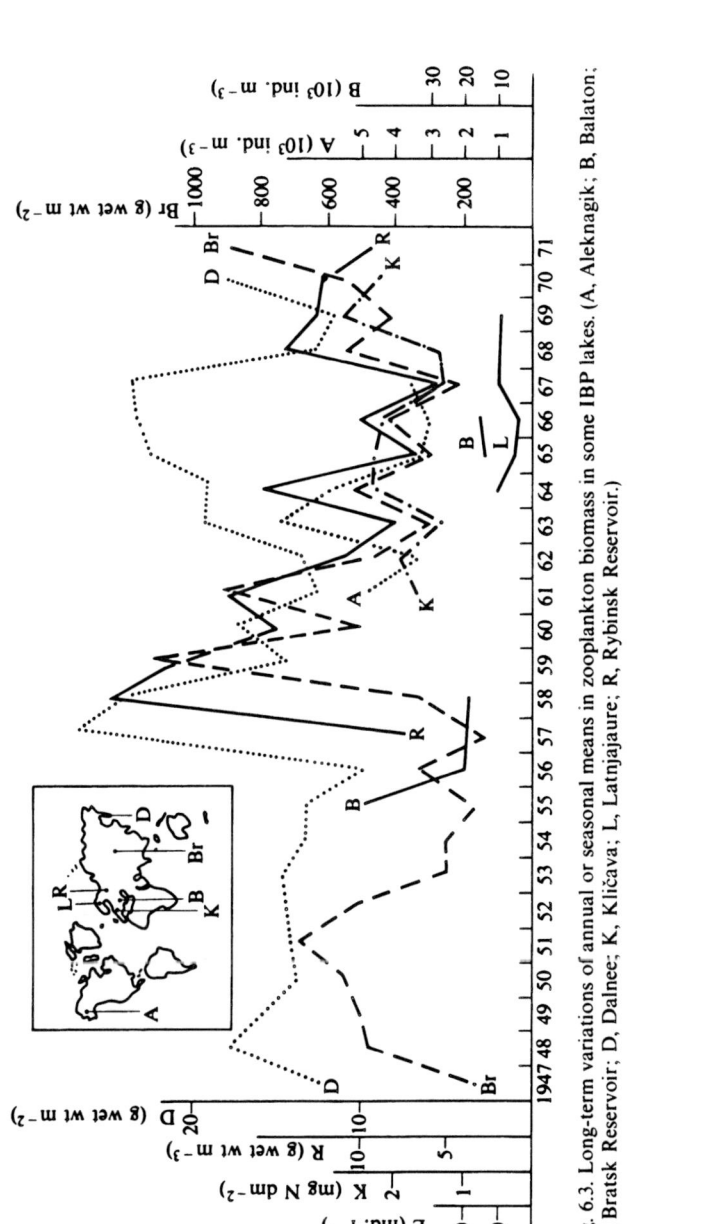

Fig. 6.3. Long-term variations of annual or seasonal means in zooplankton biomass in some IBP lakes. (A, Aleknagik; B, Balaton; Br, Bratsk Reservoir; D, Dalnee; K, Kličava; L, Latnjajaure; R, Rybinsk Reservoir.)

common long-term fluctuations can be traced, at least in certain major areas of Europe and Asia, which indicate a possible relationship to climatic change. For instance a period in the middle of the 1960s is characterised by low values in many European and some northern Asian lakes. Numerous IBP projects carried out at the end of the 1960s and the beginning of the 1970s show higher biomass and production values in later years of their study than in earlier ones.

Against this background a more detailed comparison of lakes seems possible only on the basis of several years' data. Differences from one year to another are interesting *per se*, but as long as they are not extremely dramatic, say of an order of magnitude in either direction, it is hardly possible to distinguish between climatic factors, previous biological conditions, and human activities like eutrophication, fertilisation or restorative measures as the causes of observed differences. In IBP lakes ELA-239 (fertilised) and Trummen (restored), however, changes are so big that it cannot be doubted that they are caused by these operations.

6.2.5. Production

6.2.5.1. Methods of calculation

Basically, there are two methods used for the assessment of production of zooplankton. The first can be called the 'recruitment-time' method, and was originally developed by Edmondson (1960, 1965), and modified among others by Galkovskaya (1968). This method allows the calculation of loss of biomass with time as well as production. It is mainly used for rotifers but has been adopted for calculation of crustacean production by authors such as Stross, Neess & Hasler (1961), Hall (1964) and Wright (1965). In IBP projects, the method has been used for crustaceans (e.g. in Lakes George and Wingra, Wisconsin (McNaught, DR 23) and in Lake George, Uganda (Burgis, 1971)). Using this method, production is calculated as the product of the biomass of a population and its recruitment time, (generation time from egg to egg).

The second method is known as the 'growth-increment' method (Winberg, Pechen & Shushkina, 1965) and consists of the integration of successive biomass differences. In its simplest application it is basically similar to the cohort method. This method has mainly been used by Soviet and Polish investigators and is the most common one for calculation of crustacean production.

Comparing the two methods, Stepanova (1971) and Ivanova (1973) found comparable results for production per unit biomass over long periods of time or for steady-state conditions in zooplankton populations. However, considerable differences appeared in species with periodic changes in age composition of the population, as in *Eudiaptomus graciloides*, where production was found to be largest in spring using the 'recruitment-time' method, but largest in late summer using the 'growth-increment' method. It is likely that short-term estimates tend to result in unrealistic values for species with strong seasonal patterns. This seems to be particularly the case when applying the 'recruitment-time' method to copepods with distinct and separate generations.

Modifications of the 'recruitment-time' method have also been compared by Hillbricht-Ilkowska & Pourriot (1970) and by Hillbricht-Ilkowska & Weglenska (1970) for different rotifers. In general, the highest figures and widest variations were obtained when assuming exponential growth and the lowest figures when assuming linear growth.

6.2.5.2. Generation times

The determining mechanisms for zooplankton production are inherent in the organisms. Every species faces its environment with a set of tools which enables it to survive and to reproduce and to maintain a stock. Common mechanisms are frequently found in groups of animals living in similar habitats, but it is the differences between species which allow them to exist together. Species with similar environmental demands often occur as competitors and can replace each other in similar biotopes. One of the most effective tools in this respect is mode of reproduction. It is an advantage to reach maturity quickly in a nutrient-rich environment. Parthenogenetic reproduction can speed up population growth. Reducing the number of larval stages, and giving birth to sexually mature progeny, as in rotifers and occasionally in cladocerans can speed up population growth still more. A disadvantage, in such cases, is if young animals immediately become their parents' competitors.

Differently from cladocerans and rotifers, in copepods population growth is much more regulated by the longevity and survival rates of different larval stages than by egg production. Young animals are not generally competing with their parents, as the food particles eaten belong to different size groups or feeding habits are different. The number of generations per time unit is thus smaller in copepods, and their ecological efficiency lower because they are bound to sexual reproduction, which means that only half of the adult population can produce progeny. Because of slower development and because the first clutch of eggs which is often the only one, or the only fertile one, is mainly produced from reserves stored during copepodite stages, the copepods depend less on the food conditions during the adult stage for reproduction than do rotifers and cladocerans. The number of generations produced per time unit becomes relatively more dependent on temperature. The time from an individual's birth until its first progeny is an important factor for calculation of production according to the 'recruitment-time' method.

Usually, generation time, defined as the sum of egg development time and post-embryonic development up to the first egg being laid, is found experimentally under laboratory conditions. Production of the natural population is then calculated for a given temperature by multiplying this value with the actual stock of eggs. Galkovskaya (1968), proposing the method, relates the value only to temperature according to Krogh's curve. Bottrell (1975) and others have shown that errors are involved in such a procedure.

Generation times found experimentally were similar for many of the common plankton rotifers (Pourriot & Hillbricht-Ilkowska, 1969; Pourriot & Deluzarches, 1971) ranging from 90–170 hours at 10–12 °C and from 27–72 hours at 20–25 °C. Galkovskaya (1968) obtained approximately twice as long times when not using 'optimal' concentrations of culture algae but keeping animals in filtered lake water.

According to Edmondson's method, mean generation times during summer in the epilimnion of Polish lakes were from less than 1 week to 2 or 3 weeks for *Keratella cochlearis*, *K. quadrata* and *Kellicottia longispina*, somewhat shorter for *Pompholyx sulcata* and somewhat longer for *Filinia longiseta*. This allows at least eight successive generations during the summer season. In Lake Chad, rotifer generation time is estimated as 2 days at 30 °C. In Ekmanjaure at temperatures not exceeding 2 °C, *Keratella hiemalis* still has three generations per year (Nauwerck & Persson, 1971). A similar number of generations was found by Amrén (1964) in Spitsbergen ponds, using a variation of the cohort method on *Keratella quadrata*.

Since with increasing age post-embryonic growth in most crustaceans is increasingly dependent on food, generation time cannot be related so closely to temperature as in rotifers. Nevertheless, knowledge of the development rate and longevity of different stages is necessary for the judgement of sampling frequency if the growth increment method is used. The generation time of cladocerans in Lake Chad is stated as 4–8 days, allowing 50–100 generations per year. Five days have been calculated for *Daphnia* and *Moina* in Polish fishponds under summer conditions, so that during the summer season 15–30 generations ought to be possible. In Latnjajaure, during a growing season of 3 months at 3–8 °C, three generations of *Bosmina* can still succeed.

In Lake Chad *Tropodiaptomus incognitus* takes approximately 1 month for a generation and thus can produce twelve generations per year. In Polish fishponds *Diaptomus zachariasi* has a generation time of 12–23 days during summer which theoretically allows approximately 12 generations per year. This is an infrequent species in temperate Europe living in water bodies of limited duration, hence the reproduction rhythm may be accelerated. According to Weglénska (1971), *Eudiaptomus graciloides* appears to have at least four generations per year, consisting of three shorter summer generations and one longer winter generation. In Lake Krasnoye this species has three generations per year, in Lakes Krivoye and Krugloye it has two and in lakes

Zelenetskoye and Akulkino it has one generation per year. One generation per year is also reported from Stugsjön/Hymenjaure and numerous lakes in northern Sweden (Ekman, 1964). There is evidence for 2-year cycles in certain lakes of this region (A. Nauwerck, unpublished data). Compared to the above figures, a theoretical number of twenty-eight generations per year for *Diaptomus minutus* in Lake George, Wisconsin, seems unrealistic. The same species was found to have only one generation per year in Clear Lake, Ontario by Schindler (1972).

Limnocalanus macrurus appears to have an endogenous rhythm. The generation cycle is 1 year (the life span of females somewhat longer) practically independent of temperature and trophic state of the lake. Development is the same in arctic Char Lake with temperatures between 0 °C and 2 °C, in eutrophic Ekoln with summer temperatures up to 15 °C in the hypolimnion, and in humic-oligotrophic northern Scandinavian lakes. The generation starts with nauplii during December until May, continues with copepodites, during a relatively short period, and ends with an adult maximum in August and sporadic survival until May the following year.

Heterocope (*H. saliens* in Øvre Heimdalsvatn, *H. appendiculata* in Stugsjön/Hymenjaure) has also a 1-year fixed cycle with only resting eggs being produced.

The widest variation can be found in cyclopoids. With a development time of less than 72 hours in African rain pools (Rzóska, 1961), *Metacyclops minutus* would be able to produce more than 100 generations per year if this figure can be extrapolated to permanent water bodies. Direct estimations from larger tropical waters are not available, but M.J. Burgis (personal communication) calculates the generation time of *Thermocyclops hyalinus* in Lake George, Uganda as 20–30 days at temperatures of 25–30 °C. This would give about fifteen generations per year. From Tsimlyansk Reservoir, *Cyclops vernalis* is reported to have 5-8 generations per year. In Char Lake the same species has one generation. In Lake Krasnoye *Mesocyclops leuckarti* and *M. oithonoides* have two generations. *Cyclops tatricus abyssorum* has a 1-year cycle in Gossenköllesee (Eppacher, 1968). *Cyclops scutifer* also needs 1 year in Øvre Heimdalsvatn, Lake Krivoye and Stugsjön/Hymenjaure, but the same species needs 2 years in Zelenetskoye, Akulkino and 3 years in Latnjajaure and in high-arctic Canadian lakes (McLaren, 1964). Cycles of 2 and 3 years are also reported from Norwegian lakes Gjende and Bessvatn (J.A. Eie & K. Elgmork, personal communication).

6.2.6. Production per Biomass

6.2.6.1. Total zooplankton and main groups

A large zooplankton standing crop in a water body is also roughly an expression of high zooplankton production. Certainly, a small biomass may reproduce quickly and a large one slowly, depending on temperature and food limitations, but usually this first becomes discernible in extreme situations. This general statement is in agreement with Brylinsky's findings (Chapter 9) based on material collected during IBP.

The table comparing P and B values as used by Brylinsky has been completed with a few more figures which were calculated from additional IBP

Table 6.2 *Zooplankton production characteristics in different lakes. Data as used by Brylinsky (Chapter 9) except IBP contributions 5, 13, 17, 29, 32 and non-IBP contributions 25, 28, 33. (HZP, herbivore zooplankton production; HZB, herbivore zooplankton biomass; CZP, carnivore zooplankton production; CZB, carnivore zooplankton biomass)*

		HZP (J m^{-2})	HZB (J m^{-2})	HZP/B	CZP (J m^{-2})	CZB (J m^{-2})	CZP/B
1	Flosek	1755.6	58.8	30.0	71.4	5.88	11.8
2	Dalnee	1638.0	239.4	6.8	113.4	18.90	6.0
3	Mikołajskie	1596.0	84.0	19.3	399.0	53.80	8.3
4	Tałtowisko	1495.2	133.4	15.8	147.0	7.14	20.3
5	George, Uganda[a]	994.1	10.1	—	—	—	—
6	Drivyaty	630.0	25.6	98.6	—	—	—
7	Chedenjarvi	443.5	27.7	16.0	35.7	3.78	9.5
8	Warniak	420.0	12.6	33.0	63.0	2.52	25.0
9	Batorin	420.0	18.9	22.0	63.0	4.20	15.0
10	Myastro	378.0	35.3	11.0	71.4	5.88	12.0
11	Krasnoye	371.3	21.0	18.0	41.6	2.85	15.0
12	Sniardwy	352.8	33.6	12.3	67.2	5.04	12.8
13	Baykal	338.1	25.2	13.5	34.0	7.60	1.5
14	Kiev Reservoir	231.0	27.3	8.5	10.1	0.84	12.0
15	Kurakhov	226.8	7.6	30.0	157.5	10.5	15.0
16	Rybinsk Reservoir	184.8	4.2	44.0	27.1	1.84	14.6
17	Sibaya	183.1	1.4	132.1	—	—	—
18	Karakul	171.4	7.6	22.6	13.0	0.42	30.4
19	Naroch	168.0	16.8	10.0	81.9	6.72	12.0
20	Pääjärvi	126.0	14.3	9.0	21.0	2.52	8.0
21	Bratsk Reservoir	79.0	8.0	10.0	58.8	5.46	10.8
22	Krivoye	71.4	5.5	13.0	10.7	2.14	5.0
23	Krugloye	46.1	3.7	12.4	8.5	0.75	11.2
24	Ø. Heimdalsvatn	21.0	2.5	8.5	—	—	—
25	Hymenjaure	20.4	3.6	5.3	3.5	0.37	9.2
26	Bolshoy Kharbey	13.9	2.9	4.7	1.5	0.84	1.8
27	V. Finstertaler See	10.1	2.3	4.4	2.7	0.84	3.2
28	Stugsjön	6.5	2.5	2.6	2.8	0.21	13.4
29	Char Lake	4.6	1.6	2.9	—	—	—
30	Zelenetskoye	1.5	0.4	3.5	0.1	0.01	7.9
31	Aleknagik	0.5	1.0	0.5	1.7	0.17	10.0
32	Tundra Pond	0.3	0.1	2.8	—	—	—
33	Latnjajaure	0.3	0.2	1.3	—	—	—

[a] *Cyclops* only (= > 90% of zooplankton population).

reports or other studies with the help of conversion factors recommended for IBP (Winberg, 1971a). All values are listed in table 6.2. They scatter widely, and hardly show conspicuous trends. Generally, it can be stated that P values of non-predators range from less than 1 J m^{-2} in arctic waters to more than 1700 J m^{-2} in temperate waters in Central Europe. Predatory production usually is between 10% and 20% of non-predatory production but ranges from less than 5– > 340%. The latter figure, of course, cannot reflect true planktonic conditions, but is probably only the result of the authors' decision on what to include as non-predators.

Relationships between biomass and production and between P/B ratios of different trophic groups are of primary interest. If all P and B values are plotted against each other, a linear relationship can be obtained. The relationship $P = 9.097 B^{1.237}$ for predators and non-predators together can be compared with the relationship of $P = 2.34 B^{1.62}$ found by Shushkina (1966) for one month (August) in Karelian and Byelorussian lakes. If separated, the correlation is weaker for predators and non-predators respectively. This again may be a matter of how groups were divided but also may be caused by the fact that the relation between P and B, in reality, is not a linear one.

If the values are grouped according to B or P/B, it becomes evident that highest P/B ratios for non-predators go together with moderately high B values. With increasing as well as with decreasing B the P/B ratio decreases. Low or relatively low B values, on the other hand, belong not only to the coldest but also to some of the warmest lakes. Thus, in both very cold and very warm lakes, the productive efficiency of the biomass of herbivores present is lower than at moderate temperatures, which means production conditions are suboptimal to the animals in both cases. A reasonable assumption is that temperature itself must be limiting in cold waters, i.e. low metabolic rates prevent the animal from using the food available efficiently, while food supply may be limiting in warm waters, i.e. high metabolic rates would allow higher animal production if more food were available. Zooplankton has been found under semi-starvation conditions, e.g. in Queen Mary Reservoir and in Saidenbach Reservoir during warm summer months. At comparable temperatures in August, Shushkina (1966) found 3–4 times higher P/B values for herbivorous zooplankton in shallow, eutrophic lakes than in deeper, more oligotrophic ones.

Hillbricht-Ilkowska (1972), comparing herbivorous production with primary production found a similar optimum curve, with maximum zooplankton production between primary production values of 2500 and 8500 J m^{-2}, which may indicate changed food quality at very high levels of primary production.

If the same grouping is done for carnivorous zooplankton, similar principles seem to be valid. The P/B ratio is at its highest at moderate B values, when B

increases, P/B decreases, and when B decreases to very low levels, P/B decreases again. However, the picture is less clear owing to among other things difficulty in separating carnivores from herbivores.

A comparison of carnivore data with herbivore data nevertheless indicates an optimum curve, with highest carnivore P/B ratios within a field of maximum P/B and maximum B of herbivores. Yet, the quality of the data does not allow us to distinguish whether carnivore P/B primarily depends on food supply (herbivore biomass) or on food replenishment (herbivore P/B). The general picture is that high B of herbivores is usually followed by high B of carnivores, but that high B of carnivores also restricts P/B of herbivores.

From data provided by M.B. Ivanova (personal communication) it appears that carnivorous P/B is relatively higher at high herbivorous P/B values, and may even surpass them in warm and eutrophic lakes. A comparison of different taxonomic groups shows increasing P/B values from cold-oligotrophic to warm-eutrophic lakes. The increase is strongest in cladocerans and least pronounced in copepods, but rotifers' P/B seem to depend mainly on temperature while P/B values of the other groups show dependence on both primary production and temperature. Relationships, however, can be shown only within broad limits and are obscured by the individuality of lakes even within restricted geographical areas.

Against this background P/B variations between seasons and between different years can be examined. Calculations of P during a long series of years is available from Rybinsk Reservoir (fig. 6.4a). Obviously, in this case zooplankton biomass is negatively related to its P/B ratio. For Clear Lake, Ontario, Schindler (1972) shows a clear connection between high levels for daily herbivore P/B ratios and high summer temperature, in spite of low

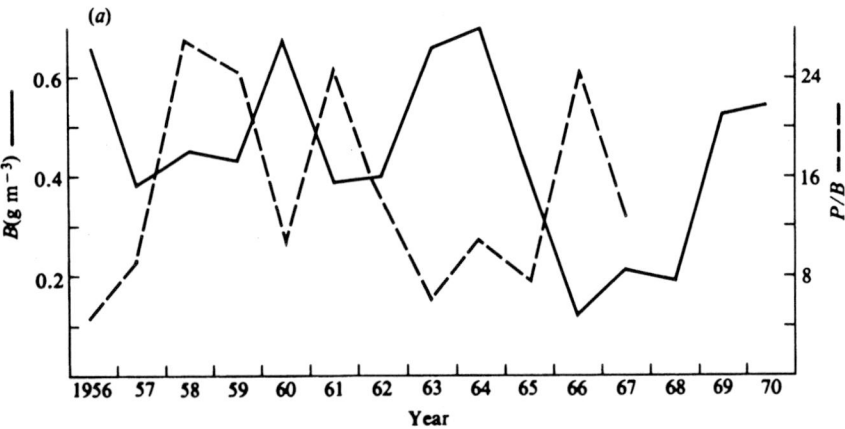

Fig. 6.4a. Long-term variations in zooplankton biomass and P/B ratio in Rybinsk Reservoir. Annual values.

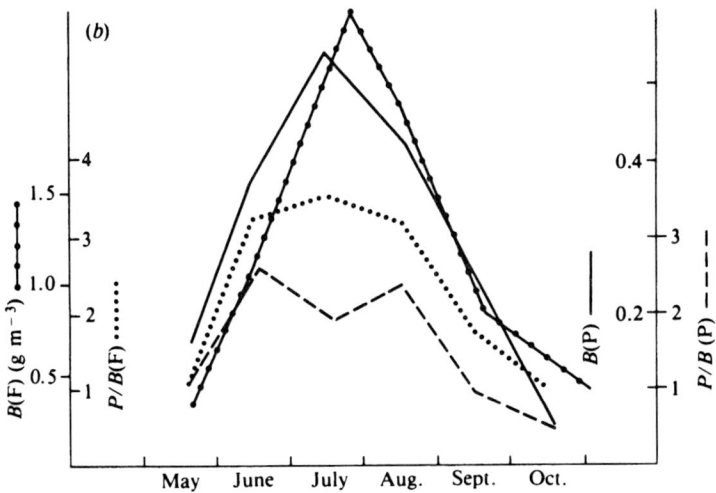

Fig. 6.4*b*. Seasonal variations in zooplankton biomass and *P/B* ratio in Lake Krasnoye. Monthly values (F, filter feeders; P, predators).

primary production at the same time. Patalas (1970) also points out the dependence of *P/B* ratios on temperature. Comparing artificially heated with unheated lakes in the middle of summer he finds a doubling or more of the *P/B* ratios of rotifers, non-predatory and predatory crustaceans. His lists of *P/B* values from different types of lakes also indicate a positive correlation with primary production. The same is evident from Lake Krasnoye (fig. 6.4*b*), where herbivorous *B* and *P/B* on the whole follow each other and are highest during summer. The values for predators practically follow the same pattern with a slight depression of *P/B* at maximal *B*.

Neither temperature nor food relationships to the overall *P/B* values are very obvious in lakes Mikołajskie, Sniardwy, Tałtowisko and Flosek (Hillbricht-Ilkowska, Spodniewska, Weglerńska & Karabin, 1972), but the modifying role of species composition in the zooplankton community is demonstrated; e.g. *P/B* values for non-predators are usually completely dominated by very high rotifer values. Thus, *P/B* ratios of heterogeneously and subjectively classified groups of animals are not suitable for more subtle comparisons of production processes in different lakes.

6.2.6.2. Individual species

Even in single species, production is directly proportional to population biomass within broad limits (fig. 6.4*c*). However, considerable variations can take place seasonally (fig. 6.4*d*). Sometimes *P/B* values roughly follow temperature change in the water (fig. 6.4*e*), but in many cases show considerable change within short periods of time. A general correlation with food

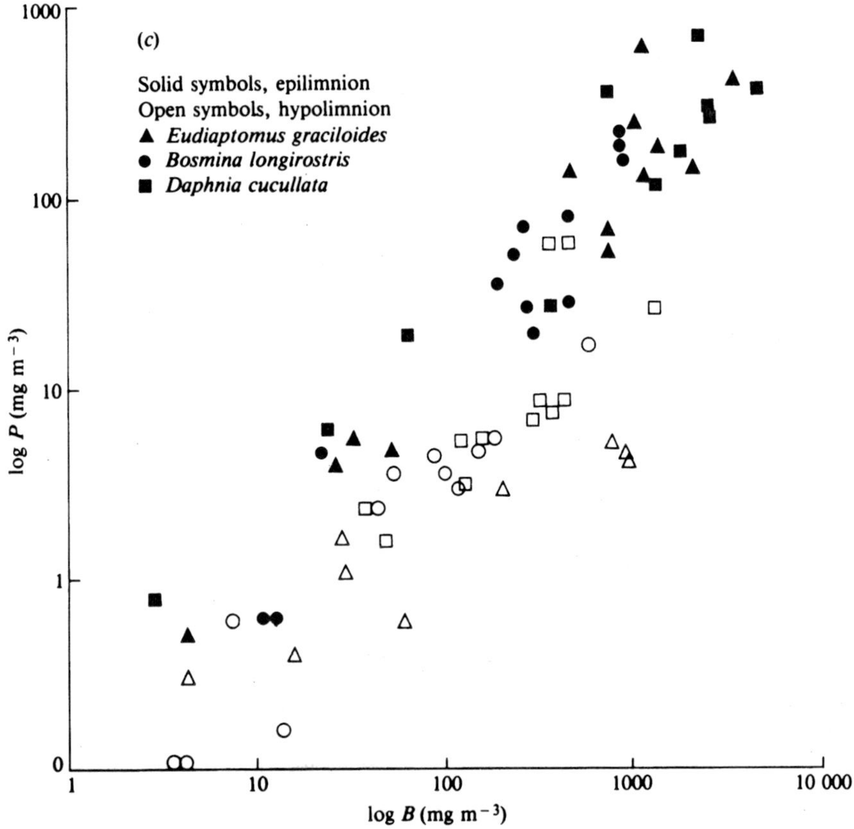

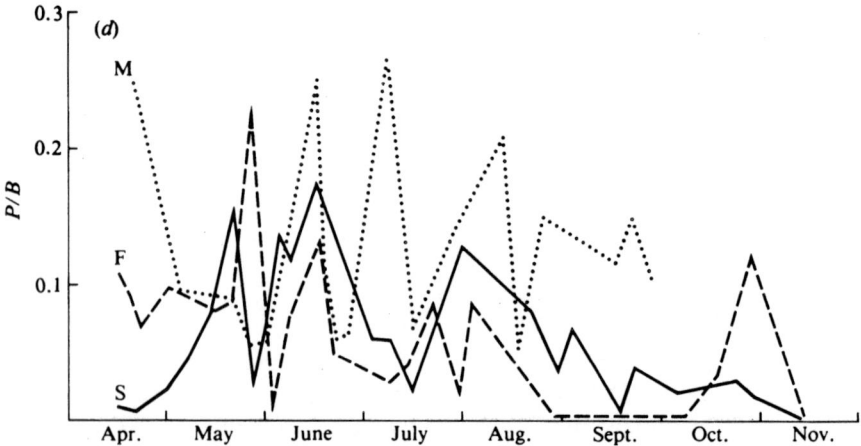

Fig. 6.4c. P/B relationships in different herbivorous zooplankton species in Lake Tałtowisko. Original from Wegleńska.

Fig. 6.4d. Seasonal variations of daily P/B ratios of *Eudiaptomus graciloides* in different Polish lakes. Original from Wegleńska. (M, Mikołajskie; F, Flosek; S, Sniardwy. All data from 1966.)

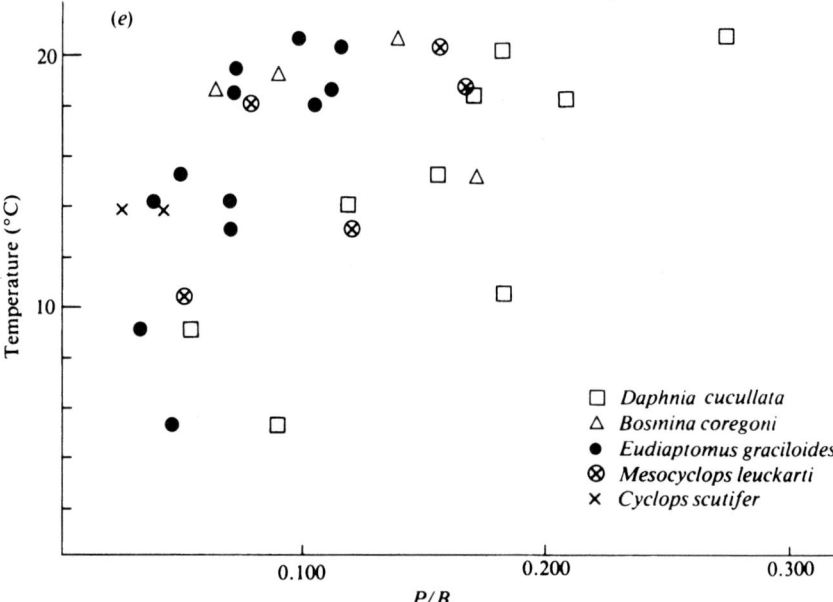

Fig. 6.4e. Daily *P/B* ratios of different zooplankton species at different temperatures. From different Soviet authors after Ivanova.

supply is seldom obvious but has been proved experimentally (Wegleńska, 1971) and statistically (Lewkowicz, 1971).

As expected the highest *P/B* ratios occur during the log phase of growth of a population. This becomes particularly clear in cases of species with marked annual cycles and synchronised development (figs. 6.4*f* and *g*).

In Øvre Heimdalsvatn *Heterocope saliens* shows to what a small extent food conditions affect the development of the population of a species which can

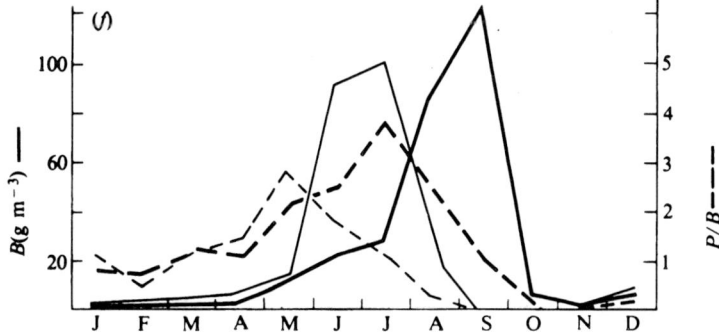

Fig. 6.4*f*. Seasonal variations of biomass and *P/B* ratio of *Eudiaptomus graciloides* (thick lines) and *Epischura baicalensis* (thin lines) in Bratsk Reservoir 1970.

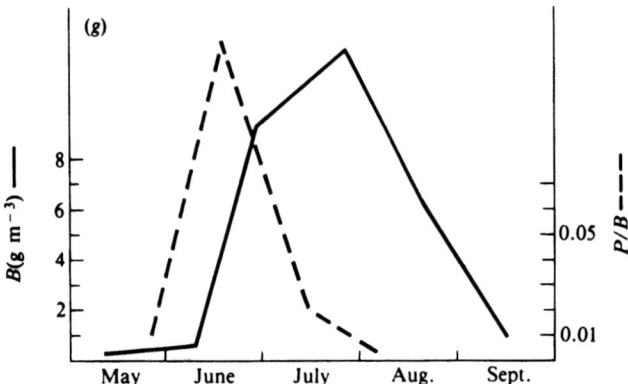

Fig. 6.4g. Seasonal variations of biomass and P/B ratio of *Heterocope saliens* in Øvre Heimdalsvatn 1971.

only produce resting eggs. The same has been indicated for *Limnocalanus macrurus* in Char Lake (Rigler, MacCallum & Roff, 1974). On the whole, in cold lakes of arctic–alpine type, hatching and survival of resting stages is the main restriction on much of zooplankton production (see also Persson, 1973). This, together with the fact that zooplankton often develops very slowly in such lakes, makes the relationship between primary production and zoo-plankton production considerably less intimate than in warmer lakes, where fast growing species can respond quickly to increases in primary production.

On a geographical scale, P/B ratios of single species seem to be primarily a function of temperature, and at the same temperature, a function of food availability. For instance, the average daily P/B values for *Bosmina coregoni* sensu lato in Kiev Reservoir (eutrophic, summer-warm), Lake Krasnoye (mesotrophic, summer-cool), Stugsjön (oligotrophic, summer-cool) and Latnjajaure (extreme oligotrophic, summer-cold) are in the approximate proportions 7 : 4 : 2 : 1. In absolute terms this means up to 3–4 times as many generations per growing season.

For *Eudiaptomus graciloides* average daily P/B ratios vary between 0.003 (Stugsjön) and 0.1 (Lake Warniak), most values recorded from other places being between 0.02 and 0.05 (see also Winberg, 1972a). During short periods of time (1–3 months) daily averages of close to 0.3 are reported from shallow, eutrophic Lake Warniak. Other calanoids show values within the same order of magnitude as *Eudiaptomus: Limnocalanus macrurus* (Char Lake) 0.006, *Diaptomus minutus* (Clear Lake, Ontario) 0.008, *Heterocope saliens* (Øvre Heimdalsvatn) 0.009, *H. appendiculata* (Lake Krasnoye, Stugsjön) 0.017–0.034, *Epischura baicalensis* (Bratsk Reservoir) 0.045, *Arctodiaptomus spinosus* (Neusiedlersee) 0.055 and *Tropodiaptomus incognitus* (Lake Chad) 0.064. A comparatively high value is attributed to *Pseudodiaptomus hessei* (Lake Sibaya) which has an annual average daily P/B ratio of 0.111, corresponding to a turnover rate of biomass of about 40 times per year. Remarkably high values in respect to ruling water temperatures occur for *Epischura*. They also keep at approximately the same level in winter and summer. In other species averages during the growing season are found to be 2–3 times higher than the annual averages and maximum values are 10 times higher. These proportions are normal for most of the other groups as well.

Values for most cyclopoids are similar to those for the calanoids or somewhat higher, obviously partly because most of them live in warmer waters. However, maximum values recorded for some calanoids are higher than the highest cyclopoid values. *Thermocyclops hyalinus* (Lake George, Uganda) has an average daily P/B ratio of 0.080. For *Cyclops kolensis* (Bratsk Reservoir) an average of 0.062 has been calculated. Values for *Mesocyclops leuckarti* vary between 0.017 (in relatively cold Lake Krasnoye) and 0.045 (in warmer Lake Karakul). The lowest value is for *Cyclops scutifer* from arctic water: 0.002 (Latnjajaure). Winberg (1972a) reports higher values for the same species from Karelian Lakes: 0.01–0.02 (Lake Krivoye, Krugloye). *Mesocyclops leuckarti* and *M. oithonoides* were also found to have relatively high values in Polish and Byelorussian lakes (Winberg, 1972a; Zaika, 1972), namely 0.01–0.08. The latter figure is the same as that for *Thermocyclops* from Lake George.

Cladocerans have generally higher values, even in colder waters. The highest daily average, *ca.* 0.3 is for *Daphnia hyalina* (Queen Mary Reservoir), but this value shrinks to about half if a few extreme daily figures are excluded. The highest value found for one day surpasses 4.0, which would mean a quadrupling of the biomass during 24 hours. Another very high value of 0.15 is reported for *Daphnia cristata* in cold Lake Krivoye (Winberg, 1972a). Other cladocerans are mainly around 0.05–0.10, except in the most oligotrophic and cold lakes where they can be as low as 0.03 (*Holopedium gibberum*, Clear Lake, Ontario) and 0.007 (*Bosmina longispina obtusirostris*, Latnjajaure). The few data reported for bigger cladocerans like *Leptodora* are between 0.04 and 0.08 (Lebedev & Mal'tsman, 1967).

For rotifers mainly overall estimates for the whole group are recorded. Where single species have been studied (Lewkowicz, 1971; Doohan, 1973) recorded values are very similar, but results from Polish lakes indicate slight differences between smaller cold-water species (*Keratella, Kellicottia, Filinia*) and larger summer species (*Pompholyx, Polyarthra*), the latter having a third to a half higher P/B ratio. Values for the whole growing season only, vary around 30, which compared to an annual average of 0.04 per day if no production occurs outside this 'growing season' or 0.08 if winter production is of the same size as summer production. Values given in IBP data reports differ between 0.022 (Zelenetskoye) and 0.17 (ELA 239). Higher values have been recorded for *Asplanchna priodonta* (Bregman, 1968; Galkovskaya, 1968) and annual averages range from 0.1 to a maximum of 0.65 (in eutrophic lowland lake,

Drivyaty), but in the IBP material only lie between 0.08 and 0.11. For smaller species (*Keratella, Synchaeta, Polyarthra*) from arctic and alpine waters ratios are only 0.007–0.018.

Generalisations are that firstly, values from all groups are within the same order of magnitude, and secondly, temperature and food differences seem to be the main reasons for variation. One would expect a more obvious correlation between animal size and *P/B* ratios, with higher *P/B* ratios for smaller organisms, but this only shows weakly when comparing rotifers with crustaceans and can be directly contradicted within one and the same group of animals. Among cladocerans only the small *Chydorus sphaericus* (Kiev Reservoir, Lake Krasnoye) appears clearly more efficient a producer than other and bigger species.

It is very difficult to draw conclusions from the data available. The *P/B* ratios referred to by IBP teams are in many cases results of rather crude estimations, sometimes they are taken from other authors for the same species or are even assumptions from data for other species. In any case, rotifer values must be taken *cum grano salis* as sampling frequency and methods are not always satisfactory.

6.2.7. Factors controlling production and biomass

6.2.7.1. Temperature

P/B ratios can be regulated by changes in *P* or in *B*. Productivity, in the sense of an ability to produce, by body growth or by reproduction, is related to specific physiology, and within a species changes with age. Production of organic matter during the life span of an animal under natural conditions is usually limited by food, production per time unit being more or less dependent on temperature. Field evidence as well as laboratory proof are ample. The role of temperature as a main regulator of field *P/B* ratios is obvious and has been touched upon sufficiently already.

6.2.7.2. Food and feeding

Successful growth of a zooplankton population can proceed only if there is enough food of suitable quality. Food particles must also be of suitable size. Changes in food size, quality, composition and concentration necessarily favour at times one species and at other times another (cf. pp. 255–8).

Detailed experiments on food intake and utilisation in plankton animals have been carried out by Monakov & Sorokin (1972) and by Gliwicz (1969a, b). As well as Gliwicz, Burns (1968) and Rigler (1972) have given special attention to size selection in filter feeders.

Size range and maximum size of particles taken is rather directly related to the size of the feeder. This is shown in table 6.1 for taxonomically fairly different groups but is particularly true for different age classes of the same filter feeders.

In spite of the fact that upper particle size limits are clearly different for animals of different size, it is notable that the optimal particle size group (i.e. the group of particles most frequently taken by, or found in the gut of the animal), is always considerably below the maximum size, and is very similar for many filter feeders. Also, within a wide range, different species and different age stages of one species overlap in particle size selection.

It is logical that, if so many zooplankton species have similar demands concerning particle size, they must become food competitors unless differences in feeding habits allow them to coexist. Such differences could be the selection of particles of different quality within one size group, or the different digestibility of certain particles by different species. Another possibility is to avoid competition by separating grazing in space and time.

Normally, optimal food conditions for filter feeders may be expected, if suitable concentrations of particles belonging to the optimal size group are available. Yet, even if small particles may be more frequently taken by an animal, the bigger ones still may be the more important food source in terms of volume. Small particles from 1 μm to 2 μm are mainly detritus and bacteria. Nannoplanktonic algae are most important in a range of 5–25 μm. Even in this size group particles of detritus may be represented, but their proportion is usually low. Larger phytoplankters and colonial forms are in order of magnitude 50–150 μm. Detritus particles of that size are rare. Large colonies of blue-green algae occasionally can be in the millimetre class.

In oligotrophic lakes the nanno-forms are normally dominant in the phytoplankton, while detritus and bacteria occur in relatively small amounts. In eutrophic lakes large forms and colonies of phytoplankton are usually much more abundant but because of more intense biological activities in such lakes, high amounts of decomposition products such as organic detritus, and consequently also of bacteria are available. In shallow, wind-exposed waters or waters with a strong through-current, bottom material can also be washed up into the open water. Even littoral vegetation, both terrestrial and aquatic, can at times contribute to the food of pelagic animals. In special cases windborne material can be an additional food source for zooplankton.

Indirectly, it is shown very clearly by a zooplankter's response by egg production etc., that naked flagellates like small Volvocales, Crypto-monadales and particularly Chrysomonades are of foremost nutritive value among algae (Saidenbach Reservoir, Ekoln etc.). Blue-green algae have for a long time been considered to be avoided by zooplankton and even to be toxic. In certain cases this assumption has been proved, but contrary evidence of their nutritive value has accumulated (Monakov & Sorokin, 1972; Infante, 1973). Highly significant evidence of blue-green algae forming zooplankton food has been reported, among others, from Lake George, Uganda.

The more zooplankton species there are present the more efficient is the breakdown of the larger particles into smaller ones. Much of the food ingested returns as faeces to the standing crop of detritus and can be taken up again by filter feeders. Naturally, decomposing faeces also favour the growth of bacteria in several ways. *Asplanchna* can even disgorge food particles (Karabin, 1974) which are then more readily broken down by other species.

In raptors, digestion may be more efficient than in filter feeders (cf. pp. 283–4) but the use of food is still less economical. For instance in carnivorous *Cyclops*, attacks on prey are seldom successful, but often result in damage to the prey organism which dies without having been eaten by the raptor. Even successfully caught animals are eaten very incompletely and at least two-thirds of the food is lost (Z. Brandl, personal communication; A. Nauwerck, unpublished). Similar losses are possible in *Leptodora* which only 'sucks' its prey (A. Karabin, personal communication). In this way the predators can contribute considerably to the support of the filter feeders. In the same way part of the larger phytoplankton can be made available to smaller animals.

Another way for micro-filtrators to share the production of larger algae is to graze upon their reproductive stages. Colonial forms lose single cells or detach them during division. Big, thick-walled peridinians or Protococcales can produce naked swarming spores in the size class of nannoplankton. Filtering zooplankton can steadily harvest such algal production, which would not be visible from population growth changes in the algae or from algae remainders in the gut of the animals.

Aggregation in swarms, both by phytoplankton and zooplankton, is probably an important factor affecting zooplankton feeding which has not been sufficiently studied.

In the past, filtration rates per unit time have been measured to evaluate the food uptake of filter feeders. Filtration rates, though dependent on food concentrations within certain limits, roughly follow body size of the animals, but are generally higher in cladocerans than in calanoids of comparable size. Animals having low filtration rates, that is small or inefficient filter feeders, are thus precluded from very dilute particle concentrations unless they are able to survive temporarily poor food conditions by other means. There is a clear trend towards larger cladocerans in oligotrophic waters. Calanoids with low filtration rates in such waters are favoured by their lower metabolic rates and reserve storing abilities. The apparent paradox that rotifers can survive and sometimes are the only zooplankton representatives in extremely oligotrophic waters can be explained by well-developed food-searching abilities.

According to Monakov & Sorokin (1972), the optimal particle concentration for filter feeders (*Daphnia longispina, Diaptomus gracilis*) in Rybinsk Reservoir is obtained at seston freshweight of $3-4$ g m^{-3}. Normal weights of bacterioplankton and phytoplankton range around 1 g and 2 g respectively, which means generally close to optimal conditions. 'Surplus feeding', that is to say decreasing food utilisation at very high food concentrations because of quick passage through the digestive tract, does occur at *ca* $8-10\,\mathrm{g\,m^{-3}}$. Similar weights for bacteria and phytoplankton are given as summer averages for surface layers of Polish lakes, but total edible seston is calculated as $4-14.5\,\mathrm{g\,m^{-3}}$.

6.2.7.3. Predation upon zooplankton

Predacious species of zooplankton, can undoubtedly play a role in controlling herbivores, or each other. But whilst large numbers of experiments have proved that this can happen, there is no clear-cut example of predatory control of zooplankton by zooplankton under natural conditions.

Anderson (1970) found experimentally, that carnivorous *Diaptomus shoshone* can eat up to 12 or more smaller cyclopoids or diaptomids per day. *Cyclops vernalis* and *Cyclops bicuspidatus thomasi* can eat 6 or more prey animals per day. Maximum numbers of small crustaceans eaten daily by *Leptodora* and *Bythotrephes* according to Monakov (1972) is around 10-30. According to Karabin (1974) *Leptodora* can eat or damage 10-15 cladocerans per day. The daily food ration for *Leptodora* ranges around 30-40% of the animals' own weight, and for carnivorous *Cyclops* around 40-80% (Monakov, 1972).

Circumstantial field evidence for the controlling effect of planktonic predators is shown for instance from Lake of Constance, (Elster, Einsle & Muckle, 1968). A direct relationship has been observed in this lake between the mass development of *Cyclops vicinus*, and the decline of the formerly dominant *Heterocope borealis*. *Cyclops* is likely to be the main predator of the young nauplii of *Heterocope*.

Cases where predation is lacking are surprisingly common particularly in arctic–alpine waters. *Limnocalanus macrurus* in Char Lake does not seem to be preyed on at all and dies of old age or from fungal infection (J. Roff, personal communication). Practically no predation on zooplankton occurs in tundra ponds (Hobbie *et al.*, 1972) and other arctic tarns (Hellström & Nauwerck, 1971). Mass death of calanoids due to infection by parasites (Eckstein, 1964) or simply by starving (Nauwerck, 1963) is known.

The role of fish in controlling zooplankton is, however, significant and is illustrated in the section on interrelationships (section 6.5.2). There is also evidence of the effect of benthic animals on the pelagic community. However, from the examples given in section 6.5 one can infer that predation by fish is likely to be the most important factor regulating zooplankton biomass, probably more important than predation by zooplankton and zoobenthos together. Occasional harvesting of zooplankton by larger, migratory organisms should also be mentioned. This may become important particularly in minor waters lacking their own predators, where amphibia (Dodson, 1970) or birds (Hellström & Nauwerck, 1971) can sometimes radically change the balance of the herbivores.

6.2.7.4. Other factors

Biomass is affected positively primarily by growth, but for certain biotopes, also by immigration. Negative effects are by losses, caused by 'natural' death, by predation, or by emigration out of the biotope. Emigration, gross or selective, can be caused by wash-out effects due to through-running streams, (Straškraba & Javornický, 1973) or by the life habits of a species changing during its life cycle. Switching of certain age classes from planktonic to benthic life, as in *Cyclops* (diapause) not only changes the biomass of the population in the plankton but also affects the P/B of the remaining part of the population in the plankton by leaving behind age classes with different reproductive or growth capacities. The same thing happens if selective predation occurs on certain age or size groups of animals or, if the biomass of the total zooplankton community is governed by predation on certain species only. Fig. 6.5 for *Cyclops scutifer* from Øvre Heimdalsvatn shows how the combined effect of different factors changes biomass and production during the season. In spring, the return of copepods from the bottom increases B quickly, while P remains low at low temperatures with little food. Increasing P in early summer

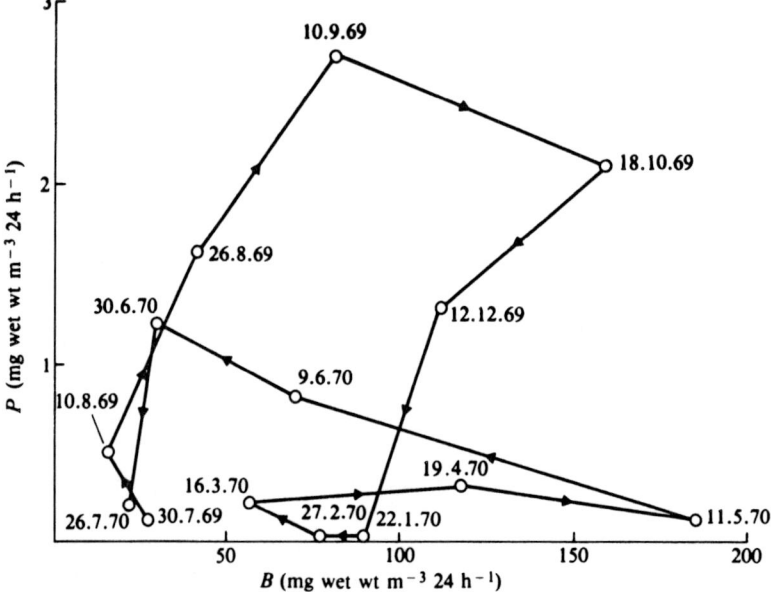

Fig. 6.5. Seasonal variation in P/B relationships of *Cyclops scutifer* in Øvre Heimdalsvatn 1969–70. Original figure from Larsson.

coincides with heavy losses of B due to wash-out through the outlet of the lake. When the population reaches adult stage, natural death (or predation) reduces B. Maximum P at high B is reached during late summer at maximum temperature and good food conditions. In autumn, growth decreases, and during early winter the population again leaves the plankton reducing B, and P becomes almost zero.

6.2.8. Ecological efficiencies

Although questions concerning food use and energy budgets of zooplankton are treated in other chapters of this book, brief mention will be made here.

Production of biomass (as body growth or population growth) in terms of energy usually represents a small fraction of energy used in respiration. Nevertheless the proportions contributing to growth and metabolic expenses can differ essentially under various conditions. For *Limnocalanus macrurus* in Char Lake, the ratio of production to assimilation is 0.074. According to Hellström & Nauwerck (1971), *Polyartemia forcipata* spends 95–98% of the energy assimilated during its life in respiration, burning up daily about the amount equivalent to the amount of carbon contained in the animal's body. Annual variations of oxygen use per day are given by Cremer & Duncan (1969) and Duncan, Cremer & Andrew (1970). The rates vary between 0.1 and 0.6 mg O_2 mg^{-1} dry weight of crustaceans.

In a number of Soviet lakes, in a series from colder, more oligotrophic waters to warmer, more eutrophic waters, M.B. Ivanova (personal communication) found no clear trends in respiration per unit biomass produced. This ratio differs from 3.6 to 46.0 and is sometimes higher for herbivores, sometimes for carnivores. The ratio of respiration to production is assumed to be approximately 1.5 for both trophic levels but in some cases is estimated as up to 8.0 for herbivores and from 0.8 to 5.7 for carnivores. The need for more direct respiration measurements is obvious.

Food utilisation in terms of food uptake by the animals is usually much higher than food use in terms of energetic benefit. In filter feeders the waste is primarily in inefficient assimilation of the food eaten. In raptors waste is primarily in inefficient food uptake and food damage whereas assimilation of the food eaten seems to be higher. According to Z. Brandl (personal communication) in *Cyclops vicinus* 80% of the *Ceriodaphnia* swallowed is assimilated.

Daphnia longispina in Queen Elizabeth II Reservoir has daily feeding rates of 0.2 to 19.2 but on average during the season stays below 4 µgC animal^{-1} (A. Duncan, personal communication). From Lake Biwa summer and winter values for the same species are between 8.6 and 2.1 µgC animal^{-1} day^{-1}, values for *Eudiaptomus japonicus* being about half that size, and egg-carrying individuals having the highest values.

The crustaceans of Lake Biwa have a daily consumption of more than their own energy content in summer and still more than half of that in winter. This would correspond to a reduction of the standing stock of phytoplankton by approximately 100% in summer and some 10% in winter, which is several times the daily primary production and in practice means another additional food source is needed. Results from London reservoirs are similar, and detritus and bacteria are the main additional food sources to algae. Dissolved organic matter is known to be used as food by cladocerans. It has been shown recently (Sorokin & Wyshkwarzew, 1973) that marine copepods at least are able to assimilate this type of food as well.

The ecological efficiency of zooplankton can also be expressed in terms of the ratio food production : feeder production in a lake. Ratios recorded from Soviet waters, table 6.3 (Ivanova, personal communication) give the following indications: as an average over the growing season, filter feeder production is always much lower, mainly less than 15% below algal production, without clear differences between different lake types. Utilisation of food by predators seems to be more efficient, up to 2–3 times higher than for filter feeders. A trend towards higher consumption rates in oligotrophic lakes is discernible, particularly for the predators. This is in agreement with the general observation that there is a relatively high proportion of algae in the seston of oligotrophic lakes and that the proportion of zooplankton is usually higher in such lakes compared to the biomass of phytoplankton.

Hillbricht-Ilkowska (1972) has compiled ratios of consumption : primary production for the plankton of various lakes. By grouping lakes roughly into oligotrophic, mesotrophic and eutrophic ones, with an 'algal', a mixed, and a 'detritus' food regime, Ilkowska comes to the following conclusions: in

Table 6.3 *Zooplankton ecological efficiencies in different lakes. Lakes arranged from cold oligotrophic to warm eutrophic. (P_f, production filter feeders; P_a, production algae; P_p, production predators; C_f, consumption filter feeders; C_p, consumption predators)*

	P_f/P_a	P_p/P_f	C_f/P_a	$C_p/P_p + P_f$
Zelenetskoye	0.15	0.11	0.68	0.32
Krivoye	0.13	0.15	0.51	1.09
Krugloye	0.31	0.18	1.19	0.62
Bolshoy Kharbey	0.02	0.11	0.07	0.31
Krasnoye	0.08	0.11	0.25	0.32
Naroch	0.11	0.35	0.48	0.81
Myastro	0.07	0.38	0.31	0.86
Batorin	0.08	0.39	0.33	0.88
Drivyaty	0.10	0.26	0.31	0.56
Karakul	0.07	0.08	0.23	0.22
Rybinsk Reservoir	0.16	0.08	0.54	0.22
Kiev Reservoir	0.21	0.03	0.16	0.10

oligotrophic lakes the larger filter feeders consume 60–100% of the phyto-plankton production, the filter feeders' production being 10–20% of that of primary production. In eutrophic lakes with an increasing number of larger species in the phytoplankton, detritus/bacteria are more important as food for the predominating smaller filter feeders. Consumption is mainly around 30% of the phytoplankton production whereas zooplankton production is mainly below 10% of it. Gak *et al.* (1972) found that almost 100% of the bacterial production was consumed by zooplankton in Dnieprowskie Reservoir and the phytoplankton was not eaten directly.

Higher levels in the food chain are stated by Ilkowska to be more efficient. Communities of invertebrate predators consume 30–100%, on average 60%, of the filter feeder production, their own production being about 20% of that of the filter feeders. This has also been found by Winberg (1972a). The production of predators is on average 2% of the net phytoplankton production.

6.3. Zoobenthos (Kajak, Bretschko, Schiemer, Lévêque)

6.3.1. Introduction

Quantitative knowledge of zoobenthos is considerably less than that for zooplankton. The comparability of quantitative benthic data is very often questionable, impeding generalisations. This is mainly because of problems of sampling methods. While sample-series from the middle of the lake may be taken as representative for the plankton of the total water body, benthic environments and communities are highly diversified, requiring numerous and properly distributed samples as well as different types of sampling gear with different sampling efficiency, and the sampling efficiencies are very often not tested properly. In addition the processing of benthos samples is very time-consuming, so that the number of samples which can be handled per person is usually a limiting factor.

The benthic environment is also much more difficult to simulate experimen-tally, to determine the rate of development, feeding etc., in near natural conditions.

These problems have to some extent limited the advancement of knowledge of benthos within IBP.

It has not been possible to touch upon all the IBP findings in this chapter because of lack of space and so we have mainly restricted ourselves to biomass, production and the factors affecting them. On the other hand some non-IBP results, where closely related, have been included.

6.3.2. Description of the lakes studied

IBP work on benthos has been carried out chiefly on standing waters and only a few running waters have been studied. The natural lakes may most conveniently be grouped into arctic–alpine, temperate and tropical.

Five arctic and three alpine lakes, situated above the tree line, scattered between latitudes 43° and 75° N were studied. All these lakes are relatively small (0.1–0.8 km²), half of them shallow, with maximum depths 4–20 m, and the others moderately deep with maximum depths of 30–40 m. They include monomictic, dimictic and polymictic lakes. In two lakes macrophytes were negligible and in four, macrophytes covered about half of the bottom. Arctic–alpine lakes are characterised by the lack of diversity of their substrates and here the growth of epipelic algae throughout the littori-profundal (Hutchinson, 1967) is extremely important for primary production. Production of phytoplankton ranges from 88 to 1100 kJ m^{-2} a^{-1}. In all the light intensity at the deepest point is sufficient to support at least epipelic algae.

Altogether thirty-four temperate lakes were examined lying between latitudes 36° and 61° N with one at 38° S. Seven are shallow homothermous lakes, three of which are small, 0.2–1.4 km², and macrophytes may cover up to 95% of the bottom, three are over 20 km² and macrophytes are less important (maximum coverage 12%) and one is a highly saline lake from Australia. Thermal stratification occurred in the other twenty-seven lakes, twenty-one of which have a mean depth < 15 m. In surface area they range from 0.6 to 385 km² and range widely in primary production level of the phytoplankton from 700 to 30000 kJ m^{-2} a^{-1}. Generally macrophytes decrease in importance the bigger the lake. The six deep lakes, ranging from mean depths of 22–175 m, were all monomictic.

Three tropical lakes, plus a series of backwaters of the White Nile, lying between latitudes 0° and 15° N, two lakes and a series of backwaters in central Amazonia, lying between 0° and 18° S, and another lake astride the equator were examined. Two reservoirs are included in those totals which have mean depths of 9 and 19 m but the natural lakes are shallow ($\bar{d} < 4.5$ m). The areas vary from 26 to 20000 km² and the annual range of water temperature is 16–36 °C. In lakes George and Chad the whole of the bottom outside the reed fringe is devoid of macrophytes whereas in the others large quantities of submerged and floating vegetation may develop.

Nine running waters studied within the programme ranged in size from small mountain streams to the Danube. All were in the temperate zone but they extended up to 2180 m in altitude and showed great variation in other physical characteristics. Most were small upland streams.

6.3.3. Main faunal components

Most of the studies deal with macrobenthos only and most of this work has been on areas outside the littoral. In all lakes (except extremely cold ones where Chironomidae do not occur (Borutskii, 1963)),Chironomidae and Oligochaeta are quantitatively important and in some circumstances

Mollusca and Crustacea make a major contribution to production. A few generalisations, characteristic of the three climatic groups of lakes, can be stressed: arctic–alpine lakes are characterised by the paucity of Mollusca, the only molluscan group of possible importance being Sphaeriidae, Odonata have not been reported and Amphipoda, Sialidae, Ephemeroptera and Trichoptera are restricted to the arctic–alpine lakes with summer temperatures above 14 °C. In stratified eutrophic temperate lakes with a mean depth of less than 15 m chironomids predominate in the profundal. In the profundal of deeper lakes Oligochaeta (most European lakes) or Amphipoda (Great Lakes of North America) predominate. In warm temperate lakes of the Soviet Union Chironomidae and/or Mollusca predominate (A.F. Alimov, personal communication). Most of the tropical lakes were dominated by Chironomidae but in Lakes Chad and Léré, in the 'sahel' zone of Africa, Mollusca were quantitatively the most important group. *Chaoborus* spp. were usually characteristic of the central zone of tropical lakes whereas amphipods were not recorded. In macrophyte-covered zones of temperate and tropical lakes other insect groups, as well as Chironomidae, Crustacea and Mollusca may predominate. Whenever meiobenthos was studied Nematoda and small Crustacea were the predominant components. In upland streams the dominant group in terms of biomass was Ephemeroptera or Amphipoda. Chironomidae were important numerically. Oligochaeta were the main group in the biomass of a small lowland stream, the Bere Stream, and Mollusca were clearly dominant in the two lowland rivers, the Thames and Danube. In Deep Creek just over 50% of the biomass was Crustacea and Mollusca. The available studies support the general conclusion that rivers progress from swift-flowing upland areas, with a fauna dominated by Ephemeroptera and Amphipoda, to slower-flowing lowland areas characterised by Oligochaeta and Mollusca.

6.3.4. Biomass*

Evaluation of biomass data is hampered by methodological difficulties which have been ignored in many studies. Therefore the accuracy and significance of the determinations are often most doubtful. These errors obviously limit the value of derived data like production.

6.3.4.1. Mean annual biomass

The range (fig. 6.6) in most arctic–alpine lakes is 0.2–0.5 g dry wt m^{-2} but Øvre Heimdalsvatn, which has exceptionally high summer temperatures, has a

---

* For conversion purposes dry weight has been taken as 5% of wet weight for *Dreissena* and other bivalves and 15% of wet weight for all other groups. The biomass of Mollusca is without shells. 1 g dry wt = 19.7 kJ.

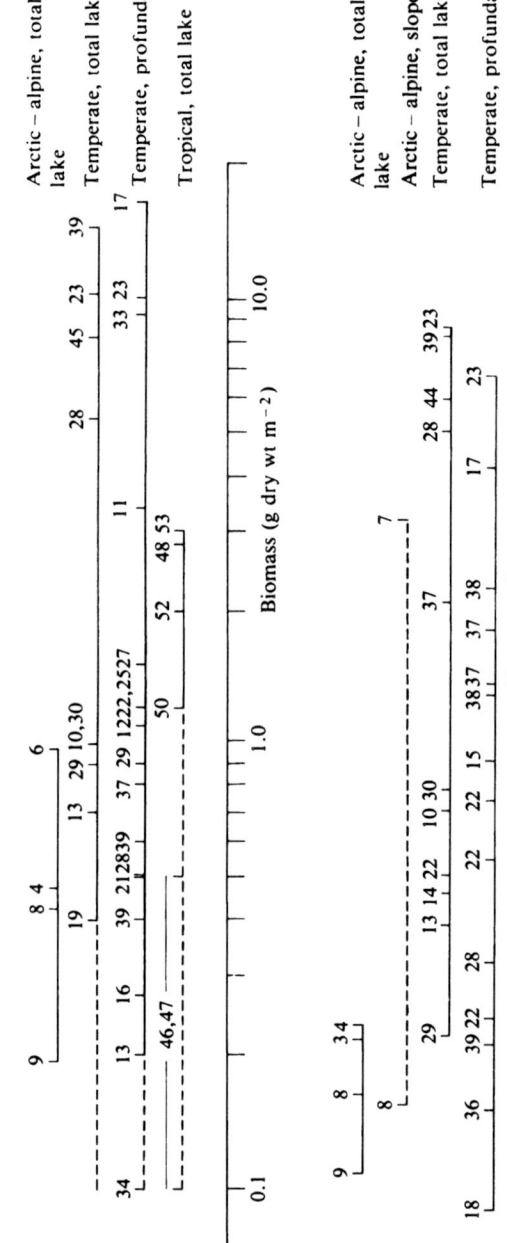

Fig. 6.6. Ranges of mean annual biomass and annual production of macrobenthos in different climatic zones. (Solid line, area where the bulk of the figures lie,; broken line, extreme range within the zone.)

Arctic–alpine lakes: 1, Char, 2, Charbiej; 3, Krivoye; 4, Krugloye; 5, Latnjajaure; 6, Øvre Heimdalsvatn; 7, Port-Bielh; 8, Vorderer Finstertaler See; 9, Zelenetskoye.

Temperate zone, lakes and reservoirs: 10, Batorin; 11, Bolsena; 12, Bracciano; 13, Chedenjärvi; 14, Drivyaty; 15, Dusia; 16, Erlaufsee; 17, Esrom; 18, Galstas; 19, Karakul; 20, Kiev Reservoir; 21, Konnevesi; 22, Krasnoye; 23, Loch Leven; 24, Lough Neagh; 25, Maggiore; 26, Marion; 27, Mergozzo; 28, Mikołajskie; 29, Myastro; 30, Naroch; 31, Neusiedlersee; 32, Obelija; 33, Ohrid; 34, Pääjärvi; 35, Rybinsk Reservoir; 36, Slavantas; 37, Sniardwy; 38, Suwa; 39, Tałtowisko; 40, Tatsu-numa; 41, Tjeukemeer; 42, Varese; 43, Vico; 44, Warniak; 45, Wewowrap.

Tropical zone, lakes, reservoirs and backwaters: 46, backwaters in Central Amazonia; 47, backwaters of the White Nile; 48, Chad; 50, George; 52, Léré; 53, McIlwaine.

mean biomass of nearly $1.0 \, \mathrm{g \, m^{-2}}$. For temperate lakes, weighted means for the total lake are scarce and they range between 0.1 and about $13.0 \, \mathrm{g \, m^{-2}}$. The few data available show strong variations: data (in $\mathrm{g \, dry \, wt \, m^{-2}}$) are 0.4 for shallow unstratified L. Karakul and 0.7 for Chedenjarvi. In Batorin, Myastro and Naroch values ranged from 0.1–1.0; 0.2–0.9 and 1.0–1.8 respectively for the period from 1959–71. Distinctly higher values have been obtained for Loch Leven (10.3), Mikołajskie (5.3) and Tałtowisko (12.5). In Loch Leven this is mainly due to a generally dense chironomid population, in the latter lakes to high animal densities (e.g. *Dreissena*) in the littoral and littori-profundal. Four tropical lakes have a mean annual biomass of $1.0–3.0 \, \mathrm{g \, m^{-2}}$. In a limited survey of Sudanese backwaters of the White Nile figures between 0.1 and $2.0 \, \mathrm{g \, m^{-2}}$ were obtained whereas in the low conductivity 'Schwarzwasser' blackwaters of the Amazon estimates were 0.02–0.17 and in 'Weisswasser' backwaters they could rise to $1 \, \mathrm{g \, m^{-2}}$. In arctic–alpine and temperate lakes the total biomass generally decreases with depth.

Biomass in the whole littoral zone of temperate lakes with submerged macrophyte cover ranged from 0.4 (Karakul) up to 33 (Esrom, including macrophyte fauna). In the littori-profundal and profundal of lakes with a mean depth $< 15 \, \mathrm{m}$ and in the upper profundal zone of deeper lakes the values obtained were generally lower than $1.5 \, \mathrm{g^{-2}}$ (Bracciano, Neagh, Maggiore, Mergozzo, Mikołajskie, Päajärvi, Sniardwy, Tałtowisko, Vico). The lowest values have been recorded from polyhumic, oligotrophic Päajärvi (0.1) and from the greatest depth (36 m) of mesotrophic Tałtowisko (0.01). Distinctly higher values have been obtained from the upper profundal zone of the warm monomictic L. Bolsena (1.7–3.4), the littori-profundal of Loch Leven (10.0) and the profundal of Lake Esrom (16.5). The high mean biomass in the profundal of Esrom is a result of high population densities of *Chironomus anthracinus* and a prolonged fourth instar larva, mainly due to temporary oxygen depletion. Few oligotrophic temperate lakes have been included in IBP but other survey work has shown that the range of *B* is similar to that for arctic–alpine lakes (Morgan, 1966). The data on benthos biomass from the literature for lakes and ponds (Deevey, 1941, 1957; Zhadin, 1950; Winberg & Lyakhnovich, 1965) fits into the range of biomass and production shown in fig. 6.6.

Usually the biomass is higher in the littoral than in the profundal zone. However it is worth stressing that whilst the biomass in the littoral of very old lakes is similar to that in more recently formed lakes, that of the profundal is higher, e.g. about $10 \, \mathrm{g \, m^{-2}}$ in 200 m in Ohrid, and about $0.5 \, \mathrm{g \, m^{-2}}$ in deeper than 500 m in Baykal (Stankovič, 1960; Moskalenko & Votinsev, 1972). This may be because the profundal fauna of these lakes has had longer to adapt to the conditions.

The biomass of the macrophyte fauna also fits into the range $1–10 \, \mathrm{g \, m^{-2}}$, (J. Kořinkova, personal communication) usually exceeding the biomass of the

Table 6.4 *Comparison of meiobenthos biomass*

Water body, stratum	B (mg dry wt m^{-2})	% of total benthic biomass	Comments
Char lake, total	90	50	B = annual mean, Nematoda, Crustacea
Zelenetskoye, total	20	11	B = annual mean, Nematoda, Crustacea
V. Finstertaler, total	40	10	Nematoda, Ostracoda, Enchytraeidae
'slope'	20	7	
'flat bottom'	70	17	*Tobrilus B* = 20 mg
Päajärvi, 0–2 m	710	17	Mean annual B, all groups considered
2–5 m	200	29	
5 m	80	61	
Mikołajskie 4–24 m	855	< 10	Mean for all zones for different seasons, all groups including first instar Chironomidae
Tałtowisko 4–36 m	435	< 10	Mean for all zones for different seasons, all groups including first instar Chironomidae
Sniardwy	225	25–30	Mean for all zones for different seasons, all groups including first instar Chironomidae
Balaton	5–10 (max 61)		Nematoda only, whole lake survey, late summer
Neusiedlersee inshore	90–360		Predatory Nematoda only, annual range of B

fauna in the sediment at the same station (Zimbalevskaya, 1966; Kořinkova, 1971). This means that the presence of macrophytes can at least double the biomass of invertebrates per unit area of bottom in the littoral. An extreme value of 40 g m^{-2} is reported from *Rorippa nasturtium* in a chalk stream (Westlake *et al.*, 1972).

Information on meiobenthos is scarce but the order of magnitude of the biomass of the meiobenthos for a number of lakes is shown in table 6.4. This indicates that under extreme oligotrophic conditions the meiobenthos may form up to 60% and in other situations up to about 20% of the total benthic biomass. Because of the probable higher turnover rate of meiobenthos it may form a higher proportion of benthic production.

Biomass data for running waters are available in four cases. The mean annual biomass at Deep Creek, USA was 5.6 g dry wt m^{-2}, at Bere Stream, England, 17.0 g dry wt m^{-2}, in the River Thames 17.0 g dry wt m^{-2} and in the River Danube 1.7 g dry wt m^{-2}. In the Bere Stream oligochaetes made up 55% of the total biomass, in the Thames Mollusca made up about 90% of the total biomass, and in the Danube about 65% of the total. The Thames and Danube Mollusca are mainly Unionidae. These running-water biomass figures fit into the same range as the biomass for standing waters.

6.3.4.2. Maximum biomass

The potential of different environments is shown by comparing the maximum biomasses recorded. In arctic–alpine lakes the maximum recorded was

3 g m^{-2} and in tropical lakes 10 g m^{-2}. The maximum biomasses observed in temperate lakes (except in situations with high *Dreissena* biomass, see section 6.3.4.1) were 24 in the profundal of Esrom Lake and 29.5 in 0–3 m at Lough Neagh, for Chironomidae only. Where *Dreissena* predominated in the littoral and littori-profundal zones biomasses as high as about 170 g m^{-2} were found (Esrom & Taltowisko). On the sloping bottom of arctic–alpine lakes maximum biomasses as high as those in that zone of temperate lakes or for the total lakes in the tropics (where extensive sloping zones were not present), were recorded.

6.3.4.3. Seasonal pattern

In arctic–alpine lakes the maximum biomass occurred between November and January inclusive with the exception of Øvre Heimdalsvatn where *Gammarus lacustris* caused a peak in July.

In the sublittoral and upper profundal of stratified temperate lakes maxima occur mostly in spring (April, May) and autumn (October). In shallow unstratified lakes additional maxima occur in summer, mainly in July.

In the tropical Lake Chad there is a clear regular maximum of Oligochaeta and Chironomidae in February, corresponding with the cold season, whereas in Lake George, which is situated on the equator, there is no seasonal variation. In the Amazon back-water lakes the maxima occur during a prolonged period from August to January varying in different lakes in relation to the recession of high river levels.

In fish-ponds the maximum biomass is normally reached 1–2 weeks after they fill with water, independent of season. The same is true for the 'flooded' zone in reservoirs.

As for the macrobenthos, the time of the meiobenthos maximum in arctic–alpine lakes was in winter with an autumn minimum, and in the shallow temperate Neusiedlersee the predominant nematodes showed a late summer minimum and a maximum from January to June; in eutrophic Masurian Lakes meiobenthos maxima varied.

6.3.4.4. Differences between years

Mean biomass and annual production may change by a factor of roughly three times between years (Lakes Krasnoye, Myastro, Naroch, Esrom, Suwa, Loch Leven). Oscillations of up to seven times are reported for the eutrophic Lake Batorin. A.F. Alimov (personal communication) believes that the variations increase with increasing trophy. Changes in the abundance of dominant species are commonly observed, as noted for two chironomid species by Morgan (1972), for *Chironomus anthracinus* by Jónasson (1972), and for *Dreissena polymorpha* by Stańczykowska (personal communication). In reservoirs the oscillations are usually high (6–8 times) for several years after filling, owing to changes in environmental conditions and the ecological succession (Mordukhai-Boltovskoi & Dzyuban, 1966; Sokolova, 1970).

6.3.5. Generation times

Determination of generation times and rate of growth, essential for production estimates, is easier and more reliable in situations where clear cohorts are discernible, which is usually for longer-living animals. Other methods of estimating are generally subjective and questionable.

In alpine lakes Chironomidae are usually univoltine as at Vorderer Finstertaler See and Lac de Port-Bielh, but may take 2–3 years in arctic lakes such as Char and Latnjajaure, to complete their life cycle. The bigger animals, like *Mysis relicta* in arctic lakes and *Sialis* in Lac de Port-Bielh, take three years. Nematodes are polyvoltine in Vorderer Finstertaler See, but in Char Lake the generation time is assumed to be one or more years (K. Prejs, personal communication).

In the profundal of temperate lakes Chironomidae are generally univoltine, but occasionally bivoltine or semivoltine. *Chironomus anthracinus* is normally univoltine, but in Lake Esrom the main part of the population is semivoltine. Where the same species occurs in different zones of a lake or in different geographical areas, the number of generations may differ accordingly. Thus within the same lake *Chironomus plumosus* has two or three generations per year in the shallows, whilst in the profundal it has one. The number of generations also varies with latitude; in shallow waters (0–2 m) in the tropics *Chironomus* spp. have up to twelve generations but the number of generations decreases with increasing latitude, altitude and depth. In shallow waters most species are polyvoltine, and the generation time may be as short as 11-15 days (Sadler, 1935; Konstantinov, 1958; Tubb & Dorris, 1965; Wójcik-Migała, 1965; Zięba, 1971; Tsuda, 1972).

Potamothrix hammoniensis takes 3–4 years to develop in the profundal (Jónasson & Thorhauge, 1972), *Tubifex tubifex* probably takes 3 years in Vorderer Finstertaler See (G. Wagner, personal communication) and *Pisidium* spp. one or more years in the littoral (Meier-Brock, 1969) but several years in the profundal of Esrom Lake (Jónasson, 1972). *Chaoborus* spp. are generally univoltine in lakes but may have several generations yearly in shallow water bodies. *Sialis lutaria* takes 1–2 years to develop, *Dreissena polymorpha* lives 4–5 years, and Unionidae 8–9 (Stańczykowska, 1964; Tudorancea, 1972); *Dreissena* starts to breed when 1–2 years old, and Unionidae when 2–4 years old.

In tropical water bodies Chironomidae and Chaoborinae took 2 to 8 weeks, and Mollusca 3 months to 1 year to complete a generation (Lévêque *et al.*, 1972).

Generation time is governed by body size and such ecological factors as temperature and food supply. The effect of temperature becomes evident by comparing species with similar ecology, e.g. *Chaoborus, Chironomus*, in different climatic zones (Borutskii, 1963). Food consumption depends on the amount present in the environment and the feeding ecology of a species (Konstantinov, 1958; Kajak, 1968; Azam & Anderson, 1969; Jónasson, 1972). The long generation times of species which feed below the mud surface (Tubificidae, *Pisidium*) compared with surface feeders (Chironomidae) may result from the different quality of food. Tubificidae however can take only one year to reach sexual maturity and breed (T.L. Poddubnaya, personal communication) but individual variation as well as differences between years is large (Wagner, personal communication).

6.3.6. Production

6.3.6.1. Validity of estimates

The validity of production estimates is mainly dependent on the accuracy of determination of biomass and growth rate. The use of generation times obtained from laboratory cultures is questionable since significant differences from natural growth rates occur, even under close to natural conditions (Lukanin, 1957; Konstantinov, 1960; Kajak, 1964; Kimerle & Anderson, 1971). In situations where cohorts overlap and are not clearly separated and the generation time is short, the latter is difficult to assess. Yearly production estimates for the total benthos are often too low ($P/B = 2-4$) because a long generation time of 1 year or more is assumed (many papers in Kajak & Hillbricht-Ilkowska, 1972; data from Zaika, 1972). As shown above this time scale is commonly true for the profundal but doubtful for shallow water. This becomes evident when production estimates obtained from measurements on cohorts are compared with those based on consumption of benthos by fish or other predators.

The average daily production in situations when the benthos is heavily eliminated by predators is about 17% of the biomass (table 6.5).* On the other hand, if one assumes only one generation per year, the daily P/B ratio would be about 2%, and accordingly 4% for two generations per year. For a cohort lasting 1 year, the yearly production, according to Waters (1969), would be 300–400% of its mean biomass; if the growth only occurs during 180 days, 300–400% divided by 180 gives about 2% daily. This is considerably less than that estimated by elimination. Thus benthos production is commonly either underestimated or overestimated depending on the accuracy of the methods and more precise studies are required.

* Assuming that elimination = production, which to some extent underestimates the production.

Table 6.5 *Turnover times of benthos, as estimated from fish production and their food requirements or from the exploitation of benthos by fish or invertebrate predators (enclosure method) (recalculated from the original data)*

Fauna	Habitat	Turnover time (days)	Daily increase of biomass %	Author	Remarks
Benthos, small Chironomidae dominant	eutrophic lakes	8.3 (5.0–25)	12.0	Kajak (1972)	8 series 6–12 days long, sampling every 3 days
Chironomidae on aquatic plants	fish-ponds, rather low fish stock	5.5 (1.6–125)	18.2	Maksimova (1961)	average from 6 sampled every 2 weeks all summer
Chironomus plumosus	fish-ponds, rather low fish stock	17.0[a]	6.0	Maksimova (1961)	average from 6 sampled every 2 weeks all summer
Annelida, Arthropod, small molluscs	River Thames	18.0	5.6	Mann (1965)	whole year
Insecta, Amphipoda	fish-ponds	4–10	25–10	Welch & Ball (1966)	enclosures in fish-ponds
Mostly Chironomidae	fish-ponds	5–17	20.0–6.0	Hayne & Ball (1956)	growing season; reversal of stocked and control parts of the ponds
Chironomidae	Marion Lake – 0.5 m	7	14.3	Kajak & Kajak (1975)	2 days' experiment, small enclosures
Insects	experimental	2–2.5[b]	50.0–40.0	Warren et al. (1965)	74 days, 5 streams
Mostly insects	Dartmoor stream	14–41	7.1–2.7	Horton (1961)	whole year
Mostly insects	Horokiwi stream	2.4–9.0	41.7–11.1	Allen (1951)	whole year
Mostly Chironomidae	carp ponds	6	16.7	I. Wójcik-Migaļowa (unpublished)	growing season
Mostly Chironomidae	carp ponds	5	20.0	Wolny & Grygierek (1972)	for the period end of June/beginning of July

[a] Range of exploitation by fish – 0–100% daily.
[b] Unclear calculation.

The prolonged appearance of young individuals (Kajak, 1968), the production of exuviae, mucus or silk (Kimerle & Anderson, 1971), and the production of Oligochaeta due to the regeneration of their tails, removed by grazing, (Poddubnaya, 1962; Kajak & Wiśniewski, 1966) are commonly ignored, but can be important for production.

In most cases only some of the dominant components of the macrobenthos have been investigated. Meiobenthos (section 6.3.4.1) and macrophyte fauna production may be substantial but were very seldom evaluated during IBP and then mostly based on rough assumptions.

6.3.6.2. Annual production

Arctic–alpine lakes are usually much less productive than tropical lakes but on the sloping bottom of Lac de Port-Bielh annual production is as high as $300 \, kJ \, m^{-2} a^{-1}$, which is of the same order as the most productive tropical lakes (fig. 6.6). On the other hand the range of production in temperate lakes was greater than the total span for arctic–alpine and tropical lakes.

Annual production for the whole lake for natural lakes ranged in the temperate zone from $20–900 \, kJ \, m^{-2} a^{-1}$. All the Russian lakes studied have a production below $110 \, kJ \, m^{-2}$. Shallow, polymictic water bodies (Leven, Warniak) and dimictic lakes with a high production in the littoral and littori-profundal (Tajtowisko, Mikojajskie, Esrom) yielded values in the range of $500–900 \, kJ \, m^{-2}$. In the littoral of these lakes annual production can be as high as $900–1500 \, kJ \, m^{-2}$. A value of $1300 \, kJ \, m^{-2}$ has also been obtained from the shallow, unstratified L. Wewowrap (Australia). On the other hand much lower values for littoral benthos have been found, e.g. Karakul (110) and Krasnoye (50).

The share of the total production of the littoral benthos contributed by the macrophyte fauna differs greatly, from significantly lower to much higher figures than those for the fauna in the bottom sediments. In shallow exposed lakes the production close to the reed beds was 160 at Tjeukemeer and $30 \, kJ \, m^{-2} a^{-1}$ at Neusiedlersee, but negligible in the open water. Annual production of the profundal zone of temperate lakes yielded values below $100 \, kJ \, m^{-2}$ (Mikojajskie, Tajtowisko, Galstas, Slavantas, Dusia, Krasnoye) with the exception of Esrom with a value of 420. The other values above $100 \, kJ \, m^{-2}$ in fig. 6.6 for the temperate profundal are obtained from littori-profundal strata.

Although energy budgets for benthic species and communities are difficult to obtain in natural conditions, they have been estimated for some water bodies (see table 6.6). Respiration values have been taken from laboratory measurements and recalculated for the temperature in the lake. Various communities appeared to produce a given amount of biomass at variable respiratory expenditure. In Lake Zelenetskoye, the zoobenthos is dominated by Chironomidae, but in Krivoye by Ephemeroptera and Crustacea which are more primitive groups with a longer life cycle. Alimov suggests that the latter probably require more energy for their development than the more recently evolved, fast-cycling Chironomidae.

The relationship between primary production and benthos production is examined in fig. 6.7 both for the total lake and for the profundal. The scatter of points is rather high but there is a distinct trend towards higher zoobenthos production at higher phytoplankton production. The proportion benthos/primary production tends to be higher in more oligotrophic lakes, up to 9% in polar ones, and can be high in eutrophic ones, as in Kiev Reservoir –

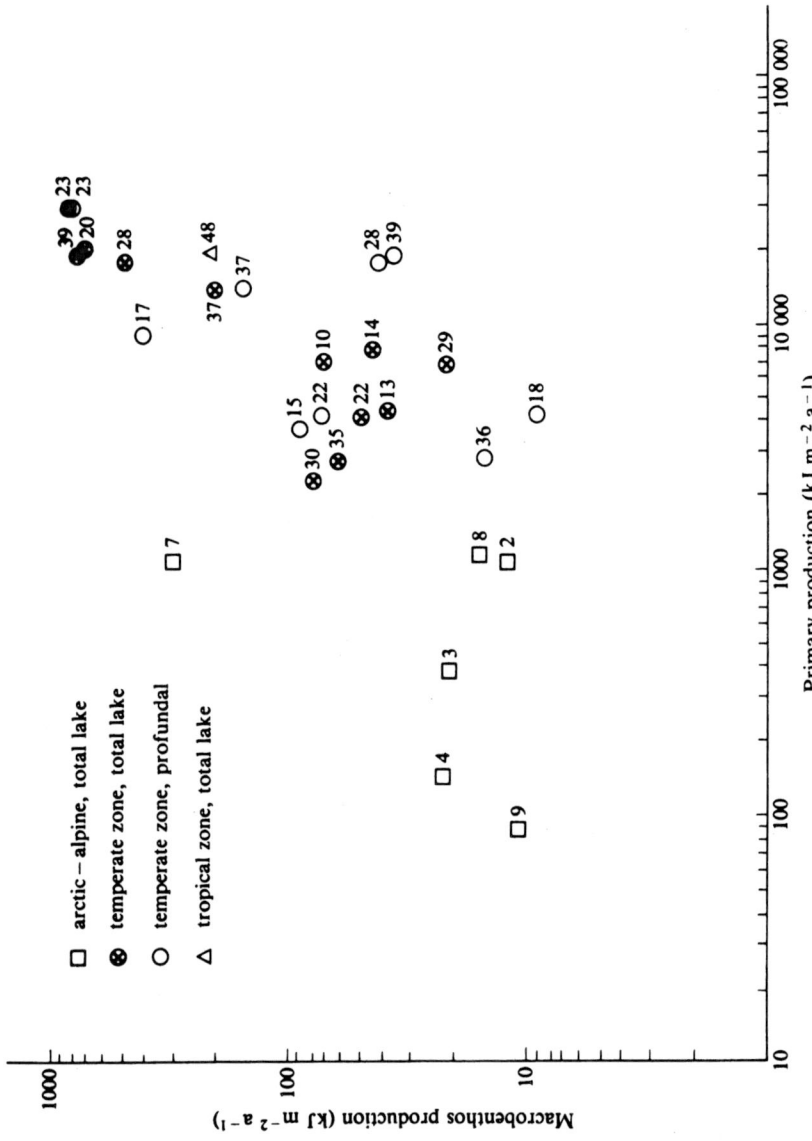

Fig. 6.7. Relationship between primary production and macrobenthos production in water bodies from different climatic zones. The numbers refer to the lakes listed in fig. 6.6

Table 6.6 *Bioenergetic calculations for several benthic communities for the ice-free period*

Community	Number of species	Depth (m)	Biomass (kJ m^{-2})	Production (P) (kJ m^{-2})	Respiration (R) (kJ m^{-2})	P/R (%)
Lake Zelenetskoye						
Apatania sp.						
— *Smittia septentrionalis*	54	0–2	8.8	21.3	54.7	39.0
Stictochironomus histrio						
— *Heterotrissocladius*	37	2–10	3.3	6.3	14.3	44.1
Tubifex tubifex						
— *Prodiamesa bathyphila*	19	10–25	5.4	11.7	24.7	47.0
Tanytarsus gregarius						
— *Pagastiella orophila*	20	1.5	8.4	13.4	36.8	36.0
Lake Krivoye						
Gammarus lacustris						
— *Ephemera vulgata*	41	0.5–4.0	16.7	4.2	142.9	3.0
Sphaerium suecicum						
— *Ephemerata vulgata*	30	4.0–7.0	15.9	10.0	118.3	8.5
Pontoporeia affinis	31	7.0–18.0	4.2	2.9	35.5	8.2
Sergentia caracina						
— *Pontoporeia affinis*	12	18.0–30.5	4.6	10.4	48.9	22.0

After A.F. Alimov (personal communication).

4.6%, although often below 1% (A.F. Alimov, personal communication and fig. 6.7); see also Brylinsky (chapter 9). It can be much higher for littoral, sublittoral and other productive environments where it can reach levels similar to those obtained for zooplankton of 20% and more (Hall, Cooper & Werner, 1970).

6.3.6.3. *P/B* ratios

P/B ratios are widely used to calculate production from biomass data, particularly when reproduction is continuous and cohorts are not discernible. They are however often arbitrary and tend to underestimate production. As they are very dependent upon conditions and on species (Winberg et al., 1973) they should not be generalised, and applied to other situations without care. This is clearly shown by the ratios listed in table 6.7, which show considerable variations, even for the same species in a particular lake, and between different years. Even so, the table does not contain ratios for the most difficult situations as with cohorts overlapping in species with life cycles of a few weeks' duration.

Most of the production calculations agree well with the excellent theoretical calculation by Waters (1969) that *P/B* for a single cohort is usually 2.5 to 5.0, the extreme values being 1.5 and 7.8 with a mode of about 3.5, depending on the pattern of elimination and type of growth. However, most benthos production data should be treated as approximate. The various methods applied to calculate the production do not give seriously different results if the

Table 6.7 *Chosen examples of the production to biomass (P/B) ratios based on annual production and mean annual biomass*

	Gener-ation time (years)	Water body	P/B per year
Crustacea			
Gammarus lacustris	2	Krivoye	0.6
Pontoporeia affinis	2	Krivoye	
3–6 m			1.5
7–9 m			0.9
12–30 m			0.2
Candona candida	0.3	Krivoye	3.2
Chironomidae			
Chironomus anthracinus	1	Loch Leven	$\begin{cases} 1.1\text{--}1.7 \ (1970\text{--}71) \\ 2.4\text{--}4.1 \ (1971\text{--}72) \end{cases}$
Chironomus plumosus	0.3	Karakul Lake	15.8
Chironomus plumosus		Uchinsk Res.	1.9–3.7
Chironomus plumosus	0.5	Jeziorak Lake	10.8
Glyptotendipes paripes	1	Loch Leven	5.5
Glyptotendipes gripekoveni	0.2	Jeziorak Lake (*Stratiotes aloides* beds)	15.5
Chironomus anthracinus	2	Esrom Lake	3.8(0.4)
Psectrocladius sordidus	1	Lac de Port-Bielh	4.1
Zavrelymia melanura	1	Lac de Port-Bielh	3.1
Lauterbornia coracina	1	Finstertaler See	4.5
Heterotrissocladius marcidus	1	Finstertaler See	6.6
Cricotopus sp.		River Thames (*Nuphar lutea* beds)	11.0
Tanytarsus gr. *mancus*	0.2	Krugloye	4.5
Procladius nigriventris	0.3	Krugloye, Krivoye	4.5
Sergentia coracina	0.6	Krivoye	2.9
Other Insecta			
Chaoborus flavicans	1	Esrom Lake	1.7
Sialis lutaria	3	Lac de Port-Bielh	1.3
Simuliidae		Bere Stream	7.7
Ephemera vulgata	2	Krivoye, Krugloye	1.2
Tubificidae			
Potamothrix hammoniensis	3–4	Esrom Lake	0.7
Isochetides nevaensis	1	Rybinsk Reservoir	4.0
Limnodrilus hoffmeisteri	1	Ivankovsk Reservoir	3.3
Tubifex tubifex	1	Bratsk Reservoirs	2.5
Mollusca			
Melania tuberculata		L. Chad	4.4
Bellamya unicolor		L. Chad	5.8
Cleopatra bulimoides		L. Chad	2.6
Corbicula africana		L. Chad	2.6
Pisidium casertanum	several	Esrom Lake	0.2
Unionidae		Thames	0.3
Sphaericum suecicum	1.5	Krugloye	1.5
Pisidium crassum	1.5	Krugloye	1.4
Pisidium lilljeborgi	1	Zelenetskoye	1.5
Pisidium nitidum	1.5	Zelenetskoye	1.1

From **IBP** data.

generation time and rate of growth are properly estimated (Konstantinov, 1960; Sokolova, 1971).

6.3.7. Feeding of non-predatory zoobenthos

The following discussion relates mainly to animals living on soft bottoms. For an extensive review of animals feeding on aquatic macrophytes, the reader is referred to Gaevskaya (1966). Trophic relationships of mainly running-water species of aquatic insects have been recently reviewed by Cummins (1973).

6.3.7.1. Method of feeding

Epipelic and sedimenting planktonic algae, detritus and bacteria form the main food items of benthic animals. Many of them feed at the sediment surface, even those building tubes in the deposits, either filter feeding or collecting food particles from the surface. Both feeding methods may occur within one species depending on external conditions (review in Oliver, 1971; see also Iovino & Bradley, 1969; Kajak & Rybak, 1970; Jónasson, 1972; Konstantinov, 1972). This group consists mainly of chironomids.

Representatives of the Tubificidae, Sphaeriidae and Nematoda, however, feed deeper in the sediments and thus exclude algae from their diet. Tubificidae feed several centimetres below the sediment surface (Poddubnaya, 1962), whilst most species of *Pisidium* live in a shallow burrow a few millimetres deep, filtering food particles from the water and mud (Meier-Brock, 1969). In the meiofauna, a few species of Nematoda have been found to live in deeper substrate layers and bacteria are believed to form their main food (Ott & Schiemer, 1973).

6.3.7.2. Nature of food

The value of different groups of algae as food for benthos is different. Diatoms form good food, while green algae are hardly digested. The nutritive value of blue-greens has been questioned but most authors now think that they are readily digestible (Borodičova, 1962; Kajak & Warda, 1968; Iovino & Bradley, 1969; Konstantinov, 1972; and others). It has been pointed out (Sorokin & Meshkov, 1959; Sorokin, 1966; Karzinkin, 1967) that digestibility very much depends on the physiological state and age of the cells. Decaying organic material is generally considered to be of low nutritive value compared with living cells. Its importance within the food chain is probably higher as a substrate for bacterial growth, rather than as a food item itself. This may also explain a time-lag of several weeks to a month which occurs after high input of allochthonous matter before maximal development of benthos takes place in high alpine lakes (Bretschko, 1974). Such periods seem to be necessary for adequate bacterial colonisation and decomposition. The role of bacteria is still not well known but they probably form the main food supply for many benthic species. According to the review by Kuznetsov (1970), bacteria form on average about 5% of the organic matter in the mud. Their biomass is up to 300 g fresh weight m^{-2} in oligotrophic lakes, Finstertaler See (Tautermann, personal communication) and up to 1500 g m^{-2} in eutrophic lakes, thus significantly exceeding the biomass of sedimented algae.

6.3.7.3. Food selection

A very distinct preference for algae as opposed to detritus has been shown by several chironomids and its proportion in the gut contents can be considerably higher than in the surface layer of the mud (Kajak, 1968). Selection within the algae usually seems to be small (Kajak & Warda, 1968), but some exceptions have been found: *Trissocladius grandis*, was found to live exclusively on diatoms (Mozley, 1970). Tubificidae select small mud particles as food (Poddubnaya, 1962) and show a clear preference for certain types of organic material (Chua & Brinkhurst, 1973).

The examination of gut contents, especially in aquatic insects, has shown that food composition is highly variable and strongly dependent on its availability. For example, strong annual differences in the gut contents have been observed (Kajak & Warda, 1968 in chironomids; for other insects see Mecom, 1972; Cummins *et al.*, 1966; for Viviparidae, Stańczykowska, 1969; Stańczykowska, Plinski & Magnin, 1972).

6.3.7.4. Food consumption

Most food ration estimates have been obtained in the laboratory, during short experiments with high food concentration and optimum availability, so they are obviously overestimated. Data from close to natural conditions range from a few per cent to 15–50% of the body weight estimated roughly according to the time of food passage through the gut (Kondratev, 1969 – Unionidae; Izvekova & Lvova-Kachanova, 1972; Mikheev, 1967 and others – *Dreissena*). Times of 3–4 hours have been found for Chironomidae and 2 hours for Amphipoda (Sorokin & Meshkov, 1959; Borodičova, 1962; Kajak & Rybak, 1970; Hargrave, 1970, 1972b). Very high figures for Tubificidae, up to 800% (Poddubnaya, 1962) and 1500–2000% (R. Wiśniewski, personal communication) probably result from only a small part of the material swallowed being digestible and having a high nutritive value. Although Unionidae and *Dreissena* have a small daily food ration, they filter a very much larger quantity of food particles in obtaining the food which they ingest.

Sometimes the food ration increases when the value of the food deteriorates, and, to some extent with increase of temperature (review by Levanidov &

Kurenkov, 1973). Assimilation of food differs significantly (40–99% according to the review by Cichon-Lukanina & Soldatova, 1973) being highest for animal food.

6.3.7.5. Food abundance as a limiting factor

To what extent does food limit benthic production? Experimental work on this problem is scarce. Schiemer, Duncan & Klekowski (1980; and see Duncan, Schiemer & Klekowski, 1974) showed a strong effect of bacteria concentrations on the growth and reproduction rates of a benthic nematode. The range at which growth and reproduction took place was close to the bacterial densities known from sediments of meso- and eutrophic lakes. Eisenberg (1966) demonstrated that the addition of spinach had an enormous influence on the fecundity and abundance of *Limnaea elodes*. Although more complicated to interpret, a number of field studies indicate the importance of food supply for benthic production. The availability of food to profundal species during the annual cycle is strongly linked to the annual rhythm of the whole lake system. Jónasson (1972) and Sokolova (1971) found that in the profundal zone Chironomidae only grow during short periods, in spring and autumn, when both a high supply of sinking algae and good oxygen conditions coincide. The supply of diatoms in the spring allows high accumulation of fat (up to 20% of body weight, compared with almost zero in other seasons) and subsequent emergence.

Changes in the biomass and production of benthos, other than Mollusca, correlated well with the course of tripton sedimentation in the sublittoral of the eutrophic Lake Mikołajskie (Kajak, Hillbricht-Ilkowska & Pieczyńska, 1972). Only bigger and heavier phytoplankton species have the chance to reach the deeper sediments during stratified conditions as they are less likely to be eaten by zooplankton or decomposed in the water column (Jónasson, 1972). Increasing water temperature and water depth will reduce the amount of sedimenting detritus of high nutritive value. This is probably the main reason for the low ratio of benthos/primary production in the tropics (fig. 6.7) and in lakes with deeper epilimnions (Hargrave, 1973; Brylinsky, Chapter 9).

Natural or experimental improvement of the food supply as a rule stimulates development rates and increases the abundance of benthos. Kajak (1968), Hargrave (1970) and others proved this by addition of different kinds of food during experiments *in situ* in lakes and similarly Egglishaw (1964) obtained increased benthos abundance by addition of detritus in running waters. The annual pattern of zoobenthic biomass and production in Vorderer Finstertaler See is regulated by the availability of appropriate food. The main energy resource is allochthonous organic material which enters the lake mainly during ice-thaw (June). It serves first as a substrate for microbenthos, which grow slowly because of low phosphorus and nitrogen concentrations.

The best feeding conditions for meio- and macrobenthos do not occur before autumn when this food is in suitable form (G. Bretschko & Vivl, personal communication).

Manuring of fish-ponds with organic material increases benthos abundance and the population dynamics (period of occurrence and size of the population) is strongly dependent on the amount of organic matter added (Zięba, 1973). The same phenomenon can be observed in polluted rivers (Edwards, Egan, Learner & Maris, 1964; review by Sivko & Lyakhnovich, 1967; Kajak, 1966, 1968). The fertilisation of ponds (Winberg & Lyakhnovich, 1965) and lakes (Petrov, 1972) always leads to increased fish production, due to a higher production of benthic food, although this is not always expressed in an increased benthos biomass. Morgan (1966) found that in small Scottish oligotrophic lakes benthos biomass increased 16–25 times following the addition of inorganic fertilisers.

Refilling of fish-ponds and flooding of dried up zones in reservoirs has a temporarily stimulating effect on benthos growth, largely due to decomposition of accumulated organic material (Mordukhai-Boltovskoi, 1955; Kajak, 1962; Winberg & Lyakhnovich, 1965; Yount, 1966).

We can conclude that quality and quantity of food are the main factors limiting benthos production. This raises the problem to what extent the food supply is exploited?

6.3.7.6. Food utilisation

Kajak & Warda (1968) found that about 7% of the algae from the upper centimetre of the sediment disappears in one day under natural benthos densities. Similarly Hargrave (1971) found that *Hyalella azteca* in Marion lake consumes about 7% of the benthic algae present daily. There seems to be a delicate balance between food supply and its rate of utilisation. On the other hand experiments with the acclimatisation and introduction of *new* species (Crustacea, Mollusca, etc.) prove that significantly higher levels of biomass and production may be obtained (Karpevich & Mordukhai-Boltovskoi, 1966; Ioffe, 1972). The same is true for spontaneous immigration – e.g. the appearance of *Dreissena polymorpha* in many European waters greatly increased the total benthic biomass, usually changing the composition, but not diminishing the biomass of other benthic organisms.

6.3.8. The role of benthic invertebrate predators

Information on predator–prey relationships is based on four types of observation: (1) relationship between predator and prey abundances and their pattern in time and space; (2) gut analysis data; (3) laboratory experiments; (4) experiments in close to natural conditions.

6.3.8.1. The proportion of predators in the zoobenthos

Quantitatively reliable investigations are scarce and often impeded because the zoobenthos is studied only in part. Most studies deal with predatory chironomids, which very often form a high percentage of the total macrobenthos. In many situations they are the dominant or even the only constituent of the macrobenthos, especially when *Procladius* spp. occur. Based on annual mean biomass data, predatory invertebrates do not usually exceed 20% of the total benthos in arctic–alpine and temperate lakes. Usually they are below 10% but in certain cases they may exceed 50%.

6.3.8.2. Food composition

Chironomidae/Tanypodinae prefer small chironomid larvae as food but Copepoda, Cladocera, Protozoa, Tubificidae and large algae are found regularly in their guts and other items sporadically (Belyavskaya & Konstantinov, 1956; Kajak, 1958; Konstantinov, 1961; Luferov, 1961; Izvekova, 1967; Roback, 1969; Tarwid, 1969; Kajak & Dusoge, 1970). All the species of the genus *Cryptochironomus* which have been studied (*C. defectus, C. pararostrasus, C. fuscimanus, C. camptolabis, C. monstrosus, C. rolli*) appear to be predators, preferring Chironomidae and Tubificidae to Crustacea (Konstantinov, 1961). Many species of the genera *Cricotopus, Psectrocladius, Endochironomus* and even *Chironomus* are facultative predators, taking animal food as an additional diet (Konstantinov, 1961; Izvekova, 1967; Jónasson, 1972), but the full extent to which this happens in the field is not known. *Protanypus forcipatus* feeds on small chironomid larvae but mainly on small benthic Crustacea. *Heterotrissocladius subpilosus* and *Pseudodiamesa* sp. feed on planktonic Crustacea by migrating to the water surface during the night and eating the plankton caught in the surface film (A. Nauwerck, personal communication). Although young chironomid larvae are very often a preferred food, cannibalism is seldom observed.

Algal food is taken regularly by Tanypodinae, but it is assumed that it is only an important and complete food for young instars. For older instars it serves as an additional food only. Some species can survive on a pure algal diet but cannot pupate without animal food. In contrast, *Psilotanypus imicola*, cannot complete its larval development on animal food alone. Heleidae and Hydracarina are supposed to feed mainly on benthic Crustacea but they attack Chironomidae and other groups when they are available (Pchelkina, 1950; Kajak & Pieczyński, 1966). Triclads prey efficiently on *Asellus*, Oligochaeta and Gastropoda (Reynoldson & Davies, 1970). Predators feeding on triclads are described by Davies & Reynoldson (1971).

Very little is known about the feeding behaviour of meiobenthos, although some components are said to be predators. Predatory instars of *Cyclops tatricus abyssorum* have been observed feeding on small nematodes (G. Bretschko, personal communication).

The composition and amount of the food may differ significantly depending on the conditions, especially the availability of food (Luferov, 1961; Kajak & Dusoge, 1970). Many authors stress the dependence of the food consumption on the conditions of the experiment; e.g. the presence of mud or algae, the choice of prey, the exposure time (Konstantinov, 1961; Kajak & Pieczyński, 1966; Izvekova, 1967; Kajak & Dusoge, 1970). According to Izvekova (1967) *Cryptochironomous defectus* only feeds on Tubificidae in natural conditions but not in laboratory ones.

6.3.8.3. Feeding rates

Because of the highly opportunistic feeding behaviour of invertebrate predators, and their strong dependence on the amount and availability of food (Kajak, 1968), feeding rates may be estimated realistically by experiments in close to natural conditions only. Feeding rates of 10–20% of body weight per day have been estimated in this way for Tanypodinae, Heleidae and *Crangonyx* (Kajak & Dusoge, 1970; Kajak & Kajak, 1975). Under more artificial conditions, feeding rates of predatory Chironomidae and *Erpobdella* oscillate between 50 and 150% with extremes of up to 200% (Konstantinov, 1961; Luferov, 1961; Izvekova, 1967). As for food composition, the amount of food depends mainly on the abundance and availability of prey organisms and the amount of food consumed influences in turn the rate of development (Azam & Anderson, 1969; Kajak & Dusoge, 1971; etc.). Various species, especially in the genus *Procladius*, are able to survive long periods of complete starvation, up to several months depending on prevailing temperatures (Luferov, 1961).

6.3.8.4. Effect on prey

Since feeding rates of invertebrate predators are mostly based on laboratory experiments, their importance in the consumption of benthos is often overestimated. In spite of this they still play a significant role in the ecosystem. Tanypodinae have been found to reduce significantly the abundance of their prey organisms and consequently influence biomass and production of the prey population (Belyavskaya & Konstantinov, 1956). The same has been shown for stoneflies and *Sialis* (Warren et al., 1965; Azam & Anderson, 1969). Experimental investigations (R. Wisniewski, personal communication) show that the cropping of tails of Tubificidae is done mainly by invertebrate predators (Tanypodinae, *Cryptochironomus*) and less by fish. A less obvious effect is the strong exploitation of young prey organisms, mainly chironomids, which possibly reduces their abundance to the capacity of the environment by removing the surplus. Furthermore, invertebrate predators are able to adjust quickly to changing situations because of their mobility, broad food spectrum and ability to survive periods of starvation. The regulating effect of invertebrate predators may well be substantial even in cases where they form a low proportion of the total biomass.

6.3.9. Effects of fish on benthos

Fish predation has often a very profound effect on the qualitative and quantitative composition, size structure and temporal patterns of the zoobenthos. More detailed discussion is given in section 6.5.

6.3.10. Competition and other relationships

Many examples of intra- and interspecific competition are described in the literature but the mechanisms involved are poorly understood. Only a few selected examples are given here.

6.3.10.1. Intraspecific competition

It is believed that competition for food is the most important consequence of increasing population densities. At high densities, the average body size of *Dreissena* is smaller (Stańczykowska, 1964), survival rates and fecundity in *Limnaea* decrease (Eisenberg, 1966) and feeding and growth rate decline in Chironomidae (Kajak, 1968).

In Chironomidae older instars may suppress the younger ones. The best-known example is that of the profundal population of *C. anthracinus* in Lake Esrom (Jónasson, 1972). The majority of the population has a generation time of two years. In alternate years the newly laid eggs are nearly completely eliminated by older instars creating a clear two-year rhythm of emergence. Only in years of low abundance, caused by other factors, is the alternate-year generation able to survive.

In experiments with *C. plumosus*, Kajak (1968) showed that the growth of younger instars is suppressed by both high densities of older instars and by periodical stirring of the mud. This may be explained by the mechanical disturbance of larvae by the tube building activity of the older ones.

6.3.10.2. Interspecific competition

The mechanisms of interspecific competition are even more complicated to assess, but some clear examples can be given. The effect of the density of *C. plumosus* on its own survival and that of *C. anthracinus* is shown in table 6.8 (Kajak, 1968), *C. anthracinus* only appeared in the less crowded cages. The

Table 6.8 *Influence of artificial crowding of* Chironomus plumosus *larvae on the numbers and growth of themselves and the related* Chironomus anthracinus (*Lake Sniardwy, 8 m*)

Density of C. plumosus	Numbers after 1 month (m^{-2})		Length after 1 month (mm)	
	C. plumosus	C. anthracinus	C. plumosus	C. anthracinus
1 × natural density	230	210	27.3	14.4
3.5 × natural density	270	90	25.1	10.0
8.0 × natural density	520	0	24.8	—

After Kajak (1968).

The numbers of C. plumosus slightly increased in the control during a 1-month exposure of experimental cages (from 150 to 230 ind. m^{-2}).

exclusion of other Mollusca from dense aggregations of *Viviparus fasciatus* (Stańczykowska, 1960), the mutual avoidance of the predatory Heleidae and the amphipod *Crangonyx* (Kajak & Kajak, 1975) and the mutual exclusion of Tricladida species (Reynoldson & Davies, 1970) are examples studied in the field. The coexistence of species with similar feeding ecology is explained by very specific food links as in *Pisidium* (Meier-Brock, 1969). Kajak (1968) demonstrated that meiobenthos abundance decreases with increasing abundance of macrobenthos and suggested that this is due to the mechanical impact of the latter. Jónasson (1972) found in Lake Esrom that the fecundity of *Potamothrix hammoniensis* is reduced when the density of *C. anthracinus* is high. Chua & Brinkhurst (1973) have shown strong interspecific interaction between Tubificidae species, resulting in significantly higher respiration and growth rates in mixed populations compared with single species situations.

6.3.10.3. Other relationships

Competition among non-predatory benthos is likely to be weakened when predation pressure is high (Hall, Cooper & Werner, 1970; Kajak, 1968). In fish-ponds the number of chironomid species is higher under heavy fish pressure (I. Wójcik-Migała, unpublished). The effect of predators on non-predatory benthos is greater when the latter is overcrowded, and in a highly competitive situation (Kajak, 1968, 1972).

Predators seem to play the controlling role in adjusting the level of benthos abundance to the environmental possibilities. Invertebrate predators can be substituted for fish since their proportion in the total benthos is always greater when fish pressure is low (Wolny, 1962; Kajak, 1968, 1972; B.E. Wasilewska, personal communication). In the absence of predators, competition seems to adjust benthos abundance. It has been shown experimentally that natural abundance could never be seriously overpassed, except in situations where the natural tendency is for an increase in numbers (Kajak, 1968; Hall, Cooper & Werner, 1970). Hargrave (1970) has shown that when *Hyalella azteca* is the dominant species it can have a strong influence on the production of benthic algae and the respiration of bacteria, stimulating these at some densities of *Hyalella* and depressing them at higher densities, within the natural range of densities.

6.3.11. Transformation of energy in the environment

The role of benthos in the exchange of chemicals between mud and water by stirring the mud and pumping water through their burrows must be significant although little work has been done on the quantitative aspects (Tessenow, 1964, 1972).

Benthic filter feeders may be very important in removing the seston from the water, and depositing it at the bottom. *Dreissena*, at densities of 1000–2000 ind. m^{-2} may filter the whole epilimnetic water during a period of several to 50 days (Stańczykowska, 1968; Lvova-Kachanova, 1971; Wiktor, 1969) and the amount of matter deposited may reach several scores, to more than 100g fresh weight m^{-2} per 24 h. The role of filtering Chironomidae, mostly in shallow depths, may be of the same order (Izvekova, 1971a, b).

The high nutritive value for Chironomidae of the matter sedimented by *Dreissena* has been proved by Lvova-Kachanova & Izvekova (1973). In Wolgogradskoje Reservoir, *Dreissena* metabolises more than 200 g of organic matter, which equals roughly 3400 kJ, and forms a substantial amount of the primary production of the reservoir.

6.4. Fish (Backiel, Thorpe, Kitchell)

6.4.1. Introduction

Within the framework of IBP studies, freshwater fish are considered mainly as a component of an ecosystem. However, this component is different from any other product of aquatic ecosystems because, since time immemorial, it has been the chief or only utilisable 'product to interest man' (Ivlev, 1945). For this reason it deserves special attention when considering the functioning of ecosystems.

Fish have been studied for many decades from many different points of view, and some workers have touched on productivity problems. With this in mind we should handle the IBP contribution on fish with modesty. It would not be difficult to prove that many points demonstrated here have been in one way or another dealt with before or concurrently with IBP. On the other hand, there is still rather scanty factual information on basic parameters like biomass and production of fish in fresh waters and even less on food consumption by fish and their role in the ecosystem. Thus, the information which has come out of or was stimulated by IBP deserves presentation.

Although based mainly on IBP work, this presentation does not aim to summarise or synthetise these studies. Instead, examples have been selected from the available data in order to demonstrate a few ideas that have occurred to the authors while considering the achievements and failures of the Programme.

6.4.2. Distribution of waters studied

Most standing waters considered in this chapter are located in the north temperate zone. The highest latitude represented is Lake Bolshoy Kharbey – 68° N. Reference is also made to three tropical lakes including Lake George, Uganda, on the equator. The running waters lie between 39° N in Japan and 57° N in Scotland, and reference is also made to two studies from the south temperate zone; in South America and New Zealand.

6.4.3. Taxonomic variety

The well-known fact that the number of fish species in fresh waters decreases with increasing latitude and altitude is also illustrated by the IBP material. Northern and high altitude (alpine) lakes are inhabited by two to four species (with salmonids dominating), temperate zone waters of the European plains have up to fifty species (with cyprinids dominating) and in large tropical lakes, e.g. Lake Malawi, several hundred fish species (mostly cichlids) have been identified. This is also true with respect to running waters. The number of fish species increases with size and habitat diversity of water bodies; e.g. large Lake Aleknagik, Alaska, although situated at 59° N contains twenty-three species but the small alpine Finstertaler See has three only. It is likely to hold true for the tropical zone when one compares Lake Malawi with Lake George, Uganda, in which 'more than 29 species have been identified' (Burgis *et al.*, 1973).

From the productivity approach the number of species does not provide much information. The share by important species in the total fish biomass and/or production determines not only the role of these species in an ecosystem but also provides the basis for the research strategy. Out of twenty-three species listed for Aleknagik two, *Oncorhynchus nerka* and *Gasterosteus aculeatus*, form the bulk of the pelagic fish biomass. In Lake George, Uganda, two herbivores, *Haplochromis nigripinnis* and *Tilapia nilotica*, make up about 60% of the total ichthyomass. The riverine populations show a similar structure; e.g. in the River Thames three species, *Alburnus alburnus*, *Gobio gobio* and *Rutilus rutilus*, contribute 86% to the total fish production (Mathews, 1971), although many more occur. Hence, in spite of the sometimes great number of species, most aquatic habitats, especially those of fairly uniform character, do not have a great number of important species and a study of these may suffice to appraise the role of fish in an ecosystem.

The dominating species (in biomass and/or production) of the Salmonidae have been studied best in relation to productivity. Next come some of the Cyprinidae, Percidae and Centrarchidae (N. America), but very little has been done about the Cichlidae which is the dominant group in African waters. Production problems of the rich ichthyofaunas of South America, and South-Eastern Asia have hardly been touched upon.

6.4.4. Methods of quantitative study

6.4.4.1. Population estimation

Although the theory of estimation of population parameters has been well developed and applications outlined (e.g. Ricker, 1968), there are difficulties in following these recommendations. The amount of effort necessary to sample fish properly is roughly proportional to the area of the water body and its environmental diversity. Hence fish populations in small water bodies have been studied in much more detail than those in large lakes, reservoirs and rivers. Relatively few of the IBP data reports include information on the statistical errors involved in the estimation of production, biomass, etc.

Estimated numbers, mortalities and growth rates of fish populations were usually used for calculating biomass and production. Population number and mortality have been most frequently estimated by sampling from a known area swept by a fishing gear or by means of the mark-recapture method and using data on age composition of the stock of fish. These methods produce estimates of the "catchable" stock only, no matter whether they are caught with special sampling gear or with ordinary fishing gear (Kipling & Frost, 1970; Beattie *et al.*, 1972; DR 1, 2, 3, 84*). These have usually been considered as adult fish and comparisons of such estimates can be misleading without information on what fraction of the populations they concern. In general the younger age groups have been inadequately sampled or omitted altogether.

In several papers (Winberg *et al.*, 1972; DR 68; F.V. Krogius, personal communication) reference is made to Tyurin's method which has been used for assessment of natural and total mortality of fish in a number of lakes in USSR. In this the ultimate age of a fish species and the maximum age in samples from an exploited population are used. It involves many assumptions and, from the point of view of statistical adequacy, can be easily criticised.

The limitations regarding the inadequately sampled fractions of populations are not overcome when the method of successive catches (e.g. Dedury's method, see Gerking, 1967) is applied for estimation of fish numbers. Recently this method has been most extensively used in streams (tables 6.9 and 6.12 for references).

Even when adequate methods have been applied, they have rarely been

* See list of data reports (DR) given in Appendix I.

Table 6.9 *Fish production and biomass ($g\ m^{-2}\ a^{-1}$)*

No.	Water bodies (latitude° N)	Original estimates: production			Corrected[a] total production ($P_{tot} = 116$) P	Correction factor $P : a$ d	Biomass B	P/B	Notes
		Total a	Piscivores b	Ratio $a : b$					

Part A. Data used for correcting P estimates

No.	Water bodies (latitude° N)	Total a	Piscivores b	Ratio $a : b$	Corrected P	$P : a$ d	Biomass B	P/B	Notes
	Lakes								
1	Bolshoy Kharbey (68°)	1.72	0.32	5.4	3.52	2.1	4.0	0.9	Coregonines mainly
2	Chedenjarvi (62°)	3.6	0.6	6	6.6	1.8	7.2	0.9	*Coregonus peled, Perca fluviatilis, Rutilus rutilus* dominate
3	Drivyaty	9.0	1.5	6	16.5	1.8	—	*ca* 1	Cyprinids dominate, 8 species estimated,
4	Naroch	4.4	0.9	4.9	9.9	2.2	8.1–12.3		22 species occur
5	Myastro	6.4	1.4	4.6	15.4	2.4	11.3–16.9	*ca* 1	
6	Batorin	8.3	1.4	6	15.4	1.8	11.7–23.5	*ca* 1	
7	Karakul (43° 50′)	4.7	1.2	3.9	13.2	2.8	5.8	2.3	*Cyprinus carpio* and *Perca schrenki* dominate, 10 species occur
	Reservoirs								
8	Rybinsk (58° 30′)	24.4	3.7	6.5	40.7	1.8	*ca* 28	*ca* 1.4	Cyprinids dominate
9	Tsimlyansk (48°)	8.0	2	4	22	2.7	*ca* 24	0.9	Cyprinids dominate, 7 species estimated, 44 occur

(Drivyaty and Naroch marked "(ca 55°)")

Part B. Original or corrected estimates

	a	b	c	d	e	f	
Lakes							
10 Dalnee (53°) (after 1958)	23.0	11.5	23.0	—	ca 11	2.0	*Oncorhynchus nerka* and *Gasterosteus aculeatus* dominate
11 Øvre Heimdalsvatn (61°)	2		2	—	1.7–2.3	ca 1	*Salmo trutta* and *Salvelinus alpinus*
12 Aleknagik (59°)	—		—	—	0.09–1.29 (mean = 0.68)	—	Range for 17 yr of 2 dominant species
13 Loch Leven (56)			44.2	—	ca 25.0	1.8	*S. trutta, P. fluviatilis, Esox lucius*
14 Baykal (54°)	3.8	—	7.6	2	7.1	1.1	Endemic *Cornephorus* sp. dominate
15 V. Finstertaler (47°)			1	—	2.3	ca 0.5	*S. trutta* and *S. alpinus*
16 Chad (13°)			—	—	5–50	—	Herbivores dominate
17 George (0°)			—	—	22.7	—	Herbivores dominate
18 Tjeukemeer (53°)			ca 71	—	—	—	Cyprinids dominate
Running waters							
19 Small streams in England (51°–54°)			3.0–72.7	—	0.6–26.7	1.3–3.7	1–5 species
20 Trout streams in Normandy, France (ca 49°)					mean 63.5 max. 161.1		*S. trutta, Anguilla anguilla, Cottus gobio*
21 Chigonosawa Brook, Japan (ca 39°)			5.9–14.1[b]	—	4.8–9.3[b]	1.2–1.6	*Salvelinus pluvius*
22 River Thames (ca 51°)			197[b]	—	120	1.5	Cyprinids and *P. fluviatilis* dominate

Sources: nos. 1–7, 10, columns a & b from Krogius (manuscript); no. 8, Sorokin (1972); no. 9 (DR 84); no. 11 (DR 2); no. 12 (DR 60): no. 13, Thorpe (1974a): no. 14, Moskalenko & Votinsev (1972) corrected by a factor of 2; no. 15 (DR 1) *P* corrected for juveniles by a factor of 2; no. 16, Lévêque et al. (1972); no. 17, Burgis et al. (1973); no. 18, Beattie et al. (1972); no. 19, Le Cren (1969), Mann (1971); no. 20, Cuinat (1971); no. 21 (DR 19); no. 22, Mann et al. (1972).
[a] Mean correction factor = ca 2.
[b] Original data in kcal roughly equivalent to somatic *P* in grams.

employed to study all fish species where more than a few occurred. The use of specific poisons, such as rotenone or toxaphene, makes it possible to assess the standing crop of all fish species but, again, fry and small fish are difficult to collect (Sumari, 1971) and thus their biomass is underestimated.

Estimations of numbers of fish smaller in size than those sampled have been made on the basis of the numbers of eggs laid by the mature fraction of the population and of the numbers of the youngest fish caught (Le Cren, 1962; Kipling & Frost, 1970; Mathews, 1971). These two anchor points have been supplemented with some observations on egg and fry mortalities in the case of a pike population (Kipling & Frost, 1970) and of brown trout (Thorpe, 1974a), but even in these cases the estimates involve several assumptions which are very difficult to prove.

6.4.4.2. Correction factor for adult production

Available data on juvenile and adult production (see section 6.4.8 of this chapter) suggests that production of the entire fish population is roughly twice that of the adult population.

The team working on Lake Karakul (DR 63) utilised information on cannibalism in perch, *Perca schrenki* Kessler, for the estimation of juvenile production. This approach can be developed further. Studies on consumption in piscivorous fishes have shown that they require from about three to ten units of food to produce one unit of their own body weight (Backiel, 1971). Obviously, production of non-piscivores must be considerably greater than the total amount consumed by predators because the latter prey selectively and there are other causes of mortality than predation. Thus, multiplying piscivore production by ten would still give a conservative estimate of production of the prey fish population. Extensive IBP studies carried out in the USSR supply relevant data (table 6.9) which show that the ratios of estimated total production, based on the above approach, to the original estimate vary little and on average equal two. This correction factor has incidentally the same value as that arrived at to correct for the fraction of total fish production contributed by the juvenile stages. Obviously, such a way of correcting production estimates cannot be applied to every situation. In Lake Dalnee (table 6.9) the piscivores contributed a much smaller fraction to the entire fish production during the period between 1937 and 1957 than 1 : 11 and that holds true in Lake George, Uganda, and the River Thames. However, in a number of cases correction of the original estimate by doubling it is justified on the grounds of both the ratios: predator/prey and adult/juvenile production.

6.4.4.3. Food consumption

The guidelines for studies of food consumption were set out by many of those participating in the organisational phase of IBP (Gerking, 1967; Ricker, 1968).

The techniques for study currently available include two general approaches. One is based on calculation of food consumption rates determined from calculations of energy requirements using bioenergetic studies. This method was used in several IBP studies (e.g DR 58, 64, 68, 84) and is not treated here as it has been well described in summaries by Winberg (1960), Mann (1967, 1969) and Warren (1971).

A more immediate and direct approach is that of calculating daily rations based on measurements of gut content and knowledge of rates of digestion or evacuation. Although not in common use, it is perhaps less subject to inherent errors than the application of laboratory data to natural systems (Healy, 1972; Solomon & Brafield, 1972), and was employed by several IBP workers.

6.4.5. Biomass

As stated in the previous section assessment of biomass and production of the entire fish population has seldom been attempted. Production figures taken from selected IBP studies could have been corrected either as in table 6.9, or by some other method (see sections 6.4.8 and 6.4.9). It is believed that biomass has been estimated with much less bias than production. Thus, table 6.9 includes original biomass estimates but corrected production figures for lakes with the exception of Loch Leven, Lake Dalnee and Tjeukemeer.

It can be seen that the biomasses of fish for lakes and reservoirs listed range from less than 0.1 $g\,m^{-2}$ in Lake Aleknagik, USA, up to 25 $g\,m^{-2}$ in Loch Leven, Scotland. Other data published recently somewhat exceed this range. Anwand (1968) reported biomasses from 0.7 to 32.3 $g\,m^{-2}$ in small lakes in East Germany. For three sewage lagoons in Great Britain the range was 18.2–38.3 $g\,m^{-2}$ (White after Burgis *et al.*, 1973) and Holčik (1972) found biomasses of up to 80 $g\,m^{-2}$ for some small ox-bow lakes in the Danube valley.

These maxima for standing waters in the north temperate zone do not appear to be exceeded by those for tropical lakes (table 6.9), though in certain areas of Lake George up to 100 $g\,m^{-2}$ were frequently recorded (Burgis *et al.*, 1973).

For running waters recent studies encompassing the entire fish population have been carried out in small trout streams and one fair-size lowland river, the River Thames near Reading, all from the north temperate zone. Even so, the range of biomass is a little greater than that for standing waters, the maximum recorded being 160 $g\,m^{-2}$ for one site in a trout stream in France.

6.4.6. Production

The entire fish production in lakes and reservoirs ranges between 1.0 $g\,m^{-2}\,a^{-1}$ in an alpine lake and 71 $g\,m^{-2}\,a^{-1}$ (table 6.9). The highest level, which was probably overestimated, was recorded in a Dutch polder lake.

Minima must be much lower than $1 \, g \, m^{-2} \, a^{-1}$ considering the low biomass in Lake Aleknagik where production was not estimated.

An interesting example of high fish production in the tropics is Laguna de Bay, a large, $900 \, km^2$, shallow lake in the Philippines, (FAO/UN 1966, Delmendo, 1966). Extensive survey of the fishery in 1962/63 resulted in an estimate of fish *yield* of about $90 \, g \, m^{-2} \, a^{-1}$. The main species caught (*Therapon plumbeus percoidea*) is a small fish with a high reproductive rate. Therefore production of fish in that lake must have been substantially higher than $100 \, g \, m^{-2} \, a^{-1}$. A very high fish yield of $220 \, g \, m^{-2} \, a^{-1}$ is reported by Sreenivasan (1972) for Chetpat Swamp in India, which indicates a much higher production, approaching that of well-managed fish-ponds. Carp ponds in central Europe can produce up to $30 \, g \, m^{-2} \, a^{-1}$ (season^{-1}) without extra feeding. The production may reach $500 \, g \, m^{-2}$ with well prepared feeds but, then, oxygen and other factors limit the density of fish. In Israeli ponds with well balanced populations of mixed species the biomass at the end of the growing season can be twice as much as in Europe (Yashouv, 1969, 1972). These figures do not differ much from biomasses of fish cultivated in well-managed ponds in the Indo-Pacific region (Hora & Pillay, 1962). With low mortality and relatively small initial biomass the biomass at the end of the growing season in ponds shows the order of magnitude of the season's production. Hence, the largest production reported is about $1000 \, g \, m^{-2} \, a^{-1}$ in Israel carp ponds with extra feeding. In Indian ponds with mixed populations of herbivores and a little supplementary feeding, production can exceed $600 \, g \, m^{-2} \, a^{-1}$ (Ganapati & Sreenivasan, 1972). A. Nauwerck (personal communication) found a pond in Africa situated near a chicken farm in which the yield of *Tilapia* sp. feeding on a constant bloom of algae was of the same if not of a greater order of magnitude.

Available data for fish production in small trout streams show a similar range as for standing waters of the same climatic zone with a maximum of *ca* $60 \, g \, m^{-2} \, a^{-1}$ in a small English stream. The lowland River Thames with its fish production of up to *ca* $200 \, g \, m^{-2} \, a^{-1}$ may well be an extreme case; there were practically no piscivores and fish larger than 100 g were exceptional.

6.4.7. *P/B* ratios

6.4.7.1. Range

The *P/B* ratios given in table 6.9 can be considered as rough order-of-magnitude estimates. The range of this ratio for standing waters is from 0.5 for Vorderer Finstertaler See to 2.3 for Lake Karakul. We do not have any figures for tropical waters but one can assume these are larger than 2. The sequence seems logical from a low *P/B* in an alpine, high altitude lake (V. Finstertaler), then subarctic Bolshoy Kharbey and high latitude (*ca* 62° N) Chedenjarvi

(0.9), followed by a number of north temperate zone lakes with a P/B around 1 to Lake Karakul, at *ca* 43° N, with 2.3.

In the running waters listed in table 6.9 the P/B ratio is greater than 1 and reaches a value of about 4. Chapman (1967) compiled data on P/B ratios ranging from 1 to 2.5, the highest figure being found in populations of juvenile *Oncorhynchus nerka* and *O. kisutsch*.

6.4.7.2. Comparison of *P* and *B*

With reservations the scanty material available enables some comparisons to be made. Two lakes; Aleknagik, Alaska, and Dalnee, Kamchatka, from similar climatic zones and with similar dominant fish species (*Oncorhynchus nerka* juveniles and stickleback, *Gasterosteus aculeatus*) had very different fish biomasses; 0.68 g m^2 and *ca* 20 g m^2 respectively. Four lakes with similar mean biomasses of fish: Dalnee, Batorin, Loch Leven and the equatorial Lake George, Uganda, differed in many other features including climatic zone, fish species composition and diversity, and the presence or absence of phytoplankton-feeding fish. Hence, the climate itself does not seem to be the major factor controlling the biomass of fish. This is also likely to be true for the production of fish although, with faster growth of individuals and longer growing season in the tropics, than, say in the subarctic region, it can be greater for the same biomass. If we include information on the yield to man, e.g. from Laguna de Bay, the range of production estimates is wider than that of biomass.

Comparisons of fish biomass and production on a unit area basis between standing and running waters have limitations. This subject is discussed later (section 6.4.9) with respect to trout populations. However, we can infer from the available data that running waters are likely to support denser populations than standing waters of the same climatic zone.

6.4.8. The role of juvenile production

6.4.8.1. Proportion of total fish population

Allen (1951) and Le Cren (1962) drew attention to the fact that in some fish populations production of gonads and of juvenile fish was much greater than that of adults. Data compiled in table 6.10 show that in extreme cases the latter can be less than 20% of production of the entire population. In most fishes studied the adults' somatic growth contributes less than one-third and seldom more than half to the total production.

The gonad production varies greatly if considered as a fraction of *adult* production; the maximum recorded is for perch in Lake Windermere (87%) and the minimum of *ca* 10% for the Horokiwi trout. This is closely related to

Table 6.10 *Gonad and juvenile production in fish*

Species and location	Age groups (years)	% of total production gonad	juvenile	Unit	References and notes
Salmo trutta					
Horokiwi Stream	all	2.3	*ca* 75	wet wt	Allen (1951)
	adults	10			
Three Dubs Tarn	all	3.15	*ca* 82	wet wt	Le Cren (1962)
Kingswell Beck ⎫					
Dockens Water ⎬	all	—	30–48	wet wt	Le Cren (1969), DR 40
Bere Stream ⎭					
Loch Leven	adults	31.2		wet wt	Thorpe (1974a)
	all	9.1		wet wt	
Øvre Heimdalsvatn	adults	7.0	—	energy	DR 2 (energy loss on spawning, migration and gonads)
Loučka Creek	2 +, 3 +	15.8	—	energy	Libosvarsky & Lůsk (1970)
Brodská Brook	all	4.5–9.4	—	energy	DR 18 (age group 0 + underestimated)
Salvelinus fontinalis					
Lawrence Creek	all	—	43–57	wet wt	Hunt (1966)
Salvelinus pluvius					
Chigonosawa Brook	all	5.5–11.4	28.5–55.1	energy	DR 19 (estimates for 7 years)
Perca fluviatilis					
Windermere	all	19	59	wet wt	Le Cren (1962) (his
	adults	87	—	wet wt	table 4)
River Thames	1 + on	*ca* 16.6	—	energy	Mann (1965)
Vistula River	4 + to 13	*ca* 25	—	energy	Backiel (1971)
Alburnus alburnus					
River Thames	all	6.7	62.3	wet wt	Mathews (1971)
	adults	14	—	wet wt	Mathews (1971)
Gobio gobio					
River Thames	all	11.5	60.5	wet wt	Mathews (1971)
	adults	29	—	wet wt	Mathews (1971)
Leuciscus leuciscus					
River Thames	all	7.7	65.3	wet wt	Mathews (1971)
	adults	22	—	wet wt	Mathews (1971)
Leuciscus cephalus					
Vistula River	3 + to 13	*ca* 21	—	energy	Backiel (1971)
Rutilus rutilus					
River Thames	1 + on	16	—	energy	Mann (1965)
	all	1.8	64.2	wet wt	Mathews (1971) (from his table 17)
Tjeukemeer	2 to 10	*ca* 30	—	wet wt	DR 21
Stizostedion lucioperca					
Vistula River	1 + to 10	*ca* 15	—	energy	Backiel (1971)
Tjeukemeer	all	—	*ca* 53	wet wt	DR 21
Esox lucius					
Windermere	all	*ca* 10	39	wet wt	Kipling & Frost (1970)
	adults	16	—	wet wt	(from their table 19)
Vistula River	1 + to 9	12–15	—	energy	Backiel (1971)
Entire Fish Population					
River Thames	all	—	70	energy	Mann *et al* (1972)
Rybinsk Res.		—	29	energy	Sorokin (1972) (his term 'fish larvae' taken as juveniles)

the growth rates of mature fish and probably does not depend much on variation in the gonad growth in each species. The gonads can contribute as little as about 2% to the *total* population production as in Horokiwi trout and River Thames roach, and as much as 19% as in Windermere perch.

The juveniles, or more precisely the 0 + age group, produce from about 30 to about 80% of the entire population production and, excluding the salmonids, the range is from 40 to 70%. For those estimates of production based on sampling adult fish the figures could be from 1.5 to 3.3 times smaller than the true estimate of total fish production. This tentative conclusion requires verification by further studies but for the time being a rough correction factor of 2 can be applied to production estimates where juveniles were excluded, if other ways of correction are not available.

6.4.8.2. Growth and mortality

One might expect that juvenile production in slow-growing fish with short life spans would be relatively greater than in fast-growing large fish. The data in table 6.10 do not support this. In the population of gudgeon (*Gobio gobio*), a small fish in the River Thames, the juveniles contributed 60% to the total population production while in pike and pike-perch, both large and fast-growing fish, this contribution was *ca* 40–50%, which is not strikingly different.

According to Le Cren (1962) and Mathews (1971) high mortality and fast growth of the young fish are responsible for the considerable share made to the total population production by juveniles. In four cyprinids in the River Thames, Mathews (1971) found mortalities during their first 2–3 months of life to be between 0.9814 and 0.9993 with corresponding instantaneous rates (Z) of 3.98 and 7.26. Zuromska (1967), in her extensive study on eggs and larvae of roach in lakes, estimated that up to 10%, but often only 1%, of actively swimming fry at several days old survived from the eggs laid. The same author (quoted in Zawisza & Backiel, 1970) found similar mortality rates among eggs of coregonids laid in lakes. Kipling & Frost (1970) assumed a 70% survival of pike eggs to hatching and 22% of larvae to the start of feeding.

However, even with a small biomass of fry the juvenile production is high because of very fast growth of the young individuals. The weight of a newly hatched fry is of the order of few milligrams in cyprinids and percids, about 10 mg in pike and about 100 mg in salmonids. Within several weeks its weight increases by several tens up to even a thousand times. The instantaneous growth coefficient (G) is therefore of the order of 4 to 7 which means that production in that short period is 4 to 7 times the average biomass. Hence, with Z between 4 and 7 as was found for cyprinids in the River Thames (Mathews, 1971) and G as above there is little change in biomass during the short juvenile period, but production must be high. At the end of that period biomass is low which means little recruitment to the 'adult' population.

Table 6.11 *Growth and survival of juvenile (O +) pike-perch in Lake Balaton*

Parameters	June	July	August	September	Whole period
Weight (g)	0.0125	1.09	2.64	3.54	—
Growth coefficient (G).		4.4682	0.8846	0.3018	5.6546
Number per 50 m²	1331	674	15	8	—
Mortality coefficient (Z)		0.6797	3.8051	0.6287	5.1135
Biomass change: inst. rate (G − Z)		+ 3.7895	− 2.9205	− 0.3269	+ 0.5411
exp (G − Z) = B_0 : B_t					1.72

N.B. Data of Biro, DR 22, recalculated.

Backiel (1971) noticed that the gonad production of six predatory fish was of the same order of magnitude as the biomass recruited to the adult population.

Data from Lake Balaton, Hungary, on juvenile pike-perch (*stizostedion lucioperca*) provide another example (table 6.11). Estimates of numbers of fish are inevitably unreliable due to difficulties of sampling, although the order of magnitude of the mortality coefficient for the pike-perch ($Z = 5.11$) is within the range given by Mathews (1971). The growth coefficient for the whole summer period ($G = 5.65$) is only slightly greater than the calculated Z, which results in small change of biomass from June to September (see $\exp(G-Z)$ in table 6.11).

The role of juvenile mortality and growth can be illustrated most reliably from data on common carp during the first four weeks of life in pond culture. The data of Wolny (1970) show that instantaneous growth rate G was about 7.8 while mortality coefficient Z was 0.28–0.69, resulting in an increase of biomass by more than 1000 times the initial one. The final biomass gives a good start to the production of older fish unlike that in natural populations. These data concur with information on carp farms where juvenile production is about 20% of the total farm production. Similar proportions are observed in rainbow trout farms and in both situations this has been achieved by creating the best possible conditions for survival of eggs, larvae and fingerlings.

The data illustrating juvenile production, discussed above, are still preliminary but there is little doubt that the share of the total production contributed by juveniles is usually very high.

6.4.8.3. Comparison of natural and fish farm population

Comparison between natural populations and those grown in the somewhat controlled conditions of fish farms indicates the significance of the juvenile production process. The classical treatment of the dynamics of exploited fish populations usually begins at recruitment to the exploited phase. The number recruited has been related to numbers of spawners or of eggs laid but little attention has been paid to the cost of producing recruits. The cost, to the ecosystem of producing utilisable stocks is proportional to the consumption of food by all fish which, in turn, is directly related to production. The cost is higher the greater the share of pre-recruit/juvenile production to the total production. It is fairly low in fish farms but may be very high in some natural populations where pre-recruits contribute a greater part to the total production. In the latter case a high mortality is the primary cause of relatively high juvenile production.

Backiel & Le Cren (1967) suggested that during the juvenile stages fish mortality is density dependent whilst growth rate is not. Although the examples discussed above have not confirmed the effect of density on juvenile mortality they have shown that survival during the first few months of life may be less than 0.1% or as great as 75% (in common carp fry) with little effect on growth rate.

6.4.9. Production of a fish population: brown trout

The limited information on factors affecting fish populations may be supplemented by data on brown trout (*Salmo trutta* L.), which have been studied both in running and in standing waters.

6.4.9.1. Streams

Le Cren (1969) reviewed the data on annual production of salmonids in small streams in England and found values ranging from $2-12$ $\mathrm{g\,m^{-2}\,a^{-1}}$ without direct correlation with general stream productivity (as p.p.m. calcium) or population age structure, growth rate, or migratory tendencies. Limitations on production were probably numerical early in life, being governed by available territory in the less productive streams and by low egg survival in the silty spawning grounds in the more productive chalk streams, and at a later age by food supply.

Within streams production varied between contiguous reaches, being influenced by stream morphology, depth, weed growth and the age groups of trout and salmon present. He stressed the need for an adequate series of samples from individual sites, as did Mann (1971), in order to improve the accuracy of estimates and to provide measures of year-to-year variability. Egglishaw (1970) demonstrated annual variations in production of salmon and trout in the Shelligan Burn, Scotland, and attributed these to variations in availability of food and of feeding sites through changes in stream discharge between years. With lower water flow the depth of the stream was reduced without appreciable diminution of the bottom area, and trout production was decreased but salmon production was not. Annual production values ranged from $7.7-12.3$ $\mathrm{g\,m^{-2}}$ for trout and $6.6-11.1$ $\mathrm{g\,m^{-2}}$ for salmon, which are of the same order as those reviewed by Le Cren.

Libosvarsky & Lusk (DR 18) and Lusk & Zdrazilek (1969) found that trout production in two tributaries of the Bečva River, Czechoslovakia, over 5 years varied little, being $8-9\,\mathrm{g\,m^{-2}a^{-1}}$ in the Brodska Brook and $7.5-7.7\,\mathrm{g\,m^{-2}a^{-1}}$ in the Lusova Brook. They concluded that production depended primarily on the number of trout present in the first part of the growing season (April–June) when the greatest growth takes place. This number was determined by space and food requirements. Le Cren (1969) has suggested that Allen's (1951) production estimate of $54.7\,\mathrm{g\,m^{-2}a^{-1}}$ for trout in the Horokiwi stream in New Zealand may be too high since density-dependent mortality during the first 3 months of life may have been more severe than Allen assumed. Le Cren recalculated production in the Horokiwi at $38\,\mathrm{g\,m^{-2}a^{-1}}$, a figure only equalled by trout populations maintained at high density (initially 100 per m^2) in artificial stream channels in Dorset, England. Kalleberg (1958) showed that densities of salmonid fry in a stream tank could be increased without increased aggressive activity among them by increasing the complexity of the stream bed and thereby the visual isolation of individual fish. Hence it is possible that differences in production between streams may depend partly on the morphology of the stream bed. The mean values for annual production of trout populations in streams are summarised in table 6.12.

6.4.9.2. Lakes

Trout production was estimated during IBP studies at Vorderer Finstertaler See, Austria (Pechlaner *et al.*, 1972a, b), Øvre Heimdalsvatn, Norway (K.W. Jensen & L. Lien, unpublished; Lien, 1979) and Loch Leven, Scotland (Thorpe, 1974a) and there appear to be no other published estimates of production by *S. trutta* in lakes. Table 6.13 summarises data from these studies together with some approximate values derived from accounts of trout populations in England, at Yew Tree Tarn (Swynnerton & Worthington, 1939) and Three Dubs Tarn (Frost & Smyly, 1952; Le Cren, 1962).

6.4.9.3. Factors affecting production

By measurements of production per unit area of lake surface, lake populations are only about one-tenth as productive as stream populations. In terms of ecosystem productivity this comparison has meaning, but in terms of a population, expression of production per unit area gives little biological insight into factors affecting that production. Most of the stream studies have been made on small systems, where trout made use of most of the bottom area included in the measurements: in the lake studies this is not the case, except probably in the smaller systems, Three Dubs and Yew Tree Tarns. Hatch & Webster (1961) reported that rainbow trout (*Salmo gairdneri* Rich.) were found chiefly in the littoral zone of four lakes studied in the Adirondack mountains,

Table 6.12 *Annual production of* Salmo trutta *L. in streams*

Country	Stream (reference)	Production (g m^{-2} a^{-1})
Czechoslovakia	Lušová Brook (Lusk & Zdrazilek, 1969)	7.6
	Brodská Brook (Libosvarsky & Lusk, DR 18)	8.5
	Loučka River (Libosvarsky, 1971)	10.4
England	Appletreworth Beck (Le Cren, 1969)	3.0
	Devil's Brook (Mann, 1971)	4.8
	Bere Stream (Mann, 1971)	6.2
	Hall Beck (Le Cren, 1969)	5.2
	Nether Hearth Sike (Crisp, 1963 in Le Cren, 1969)	5.0
	Kingswell Beck (Le Cren, 1969)	7.4
	Black Brown Beck (Le Cren, 1969)	10.0
	River Tarrant (Mann, 1971)	12.0
	Dockens Water (Mann, 1971)	12.1
	Walla Brook (Horton, 1961)	12.6
Scotland	Shelligan Burn (Egglishaw, 1970)	10.1
New Zealand	Horokiwi Stream (Allen, 1951)	54.7
	Range	3.0–54.7
	Mean	11.3

Table 6.13 *Annual production of* Salmo trutta *L. in lakes*

Lake	Area (ha)	Nursery stream area (ha)	Ratio Stream area/Lake area	Ca Co$_3$ (mg l^{-1})	P (g m^{-2} a^{-1})	P/B
Vorderer Finstertaler[a]	15.8	0	0	0.0–6.5	0.15	0.24
Three Dubs Tarn[b]	1.6	0.003	0.0019	0.7–7.5	0.76	0.89
Yew Tree Tarn[c]	4.0	ca 0.15	0.0375	0.7–7.5	1.2	0.54
Øvre Heimdalsvatn[d]	77.5	0.98	0.0126	2.5–7.5	2.0	1.03
Loch Leven[e]	1330	54	0.0406	30–70	3.9	0.93

[a] Data from Pechlaner (1966) and Pechlaner *et al.* (1972a, b)
[b] Data from Macan (1949), Frost and Smyly (1952) and Le Cren (1962). P estimated from the two latter sources.
[c] Data from Swynnerton & Worthington (1939) used for estimates of P and B. Also data from 1-inch Ordnance Survey map, Sheet 88, 1948.
[d] Data from DR 2 and Lien, personal communication. P corrected for juveniles.
[e] Thorpe (1974a).

USA, and Fish (1963) found the same distribution in New Zealand lakes. In the three IBP lakes, the littoral areas are probably of primary importance to the adult trout during the summer growth period, but within the littoral, substrate may determine the relative use of different areas through the quality, quantity and availability of food organisms there. When comparing the standing crops of rainbow trout in five New Zealand lakes, Fish (1968) used biomass per kilometre of shoreline as his unit, arguing that this was more meaningful than a unit based on total area since the fishes' food resources appeared to be concentrated near the shoreline. However, Fish restricted his study to trout of at least 20 cm in length, and no detail is given of the habitat of the smaller fish. In Loch Leven, Thorpe, (1974b) found that juvenile brown trout were offshore at all times during their first 2 years of loch life. Since production during this interval was 75% of the total during the loch life of a year-class (Thorpe, 1974a), production in kilograms per kilometre of shoreline is here an unrealistic unit for total trout production. Until the precise boundaries of the productive area of lakes for juvenile brown trout can be defined production per unit area of the whole system is probably the best unit for comparing productivity of trout between lakes, but as Le Cren observed for the stream populations, it will tend to show values lower than those which are achievable by the species.

Within these limitations some generalisation is still possible. As in the stream populations production in lakes is limited initially by number, since the lake populations depend on immigrating young trout whose numbers depend on the success of the fry stage in the streams (cf. Le Cren *et al.*, 1972). Thereafter density may be influenced by predation and by disease (Thorpe & Roberts, 1972).

Territoriality may occur in lakes, but although restricted range of movement has been shown (K.W. Jensen, personal communication; Holliday, Tytler & Young, 1974; Thorpe, 1974b) active defence of that range has not been shown. Limitation of production via food intake may occur later, as negative production occurs regularly in winter in Loch Leven (Thorpe, 1974a). Nilsson (1965, 1967, 1972) showed that production of trout in lakes was reduced when arctic char (*Salvelinus alpinus* L.) were introduced there, since with interactive segregation their numbers and growth rate decreased. Limitation of production through interspecific competition of the kind studied by Nilsson can be demonstrated in Vorderer Finstertaler See (Pechlaner *et al.*, 1972a). In Loch Leven the principal components of the diet of trout and perch (*Perca fluviatilis* L.) were young perch which amounted to 29.9 and 13.5% of the trout and perch diets respectively during the period June–September (Thorpe, 1974a). Shortly after hatching, the 1969 brood of perch failed and during the summer of 1969 somatic production by the adult trout was negative. This limitation of production was not due to factors affecting adult stock density, which was stable from April 1968 to April 1970, but to factors affecting growth. The failure of a key component of the diet of two potentially competing species must increase the probability that they compete for other shared components, in this case *Asellus* and chironomids which together formed 39.3 and 46.9% of the trout and perch diets respectively in 1971. It is at least suggestive that interspecific competition could have restricted trout production in Loch Leven in 1969. Le Cren *et al.* (1972) found that when the perch population of Windermere was reduced by trapping to 20% of its former density between 1941–64, the char (*Salvelinus willughbii*) increased in number five-fold. However, no drastic changes in the trout population seem to have occurred and he suggested that the lake stock may be limited by the space available in the streams for rearing the fry. Table 6.13 provides some evidence in support of this hypothesis as annual production in other lakes is correlated with the ratio of nursery area to lake area ($r = 0.711$).

As in stream populations year-to-year variability of production occurs in the Loch Leven population of adult trout. Production from April to October varied by 11 times between extremes over 4 years. If the extreme low of 1969 is omitted the variation was 2 times. Some major environmental changes can be linked with the extreme values of somatic production, namely food shortage in summer 1969, followed in 1970 by a spectacular increase in production at a time of unusually high mortality and increased growth rate of survivors. During winter assimilation of food is usually inadequate to meet metabolic demands, and negative production occurs (cf. Morgan, 1974) but in autumn 1969 unusually high water levels allowed the fish access to terrestrial food on flooded shores, and positive production took place.

Fluctuating water levels in reservoirs have been shown (Grimas, 1961; Aass, 1969) to influence trout density and growth rate adversely (and thus, by implication, trout production) through elimination of important food organisms. In shallow systems such as Loch Leven, a drop of 50 cm in water level excludes trout from about 200 ha of the littoral (see Smith, 1974). Such a reduction in July and August contributed to the poor performance of the trout during 1969.

Indirect effects on salmonid production through direct effects of eutrophication on benthic Crustacea were implied by Colby *et al.* (1972). Nümann (1972) gave instances of eutrophication in the Bodensee affecting fish production through increased parasite abundance, where changing physicochemical conditions had favoured the parasites' intermediate hosts. Similarly Grimaldi *et al.* (1973) have found widespread infection of the fungus *Branchyomyces* sp. on the gills of many species of fish in eutrophicated lakes in northern and central Italy and northern Switzerland. In the three IBP lakes where trout were studied the parasite burden was low; L. Lien, R. Pechlaner (personal communication). The production of the cestode *Eubothrium crassum* Bloch in trout in Loch Leven has been estimated at about 1% of that of its host (Thorpe, 1974a; A.D. Campbell & J.E. Thorpe, unpublished data) where levels of infestation have been low, ranging from 0.1 to 6.4 worms per fish (Campbell, 1974). As no data is available on the conversion efficiency in *Eubothrium*, or on the tapeworm's biochemical demands on its host, the true impact of this low level of cestode production on that of its host is not known. Ingham & Arme (1973) have shown that in rainbow trout carrying 1–5 *Eubothrium* per fish these cestodes had no effect on either nutrient absorption or fish growth under culture conditions. However, since half the parasite's production occurs in autumn at a time of low food intake and low conversion efficiency in the trout (Brown, 1946) it must aggravate tissue wastage at a time of regular negative production.

Close comparison between the five lake populations is not yet very profitable. However the range of production is narrow and appears to depend first of all on the ratio of the nursery area available to fry to the area of the lake to which they migrate (table 6.13). This relationship expresses the effect of initial numerical restraint on production (similar dependence of production on recruitment occurs in the Loučka River population). Limitations through growth will depend on properties of the lake itself; in the present case, production P correlated well with calcium carbonate content of the water which is a rough lake productivity index.

6.4.10. Food consumption

Studies of food consumption by fish serve as a focal point for both basic and applied interests in fish growth processes. An extensive literature on food habits of fish has, unfortunately, provided insufficient insight about the rates of consumption by fish in nature (Mann, 1969) which must be measured directly to evaluate effectively the role of fish in ecosystem dynamics or to optimise management strategies.

6.4.10.1. Temporal patterns of food consumption

Quantitative and qualitative monitoring of the stomach contents over a period of time provides an immediate measure of the selectivity and efficiency of the food consumption process. Variation in the average gut content tends to suggest both a short-term or diel component and a long-term or seasonal change associated with relative food abundance and feeding behaviour.

Since considerable individual variation and opportunism are the rule (Keast & Welsh, 1968; Biro, 1969) knowledge of the diel periodicity in feeding intensity of an adequate sample is highly desirable. In fact, Davis & Warren (1968) found this to be the major deficiency of daily ration studies as a measure of food consumption rates and few investigations had been based on the 24-hour sampling schedule necessary to solve the problem.

Within the framework of IBP, three major studies have coupled laboratory and in-situ measurement of gastric evacuation rates with intensive diel sampling to provide direct measures of daily rations (Magnuson & Kitchell, 1971; Moriarty et al., 1973; Thorpe, 1974a).

Moriarty et al. determined rates of food consumption for the herbivorous *Tilapia nilotica* and *Haplochromis nigripinnis* in Lake George, Uganda.

Quantities of phytoplankton consumed per day per gram of fish were inversely related to fish size for *Tilapia* but relatively constant over all sizes for *Haplochromis*. For example, 10-g *Tilapia* and *Haplochromis* had daily rations of 40% and 19% respectively, while 100-g fish of each species had rations of 16% and 22%. The lower rations for small *Haplochromis* were compensated for by a higher assimilation efficiency (66%) than for *Tilapia* (43%). Moriarty *et al.* (1973) suggest that the grazers, including zooplankton, in Lake George are not

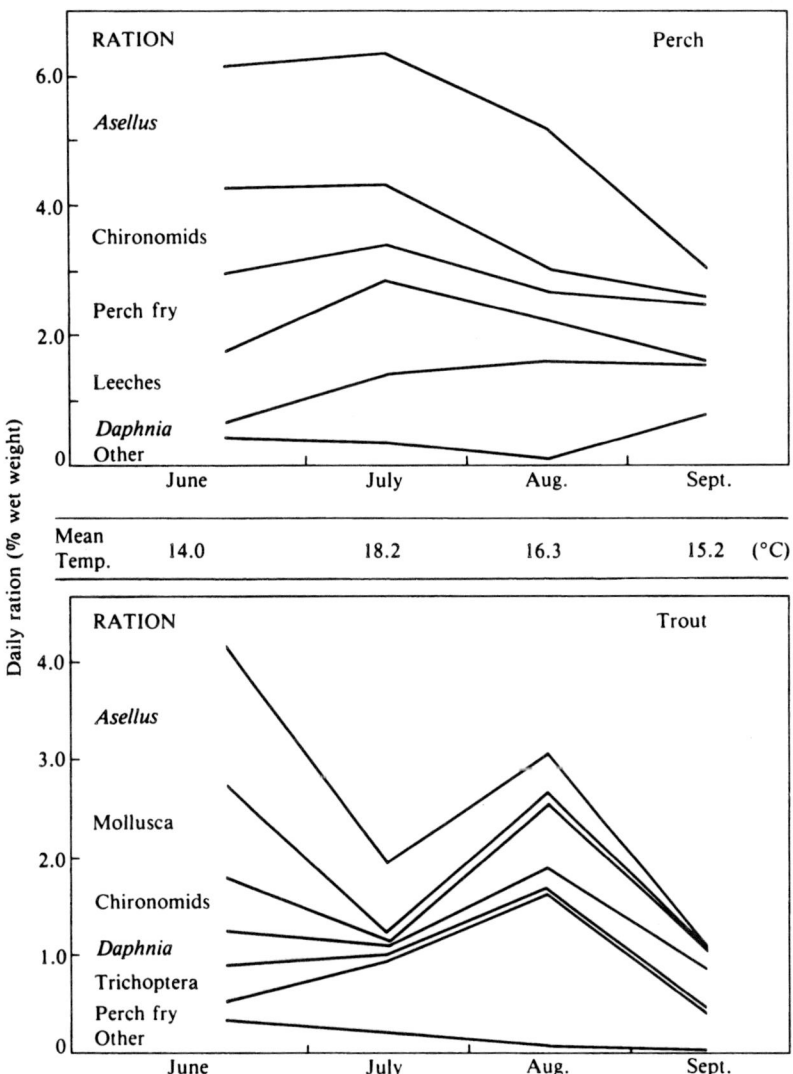

Fig. 6.8. Daily rations and principal diet components of adult perch (*Perca fluviatilis*) and brown trout (*Salmo trutta*) in Loch Leven, Scotland, during June–September, 1971 (after Thorpe, 1974a).

food limited and can be assumed to obtain their maximum requirements. Backiel (1971) also reported high daily rations (12–18%) for chub, *Leuciscus cephalus*, and noted that algae (*Spirogyra*) were dominant in the diet.

Daily rations of many carnivores are generally lower than those for herbivores. Thorpe (1974a) determined rates of food consumption for adult perch, *Perca fluviatilis*, and brown trout, *Salmo trutta*, in Loch Leven. Daily rations of the principal diet components are given in fig. 6.8 for the months of June to September, 1971. Rates of food consumption generally decrease from late spring to early autumn. High water temperatures during July may restrict habitat utilisation by brown trout and be responsible for the reduced rations observed. Brown trout also demonstrated a major increase in predation on perch fry during late summer (fig. 6.8). High growth rates, daily rations and condition factor values for adult Loch Leven perch all suggest little evidence of food limitation during late spring and early summer.

As evidence of the converse extreme, Magnuson & Kitchell (1971) and El Shamy (1973) found adult bluegill, *Lepomis macrochirus*, in Lake Wingra, Wisconsin, USA, to be severely food limited. Daily rations were computed for three size classes of bluegill for the period April–November, 1970. The size classes were arbitrarily stratified as small, medium and large but broadly correspond to age classes II, III and IV–V; with weights of 10–20, 20–35 and 35–55 g respectively. In general, daily rations were inversely related to size for all months (fig. 6.9) averaging 1.69, 1.04 and 0.82% for small, medium and large fish respectively.

The importance of zooplankton in the diet was also inversely related to size. Rations for the smallest size group generally reflect the population dynamics of their major food item with spring and early autumn peaks in food consumption. Larger food items, i.e. insects, molluscs, terrestrial drift, plant material and some small fishes, were major dietary components of larger fish. Daily rations for all size groups were higher during spring than at comparable temperatures during autumn months (e.g. April–May compared with September–October) but were generally lower than those for bluegill in Minnesota lakes (Seaburg & Moyle, 1964).

El Shamy (1973) compared daily rations, food abundance and growth rates of bluegill in Lake Wingra with those of nearby Lake Mendota. In general, rates of food consumption and growth were comparable for yearling fish. However, differences between the populations increased with age and size of fish. Adult bluegill in Lake Mendota had daily rations 2–3 times those of adult bluegill in Wingra and weighed approximately 3 times as much by age 5. There were equally significant qualitative differences as the diet of adult bluegill in Lake Mendota was almost exclusively composed of large invertebrates (e.g. *Hyallela azteca*, Diptera, Ephemeroptera) and young fish. Further, the diversity and abundance of these macrofood items was substantially greater than that for Lake Wingra.

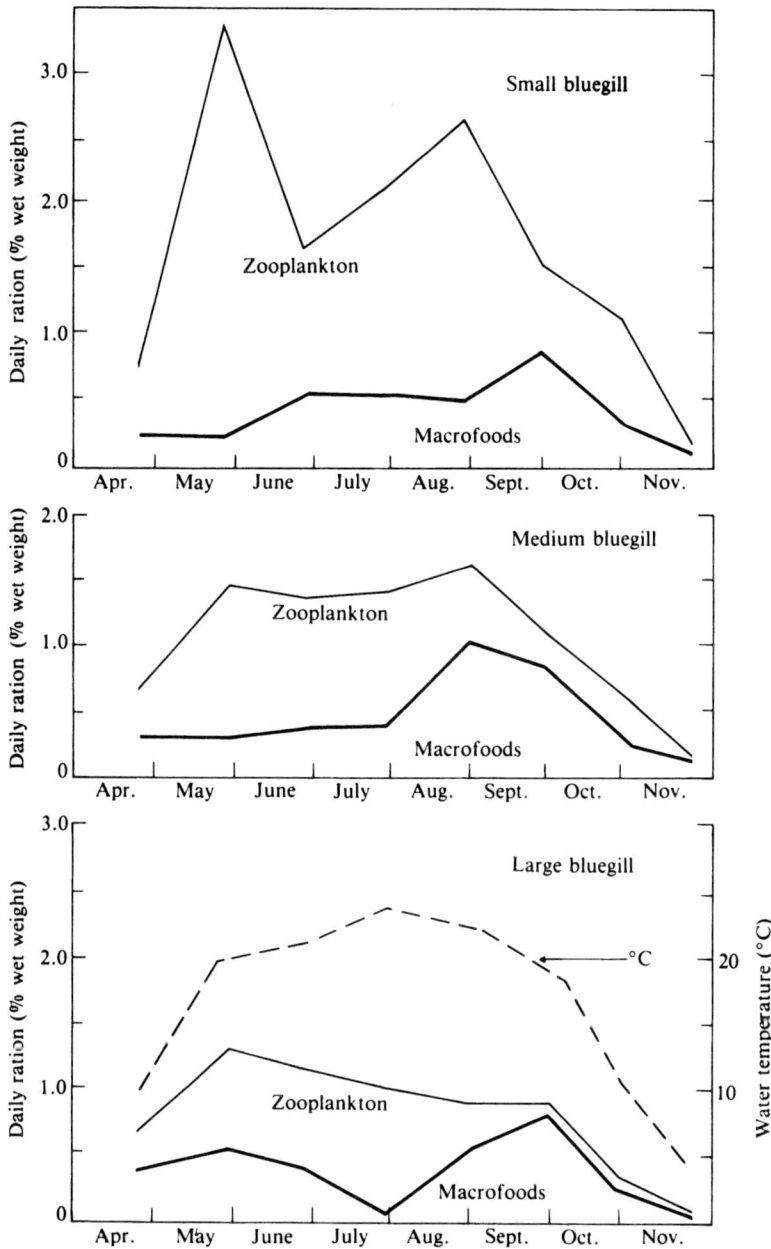

Fig. 6.9. Daily rations (% wet weight) for three size classes of bluegill (*Lepomis macrochirus*) in Lake Wingra, Wisconsin, USA, during the period April–November, 1970. Size classes small, medium and large broadly represent fish of 10–20, 20–35, and 35–55 g wet weight, and generally correspond to age groups II, III and IV–V, respectively. Dietary components are represented as zooplankton and macrofoods, the latter was largely composed of insects but included molluscs, terrestrial drift, plant material and some small fishes. (After Magnuson & Kitchell, 1971 and El Shamy, 1973.)

Examples of recent daily ration studies on piscivores are those by Biro (1969) and Backiel (1971). Both applied the Bajkov (1935) method. Although neither was developed around stratified diel sampling, the data base was large and the seasonal changes in calculated rations of general comparative interest.

Daily rations of 300–500g pike-perch *Stizostedion lucioperca*, in Lake Balaton, Hungary, ranged from 0.23% during March to 1.17% during July and averaged 0.87% for the period March–August (Biro, 1969). Daily rations during summer months were comparable with those of the walleye, *Stizostedion vitreum* (Seaburg & Moyle, 1964). The observed rations were estimated to be one-third or less of maximum rations and Biro concluded that the fish were substantially resource limited.

Backiel (1971) calculated daily rations for several predatory fish of the Vistula River, Poland. He presented a range of estimates based on a thorough review of available information. In general terms, he determined that rations were inversely related to fish size. Again, rations during spring were greater than those at comparable thermal conditions in autumn. Even the conservative estimates for pike, *Esox lucius*, and pike-perch, *Stizostedion lucioperca*, indicate relatively high rates of food consumption for these piscivores: two to three times those recorded by Biro (1969) and Seaburg & Moyle (1964). Comparison of Winberg's (1960) approach with the daily ration method indicated tentative agreement of consumption estimates given that either or both methods had an assumed bias of $\pm$ 30%. It is of interest to note that Solomon & Brafield (1972) also found agreement with Winberg's statement that metabolic rates in nature are twice those of the standard metabolism in the laboratory, but only when rations were maintained at high levels. Healy (1972) found that metabolic rates in nature were poorly estimated by the Winberg method when feeding rates were less than maximal.

6.4.10.2. Energy budgets

Few studies have sought to measure all the components of an energy budget for fish in natural or laboratory environments. Recent studies in the field (Healy, 1972) and laboratory (McComish, 1970; Solomon & Brafield, 1972) suggest that the bias introduced by applications of laboratory-determined respirometry to estimates of food consumption are more severe than those derived from laboratory data on food passage rates properly employed in determining daily rations. The fundamental criticism of Winberg's approach appears to be associated with the variation of specific dynamic action (Warren & Davis, 1967) which is a function of feeding level. To estimate food consumption rates based on laboratory measurements of respiration, experimental information must be compiled on temperature, body size, activity and feeding level effects. Moreover, respiration rates are significantly responsive to seasonal hormonal controls (Gerking, 1966). Combining all variables for application to nature still requires some direct measure of food consumption rates to estimate specific dynamic action levels. Of course, the best approach would involve measures of all process rates. However, given finite time and resources it seems logical that the evaluation of energy budgeting processes in nature might be better served by direct measures of daily rations and the remaining terms of an energy budget – excretion, growth, gamete production and mortality – and determining respiration levels by difference. When coupled with stomach analyses this approach would provide a seemingly more accurate and direct assessment of the quantitative and qualitative role of food consumption by fishes in the production dynamics of aquatic ecosystems.

6.4.11. Transfer efficiency

Brylinsky (Chapter 9) calculated the ratios of consumer production to gross primary production and called these ratios transfer efficiency. He distinguished four groups of consumers consisting of herbivores and carnivores in the zooplankton and benthos respectively. The entire fish population in an ecosystem does not form one particular trophic level. Even a single species requires a certain diversity of food items to survive; fry feed on small, easily digestible organisms only, the proportion of larger particles increases as the fish grow and the diet can shift either to plant material or to large benthos or pelagic animals including vertebrates.

We have demonstrated that there are situations in which production of fish is food limited, e.g. in Lake Wingra (page 326) or the 1969 summer for the trout population in Loch Leven (page 322), but it has also been shown that trophic relations are perhaps not the most important factor controlling trout populations in lakes. However, where fish reproduction is not limited and physical and chemical conditions do not preclude a fish existing, then their production depends on food supply. This, in turn, is directly or indirectly related to the primary sources of organic matter, primary production and influx of allochthonous organic matter (see Chapter 7), and can be best illustrated by data on fish-ponds where the number of fish is initially controlled.

Müller (1966) found a significant correlation between the average daily gross production of phytoplankton and the production of common carp in ponds ($r = 0.72$), but no correlation with net primary production. Hrbáček (1969) and Winberg & Lyakhnovich (1965) demonstrated similarly high correlations ($r = 0.85$) in carp ponds of Byelorussia. Wróbel (1970) estimated the same correlation ($r = 0.89$) and Wolny & Grygierek (1972) found that this coefficient for carp fry ponds was as high as 0.98. However, Kořínek (1972) reported a much weaker relationship ($r = 0.54$) in a series of Czechoslovakian ponds. He pointed out that in waters stocked with carp of older age groups which feed mainly on bottom fauna the relationship may not be as simple as in the above quoted cases. This conclusion concurs with the finding of Brylinsky (Chapter 9) that benthos production in lakes is less correlated with primary production than is zooplankton production.

Some IBP data on primary production and production and yield of fish are presented in table 6.14. They should be treated with due reservation because the estimates of the basic parameters are often rough and preliminary. Data on fish production and yield to man are taken from tables 6.9 and 6.13 for most sites. Data on gross primary production came from the sources referred to in those tables except for a few cases discussed below.

From among three reservoirs in USSR (Kiev Res., location *ca* 50° N, DR 66; Rybinsk Res., 58° 30′ N, table 6.9; and Tsimlyansk Res., 48° N, table 6.9) and one polder reservoir in Holland (Tjeukemeer, table 6.9), the latter and Rybinsk occupy an outstanding position in fig. 6.10. Sorokin (1972) reported that the influx of allochthonous organic matter to Rybinsk is about twice as much as

Table 6.14 *Primary and fish production in ponds*

Ecosystem	Range of primary production, P_1 (kJ m^{-2})	Range of fish production P_{fish} (m^{-2})	$(P_{fish}/P_1) \times 100$ %	References and notes
Carp ponds (19 measurements, *ca* 7-month period, Czechoslovakia)	2 008–10 460	18.0–47.0 g	1.7–8.1	Kořínek, 1972 (his tables 9 and 10, carp of age 1 + and 2 +)
Carp fry ponds (46-day period, 6 ponds, Poland)	1 740–4 351	59.4–150 kJ	2.9–3.6	Wolny & Grygierek, 1972 (juvenile carp and Chinese carps, original data recalculated)
Carp fingerling ponds (120-day period, 3 ponds, Poland)	2 100–5 715	23.5–52.7 g	*ca* 3.9–4.7	Wróbel, 1972 (carp of age O +)
Carp ponds (16 measurements, 165-day period, Poland)	2 297–11 606	52.3–215 kJ	1.46–2.89	Wróbel, 1970 (carp of age 1 +, average efficiency = 2.06%)
Sturgeon pond (lower Don region, USSR; 25–40 day period)	1.6 g O$_2$ m^{-2} day^{-1} net, hence, *ca* 586–938 kJ m^{-2} per period	12–18 g	8.0–8.6	DR 73 (recalculated; efficiencies related to net P_1)
Ponds in Malacca (*ca* 6-month period, Malaysia)	55 187–97 696	28.6–43.9 g	1.02–1.79	Prowse, 1972 (*Tilapia* sp, *Puntius* sp, grass carp; efficiencies of Prowse related to net P_1)
Various ponds with herbivore fishes (India)	13.07–29.5 g O$_2$ m^{-2} day^{-1} (annual averages)	90–618.3 g	0.4–0.78	Ganapati & Sreenivasan, 1972, their tables 3 and 6 (efficiencies as per cen of conversion of carbohydrates to edible fish flesh)

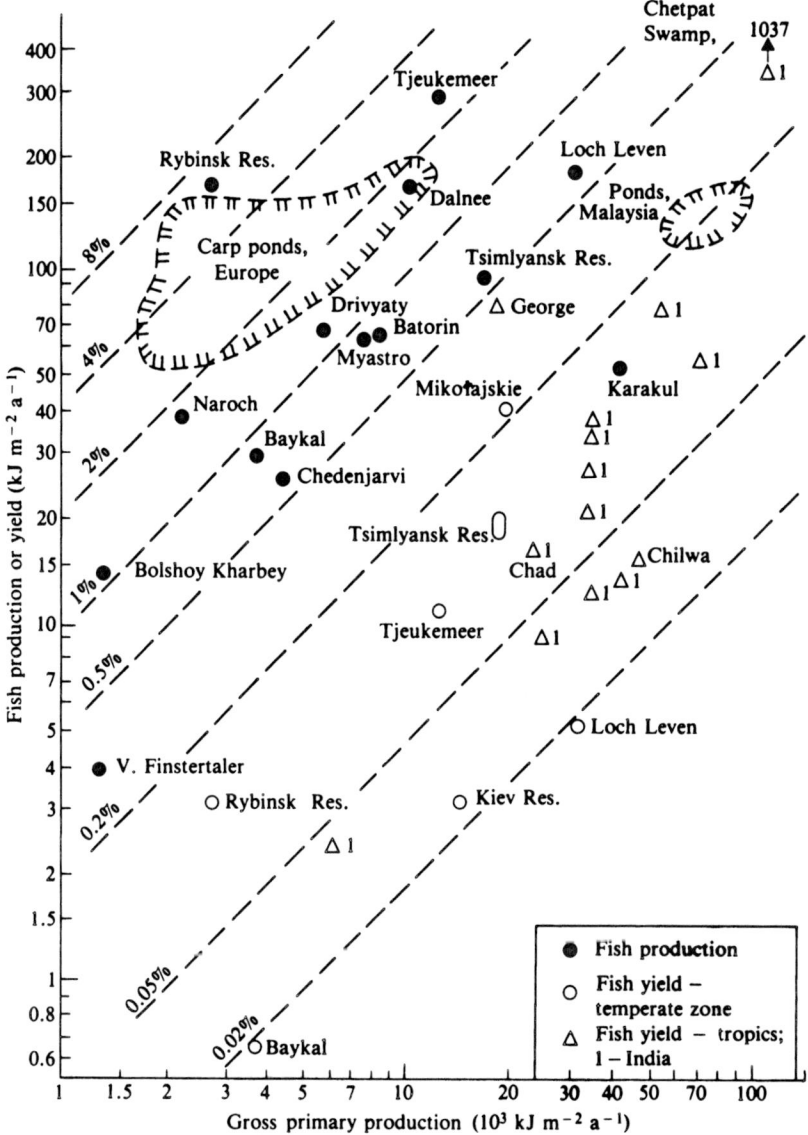

Fig. 6.10. Fish production (P_f) and fish yield to man versus gross primary production (P_1), both expressed in kJ (assuming that 1 g wet weight fish and 1 kcal of P_f and P_1 are equivalent to 4.184 kJ). Diagonal lines indicate efficiencies of transfer from P_1 to P_f or to yield. For explanations and references see text and tables 6.9 and 6.14.

the primary production (see Chapter 7). This can also be true for Tjeukemeer where, in addition, immigration of fish can contribute to the overestimation of the transfer efficiency related to autochthonous production only (Beattie et al., 1972).

Low efficiency in Lake Karakul (0.12%) is because the macrophytes contribute about 90% to the primary production (DR 63). In European and Malaysian ponds primary production (table 6.14) was estimated by the light and dark bottle method and these estimates seem fairly comparable. The areas encircled by a hatched line in fig. 6.10 encompass all plots of primary and fish production found in the references quoted in table 6.14. If the two reservoirs showing apparently very high efficiency are excluded then maximum fish production is *ca* 1.6% of gross primary production in lakes and reservoirs and the minima recorded are *ca* 0.2% (V. Finstertaler) or even less, e.g. Lake Karakul.

The efficiencies in carp ponds are usually higher. Hepher (1962a) estimated that daily fish production in Israeli carp ponds was on the average 1.6% of the primary production which ranged from 57 to 175 kJ m^{-2} day^{-1}. Winberg & Lyakhnovich (1965) showed that this ratio ranged between 2.2 and 12.9% of *net* primary production. In fingerling carp ponds in Byelorussia fish production was from 6.1 to 8.2% of gross primary production. The highest efficiency reported in table 6.14 is 8.1% thus about the same as the maximum reported in Winberg and Lyakhnovich. However, most ratios lie between 1 and 4%. The maxima reported (*ca* 8%) are surprisingly high considering the feeding habits of carp in Europe. It is possible that the low level of primary production given in Kořínek (1972) is underestimated resulting in an apparently high transfer efficiency. Another possibility is that organic matter accumulated in one growing season is effectively utilised by the pond biocoenosis in the next season (Wróbel, 1970). This observation is similar to the delayed effect of pond fertilisation observed by a number of authors (Winberg & Lyakhnovich, 1965). Such delayed recycling of organic matter complicates the interpretation of the transfer efficiency ratio.

The transfer efficiencies reported for tropical ponds (fig. 6.10 *ca* 0.2%) are of the order of 'inefficient lakes' in spite of the fact that a mixed fish population including the herbivorous *Tilapia* sp. was grown there, Prowse (1972). Wróbel (1970) noted that ponds situated in higher latitudes showed higher efficiencies and the data in fig. 6.10 seem to confirm his view. Apparently fish production is not proportional to the increased photosynthesis in the tropics.

Another set of data in fig. 6.10 concerns the ratio of fish *yield* to primary production. The highest ratio for the north temperate zone is represented by the data for Lake Mikołajskie, 53° 45′ N (primary production from Kajak *et al.*, 1972, fish yield from Inland Fisheries Institute data register). It can be explained by intensive fishing for plankton-feeding fish and by possible immigration of those from the nearby lakes. However, similar yields of *ca*

$10 \, \mathrm{g \, m^{-2} \, a^{-1}}$ have been recorded from some other Polish lakes. These and other data (fig. 6.10) show that man utilises not more than *ca* 0.2% of the gross primary production of lakes and reservoirs as fish yield. Winberg (1972b) stated that lake fisheries in the European part of the USSR yielded about 0.2% of net primary production but this efficiency seems to be on the high side of the recorded cases.

Data for tropical waters presented in fig. 6.10 also require some explanation. The value of primary production in Lake George, Uganda, has been derived from estimates of consumption of phytoplankton by herbivorous fish and zooplankton at $1240 \, \mathrm{mg \, C \, m^{-2} \, day^{-1}}$ (Moriarty *et al.*, 1973). It is considered to be the minimum *net* production roughly equivalent to $52 \, \mathrm{kJ \, m^{-2} \, day^{-1}}$ and *ca* $18\,800 \, \mathrm{kJ \, m^{-2} \, a^{-1}}$. Whatever the bias the high efficiency of fish yield in Lake George – *ca* 0.4% – is because the exploited species feed on phytoplankton and fishing is fairly intensive. Primary production in Lake Chilwa was estimated by standard methods (DR 5) and the value used is for the open lake area, the surrounding swamp being disregarded. Here the exploited fish are also mainly herbivorous.

Indian waters are represented by six reservoirs, three small lakes, one 'tank' and the Chetpat Swamp (Sreenivasan, 1972). The latter is an extreme case, but no explanation can be offered for both the very large primary production and fish yield and the highest efficiency of utilisation (*ca* 1%).

For running waters the index of transfer efficiency has little meaning: these ecosystems are heterotrophic and rely heavily on allochthonous resources. As shown by Westlake *et al.* (1972) and in DR 40, in a tentative budget of organic matter for a chalk stream, the dissolved and suspended organic matter reaching and leaving the stream section was many times larger than primary production. The fish production of $146 \, \mathrm{kJ \, m^{-2} \, a^{-1}}$ corresponded to several hundreds of thousands $\mathrm{kJ \, m^{-2} \, a^{-1}}$ of organic matter sources, an efficiency of the order of 0.001%. Quite a different picture is obtained from the data on the River Thames (Mann *et al.*, 1972; Chapter 7, p. 388) where the ratio is of the improbable order of 10%. This large difference obviously derives mainly from the different methods of estimating the influx of allochthonous matter.

It is noteworthy that in the River Thames roach utilised detritus directly as a food component which may to some extent explain the high transfer efficiency as the detritus pool represents both allochthonous and autochthonous production. Similarly high efficiencies can be expected in the South American waters where mud-eating (iliophagous) fishes are fairly abundant. Bonetto *et al.*, (1969) reported that in lentic environments of the Paraná River valley the iliophagous *Prochilodus platensis* amounted to more than 60% of the total fish biomass which reached *ca* $100 \, \mathrm{g \, m^{-2}}$.

Brylinsky (Chapter 9) discusses a number of papers dealing with possible correlations between fish yield or standing crop and abiotic factors or indices. He shows that they have some predictive value for certain ecosystems. This is also true with regard to the P_f/P_1 relationship in simple fish-pond systems. We cannot draw similar conclusions from the order of magnitude estimates discussed above. Tentative conclusions are that fish production varies between 0.1 and 1.6% gross primary production in lakes and reservoirs and that it can reach *ca* 4% in fish-ponds. The fish yield is likely to be an order of magnitude lower. Hence the cost to the ecosystem of producing fish is the reciprocal of the transfer efficiency which is somewhere between 62 and 1000 energy units of gross primary production for one unit of fish. The cost of producing edible fish is probably 10 times as much, but in well-managed ponds it may fall down to about 25 energy units. In tropical ecosystems the cost is probably higher than in temperate zones. These and other questions require further thorough consideration of both the theory and methods of obtaining reliable data.

6.5. Interrelationships between secondary producers (Morgan, Backiel, Kajak, Nauwerck, Kitchell, Larsson)

6.5.1. Interrelationships at lake outfalls

The outlets from lakes form a focal point for the interaction between living plankton, benthic communities and fish since the zooplankton forms a continuous stream of potential food flowing past the benthos and fish. Filter-feeding benthos, such as the net-spinning caddis, are particularly adapted to exploit this food supply. There may also be a two-way flow of energy between the lake and the river. This phenomenon was only investigated in Øvre Heimdalsvatn (Larsson, manuscript) where energy leaves the lake as zooplankton drift and the eggs of spawning brown trout and is partly returned in the form of recruits to the trout population of the lake. Unfortunately quantitative estimates are only available for August when 770 g dry wt 24 h^{-1} of zooplankton ran out of the lake of which 747 g dry wt 24 h^{-1} disappeared in the first 500 m of the outflow river mainly as food of trout and the net-spinning caddis, *Polycentropus flavomaculatus*. The movement of trout from the river to the lake averaged 107 g dry wt 24 h^{-1}. Of the trout guts 50% contained zooplankton, which may be as important in trout food as is the river benthos. *Polycentropus* congregated below the lake outlet where they dominated the benthos in terms of biomass. Chironomidae larvae, which could have also drifted from the lake, were the most important item in their food, but small crustacea occurred in 29% of the guts so that the caddis was also a significant predator on the drifting zooplankton.

An interesting feature was that although the filter feeding *Polycentropus* formed 50% of the biomass of total zoobenthos none were eaten by the trout. *P. flavomaculatus* is known to be fed on by trout and it is possible that at other times of year this may form an important link between the zooplankton and trout. Also *P. flavomaculatus* may feed more extensively on zooplankton at other times.

6.5.2. Influence of fish on zooplankton

Examples from Polish carp ponds (Grygierek, 1973) show how particularly the larger cladocerans are reduced in numbers and change in composition and size when the fish stock is increased. The larger the fish stock, then the smaller is the mean size (age) and the higher the number of eggs per clutch of cladocerans, and *Daphnia* may disappear completely. A higher egg production of *Daphnia pulicaria* is also recorded from Jezárko pond in Czechoslovakia (DR 10).

The impact of fish on zooplankton becomes even more pronounced when several species graze upon it. Grygierek (1973) found distinct differences between ponds with common carp only and ponds stocked with the latter plus silver carp, *Hypophthalmichthys molitrix*, grass carp, *Ctenopharyngodon idella*, and big-headed carp, *Aristichthys nobilis*. The main difference was that larger species – *Daphnia* sp., *Moina* sp. – were relatively more abundant in ponds with mixed fish populations than in those with common carp only. There were also significant changes in the species composition of the phytoplankton. These phenomena, according to Grygierek, resulted not only from the utilisation of phytoplankton by fish but from complex interactions between various grazers and the food resources. Unlike common carp, silver and big-headed carp filter out almost all sizes of plankton both small and large crustaceans and larger specimens of phytoplankton. The latter are not grazed upon by herbivorous zooplankton. Removing large phytoplankters creates better conditions for small ones. Increased grazing on phytoplankton results in more abundant particulate organic matter in the faeces, hence, more abundant bacteria which, in turn, are a food source for crustaceans (e.g. *Moina* sp.). With a greater amount of food the crustaceans reproduce more quickly and therefore the effects of grazing by fish on the zooplankton may be smaller.

Changes of species composition in lakes in favour of smaller species after introduction of planktivorous *Alosa* has been demonstrated by Brooks & Dodson (1965) and Brooks (1969).

In Lake Warniak, as a result of successive introductions of carp and bream (Kajak *et al.*, 1972; Hillbricht-Ilkowska & Węgleńska, 1973; Hillbricht-Ilkowska *et al.*, 1973) the following changes in zooplankton were observed:
(1) Decrease in numbers, biomass and production of the bigger zooplankton components (*Eudiaptomus graciloides*, *Daphnia cucculata*, *Diaphanosoma brachyurum*, *Asplanchna priodonta*), mainly by size selection by fish.
(2) Increase in the same parameters of the smaller components (*Bosmina longirostris*, *Ceriodaphnia quadrangula*, small rotifers).
(3) Increase in fecundity and size of all crustaceans, the latter being explained by indirect effects like increase of nutritive detritus and changes in the phytoplankton caused by fish.

Nilsson & Pejler (1973) show evidence for *Coregonus* affecting *Heterocope* species and daphnids. In the Vorderer Finstertaler See large *Daphnia*, which, based on remains in the lake sediments must have been present in the lake

earlier, have not been found in recent times and probably disappeared after the
introduction of char (*Salvelinus alpinus*) into the lake in the sixteenth century
(R. Pechlaner & G. Bretschko, personal communication). A similar example is
mentioned by Larson (1973). Generally large *Daphnia* species and large
individuals of other zooplankton species cannot usually survive together with
fish (Hrbáček, 1962). Alternatively, at Queen Mary Reservoir, heavy fish
mortality was followed by a big increase in the zooplankton stock (cladocerans mainly), which in turn was followed by a decrease in phytoplankton.

6.5.3. Influence of fish on the zoobenthos

The biomass of benthos may be reduced to about half or a third by fish
predation (Kajak & Zawisza, 1973; review by Kajak, 1968). Providing the fish
pressure is not too strong, and the decrease of biomass not too great this may
stimulate the production of zoobenthos keeping biomass at the optimum level
for intensive growth by reducing competition within the benthos, removing
invertebrate predators, and favouring younger stages and those species with a
shorter life cycle (Hayne & Ball, 1956; Wolny, 1962; Kajak, 1972). Invertebrate
predators often replace fish in this respect (Wolny, 1962; Kajak, 1972). The
highest biomass of benthos and probably the highest production in spite of
strong elimination, was observed at moderate stocking rates of fish in fish-
ponds (Wójcik-Migałowa, 1965) owing to manuring of the bottom by fish. The
seasonal pattern in numbers and biomass remains unchanged until a very high
level of fish predation is reached and, unless this is reached, factors other than
fish feeding are probably decisive for seasonal changes in benthos biomass.
Under very strong predation which lowers the biomass level very early in the
season, change in the seasonal biomass curve, species composition and size
structure may occur (Grygierek & Wolny, 1962; Wójcik-Migałowa, 1965).
Heavy fish predation can be especially important if it occurs during the
breeding period of the benthos (Kajak & Wiśniewski, 1966, for Tubificidae;
Berglund, 1968, for *Asellus*).

   After lowering the benthic biomass significantly fish move to other feeding
places (Kajak, 1972), or to other food such as zooplankton which allows the
benthos to recover (Gurzeda, 1965). Prolonged input and hatching of the eggs
of temporary fauna, mostly Chironomidae, seems very important for the
restoration of the benthic biomass (Lyakhnovich, 1965; Kajak, 1964, 1968).
Increase in the benthos biomass after fish pressure stops is due to both increase
in numbers and of the size of individuals (Kajak, 1972). The biomass rises to its
'normal' level very quickly, often within a few days. Bigger, older specimens of
the usual species present as well as new big species often develop when fish
pressure is low (Assman, 1958, 1960a, b; Hruška, 1961; Wolny, 1962; Kajak
& Zawisza, 1973).

Obviously the impact of fish is more complex than just by grazing as was briefly mentioned with respect to zooplankton in ponds.

The size and activity of benthos is probably important for their availability to fish (Morgan, 1956; Luferov, 1963). In some situations fish influence the benthos by destroying the filamentous algal layer (Wolny, 1962) and in others by stirring the bottom and thus increasing the amount of seston, which may improve the feeding conditions for benthos (Kajak *et al.,* 1972). Various groups of fish feeding on benthic algae, macrophytes, plankton etc. produce different benthic biomass patterns in fish-ponds by indirectly changing the food and living conditions for benthos (Lellak, 1966a, b).

There is commonly a greater proportion of invertebrate predators in the benthos biomass when fish grazing is small (Wolny, 1962; Kajak, 1968, 1972; B.E. Wásilenska, personal communication) which indicates selective grazing of predatory benthos by fish.

Studies by Hurlbert *et al.*(1971) demonstrated the role of fish in determining not only the abundance of other species but both direct and indirect effects on the chemical composition of waters. Nutrient cycling and remineralisation through excretion processes were substantially altered in those ponds with fish present. Many other studies (e.g. Wolny & Grygierek, 1972) demonstrate the dramatic impact even of a single fish species on biological, chemical and physical features of ecosystem dynamics.

6.5.4. Influence of zooplankton and benthos on fish

The most dramatic effect of invertebrate fauna on fish has been found on spawning grounds in lakes. Eggs and larvae of cyprinids and coregonids are often heavily preyed upon by predatory Cyclopidae and larvae of several insect species (Slack, 1955; Zuromska, 1967; Zawisza & Backiel, 1970). The limiting role of food organisms on fish populations is known from the well-known evidence from fish culture practice showing their importance for growth and survival rate of fish; e.g. in central European ponds carp, if not given artificial feeding, cease to grow in August when benthos and large forms of zooplankton are very scarce.

To date, quantitative measures of food consumption by fish in natural waters have rarely been effectively coupled with studies of food species production in ways that allow direct evaluation of interrelationships. There is, however, little doubt that the greater the monthly food resources available to the fish, the greater the fish production and vice versa. This trivial statement can be illustrated by a historical perspective indicating perhaps long-range feed-back mechanisms.

Through a rather complex series of physical and biological changes in Lake Wingra over the last 100 years, the bluegill has become the dominant fish

(Baumann *et al.*, 1974). Growth rates of bluegill have been reduced to one-third of those recorded three decades ago. As demonstrated by Hall *et al.*, (1970) the bluegill is an intensively size-selective and very efficient predator. Although once the most abundant large invertebrate in Lake Wingra, *Hyallela azteca*, a favoured food item, has declined to virtual extinction with the growth of the bluegill populaton. The number of species of Cladocera has been reduced to one-half that recorded by Birge (1891) and the majority of those which have been lost are large forms (*Daphnia* spp. and *Ceriodaphnia* spp.). Daily rations and growth rates of bluegill have declined to that of a population living at or about its dietary minimum in a closely coupled trophic exchange system (Baumann *et al.*, 1974). Over the last two decades, intensive predation by fish has brought about the same magnitude of change in community structure as that recorded in experimental ponds (Hall *et al.*, 1970).

6.5.5. Influence of benthos on zooplankton

Besides self-predation and fish predation, zooplankton can also be controlled by meroplanktonic members of the benthos or other nektonic species. In Latnjajaure dytiscid beetles live in the open water. In Esthwaite Water, Smyly (1972) has demonstrated a negative correlation between the relative abundance of *Cyclops strenuus abyssorum* and *Chaoborus*. Dodson (1970) has proved selective feeding on *Daphnia* by *Chaoborus*. In Latnjajaure there is predation on *Cyclops scutifer* by some species of chironomid larvae (*Pseudodiamesa, Heterotrissocladius*) partly close to the bottom but also during vertical migration of the chironomids to the surface at night, in summer, when the zooplankton are caught in the surface film (A. Nauwerck, personal communication). The role of zoobenthos, e.g. tube-building chironomids, in releasing nutrients from the mud indirectly produces more food for zooplankton.

Finally the possibility of occasional harvesting of zooplankton populations by larger, migratory organisms should be mentioned. This may be important particularly in minor waters lacking their own predators, where Amphibia (Dodson, 1970) or birds (Hellström & Nauwerck, 1971) can sometimes radically change the balance of the herbivores in the zooplankton.

6.6. Discussion

Within the natural lakes included in IBP the range of biomass is similar for zooplankton and zoobenthos but as might be expected fish biomass does not reach such high levels. The ranges are 0.01–100 g wet wt m^{-2} for zooplankton, 0.7–100 for zoobenthos and 0.1–25 for fish. It is not easy to compare zooplankton production with the other groups as it has been estimated on a daily basis only. However assuming the zooplankton produces at a high rate for half the year figures of $0.2-310 \text{ kJ m}^{-2}\text{a}^{-1}$ can be computed, compared with 10–1000 for zoobenthos and 4–300 for fish. The figures for zoobenthos and fish differ by the same proportion as for biomass but the computed figures for zooplankton are probably too low. There is also the problem that production may vary greatly between different zones in a lake* P/B values for zoobenthos and fish also vary by the same proportion, 1.5–7.8 compared with 0.5–2.3. As might be expected the ecological efficiency decreases from zooplankton through zoobenthos to fish, zooplankton being 2–30%, zoobenthos 1–9% and fish 0.2–2%. Although P/B ratios have been used to provide production estimates when only biomass data are available the IBP studies have shown that this can be very variable even for the same species within the same lake and P/B ratios from elsewhere should only be applied cautiously to indicate the possible range of production. Clearly, annual P/B is a very different measurement when calculated for an arctic lake, with a short growing season, and for a tropical lake with a continuous growing season. The truest measure of production per biomass can only be obtained by measurement for a single generation of each species. P/B has however a handy rule of thumb application to the whole of a component such as benthos but the wider the number of species covered the less meaningful the ratio.

The interrelationships between zooplankton, zoobenthos and fish are delicately balanced so that changes in the composition of one may initiate responses in the composition of the others. Within the secondary producers fish are the most important for the welfare of man as a food supply. Fish production may be manipulated by man either through factors affecting their food supply or through factors influencing the fish directly. Either way this involves changes in the whole ecosystem which must be understood for the most effective management. The same applies if ecosystems are being viewed from the viewpoint of conservation of natural habitats. From the data presented here, as well as that from other sources, we can infer that although there is a web of trophic interrelationships some pathways are more efficient in transferring matter/energy; e.g. zooplankton–fish, than others, e.g. benthos–fish or zooplankton benthos. This may be related to the time scales of the life cycle in the three groups as well as the choice of index selected for measurement of matter. Once the objectives of a study are elaborated and a possible web of trophic interrelationships drawn up, only those pathways which are considered to be important need be studied initially to exploit the food supply for man.

In ending this chapter it is worthwhile reiterating the need for more precise and sophisticated techniques for measuring food uptake and utilisation by

---

* The ranges given here are for total water bodies or the profundal zone; for some zones maxima are much higher. (Note that these figures are based on dry weights, wet weights would give values at least 7 times higher, and more for most bivalves in which wet weight is 20–25 times higher than dry weight.)

secondary producers in order to determine their energy demands on the ecosystem. This must be coupled with improvement in quantitative sampling strategies and techniques in order to determine population size, growth and mortality of secondary producers if we are to improve our knowledge of the delicate interrelationships which determine production to a level whereby we can manipulate them with some precision.

The IBP programme has highlighted the difficulties of constructing simple trophic relationships particularly for the zooplankton and zoobenthos and pointed the way for more studies of the functional relationships of organisms within the ecosystem.

The authors wish to express their appreciation to Dr F.V. Krogius for her concise review of fish production studies made in the USSR. Dr J. Rzóska assisted this work in many ways which has been highly appreciated during every stage of preparing this contribution.

7. Organic matter and decomposers

Coordinator: G.W. SAUNDERS

7.1. Introduction

It has been intuitively obvious for a long time that non-living organic matter plays a very important role in the structure and function of aquatic ecosystems. Both dissolved and non-living particulate organic matter tend to accumulate in aquatic systems until a quasi-equilibrium is attained, so that quantitatively they dominate these environments compared to the living components. Of course, this may not always be true but on the average it seems to hold. Therefore, organic matter functions as an energy reservoir, which may be more rapidly or more slowly utilized depending on the functional requirements of the system.

Organic matter is in the reduced state compared to carbon dioxide and therefore it has an energy content part of which can be coupled to do work. Most of the work that is done with this available energy is concerned with maintenance, growth, development and reproduction of organisms.

Organic matter not only contains a store of free energy but it also contains the elements that enter into the construction of living protoplasm. When it is decomposed these elements are released and become available for storing energy in the form of new protoplasm. Traditionally the ultimate decomposition of organic matter has been attributed to microbes, in particular the bacteria. However, other organisms, the zooplankton and zoobenthos, may also generate smaller or larger short cycles, dissipating energy and releasing nutrients in intervening steps between the production of new organic matter and its decomposition by bacteria.

Although quantitatively very little is known about non-living organic matter in aquatic systems and the processes of its transformation and decomposition, it is obvious that the quantity in a system and the rate at which it is decomposed is some function of the rate at which energy enters the system. The two forms in which energy may enter an aquatic system are: the kinetic energy of solar radiation, which has yet to be stored in organic matter by photosynthesis, and the potential energy of organic matter formed outside the aquatic system, i.e., allochthonous material derived from the catchment area. The form of the predominant energy input must have profound influence on the structure of an aquatic system.

The perspective developed here relates to what happens after organic matter

Contributors: G.W. Saunders, K.W. Cummins, D.Z. Gak, E. Pieczyńska, V. Straškrabová & R.G. Wetzel.

enters or is produced within the system. Some data will be derived from IBP
studies. However, since this subject area was not intensively studied within
IBP much information will be reviewed from other studies. We will attempt to
develop an integrated picture of this most important problem area and suggest
subjects in need of study and strategies of analysis that should prove
rewarding.

7.2. Sources of organic matter

7.2.1. Allochthonous organic matter

Compared with autochthonous primary production, there are few data
concerning the amount, forms and fate of allochthonous organic matter in
water bodies. The importance of allochthonous organic matter varies in
different types of water. Various systems of lake typology have attempted to
consider this importance (Aberg & Rodhe, 1942; Elster, 1963; Odum, 1971) in
undisturbed lakes; problems concerned with allochthonous inputs and
typology of polluted waters have been discussed in detail elsewhere (Caspers &
Karbe, 1967; Rodhe, 1969; Olszewski, 1971; Rossolimo, 1971). These schemes
are mostly qualitative because of the difficulties in quantitatively estimating all
the input rates associated with any single water body. However, as a general
rule, the relative importance of allochthonous organic matter tends to increase
as the ratio of the area of the water body to length of shoreline decreases. This
would appear to be true for streams as well as lakes.

The most important vehicles of input of allochthonous organic matter into
water bodies are streams, surface runoff, ground-water, shore erosion, litter
fall, municipal and industrial wastes, and atmospheric precipitation
(Pieczynska, 1972a). The proportion of organic to inorganic matter varies
from a few per cent of dry weight in the case of most surface drainages to more
than 90% for litter fall. The allochthonous organic matter may enter a system
as large particles, such as leaves, or as fine particles that have been partially
processed in terrestrial systems, either mechanically or biologically, or as
dissolved material. In undisturbed open aquatic systems the main constituents
of dissolved organic matter are considered to be humic substances. It is not
absolutely clear that this is true. During this era of increased industrialization
and intensification of agriculture many water bodies are affected by such
human activities and are supplied with a diversity of fertilizing and toxic
substances. From an ecological point of view water bodies supplied with
allochthonous substances not only contain an increased amount of organic
matter, but because of its specific nature in different regions, this organic
matter may function in very different ways so that the normal operation of
such waters is drastically modified by these additions.

Most inventories of allochthonous input to water bodies have been only partial because selected sources have been estimated or have been presented in the form of proximate indices of area, or indices of fertility of the drainage basin, or as total nitrogen or total phosphorus inputs. The latter are considered in Chapter 4. The units of measurement have differed as well as have the time scales for the budget, so there is little comparability. Conversion factors are often lacking. Thus it is possible to present only a very general and cursory summary of the importance of allochthonous organic matter to freshwater systems.

Detritus in the form of dead leaves is the most conspicuous and the most frequently studied input of allochthonous particulate organic matter to streams and to some lakes. In the case of streams this leaf litter may be quantitatively and qualitatively the most significant input but it is by no means the only one. Leaves from riparian woody plants are not only significant in woodland biomes but also in grasslands where trees and shrubs frequently border the water courses. Leaf litter input is pulsed in the autumn and spring in the temperate zone (Minshall, 1967; Fisher, 1971; Fisher & Likens, 1972). Autumnal abscission, the rule among deciduous species, together with autumn runoff maximizes the introduction of leaf litter, fruits and seeds. The annual input for a small woodland stream averages between 1 and 5 g m^{-2} day^{-1} in dry weight (Petersen & Cummins, 1974). For streams in general the range is probably greater than this. Input is reduced in winter, particularly at latitudes where snow and stream ice cover restrict the movement of litter. Spring runoff and spring-shed leaves (e.g. beech and oaks) produce a second pulse. Spring and early summer are marked by introductions of bud scales and flowers and some of the vegetation sheds leaves throughout the summer period, e.g. willow and alder, and summer rains continue to introduce partially processed terrestrial litter. Terrestrial insect faeces from the forest canopy also may constitute a significant input into some streams and the margins of standing waters.

Other particle sizes of allochthonous detritus are derived primarily from fragmentation of leaf litter, trees and shrubs, and other vegetation by mechanical forces or biological activity and from faecal output of herbivores.

In the case of lakes the amount of leaf litter dropping in autumn ranged from 100 to 500 g dry weight per metre of shoreline (Levanidov, 1949; Szczepanski, 1965; Pieczyńska, 1972a; Gasith et al., 1972). Pieczyńska (1972a) has shown that in Mikołajskie Lake the leaf fall from September–November is 70% of the total annual leaf drop. Reimers (1954) showed that leaf fall is the major source of allochthonous organic input to a small forest pond.

However this source of allochthonous matter is frequently only a small part of the total annual organic input in lakes. In Mikołajskie Lake, for example, it is less than 0.2% of primary production and only 2.7% of the total allochthonous organic input (Pieczyńska 1972a). Leaf drop is not an important source of organic matter in Lawrence Lake, an unproductive marl lake surrounded by marsh vegetation (Wetzel & Otsuki, 1973) or in Lake Wingra (Gasith *et al.*, 1972).

Wind-blown material may be an important source of nutrients especially in heavily forested areas containing small ponds. Sorokin (1967) has studied the meromictic lakes Belovod and Geh-Gel in which the zooplankton depends mainly on the allochthonous matter decomposed by bacteria. On the other hand Wetzel *et al.* (1972) have shown that wind-blown loess is insignificant in the energy budget of Lawrence Lake. Smirnov (1964) has suggested that allochthonous organic matter may have special properties. The pollen and spores entering Rybinsk Reservoir amount to 6×10^{-4} kg m^{-2} a^{-1}. In spite of its small amount this material is an important source of vitamins for animals.

The amount of water-borne allochthonous organic matter entering lakes is variable but sometimes quite important. Surface runoff supplied Mikołajskie Lake with 34 tonnes dry weight per year, which is several times greater than

Table 7.1 *Annual input and output of organic carbon in Lawrence Lake, Michigan, 1971–2*

	g C m^{-2} a^{-1}
Inputs	
Inflow	
Particulate	
Inlet 1	1.99
Inlet 2	0.96
Ground-water and seepage	1.15
Wind-blown	0.00
Dissolved	
Inlet 1	7.00
Inlet 2	7.94
Ground-water	6.01
Biological Production	
Particulate	75.10
Dissolved	20.20
	120.35
Outputs	
Outflow	
Particulate	2.80
Dissolved	35.82
Sedimentation	23.70
Respiration	55.30
	117.62
Net Increment	2.73

After Wetzel & Otsuki (1973).

direct leaf fall. The second, but much less important source of allochthonous matter in Mikołajskie Lake is shore erosion. This amounts to 2 tonnes a^{-1} or about 0.8% of total allochthonous input. Efford (1972) in studies of Marion Lake, which is part of a stream system, has estimated the input of fine particulate carbon from inlets to be $50 \, g \, C \, m^{-2} a^{-1}$. This is a small value compared to the input of dissolved organic carbon or other sources of particulate carbon. The input of dead leaves is of considerable importance in this system although this source has not been studied in detail. Wetzel & Otsuki (1973) have estimated the annual input of particulate allochthonous carbon to Lawrence Lake (table 7.1).

The introduction of particulate organic matter of domestic, industrial, or agricultural origin may often be qualitatively similar to natural detritus, at least with regard to the variety of substrates made available. Quantitative differences often exist depending upon such factors as the magnitude of natural inputs over which man-engendered effluents are superimposed. Such wastes are usually introduced at continuous high levels or short pulses rather than in the seasonally pulsed fashion typical of undisturbed streams and standing waters. A source of large particulate organic matter not usually considered is via storm sewers. For Lake Wingra this amounts to more than 0.5 tonnes organic matter in one season (Gasith *et al.*, 1972).

The structural and functional properties of lake and pond ecosystems must vary considerably in relation to the amount of particulate allochthonous organic input as well as to its qualitative content and the temporal distribution of its input. However, no detailed studies attempting to elucidate the functional importance of allochthonous organic inputs to lakes have yet been conducted.

The metabolic pattern of streams tends to follow a continuum proceeding downstream. Small streams and head waters are usually natural heterotrophic detritus based systems. Intermediate size streams tend to shift to autotrophy because of increased incidence of light and increased nutrient loading. Very large and deep streams tend to become less autotrophic or even heterotrophic because of limitations of light penetration and increased organic loading from upstream sources. Agricultural practices tend to intensify autotrophy by clearing cover from streams and increasing inorganic nutrient loading. Heterotrophy is intensified downstream by increased organic loading. There are very few intermediate and large streams that are not modified by such human activities.

The role of large particulate allochthonous organic matter in running water may be quite significant, especially in small forest streams (Kaushik & Hynes, 1971). In larger streams the importance of this kind of material may not be so great. Westlake *et al.* (1972; DR 40*) estimate the allochthonous organic input to Bere Stream to be as follows (metric tonnes organic matter per year): leaf fall

* See list of data reports (DRs) given in Appendix I.

2, marginal plants 1, cress-beds (particulate and dissolved) 38, spring water 17, drains and minor effluents 81, total 139.

In the case of the River Thames at Reading (Berrie, 1972b) the sewage input in summer provides $2.4 \, \text{g m}^{-3}$ suspended material in comparison to $5-10 \, \text{g m}^{-3}$ total suspended material in the river. Therefore, in this case, sewage is an important source. Mathews & Kowalczewski (1969), in the same stretch of river, estimate leaf fall to be $28.4 \, \text{g dry wt m}^{-2} \text{a}^{-1}$. The units are different but in comparison to values given above for leaf litter inputs this estimate is small. Mann *et al.* (1972) suggest that terrestrial insects, bread, and bait due to angling are a significant allochthonous input to this same region of the River Thames.

It has proved useful in comparing selected woodland streams in the USA to follow the annual cycle of the ratio of coarse particulate organic matter to fine particulate organic matter (CPOM/FPOM) in the sediments. Coarse particulate organic matter is defined as detrital particles larger than 1 mm in size. This is the approximate size at which the processing of particulate matter shifts from a dominance of fungal–bacterial colonization and shredding action by invertebrates to bacterial colonization and fine particle collector detritivores (Cummins, 1974). The CPOM/FPOM ratio ranges between 0.4 and 0.6 during the autumn and spring and between 0.1 and 0.3 in the winter and summer.

Allochthonous water-borne particulate organic matter of natural streams is on the average 10–20% of the amount of dissolved organic matter. Large fluctuations of allochthonous particulate organic matter of streams are found in relation to precipitation, velocity of stream flow and scouring, and terrestrial vegetation of the watershed (Fisher & Likens, 1973; Hobbie & Likens, 1973; Wetzel & Manny, 1977). In a cold woodland stream fine particulate organic matter maxima were generally found in the period of maximum terrestrial growth during the summer; constant levels of fine particulate organic matter were characteristic of autumn and winter months (Wetzel & Manny, 1977).

Diel fluctuations of particulate organic matter in transport can be marked in streams; levels in a natural woodland stream were consistently higher during dark periods than during daylight, presumably related to increased activity of benthic fauna and fish during the night (Manny & Wetzel, 1973).

Minshall (DR 30) has estimated that for Deep Creek Stream wind-borne allochthonous detritus input ranges from $0.5-6.88 \, \text{mg m}^{-2} \text{d}^{-1}$ organic matter as insects and $3.52-132.2 \, \text{mg m}^{-2} \text{d}^{-1}$ for plant material.

It is obvious that there is no consistent quantitative pattern of allochthonous particulate inputs to lakes and streams. Nor is any pattern likely to be discovered until budgets for many more water bodies have been established. It is also obvious that particulate allochthonous inputs may be important in many lentic and lotic systems but especially in the smaller systems that are shaded and more contiguous to terrestrial ecosystems.

The main sources of allochthonous dissolved organic matter are surface runoff, stream inflow, ground-water seepage, sewage and leaching from particulate matter, particularly dead leaves. Leaching of dead leaves is a relatively rapid process and has been recorded by several authors. The rate of leaching is highly specific, depending upon plant species, water temperature and other factors. It usually varies from a few per cent up to 30% of the dry weight per day (Kaushik & Hynes, 1971; Bretthauer, 1971; Lush & Hynes, 1973).

In an arctic tundra pond, the runoff following the summer thaw provided a large percentage of the total input of dissolved organic carbon into the pond (Hobbie *et al.*, 1972).

In a small hard-water Michigan lake (Wetzel *et al.*, 1972) the allochthonous dissolved organic input was an order of magnitude greater than the input of particulate organic matter. Surface runoff was minimal during summer months in which rainfall was reduced (Wetzel & Otsuki, 1973). This is also a period of maximum growth of marsh plants which otherwise might release greater amounts of dissolved organic matter into the drainage water. The concentration of dissolved organic carbon in ground-water was low but the total input from this source amounted to about one-third the total annual input of allochthonous dissolved organic carbon (table 7.1).

The proportion of humic fractions in the inflow, lake water and outflow, were similar, suggesting that the inflow contained mostly refractory dissolved organic substances. A sedimentation of humic materials to the hypolimnion was observed in this hard-water lake, in part due to absorption on $CaCO_3$ precipitating during epilimnetic decalcification (Otsuki & Wetzel, 1973; Wetzel & Otsuki, 1973). Variation in the input of dissolved organic matter during the year has also been observed for Lake Wingra (Gasith *et al.*, 1972). They also observed a high concentration of dissolved organic carbon in rainfall, $6-11$ $mg\,l^{-1}$. Efford (1972) has estimated an allochthonous inflow to Marion Lake of $467\,g\,C\,m^{-2}\,a^{-1}$.

Generally the concentrations of dissolved organic matter in unpolluted streams are low, of the order $6-20$ $g\,m^{-3}$ (Fisher, 1971; Fisher & Likens, 1972, 1973). For short periods organic matter leached from particles that fall into streams may be significant. From 5 to 30% of the dry weight of leaves is lost through leaching in the first 24 hours after wetting. Experimental data (Cummins *et al.*, 1972) indicate a rapid processing of high levels (35 $g\,m^{-3}$) of organic carbon leached from leaves, and thus the concentrations measured in monitoring studies probably represent mainly extremely resistant organic substances. The usual sampling frequency applied in stream studies probably does not permit shorter-term pulses of rapidly metabolized dissolved organic matter to be observed. However, there will be a small pool of labile dissolved organic matter which is maintained continuously by long-term gradual leaching of particulate organic substrates, particularly invertebrate faeces, and extracellular release from producer and decomposer organisms, as well as hydrolysis by extracellular enzymes. In the case of a stream passing through an

Table 7.2 *The relation of allochthonous and autochthonous organic matter input to various types of water bodies*

Water body	Allochthonous matter		Net primary production (kJ m^{-2} a^{-1})	Author
	total input (kJ m^{-2} a^{-1})	composition		
Silver Springs (USA)	2 033	detritus	36 957	Odum (1957b)
Root Spring (USA)	10 586	detritus	2 740	Teal (1957)
River Thames (below Kennet, England)	1 363	terrestrial insects, bread, bait, leaves, upstream sewage	8 163	Mann *et al.* (1972)
Bere Stream (England) (40 ha part of river)	139 (tonnes org. matter a^{-1})	trees, cress-beds, spring water drains, minor effluents	19 (tonnes org. matter a^{-1})	Westlake (DR 40)
Mikolajskie Lake (Poland)	711	sewage, litter fall, surface runoff, shore erosion	14 393	Kajak *et al.* (1972)
Rybinsk Reservoir (USSR)	5 356 (kJ m^{-2} season^{-1})	surface runoff, precipitation	2 205 (kJ m^{-2} season^{-1})	Sorokin (1972)
Chedenjarvi Lake	149.4 (g C m^{-2} a^{-1})?	surface inflow, rainfall, flood, runoff	88.2 (g C m^{-2} a^{-1})	Romanenko (1973) Gorbunova *et al.* (1973)
Marion Lake (Canada)	525 (g C m^{-2} a^{-1} + great amount of allochthonous material)	inlet/dissolved and fine particular material	70 (g C m^{-2} a^{-1})	Efford (1972)

	(g C m^{-2} a^{-1})		(g C m^{-2} a^{-1})	
Lawrence Lake (USA)	25.1	inlet, ground-water	171.12	Wetzel et al. (1972)
Kiev Reservoir (USSR)	16 736	inlet	11 506	Gak et al. (1972)
Red Lake (USSR)	1 402	inlet	5 084	Andronikova et al. (1973)
Augusta Creek (USA)	9 678	particulate detritus	79	Cummins, Petersen, King (unpublished data)
	8 389	leaf litter and 'non-resistant' detritus		
	1 289	'resistant' detritus, branches, bark, roots		
Total	19 356			
Bear Brook (USA)	13 493	particulate detritus	40	Fisher & Likens (1973)
Watershed 10 (USA)	8 962	particulate detritus	94	Sedell et al. (1974)

extensive marsh area the dissolved organic matter increased markedly, as much as doubling (Wetzel & Otsuki, 1973; Manny & Wetzel 1973; Wetzel & Manny, 1977).

It is obvious that the amount of allochthonous material entering a water body is some function of the nature of the drainage basin. A number of indices have been developed to express the relationship between production in lakes, and the size and nature of the drainage basin (Patalas, 1960b; Ohle, 1965; Vollenweider, 1971; Schindler, 1971b). However, Brylinsky & Mann (1973; Chapter 9) found that Schindler's index did not correlate for those lakes studied in the IBP/PF Programme; nor did Ryder's index (Ryder, 1965). It is probable that the indices are applicable within fairly homogeneous ecological regions. However, the indices fail when applied to a diversity of geographical regions because other factors may be as dominant in controlling production as the size and nature of the drainage basin. Increase in the fertility of lakes has been correlated with increased population densities, industrialization, tourism, or intensification of agriculture (Caspers, 1964; Zahner, 1964; Edmondson, 1968, 1970; Vollenweider, 1971; Ohle, 1972) all of which imply increased allochthonous input. There have been two classic examples of the impact of allochthonous inputs on lake ecosystems. They involve the restoration to a previous less productive state by diversion of sewage (Edmondson, 1968, 1970) and by removal of bottom sediments accumulated over many years (Björk, 1972).

The relative importance of allochthonous organic matter in water bodies depends on the size and character of the drainage basin and factors controlling autochthonous production. Allotrophic and autotrophic inputs to certain aquatic systems are compared in table 7.2.

There are so few data available that it is not possible to observe any regular patterns concerning the importance and the processing pathways of allochthonous organic matter in natural waters. There are great differences between even such special systems as springs. Root Spring (Teal, 1957) is dominated by allochthonous detrital input while Silver Springs (Odum, 1957b) is dominated by an autochthonous primary production input. Meaningful generalizations cannot be made until additional comparisons become available. However, in most cases allochthonous inputs are probably underestimated because usually not all sources are measured in a single system. Therefore, the significance of allochthonous matter is probably greater than indicated in table 7.2.

7.2.2. Autochthonous particulate organic matter

7.2.2.1. Introduction

The particulate non-living organic matter of autochthonous origin must be derived from the death of living organisms. Death may be due to natural mortality or to feeding by herbivores and carnivores. In lakes, emphasis has been placed on estimating zooplankton mortality because methods are available and sampling variance is relatively low compared to sampling benthos or higher trophic levels (Edmondson & Winberg, 1971). It is obvious, however, that bacterial and algal mortality cannot be accounted for solely on the basis of feeding by animals (Gak, 1967; Winberg, 1971c). There are many fewer animals adapted for feeding on living vascular hydrophytes than those which graze on living phytoplankton. There are many examples in which the population development and decline of phytoplankton are out of phase with zooplankton population oscillations, suggesting that other factors cause the mortality of the phytoplankton (Nauwerck, 1963; Saunders, 1971; Haney, 1973). Natural mortality has been very difficult to estimate so that no broad studies have concerned themselves with the mechanisms of mortality relative to ecosystem structure. The following presentation is only very general. Greater detail concerning plant production and destruction can be found in Chapter 5 and animal production in Chapter 6.

7.2.2.2. Algae

In large lakes phytoplankton must be a dominant source of organic detritus in the pelagic zone (Wright, 1959; Wetzel *et al.*, 1972). Production of detritus fluctuates irregularly and seasonally (Pieczyńska, 1972b; Wetzel *et al.*, 1972). Seasonal fluctuations also occur in tropical lakes where environmental conditions appear to be much more regular than in temperate regions.

Large masses of benthic algae produce organic detritus periodically and probably continuously in the littoral zone of lakes. Very few estimates of the rate of production of detritus from algae are available. However, epibenthic algae constitute half of the primary production of a large shallow lake in northern California (Wetzel, 1964). Epilithic algae are an important component of production in Char Lake in the Canadian Arctic (Kalff & Wetzel, 1978). Epiphytic and epipelic algae are also important components of production in an oligotrophic Michigan lake amounting to about 23% of total annual primary production. This is roughly equal to phytoplankton production in this lake.

7.2.2.3. Macrophytes

Macrophytes may predominate the plant biomass in small shallow lakes and *Chara* sp. may be very important in large lakes where light penetration is deep and a productive littoral and sublittoral zone develops. In an arctic tundra pond emergent macrophytes and epibenthic algae dominate the sources of particulate organic carbon (Hobbie *et al.*, 1972). Perennial mosses are of major importance in a Swedish Lapland lake (Bodin & Nauwerck, 1968). Macrophytic production dominates an oligotrophic hard-water lake in

Michigan. Macrophytes may produce detritus throughout the year but the major input of dead particulate matter occurs with the change of conditions in the autumn in temperate regions and is mainly related to the life cycles of these plants.

As emergent macrophytes assume greater dominance and eventually encompass most of the lake basin, an exceedingly productive combination of littoral macrophytes and attendant microflora develops (Wetzel & Hough, 1973). Ultimately all of this production must die and be processed in the decomposition pathway.

7.2.2.4. Animals

Feeding by herbivores and carnivores results in production of detrital particles by fracturing of food particles prior to ingestion, and through egestion. Most animal feeding produces detritus as faecal material which may be dispersed as fine particles or it may be surrounded by an envelope and packaged as a pellet. Thus, pelletizing may produce larger more concentrated food sources which are easily collected by detritivores. Digestion in the gut may result in a rapid breakdown and transformation of detritus into a more easily assimilable form of detritus or into bacterial protoplasm. Bacteria usually are not as concentrated as detritus either in suspension or in the sediments. However, they appear to be more assimilable than detritus by animals in general (Saunders, 1969, 1972a). Animal feeding can produce various kinds of particles that may be used directly by other animals or indirectly by the production of assimilable bacteria as an intervening step.

Animals that graze on algae are always present in non-polluted streams, but they are not obligate herbivores, often feeding on significant amounts of detritus (Cummins, 1973). As in lentic systems, the tissues of vascular hydrophytes tend to enter the processing cycle after death rather than through the grazing activity of stream herbivores.

Materials such as animal faeces, often considered to constitute autochthonous inputs, are probably primarily conversions of detrital particles of terrestrial origin. The fine detrital particles are surface colonized by bacteria. The ingestion–digestion activity of fine-particle-feeding detritivores serves to strip off the surface colonizers which constitute the basic source of nutrition for the animals. The surface cleaned material in the hind gut is recolonized by resident anaerobic gut flora. The egested faeces represent sites for new microbial colonizers and particles for ingestion by other (or the same) detritivores (M.J. Klug, personal communication).

7.2.2.5. Resuspension

Resuspension of organic particles from the sediment associated with turbulence probably occurs more or less continually in lakes (Wetzel *et al.*, 1972) and particularly in streams. In lakes turbulence is mainly due to wind but in the winter in ice-covered lakes it may be associated with density currents generated by differential heating in the littoral zone. There may be a periodic resuspension, lateral transport, sorting and deposition into the pelagic region of lakes (Wetzel *et al.*, 1972). There are also more major resuspension periods in lakes with mass overturns, as exemplified by vernal and autumnal mixing in lakes of temperate and higher latitudes.

Major resuspension in streams is associated with high water level. This may be caused by a periodic rainfall or may be seasonal as a result of melting snow in north temperate regions or prolonged rainy seasons in other areas. The result is transport downstream of particles which may be important to filter-feeding detritivores. Jónasson (1972) has demonstrated that growth of benthic filter feeders is correlated with sedimentation of detritus resulting from high phytoplankton production in the spring and autumn in Lake Esrom.

7.2.2.6. Flocculation

Flocculation of dissolved organic matter on bubbles that occur in water has been proposed as a mechanism for producing particles of organic matter. This may occur regularly in the ocean (Riley, 1970, 1973) but it may not be an important process in fresh water (Saunders, 1969).

Adsorption of dissolved organic matter to detrital particles may be a mechanism for scavenging and concentrating substrates on particle surfaces (Lush & Hynes, 1973) although this process has not been quantitatively estimated in lakes. Adsorption of dissolved organic matter to precipitated calcium carbonate has been documented in experiments and in lakes. This phenomenon is more intensive during the summer and in productive lakes. The quantitative significance has not been studied in a spectrum of lakes. It appears to be quantitatively small but qualitatively important in a marl lake (Wetzel *et al.*, 1972) and probably also in streams (Lush & Hynes, 1973).

7.2.3. Autochthonous dissolved organic matter

7.2.3.1. Algae

Phytoplankton and benthic algae are known to release dissolved organic matter during the course of photosynthesis and in some cases during dark metabolism. In the latter case there appear to be at least two general mechanisms of production of dissolved organic matter. One mechanism is associated with the general metabolism of the plants; the second occurs during dark fixation of carbon dioxide.

In general, release of dissolved organic matter is quantitatively small, amounting to 1–10% of total photosynthesis (Nalewajko, 1966; Watt, 1966;

Nalewajko & Marin, 1969; Anderson & Zeutschel, 1970; Fogg, 1971; Miller, 1972; Wetzel *et al.*, 1972; Saunders, 1972c). However, there are reports of very high relative extracellular release rates which range from 30–95% of photosynthesis (Allen, 1972; Fogg, 1971).

No correlation of high extracellular release rates with any particular species or community of algae has been observed but the absolute rate of extracellular release is correlated with high productivity rates. Fogg (1971) reports that percentage extracellular release is positively correlated with low phytoplankton concentrations and oligotrophy. At light intensities that inhibit photosynthesis the relative rate of release may be very high. More recent data do not support the last two generalizations so that the relation of photosynthesis and extracellular release is still very much an open question.

Quantitatively, the daily release of extracellular dissolved organic matter in the photic zone is likely to be relatively unimportant compared to other sources of dissolved organic matter. However, these substances may sum their daily increment until very high absolute concentrations of organic matter are attained and a quasi-equilibrium between absolute release rate and utilization rate is developed. If the molecules released are low molecular weight substrates in general, then extracellular release by algae may be quantitatively more important in the metabolism occurring in the photic zone. Otherwise it is unimportant except in the special case where very high relative release rates occur.

Epiphytic and epipelic algae are reported to have mean release rates of 5% of photosynthesis.

7.2.3.2. Macrophytes

Very little work has been done on extracellular relase by macrophytes and this has been mostly in pure or axenic culture (Allen, 1971a; Khailov, 1971; Hough & Wetzel, 1972; Wetzel & Manny, 1972a, b; Wetzel *et al.*, 1972). Release rates range from less than 1% to greater than 50% of photosynthesis. Most rates were less than 10% of photosynthesis. In a marl lake mean release by macrophytes was 5% of photosynthesis. This is comparable to values for phytoplankton.

7.2.3.3. Animals

No information is available concerning direct release of dissolved organic matter by animals. However, Krause (1959) has shown that zooplankters lose about 25% of their body weight immediately upon death. Presumably this is due to disruption of membranes and lysis.

Release of dissolved organic matter by rupture of food cells by grazing zooplankton has been reported as insignificant in natural lake water under normal zooplankton concentrations (Miller, 1972).

7.2.3.4. Bacteria

It is well known that bacteria release exoenzymes, vitamins and other substances in culture. Quantitative estimates of extracellular release by bacteria in natural waters are lacking with the exception of one set of experiments. Clear demonstration that bacteria release extracellular dissolved organic matter in lake water has been obtained through kinetic studies. When fed glucose and glutamic acid in tracer amounts 5–10% of the original substrate appeared in another organic form in the dissolved state. These values represent minimal estimates of release by bacteria because the calculation assumed that the radioactive tracer was at isotopic equilibrium with the intermediary pool releasing the product (Saunders, unpublished).

7.2.3.5. Comparison of detrital decomposition with plant secretion

Saunders (1972a) has reported that planktogenic detritus undergoing decomposition lost about 1% of its carbon per day as dissolved organic matter. This amounted to roughly 2–6 times the dissolved organic matter released by phytoplankton during the same period. Whether release of dissolved organic matter from detritus was due to hydrolysis by enzymes dispersed in the lake water or due to diffusion from aggregates being hydrolysed by bacteria or due to extracellular release by bacteria could not be distinguished.

7.3. Fates of primary sources of organic matter

The possible fates of organic matter are:
1. Leaching of soluble compounds.
2. Decomposition in microbiological processes.
3. Fragmentation by mechanical forces.
4. Direct feeding by animals.
5. Flocculation.
6. Sedimentation.
7. Export beyond the ecosystem.

7.4. Distribution of microorganisms

7.4.1. Introduction

Convention has dictated that the description of distributions of microorganisms in aquatic ecosystems be approached from two points of view. One is to isolate and identify species occurring in these systems. Success has been limited with this approach because of the inability to culture the majority of microorganisms and therefore most of the microorganisms that occur in fresh water have not yet been identified.

The second approach has been to enumerate those particles identifiable as microbes by direct microscopic observation and by estimating numbers using various selective culture methods. The selective culture methods suffer the disadvantage that only a fraction of the microbial community can be cultured and the results can apply only to rather special questions. Direct microscopic enumeration has employed several variations of specific method, including the promising method of fluorescence microscopy. Discrimination of particles and determination of viability are the more serious problems associated with direct microscopic counting. The methodology of microbial enumeration has been summarized by Rodina (1965, 1971) and Sorokin & Kadota (1972). The general results of aquatic microbiology over several decades have been summarized by Kuznetsov (1959, 1970) and Zobell (1946).

The most comprehensive microbiological studies of the IBP have been conducted in the USSR where nine lakes and seven reservoirs have been investigated. Brief descriptions of these lakes and reservoirs have been given by Winberg (1972b and in Chapter 11).

The more important variables estimated include number of bacteria per unit volume, biomass of bacteria, rate of increase of bacteria and bacterial production. In a few cases dissolved and/or particulate organic matter was estimated. Most of the methods contain some broad assumptions and methods are not always strictly comparable when applied to lakes outside the USSR. Since space does not permit presentation of specific methods the results must be considered as approximate. The original publications or data lists should be consulted in order to make more specific comparisons.

7.4.2. Comparison of bacterial concentration in lake types

Earlier work has demonstrated a tendency for concentration of bacteria to increase with increasing inorganic and/or organic nutrient concentration (Kuznetsov, 1959, 1970). This correlation is confirmed for lakes investigated in the IBP (table 7.3). Most of the Russian lakes were studied during the warmer vegetative season. From 20 to 100 enumerations and from 20 to 50 estimates of bacterial production were made on each lake during this season. This correlation is only very general as evidenced by the wide range of values among lake types and overlap between lake types. However, it is clear that oligotrophic lakes have fewer bacteria and lesser biomass of bacteria than ponds and eutrophic lakes. Mesotrophic and dystrophic lakes tend to be intermediate and reservoirs compare with eutrophic lakes, probably reflecting the fact that these large impoundments tend to occur on rivers that receive both industrial and municipal wastes.

Direct microscopic counts are fairly reliable when used by experienced

Table 7.3 *The number, volume, and biomass of bacteria in lakes of different trophic state*

Lake	Year studied	Number (10^6 ml^{-1})	Volume (μm^3)	Biomass $(g \text{ m}^{-3})$
Oligotrophic lakes				
Zelenetskoye[a]	1970	0.09	0.265	0.03
Zelenetskoye	1971	0.26	—	0.09
Krivoye[a]	1968	0.84	0.43	0.26
Krivoye	1969	0.50	—	0.16
Ladozhskoe	—	0.2–0.5	—	—
Onezhskoe[b]	—	0.2–0.3	—	—
Baykal[b]	—	0.20	—	—
Konchozero[b]	—	0.17	—	—
Pertozero[b]	—	0.13	—	—
Oligo-dystrophic lakes				
Akul'kino[a]	1970	0.12	0.24	0.03
Akul'kino	1971	0.34	—	0.09
Krugloye[a]	1968	0.87	0.38	0.33
Krugloye	1969	0.87	—	0.33
Mesotrophic lakes				
Krasnoye[a]	1964–70	0.70	0.43	0.30
Naroch[a]	1968–70	0.64	0.50	0.32
Dalnee[a]	1970–1	1.50	0.67	1.00
Sevan	1952	0.32	1.00	0.31
Seven	1962	0.29	1.20	0.35
Sevan	1966	0.56	1.15	0.64
Glubokoe[b]	1932	1.0–1.4	—	0.97
Reservoirs				
Rybinsk[a]	1964–8	1.70	0.60	1.00
Bratsk[a]	1965–72	0.85	0.90	0.77
Klayaz' mёnskoe	1945–6	0.5–0.9	—	0.4–0.9
Kiev[a]	1967	3.50	0.89	3.30
Kiev	1968	4.70	0.79	3.40
Kremenchug	1968	3.50	0.60	2.10
Kakhov	1968	4.00	0.47	1.90
Dneprodzerzhin	1968	3.40	0.65	2.20
Zanorozh	1968	3.60	0.52	1.90
Eutrophic lakes				
Drivyaty[a]	1964	1.84	0.76	1.40
Myastro[a]	1968–70	2.20	0.50	1.10
Batorin[a]	1969–70	6.40	0.50	3.20
Vyrts'yarv	—	4.40	—	0.2–1.8
Beloe[b]	1932	2.23	—	—
Chernoe[b]	1932	4.00	—	—
Ponds				
Kramet-Niyaz				
without fertilization		2.0–6.0	—	2.0–6.0
mineral fertilizer		5.0–20.0	—	5.0–25.0
Dystrophic lakes				
Piyavochnoe		0.43	—	—

(contd.)

Table 7.3 (*contd.*)

Lake	Year studied	Number (10^6 ml^{-1})	Volume (μm^3)	Biomass (g m^{-3})
Chernoe		1.07	—	—
Melnezers		1.00	—	—
Serpovidnoe		0.1–0.5	—	—
Average values				
Oligotrophic		0.50	0.2–0.4	0.15
Mesotrophic		1.00	0.4–1.2	0.70
Eutrophic		3.70	0.5–0.9	2.30

[a] Lakes investigated under IBP.
[b] Kuznetsov (1970).

persons. Counts of the same sample, performed in different laboratories, usually vary less than $\pm 20\%$ of the mean.

Total direct counts of bacteria are correlated with the amount of easily decomposable organic matter (fig. 7.1). The latter is estimated indirectly by measuring biological oxygen demand, BOD$_5$. The relation is true over a broad

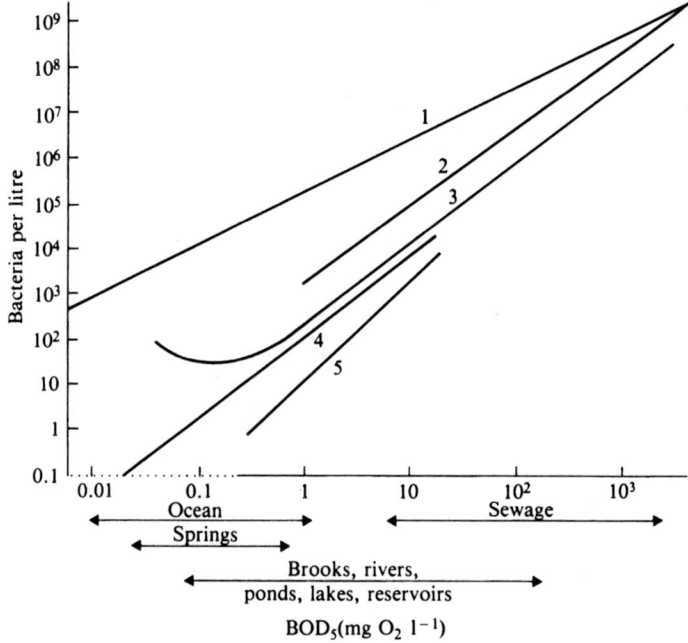

Fig. 7.1. The relation between concentration of bacteria and trophic state of water body types as measured by the five-day biological oxygen demand. Curve 1, direct microscopic count; 2, ten-day beef peptone agar plate for lotic waters; 3, two-day beef peptone agar plate for lotic waters; 4, ten-day beef peptone agar plate for standing waters; 5, two-day beef peptone agar plate for standing waters.

range of trophic states, varying from very unproductive oceanic regions to water bodies with heavy sewage input (Straškrabová & Legner, 1969; Straškrabová, 1973).

There is also a correlation between microscopic counts of bacteria able to grow on beef peptone agar plates and easily assimilable organic matter as estimated from BOD_5 (fig. 7.1). However, total counts and plate counts differ in unpolluted waters by four to five orders of magnitude and converge only at high levels of sewage input. Bacteria in unpolluted water are probably living under starvation conditions and are inhibited by higher substrate concentrations whereas bacteria in polluted waters may be fully adapted to high organic nutrient conditions.

In surface waters with moderate BOD_5 values, agar plate counts tend to be an order of magnitude greater in flowing waters than in static waters. This is probably due to greater turbulence in streams effecting suspension of bacteria from the superficial sediments. This general picture may be somewhat distorted if organisms feeding on bacteria are present. Concentrations of bacteria are greater over the whole range of BOD_5 in the absence of bacterial feeders.

Estimates of bacterial biomass are based mostly on direct microscopic counts and cell volume measurements. The variance of the latter measurement is greater than for cell counts. In spite of this, bacterial biomass is correlated with increasing concentration of easily assimilable organic matter (fig. 7.2), (Winberg & Lomonosova, 1953; Uhlmann, 1958; Straškrabová, 1968, 1973; Straškrabová & Legner, 1969). In order for this to be true the mean volume of

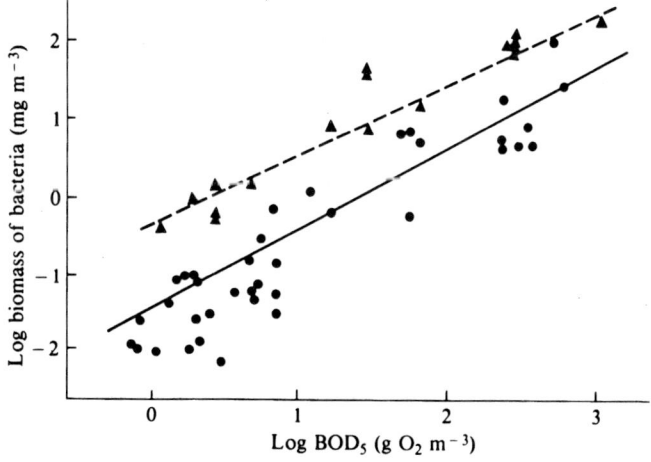

Fig. 7.2. The relation between bacterial biomass and trophic state of the water system as estimated from the five-day biological oxygen demand. Broken line, experimental systems without bacterial consumers; unbroken line, natural and sewage systems with bacterial consumers.

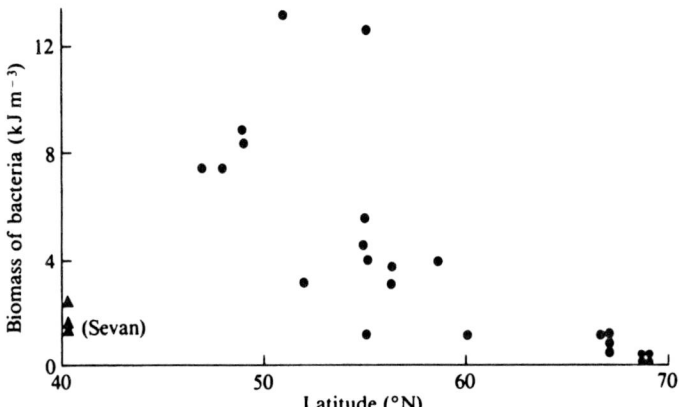

Fig. 7.3. The relation between bacterial biomass and latitude for lakes studied within the IBP/USSR.

bacterial cell cannot be substantially affected by the concentration of easily decomposable matter. The evidence for this is not conclusive and requires further investigation.

Planktonic bacterial biomass tends to decrease with increasing latitude, with the exception of Lake Sevan (fig. 7.3). Lake Sevan is a large high-altitude unproductive lake. Therefore, a general correlation analogous to that observed by Brylinsky & Mann (1973) between latitude and primary production or chlorophyll concentration is obtained. One would expect bacterial biomass to be generally correlated with the intensity of nutrient supply in the sense that more productive systems usually have greater standing crops of organisms than unproductive systems.

7.4.3. Growth rate and production of bacterial biomass

Production of bacterial biomass, P, and its relative rate of increase P/B (production/biomass) are presented in table 7.4. Values of bacterial production per day are correlated highly with trophic type of water body. In oligotrophic lakes values range from $0.02-0.67$ kJ m^{-3}; in mesotrophic lakes, $0.50-1.80$; in eutrophic lakes, $1.67-6.07$ (averaged over the whole growing season). Bacterial production per unit area of lake surface integrated for the growing season varied from 11 kJ m^{-2} in subarctic Lake Akul'kino to 7950 kJ m^{-2} in eutrophic Zaporozhsky Reservoir. The length of the growing season is generally correlated with the trophic state.

The greatest number of turnovers of bacterial biomass per season occur in the more southern water bodies, mainly due to the longer growing season. If relative daily bacterial production, P/B, is compared the values range from $0.2-0.6$; such relative rates are observed in oligotrophic as well as eutrophic

Table 7.4 *Estimates of average daily bacterial production, total bacterial production per growing season, and the ratio of bacterial production to bacterial biomass in different types of water bodies of the USSR*

Water body	P (kJ m^{-3} d^{-1})	P/B per day	P (kJ m^{-2} season^{-1})
Oligotrophic			
Zelenetskoye Lake	0.02	0.20	21
Akul'kino	0.04	0.23	11
Krivoye Lake	0.17	0.26	222
Krugloye Lake	0.67	0.62	205
Mesotrophic			
Krasnoye Lake	0.50	0.34	879
Naroch Lake	0.59	0.58	1 004
Dalnee Lake	0.50	0.20	2 444
Rybinsk Reservoir	1.67	0.30	1 464
Bratsk Reservoir	1.80	0.55	8 954
Eutrophic			
Myastro Lake	1.67	0.50	1 716
Batorin Lake	3.35	0.38	1 841
Kiev Reservoir	3.77	0.22	3 013
Kremenchug Reservoir	3.35	0.38	4 017
Kakhov Reservoir	2.30	0.33	3 891
Dneprodzerzhin Reservoir	6.07	0.65	5 272
Zaporozhsky Reservoir	5.02	0.60	7 950

water bodies. This means that the specific rate of metabolism of bacteria is closely regulated in the dilute media that normally is constituted in lakes. If nutrient levels vary in lakes it must be the concentration of bacterial biomass that responds to this over a wider range rather than the relative growth rate, although there may be brief transient conditions during which relative growth rates might be very high. These high rates would not be observed readily at the sampling frequencies normally used to estimate bacterial production rates or generation times.

It must be clear that the results presented above are only broadly correlative and cannot be used for detailed functional analysis. Bacterial production, P, and bacterial biomass, B_B, as well as bacterial numbers, N_B, are all correlated inversely with latitude. While there may be a tendency for high-latitude lakes to be unproductive and low-latitude lakes to be very productive, not all lakes at a latitude are in the same trophic state. The analysis presented here cannot distinguish between oligotrophic lakes and eutrophic lakes at the same latitude.

Lake Sevan (fig. 7.3) is an exception with regard to bacterial biomass. This may be an artifact of small sample size. It may also suggest that a better correlation of bacterial variables is with trophic state of lake and that latitude is only indirectly relevant. Or Lake Sevan may have unique properties. The problem may be resolved partially by increasing sample size or increasing sampling frequency. However, the detail produced by increased sampling is not likely to provide much more information. This suggests that, while enumeration of bacteria is necessary, greater advances in understanding the role of bacteria in the structural and functional operation of aquatic ecosystems are likely to be made by developing other approaches and examining other aspects of the activities of bacteria in these systems. It would be more interesting to determine why Lake Sevan is so discrepant from the extrapolation in fig. 7.3 than to determine that other low-latitude lakes of the USSR fall on that extrapolation.

The generation time of bacteria is an index of their metabolic activity in water bodies. Several methods, not all of which may be strictly comparable, are available for estimating the generation time of bacteria (Winberg, 1971c; Sorokin & Kadota, 1972). Growth rate and generation time are conceptual reciprocals. Because of slight differences in the manner in which they are determined this is not exactly true in practice. The relative growth rate is calculated as the ratio of production to bacterial biomass, P_B/B_B. The generation time is calculated as the time it takes the bacterial community to double, assuming exponential growth.

The bacterial generation times for a number of water bodies are presented in table 7.5. There is a tendency for the generation time to decrease with increasing trophic state of the water body. In a few cases the generation times

Table 7.5 *The range in generation time, g, of bacteria in lakes and reservoirs of the USSR*

Name	Type[a]	g (hours)[b]
Lake Baykal	O	72
		218
Sevan	O	20–933
Irkutsk Reservoir	O	12–301
Glubokoe Lake	M	72–218
Lake Naroch	M	61–103
Lake Krasnoye	M	15–200
Lake Ritza	M	3.6–50
Rybinsk Reservoir	M–E	5–36
Gorkovsky Reservoir	M–E	8.8–120
Kiev Reservoir	M–E	4.6–151
Mingechaursky Reservoir	M–E	9–65
Tkibul'skoe Reservoir	M–E	3–50
Lake Batorin	E	21.7–57.5
Bay of Kurshu-Mares	E	1.8–50.7
Ponds of Volga Delta	E	14–35

[a] O, oligotrophic; M, mesotrophic; E, eutrophic.
[b] Average annual generation time.

were of the order of 2–4 hours. At the other extreme the generation time was several days to as long as one month. However, on the average the generation time was of the order of one or two days. It is not clear that the generation time is a reliable index of trophic state of a water body and only accumulation of more data on lakes of different type will resolve this question.

It must be noted that the P/B per day data from table 7.4 are not consistent with the data for generation times given in table 7.5. Generation time is clearly inversely correlated with trophic state. The ratio P/B has no significant trend. It is not possible for the relative growth rate to be constant over all trophic states and for the generation time to decrease with increasing trophic state. It is not known whether this contradiction is an artifact due to the small sample size and a high variance of the methods used or whether there is some consistent bias within one set of data. It is reasonable to expect the generation time to be shorter with increasing nutrient supply.

The generation time exhibits a seasonal trend in Kiev Reservoir. In the late spring it averages 15.5 hours, in summer 24.2 hours, and in autumn 50.0 hours. A comparable range in generation time, 2.5–41.6 hours has been estimated in two reservoirs (Straškrabová-Prokesova, 1966). An increase in generation time has been noted by other authors during the summertime development of phytoplankton (Novozhilova, 1957; Drabkova, 1965). They attribute this decrease in growth rate to a toxic effect of the algae but proof of this effect is not presented.

In Kiev Reservoir at the time of the spring flood bacterial reproduction proceeds the most irregularly; in certain sections a rapid growth of bacterial biomass occurs whereas at the same time in other sections a dying-off of the flood water allochthonous microflora is observed. In summer the growth rate is reduced in comparison to springtime. This occurs over the whole reservoir. A decrease in the growth of bacteria is sometimes observed at definite stages in the development of blue-green algal blooms. Bacterial growth occurs in the upstream shallow waters of the reservoir but is insignificant in the lower reservoir where a blue-green bloom is present. In autumn the growth rate of the bacteria is very low and in many sections of the reservoir the lysis of bacterial cells is observed. In winter there is essentially no bacterial growth except in the 10-cm layer of water near the bottom, where division is observed.

Bacterial concentrations fluctuate in all water bodies. The generation time is calculated for an increase in bacterial biomass assuming exponential growth. Growth is not necessarily always exponential because the nutrient source may be limited and there may be death or physical transport from the system. Sometimes there must be a decrease in bacterial concentration. This is analogous to a negative generation time or a half-life and has been discussed by Gak (1967) and Straškrabová-Prokesova (1966). In Kiev Reservoir, the half-life of the bacterial community has the same range in hours as the generation time. This should result in a quasi-stable community that oscillates about some mean level most of the year. Changes in the amplitude of these oscillations would reflect differences in such factors as nutrient inputs and invertebrate feeding.

It should be obvious that generation time is a variable of limited usefulness. It may provide some index of bacterial activity but over a year, as there is no net accumulation of bacteria in an aquatic system, it must be balanced by negative generation time. At certain times the reproduction rate may be balanced by the removal rate, due to death or predation or sedimentation. The generation time would be indeterminate yet the bacterial community could be highly active metabolically.

7.4.4. Vertical distribution of bacteria in lakes

Few data concerning the vertical distribution of bacteria in lakes are available and these are briefly summarized by Kuznetsov (1970). In deep oligotrophic Lake Baykal concentrations of bacteria are greater in the near-surface photic zone decreasing to low concentrations below the photic zone by a factor of ten and remaining more or less uniform to depths greater than 1600 m. This appears to be an analogue of bacterial distribution in the oceans. In the shallower oligotrophic Lake Dalnee the pattern is similar but more irregular. In mesotrophic Lake Glubokoe the vertical distribution is uniform with minor irregularities. In eutrophic Beloe Lake there is a tendency for bacterial numbers to increase with depth but the pattern is very irregular seasonally and contains extreme degrees of stratification in bacterial numbers, probably associated with highly stratified availability of particulate and dissolved organic matter. In a detailed study of a small eutrophic lake (Saunders, 1971) there is a tendency for bacterial numbers to increase with depth. Sharp increases occur in the hypolimnion both in winter and summer associated with anaerobic conditions. This may reflect reduced grazing due to exclusion of zooplankton from the anaerobic zone or it may reflect greater concentrations of easily assimilable organic matter under these conditions. These general tendencies can be completely inverted and reverted in periods of two to five days. Very rapid changes in the vertical distribution of bacteria in thermally stratified lakes have also been observed by Rasumov (1962).

This suggests that the true dynamics of bacterial populations cannot be observed without very frequent short-term detailed sampling and that causal interactions cannot be revealed under conventional modes of sampling.

7.4.5. Temporal distribution of bacteria in lakes

Earlier studies of the seasonal distribution of bacterial numbers in lakes have shown that the time and magnitude of maximum development of bacteria differs between lakes and may vary from year to year within a lake (Saunders,

1971). More recent investigations of direct microscopic counts confirm this conclusion (Gambaryan, 1968; Drabkova, 1971; Tseeb & Maistrenko, 1972).

Saunders (1963, 1971) has shown that pulses of bacteria in a small productive lake may be broadly correlated with movement of nutrient-bearing littoral water into the limnetic zone, oxidation–reduction conditions, death and decomposition of higher aquatic plants, death and decomposition of phytoplankton, higher summer productivity, precipitation and runoff, and zooplankton dynamics. During successive winters the distributions of bacteria were quite different, as was the weather. Very large short-term fluctuations in bacterial numbers occurred for which no correlations were observed.

This suggests again that bacterial population dynamics cannot be predicted with accuracy and that direct causal interactions cannot be identified without more detailed analytical approaches.

7.4.6. Succession of microbial communities

In lakes from 75–99% of the organic matter produced in the pelagic photic zone is decomposed in the water column (Kleerekoper, 1952; Ohle, 1956; Deevey & Stuiver, 1964). In the aerobic water column the bacteria are very small and not easily identifiable so that species succession of bacteria is difficult to observe.

In the microaerobic and anaerobic zones of lakes there is clearly a stratification of morphologically distinguishable bacterial types most of which are unidentified as to species. This vertical stratification suggests a vertical succession of bacteria which results from the change in composition of organic materials being decomposed as they rain down from the near-surface water.

The fact that the annual cycles of particulate organic matter, dissolved carbohydrate, dissolved organic nitrogen and dissolved organic carbon all fluctuate and are out of phase with one another (Saunders, 1971) suggests that the bacteria producing these cycles or resulting from them must vary in their biochemical attributes during the year and thus reflect the succession of bacterial species populations during the year. However, essentially nothing specific is known about the succession of species populations and metabolic capabilities of the bacteria processing dissolved organic matter in lakes in general.

Somewhat more is known about the processing of particulate organic matter. Most of this information has been derived from studies of leaf decomposition in streams (Kaushik & Hynes, 1971). Although one would expect the same sequence of events to occur in lakes, the sequence of microbial development on leaf litter in lakes has not been studied sufficiently to verify the specific bacterial and fungal successional patterns.

The most analogous comparison of leaf litter decomposition is between stream riffles and the littoral zones of lakes – particularly those with coarse sediment and wave-swept shores (erosional habitats). These environments are dominated by aerobic processing. In depositional habitats anaerobic events occur through a significant portion of the year. The homogeneity of sediments in depositional zones allows for only a thin veneer of aerobic conditions, even when the overlying water is oxygenated. Comparison of leaf litter processing in riffles of a small woodland trout stream and the wave-swept shore of a trout lake in Michigan demonstrate the rate of leaf litter processing in the lake to be only about 25% of the stream (K.W. Cummins, personal communication). Hodkinson (1975) has observed that the decomposition of plant material in a North American beaver pond varied over a wide range, but was much slower than that observed in streams. *Picea* sp. bark decayed at about one-fifth the rate of leaf and stem materials. *Populus* sp. logs decayed one to two orders of magnitude more slowly than various leaf litters.

When leaves fall into streams and are wetted a very rapid leaching of soluble organic matter occurs. This leaching is probably due to diffusion of those small water-soluble molecules present in the protoplasm of the original leaf plus molecules derived from lysis of higher molecular weight organic matter by residual enzymes during the period up to abscission of the leaf. Leaching is usually essentially complete within 12–48 hours. The percentage of the original leaf material leached varies according to the leaf species and pH of the water (Lush & Hynes, 1973) and ranges from 5 to 30% in the first 24 hours. The leachate is then processed within the flowing stream or exported (Fisher & Likens, 1973) but a significant fraction appears to be flocculated by physical as well as biotic mechanisms (Cummins *et al.*, 1972; Wetzel & Manny, 1972a; Lush & Hynes, 1973; Manny & Wetzel, 1973). This particulate organic matter is rapidly removed from the water phase presumably being adsorbed or filtered by the sediment surface where it is further processed by animal and microbial action, including both bacteria and fungi.

The remaining leaf is decomposed at a slower rate. Within a few days after immersion in the stream the litter is colonized by bacteria and aquatic hyphomycete fungi. The invasion by fungi, which may be dominant or sub-dominant depending on the stream ecosystem (Iverson, 1973) is critical since they penetrate the matrix of the leaf through hyphal development while the bacteria are predominately surface colonizers. The rates of microbial succession and leaf litter processing are specific to leaf type but do not involve different microbial forms.

During this initial period of microbial colonization and growth there is an increase in the nitrogen of the leaf litter (Kaushik & Hynes, 1971). The increase in nitrogen, attributed to increase in microbial protein in the leaf material is derived mostly from inorganic nitrogen in the overflowing water. It is not until the leaf litter has been conditioned by invasion and growth of microbial biomass that invertebrates begin to feed on the leaf litter. The evidence suggests that this feeding is selective for the microbial protoplasm in the litter rather than the leaf detritus itself (Triska, 1970; Cummins, 1973; Mackay & Kalff, 1973). In some cases there appears to be a very specific relation between the fungal species on the leaf litter and the invertebrate species consuming that material (Bärlocher & Kendrick, 1973a, 1973b).

Invertebrates feeding on the large leaf particles shred the material, breaking it into smaller pieces, and pass it through the gut, modifying it for subsequent utilization by other microbial species which in turn may be fed upon by gathering and filter feeding detritivores. In this manner, the important large-particle detritus is continuously and sequentially reduced in size, and oxidized as it is passed on through a set of trophic steps specialized for utilizing that particle size and quality of food, until the energy stored in the leaf material is completely dissipated. Fragmentation due to non-biotic forces may also be significant, but has not yet been studied.

It is clear that most lower order (headwater) streams are heterotrophic (Hynes, 1970; Fisher and Likens, 1972; Straškraba, Chapter 3) in character and depend principally on inputs of organic matter; the particle sizes introduced to such streams can influence the structure of the animal communities in them (G.W. Minshall, personal communication). While it is obvious that detritus, including the associated microorganisms, is most important in the energetics of stream invertebrates (Egglishaw, 1968; Madsen, 1972) it is not clear whether this organic matter is mainly passed through the guts of invertebrates as the major functional processors in stream systems or whether it is processed principally by microbes. In the River Thames, it is estimated (Mathews & Kowalczewski, 1969; Kowalczewski & Mathews, 1970) that the utilization of particulate detritus by invertebrates is insignificant in the total energetics of this system. However, in a small woodland stream, animal processing accounts for 30% of the leaf litter conversion (fig. 7.4).

As for organic particles in lake water, Olah (1972) has demonstrated that ground up *Phragmites* particles are leached and then a succession of bacterial types occurs as these particles are decomposed. This succession is analogous to that which occurs on leaf litter in streams with the exception that fungi do not appear to participate in the decomposition process, probably because the *Phragmites* was ground into small particles initially, preventing fungal hyphal development. The *Phragmites* particles ultimately become very small and populated by very few bacteria suggesting that they are highly degraded and no longer nutritious to bacteria and in turn would not provide a food substrate for benthic invertebrates. However, this latter possibility was not studied by Olah.

Phytoplankton and zooplankton appear to undergo an immediate lysis upon death (Krause, 1959; Golterman, 1972). Subsequent decomposition is mediated by bacteria.

Decomposition of particulate organic matter is very important in the littoral zone of lakes. The organic matter may be brought in from the drainage

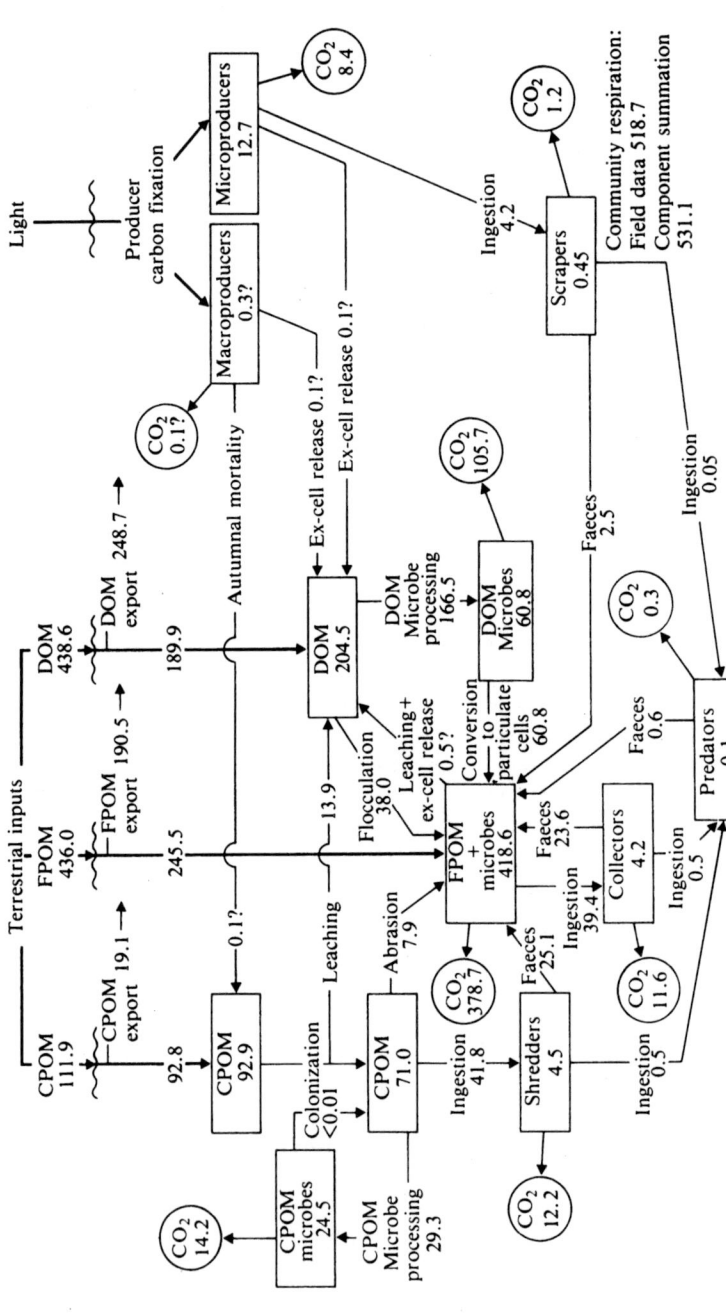

Fig. 7.4. A generalized model of a stream ecosystem organic budget showing the dependence upon particulate and dissolved organic matter of terrestrial origin and the dominance of the functional groups which process such allochthonous organics. The model is based on a first-order stream in the Augusta Creek watershed (Kalamazoo County, Michigan, USA). All values in grams ash-free dry weight m^{-2} a^{-1}. Squares represent pools of organic matter in various states; arrows represent transfers (inputs and outputs) and circles represent respiratory consumption of organic matter. CPOM = coarse particulate organic matter (> 1 mm), e.g. leaf litter; FPOM = fine particulate organic matter (< 1 mm), e.g. animal faeces, CPOM fragments and bacteria; DOM = dissolved organic matter (< 0.5 μm), e.g. leachate, microbe and producer extra-cellular release (Ex-cell release) and animal excretion (not shown; animal excretion is assumed to be < 0.05 g m^{-2} a^{-1}). Aquatic hyphomycete fungi are a dominant group of CPOM microbes. Shredders = CPOM feeding detritivores; collectors = FPOM feeding detritivores; scrapers = periphyton grazers; predator = fish and certain macroinvertebrates.

basin, may be derived from macrophytes growing in the littoral zone, or may result from algae being washed into the littoral zone by water movement and wave action. In the eulittoral the animals are predominantly detritus feeders but decomposition of organic matter is mainly due to bacterial action (Pieczyńska, 1972a, 1972b). This is because the large particles of macrophytes and allochthonous matter are not directly available to the dominant fauna because of their size.

7.5. Decomposition of organic matter

7.5.1. Introduction

Decomposition of organic matter is viewed as the reduction of higher molecular weight organic compounds to lower molecular weight compounds by enzymatic mechanisms leading to the ultimate transformation of such organic substrates into the inorganic building blocks of protoplasm. Mechanical, biochemical and microbial attrition of particulate material as well as attrition by feeding animals may help to prepare the organic surfaces for digestive attack or for the exocellular enzymes produced by bacteria.

Rates of decomposition are usually measured indirectly by estimating oxygen consumption or carbon dioxide evolution. They may also be estimated by weight loss or some other measure of decrease in concentration in time, such as loss of radioactive carbon. Whether the organic matter is particulate or dissolved, decomposition can be considered as the interaction between the organic substrate and the bacteria, and the rate of change of substrate concentration can be expressed as a bimolecular second-order reaction (Saunders, 1972b):

$$\frac{-d[S]}{dt} = c_1[S] \times c_2[B]$$

Where $[S]$ = concentration of substrate
$[B]$ = concentration of bacteria
$[N]$ = concentrations of nutrients
$[O]$ = concentration of oxygen
T = temperature
$c_1 = f(m_1, m_2, m_3 \ldots m_m)$
$c_2 = f(T, [N], [O], x_1, x_2, x_3, \ldots x_m)$
m_1, m_2, x_1, x_2, etc. = other variables

The rate of decomposition of organic substrate is a function both of the concentration of substrate and of the microbes dispersed in the water or attached to surfaces of particles. The microbes represent integrated enzyme systems.

In oligotrophic systems decomposition rates would be expected to be low, as both substrate concentrations and bacterial numbers are known to be low. In hypereutrophic systems the rates are very high; and in mesotrophic and eutrophic systems intermediate.

The rate of decomposition will depend not only on the substrate concentration but also the nature of the substrate, i.e., it will make a difference whether or not the substrate is a polymer and what kind of polymer. If the bonds in one polymer are more difficult to hydrolyse than in another, the reaction rate will be slower for the same molecular concentration.

On the other hand the enzyme concentration represented [B] is not usually a single enzyme but a set of enzyme systems, the activity of which is some function of the number or biomass of bacteria as well as other qualities of the bacteria. If the substrate is organic detritus or colloidal or dissolved organic polymers, exoenzyme must serve to hydrolyse the polymer, the hydrolysis product must collide with the bacterium at a site for active enzymatic transport into the cell and then a sequence of enzymatic reactions must process this substrate toward some end product which may result in growth or in extracellular release and in release of respiratory carbon.

Therefore, the decomposition rate is a function not only of the substrate concentration and the enzyme concentration but also a function of the chemical nature of the substrate and the nature of the total enzyme system acting.

In addition other factors will operate to modify the basic reaction rate. These are variables such as temperature, oxygen concentration, nutrient concentration, particle size, nutritional status of a particle (faecal pellets, age of detritus), adsorptive properties of surfaces and colloids, colonization sequence for microbes, turbulence, etc. The effects of these qualitative factors appear in the coefficients c_1 and c_2, the values of which must be determined empirically. These factors are the control operators for the decomposition reaction system.

This formulation is a generalized description of the rate of decomposition of organic matter in aquatic systems. The coefficients c_1 and c_2 are continuously varying coefficients which are a function of other environmental control variables and representative of the dynamic nature of any natural ecosystem or subsystem. The fundamental reaction is a simple one between two components and may be considered a primary reaction system. However, this basic reaction system is controlled by a host of operator variables which are finely tuned and provide second-order control over the performance of the fundamental decomposition reaction.

The rate of addition of nutrients to a water basin will in general dictate the intensity of production of organic matter and consequently the intensity of decomposition of that organic matter. We would predict that oligotrophic lakes will have lower concentrations of organic substrate and bacterial concentrations than eutrophic lakes. In fact experience bears this out. The rate of decomposition in oligotrophic water should be slower than in eutrophic waters. We will see later that this prediction is verified.

Within a single lake type, however, decomposition rates will vary continuously as the system changes its structure and as the secondary control variables change in the intensity of their operation. Although the general level of metabolic intensity that occurs in a water body is of interest, it is the shorter-term fluctuations about this general metabolic level that attract attention and focus on the phenomena that man considers interesting and important and that are particularly graphic as problems. In fact the fluctuations, and not the average level of metabolic activity, represent the true performance of the system. The inputs and outputs are controlled by environmental operators. It is not until the control operators are identified and the intensity with which they operate is measured that we can begin to understand how a system functions and what controls its function.

Therefore, it seems appropriate to emphasize that it is feasible and obviously more interesting to examine the variables that control decomposition processes in aquatic systems. Study of control should produce much greater dividends toward understanding the dynamics of aquatic systems and the role decomposition plays in these systems than the more conventional descriptive studies that have been conducted to date.

Earlier the general progression of decomposition in water bodies was described. Some investigation concerning rates of decomposition of classes of organic substrates have been performed. No studies of control of decomposition have been conducted in natural systems, either within the IBP or within the field of aquatic science in general. The next section presents some of the results concerning decomposition rates of classes of organic matter.

7.5.2. Decomposition rates

7.5.2.1. Introduction

In any aquatic system allochthonous inputs of organic matter plus autochthonous primary production must on the average be balanced by decomposition processes plus other removal mechanisms such as outflow or sedimentation. Over shorter time intervals, i.e., hours, days, or weeks, decomposition and production rates for organic matter may be quite out of balance.

One would expect to find in aquatic systems all the species of organic compounds that occur in protoplasm and those species should range in physical form from simple organic molecules to dissolved polymers, to colloids, to fine heterogeneous detrital particles, to parts of higher plants and animals. On the average each category of organic material should accumulate in an aquatic system until the input rate is essentially balanced by the decomposition rate where decomposition is the major removal process.

Therefore, if something is known about decomposition rates and the concentrations of those substances being decomposed we can begin to infer something about the dynamic structure of an aquatic system. If we can learn something about the factors that control decomposition rates we can begin to understand how aquatic ecosystems function and what controls this function. As the last has not yet been attempted on any significant scale and since only the most proximate estimates of concentrations have been made to date, we are left with examining what is known about decomposition rates.

If organic matter is added instantaneously to an aquatic system and the activity of microorganisms decomposing this material remains constant, this organic matter will decompose in a way such that its concentration in time will be described by a simple exponential decay term, if the substrate is homogeneous, or by the sum of a number of exponential terms if it is heterogeneous (Saunders, 1972a). Estimates of such curves have been made by weighing litter bags, by biochemical methods, or by determining the residuum of radioactivity in labelled substrates. The latter method is basically a batch culture technique and suffers the limitation of that approach. At the end of a relatively long period of time there is a residuum of material that amounts to about 20% of the original material. Actually the range of this value varies from about 5 to 50% depending on the quality of the original substrate. Presumably this residuum continues to undergo decomposition at some very low rate. Natural systems, however, are not batch cultures. They are continually changing structures and are continually being replenished in such a way that the rate of decomposition of the residuum may be much more rapid than indicated by the batch culture. The initial rate of decomposition should, however, be a reasonable estimate of the true rate of decomposition of the material being studied (Saunders, 1972a).

The vectors of decomposition in aquatic systems conventionally have been considered to be bacteria and in more specialized cases fungi. However, animals may also be considered as decomposers. Zooplankton do respire carbon dioxide and excrete ammonia and phosphates. In some lake waters and in regions of the oceans zooplankton appear to consume most if not all the phytoplankton production, at least for short periods of time.

7.5.2.2. Particulate organic matter

Most of the estimates of decomposition of particulate organic matter are from work on streams and particularly with leaf litter. When an organism dies or when a dead organism falls into the water part of the organic matter is lost immediately and part more slowly. The percentage loss per day due to leaching varies with the nature of the particulate material from 8% per day for *Fagus sylvatica* leaves and copepods to 68% per day for *Tubifex* sp., (table 7.6).

Table 7.6 *Percentage of initial dry weight leached per day*

Substrate	% day^{-1}
Copepod	8
Fagus sylvatica leaves	8
Fraxinus americana leaves	11
Pinus sylvestris needles	11
Decodon verticillatus leaves	12
Alnus glutinosa leaves	13
Picea abies needles	14
Quercus robur leaves	14
Planktonic diatoms	15
Betula verrucosa leaves	16
Callitriche hamulata	16
Populus tremuloides leaves	19
Fraxinus excelsior leaves	25
Salix lucida leaves	23
Cornus amomum leaves	27
Small fish (*Lebistes* sp.)	28
Cladocera	29
Planktonic green algae	33
Scirpus terminalis	35
Mixed zooplankton	52
Tubifex sp.	68

After Krause (1959, 1962), Nykvist (1963), Otsuki & Wetzel (1974) and Petersen & Cummins (1974).

In general this leaching loss is of the order of 10–30% per day. This leachate provides a significant source of organic material that is defined operationally as 'dissolved' organic matter, usually < 0.5 μm diameter.

In the case of leaves in stream water, the residual 70–90% of leaf biomass is decomposed at a relatively slow but finite rate. Leaf decomposition rates have been classified as slow, medium and fast, (table 7.7). The rates of decomposition are the order of 0.1 to 1.7% per day. Therefore leaf fall provides a relatively large and long-lived substrate for microorganisms and invertebrates lasting on the average from 50 to 400 days depending upon the species of leaf. In general most of the input from one year is completely decomposed within that year. Most leaf fall occurs in the autumn in temperate latitudes, but certain species absciss in spring or even continuously. Colonization of leaf material by microorganisms may occur roughly in proportion to the time that the leaf material persists in each system; more slowly decomposed leaves last longer and are colonized later than rapidly decomposing leaves. Whether invertebrates feed directly on the dead leaf material or the microbes growing on this leaf material, dissipation of the energy in the leaf material occurs over a relatively long period of time and in a drainage system containing a complex community of trees and shrubs, leaf fall provides a stable and enduring source of substrate for both microbial and invertebrate communities.

Table 7.7 *Average decomposition rate of particulate leaf organic matter in stream water*

Substrate	% day^{-1}	Author
Leaves – slow		
Acer macrophyllum	0.25–0.70	Sedell *et al.* (1974)
Populus spp.	0.33–0.46	Petersen & Cummins (1974)
Carya glabra	0.3	Petersen & Cummins (1974)
Fagus grandifolia	0.25	Petersen & Cummins (1974)
Quercus borealis	0.27	Petersen & Cummins (1974)
Ulmus americana	0.05–0.15	Petersen & Cummins (1974)
Pseudotsuga menziesii ⎱ *Tsuga heterophylla* ⎰	0.20–0.75	Sedell *et al.* (1974)
Leaves – medium		
Salix lucida	0.62–0.78	Petersen & Cummins (1974)
Juglans nigra	0.70	Petersen & Cummins (1974)
Betula lutea	0.5–1.2	Petersen & Cummins (1974)
Carpinus caroliniana	0.83	Petersen & Cummins (1974)
Alnus glutinosa	0.75	Petersen & Cummins (1974)
Acer saccharum	0.33–1.07	Petersen & Cummins (1974)
Acer platanoides	0.76	Petersen & Cummins (1974)
Acer circinatum	0.5–1.9	Sedell *et al.* (1974)
Leaves – fast		
Tilia americana	1.75	Petersen & Cummins (1974)
Cornus amomum	1.15	Petersen & Cummins (1974)
Decodon verticillatus	1.01	Petersen & Cummins (1974)
Fraxinus americana	1.20	Petersen & Cummins (1974)
Alnus rubra	1.2–1.6	Sedell *et al.* (1974)

The exuviae of chironomid pupae from streams sink after about 2 days in water at room temperature (20° C). After 6–14 days the water-insoluble material is no longer intact (Coffman, 1973). Thus the first-stage breakdown of this material occurs at an average rate of the order of 5–20% per day. This breakdown is in part physical but it suggests that decomposition of this type of organic material is comparable to that of organic particulate matter listed in table 7.8.

In lake water phytoplankton and zooplankton exhibit an immediate loss of organic matter upon dying. In phytoplankton this is of the order of 5–10% of dry weight (Golterman, 1964) and for zooplankton an average of 26% dry weight (Krause, 1959).

Subsequent to leaching, the residual particulate organic matter is decomposed. For phytoplankton, zooplankton and detritus the relative rate of decomposition of a major component of this particulate material is of the order of 10% per day (table 7.8). There is a smaller component of this material that decomposes at a much slower rate, of the order of 1% per day (table 7.8; Krause, 1959). There is undoubtedly a very small component of this material

Table 7.8 *Decomposition of fine particulate organic matter in lake water*

Substrate	% day^{-1}	Author
Phytoplankton	14	Saunders (1972a)
Mixed dead phytoplankton and detritus	5–20	Saunders (1972a)
Detritus	8	Saunders (1971)
Cladophora glomerata	7	Pieczyńska (1972b)
Gloeotrichia echinulata	1–7	Pieczyńska (1972b)
Zooplankton	5–33	Krause (1959)
Zooplankton	1	Krause (1959)

that has a much slower rate of decomposition, perhaps the rate of 1% per month or per year.

Studies of decomposition of particulate matter have been conducted under more or less eutrophic conditions. However, this may not be particularly important even for ultra-oligotrophic lakes since particles constitute local highly concentrated sources of organic matter. Such organic matter may be populated very rapidly by bacteria so that rates of decomposition of this matter in oligotrophic water bodies may not be fundamentally different from that in eutrophic water bodies as substrate concentration might effectively control the rate of decomposition. As oligotrophic lakes tend to be somewhat cooler than eutrophic lakes, temperature may exert greater control over decomposition rate than may particle size and/or surface available for bacterial attachment (Floodgate, 1972). However, in streams, under aerobic conditions, particle size seems to be more important than temperature, decomposition being dominated by temperature-adapted forms.

Pieczyńska (1972a, b) has investigated decomposition of plant material in the eulittoral zone of Mikolajskie Lake. Parts of dead plant material are frequently deposited in the littoral zone and washed up on the shore. Using *Potamogeton*, *Phragmites* and *Salix* leaves, she found that the average percentage decomposition per day for these leaves was 0.5 in the emergent zone one metre above the water line, 3.5 under heaps of reeds, 6.4 in small puddles at the shore line but isolated from the main water body, and 0.8 in the submerged littoral at 0.5 m depth. The small puddles were warmer than the lake water. Therefore the relative rate of decomposition appears to be correlated both with moisture content and with temperature in this series of habitats. Masses of *Cladophora glomerata* and *Gloeotrichia echinulata* washed into the eulittoral zone had relative decomposition rates per day of 7, 20, 25 and 7 for the four zones respectively. The differences in the rates of decomposition of higher plant and algal organic matter may result because of differences in biochemical quality of the matter, differences in the mechanical structure of these two classes of organic matter, and differences in mean particle size.

Thus, dead particulate organic matter exhibits an immediate loss of soluble organic matter amounting to 10–20% of dry weight. The residual particulate organic matter subsequently decomposes at rates that can be classified as in the order of 10, 1, and lesser percentages per day. Undoubtedly, if organic matter decomposition could be examined in greater detail a more continuous yet discrete distribution of classes of decomposition rates would be observed. This would appear to have important implications concerning the structure of aquatic ecosystems and the recycling of mineral nutrients.

7.5.2.3. Dissolved organic matter

Dissolved organic matter is operationally defined as that organic matter which passes through a filter of very small effective pore size usually of the order of 0.5 μm but sometimes smaller. It is generally considered that a small fraction of this dissolved organic matter, perhaps of the order of 10%, consists of small molecules such as sugars, amino acids and carboxylic acids. The remainder consists of higher molecular weight substances that are truly dissolved, colloidally suspended, or small particles. The smaller pool of dissolved simple molecules turns over relatively rapidly while the much larger pool of higher molecular weight material is relatively resistant to microbial attack and turns over very slowly. The total pool of dissolved organic matter is of the order of ten times that of the particulate matter suspended in lake water or stream water. This total pool has been characterized physically as labile or refractory depending on the rate with which it decomposes into inorganic constituents under high-intensity irradiation with ultraviolet light. However, the biological importance of this distinction in natural waters has yet to be demonstrated.

Measured concentrations of simple sugars, amino acids and carboxylic acids dissolved in natural waters are of the order of 10^{-4} to 10^{-5} mol m^{-3}. Therefore, the total pool of these simple molecules appears to be a small fraction of the total dissolved organic pool, which may range from 1–20 g C m^{-3} or more. There are not sufficient estimates of concentrations to suggest major differences in the amounts of these simple substrates in oligotrophic and eutrophic lakes or in the oceans. However, the substrate concentrations should vary somewhat depending on the dynamics of each ecosystem.

The turnover time for glucose and other simple substrate pools, i.e., the time required to completely replace the amount in the pool, has been determined in lake water from different trophic lake types. Data for glucose turnover time are presented in fig. 7.5. There is an obvious general correlation between the rate of glucose turnover and the rate of primary production, which is generally correlated with nutrient level. The correlation has little power to predict the metabolism of simple organic substrates at any time in any lake. It is clear that

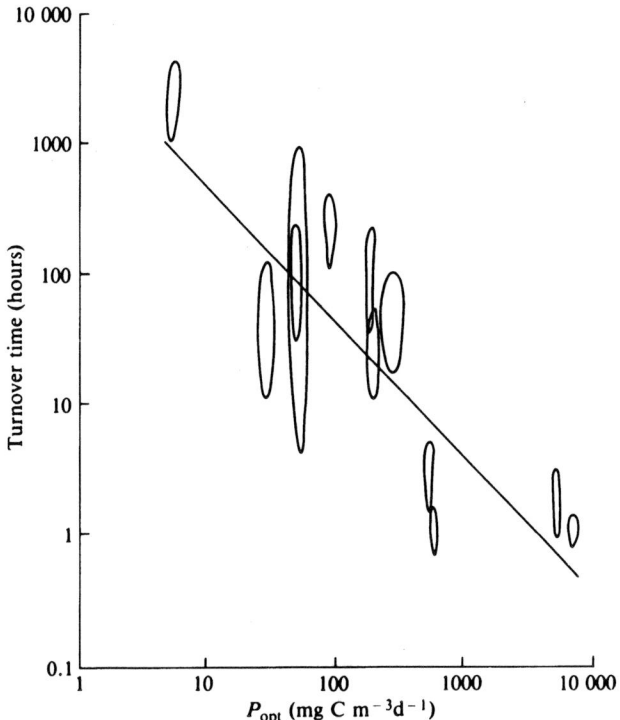

Fig. 7.5. The relation between turnover time of glucose and trophic state of a standing water body, as estimated from optimal photosynthesis (after Hobbie & Wright, 1965; Rodhe, 1965; Wetzel, 1967; Vaccaro *et al.*, 1968; Allen, 1969; Hobbie *et al.*, 1972; Miller, 1972; Seki *et al.*, 1974; G.W. Saunders, personal communication).

the turnover time of glucose varies within one to several orders of magnitude in any single lake system. It is of limited value to be able to say that simple organic substrates turnover more rapidly in productive lakes than in unproductive lakes. The data do not indicate what the pattern of variation may be. Thus nothing specific about the internal dynamics of decomposition of small molecules in any single lake can be inferred. It is this dynamic pattern that it is necessary to know as well as the variables which control the pattern, if the dynamics are ultimately to be understood.

The relative rate of decomposition of higher molecular weight dissolved organic matter is somewhat slower than for simple molecules. It has been suggested that phytoplankton predominantly release organic material extracellularly that is of higher rather than lower molecular weight, i.e., the molecules are not monomers (Saunders, 1972c). A few average relative rates of decomposition of extracellular dissolved organic matter are given in table 7.9. Frains Lake is a small eutrophic lake, Sanctuary Lake a hypereutrophic lake, and the Indian ponds are quite anomalous in being hypereutrophic but having

Table 7.9 *Average decomposition rate of total dissolved organic matter released extracellularly by phytoplankton*

Lake	Relative decomposition rate ($\% \, h^{-1}$)
Frains (March to November)	0.6–1.2
Sanctuary (23 June 1970)	7.4
Microcystis Pond (India)	76
Mariamman Tank (India)	76

an exceptionally large amount of bacteria, perhaps because of human usage.

The relative decomposition rate of dissolved organic matter released extracellularly by bacteria in Frains and Wintergreen lakes in Michigan is 1.0 and 1.8% h^{-1}, respectively, (G.W. Saunders, unpublished). This is comparable to the rates of decomposition of algal extracellular release. There is a component of the dissolved organic matter released by algae that is decomposed at a much slower relative rate, of the order of 1 to 4% per day (Storch, 1971).

Most of the total dissolved organic pool in lake waters in considered to be more resistant to decomposition. Estimates of the percentage decomposition per day of this pool in Frains Lake and a tundra pond are 0.7 and 0.8 respectively (Saunders, 1971; Hobbie *et al.*, 1972). However, the percentage decomposition of the interstitial dissolved organic matter of the tundra pond was 50% per day (Hobbie *et al.*, 1972).

There is sufficient evidence to indicate that this total dissolved organic pool in lakes is not completely resistant to decomposition and that in most lakes it is potentially susceptible to being completely metabolized during any single year. Certainly the annual input of dissolved organic matter is essentially completely decomposed. Usually it is not possible to measure the rate of change of concentration per day of this pool because the change is very small and lies within the sampling variance. But average trends over some period of time can be determined. Small relative changes that are prolonged over a long period of time will produce significantly large changes in concentration, (table 7.10). In the case of Frains Lake the total pool of dissolved organic nitrogen was reduced to 25% of its initial value at an average net rate of change of 0.5% per day.

Less detailed data for several Michigan lakes (table 7.11) indicate that even though the total dissolved organic pool appears to be more or less stable, minimum concentrations are reached during the whole year or during the summer such that from 50 to 78% of this pool is decomposed at some depth in these lakes. The input to subcompartmental pools due to biological or physical dispersion mechanisms is not likely to be zero. Therefore, these relative decomposition rates represent net changes and are likely to be minimum values.

This suggests that the total dissolved organic pool in lakes is a dynamic pool and if there is a very refractory component it is likely to be a very small fraction of the total pool. Most of the total dissolved organic pool is potentially susceptible to relatively rapid decomposition. In many cases decomposition is not so rapid. This should reflect the fact that variables other than the chemical nature of the substrate control the decomposition rate,

Decomposition of dissolved organic matter in streams has been little studied. Manny & Wetzel (1973) suggest that this material is more mobile than might be expected. Loss rates of dissolved organic carbon leached from leaves placed in an experimental stream have been studied (Cummins *et al.*, 1972; Wetzel & Manny, 1977). There appeared to be two organic pools in this material, one about 28 g C m^{-3} decomposing at a rate of 20% per day and a smaller 8 g C m^{-3} pool decreasing at a rate of 1.4% per day.

7.5.2.4. Community oxygen consumption

Total oxygen consumption in the dark has been used as a measure of total decomposition in subsamples of lake water and sediments. There is some ambiguity to the meaning of this definition of decomposition as the respiration of photosynthate by the primary producers is included in the estimate. In addition there are technical difficulties in partitioning microbial, animal and plant respiration. This manner of viewing decomposition processes is presented more fully in Chapter 5 (primary production).

7.5.2.5. Decomposition in sediments

Attempts have been made to estimate decomposition in sediments (Rybak, 1969; Harrison, Wright & Morita, 1971; Hall *et al.*, 1972; Hargrave, 1972a; Wetzel *et al.*, 1972). There are technical difficulties, also major difficulties in interpreting the results. However, it does appear that the intensity of decomposition in the sediments is several orders of magnitude greater than in the open water. There are horizontal differences in the intensity of decomposition in sediments and the proportion of total decomposition that occurs in the sediments will vary depending on the depth and the morphometry of the basin.

7.5.2.6. Co-metabolism

Lakes and some streams (Crisp, 1966) in peaty drainage areas receive large amounts of allochthonous particulate and dissolved humic compounds and humic compounds are probably produced to some extent in other lakes *in situ*.

Table 7.10 *Decomposition of total dissolved organic matter in lake water and some proximate organic substrates*

Lake	Average daily percentage decomposition	Duration of decrease in days	Final concentration / Initial concentration
Frains			
Dissolved organic carbon	2.7	10	0.73 Storch (1971)
Feb. 1970	3.1	8	0.75 Storch (1971)
Dissolved organic carbon	4.4	10	0.56 (G.W. Saunders, unpublished)
April 1960	3.1	13	0.59 Saunders
	6.9	10	0.31 Saunders
Dissolved carbohydrate 1960	0.38	195	0.25 Saunders
Dissolved organic nitrogen 1960	0.5	150	0.25 Saunders
Dissolved carbohydrate 1961	0.45	125	0.44 Saunders

Table 7.11 *Maximum and minimum concentrations of dissolved organic carbon in four Michigan lakes*

Lake	Dissolved organic carbon ($mg\,l^{-1}$) Maximum concentration	Minimum concentration	Min. conc. / Max. conc.
Section Four 1970 (unproductive marl)	6	2	0.33
Hemlock 1970 (productive)	10	6	0.60
Lawrence Lake 1967	5	2	0.40
(unproductive marl) 1968	9	2	0.22
1969	9	2	0.22
1970	8	3	0.37
1971	6	3	0.50
Duck Lake 1968	16	7	0.44
(productive)			

After Fast *et al.* (1973).

These compounds are supposed to be refractory. The evidence for the microbial decomposition of humus in Finnish lakes and the isolation of a bacterium (De Haan, 1972) able to grow slowly on humus support the view that these compounds may not be as refractory as is usually believed.

De Haan (1974) reported that an isolate of *Pseudomonas* sp. from Lake

Tjeukemeer does not grow on a fulvic acid medium but it does produce a greater yield of cells upon growth in fulvic acid/lactate medium in comparison with lactate medium alone. As co-metabolism of many refractory organic substances is known (Horvath, 1972), co-metabolism of humic substances in natural waters may be an important mechanism of decomposing these refractory substances in lakes and streams and this mechanism is worthy of more intensive study. It seems almost certain that fulvic acids are completely decomposed by mixed populations of physiologically different bacteria whether this decomposition is due to a priming effect or to co-metabolism (De Haan, 1975).

7.6. Non-living organic matter and bacteria as a source of food for animals

7.6.1. Detritus

Naumann (1921) stated that detritus and bacteria in lakes must be the most important food sources for many animal forms. Since that time many authors have implicated organic detritus as a dominant source of food for organisms. In many natural systems it is obvious that this must be so (Ladle, 1972). However, few conclusive data are available to support this argument. Since bacteria are closely associated with many organic detrital particles and since it is virtually impossible to separate the organic particles from bacterial particles, detritus has been defined operationally as this conglomeration of two kinds of organic particles (Odum & de la Cruz, 1967). In this case it is not clear whether organisms ingesting such particles derive energy directly from the dead organic particles or from the bacteria which have transformed a part of the detrital particle into protoplasm. The bacteria growing on the particles may be the main source of food for organisms ingesting these particles. In this case the particles are hydrolysed partially and bacteria develop, only to be stripped from the particle when it passes through the digestive tract. The surface of this particle is thus prepared for recolonization by a new set of bacteria. For one stream detritivore, colonization by hind gut anaerobes prior to faecal release has been demonstrated (M.J. Klug, personal communication). The process of stripping and recolonization upon passing through the digestive tract of animals is repeated until the organic particle disappears or until the particle is reduced to a residue of metabolically inert organic substrates such that they cannot support significant numbers of bacteria during their time of suspension in the pelagic region of lakes or the turbulent zones of streams.

In the special case of leaf detritus in streams it is clear that feeding by invertebrates on these leaves is closely correlated with the development of hyphomycetous fungi on and in the leaf tissue. The correlation suggests that the animals will not feed until the leaf is modified and that the animals in fact depend mainly on the fungi (Bärlocher & Kendrick, 1973a, b; Mackay & Kalff, 1973).

After passing through the anterior portion of the digestive system, a specialized microbial flora may develop in the hind gut on the fine detrital leaf particles produced by mastication. This microbial flora may produce dissolved organic molecules by hydrolysis that are absorbed through the gut or the microbes may be stripped from the particles by digestive enzymes secreted in the gut. According to the generalized stripping hypothesis fine detrital particles along with their special microbial flora may be packaged in faecal pellets to be defecated, recolonized by stream aerobes and processed by coprophagous organisms.

Using artifically generated detritus which contained no bacteria, it has been shown that zooplankton can assimilate organic detritus directly and that energetically this detritus can be as important as phytoplankton to the organism (Saunders, 1969, 1972a).

However, in general it is debatable whether the mechanism of obtaining energy from detrital particles is direct or indirect, and it seems obvious that this difficult area of research on feeding requires some careful and incisive study.

7.6.2. Bacteria

A number of authors have stated that bacteria are an important food source for zooplankton and for other animals. This subject is reviewed in Chapter 6 (Section 6.2.3.2). It is clear that bacteria can be a major source of food if their biomass is of the same order or greater than the biomass of nanno-phytoplankton. The biomass concentration may be greater than that of nannophytoplankton if a major source of organic substrate for bacteria is derived from net phytoplankton which is not grazed by zooplankton or if there is a large source of allochthonous organic substrate. If bacterial biomass is small relative to the nannophytoplankton biomass, bacteria cannot be a major source of food unless they are selectively grazed but there is no evidence to support the latter possibility.

7.6.3. Fungi

As in the soil community (Whitcamp & Olsen, 1963; Whitcamp, 1966; Whitcamp & Crossley, 1966), the fungi constitute extremely important decomposers of coarse particulate detritus, such as leaf litter, in streams. Selective feeding by stream detritivores on different leaf species or litter of different ages is strongly dependent upon the presence of fungi (Mackay &

Kalff, 1973). Bärlocher & Kendrick (1973a, b) have demonstrated that the feeding preference of *Gammarus* can be altered by manipulation of the fungal species colonizing the leaf litter food. In an experiment with a variety of detritivores (K.F. Suberkropp, personal communication), sterile leaves were not fed on at all, bacteria colonized leaves to a small extent and fungi colonized leaves to a highly significant degree when these three foods were presented singly and in combination.

7.7. Ecosystem structure

Bacteria in lakes and streams can function as primary producers; bacteria and fungi as decomposers or as food for other organisms. The first functional role is of limited importance and not the subject of this chapter. The third role has been mentioned briefly in this chapter and is discussed more extensively in Chapter 8.

The dominant role of bacteria in aquatic systems has conventionally been considered to be that of decomposers and mineralizers. Fungi have been relegated to a secondary role in decomposition with the exception of coarse particulate detritus in the sediments, especially leaf detritus.

However, animals of the plankton and benthos may also be considered as decomposers. They assimilate both living and dead organic matter and excrete ammonia, urea, carbon dioxide and phosphate, all of these substances being building blocks recycled into photosynthetic production. Grazing by zooplankton in Lake George, Uganda, may recycle more nitrogen and phosphorus than enters the lake from the drainage basin per year (Ganf & Viner, 1973). Although the relative rate of grazing is very low it is the same percentage order as the turnover of phytoplankton due to net primary production (Moriarty *et al.*, 1973). This suggests that zooplankton grazing is a major mechanism of decomposition in this lake. Haney (1973) has shown that zooplankton in a productive lake may graze more than 100% of the water volume per day for short periods of time. For several months from 30–60% of the water volume per day is grazed. If the assimilation efficiency of the zooplankton for phytoplankton is about 65%, the zooplankton should occupy a dominant role as decomposers during these periods of very high clearing rates.

A number of attempts have been made to describe the compartmental biomass structure and material flow in standing waters (Saunders, 1971, 1972a; Alimov & Winberg, 1972; Hobbie *et al.*, 1972; Kajak *et al.*, 1972; Moskalenko & Votinsev, 1972; Sorokin, 1972; Sorokin & Paveljeva, 1972; Wetzel *et al.*, 1972; Winberg *et al.*, 1972; Andronikova *et al.*, 1973; Tseeb *et al.*, 1973; Umnov, 1973) and in streams (Odum, 1956; Nelson & Scott, 1962; Minshall, 1967; Tilly, 1968; Coffman *et al.*, 1971; Cummins, 1972; Cummins *et al.*, 1972; Fisher & Likens, 1972, 1973; Mann *et al.*, 1972; Westlake *et al.*, 1972;

Manny & Wetzel, 1973; Boling *et al.*, 1974a; Boling, Peterson & Cummins, 1974b). Most of these budgets are incomplete and gross assumptions have been made that can provide only very approximate estimates. However, there are some very general statements that can be made concerning the structure of these ecosystems.

In lake and stream ecosystems the non-living organic matter is greater than the living matter by one or two orders of magnitude. The dissolved organic matter is about ten times that of the particulate organic matter and the organic detritus is five to ten times greater than the living matter. How does such structure come to exist and what is its significance?

Particulate and dissolved organic matter occur in water bodies as classes of organic substrates. Although the total kinds and amounts of organic substances have never been determined for even a single water body, it is clear that general and proximate types of organic substrates decompose at different relative rates. The relative rates of decomposition of classes of organic substrates in productive type water bodies are given in table 7.12. The same substrates in less productive water bodies should decompose at relatively slower rates. However, dissolved substrates in general tend to decompose somewhat more rapidly than particulate substrates. There is a class of dissolved substrate that decomposes at very slow relative rates as carbon-14 dating has revealed organic matter which is apparently a few thousand years old in sea water and a few hundred years old in soils.

If a substrate decomposes at a rate of 10% per day, it must accumulate in the system until on the average its absolute rate of decomposition is equal to its absolute rate of supply to the system. In this case the concentration in the water should be about ten times its daily supply.

Table 7.12 *The relative rates of decomposition of organic substrates in productive lake and stream systems*

	Decomposition $(\% \, d^{-1})$	Type of substrate
Dissolved	500–1000	simple small molecules
	25	extracellular algal and bacterial products; leaf leachate (labile fraction)
	2.5	refractory component of extracellular products
	1	refractory component of leaf leachate
Particulate	10	phytoplankton, zooplankton, detritus, insect exuviae, algae in littoral zone
	1	leaves in streams, residual component of zooplankton, leaves in littoral zone
	0.03	small secondary residuum of zooplankton

The absolute concentration of dissolved organic matter is therefore a function of its relative rate of decomposition and its absolute rate of supply. The absolute rate of supply must be some function of the quality of organic protoplasm and of the metabolites released by organisms and transported into water bodies from drainage basins.

As dissolved organic matter is about an order of magnitude more concentrated than particulate organic matter in natural water bodies, this should mean that it is more resistant to decomposition than particulate matter or that it is supplied to the water body at a higher absolute rate. The available evidence concerning extracellular release of dissolved organic matter by organisms or through hydrolysis of particles does not yet support the latter possibility. As shown earlier, 50–80% or more of the total dissolved organic pool in productive systems is potentially susceptible to decomposition at rates of a few per cent per day or per week. This does not suggest that this organic matter is highly resistant to decomposition and as long-lived as carbon-dating implies. From 80–99% or more of the organic matter produced in or entering an aquatic system per year must have a longevity in that system of less than a year.

Although the data are minimal, concentrations of glucose and amino acids in aquatic systems are only a few micrograms per litre and appear to vary within an order of magnitude. These considerations suggest that factors other than the molecular structure of the organic substrates are highly significant in controlling the general structure of aquatic ecosystems and the functional operation of decomposition processes in these systems.

The fact that organic substrates can be placed into classes according to decomposition rate suggests that these pools are chemically different and may exert different types of environmental control on these organisms functioning as primary producers as well as the decomposers.

Aside from their qualitative role in controlling ecosystem function (Saunders, 1957) these organic pools are important from the point of view of energetics. They provide stability in the general operation of any system. The more slowly metabolized organic pools will build up concentrations and by virtue of their longevity will be dispersed downstream in rivers or into the deeper waters of lakes due to settling in the water column or by turbulent mixing. In this manner the volume of water in which living organisms may exist is extended beyond the photic zone and dilution due to mixing will dampen the relative rate of utilization of substrate and in these senses, ecosystem metabolism is buffered. As the small more rapidly metabolized pools are used up the large more slowly metabolized pools will become the major sources of substrates released at slower, yet significant rates. This will maintain yet dampen primary production. As inorganic nutrient release increases the larger organic storage pools will increase tending to make inorganic substrates less available and decrease the intensity of primary production or microbial activity. Thus the quantitative release of organic and inorganic nutrients to heterotrophs and autotrophs from the various organic storage pools will stimulate or depress their activities in such a way that oscillations between extremes will tend to be minimized (Saunders, 1969, 1972a).

It is clear that the decomposition subsystem is a highly sophisticated and subtle functional system. It would seem that it is worthy of a much more sophisticated analytical approach if the subtleties of its operation are eventually to be placed in perspective and properly understood. Why does glucose in the unproductive oceans have a turnover time of the order of 10 000 hours and in hyper-productive lakes a turnover time of the order of 1 hour? Certainly glucose in the oceans is not a resistant chemical molecule and in lakes it is a chemically labile molecule. Is there something about the structure of these systems, other than concentrations, that is very different and if so what is it that brings these differences about? Aquatic microbiologists have not yet begun to address themselves to such questions.

The microbes and animals in aquatic systems are functional analogues as decomposers, although the physical manner in which they accomplish decomposition is quite different. Attempts have been made to estimate the relative roles of bacteria and zooplankton as decomposers in lakes (table 7.13). A list of references has been given earlier in this section. All of the budgets are incomplete, some may be in error and they have been calculated on different bases: daily, monthly, seasonal and yearly.

If zooplankton clear nearly 100% of the water volume per day (Haney, 1973)

Table 7.13 *The relative roles of bacteria and zooplankton in the metabolism of lakes of different types*

		Decomposition (%)	
Lake	Type	Bacteria	Zooplankton
Krivoye	oligotrophic	68	32
Krugloye	bog	61	39
Krasnoye	mesotrophic	80	20
Mikołajskie	eutrophic	93	7
Tundra pond	oligotrophic	98	2[a]
Baykal	oligotrophic	55	45
Frains	eutrophic	76	24
Rybinsk	meso-eutrophic	91	9[a]
Dalnee	eutrophic	76	24[a]
Kiev	eutrophic	79	21[a]
Lawrence	oligotrophic	79	negligible
Myastro	slight eutrophic	71	29
Zelenetskoye	oligotrophic	63	37

[a] Includes much allochthonous material (71% of total input in Rybinsk, 35% in Dalnee).

and if their assimilation efficiency for algae is 65%, on the average, this latter value should be roughly the upper limit for the role of zooplankton as decomposers. If the zooplankton recycle the organic detritus defecated and if they digest and assimilate bacteria that develop on this detritus, the upper limit for the role of zooplankton in decomposition should be somewhat higher than 65% of primary production. The lower limit should be near zero. There is no obvious pattern among the estimates given in table 7.13. In general, bacteria occupy a greater role in decomposition than zooplankton, their greatest dominance occurring in lakes receiving large amounts of allochthonous material. The greater dominance of zooplankton in decomposition occurs in lakes that are unproductive but it is also high in productive lakes and intermediate in some lakes with high input of allochthonous matter. These results may be the artifacts of errors in assumptions and estimates of lake budgets or the estimates may be reasonable. The dominance of zooplankton as decomposers in unproductive lakes is in contradiction with the theory that zooplankton is relatively ineffective in grazing phytoplankton in unproductive lakes (Chapter 5, p. 202; Haney, 1973).*

Table 7.14 *Annual carbon budget for Lawrence Lake, Michigan*

	$(\text{g C m}^{-2} \text{a}^{-1})$
Inputs	
Macrophytes	87.9
Epiphytic algae	37.9
Epipelic algae	2.0
Heterotrophy	2.8
Algal secretion and autolysis	14.7
Phytoplankton	43.4
Allochthonous particulate	4.1
Allochthonous dissolved	21.0
Littoral plant secretion	5.5
Bacterial chemosynthesis	7.1
	226.4
Outputs	
Benthic respiration	117.5
Outflow dissolved organic matter	35.8
Outflow particulate organic matter	2.8
Permanent particulate sedimentation	14.8
Co-precipitation with Ca CO_3	2.0
Bacterial respiration dissolved organic matter	20.6
Bacterial respiration particulate organic matter	8.6
Algal respiration	13.0
	215.1

After Wetzel *et al.* (1972).

* But see also p. 284/5 (eds.).

A detailed carbon budget for a marl lake is presented in table 7.14. In this case the budget is within 5% of being balanced. The number of processes estimated is remarkable yet details are missing and the whole animal kingdom is considered to be negligible in its role in carbon flow in this system. The budget does not contain any information as to why this system has this particular structure and what factors control the development of this structure. This most graphically illustrates the point that analysis of function in any natural ecosystem is a tremendously large undertaking in which ecologists have not yet been successful. This is not to deprecate the accomplishment that has been made but merely to point out that while the descriptions of ecosystem structure and dynamics are necessary to any analysis they are not sufficient to complete analysis from a functional point of view.

The structure of streams is much less well documented. A general comparison of the organic matter inputs for three woodland streams is given in table 7.15. Primary production constitutes less than 1% of the organic matters input for each case, demonstrating the dependence of such waters on allochthonous organic matter. Respiration accounted for one-third to one-half of the total organic matter export. A more detailed budget of the average daily biomass flow in Augusta Creek has been estimated by Cummins (1972) (fig. 7.4). In this case 25–30% of the particulate biomass input flows through the animals, assuming an average assimilation of 40% for non-predators, and is decomposed in this manner. Allochthonous input of dissolved organic matter is rapidly converted to particulates by physical flocculation or microbial metabolism. On the other hand less than 1% of detrital inputs to a small New England mountain stream were calculated to be processed by macroinvertebrates (Fisher & Likens, 1972, 1973). For the River Thames, Berrie (1972b) and Mann et al. (1972) have shown that detritus is a dominant source of food for many fishes and suggest that it is probably a dominant source of food for macroinvertebrates. Unfortunately no quantitative estimates of the role of invertebrates in this pathway of decomposition were made for Thames, although Mathews & Kowalczewski (1969) believe that it is small for leaf litter.

Insufficient analysis has been made of stream structure and function to facilitate any broad comparison of decomposition in streams. However, it is obvious from the examples given above that the roles of microbes and of animals in the decomposition process must be distributed over a fairly wide spectrum of relative dominance.

In the processes of decomposition, microbes and animals, both zooplankton and zoobenthos, operate as functional analogues in determining the physical size and availability of particulate food for other animals and microbes as well. Very large detrital particles are made smaller by the mechanical shredding action of feeding animals and by the mechanical disruption following enzymatic hydrolysis due to microbes. This reduction in size makes food

Table 7.15 *Organic matter budgets for three North American streams (POM = particulate organic matter; CPOM = coarse particulate organic matter; FPOM = fine particulate organic matter; DOM = dissolved organic matter)*

Parameters	Bear Brook USA[a]		Watershed 10 USA[b]		Augusta Creek USA[c]	
	(kJ m⁻² a⁻¹)	%	(kJ m⁻² a⁻¹)	%	(kJ m⁻² a⁻¹)	%
Inputs						
POM	13 495	53.5	8 962	53.7	9 695	53.8
CPOM	(11 435)	(45.3)	(3 477)	(20.8)	(1 486)	(8.2)
FPOM	(2 060)	(8.2)	(5 485)	(32.9)	(8 209)	(45.5)
DOM	11 697	46.3	7 635	45.7	8 258	45.8
Primary Production	40	0.2	94	0.6	79	0.4
Total	25 232	100.0	16 691	100.0	18 032	100.0
Outputs						
POM	5 024	19.9	2 015	13.3	3 587	19.9
CPOM	(3 876)	(15.4)	—	—	—	—
FPOM	(1 148)	(4.5)	—	—	—	—
DOM	11 701	46.3	7 635	50.3	4 679	25.9
Respiration	8 512	33.8	5 535	36.5	9 766	54.2
Total	25 237	100.0	15 185	100.0	18 032	100.0

[a] Fisher & Likens (1973).
[b] Sedell et al. (1974).
[c] Cummins (1972).

particles available to other organisms whose feeding apparatus is smaller in dimension and also provides greater surface area for attachment of bacteria. There is clearly a correlation between microbial activity (Hargrave, 1972a; Olah, 1972) and particle size down to a minimum size below which microbial activity decreases again. It is also well known that zooplankton and benthic invertebrates are effective in filtering particles over specific size ranges which bear some general relation to body size.

On the other hand, bacteria may assimilate dissolved organic molecules and grow to produce larger particles. In turn these bacteria may aggregate and there is a tendency for some bacterial aggregations to be associated with flocculent organic detrital material producing even larger aggregates. Invertebrates will ingest particles and reduce their size during digestion in the gut. Associated with this may be the production of a relatively large biomass of bacteria that are indigenous to the gut of the invertebrate species feeding. However, in defecating this material often it is packaged in a faecal pellet that is the same size or larger than the particles ingested.

Thus in the decomposition process there is a tendency to produce particles of an intermediate size range. This size range may be specific for a particular ecosystem type. The manner in which these particles are produced probably has a regulatory function concerning community structure in both stream and lake ecosystems. It should also have survival value if and when food supply is decreased toward some extreme limiting abundance. This aspect of decomposition is only intuitively understood. The details are undoubtedly quite subtle and are worthy of much more intensive study.

Another important aspect of ecosystem structure relates to dissolved organic matter. This is an operationally defined constituent that probably contains a high percentage of colloidal material, as has been demonstrated for sea water by Sharp (1973). Colloidal micelles are potentially highly labile in a physical–chemical sense, their structures being dependent on such variables as pH, temperature and ionic strength of the medium. Colloids may aggregate to form particles or they may disassociate to form smaller micelles and dissolved molecules. The colloidal fraction of the 'dissolved' organic phase must exert both qualitative and quantitative control on the biochemical reactivity of the molecules of which the micelles are constituted as well as affect the absolute rates of decomposition of the simple molecules with which the micelles are in equilibrium.

While this is a technically difficult area of investigation, it is of fundamental importance to approach this problem from this perspective if a functional analysis of the 'dissolved' organic pool is ultimately to be obtained.

7.8. Concluding remarks

Any proper discussion concerning organic matter and decomposers should contain some evaluation of that discussion. Much of the information presented in this chapter has not been obtained within the formal framework of the IBP. This is not a condemnation of IBP but is representative of the fact that the scientific community has chosen not to investigate intensively this difficult problem area. Many of the data are highly selected from the work of a few individuals. Since few individuals have conducted research in these areas this should not be too objectionable. Much of this research is unique and from that point of view is at least interesting. Whether or not it is accurate awaits the tests of future research. No attempt has been made to evaluate precision of methods and accuracy of the estimates for the purposes of this review. Considering the relative paucity of information available and the limitation imposed upon the chapter, such an analysis was not warranted.

The general correlations presented here are undoubtedly true but have little power of resolution. Although bacterial biomass is inversely correlated with latitude it is not possible to distinguish the relation between bacterial biomass and trophic states of lakes at the same latitude. The turnover time for glucose is inversely correlated with the intensity of optimal photosynthesis. However, the turnover time can vary through several orders of magnitude, within a single lake system. It seems obvious that the question that asks what it is that controls the turnover time of substrate in a water body will result in greater information content than the question which asks only what is the turnover time.

Some of the results are conflicting. The generation time of bacteria is correlated with trophic state of a lake but the production biomass coefficient of bacteria is the same over all trophic states. Both relationships cannot be true. The idea that zooplankton are relatively less able to graze phytoplankton in oligotrophic lakes (Chapter 5, p. 202)* and that zooplankton are relatively dominant in decomposition processes in oligotrophic lakes (table 7.13) seem inconsistent. Such conflicting points of view cannot be resolved unless accurate methods are used and unless a truly representative spectrum of ecosystems is examined. Clearly much more work is needed here.

Many interesting details are not included in this chapter because of space limitations or because they have not yet been examined.

It must be abundantly clear to persons reading this chapter that our understanding of decomposers and decomposition processes in aquatic systems is in a very immature state. Decomposition, whether it is carried out principally by animals or by microbes must be at least equal in magnitude to the processes of photosynthesis in freshwater systems. Clearly this half of biologically mediated energy flux has been sadly neglected. This is not to say that the information summarized in this chapter is not necessary but merely that it is not sufficient.

If the information we have concerning decomposition is not sufficient what then is necessary to provide sufficient understanding of decomposition processes in aquatic ecosystems? We have described the sequence of events in

* But see also p. 284/5 and p. 478 (eds.).

several decomposition processes so that we do have fundamental notions concerning those pathways. However, to know the turnover time or the relative rate of decomposition of a substrate does not permit us to know the relationship between the several decomposition pathways operating in a system, except in the most general manner. However, if the absolute rates of decomposition can be estimated, the decomposition pathways in any system can be proportionated, whether they are mediated by microbial or by animal activities.

As the chemical nature of the substrate will effect one primary control over the reaction rates in decomposition processes, major emphasis needs to be placed on qualitative identification of particulate and dissolved organic substrates as well as quantitative estimation of the substrate concentrations. Other factors such as temperature and inorganic nutrients may exert secondary control over decomposition processes. Therefore, additional emphasis needs to be placed on developing experimental approaches to determine the intensity with which environmental variables control the various decomposition pathways. Not until such approaches are initiated on a spectrum of type systems and on a scale that is adequate can we hope to develop a more definitive and fundamental understanding of the manner in which stored chemical energy is dissipated in aquatic ecosystems.

8. Trophic relationships and efficiencies

P. BLAŽKA, T. BACKIEL & F.B. TAUB

8.1. Introduction

In most ecosystems organic matter is formed by primary producers (auto-trophic organisms) and is degraded to carbon dioxide and nutrients by heterotrophs, mainly animals and bacteria. Degradation may be quite straightforward, for example in an association of algae and bacteria, or it may proceed through several steps. Formation and degradation pathways are called 'chains' or 'webs' according to their complexity. In any system, the flow of energy coming from the sun's radiation is unidirectional and most of it is transformed to heat within the system. Nutrients such as carbon, nitrogen, phosphorus and many others, brought into aquatic systems with inflowing water, rainfall, sewage, falling leaves etc., may cycle within the system before they are exported from it.

The aim of this chapter is to discuss factors which determine the complexity of ecosystems and the quantitative relationships between the major compart-ments. The term 'major', an arbitrary one, indicates relationships between two closely related species or minor compartments which influence or change the structure of the ecosystem quite considerably. Trophic relationships within the phytoplankton, zooplankton, benthos, and among microorganisms have, however, already been discussed in preceding chapters, and this chapter will be limited to a discussion of the relationships between associations.

During the activity of the IBP/PF in Europe the tacitly assumed leading idea was that of a simple chain:

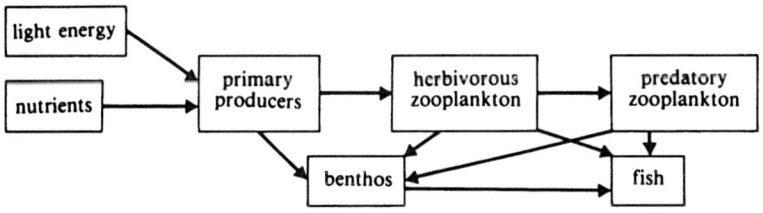

with a bacterial compartment placed somewhere in the middle of this scheme. Bacteria may also be considered as part of another pathway called the 'detritus food chain'.

It is evident that both pathways, that of grazers and that of the detritus food chain, operate in most freshwater habitats. Their mutual relationships may be visualized as two parallel 'shunts'. Their quantitative relationships can best be judged from ecosystem production budgets, therefore efficiencies will first be discussed.

8.2. Efficiencies

The above two schemes are certainly simplifications (see for example the discussion in the zooplankton section of Chapter 6), but it was found very difficult to rearrange data collected according to these schemes into some other scheme without adding a number of unproved assumptions. The discussion here of energetic efficiencies will therefore be based on the grazing food chain scheme given above (simplified yet further) assuming that it does describe the main pathway in terms of energy transformations, and that it may be useful in considerations of potential yields from freshwater ecosystems for human nutrition. This scheme has very little predictive power in terms of understanding nutrient limitations, or of the surpluses which provide the scientific background to the problems of technologically advanced areas (eutrophication, pollution, sources of drinking water etc.). A rough statistical treatment of transfer efficiencies is given by Brylinsky (Chapter 9, table 9.13).

The transfer efficiency is defined as the ratio of production of the particular step of the trophic chain to the gross primary production. Surprisingly, it is not identical with any of the production indices used by earlier students of ecological efficiencies, listed by Kozlovsky (1968), and it was evidently introduced by Lasker (1970).

8.2.1. Efficiency of secondary production

The separate listing of efficiencies for herbivorous zooplankton and benthos respectively is not strictly correct as these two compartments are not independent of one another, and some other parameters determine the partitioning of primary production between them. Moreover, separation of herbivorous and carnivorous zooplankton (and of comparable compartments in the benthos) has in many studies been rather arbitrary. For example, cyclopoid copepods have been considered as predators, although the early life-history stages of all of them, and the later stages of many of them, are herbivorous or rely on mixed food (see Chapter 6, section 6.2.3.2).

An attempt to overcome these problems has been made here by summing the production of herbivorous and carnivorous zooplankton and benthos to get total secondary production, though it is recognized that the value obtained in this way also includes a great deal of the tertiary production. On the other hand, the part of the secondary production which is due to herbivorous fish

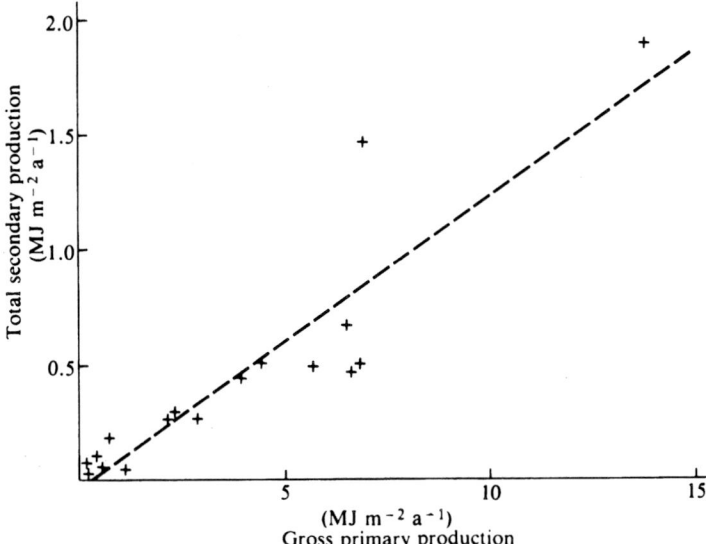

Fig. 8.1. Relationship between total secondary production (see text for explanation) (*y*-axis) and gross primary production (*x*-axis) for the 17 IBP/PF sites for which secondary production data were available (listed in table 9.1, p. 412).

feeding is not included (fig. 8.1). The resulting relationship has a higher correlation coefficient ($r = 0.906$) than any of the partial relationships listed in Chapter 9, though N is smaller. This suggests that the assumption made on partitioning was correct. The regression equation

net total secondary production $= -36.05 + 0.128$ gross primary production
$$(8.1)$$

(both production rates as kJ m^{-2} a^{-1})

suggests:
(1) that the line intercept is not statistically different from the origin,
(2) inspection of the data and high correlation coefficient (none of the non-linear transformations tested gave a significantly higher one) both suggest a linear relationship,
(3) the slope of the line $+0.128$ suggests an approximate 13% transfer of gross primary production into net secondary production. This is certainly an overestimate due to including both predatory zooplankton and benthos. Applications of an arbitrary correction of 15% to account for this would give a value of 10.9%.

In Chapter 5 it has been shown that determination of net primary production of phytoplankton is not feasible as a routine procedure. The ratio of net to gross would certainly vary within broad limits due to localities, depth, biomass, time of day, season and other variables. It is, however, likely that a *statistical* mean value would lie somewhere between 50% and 80% on a 24-hour basis. If this is true, then corrections applied by Brylinsky (Chapter 9, section 9.1.2) to make carbon-14 and oxygen determinations comparable are not large enough, and a *statistical* mean transfer coefficient between primary and secondary production would be likely to be somewhere about 15%. The earlier value of 10% suggested by Slobodkin (1959, 1960) and confirmed by a few other reports was originally obtained in the laboratory on fairly crowded *Daphnia* populations (up to 5.9 individuals/ml) and fed once in a 4-day period. The bias in the present data is not limited to primary production estimates, but also involves secondary production estimates (Hall, 1971).

Fairly strong support for a higher transfer coefficient than was hitherto assumed comes from marine studies. Steel (1978) has summarized studies from the North Sea and other areas, and Cushing (1973) from the Indian Ocean; both authors suggest transfer coefficients of between 15% and 20%.

Another point of interest is the linearity of the relationship in equation 8.1, though relationships between primary production biomass and zooplankton feeding are non-linear. Gutelmacher (1975) and Desortová (1976) studied the relationship between the specific uptake rate of carbon and cell size by autoradiography. Both found the relationship of photosynthetic activity to be inversely proportional to cell mass: to the power of -0.63 ($m^{-0.63}$) according to Desortová. Javornický (Chapter 5 and unpublished data) has summarized relationships between primary production and photosynthetic activity (PA). Biomass data were derived from mixed samples through the euphotic layer. Thus he got a relationship where effects of self-shading by increasing biomass and shading by organic matter produced by the phytoplankton were included:

$$PA[d^{-1}] = 0.24B^{-0.63}[J\ m^{-3}]$$

To arrive at the type of relationship between secondary production and food concentration we may examine some older unpublished data collected by Blažka on the growth rate of *Daphnia hyalina* at 20 °C fed on a culture of *Chlorella pyrenoidosa*, contaminated by bacteria (Taub & Dollar 1968; M. Hrbáčková, personal communication). The values measured were: time taken (days) for embryonic development (E) and for postembryonic development (P); number of young ($+ 1$) per brood and per day (I). Times when there were no eggs or young in the brood chamber of adult females (this occurred at the two lower food concentrations) and uncompleted developments were recorded and corrections applied. The ratio of adult females in the population (R) was assumed to be 0.5. Food was changed daily and dilutions were based on a chemical oxygen demand procedure similar to that of Ostapenya (1965) but standard solutions and corrections were applied to account for incomplete oxidation of algae. Growth rate (μ_z) was calculated using the expression:

$$\mu_z = \frac{I \times R}{E + P}$$

The resulting mean data can be formulated by several functions; in at least five of them the linearized function has $r > 0.95$, but two of them are of particular interest:

(i) $\mu_z = 0.0257 B_p^{0.77}$ $r = 0.973$ (8.2)

solving (8.2) using (8.1):

$\mu_z = cPA^{-0.77/-0.63}$ $\mu_z = cPA^{-1.22}$

This expression is satisfactorily close to the linear form (8.1) and explains why the relationship of primary production and secondary production is linear.

(ii) In fig. 8.2 the relationship between rate of population growth and food concentration (S) for *Daphnia hyalina* is plotted in terms of the Monod equation, suggesting the formula $\mu_{max} = 0.77 \, d^{-1}$ and $K_s = 40 \, J \, l^{-1}$. A more detailed analysis of data used for construction of fig. 8.1 would demonstrate that another set of constants for individual growth is suggested: $\mu'_{max} = 0.139 d^{-1}$; $K'_s = 0.733 \, J \, l^{-1}$. This also suggests that the rate of population growth is influenced more by variations in number of eggs than by rate of individual growth.

The feasibility of expressing the rate of population growth of a typical secondary producer by the same mathematical treatment as rate of growth of microorganisms will greatly facilitate the modelling of biological events in aquatic environments. It will also facilitate the study of interspecific

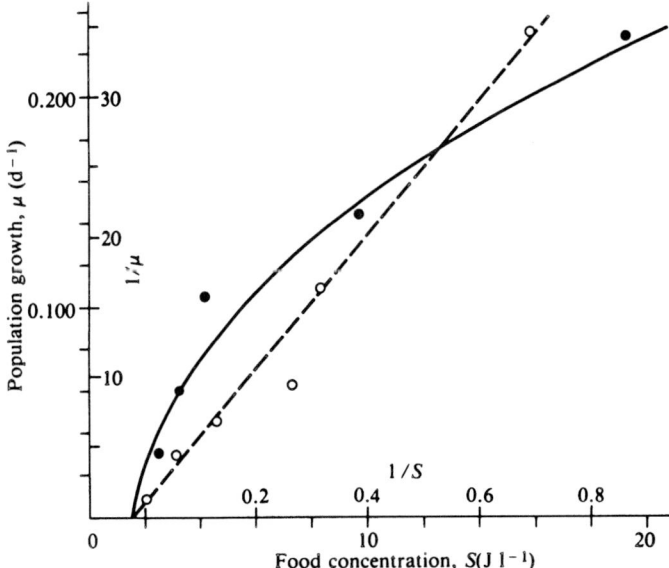

Fig. 8.2. Rate of population growth in laboratory cultures of *Daphnia hyalina* plotted against food concentration (in $J \, l^{-1}$) (——) and the Lineweaver Burk plot for the same relationship (- - - - -).

relationships and put this on a sound experimental basis (Hrbáčková & Hrbáček, 1978).

Two IBP localities were not included in the plot of fig. 8.1: Lake Sniardwy (Poland) and L. George (Uganda). While the former is out of line evidently due to some disagreement in presentation of the data (compare fig. 4 in Hillbricht-Ilkowska *et al.*, 1972 with table 3 in Kajak *et al.*, 1972), in Lake George, Uganda, the reason lies in the very low ratio of net to gross primary production estimates (*ca* 0.3) (Ganf & Viner, 1973; Moriarty *et al.*, 1973). The latter authors and Ganf (1972) were, however, able to interpret this ratio by the light and temperature regimes of the lake. Relative respiration rates (R/P_{max}) measured were about 0.05, which suggests that the ratio of bacterial and autotrophic activity was shifted much in favour of autotrophs.

The relationship between primary and total secondary production is thus linear over the whole of the small range examined, though there is a suggestion of a decline in the ratio with increasing primary production (Hillbricht-Ilkowska *et al.*, 1972).

8.2.2. Ratio of primary and fish productions

Another set of observations supporting the statement that efficiency is higher than 10% are the transfer efficiencies between primary production and fish net production or yield. Most fishes in temperate and arctic waters are carnivores and therefore represent a third or fourth link in the production chain. Assuming two equally effective steps in a simple straightforward chain (autotrophs and herbivores), the percentage transfer efficiency of the fish production would be $(0.1\,E)^2$, E being percentage transfer coefficient for primary–secondary and secondary–fish production. Thus a 4% transfer coefficient of primary to fish production suggests a 20% transfer coefficient in each step. In tropical waters the increased ratio of herbivorous fishes may permit higher fish production efficiencies to occur.

Thus the statement already made concerning the equivocal position of zooplankton and benthos species in the production chain also holds for fishes. Even a single fish species requires a certain diversity of food in order to survive: the fry start feeding by consuming small items only and the proportion of larger particles grazed increases with age and size. The composition of the diet can shift either to plant material or to large benthic and pelagic animals, including vertebrates.

Provided that the spawning of the fish and their relationships with the environment are not adversely affected by lack of suitable biotopes, that physical and chemical conditions do not preclude the existence of fish, then their production depends on food supply. This in turn is directly or indirectly related to the primary sources of organic matter. The best illustrations of this come from fish-pond data, where fish production is routinely determined with much higher precision than in any other type of water body.

Müller (1966) found a significant correlation between the average daily gross production of phytoplankton and the production of common carp in ponds ($r = 0.72$); Hrbáček (1969) and Winberg & Lyakhnovich (1965) demonstrated similar high correlations in fish-ponds in Czechoslovakia and Byelorussia. Wróbel (1970) estimated the same correlation at $r = 0.89$, and Wolny & Grygierek (1972) found that this coefficient for carp fry ponds was as high as 0.98. Kořínek (1972), however, reported a much weaker relationship ($r = 0.54$) in a series of Czechoslovakian ponds. He pointed out that in waters stocked with carp of older age groups which feed mainly on bottom fauna the relationship may not be as simple as in the above quoted cases.

Some IBP data on primary production and production and yield of fish are presented in Chapter 6, table 6.14 and fig. 6.10. The Tjeukemeer polder reservoir in Holland and the Rybinsk reservoir in USSR occupy outstanding positions in fig. 6.10. Sorokin (1972) reported that the input of allochthonous organic matter by Rybinsk is about twice as much as the primary production. This can also be true for Tjeukemeer, where immigration of fish may contribute to the overestimation of the transfer efficiency related to the primary production only (Beattie *et al.*, 1972).

Low efficiency in Lake Karakul (0.12%) is apparently caused by the very high contribution of macrophytes to the primary production – about 90% (Khusainova *et al.*, 1973).

In both European and Malaysian ponds primary production was determined by the light and dark bottle methods (O_2) and the estimates seem to be fairly comparable.

Excluding the two reservoirs with apparently high efficiency, the maximum fish production is *ca* 1.6% of gross primary production in lakes and reservoirs, and the minima recorded are 0.2% (Vord. Finstertaler, Pechlaner *et al.*, 1972a, b) and 0.12% (Lake Karakul).

As already discussed in Chapter 6 (section 6.4.11) efficiencies recorded in carp ponds are usually higher. Hepher (1962b) estimated that the daily production in carp ponds in Israel was on average 1.6% of the primary production, which ranged from 57 to 175 kJ m^{-2} d^{-1}. Winberg & Lyakhnovich (1965) showed that this ratio ranged between 1.8% and 10.4% of gross primary production. In fingerling carp ponds in Byelorussia fish production was from 6.1% to 8.2% of gross primary production. Most ratios are, however, between 1% and 4%. The maxima of about 8% are surprisingly high considering the habits of the carp in Europe, underestimates of primary production, and that unconsidered inputs of allochthonous material might have occurred.

The transfer efficiencies for tropical ponds are of the same order of magnitude as those for 'inefficient' lakes, *ca* 0.2%, despite the fact that a mixed population of fish, including herbivorous *Tilapia*, were grown in them

(Prowse, 1972). This is apparently associated with a higher respiration : primary production ratio at higher algal and blue-green standing crops (see Chapter 5).

Another set of data in fig. 6.10 concerns the ratio of fish yield to primary production (also considered in Chapter 6). The highest ratio for the north temperate zone comes from data for Lake Mikołajskie in Poland (primary production data from Kajak *et al.*, 1972, fish yields from the Inland Fisheries Institute data register). This can be explained by intensive fishing for small plankton-feeding fishes and by possible immigration from neighbouring lakes. The yield data in general suggest that man does not utilize more than about 0.2% of the gross primary production of lakes and reservoirs. Winberg (1972b) stated that a lake fishery in the European part of USSR cropped about 0.2% of the net primary production, but this value seems to be on the high side of the yield variation.

The ratio of production and fish yield reflects several factors which are of a methodological and technological nature, rather than a biological one. In well-managed fish-ponds the production of fish equals the yield. In lakes and reservoirs, however, the bulk of the production is attained by fishes in their first and second year of life which have high mortality rates. These fishes are very difficult to catch, so it is certain that most production estimates are grossly underestimated for this reason. (Holčík & Pivnička, 1972; Pivnička, 1975). For economic fishing only certain fish species are attractive, and of these only the older and larger fish. Man therefore exploits only a fraction of the actual fish production in lakes and reservoirs, though exploitation may approximate to production in fish-ponds (Holčík & Pivnička, 1972).

Indian waters were represented by studies in six reservoirs, three small lakes, one tank and the Chetpat Swamp (Sreenivasan, 1972). The name 'swamp' does not seem to be descriptive, for this last locality was actually a pond, area 73 ha, mean depth 1.0 m, in which the predominant primary producers were blue-greens, diatoms and green algae in a well-mixed suspension. No data on macrophytes were given, but it is unlikely that they would be very abundant in the thick and productive suspension of algae and blue-greens, except perhaps at the fringes of the pond. The main fish species were described as plankton-feeding. The shallow depth evidently reduces the ratio of respiration to gross primary production; net primary production may be relatively higher than in L. George (Uganda) which was studied in greater detail. The shallow depth may also help efficient fishing, and thus contribute to the highest yield per hectare of any IBP site.

As a tentative conclusion it may be stated that most fish production estimates range between 0.1% and 1.6% of gross primary production in lakes and reservoirs, but that they are about 4% in fish-ponds. It is suggested that fish production estimates in lakes and reservoirs are underestimated due to the problems of studying fish in their first and second years of life. Yield is about

10% or less of production in lakes and reservoirs, but equals production in fish-ponds. Tropical biotopes with high primary production are less efficient, despite a higher ratio of phytophagous fish. The reason for this lies in the very shallow euphotic zones of these environments which result in great respiration losses where mean depth is greater than euphotic depth. This also suggests that very productive ponds should be shallow to conform with the relationship determining euphotic depth (Chapters 3 and 5).

8.3. Relationships of the grazing and detritus food chains

The linear relationship between primary production and secondary production has one more implication: it suggests that the ratio of grazing and detritus food chains ('shunts') remain roughly constant on a round-the-year basis and within the range examined by the IBP/PF teams. If this were not so, the efficiency of secondary production would drop with increasing participation of the detritus chain and vice versa. Having bacteria as one more link in the simple chain outlined above the transfer coefficients suggested (section 8.2.1) would then become completely unrealistic.

Similar suggestions with somewhat wider variance are made in Chapter 7 when dealing with relationships between bacterial numbers and activities and primary production. This of course, is only true if allochthonous input is negligible or its nature prevents fast development of activity of bacteria and other heterotrophic microorganisms.

During IBP considerable progress was made in studies of biotopes which are too small for boundary influences to be neglected. Prominent examples are the tundra pond of Hobbie *et al.* (1972) and some forest streams. In both types the aquatic environment functions as a trap of flooded and wind-blown allochthonous matter and its input to the water body is thus greater than to the surrounding land (per unit area), and greater than primary production of these systems. Here the detritus food chain thus provides the bulk of the food for the animals and is the major pathway (Teal, 1957; Egglishaw, 1972; Ladle, 1972; Chapter 7).

The development of macrophytes imparts certain characteristics to a water body. Macrophytes limit growth of algae and blue-greens both by shading and competition for nutrients; they limit mixing so that spatial distribution of dissolved organic compounds, including organic carbon, becomes very uneven (Wetzel *et al.*, 1972). Macrophyte-grazers are less efficient than algal-grazers. They cannot control the development of higher plants, and the accumulated plants are decomposed through the detritus food chain, which evidently prevails in macrophyte-rich localities (Wetzel *et al.*, 1972; Hobbie *et al.*, 1972). This results in lower efficiency of zooplankton production and fish yield (Wetzel *et al.*, 1972, and L. Karakul, section 8.2.2). Factors which determine the prevailing primary producers (whether macrophytes or algae and blue-greens) thus also determine the prevailing food chain, whether it is the detritus chain or the grazing chain. Unfortunately these factors are only known in very general terms: shallow depth and low turbidity allowing access of light to the bottom promote macrophyte growth and vice versa.

In arctic and alpine lakes a part of the bottom, usually a belt, is covered by macrophytes, mainly mosses: in Char Lake (74° 42′ N, 94° 53′ W, Cornwallis Is., Canada) these cover 27.5% of the bottom area, in V. Finstertaler (Central Alps 2237 m above sea level) 9% of the bottom, in Bolshoy Kharbey (67° 30′ N, 64° E) 5% of the bottom (Vlasova *et al.*, 1973) and in the Lapland lake, Latnjajaure, 39% of the bottom area (Bodin & Nauwerck, 1968). The detritus chain associated with these macrophytes may be an important stabilizing factor in communities likely to be subjected to fast changes of weather during the short summer.

In Chapter 7 table 7.13 the relative zooplankton and bacterial decomposition rates are given for a number of lakes and reservoirs. The way these values were derived is not given in detail; the high transfer efficiencies from primary production to secondary and fish production would be most unlikely with a strong detritus food chain shunt.

Clearly designed experiments by Moriarty (1973) strongly suggested high assimilation ratios of *Microcystis* clumps after digestion at low pH (pH 2) in *Tilapia nilotica*. Similar assimilation ratios with the same food were found in *Thermocyclops hyalinus* (Moriarty *et al.*, 1973). This means that blue-greens need not be processed through the detritus chain, but can be used directly as food by herbivorous animals with the peptic digestion step.

At present it is probably presumptuous to suggest ratios of the two food chains. In gross primary production estimates based on the classical concept of light and dark bottles, assumptions of equal bacterial respiration rates and equal photorespiration in light and dark are included. New approaches and methodologies will be required in the near future to answer this classic limnological question. Use of antimetabolites (Golterman, 1971) may be attractive, but their side-effects must be considered at each step in considerable detail (Corbett, 1974). Short-time kinetic analysis of processes involved is likely to be less objectionable.

8.4. External influences and feed-backs

8.4.1. Catchment/water body relationship

In recent years we have begun to understand water bodies as parts of larger systems (for example the biomes of American IBP programmes) and particularly as parts of their respective catchments. Two lines of enquiry become evident, both related to speeding of nutrient cycles by man. The input of nutrients in sewage is recognized as a major source of pollution and eutrophication of lakes; the feasibility of lake restoration by sewage diversion and by sediment removal has been demonstrated by Edmondson (1972a) and Björk (1972) respectively. Modern agriculture has been identified as the major contributor of nitrate, by releasing in this way about 20–25% of the nitrogen supplied each year. Other nitrogenous compounds do not apparently escape at a comparable rate. Phosphorus does not seem to escape from fields in significant amounts – only a few per cent, or less, of the amount supplied (Procházková, 1975a, b; Gächter, 1973). A small increase in P-release would however cause major disturbance of slightly eutrophied waters (Straškraba, Chapter 3). At present the major phosphorus release seems to be through domestic and industrial sewage, though some phosphorus may be brought in by erosion from steeply sloping fields and pastures (Gächter, 1973). The concept of load – the amount of nutrients brought to a lake per unit area and time (Vollenweider, 1975; Dillon & Rigler, 1975) seems promising as a practical guide for water management.

8.4.2. Components of nutrient metabolism

Before the start of IBP a series of papers suggested limitation of production by micronutrients (Goldman, 1964, 1972). Parallel to IBP in America there was the great 'eutrophic' discussion on the relative significance of major nutrients–carbon, nitrogen and phosphorus–for limitation of phytoplankton growth and as the main cause of eutrophication (Schindler, 1975). Ideas on limitations may be briefly summarized as follows: phosphorus is apparently the most frequently limiting nutrient, though nitrogen and carbon may limit growth in some special situations. In technologically advanced areas most phosphorus is supplied in domestic sewage and phosphorus-containing detergents, very little comes from arable land. Micronutrients may limit growth of phytoplankton in some sparsely populated areas of crystalline geological formations. Their significance will decrease in the future, as man's activity is likely to circulate more and more nutrients through the biosphere.

Calculations suggest that – except in lakes with a very high and constant load of phosphorus – measured concentrations of phosphorus can sustain measured values of primary production for only very limited periods:

$$C:N:P \text{ in biomass of algae} = 100:12:1$$
$$1 g\, C\, m^{-2}\, d^{-1} \rightarrow 8.3\ mg\, P\, m^{-2}\, d^{-1}$$
$$z_{eu} = 3m; \rightarrow 2.8\ mg\, P\, m^{-3}\, d^{-1}$$
$$PO_4-P = 10\ mg\, m^{-3} \rightarrow 3.6\ days$$

Stumm & Morgan (1970) have shown that the solubility of total inorganic phosphorus is governed primarily by interactions of phosphate with Ca^{2+}, OH^-, Fe^{3+} and possibly Al and Si. Similar reactions evidently regulate leaching of phosphate from fertilized fields and retention of phosphorus in reservoirs and lakes, which is proportional to its concentration (Procházková, Straškrabová & Popovský, 1973). Rigler (1966, 1968) has demonstrated conclusively that a very small fraction of total dissolved inorganic phosphorus is available as substrate for algae; presumably only one or a few forms from the whole array occurring in natural waters may be taken up by organisms. The relationship of phosphate uptake rates and pH given by Button, Dunker & Morse (1973) for *Rhodotorula* may partly support this hypothesis. In the analytical procedure the addition of molybdate at pH *ca* 2 may cause drastic shifts in most of these ionic equilibria, and thus much more phosphate is determined as molybdenum blue reactive phosphorus than is available to microorganisms.

At least three cycles of phosphate exchange have been identified in fresh water:

(1) release of phosphate from sediments
(2) uptake and release of phosphorus by sestonic particles
(3) excretion of phosphate by zooplankton.

In the first mechanism only part of the sedimented phosphate is returned to the water column, but the bulk of sedimented phosphate is trapped permanently in the sediments. Most phosphorus is released during winter and this results in high spring concentrations, with spring algal blooms, and it characterizes the level of algal activity during the following summer (Dillon & Rigler, 1974b).

The second mechanism is the fastest one, with turnover times frequently of the order of 15 minutes (Rigler, 1964), it is however the least well understood, particularly its metabolic functions. Recently Lean (1976) suggested that fragments of larger particles (organisms) may be involved in measurement of this exchange rate.

The third and best-understood mechanism is phosphate excretion by zooplankton. It became clear that data from the upper end of the range reported in the literature – about 2–4 μg P, $(\mu$g N$)^{-1}$h^{-1} at 20 °C – may be measured by both isotopic and colorimetric methods (Peters & Rigler, 1973; Ganf & Blažka, 1974; P. Blažka & Z. Brandl, unpublished). Lower rates reported by some earlier workers (for references see Ganf & Blažka, 1974; Peters, 1975b) were evidently due to the presence of phytoplankton during measurements of excretion rates by colorimetric procedures. It was also clearly established that phosphorus released by zooplankton is available to phytoplankton and is, in fact, used instantaneously (Peters & Lean, 1973; Ganf & Blažka, 1974).

For nitrogen it has been shown that ammonia is preferred as a nitrogen source by plants and that analytical concentrations of ammonia determine the extent of uptake of other substances (Fitzgerald, 1968; Procházková, Blažka & Králová, 1970). Also nitrogen fixation by blue-greens has been demonstrated at low concentrations of other nitrogenous compounds (Horne & Viner, 1971).

Recirculation mechanisms include ammonia release from the sediments and by the zooplankton. Similar mechanisms within the seston are most likely, but have not so far been demonstrated conclusively. Zooplankton excretion rates of ammonia are about five times higher than the phosphorus rates, but are about two times lower relative to phytoplankton requirements (Ganf & Blažka, 1974). Excretion rates of ammonia increase through the production chain, as does the percentage of protein in the biomass (Blažka, 1966).

8.4.3. Relationships between inorganic nutrient input, algal density, herbivore density and residual inorganic nutrient*

The relationships between inorganic nutrient supply and the first two trophic levels have long been understood in a very approximate manner but the dynamics of such a system cannot be explored adequately by simple arithmetic when non-linear relationships exist. An exploration of such a system was accomplished by the use of a mathematical model based on experimental data from a pair of two-stage continuous cultures of the alga *Chlamydomonas reinhardti* and the herbivorous protozoan *Tetrahymena vorax*.

To interpret the data, one needs a few details of the culture method. The inorganic culture media were designed to have nitrate as the sole limiting nutrient; the phosphate concentration was varied along with the nitrate concentration in a molar ratio of 12.5 N to 1 P, and phosphate and all other requirements were assumed to be present in excess. The algal cells provided the nutritional support for the protozoa since the protozoa could not grow in the initial inorganic media nor in the cell-free, spent medium. The two-stage culture unit had an axenic algal culture in the upstream flasks which overflowed into the downstream flasks. In one of the two-stage culture units, protozoa had also been added to the downstream flask. Thus the algal cells were protected from predation in the upper flask and could be introduced into the downstream flask in high concentration. The flow served both to import nutrient and to export cells. The culture volume in each flask was 500 ml. The culture method has been described in detail in Taub & McKenzie (1973). The results of batch and continuous cultures of these organisms which formed part of the basis for this model have been published (Taub, 1969a, b, c). For reviews of continuous culture methods and the calculations for multistage continuous cultures, see Kubitschek (1970) and Herbert (1964) respectively.

One needs to understand also the major assumptions and parameters of the model. The algal growth rate was based on a modified Michaelis–Menten equation that included the effects of nitrate concentrations and the effective light concentration as calculated by Beer's law to account for self-shading. The maximal growth rate (but not a standard μ_{max}) was 4.13; the K_s was 2.25×10^{-3} mM NO_3^-; and the K_L was 5500 (arbitrary light units). The

---

* Prepared by F.B. Taub.

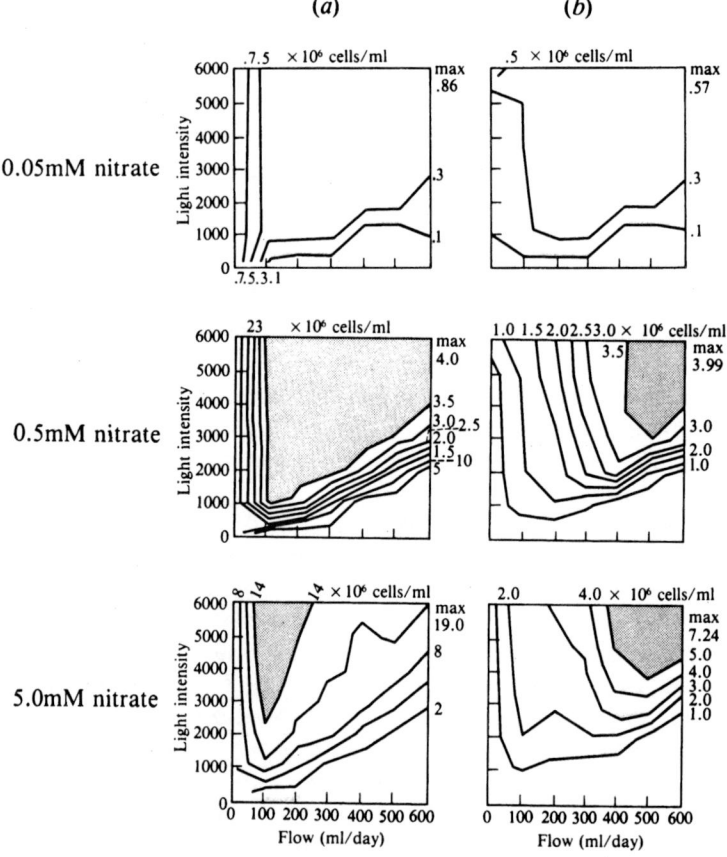

protozoan growth rate was based similarly on the Michaelis–Menten relationship with algal cell concentration as a substrate: the μ_{max} was 1.3, and the K_s was 7.2×10^5 algal cells. The resultant relationship was similar to an Ivlev saturation curve (Sushchenya, 1970). The relationships between the algae and protozoa were basic Lotka–Volterra equations modified to include the effects of inflow and washout to represent the chemostat environment. These basic relationships were modified by numerous interactions as indicated by experimental data. A portion of the algal cell protein consumed by the protozoan was assumed excreted as ammonia, and this, in turn, was assumed available for algal uptake preferentially to available nitrate. The protein of the algal cell varied with the available nitrate concentration. The per cent protein of the algal cell modified the μ_{max} of the protozoa and the number of algal cells consumed per protozoan produced. A light-induced mortality affected the protozoa. Time lags and several other minor features were included in the model. The model generally predicts a smooth approach to a steady state

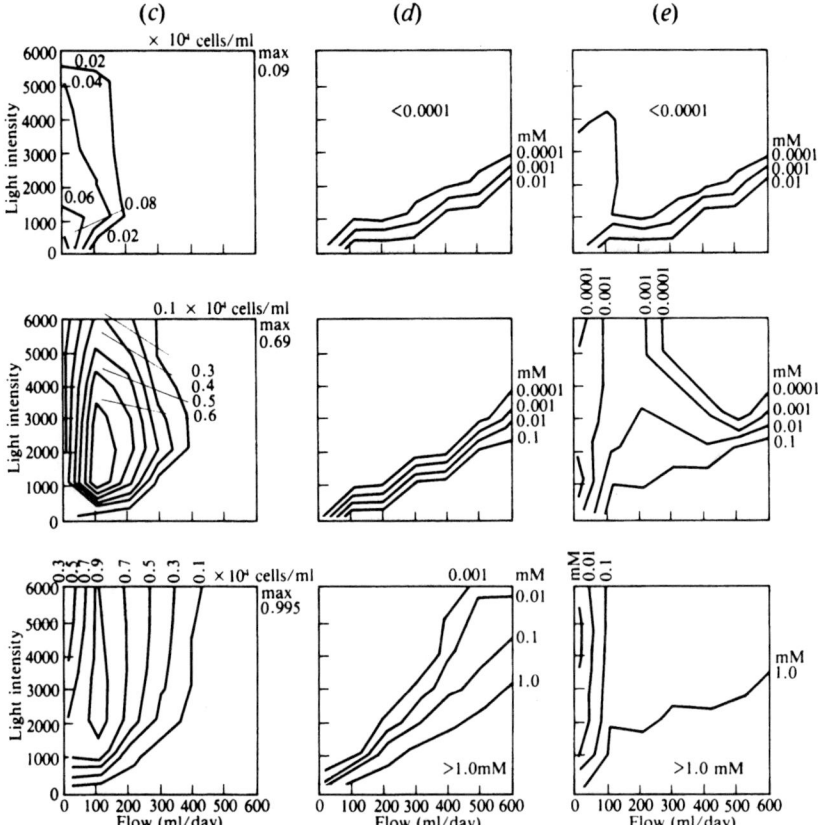

Fig. 8.3. The concentrations predicted by the model for low, moderate and high nutrient concentrations (0.05, 0.5 and 5.0 mM nitrate). The isopleths represent the concentrations at approximate steady states (24 days).

(a) Algal densities in the absence of herbivores (shown as isopleths). A high density of lines indicates instability, i.e., a rapid change in cell density due to small environmental changes, whereas an absence of lines indicates stability, i.e., a plateau over which changes in cell density would be slight despite major changes in the environmental conditions.

(b) The algal density in the presence of a herbivore.

(c) Protozoan density in the community shown in (b).

(d) The residual nitrate concentration in the axenic algal community shown in (a).

(e) The residual nitrate concentration in the algal–herbivore community shown in (b) and (c).

condition; major predator–prey cycles do not occur. Under some conditions, e.g., very slow flow rates, the true steady state is approached so slowly that a practical time limit of 24 days was used. The details of the model are described elsewhere (McKenzie, 1975).

The results, shown in fig. 8.3, should be considered as provisional model results of approximate steady states, not as experimental data. The model was run until densities appeared to be stable, maximum 24 days.

Somewhat different values would have occurred if the model had been run to the true stability, but these long time periods would be impractical to verify experimentally. The figures demonstrate how a system would respond if the organisms behaved according to the assumptions and estimated parameters of the model. Although virtually all of the assumptions and parameters were experimentally derived, they do not reflect all of the adaptive capabilities of organisms.

At the lowest concentration of limiting nutrient, 0.05 mM NO_3^-, fig. 8.3(a), it can be seen that the maximum algal concentration would occur only under the restricted conditions of extremely low flow rates, almost independently of light intensity. Although the protozoa reduced the maximal algal densities, fig. 8.3(b), they had virtually no effect on the algal concentrations under the remainder of the environmental conditions. The protozoa densities, fig. 8.3(c), were quite low, but they were greatest at extremely low light intensities and extremely low flow rates. The residual nitrate, fig. 8.3(d), was appreciable only where cells could not accumulate to use it, where light was low and flows high. The effect of the protozoans on residual nitrate, fig. 8.3(e), occurred only under restricted environmental conditions and at concentrations difficult to detect. For broad expanses of environmental conditions stability existed, i.e., concentrations remained relatively unchanged despite changes in light or flow. These results occurred because at these very low nutrient concentrations, the algae grow slowly and accumulate only at low dilution rates. Similarly, at these low algal densities, the protozoa grow slowly and cannot persist at high dilution rates.

A ten-fold increase in the concentrations of limiting nutrient to 0.5 mM NO_3^- increased the maximal algal density 4.7-fold and shifted the conditions yielding the maximal density (fig. 8.3(a)). The decrease in density at low flow rates was due to the reduced input of nutrient, and slow replacement of dead cells. At higher flow rates, the density was very responsive to changes in light intensity. At the plateau of maximal density, relative stability prevailed. The presence of the protozoans had a marked effect on algal density over much of the environmental range (fig. 8.3(b)). Comparing figs. 8.3(a) and (b), we see that although the same maximum density could be obtained, it could be maintained in the presence of the protozoa only over a much smaller environmental range: very high flow rates and very high light intensities. The protozoan density was 7.7-fold that obtained at the lower nutrient concentration and the distribution of the density relationship was shifted so that the maximum occurred at moderate light intensities and 100-ml/day flows and decreases in density occurred in all other directions. Since the protozoa were able to have a major effect on the algal density over most of the range, they were able also to have a major effect on the residual nitrate concentration, as can be seen by a comparison of figs. 8.3(d) and (e). Under environmental conditions where the protozoa had little effect on the algae, they also had little effect on the residual nitrate.

Another ten-fold increase in the limited nutrient to 5.0 mM NO_3^- resulted in a 4.8-fold increase in maximal algal density and another shift in the position of the maximum, fig. 8.3(*a*). The protozoa were able to reduce the concentrations of algae over the entire range of environmental conditions, fig. 8.3(*b*). The protozoan density distribution was very much like that at the moderate nitrate concentrations although the densities were 1.4-fold higher. The residual nitrate concentration was most dramatically increased over those ranges where the algal concentration was most reduced. Part of the shift in nitrate concentration due to protozoan feeding lies in the assumed preferential uptake of the ammonia that would arise from excretion by the protozoa and the comparatively smaller number of cells that would pick up the residual nitrate.

As can be seen from these inclusive model results (it would have taken almost 20 years of experimental data to collect enough values for these relationships to have been drawn empirically), universal general relationships could not have been obtained at any single light intensity, any single flow rate, nor any single nutrient concentration. The relationships between the concentration of nutrient and the densities of algae and protozoa are not linear. The systems tended to remain stable under changes in flow rate only as long as the organisms had unutilized capacity to grow. As they approached their maximal growth rates, further increases in flow resulted in lower concentrations of organisms and higher concentrations of residual nutrient. (Taub, 1979).

Although these data are only provisional model results based on constants derived from cultures of the single alga *Chlamydomonas reinhardti* and the single protozoan *Tetrahymena vorax*, the relationships are extremely similar to those postulated generally between an alga and a herbivore. Therefore it is felt that these results have a general significance and that the model can be used as a means of exploring the probable relationships between other kinds of organisms.

8.4.4. Relationship of load and recirculation of nutrients

Two aspects of nutrient metabolism in lakes and reservoirs have been considered in the preceding sections:

(*a*) Release of nutrients from the sediments during winter, followed by a high spring input resulting, in temperate zones, in a high spring concentration and algal blooms which are not controlled by zooplankton. This lack of control, however, means a lack of support by excretion, and the bloom eventually collapses, part of the phosphorus is immobilized and lost from the open water back to the sediments.

(*b*) In summer the recirculation mechanisms, both within the seston (the fast cycle) and between particulate matter and the animals, keep the nutrients cycling and prevent their loss from the epilimnion. In lakes with a high nutrient load – high concentration and low flow or low concentration and high flow –

the recirculation mechanisms are of less importance. At this stage the weak point of the load concept becomes evident – in the first case there is a strong algal bloom, but in the second case there need not be any.

Our knowledge is, however, limited for understanding the real relationships of both aspects – input and cycling. The study of transition periods – in late spring and autumn in temperate lakes – might therefore help to expand our understanding of nutrient cycles in general.

8.5. Conclusions

There is growing evidence for a higher efficiency (*ca* 15–20%) of transfer from primary to secondary production in most natural aquatic ecosystems, instead of the 10% hitherto assumed. Fish production in lakes and reservoirs may be comparable with that in fish-ponds (per unit area) but only a small fraction of it is of edible and harvestable fish.

In primary production studies a new methodology is required to include progress made in plant physiology in the last ten or fifteen years.

In the field of secondary production populations have so far been assumed to be in equilibrium, methodology of transient phenomena is now being developed (Canale, 1969; Curds, 1971; Legner, Punčochář & Straškrabová, 1976) and its use may expand to include metazoan populations. An alternative approach involving substrates (nutrients and organic matter) comes from models based on Monod equations (section 8.4).

Knowledge of how to supply protein food for the human population will probably need more emphasis on the population side of future production studies. Environmentally orientated research will be concerned primarily with cycles of nutrients and organic matter, but the two approaches are complementary and will necessarily be brought together (though on different sides) in general models.

More academic research will necessarily contribute to one or other side of these problems, but our budgets and the numbers of those able to continue the work will depend on our ability to answer also the practical problems of mankind.

The work reported in section 8.4.3, by F.B. Taub, was supported in part by the Environmental Protection Agency, Grant No. 16050 DXM, and in part by the National Science Foundation, Grant No. GB-20963, to the Coniferous Forest Biome, International Biological Programme. (This is Contribution No. 008 from the Coniferous Forest Biome and Contribution No. 476, College of Fisheries, University of Washington, Seattle, Washington.)

9. Estimating the productivity of lakes and reservoirs*

M. BRYLINSKY

9.1. Introduction

Since its beginning a major goal of IBP/PF has been to collect comparable information on a diversity of freshwater ecosystems. This information could then be analysed with a view to identifying those factors most important in controlling biological production. Once identified, management efforts could be directed towards manipulation of those factors appearing most important. As is exemplified by the contents of this volume, numerous approaches can and have been employed to achieve this goal. The approach adopted in this chapter has been to consider basic abiotic characteristics of a freshwater system in an attempt to identify variables having predictive value with respect to the system's biotic characteristics. The basic premise involved in this sort of approach is that useful information on the biological functioning of an ecosystem can be obtained by consideration of simple abiotic characteristics describing the structure of that system. This, of course, does not imply that the details of a system's structure, such as the kinds of organisms present, or the spatial and temporal distribution of these organisms is not important, but only that, on a different level valuable information can be obtained by comparing systems on the basis of simple characteristics.

Two popular techniques for analysing systems in this way are statistical analyses and dynamic modelling of whole ecosystems. The dynamic modelling approach is considered in Chapter 10. This chapter deals with the results of statistical analyses.

The use of statistical procedures to compare lakes and identify variables that may be useful in estimating production is not new to limnology. Many limnologists have used correlation and regression analysis to formulate models that could be used to estimate production (Deevey, 1941; Rawson, 1952; Northcote and Larkin, 1956; Hayes, 1957; Ryder, 1965; Jenkins, 1968; Schindler, 1971b; and others). Most of these analyses, however, have been confined to a relatively narrow range of lakes in terms of both location and lake type. As a result, many environmental variables, particularly those related to climate, remain constant within this narrow range and their effect cannot be adequately evaluated. Through the coordinated efforts of those involved in IBP/PF, data have become available on a diversity of lakes, in terms of lake

---

* This chapter considers the preliminary analyses of IBP data based on Data Reports received before March 1973, so figures are not always in agreement with those in other chapters based on more complete data (eds.).

Table 9.1 *Lakes and reservoirs included in the analysis*

Data Report[a]	LAKES	Phytoplankton Production (kJ m⁻² a⁻¹)	Data Report[a]	LAKES	Phytoplankton Production (kJ m⁻² a⁻¹)	Data Report[a]	LAKES	Phytoplankton Production (kJ m⁻² a⁻¹)
North America			**Europe**			**Europe (cont.)**		
11	Sammamish	6 490	1	V. Finstertaler See[b]	1 130	KD	Drivyaty[b]	6 570
11	Chester Morse	3 300	2	Øvre Heimdalsvatn	310	74	Maggiore	10 460
15	Katepwa	22 590	3	Loch Leven	29 870	—	Frederiksborg Slotsø	26 360
15	Buffalo Pound	11 300	28	Trummen	11 800	—	Furesø	14 230
15	Pasqua	17 990	32	Port-Bielh	940	—	Borresø	17 570
15	Echo	8 370	38	Mikolajskie[b]	14 080	—	Sollerød Sø	25 940
60	Aleknagik[b]	610	38	Sniardwy[b]	13 800	—	Grane Langsø	3 640
25	Lawrence	2 290	38	Tajtowisko[b]	13 280	—	Vombsjon	16 740
12,13,17,23	Wingra	35 560	38	Flosek[b]	7 730			
KD	Clear (Canada)[b]	14 640	38	Warniak[b]	9 000	**Asia**		
KD	Coon	2 260	39	Hakojärvi	210	34,48	Dal	3 190
KD	McLeod	1 360	43	Stechlinsee	3 870	34	Anchar	2 320
KD	Winnipeg	1 090	58,64	Chedenjarvi[b]	4 435	34	Manasbal	5 250
KD	Great Bear	670	61	Zelenetskoye[b]	90	85	Dalnee[b]	8 370
KD	ELA–239[b]	4 980	61	Akulkino	30	63	Karakul[b]	2 510
—	Hertel	3 600	62	Krasnoye[b]	4 020	KD	Ooty	54 180
—	Crater	418	65	Dusia	4 180	KD	Kodaikanal	5 770
—	Clear (United States)	7 320	65	Obelija	5 020	KD	Yercaud	41 340
—	Fayetteville Green Lake	12 970	65	Galstas	3 680	KD	Kasumigaura	5 860
—	Malikpuk	1 360	65	Slavantas	2 800			
—	Imikpuk	520	68	Naroch[b]	2 200	**Australia**		
—	Ikroavik	540	68	Myastro[b]	6 690	14	Coragulac	14 770
—	Sylvan	25 770	68	Batorin[b]	6 860	14	Corangamite	28 030
—	Goose	12 050	70	Krivoye[b]	380	51	Leake	3 770
—	Oliver	5 440	71	Krugloye[b]	140	51	Tooms	2 930
			72	Bolshoy Kharbey[b]	590	51	Sorell[b]	8 370
Africa			33	Esrom	8 790	51	Crescent[b]	20 500
4	Chad	4 500	50	Neusiedlersee	2 090			
5	Chilwa	6 400	82	Pääjärvi[b]	730			
			KD	Kurakhov[b]	13 600			
				Red[b]	—			

RESERVOIRS

North America		
—	Waco	8 370
Europe		
49	Lipno	7 030
—	Slapy	5 100
—	Kličava	5 130
69	Rybinsk[b]	2 930
66	Kiev[b]	5 070

RESERVOIRS

Asia		
67	Bratsk[b]	5 650
KD	Amaravathi	46 020
KD	Stanley	33 050
KD	Bhavanisager	33 250
KD	Sathanur	32 970
KD	Krishnagiri	22 800
KD	Sandynulla	33 010

[a] For Data Reports see Appendix I. – indicates lakes studies outside of IBP. KD indicates data obtained from papers presented at Kaziemierz – Dolny Symposium.
[b] Indicates secondary production data available.

type as well as geographical range, and there now exists a unique opportunity to compare lakes on a much larger scale.

9.1.1. Sites considered in the analysis

The sites included in the analysis are listed in table 9.1. The list consists of all those IBP sites for which data reports had been submitted to IBP headquarters before March 1973, and which contained, at a minimum, information on total primary production for one growing season. In addition to those studied by participants of the IBP, twenty-two other sites* which have been described in the literature were included to produce a total of ninety-three lakes and reservoirs for analysis. They are distributed over a wide geographical area (fig. 9.1), ranging from 38°S to 71°N. Most, however, are located in the northern hemisphere. The range in altitude is 0–2237 m above sea level. The variation in size is also great. The smallest has a surface area of 0.01 km^2, and the largest 24 000 km^2. Maximum depths range from 2 to 450 m. The range in productivity is equally large. The lowest phytoplankton production reported is less than 200 kJ m^{-2} a^{-1} and the highest is greater than 40 000 kJ m^{-2} a^{-1}.

An earlier analysis along the same lines (Brylinsky & Mann, 1973) had been carried out on data presented at the IBP symposium on 'Productivity Problems of Freshwaters' (Kajak & Hillbricht-Ilkowska, 1972). It used data from only fifty-five sites, obtained before 1970. The present analysis uses data from more sites, over a longer time span.

9.1.2. Variables considered in the analysis

The variables included in the analysis are listed in table 9.2. They have been classified into groups based on the role they would be expected to play in influencing lake biology. There are three major classifications; driving variables, site parameters and state variables. *Driving variables* include those factors having an effect on a system but which are not in turn affected by the system. Examples would include incident radiation, precipitation, wind and nutrient input from the drainage area. *Site parameters*, or site constants, include those variables which are specific for and do not vary within a particular site. They are considered constants because, for all practical purposes, they do not vary within any one particular site. They are, however, variables since their magnitude varies between sites. Examples of site parameters would include variables describing lake morphometry such as mean depth, surface area and volume. *State variables* are those we wish to make predictions about. Theoretically, their variations are a result of the

---

* See papers by: Wright, 1961; Goldman & Wetzel, 1963; Wetzel, 1964, 1966a, b; Sakamoto, 1966; Culver & Brunskill, 1969; Whiteside, 1970; Gelin, 1971; Howard & Prescott, 1971; Kalff, 1972; Kimmel & Lind, 1972; Larson, 1972.

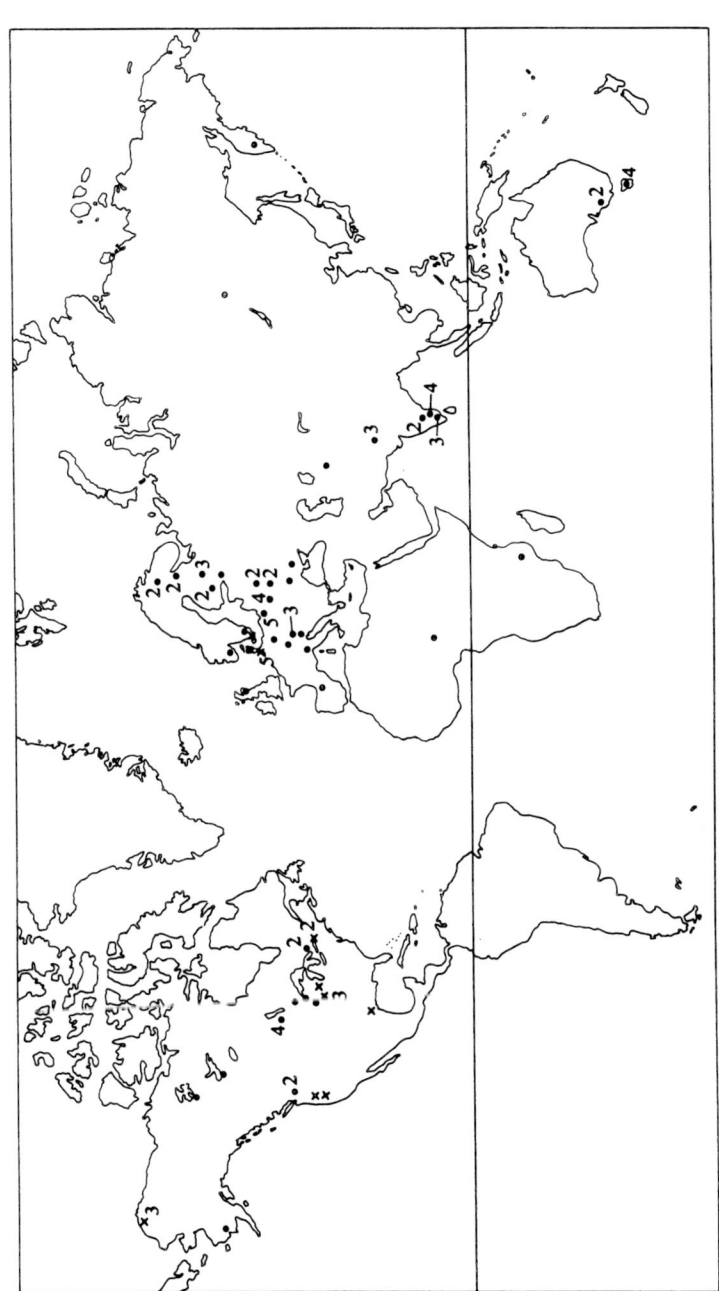

Fig. 9.1. Distribution of lakes and reservoirs studied in the analysis (listed in table 9.1). Numbers indicate several sites in the same locality. •, sites studied within IBP. ×, sites studied outside of IBP.

Table 9.2 *Variables included in the analysis*

Variable	Units
Driving variables	
Visible incident radiation	kcal m^{-2}
Daylength range	hours
Air temperature	°C
Precipitation	cm a^{-1}
Site parameters	
Latitude	degrees
Altitude	m above MSL
Mean depth	m
Maximum depth	m
Surface area	km^2
Volume	km^3
Drainage area	km^2
Development of volume (maximum depth/mean depth)	—
Specific drainage area (drainage area/surface area)	—
Drainage area/volume	m^{-1}
State variables	
Physical characteristics	
Retention time	years
Duration of ice cover	days
Duration of summer stratification	days
Maximum depth of summer thermocline	m
Mean depth of summer thermocline	m
Epilimnion mean temperature	°C
Epilimnion maximum temperature	°C
Hypolimnion mean temperature	°C
Hypolimnion maximum temperature	°C
Secchi mean depth	m
Secchi maximum depth	m
Chemical characteristics	
Total dissolved solids	mg l^{-1}
Conductivity	mohms cm^{-1}
Alkalinity	mg l^{-1}
Calcium	mg l^{-1}
pH mean	—
pH minimum	—
pH maximum	—
Total phosphorus mean	mg l^{-1}
Phosphate mean	mg l^{-1}
Total nitrogen mean	mg l^{-1}
Nitrate mean	mg l^{-1}
Epilimnion oxygen mean	mg l^{-1}
Epilimnion oxygen minimum	mg l^{-1}
Hypolimnion oxygen mean	mg l^{-1}
Hypolimnion oxygen minimum	mg l^{-1}
Biological characteristics	
Phytoplankton production/growing season	kJ m^{-2}

(contd.)

Table 9.2 (*contd.*)

Variable	Units
Phytoplankton production/growing season	kJ m^{-3}
Phytoplankton biomass – growing season mean	kJ m^{-2}
Phytoplankton chlorophyll *a* – growing season mean	mg m^{-2}
Phytoplankton *P/B* – growing season	—
Phytoplankton production/chlorophyll *a*	kJ mg chl. a^{-1}
Phytoplankton photosynthetic efficiency	—
Herbivorous zooplankton biomass	kJ m^{-2}
Herbivorous zooplankton production/growing season	kJ m^{-2}
Herbivorous zooplankton *P/B*	—
Herbivorous zooplankton efficiency	—
Carnivorous zooplankton biomass	kJ m^{-2}
Carnivorous zooplankton production/growing season	kJ m^{-2}
Carnivorous zooplankton *P/B*	—
Carnivorous zooplankton efficiency	—
Herbivorous benthos biomass	kJ m^{-2}
Herbivorous benthos production/growing season	kJ m^{-2}
Herbivorous benthos *P/B*	—
Herbivorous benthos efficiency	—
Carnivorous benthos biomass	kJ m^{-2}
Carnivorous benthos production/growing season	kJ m^{-2}
Carnivorous benthos *P/B*	—
Carnivorous benthos efficiency	—

interactions between driving variables and site constants. All of the biological variables fall into this category.

In statistical terms the site parameters and driving variables are usually the independent variables and the state variables are the dependent variables.

Most of the variables listed in table 9.2 are simple measures commonly used to describe freshwater systems and therefore require little explanation. A few, however, require some comment as to their meaning or their derivation. Daylength range was chosen as an index of the amount of variation in daylength between sites located at different latitudes. It was simply calculated as the difference between the longest and shortest day in a single year, the values of which were obtained from *Smithsonian Meteorological Tables* (Smithsonian Institution, 1966). Visible incident radiation was calculated as half of total incident radiation. Mean annual precipitation and mean annual air temperature, when not supplied in the original data, were obtained from the *Yearbook of Agriculture* (United States Department of Agriculture, 1941). Development of volume refers to the ratio of mean depth to maximum depth and specific drainage area is the ratio of the area of the drainage basin to the surface area of the lake. Mean depth of thermocline and duration of stratification both refer to summer conditions, winter stratification not being taken into account. All of the variables related to lake chemistry, with the exception of hypolimnion oxygen, are for surface waters only and represent mean values for the growing season. Growing season, when not defined in the original data reports, was taken as the time period beginning at the onset of the spring phytoplankton rise, and extending to the autumn decline.

Phytoplankton production was expressed as kilojoules per square metre of surface area and kilojoules per cubic metre of the euphotic zone, for the growing season. Where the carbon-14 method had been used, daily rates given by authors were unchanged but where the oxygen method had been used daily rates were reduced by 15% in an attempt to make them comparable with the carbon-14 figures. Integration over the growing season was done graphically, if it had not been done by the authors. For a discussion of the interpretation of primary production data see Chapter 5, section 5.6.1. In converting to kilojoules it was assumed that 1 g of carbon is equivalent to 40 kJ. When the light and dark bottle method was used to measure production and oxygen values reported, 1 g of oxygen was assumed to equal 14.6 kJ.

For secondary producers biomass was expressed as mean kilojoules per square metre in the growing season and production as kilojoules per square metre per growing season. Where conversions from other units were necessary, the factors used were from Winberg (1971a).

Phytoplankton photosynthetic efficiency was calculated as phytoplankton production divided by visible incident radiation, both expressed in kilojoules per square metre per growing season. The efficiency of each secondary producer was calculated as the ratio of phytoplankton production to the production of the appropriate secondary producer.

9.2. Methods

An analysis of this kind, because it deals with such a large number of variables, requires the aid of some sort of model describing the relationships thought to exist between the variables being considered. The types of model most often used in this analysis were hierarchical models, an example of which is shown in fig. 9.4. The models are developed on the basis of current theory describing how variables relate to one another and essentially represent a set of hypotheses that can be tested by statistical procedures. For example, in fig. 9.4(a) the variables considered to be important in affecting phytoplankton production are grouped into those related to energy availability and those related to nutrient availability. The energy related variables are then subdivided into incident radiation, cloud cover, turbidity and so on. These relationships can be tested statistically to determine the validity of the model and, if required, modified to conform with the results of data analysis. In use then, the models are continuously evolving and provide a framework to direct the analysis towards consideration of relevant associations between variables.

The statistical procedures used in conjunction with these models included simple correlation, partial correlation, multiple regression, cluster analysis and discriminant function analysis. Correlation analysis was used to determine the kind and degree of relationship existing between variables. As many of the variables in this study were correlated, partial correlation was essential to proper interpretation of the data. Multiple regression analysis served to evaluate how well the magnitude of one variable could be estimated from a knowledge of two or more other variables. Cluster analysis is a relatively sophisticated statistical procedure which served a number of purposes in this analysis. Briefly, cluster analysis places samples, or variables characterizing samples, into 'clusters' within which interaction is strong and between which interaction is weak, but not necessarily unimportant. One of its greatest advantages is that it provides a means for reducing the number of variables required for analysis since variables contained in the same cluster are highly correlated. In addition, it is especially helpful in determining whether some underlying pattern of relationships exist between variables in the data set.

Simple and partial correlation, multiple regression and discriminant function analyses were performed using programs contained in Statistical Packages for the Social Sciences (SPSS) and Biomedical Computer Programs (BMD). The cluster analyses were performed using a program developed by Dr Richard A. Park of the Geology Department at Rensselaer Polytechnic Institute, Troy, New York.

9.3. Interrelations between abiotic variables

The abiotic variables used to describe each site are composed mainly of those related to climatic factors, lake morphometry and lake chemistry. An understanding of how these variables relate to each other is important for a proper interpretation of their effect on lake biology. With this in mind, cluster analysis has been used to examine these relationships.

Fig. 9.2 presents a dendrogram summarizing the results of a cluster analysis between the major morphological and physical characteristics of each site. The scale represents the degree of similarity between variables based on Sorenson's coefficient (Bray & Curtis, 1957), those having a high degree of similarity being highly correlated with one another. There are two major clusters represented. One contains variables closely related to latitude such as water temperature and daylength range. The other is composed of variables related to lake morphometry, particularly mean depth. The majority of remaining variables fall between these two groups indicating that they are related to both latitude and lake morphometry. Thus, latitude and mean depth are probably good integrative variables in terms of their ability to summarize a number of site characteristics.

Mean Secchi depth appears to be very dependent on lake morphometry since it clusters very strongly with mean depth. This is also true of retention

Variable

Per cent similarity

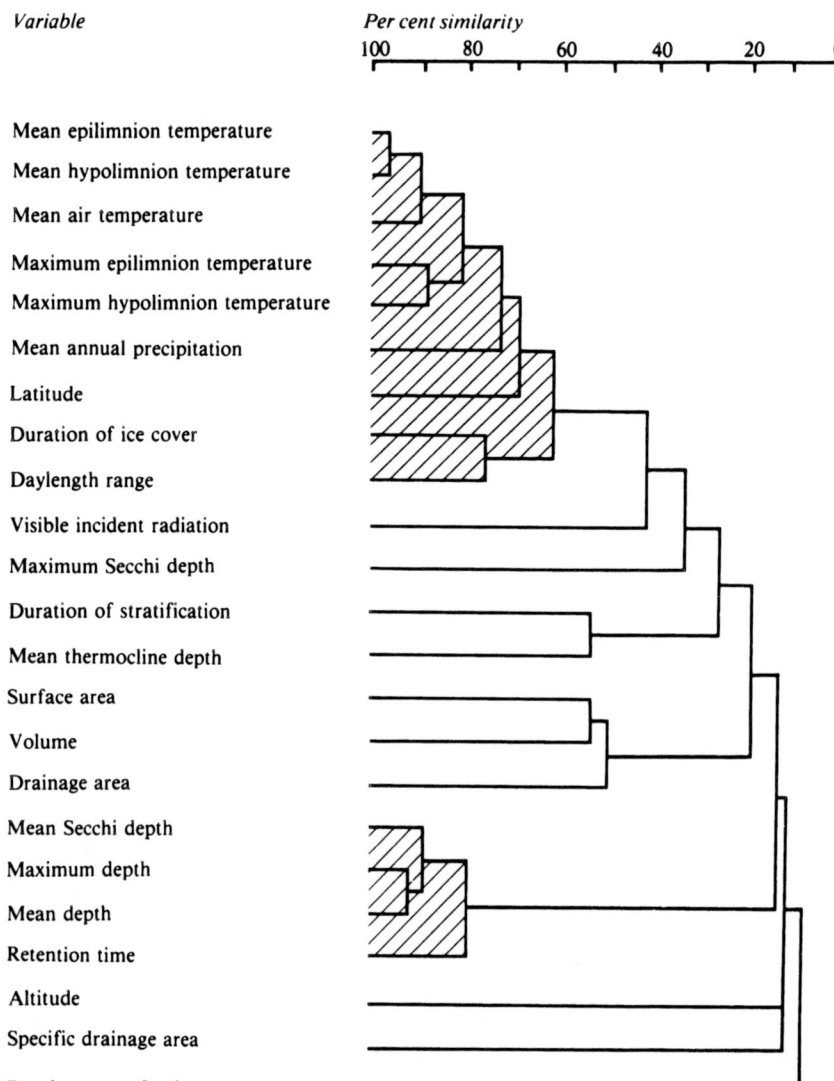

Fig. 9.2. Results of cluster analyses on morphological and physical characteristics. See text for explanation.

time. Maximum Secchi, however, seems to be also influenced by variables related to latitude. The degree of stratification, indicated by both duration of stratification and depth of thermocline, shows about equal amounts of clustering with variables related to lake morphometry and those related to latitude suggesting that it is dependent on both of these factors. Mean annual precipitation, interestingly, falls within the group of variables related to latitude. Its correlation with latitude is strongly negative (-0.4342, $N = 84$) indicating that, for the sites considered, low-latitude lakes receive greater amounts of rainfall per year.

Altitude, specific drainage area and development of volume are all relatively independent of the other variables considered. The lack of any strong relationship between lake morphometry and latitude or altitude indicates that, within the data set considered, there is no clear relationship between the morphological type of a lake and its location.

Fig. 9.3 is a result of a cluster analysis on variables related to lake chemistry together with a number of abiotic variables describing morphometry and physical characteristics. Conductivity, total dissolved solids and calcium concentration form one cluster which is in turn closely related to pH and alkalinity. Aside from this grouping, the other chemical variables appear to have complex relationships with other factors. Epilimnion and hypolimnion oxygen concentrations exhibit a strong relationship with each other but weak

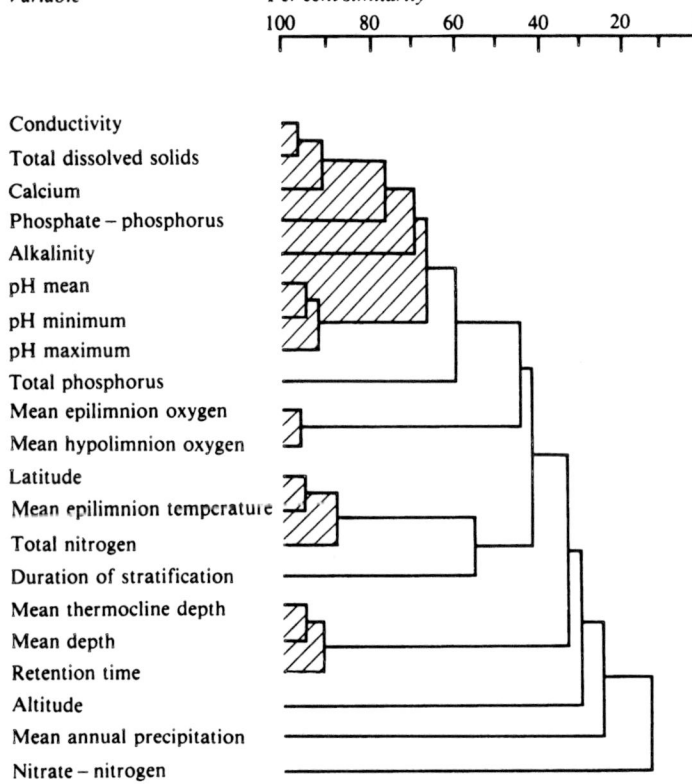

Fig. 9.3. Results of cluster analyses on morphological, physical and chemical characteristics. See text for explanation.

relationships with other variables. Total nitrogen shows a surprisingly strong relationship to latitude and water temperature, but a poor relationship to nitrate nitrogen. The latter shows no significant relation to any of the other variables considered. Total phosphorus and phosphate phosphorus are also poorly related to each other but both seem to be related to calcium and alkalinity.

Altitude, precipitation, retention time and lake morphometry, as indicated by mean depth, show no clear effect on water chemistry. As is the case with lake morphometry, there does not appear to be any clear relationship between latitude or altitude and lake chemistry.

9.4. Phytoplankton production

9.4.1. Interrelations between phytoplankton production and abiotic variables

In examining the relationship between abiotic variables and phytoplankton production the latter was considered to be dependent on two major factors: the availability of solar energy and the availability of nutrients. The models proposed to illustrate the possible relationships between phytoplankton production and abiotic factors are shown in fig. 9.4.

The simple correlations between the major abiotic variables and phytoplankton production, on both an area and volume basis, are listed in table 9.3. In general, phytoplankton production exhibits the strongest correlations with variables related to solar energy input, moderate correlations with variables related to water chemistry, and poor correlations with variables related to lake morphometry. This same trend is shown whether production is expressed on an area or volume basis. However, the correlations are greater for production per unit area.

Latitude, mean annual air temperature and mean temperature of the epilimnion are the variables correlating best with phytoplankton production. Visible incident radiation shows a surprisingly low correlation, as does altitude. If one wished to choose a simple abiotic variable to estimate phytoplankton production, latitude would be best.

The question arises as to what in particular is important about latitude that accounts for its high correlation with production. The most obvious explanation is the variation in intensity of incident radiation with latitude. However, as pointed out above, the correlation between visible incident radiation and phytoplankton production is relatively poor. Other factors that are closely related to latitude are air temperature and water temperature. Air temperature obviously cannot directly affect production and water temperature probably has little affect on adapted communities. Another possibility is that a number of factors associated with latitude may vary in such a way that

(a)

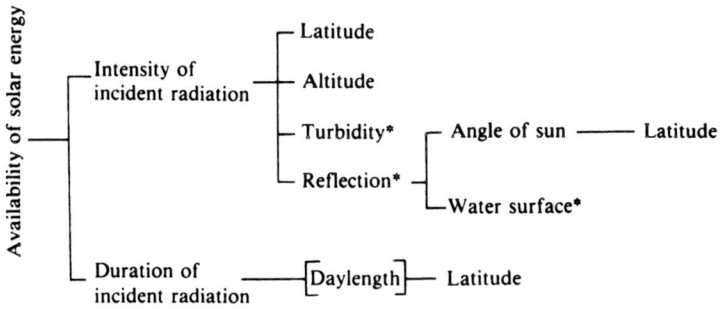

(b)

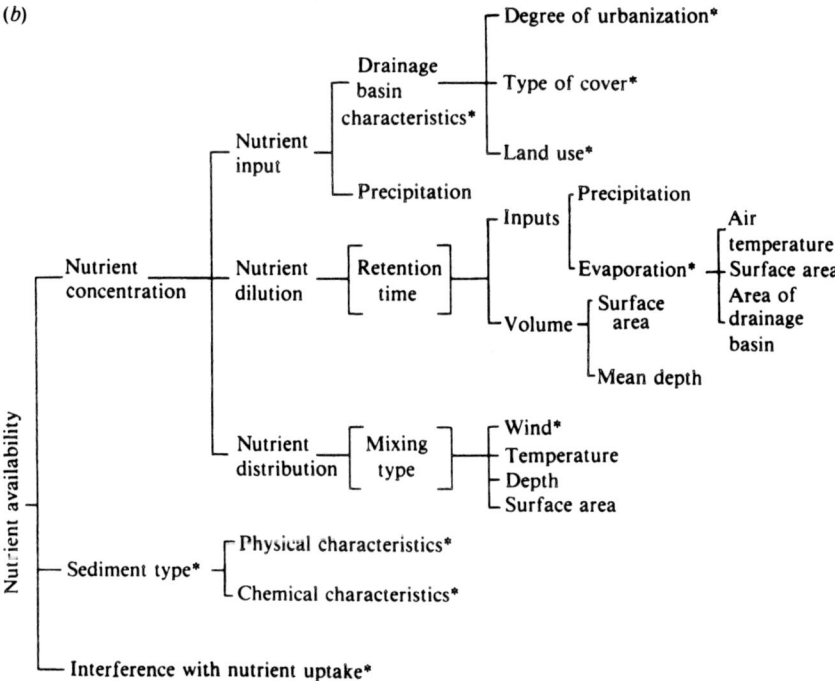

Fig. 9.4. (a) Hierarchical model of factors influencing phytoplankton production through solar energy availability. Asterisks indicate factors that are probably important but for which no data were available.

(b) Hierarchical model of factors influencing phytoplankton production through nutrient availability. Asterisks indicate factors that are probably important but for which no data were available.

Table 9.3 *Simple correlations between phytoplankton production and major abiotic variables related to solar energy input, water chemistry and lake morphometry*

Variable	N	Production m^{-2}	N	Production m^{-3}
Solar energy input				
Latitude	93	−0.6787[a]	61	−0.3345[a]
Altitude	93	0.1664[b]	61	−0.1311[b]
Visible incident radiation	93	0.3564[a]	61	0.1349[b]
Daylength range	93	−0.5219[a]	61	−0.3010[a]
Air temperature	93	0.6328[a]	61	0.3510[a]
Epilimnion temperature	40	0.6283[a]	35	0.3998[a]
Water chemistry				
Conductivity	65	0.2902[a]	46	0.1805[b]
pH mean	82	0.4570[a]	53	0.3533[a]
Total phosphorus	33	0.1732[b]	25	0.1301[b]
Phosphate phosphorus	65	0.1781[b]	48	0.2774[c]
Total nitrogen	27	0.7133[a]	19	0.7057[a]
Nitrate nitrogen	58	0.1431[b]	41	−0.0216[b]
Lake morphometry and physical factors				
Mean depth	91	−0.1673[b]	59	−0.1006[b]
Surface area	92	−0.1079[b]	59	−0.1863[b]
Volume	91	−0.1399[b]	59	−0.0554[b]
Drainage area	49	−0.0091[b]	40	−0.0023[b]
Depth of thermocline	66	−0.1461[b]	47	−0.1776[b]
Duration of stratification	51	−0.1654[b]	36	−0.1885[b]
Precipitation	84	−0.0411[b]	58	−0.0563[b]
Retention time	45	−0.2028[b]	40	−0.0838[b]

[a] Significant at 99% level; [b] not significant at 95% level; [c] significant at 95% level.

latitude acts as an integrator and combines the effects of these factors. However, identification of these factors would be difficult since, as mentioned previously, there is no indication of any consistent variation in either lake chemistry, lake morphometry, or physical factors with latitude.

One important factor not yet considered is the variation in daylength at different sites. Daylength range was the variable chosen to reflect this variation and its log is very strongly correlated with latitude ($r = 0.9634$, $N = 93$). Thus, there is a good possibility that the strong correlation between phytoplankton production and latitude is due primarily not to the variation in intensity of light but to the variation in time during which light is available.

Fig. 9.5 shows that in high latitudes the range of primary production values is small, while in low latitudes it is much greater. This may indicate that in high latitudes light is more important as a limiting factor, but in lower latitudes factors related to the availability of nutrients become more critical in determining the level of production obtained.

Most indices of water chemistry exhibit significant correlations with

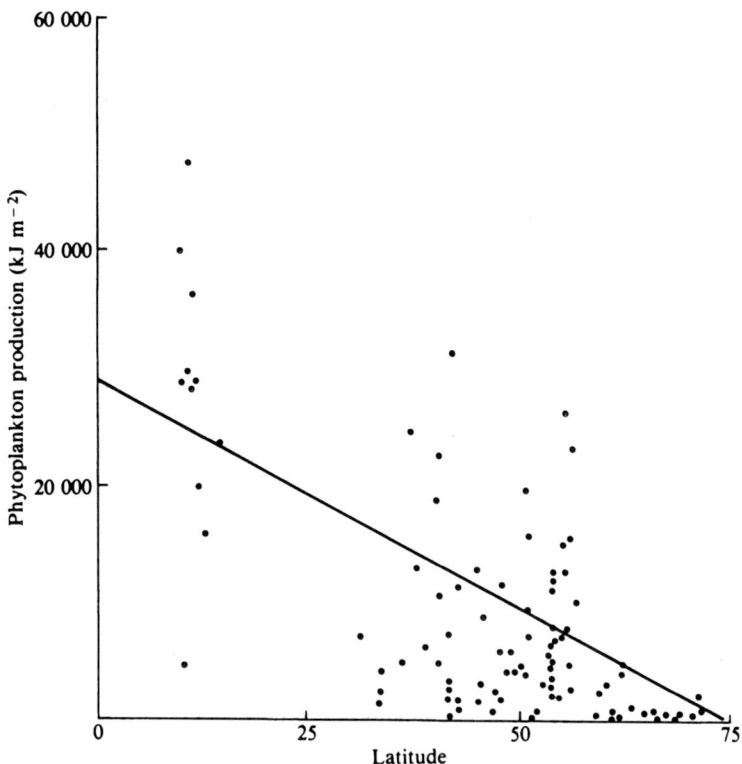

Fig. 9.5. Relationship of phytoplankton production to latitude.

phytoplankton production. Variables not showing any significance are nitrate-nitrogen and those related to phosphorus. Total nitrogen has a surprisingly high correlation with production, as does pH.

It is to be expected that production and pH are closely related since high levels of productivity would increase the pH of lake water as a result of carbon dioxide utilization. Thus in most cases, high pH is actually a result rather than a cause of high productivity. The strong correlation with total nitrogen and production, together with the poor correlation of total phosphorus and production, is not easily explained, especially since phosphorus is often considered to be the major limiting nutrient in lakes. This difference may somehow be related to the dynamics associated with these nutrients. Nitrogen has been shown to be a good indicator of phytoplankton biomass (Pavoni, 1969) and the latter is well correlated with phytoplankton production (see below). Phosphorus, however, has a very rapid turnover time in most aquatic systems (Pomeroy, 1960; Rigler, 1956). For this reason its standing stock value is probably a poor indicator of productive potential. The relationship between total nitrogen and phytoplankton production is illustrated in fig. 9.6.

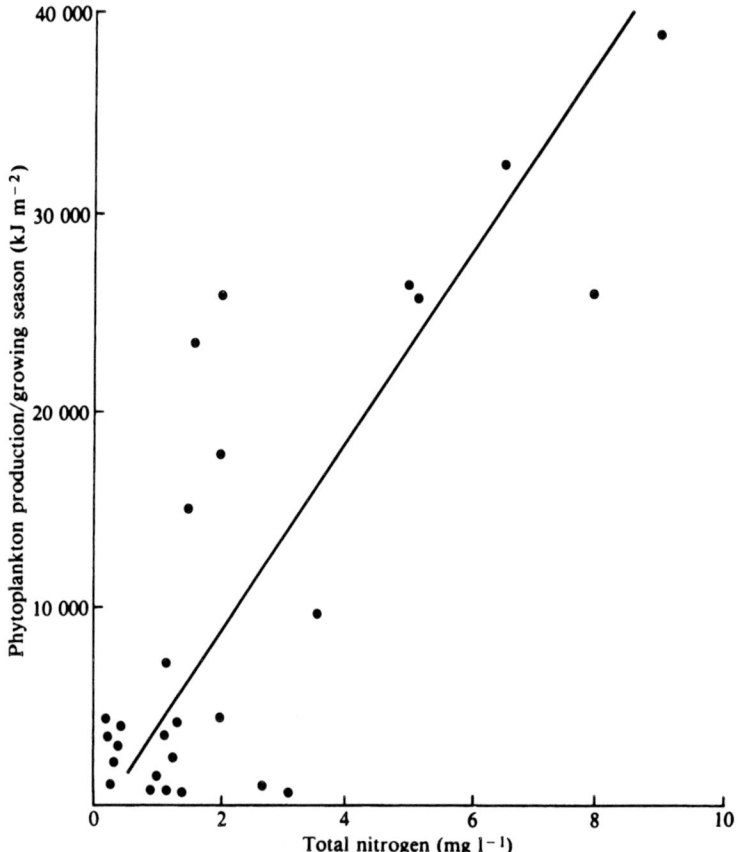

Fig. 9.6. Relationship of phytoplankton production to total nitrogen concentration.

Lake morphometry appears to have little direct influence on phytoplankton production. None of the correlations between morphological characteristics and production, on either an area or volume basis, are significant. This is not unexpected since the lakes being compared are distributed over a wide geographical range and factors related to energy availability and chemistry would certainly be more important than factors such as lake depth and surface area. The same is true of retention time and degree of lake stratification as indicated by thermocline depth and duration of stratification. These factors, when considered alone, show little correlation with phytoplankton production.

Because variations in latitude and factors associated with solar energy input exhibit a strong influence on phytoplankton production, the importance of abiotic factors that covary with latitude, in either a positive or negative manner, may be masked and their significance underestimated. To evaluate more adequately the importance of abiotic factors not directly related to solar energy input, partial correlations, after correcting for latitude, were calculated. These are shown in table 9.4.

After correcting for difference in latitude, the correlations of phytoplankton production with the variables daylength range, air temperature and water temperature become insignificant. Visible incident radiation, however, shows a significant negative correlation. This may result from a more efficient use of a given daily radiation income, if distributed over a longer daylength; i.e. light saturation inefficiency may be less at higher latitudes. Altitude assumes a greater importance and has a negative effect indicating that high altitude lakes tend to be less productive. Precipitation shows a very strong negative effect on production. If the importance of precipitation is related to nutrient input it would be expected to have a positive correlation with phytoplankton production. The fact that it is negative indicates that it is important for some other reason related, perhaps, to the extent of cloud cover and therefore light availability.

As a whole variables related to water chemistry exhibit the highest correlations with phytoplankton production after accounting for variations in

Table 9.4 *Partial correlations between phytoplankton production and abiotic variables after correcting for differences in latitude*

Variable	N	Production m^{-2}	N	Production m^{-3}
Solar energy input				
Altitude	93	-0.2317^a	61	-0.2417^a
Visible incident radiation	93	-0.2755^a	61	-0.2133^b
Daylength range	93	0.0910^c	61	-0.0726^c
Air temperature	93	0.1253^c	61	0.0912^c
Epilimnion temperature	40	0.1362^c	35	0.1119^c
Water chemistry				
Conductivity	65	0.3718^a	46	0.2432^b
pH mean	82	0.3669^a	53	0.3226^b
Total phosphorus	33	0.3081^b	25	0.2538^b
Phosphate–phosphorus	65	0.0612	48	0.1476^c
Total nitrogen	27	0.5803^a	19	0.3947^b
Nitrate–nitrogen	58	0.1394^c	41	-0.0125^c
Lake morphometry and physical factors				
Mean depth	91	-0.1434^c	59	-0.2788^b
Surface area	92	-0.1015^c	59	-0.1690^c
Volume	91	-0.0317^c	59	-0.0299^c
Drainage area	49	-0.1356^c	40	-0.0539
Depth of thermocline	66	-0.2094^a	47	-0.2390^b
Duration of stratification	51	-0.1434^c	36	-0.1656^c
Precipitation	84	-0.4708^a	58	-0.2636^b
Retention time	45	-0.1134^c	40	-0.0604^c

^{a}Significant at 99% level; ^{b}Significant at 95% level; ^{c}not significant at 95% level.

solar energy input. Only the inorganic form of nitrogen and phosphorus fail to exhibit a significant correlation. This again may be related to the importance of nutrient dynamics as opposed to nutrient standing stock values.

With the exception of thermocline depth, which shows a significant negative correlation, morphometry shows little correlation with phytoplankton production on an area basis. When considered on a volume basis however, mean depth has a negative influence on production indicating that shallow lakes have a greater production per unit volume than deep lakes. This is an obvious result and, although the correlation is significant, it is surprising that it is not much greater.

Table 9.5 lists the partial correlations after correcting for both latitude and conductivity. Conductivity was chosen as a general index to reflect differences in lake chemistry. Visible incident radiation and precipitation retain their importance. The lack of any change in the partial correlation between precipitation and production after variations in lake chemistry are accounted for further suggests that the importance of precipitation is not related to nutrient input. On the other hand, the relation between altitude and production may actually be related to chemical conditions, e.g. high altitude lakes would be expected to be poorer in nutrients than those at lower altitudes.

Table 9.5 *Partial correlations between phytoplankton production and abiotic variables after correcting for differences in latitude and conductivity*

Variable	N	Production m^{-2}	N	Production m^{-3}
Solar energy input				
Altitude	93	-0.1639^a	61	-0.1957^a
Visible incident radiation	93	-0.3286^b	61	-0.1332^a
Daylength range	93	0.0304^a	61	-0.0055^a
Air temperature	93	0.1080^a	61	0.0804^a
Epilimnion temperature	40	0.1379^a	35	-0.1473^a
Water chemistry				
pH mean	82	0.2169^c	53	0.2302^c
Total phosphorus	33	0.1288^a	25	0.1004^a
Phosphate–phosphorus	65	0.1153^a	48	0.1924^a
Total nitrogen	27	0.6073^b	19	-0.3164^a
Nitrate–nitrogen	58	0.1329^a	41	-0.0949^a
Lake morphometry and physical factors				
Mean depth	91	-0.1042^a	59	-0.3013^b
Surface area	92	-0.0724^a	59	-0.2719^c
Volume	91	0.0146^a	59	-0.1386^a
Drainage area	49	-0.1245^a	40	-0.1123^a
Depth of thermocline	66	-0.2934^b	47	-0.1987^a
Duration of stratification	51	-0.2731^c	36	-0.2647^c
Precipitation	84	-0.4402^b	58	-0.2141^c
Retention time	45	-0.1030^a	40	-0.0495^a

^{a} Not significant at 95% level; ^{b} significant at 99% level; ^{c} significant at 95% level.

Most chemical factors, of course, become less important after correcting for conductivity. An exception is total nitrogen which retains its high correlation with phytoplankton production.

Variables related to lake morphometry still show no clear relationship to phytoplankton production. The extent of stratification, however, becomes quite important indicating that, in addition to solar energy input and chemistry, the extent of mixing within a lake is an important factor in controlling production.

The fact that the simple correlation as well as the partial correlations between retention time and phytoplankton production are not significant is not so surprising as it might first appear. Theoretically, both short and long retention times would be expected to have negative as well as positive effects on production. Short retention times are associated with better mixing and higher nutrient input (although the latter is also a function of the quality of the input), but they are also associated with a greater export of biomass and thus smaller standing crops of phytoplankton. In contrast, long retention times allow for greater biomass development, but would probably not be associated with a great amount of mixing or nutrient input. These opposing factors may operate in such a way that it is difficult to illustrate a clear relationship between retention time and phytoplankton production.

9.4.2. Relationship between phytoplankton production and biotic variables

The correlations between phytoplankton production and closely related biological variables are summarized in fig. 9.7. Phytoplankton biomass and chlorophyll *a* are well correlated with each other and are about equally correlated with phytoplankton production. Thus indices of phytoplankton biomass are relatively good estimators of phytoplankton production.

Photosynthetic efficiency (production divided by visible incident radiation) shows an exceptionally high correlation with production.* This implies that lakes having a high level of production use solar energy more efficiently.

Figs. 9.8 to 9.12 illustrate the regressions between phytoplankton production and biotic variables.

9.4.3. Estimating phytoplankton production

One of the major goals of this analysis is to evaluate how well the production of a particular lake can be estimated based on varying degrees of information

---

* Statistically this is not a completely valid correlation since photosynthetic efficiency is not entirely independent of phytoplankton production with which it is being correlated. However, even when this is considered, the correlation is strong enough to suggest the existence of a real relationship.

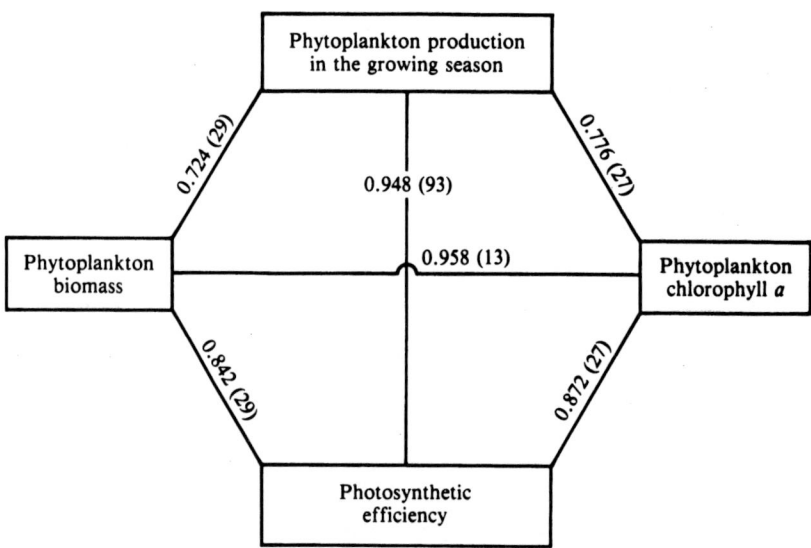

Fig. 9.7. Correlations between phytoplankton production and related biological variables. Numbers in parentheses indicate number of observations.

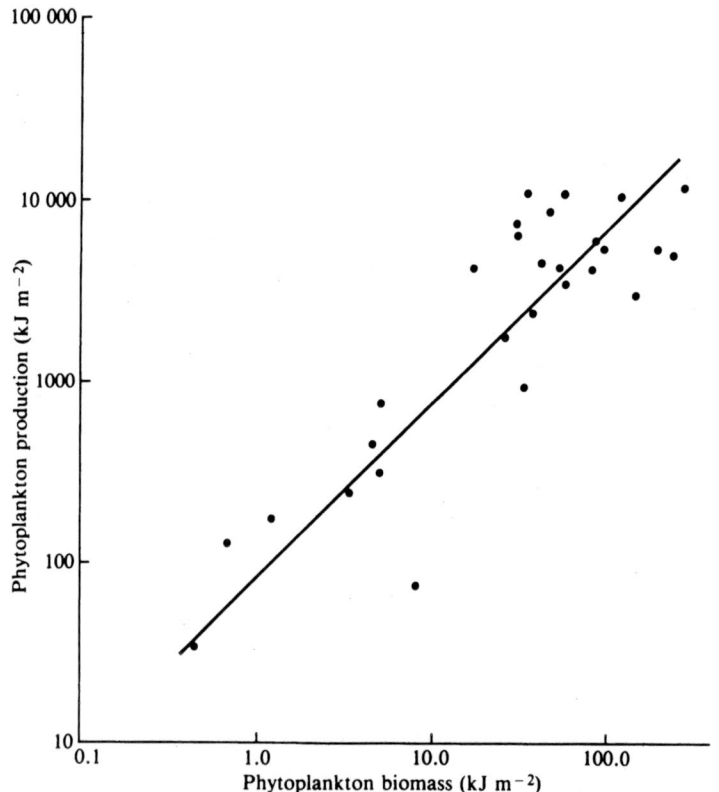

Fig. 9.8. Relationship of phytoplankton production to phytoplankton biomass.

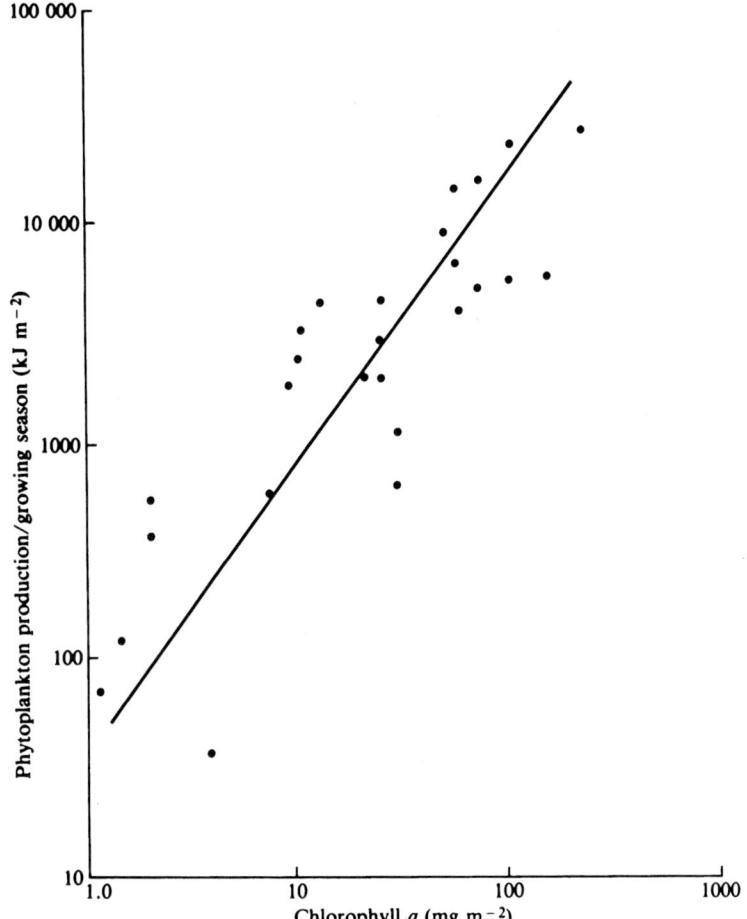

Fig. 9.9. Relationship of phytoplankton production to phytoplankton chlorophyll *a* concentration.

describing its characteristics. One method of doing this is through multiple regression analyses.

The results of estimating phytoplankton production using linear regression models involving abiotic variables are presented in table 9.6. The first regression indicates the extent to which production can be estimated on the basis of location alone. About 50% of the total variability in production is explained, most being due to differences in latitude. Altitude accounts for less than 3%.

The second entry includes location plus a number of variables related to the input of solar energy. All of this information can be obtained from published tables and graphs once the latitude and altitude of a particular site is known.

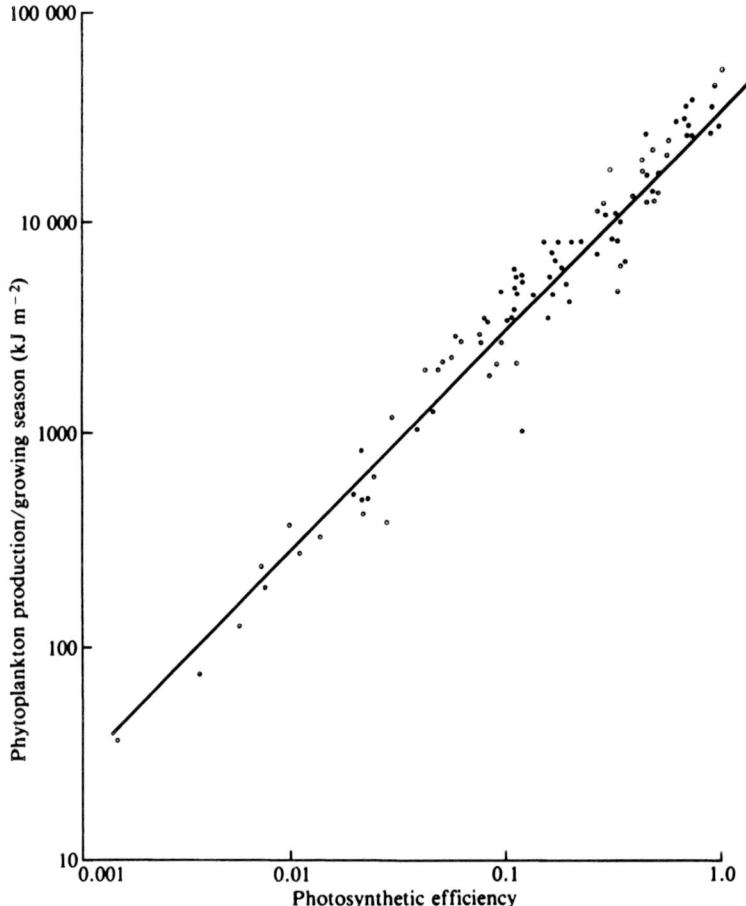

Fig. 9.10. Relationship of phytoplankton production to photosynthetic efficiency.

Of the variables included only visible incident radiation and precipitation are significant in increasing the amount of variability explained.

The third regression includes variables related to solar energy input together with some simple indices of lake morphometry. Precipitation has been retained since, as discussed previously, this may be a good indirect indicator of cloud cover. The results indicate that consideration of lake morphometry adds little towards the ability to estimate phytoplankton production.

The fourth entry includes energy related variables together with conductivity, a simple index of lake chemistry. Together these variables explain about 62% of the variability in phytoplankton production. This is relatively good considering the ease with which this kind of information can be obtained.

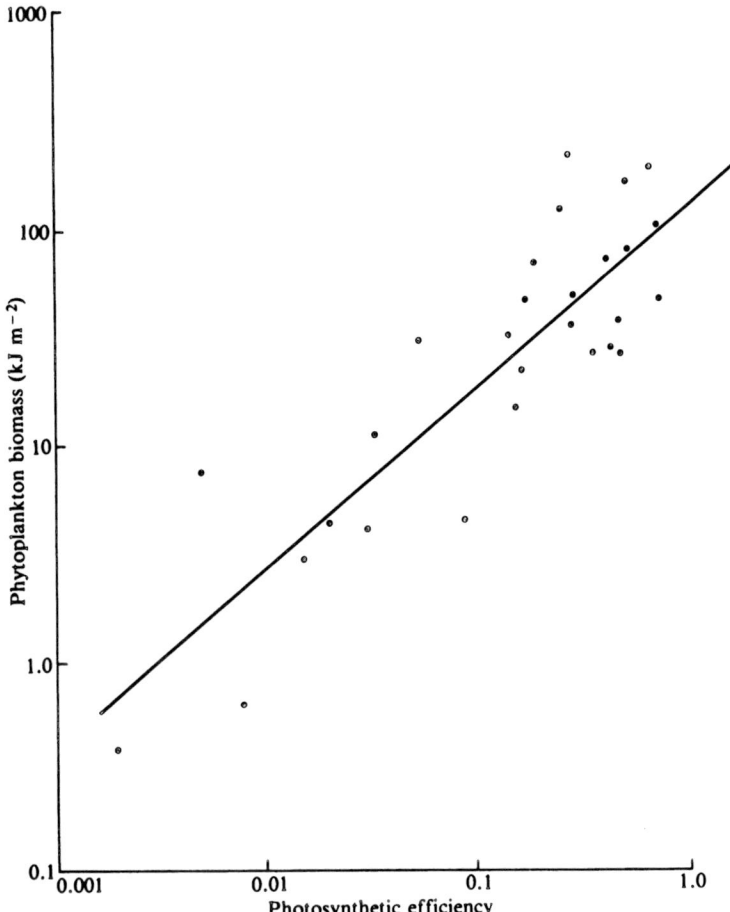

Fig. 9.11. Relationship of phytoplankton biomass to photosynthetic efficiency.

In the fifth regression more detailed indices of lake chemistry, phosphate and nitrate concentration, are substituted for conductivity. It is interesting that, although data on these variables are more difficult to obtain than on conductivity, they are relatively poor estimators. Total nitrogen and total phosphorus would have been better choices to include in the regression but data on these forms were too limited to perform comparable regression estimates.

The last regression includes indices of solar energy input and chemistry together with an index of the degree of stratification. When considered together this group of variables explains almost 70% of the variation in production.

The above regressions have considered only abiotic variables. In table 9.7

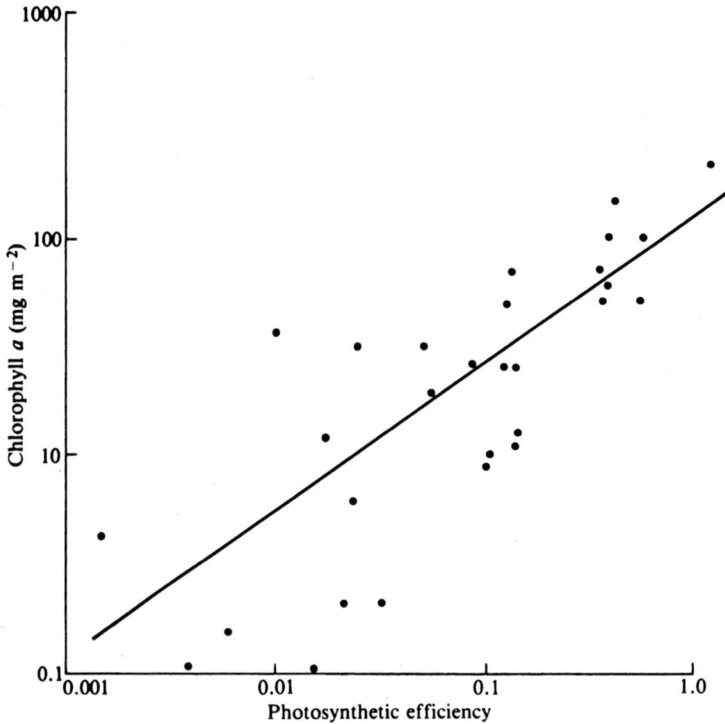

Fig. 9.12. Relationship of phytoplankton chlorophyll *a* concentration to photosynthetic efficiency.

Table 9.6 *Results of linear multiple regression analysis using abiotic variables to estimate phytoplankton production*

Variables included and percent variance explained by each	N	Per cent of total variance explained	Per cent error of prediction of mean
1. Latitude (46.1), altitude (2.9)	93	49.0	81.0
2. Latitude (43.6), altitude (0.5), visible incident radiation (2.6), daylength range (0.1), air temperature (0.2), precipitation (11.3)	84	58.3	74.2
3. Latitude (43.6), visible incident radiation (2.6), precipitation (11.3), mean depth (0.6), surface area (1.8), volume (0.3)	84	60.2	71.4
4. Latitude (42.1), precipitation (10.6), conductivity (9.7)	63	62.4	71.3
5. Latitude (44.1), precipitation (11.2), phosphate–phosphorus (1.7), nitrate–nitrogen (0.8)	54	57.8	73.6
6. Latitude (43.7), precipitation (9.9), conductivity (10.1) thermocline depth (6.2)	61	69.9	65.2

Table 9.7 *Results of linear multiple regression analysis using both abiotic and biotic variables to estimate phytoplankton production*

Variables included and per cent variance explained by each	N	Per cent of total variance explained	Per cent error of prediction of mean
1. (a) Biomass (52.4), latitude (19.6)	29	72.0	55.7
(b) Chlorophyll *a* (60.3), latitude (18.3)	27	78.6	52.4
2. (a) Chlorophyll *a* (59.7), conductivity (4.4)	20	64.1	69.2
(b) Chlorophyll *a* (60.1), nitrate–nitrogen (3.7)	25	63.8	67.4
(c) Chlorophyll *a* (58.8), phosphate–phosphorus (2.1)	25	60.9	68.5
3. Chlorophyll *a* (60.1), latitude (17.6), precipitation (12.3), thermocline depth (9.2)	23	88.4	37.7

the results of regressions using both abiotic and biotic variables are presented. The number of sites for which this kind of data is available is limited but the results presented are interesting. When either biomass or chlorophyll *a* are considered together with an index of solar energy input such as latitude, about 75% of the variation in production is explained. When a biological index is combined with some index of chemistry, little additional variance is explained by the chemical index. This suggests that the biological indices, in particular chlorophyll *a* concentration, tend to integrate the nutrient characteristics of a lake. Lund (1970) showed a relationship between maximum winter nutrient concentrations and maximum summer chlorophyll in a number of British lakes.

The best estimates of phytoplankton production are obtained when latitude, precipitation and thermocline depth are considered together with chlorophyll *a* concentration. Almost 90% of the total variation in phytoplankton production is accounted for by these variables.

Table 9.8 lists some of the more useful regression equations together with the appropriate coefficients and constants.

When estimating phytoplankton production from abiotic variables, however, there still remains unexplained variance in the estimates. Some of this is, of course, a result of imperfections in the data, and an additional amount must be attributed to not taking into account certain other variables for which no data are available. In the latter respect, it is probable that indices related to non-biological turbidity and nutrient dynamics would be helpful. In addition, a better index of the degree of mixing within a particular lake would also be helpful.

One other possibility for decreasing the unexplained variance is a better treatment of non-linear relations. The correlations and regressions discussed thus far have assumed a linear relationship between variables and it is probable that some of these relationships would be better expressed in a non-

Table 9.8 *Linear regression equations for estimating phytoplankton production*

1. Phytoplankton production = − 473.7 latitude − 3.6 altitude + 32 734.3

2. Phytoplankton production = − 535.3 latitude − 1.3 altitude − 13 736.1 visible incident radiation − 75.8 daylength range + 138.9 air temperature − 89.7 precipitation + 49 261.2

3. Phytoplankton production = − 488.6 latitude − 87.6 precipitation + 26.7 conductivity + 33 652.6.

4. Phytoplankton production = − 468.9 latitude − 78.7 precipitation + 29.0 conductivity + 96.5 thermocline depth + 33 562.9.

5. Phytoplankton production = + 1542.2 phytoplankton biomass − 289.4 latitude + 18 666.5.

6. Phytoplankton production = + 118.7 phytoplankton chlorophyll *a* − 282.3 latitude + 17 260.1.

7. Phytoplankton production = + 136.8 phytoplankton chlorophyll *a* − 391.3 latitude − 106.8 precipitation + 131.0 thermocline depth + 30 389.4.

Table 9.9 *Results of non-linear multiple regression analyses using both abiotic and biotic variables to estimate phytoplankton production*

Variables included and per cent variance explained by each	N	Per cent of total variance explained	Per cent error of prediction of mean
1. Log latitude (46.3), log altitude (0.9)	93	47.2	80.4
2. Log latitude (45.2), log precipitation (3.4), log visible incident radiation (0.5), log mean depth (0.4)	84	49.5	80.2
3. Log latitude (46.4), log precipitation (0.9), log conductivity (9.5)	63	56.7	73.6
4. Log latitude (47.2), log precipitation (0.9), log conductivity (9.7), log thermocline depth (3.2)	61	60.0	72.4
5. Log chlorophyll *a* (46.4), log latitude (15.8)	25	62.2	69.4
6. Log chlorophyll *a* (46.2), log latitude (15.3), log precipitation (6.6), log thermocline depth (0.7)	20	68.8	65.2

linear form. This is particularly true for factors that are potentially limiting, such as light and nutrients. In an attempt to evaluate the possible importance of non-linear relationships a number of multiple regressions were performed after making logarithmic transformations of the independent variables. The results, presented in table 9.9, indicate that, with few exceptions, there is little improvement over the linear models in terms of the amount of total variance explained. In fact, as a whole the logarithmic regression models are poorer than the linear regression models.

9.5. Secondary production

In contrast to the large amount of information available on phytoplankton production, data on secondary production are relatively scarce. In the case of zooplankton and benthos production, data were available from only about twenty-seven and eighteen sites respectively. In addition, most of the data are from sites located within a relatively narrow range (latitude 40° to 69 °N) and the diversity in both habitat and biological activity is not nearly as great as that represented for phytoplankton. Bacterial and fish production were not considered since information on these groups was very limited and that which was available could not easily be made comparable. As a result, no attempt was made to perform as rigorous an analysis on secondary production as on phytoplankton production. However, in the analysis which follows the aims are similar in that an attempt has been made to identify some of the more important variables, both biotic and abiotic, affecting secondary production.

In keeping with the general convention of IBP/PF studies, the secondary producers were divided into four major groups; herbivorous zooplankton, carnivorous zooplankton, herbivorous benthos and carnivorous benthos. The classification into trophic types was made by the authors. For a discussion of the difficulties and uncertainties of this procedure, see Chapter 8.

9.5.1. Relation to phytoplankton

The correlations of the secondary productions with phytoplankton biomass and production are shown in table 9.10. In all instances secondary production correlates better with production than with biomass of phytoplankton. Figs. 9.13 and 9.14 show the regressions of zooplankton production on phytoplankton production. Both herbivorous and carnivorous zooplankton are surprisingly well related to phytoplankton production. Total zooplankton production also shows a significant correlation with primary production (see Chapter 10). Benthos production, however, shows a comparatively

Table 9.10 *Simple correlations between secondary production and phyto-plankton biomass and production* (*in parentheses, number of observations*)

	Phytoplankton biomass	Phytoplankton production
Herbivorous zooplankton	0.4633[a] (22)	0.6939[b] (27)
Carnivorous zooplankton	0.5154[a] (21)	0.6731[b] (24)
Herbivorous benthos	0.1113[c] (16)	0.4956[a] (18)
Carnivorous benthos	0.2521[c] (15)	0.4708[a] (17)

[a] Significant at 95% level; [b] significant at 99% level; [c] not significant at 95% level.

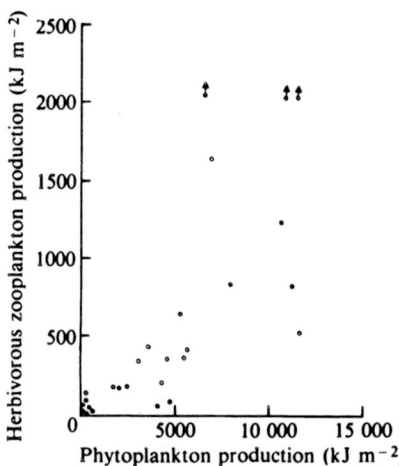

Fig. 9.13. Relationship of herbivorous zooplankton production to phytoplankton production. Points with arrows lie beyond limits of axes shown.

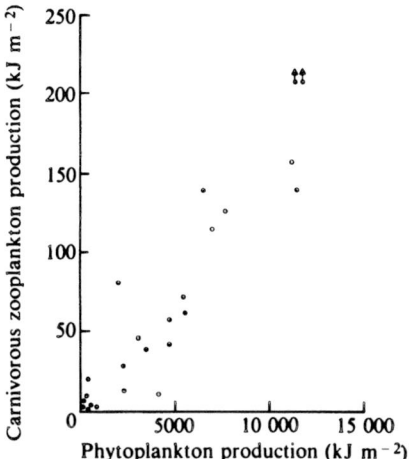

Fig. 9.14. Relationship of carnivorous zooplankton production to phytoplankton production. Points with arrows lie beyond limits of axes shown.

poor relationship with phytoplankton production (Table 9.10). This suggests that factors other than phytoplankton production have a strong influence on benthic production. Hargrave (1973) has found that, in situations where phytoplankton production is the main energy input, the respiration rates of lake sediments are directly related to phytoplankton production but inversely related to the mixed-layer depth. He has also shown (personal communication) that in marine environments the biomass of benthic organisms is related to the ratio of primary production and mixed-layer depth. The proposed reason for these relationships is thought to be related to the amount of time the products of phytoplankton production remain in the water column. With increasing depth of mixing, a greater proportion of the products of phytoplankton production are decomposed in the water column. This results in less material being available to benthic organisms.

To test this hypothesis the relationship between phytoplankton production and benthos production was re-examined after correcting for the depth of the mixed-layer. Assuming that thermocline depth adequately represents the mixed-layer depth, an index of the proportion of phytoplankton production available to benthic organisms was calculated as the ratio of phytoplankton production to mean thermocline depth. This ratio was then correlated with herbivorous and carnivorous benthos production. The linear correlation coefficients are 0.6732 ($N = 18$) and 0.6394 ($N = 17$) for herbivorous and carnivorous benthos respectively. Although there is still a considerable amount of unexplained variance, the relation is greatly improved over that obtained when considering production alone. This suggests that mixing depth, which is closely related to mean depth, is an important factor influencing benthic production.

9.5.2. Relation to abiotic variables

In an attempt to discover how abiotic factors affect secondary production, partial correlations, after correcting for phytoplankton production, were computed between the various secondary productions and a number of selected abiotic variables. These are listed in table 9.11.

Because the number of observations is low, many of the correlations are not statistically significant. However, it is possible to observe certain trends. One generalization that becomes obvious is that factors having a positive effect on zooplankton production tend to have a negative effect on benthos production. This suggests that these two groups indirectly compete with each other. Mean depth and thermocline depth both have a negative effect on benthos production and, although not statistically significant, a positive effect on zooplankton production. This is expected in view of the relationship previously shown between benthic production and mixed-layer depth. Duration of stratification follows the same trend. This is also to be expected since the longer a lake remains stratified, the greater will be the proportion of decomposition by pelagic organisms in the mixed-layer zone, and the less will be the proportion of food materials available to benthic organisms.

Retention time appears to have a negative effect on all groups. Epilimnion and hypolimnion water temperatures, however, show no clear influence on

Table 9.11 *Partial correlations on secondary productions after correcting for phytoplankton production (in parentheses, number of observations)*

Variable	Herbivorous zooplankton	Carnivorous zooplankton	Herbivorous benthos	Carnivorous benthos
Mean depth	0.1556[a]	0.1611[a]	− 0.3753[a]	− 0.3725[a]
	(27)	(24)	(17)	(16)
Retention time	0.0162[a]	− 0.2507[a]	− 0.3011[a]	− 0.0437[a]
	(17)	(15)	(10)	(10)
Mean epilimnion temperature	− 0.2566[a]	0.2826[a]	0.3332[a]	− 0.2656[a]
	(16)	(15)	(13)	(13)
Mean hypolimnion temperature	0.3314[a]	0.0718[a]	0.2320[a]	− 0.2106[a]
	(15)	(14)	(12)	(12)
Mean epilimnion oxygen	− 0.1217[a]	0.4192[b]	0.1268[a]	+ 0.0844[a]
	(19)	(19)	(15)	(15)
Mean hypolimnion oxygen	0.3876[a]	0.2336[a]	− 0.4489[a]	− 0.4601[b]
	(15)	(15)	(14)	(14)
Mean thermocline depth	0.2073[a]	0.3122[a]	− 0.5243[b]	− 0.4223[b]
	(17)	(16)	(14)	(14)
Duration of stratification	0.2749[a]	0.2183[a]	− 0.2374[a]	− 0.2013[a]
	(16)	(16)	(13)	(13)

[a] Not significant at 95% level; [b] significant at 95% level.

either zooplankton or benthos production. This is possibly a result of a complexity of effects associated with water temperature. For example, high water temperatures would be expected to be associated with high production as a result of increased feeding rates, but respiration would also be increased. In addition, water temperature is related to mixing depth and thus has an indirect effect on production. Although water temperature must surely have an important influence on secondary production, any clear effect is difficult to illustrate.

Epilimnion oxygen concentration also shows no clear effect on secondary production. Hypolimnion oxygen concentration, however, shows a strong negative correlation with benthos production. Deevey (1941) has found this same relationship between benthos biomass and oxygen concentration in a statistical analysis dealing with a large number of lakes. His explanation was that lakes with low hypolimnion oxygen concentrations tend to be populated by *Chironomus* and *Chaoborus*, and therefore to be more productive in total bottom fauna than other types. An alternative explanation is that low hypolimnion oxygen concentrations are often associated with high levels of phytoplankton production. As a result, a low hypolimnion oxygen concentration may be an indirect index of high amounts of food available for benthic organisms and this may account for the negative correlation.

9.5.3. Estimating secondary production

The small number of observations on the productivity of secondary producers limits the extent to which multiple regression analyses can be used to estimate secondary production. However, for comparative purposes, the results of a few very simple multiple regression estimates, involving no more than three independent variables in each case, are presented in table 9.12.

The estimates are slightly better for zooplankton than benthos, but in no instance does the amount of variability explained exceed 50%. In the case of zooplankton production, none of the abiotic variables greatly increased the amount of variability explained above that obtained when considering phytoplankton production alone. It is interesting that carnivorous zooplankton production is estimated about equally well whether phytoplankton production or herbivorous zooplankton production is used as one of the estimators.

Benthos production is better estimated by combining phytoplankton production together with thermocline depth than mean lake depth. The combination of phytoplankton production, mean thermocline depth, and duration of stratification produces the best estimate of benthic production. This again is consistent with the idea that the mixing characteristics of a lake are important in determining the amount of food available for benthic organisms.

Table 9.12 *Multiple regression estimates of secondary production*

	Dependent variable	Independent variables and per cent variance explained by each	N	Per cent of total variance explained	Per cent error of prediction of mean
1.	Herbivorous zooplankton production	Phytoplankton production (44.1), mean depth (0.5), mean epilimnion temp. (3.7)	16	48.1	102.8
2.	Herbivorous zooplankton production	Phytoplankton production (43.2), mean thermocline depth (3.1), mean epilimnion temp. (2.2)	15	48.5	102.4
3.	Carnivorous zooplankton production	Phytoplankton production (45.3), mean depth (2.8), mean epilimnion temp. (1.1)	15	49.2	100.3
4.	Carnivorous zooplankton production	Herbivorous zooplankton production (43.8), mean thermocline depth (4.7), mean epilimnion temp. (0.2)	13	48.7	103.1
5.	Herbivorous benthos production	Phytoplankton production (20.9), mean depth (8.3), mean hypolimnion oxygen (6.1)	12	35.3	120.4
6.	Herbivorous benthos production	Phytoplankton production (20.2), mean thermocline depth (16.2), mean hypolimnion oxygen (5.4)	11	41.8	111.9
7.	Herbivorous benthos production	Phytoplankton production (24.4), mean thermocline depth (19.2), duration of stratification (6.2)	10	49.8	103.2

9.6. Efficiencies

The only efficiency that could be calculated for secondary producers was the ratio of their production to phytoplankton production, both on a growing season basis. This ratio provides an index of the amount of phytoplankton production that is subsequently incorporated into the various secondary productions. These efficiencies are somewhat analogous to the 'ecological efficiency' of Slobodkin (1960). They have been termed 'energy transfer efficiencies' and the mean and range of each is presented in table 9.13, together with phytoplankton photosynthetic efficiency (the ratio of phytoplankton production to visible incident radiation). The absolute values of these ratios should be interpreted with caution. Since they are calculated as a ratio of productions, small errors in production estimates result in relatively large errors in the corresponding efficiencies. They should therefore be considered as rough order of magnitude estimates. Their main value lies, not so much in their absolute values, but in the relative values of these efficiencies among the various secondary producers.

The mean photosynthetic efficiency of phytoplankton is about 0.34%. This value is comparable to that reported by other workers for freshwater systems (Patten, 1959; Hillbricht-Ilkowska and Spodniewska, 1969). The range in efficiencies is great and as discussed previously, is closely related to the level of phytoplankton production occurring within a particular lake.

In the case of secondary producers, zooplankton appears to have a greater mean transfer efficiency than does the benthos corresponding to the same tropic level. This suggests that for the group of lakes considered a greater proportion of phytoplankton production is utilized by zooplankton than by benthos. The means also suggest that the idea of a 90% energy loss at each trophic level (Odum, 1971) may have some validity.

One other interesting point is that although the efficiencies of all groups vary greatly, the range tends to be larger for herbivores than carnivores. The range in efficiencies for the herbivorous forms is about two orders of

Table 9.13 *Phytoplankton photosynthetic efficiency and secondary production energy transfer efficiencies expressed as percentages (see text for explanation)*

	N	Mean	Range
Phytoplankton	93	0.34	0.002–1.0
Herbivorous zooplankton	27	7.1	0.10–27.4
Carnivorous zooplankton	24	1.2	0.17–5.0
Herbivorous benthos	17	2.3	0.16–11.1
Carnivorous benthos	16	0.3	0.35–1.8

magnitude, but for the carnivorous forms it is only about one order of magnitude.

9.7. *P/B* Values

The *P/B* value (turnover rate) of an organism, or group of organisms, is the ratio of production to biomass. When calculated on a growing season basis for an entire trophic level it serves as a general index of the rate of energy flow relative to standing stock. It is thus a useful parameter for making comparisons between different trophic levels as well as between similar trophic levels under different conditions. Table 9.14 lists the mean and range of the *P/B* value for each trophic level.

In general, the means tend to decrease with increasing trophic level. In addition, the benthic groups have considerably lower *P/B* values than the pelagic groups. These trends are probably closely related to the size of the organisms. Smaller organisms would be expected to have greater *P/B* values and, since higher trophic levels tend to have larger organisms, *P/B* values tend to decrease with increasing trophic level. The lower *P/B* values for benthic organisms in comparison with those of pelagic organisms is also probably related to size since benthic organisms tend to be larger than pelagic organisms.

From table 9.14 it is also evident that the range in *P/B* values tends to be proportional to the mean, i.e. high average *P/B* ratios are associated with a high range in *P/B* values. At first it appears that the ranges are exceptionally large suggesting no close relationship between production and biomass within each group. However, these ranges must be considered together with the ranges in production of the corresponding trophic level. In general, *P/B* values range over about one or two orders of magnitude. Productions, however, range over three or four orders of magnitude. Thus, there is some consistency in *P/B* values within a particular trophic level implying that a reasonable estimate of production can be obtained from biomass estimates. This is further suggested by the relatively high correlations between the biomass and production of each group (table 9.15).

In examining the relationship between *P/B* values and other variables it was

Table 9.14 *The mean and range of P/B values for each trophic group*

	N	Mean	Range
Phytoplankton	28	113.0	8.9–358.6
Herbivorous zooplankton	26	15.9	0.5–44.0
Carnivorous zooplankton	25	11.6	1.5–30.4
Herbivorous benthos	15	3.7	0.6–12.8
Carnivorous benthos	14	4.8	1.0–25.0

Table 9.15 *Correlations between biomass and production of each group*

	N	r
Phytoplankton	27	0.7732[a]
Herbivorous zooplankton	26	0.8396[a]
Carnivorous zooplankton	25	0.8432[a]
Herbivorous benthos	15	0.6142[a]
Carnivorous benthos	14	0.7243[a]

[a]Significant at 99% level.

Table 9.16 *Simple correlations between P/B values and latitude*

	N	r
Phytoplankton	28	− 0.3729[a]
Herbivorous zooplankton	26	− 0.3734[a]
Carnivorous zooplankton	25	− 0.4422[a]
Herbivorous benthos	15	− 0.5669[a]
Carnivorous benthos	14	− 0.5328[a]

[a]Significant at 95% level.

Table 9.17 *Partial correlations on P/B values after correcting for latitude (in parentheses number of observations)*

Variable	Phytoplankton	Herbivorous zooplankton	Carnivorous zooplankton	Herbivorous benthos	Carnivorous benthos
Phytoplankton production	0.3361[a] (28)	− 0.2664[b] (26)	− 0.0061[b] (25)	− 0.2305[b] (15)	− 0.3418[b] (14)
Mean depth	0.2690[b] (28)	− 0.4980[c] (26)	− 0.3495[a] (24)	− 0.4808[a] (16)	− 0.5117[a] (15)
Retention time	0.2800[b] (19)	0.2015[b] (16)	− 0.1255[b] (15)	− 0.4839[b] (10)	− 0.0597[b] (10)
Mean thermocline depth	0.2599[b] (16)	− 0.2083[b] (15)	− 0.0113[b] (15)	− 0.1908[b] (14)	− 0.2203[b] (13)
Duration of stratification	0.3204[b] (12)	0.2549[b] (11)	− 0.2514[b] (11)	0.3041[b] (12)	− 0.2719[b] (12)

[a]Significant at 95% level; [b]not significant at 95% level; [c]significant at 99% level.

discovered that differences in latitude have a very strong influence on *P/B* values, low latitudes being associated with high *P/B* values (table 9.16). This is probably simply a result of the fact that, with decreasing latitude, there is an increase in the length of the growing season and a corresponding increase in the number of crops produced per growing season. Thus, although mean biomass may be the same in a high- and low-latitude lake, the number of crops produced and therefore the amount of production per unit of biomass tends to increase with decreasing latitude. This implies that if one wishes to make comparisons between the productivity of different sites using biomass as an index of production, it is important that allowance be made for the differences in latitude among the sites being compared.

The effect of other variables on P/B values has also been examined. Table 9.17 presents the partial correlations, after correcting for latitude, between a number of variables and P/B values. Most of these correlations are not statistically significant and, for the most part, no general trends are evident. There does, however, seem to be one consistent relationship in that all of the P/B values for secondary producers exhibit a negative correlation with phytoplankton production. Thus, organisms in oligotrophic lakes may tend to have lower P/B values than those in eutrophic lakes.

9.8. Indices

A number of limnologists have attempted to develop indices from abiotic variables that would provide estimates of the amount of production that could be expected from a particular freshwater system. Two indices that have been shown to be successful in this respect are Ryder's 'morpho-edaphic' index (1965) and an index proposed by Schindler (1971b). Ryder's index, developed primarily for use in estimating the potential fish yield from lakes and reservoirs, is calculated as the ratio of total dissolved solids to mean depth. Schindler's index is calculated as the ratio of surface area plus drainage area to volume. Both of these indices attempt to summarize the nutrient and morphological conditions of lakes in a way that indicates their relative productivities.

In examining the relationship of these indices to the productivities of the trophic groups considered in this study it was found that, in all instances, simple linear correlations were not significant. This, however, is expected since neither of these indices takes differences in location, and therefore solar energy input, into account. In an attempt to correct for this, partial correlations were calculated after factoring out the variance due to latitude. However, even in this instance no significant correlations were found. Thus, as the authors of these indices recognize, they are probably of limited general applicability unless the sites being considered occur within a narrow geographical range and are of a similar type. This is probably more true of Schindler's than Ryder's index. In using Schindler's index to estimate production it is important that the drainage basins of the lakes being compared have similar characteristics. Ryder has indicated that his index does not work well for lakes with unusual characteristics such as high turbidity and excessive water fluctuation.

9.9. Discussion

The identification and understanding of abiotic and biotic factors controlling production in freshwater systems is one of the major goals of limnology. Of particular importance is the development of simple indices useful in estimating potential production. In this respect numerous limnologists have attempted to identify important environmental variables describing the morphological, chemical and physical characteristics of lakes and use them, either singly or in combination, to estimate the production of various trophic levels.

The possible importance of lake morphometry, particularly mean depth, in determining the production of lakes was first suggested by Thienemann (1927). Although he presented numerous arguments as to why lake depth should be important, measurements of production were scarce at that time and he was unable to present quantitative evidence to support his ideas. Rounsefell (1946) was one of the earliest workers to present quantitative evidence suggesting that lake morphometry is related to production. He showed that fish production in a number of lakes was inversely related to surface area. This concept was further developed by Rawson (1961 and earlier papers). His studies indicated that the standing crop of net plankton and bottom fauna, as well as fish yield, in a number of larger Canadian lakes, showed an inverse hyperbolic relationship to mean depth.

On theoretical grounds alone, there is good reason to suspect that lake size should have an important influence on production. Hayes (1957) has pointed out that, although nutrient supply to a lake is an area phenomenon, nutrient availability depends on lake volume, and this is inversely proportional to the amount of water and therefore mean depth. Northcote & Larkin (1956) have suggested that depth is important in determining lake stratification and mixing and therefore in controlling the distribution of nutrients and dissolved gases. Others (Rounsefell, 1946; Larkin, 1964) have stressed the importance of, not mean depth, but surface area as it relates to the extent of development of the littoral zone. As lake size increases, there is a corresponding decrease in the proportion of area occupied by the littoral zone and, since this zone is often the most productive area of a lake, the average production per unit area decreases as surface area increases.

Despite these theoretical considerations, few workers have been able to demonstrate any clear relationship between lake size and production. Rawson's work, mentioned previously as supporting the concept that production is related to lake size, is somewhat atypical in that his studies were confined to a series of lakes having a very large range in depth. As Northcote & Larkin (1956) have indicated, if it were not for the inclusion of a few lakes with mean depths greater than 100 metres, the relationship of production to mean depth would not be very strong. Thus Deevey (1941), in studying the differences in standing crops of bottom fauna in 116 lakes, failed to find any significant relationship between biomass and depth. Hayes (1957) extended Deevey's analysis to 156 lakes and reached the same conclusion. Northcote & Larkin (1956), in an analysis of 100 British Columbia lakes, found no indication that the standing crops of plankton, bottom fauna and fish is related to mean depth. The present analysis also suggests that no clear relationship exists between lake size and production.

Thus, with the possible exception of very large lakes, there is little evidence to support the concept that lake size is important in determining the amount of production attainable within a particular system. This should not be interpreted to mean that lake morphometry in general is not important, but only that mean depth and morphometric factors closely correlated with mean depth such as surface area and volume, are not good indices of the amount of production that may be expected of a particular lake. Other morphometric factors may prove to be better. In particular, since thermocline depth was shown to influence primary production (table 9.10) an index relating morphometric factors to mixing type might make the relationship between lake morphometry and production more evident. An approach to this was made in Chapter 3, section 3.4.

Indices of lake chemistry have also been studied with respect of their ability to estimate potential production. Naumann (1932) was one of the first limnologists to suggest that lake chemistry, as determined by geological surroundings, has an important influence on production. However, like Thienemann, he also lacked good measurements of production and was unable to quantify his results. More recent studies have indicated that indices related to lake chemistry often have a significant relationship to production. Moyle (1956) related fish production to carbonate alkalinity for a number of Minnesota lakes. Similarly, Turner (1960) demonstrated a significant positive correlation between fish standing crop and carbonate alkalinity in a study of twenty-two Kentucky ponds. Larkin & Northcote (1958) found that total dissolved solids was more important than mean depth in explaining the variations in standing crop of plankton, bottom fauna and fish yield in British Columbia lakes. In most of these studies, however, the amount of variability that can be explained by simple chemical indices such as alkalinity and total dissolved solids is relatively low and of limited value for use in estimating production. Although it would seem probable that indices more closely related to nutrients suspected of being limiting, such as phosphate and nitrate, would have greater predictive value, the results of the present analysis indicate that the standing stock values of phosphate and nitrate are poorly related to phytoplankton production. This is probably related to the variable turnover rate of these nutrients, so that, as Schindler (1971b) has pointed out, a knowledge of their concentration is a poor indicator of their dynamics and a poor predictor of production in the growing season. Perhaps some chemical index that takes nutrient turnover times into account may have more value than nutrient concentrations in estimating production.

Greater success in estimating production has been obtained when abiotic factors are combined, either in the form of ratios or through multiple regression analysis, to produce more complex indices. Ryder (1965) has presented a 'morpho-edaphic' index that has proved relatively successful when applied to estimating potential fish yield of lakes and reservoirs. This index, calculated as the ratio of total dissolved solids to mean depth, combines a measure of lake chemistry and lake morphometry.

Multiple regression analyses have also been shown to provide reasonable estimates of potential fish yield. Hayes (1957) and Hayes & Anthony (1964) derived a 'Productivity Index' for forty-one lakes based on fish standing stock, angler harvest, and fish yield and computed a series of multiple regressions to relate this index to morphological and chemical variables. The best equation they produced accounted for 67% of the variability, 29% of which was attributed to surface area, 20% to depth, and 18% to alkalinity. More recently, Jenkins (1968) and Jenkins & Morais (1971) developed multiple regression equations on environmental factors associated with fish standing crops and angler harvest in US reservoirs. The ability of these equations to estimate these variables varied greatly depending on the type of reservoir considered but were significantly better than estimates obtained from any single index. The environmental variables that proved most useful included total dissolved solids, depth, reservoir age and storage ratio.

It is interesting that in studies dealing with fish production, mean depth appears to gain importance when considered together with some chemical index. The reasons for this are not well understood. It may be related, not to some direct biological effect mean depth has on fish production, but to the greater efficiency with which fish populations can be harvested in shallow lakes or reservoirs (Rawson, 1952; Ryder, 1965). Since in most cases the importance of mean depth has been related to fish 'yield', as opposed to actual production, it is difficult to evaluate how much of this relationship is simply due to the fact that fish populations are more concentrated in shallow waters and therefore easier to harvest, giving higher 'yield' estimates.

The present analysis differs from most of the studies referred to above in one important respect; whereas most other studies have dealt with comparisons on a regional basis, i.e., confined to sites within a narrow geographical range, this study has considered lakes distributed on a global scale. The implications of this is that a whole new set of variables must be considered, primarily those related to climatic factors such as solar energy input and precipitation.

Despite this important difference, most of the results of the regional analyses are confirmed by the present global analysis. In particular, when considered alone variables related to lake size are poor indices of production, and most chemical indices, although somewhat better, lack a strong enough relationship with production to make them useful in estimating production. Although total nitrogen concentration was found to have a relatively strong correlation with phytoplankton production, the strength of this correlation depends heavily on a small number of lakes with high total nitrogen (4–10 mg N/1). For those lakes with a lower total nitrogen, the correlation is poor.

Perhaps one of the most noteworthy findings of this study has been the dominance shown by factors related to the availability of solar energy, as opposed to those related to lake chemistry or morphometry, in determining the level of production attainable within a particular lake. This, however, is not really so surprising when one realizes that the amount of energy input to a system is important for a number of reasons. The most important reason for the large influence of solar energy input on production is, of course, that it provides the energy source for photosynthesis. But in addition to this, and not as immediately obvious, is that solar energy has a strong influence on nutrient availability. Its effect on water temperature influences the biological activity of organisms and therefore the biological aspects of nutrient cycling and, because of its indirect influence on stratification and the mixing of lake waters, it also affects the non-biological aspects of nutrient recycling. Indeed, the fact that nutrient cycling is a dynamic process and thus involves *rates* of material flow, implies that it is dependent on energy input to the system. As a result, variables related to the input of solar energy act in an integrative capacity and in practice it is quite difficult to evaluate the importance of the availability of solar energy without indirectly considering nutrient availability also.

Of those variables related to solar energy input, latitude shows the highest correlation with phytoplankton production. The importance of latitude is most probably related to the variations in daylength since daylength range, when transformed to its logarithmic form, produces about the same plot with phytoplankton production as does latitude. Visible incident radiation, on the other hand, is poorly related to phytoplankton production. In fact, after correcting for differences in latitude, the partial correlation between visible incident radiation and phytoplankton production is negative. This suggests that, beyond a certain point, high light intensities may actually result in a decrease in production.

High amounts of precipitation were found to have a strong negative influence on phytoplankton production. The reasons for this are not entirely clear. It is possible that low precipitation combined with a hot climate could lead to high nutrient concentrations as a result of low flushing and high evaporation rates. At the other extreme, high precipitation and low evaporation may lead to high flushing rates and therefore low phytoplankton production. However, the results of the data analyses on interrelations between abiotic factors indicate that these combinations do not occur. In fact, precipitation tends to be greater at lower latitudes, i.e., in warmer climates. In addition, there does not appear to be any clear relationship between precipitation and nutrient concentration. Thus, there is little evidence to suggest that the importance of precipitation is related to nutrient concentration. An alternative explanation is that precipitation may simply be a good index of the extent of cloud cover and is therefore related to the availability of solar energy. This would be consistent with its observed negative effect on phytoplankton production.

When using abiotic factors to estimate phytoplankton production relatively good estimates are obtained when a chemical factor is combined with indices of solar energy input. When factors related to mixing are included, particularly duration of stratification, phytoplankton production can be estimated quite well. Stratification has a negative influence on production indicating that well-mixed lakes are more favourable to high productivity.

The best estimates of phytoplankton production are obtained when an energy related variable is combined with an index of phytoplankton biomass. This has also been found to be true in marine systems (Ryther and Yentsch, 1957, 1958). In this respect, the combination of latitude and chlorophyll *a* concentration produces very good estimates of production.

The three abiotic factors that appear to be most important in controlling production, i.e., solar energy availability, nutrient concentration and mixing, all have management potential. The control of production through alterations in nutrient input is an obvious management technique and has been used to increase production in systems such as fish-ponds, as well as to decrease production in cases where nutrient input adversely affects water quality. The availability of solar energy can also be controlled by alterations in non-biological turbidity. Murphy (1962) has presented evidence to suggest that non-biological turbidity may be one of the primary limiting factors in many freshwater systems. He suggests that turbidity may be altered by controlling erosion from the drainage basin, fish roiling and planting fish species that do not disturb bottom sediments. Mixing also has management capabilities. An example is the work of Hooper, Ball & Tanner (1953) who showed that pumping hypolimnion water into the epilimnion of a small Michigan lake resulted in a phytoplankton bloom.

The significant positive correlations between phytoplankton production and secondary production support the idea that high levels of primary production lead to high levels of secondary production. The relative importance of the pelagic and benthic components of secondary production, however, seem to be related to mixing depth. Whereas zooplankton production appears to be related primarily to phytoplankton production, benthos production seems to be also related to the inverse of mixing depth. Thus, well-mixed lakes or those having a shallow thermocline tend to favour greater production of benthos. The reason for this may be related to the length of time the products of phytoplankton production remain in the water column, short times being more favourable to benthos as a result of lesser amounts of material being metabolized before reaching the sediments.

The fact that a significant relationship between secondary production and phytoplankton production was found is surprising considering that other energy sources, particularly allochthonous organic inputs, were not taken into account. This suggests that the importance of allochthonous inputs as an energy source for secondary producers may be somewhat overestimated. Although many investigators have found that the input of allochthonous materials may often be considerable, few have simultaneously measured both inputs and outputs to determine what proportion of the input is actually utilized. It may be that inputs often equal outputs, especially if these materials are relatively resistant to metabolism or if the lake has a short retention time. Alternatively it may be that allochthonous input is taken into account indirectly when phytoplankton production is considered. Systems receiving large amounts of organic inputs would also be expected to receive large amounts of nutrients and therefore may have relatively high levels of phytoplankton production.

Despite the relationships found to exist between phytoplankton production and abiotic factors, and between phytoplankton production and secondary production, there still remains a considerable amount of unexplained variance when using multiple regression equations to estimate production. The three most probable sources of this variability are: (1) the inadequate treatment of non-linear relationships, (2) errors within the data, and (3) the omission of factors that are important in controlling production.

Except for a few multiple regression analyses using logarithmic transformations of independent variables, little attempt has been made to evaluate fully the importance of non-linear relationships between production and the numerous variables considered in this analysis. Instead, the emphasis has been placed on identifying important variables and determining to what extent they could be used to estimate production assuming linear relationships. This is more appropriate as a first step. To evaluate adequately the importance of non-linear relationships will require a considerable amount of time and effort since the different kinds of non-linear relationships possible are quite numerous. However, this is an obvious second step and should be considered in future analyses.

The amount of residual variance, when estimating phytoplankton and secondary production using multiple regression equations, is of the order of 30 and 50% respectively. It is not at all unlikely that the experimental techniques used to measure the various productions have this much error inherent in their methodology. If this is true, one is not really justified in trying to refine these estimates prior to the development of better experimental techniques for measuring production.

However, assuming that the above is not true, it is worthwhile to consider the third most probable source of unexplained variability, i.e., the omission of additional factors important in controlling production but for which data were not available in this analysis. Most of these have already been mentioned but, because of their potential importance, deserve to be mentioned again. In relation to the availability of solar energy, information on non-biological turbidity would certainly be helpful, and an index of the amount of reflection from the surface of each site may also prove useful. Factors that relate closely to nutrient availability would probably be most helpful, especially some index of the rate of nutrient recycling and a better index of the amount of mixing occurring within a lake. In the case of secondary production, information on allochthonous organic inputs and outputs, together with some index of potential macrophyte production, such as the extent of the littoral zone, would appear to be most important.

In the context of additional factors that should be considered in future studies, one other point should be mentioned. It is unfortunate that no detailed data on drainage basin characteristics were available for the sites considered in this analysis. Alteration within the drainage basin of a lake or reservoir is one of the more common ways by which man indirectly affects these systems. In this respect, it would have been of considerable interest to attempt to evaluate how factors such as the degree of urbanization, industrialization and other forms of land use affect lake biology. Perhaps future studies will take these factors into consideration.

Financial support for this work was provided by the Canadian Committee for the International Biological Programme and the International Council of Scientific Unions. Dr K.H. Mann was responsible for directing the work and provided much valuable advice. Technical assistance was provided by Mr W. Hunt and Mr W. Cundiff. Dr Julian Rzóska arranged the funding and assisted in many other ways. Dr Barry Hargrave read the manuscript and provided helpful criticisms. Special thanks go to the workers who allowed their data, many of which are not published, to be used in this analysis. Many others, too numerous to mention, assisted in interpretation of the results. To these people I am especially thankful.

10. Dynamic models of lake ecosystems

C.J. WALTERS, R.A. PARK & J.F. KOONCE

10.1 Introduction

A major goal for the IBP has been the development of dynamic models to predict ecosystem change over time (Menshutkin, 1971; Goodall, 1972; Walters & Efford, 1972; Huff et al., 1973; McCormack et al., 1973; Bloomfield, Park, Scavia & Zahorcak, 1974; Park et al., 1974). Such models have been viewed as a tool for synthesis of the diverse data that come from cooperative projects involving numerous investigators, and as a vehicle to help ensure better communication between investigators. Outside the IBP, aquatic ecosystem models are beginning to be widely used in applied problems of water resource planning and management (Chen & Orlob, 1972; Thomann, 1972; Di Toro, O'Connor, Thomann & Mancini, 1973; O'Connor, Thomann & Di Toro, 1973). In general, the IBP models are more detailed; while simplified versions of these models can be used for management purposes (Bloomfield et al., 1974), the models also have the potential for use as diagnostic tools for better understanding of the intricacies of aquatic ecosystems.

Most of the models that have been constructed for aquatic systems have used essentially the same basic approach and assumptions. The ecosystem is usually viewed as a series of biotic and abiotic pools or compartments, with exchange among these pools dependent on pool sizes and various physical and chemical factors. It is not our intention to describe all the the details of any one of those models; presumably, as they are validated detailed descriptions of them will be published by the respective IBP groups, (e.g., Park et al., 1974). Furthermore, several of the process-level models are described elsewhere in this volume. However, we feel it appropriate to describe some of the general considerations of ecosystem modelling and to give examples of ways in which the models treat couplings among some of the ecosystem components. We will also describe the assumptions of a general biomass model based on the Eastern Deciduous Forest Biome model (Park et al., 1974) and the Marion Lake model (Walters & Efford, 1972), and compare this model with some time-series data from IBP lakes. Time and research funds have not permitted a comprehensive comparison of the model to all IBP data sets, but we can make some conclusions about the kinds of factors that would have to be considered in a really comprehensive lake ecosystem model.

10.2. Selection of model components

We are faced in any modelling exercise with the dilemma of the extremes: simple models produce trivial answers, while detailed models require impossible amounts of data. The problem is that information requirements increase geometrically with detail added. For example, consider the problem of moving from prediction of total benthic biomass to prediction of several biomass components such as oligochaetes and chironomids. If the overall biomass model had four parameters to be estimated, one might think that the more detailed model would need only 4 × (number of component pools) parameters. This unfortunately is not the case; a host of new assumptions must be made, for example, about competition among and food selection by the component pools and about selection among them by predators. These additional factors cannot be ignored in a mathematical model as they can in field research; any prediction of change requires numbers for *every* component of that change.

However, the problem of detail is not completely insoluble. Though species composition in an aquatic community may change dramatically over time, some general types of organisms are usually represented. Although trophic level may be an important element in segregating general types of organisms, greater detail may result by distinguishing growth form, habitat preference or feeding mechanisms. Thus, while phytoplankton may be considered primary producers, many morphologically or physiologically similar species may be combined into functional groups without major information loss (cf. Allen & Koonce, 1973). These general roles, rather than simply trophic levels, form the basis for most biomass pool models.

We have not considered the alternative strategy of looking at only part of the ecosystem in greater detail. Taking this strategy in model building implies the assumption that all ecosystem components not considered in the subsystem model either remain constant over time or respond with no feedback to the modelled components. This assumption does not seem reasonable. Also, an explicit goal of the IBP is to obtain overall perspectives on ecosystem functioning.

Formulation of mathematical models requires that each system component be represented in terms of quantitative *indices* (called state variables) representing its condition at any time. While several indices could be used for each component (e.g., number of organisms, mean size of individuals), many current models are based on the single index, total biomass. These models carry no information about the number of individual animals present, or other attributes that might affect flow between biomass pools. Thus, these models have the hidden assumption that either internal pool compositions do not change over time, or effects of changing composition are compensated within other pools (for example, fish eating more zooplankton individuals when zooplankton body size is smaller). No good *a priori* justification has been proposed for this assumption.

The vulnerability of this assumption has been recognized in IBP, and consequently several groups have incorporated into their respective models indices for numbers of individuals and for mean weights. Krogius *et al.* (1972) utilized numerical abundances and mean weights in the Lake Dalnee pelagic fish model, as did Walters & Efford (1972) in their model of Lake Marion consumers. Likewise, the aquatic model developed in the US Desert Biome has indices for numbers and biomass (Goodall, 1973).

An algorithm developed in the US Eastern Deciduous Forest Biome demonstrates one way in which biomass and numbers can be related (Bloomfield *et al.*, 1974). Consumer processes are divided into two groups: those that represent an overall change in biomass for the entire population (consumption, defecation, excretion, respiration and gamete production) and those that affect the numbers of individuals (predatory and non-predatory mortality, recruitment, promotion to next size class, emergence and emigration). Thus

$$\frac{dN_j}{dt} = \frac{I_j}{W_k} - \frac{P_j}{Y_k} - \left(M_j + \sum_k C_{jk} \right) \Big/ \bar{w}$$

in which

N = numbers of individuals
j = designates jth size class
I = recruitment biomass
W_k = weight of entering organisms
P = biomass lost through promotion, emergence and emigration
Y_k = weight of organisms when they leave
M = biomass lost through non-predatory mortality
C = biomass lost through predation, and
$\bar{w}$ = mean weight of group at time t

Promotion to the next size class P_j is a function of the number of cohorts N_j and of the exponent of the difference between the weight at promotion w_c and the mean weight of the cohorts $\bar{w}$ when the mean weight is equal to or greater than the threshold weight $\bar{w}_s$. Expressed mathematically

$$P_j = ew_c \quad \text{and}$$

$$e = \begin{cases} 0 & , \quad \bar{w} < \bar{w}_s \\ N_j k_1 \{ \exp[k_2(\bar{w} - w_c)] - 1 \}, & \bar{w} \geq \bar{w}_s \end{cases}$$

where k_1 and k_2 are fitted coefficients. Similarly, mean weights can be compared to threshold weights to trigger migration and, in the case of insects, emergence. The mean weight of the cohorts can also be used to correct for the influence of size effects on processes such as respiration, feeding and non-predatory mortality. In this way additional ecological realism has been imparted to the model, although at the expense of increased data requirements and greatly increased computational load because of the additional state variables.

10.3. General simulation format

Although large dynamic models may be formulated in a variety of ways (differential equations, difference equations), in practice they are usually converted into the same general form for solution. This form consists of a bookkeeping system, coupled with rules (functional relationships) for how the various components of change are related to one another. The basic format followed in all calculations is:

$$\begin{array}{ccccc} \text{new biomass} & = & \text{old biomass} & + & \text{gains} & - & \text{losses} \\ \text{value} & & \text{value} & & \text{(feeding,} & & \text{(respiration,} \\ & & & & \text{production)} & & \text{predation, etc.)} \end{array}$$

The gains and losses are assumed to change over time as biomass values and physical conditions change.

Thus to simulate the system we start with a list of biomass values (one for each functional group) and a series of concentrations of organic materials, nutrients and dissolved gases for the beginning of the simulation period. Gains and losses for each time step are calculated based on these starting values, to give new estimates for all pools. The time step may be of uniform interval, usually one day or less, or it may be variable, utilizing an iterative algorithm to achieve a minimum permissible error in the numerical analysis. The new estimates form the basis for a second projection of gains and losses, and so forth. The reason for this time-phased procedure is that we want to represent the effects of feed-back processes: changes in the system affect conditions for further change.

The procedure is repeated until values have been obtained over the full simulation period, which may range from a single growing season to several years. The key feature of the calculations is that each change in time is estimated only from system conditions at the start of that time step, according to general gain–loss rules that are assumed to hold for all time steps. Thus time-series data are not used in the model except as driving variables (e.g., light, temperature and allochthonous inputs). The general rules must be able to represent the effects on gain–loss rates of *any* combination of biomass levels and physical conditions that could possibly occur through the course of the simulation, so single point estimates of rates are of little value in establishing these rules.

10.4. Model couplings

The ways in which model components are coupled are important for they determine the validity of the model as a representation of the ecosystem. It is beyond the scope of this paper to consider all the pathways by which components can be coupled or to consider all the alternative ways in which a particular coupling can be modelled; but rather we will give a few repre-

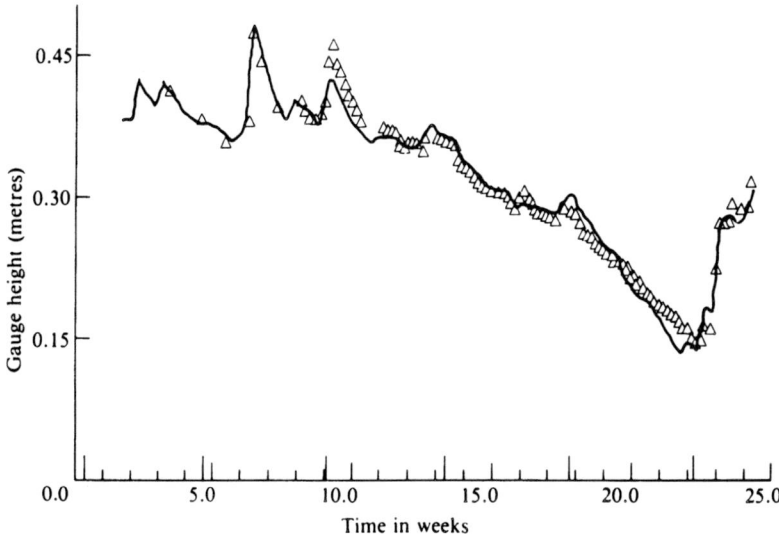

Fig. 10.1. Observed and simulated lake stages; Lake Wingra, April 10–September 15, 1970. From Huff *et al.*, 1973.

sentative examples of couplings and note how these have been treated in some of the principal IBP models, particularly the Eastern Deciduous Forest Biome (USA) model.

10.4.1. Hydrology–lake

In a study reported on by Huff *et al.* (1973) the terrestrial and aquatic components of the Lake Wingra, Wisconsin (USA), basin ecosystem were linked by a Hydrologic Transport Model (HTM). The HTM was developed by Huff (1968) as an adaptation of the Stanford Watershed Model (Crawford & Linsley, 1966) to simulate water balance and nutrient loadings in a lake basin. Climatic data, including precipitation, wind speed, air and dew point temperatures and solar radiation, in the form of time-series were used to drive the model; and from these were predicted runoff, groundwater flow, and lake evaporation, discharge and stage (e.g., fig. 10.1). By multiplying the observed concentrations of nutrients in each source by the flow from each source, nutrient loadings were calculated on a daily basis. These loadings then were used as input values in computing nutrient concentrations for the lake ecosystem simulations.

10.4.2. Abiotic factors – phytoplankton

Most models relate primary productivity to nutrient concentrations, light and temperature. The latter two factors are treated as exogenous or 'driving' variables; however, nutrient concentrations are linked dynamically in several of the models so that as growth occurs, the nutrients are depleted accordingly.

Maximum photosynthesis in phytoplankton can be symbolically scaled down by reduction factors accounting for nutrient limitations and non-optimal light and temperature. In the Eastern Deciduous Forest Biome (USA) model a construct analogous to mean resistance:

$$u_t = 4 \Big/ \left(\frac{1}{u_l} + \frac{1}{u_n} + \frac{1}{u_p} + \frac{1}{u_c} \right)$$

is used to account for the overall reduction, u_t, in photosynthetic capacity due to the simultaneous limitation effects of light, u_l, nitrogen, u_n, phosphorus, u_p, and carbon u_c (Park et al., 1974). The Desert Biome (USA) model utilizes a construct that multiplies the light and nutrient limitations (Goodall, 1973), whereas other of the IBP models, including the general version reported on later in this chapter, use the factor with the minimum value for a given date. The variety of constructs is symptomatic of our lack of understanding of how the photosynthetic process is limited.

Non-optimal temperature is usually treated by means of a multiplicative reduction factor for each physiological process. In some models a linear relationship can be assumed (e.g., Goodall, 1973). A more general construct, first proposed by O'Neill et al. (1972) can be used to represent an exponential increase in rate (the Q_{10} value) up to a maximum at optimal temperature, beyond which the rate decreases rapidly until the lethal temperature is reached. Such a formulation is more realistic and permits the model to be used under wide-ranging conditions, including situations where the effects of thermal pollution are of interest.

10.4.3. Predator – prey

Formulation of a realistic but general feeding term has received high priority in the Eastern Deciduous Forest Biome (USA) with provision being made for multiple food sources, non-linear physiological effects, and behavioural characteristics (Kitchell et al., 1974; Park et al., 1974). Of particular interest is the construct representing the complex non-linear relationship between consumer biomass and food supply, as described by Bloomfield et al. (1974). The weighted biomass on which a consumer can feed is $w_{ij}(B_i - Bmin_i)$ where w_{ij} is the food preference or capturability factor (O'Neill et al., 1972) and $Bmin_i$ is the minimum utilizable prey biomass. Preference for one food compared to all i food types is

$$\frac{w_{ij}(B_i - Bmin_i)}{\sum_i w_{ij}(B_i - Bmin_i)}$$

Minimum feeding area is represented by Q_j and density interference in feeding is represented by $r_j B_j$; $(Q_j + r_j B_j)$ can therefore be thought of as a half-saturation 'constant' with r_j being the slope. Summing over i food sources, the term becomes

$$\frac{\sum_i w_{ij}(B_i - B\min_i)}{(Q_j + r_j B_j) + \sum_i w_{ij}(B_i - B\min_i)}$$

10.5 General biomass model

Several of the IBP data sets provide an opportunity to test the hypothesis that a single general model of the biomass pool type can be applied to a wide variety of situations, with little or no modification of its structure or parameter values. If such a model can describe the data fairly well without having its parameter values estimated from that data (that is, with parameter values from independent experimental studies), we can have considerable confidence that its basic assumptions capture most of those essential features of aquatic ecosystems that are important in determining overall productivity and biomass changes. This approach should also help us to avoid falling into a common trap for the scientist: it is always possible to concoct an inductive hypothesis after the fact to explain any set of observations, but such a hypothesis cannot really be tested without further observations. The deductive hypothesis provided by a model, however, is directly testable.

10.5.1. Model structure

The roles that we have chosen to consider for the IBP data are shown in fig. 10.2 along with the basic physical factors that are represented. This choice of roles and level of detail was made by deliberately trying to set up several alternative models of increasing detail, while looking at the increase in number of assumptions (and data needed) at each step. The next step beyond the roles in fig. 10.2 would be essentially a species level model, requiring enormously more assumptions about timing of life histories, and other strategic characteristics that distinguish competitive potential and susceptibility to predation.

In the following discussion of model components, we make no attempt to cite the extensive literature sources on functional relationships and parameter estimates. Most of these sources are cited elsewhere in this volume.

10.5.1.1. Physical conditions and processes

The model considers events under one square metre of lake surface, near the lake centre. The basic physical and chemical conditions represented over time

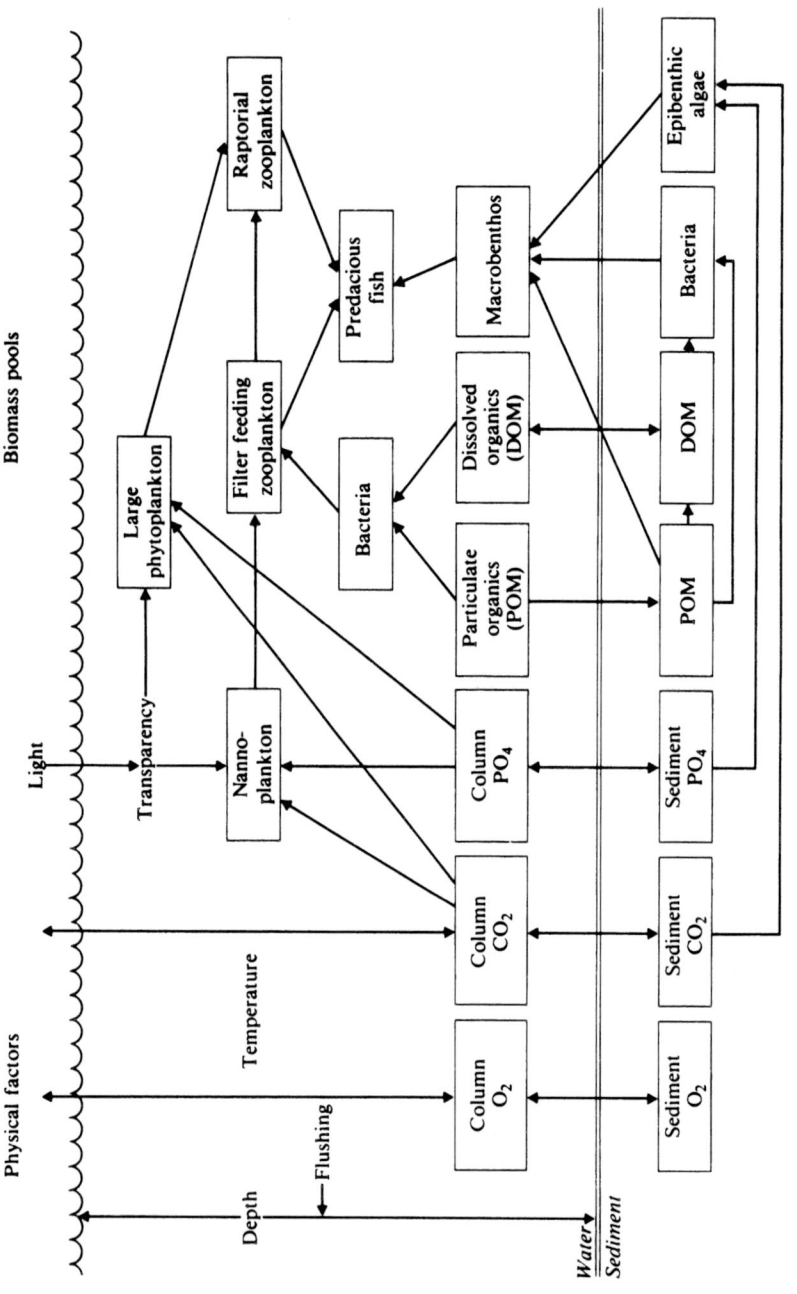

Fig. 10.2. Major components and uptake pathways in a general biomass model to compare IBP lakes. Transfer pathways that are considered in the model but are not diagrammed include: (1) production and respiratory transfers of oxygen and carbon dioxide; (2) phosphate release through respiration; and (3) releases of particulate organic matter (POM) and dissolved organic matter (DOM) through excretion, faecal production and mortality.

at this point are: (1) water depth, (2) surface water temperature, (3) incident radiation, (4) light, (5) water exchange rate, (6) water surface and sediment surface oxygen concentration, (7) water surface and sediment surface carbon dioxide concentration (total), (8) water surface and sediment surface phosphate concentration. No attempt is made to simulate temperature (by heat budget), radiation, or water exchange rate; instead, empirical values for these variables over time are provided as driving functions. Also, the model does not consider the full depth distribution of these variables. To do so would require that we replicate several of the state variables over many discrete layers of water (for examples, see Menshutkin & Prikhodko, 1971; Chen & Orlob, 1972), which would greatly increase the computational load.

In order to simulate nutrient cycling, all biomass pools are assumed to have the same carbon : oxygen : phosphorus ratio by weight (0.75 : 1.0 : 0.07), and simple stoichiometric relationships are used to keep track of uptakes and releases of these elements in production and respiration processes. Every respiration rate is assumed to be associated with a loss from the oxygen pool (32/30 mg per mg dry wt respired), and gains to the carbon dioxide and phosphate pools (44/30 mg and 0.07 mg per mg dry wt respired, respectively).

Vertical transport of carbon dioxide, oxygen and phosphate within the water column is represented with a simple diffusion equation, where the diffusion coefficient is assumed to be inversely proportional to water depth. No seasonal effect of water temperature, ice cover, or wind turbulence on transport rates is considered; only one of the test lakes (George, USA) shows significant stratification. A similar diffusion equation represents carbon dioxide and oxygen transport across the air–water interface. Inflow and outflow of these elements due to water exchange is also represented, though the inflow estimates, especially for phosphorus, were not considered reliable for any lakes used to test the model. Dissolved organic materials are assumed to diffuse vertically at the same rates as phosphorus.

Organic detritus and algae are assumed to have sinking rates that are constant over time. We would not hope with this simplifying assumption to gain more than a general picture of the rates at which organic material is made available to benthic organisms, as a function of planktonic standing crops and water depth.

10.5.1.2. Primary production

The same basic primary production rules are used for large phytoplankton (> 50 μm cell size), nannoplankton and epibenthic algae. Gross production rate is assumed to depend on light, nutrient concentration (carbon dioxide, phosphate), and algal biomass. In each time step, three estimates of potential production rate per unit biomass are made, based on light conditions, carbon dioxide and phosphate. The minimum of these three values is taken as the

Table 10.1 *Primary-production parameters used in a general biomass model for IBP lakes*

	Maximum gross P/B (per day)	Relative radiation giving max. P/B	Total CO_2 conc. giving $\frac{1}{2}$ max. P/B (g m^{-3})	PO_4-P conc. giving $\frac{1}{2}$ max. P/B (g m^{-3})	Basal respiration per biomas per day 4°C	max.	Exudation rate (proportion of gross P)	Sinking + death rate (proportion of B/day)
Large phyto-plankton ($> 30\ \mu$m)	0.7	250.	1.0	0.02	0.014	0.1	0.2	0.07
Small phyto-plankton ($< 30\ \mu$m)	2.0	250.	1.0	0.01	0.014	0.1	0.12	0.015
Epibenthic algae	2.0	200.	0.2	0.01	0.014	0.1	0.12	0.03

Additional assumptions:
(1) Minimum respiration loss is 0.2 P.
(2) Light extinction coefficient = (base for lake) + 0.017 × (algal biomass in g dry wt m^{-3}).
(3) 25°C.

realized production rate. Calculation of the potential production rate based on light conditions and depth follows the theoretical models proposed by Steele (1965) and Vollenweider (1965); the effect of algal standing crop and detritus on light penetration is included. Carbon dioxide and phosphate uptake rates are assumed to follow Michaelis–Menten kinetics, with potential biomass production per unit uptake of each determined by the stoichiometric relationships outlined in the previous section. Taken over a series of time steps, these rules can result in complicated algal responses. Nutrient limitation may occur only if uptake results in lowered nutrient concentration after several time steps; nutrient pools may remain high in the model even as algal biomass increases, if respiration rates in other parts of the system result in high nutrient release rates.

Primary-production parameter values used in tests against the IBP data are given in table 10.1. The large phytoplankton are assumed to differ from grazeable forms only in being less efficient at obtaining all limiting resources; that is, no special assumptions are made about differential ability to use certain nutrients.

Algal respiration is assumed to have two components, a dark rate dependent on temperature and an additional rate proportional to gross production rate. Dark rates (table 10.1) are assumed to follow the temperature curve proposed by O'Neill *et al.* (1972) as described above in section 10.4.2. The additional respiration rate is assumed to use up 30% of gross primary production. Algae are also assumed to have exudation and death rates (table 10.1) that depend on gross production rate.

10.5.1.3. Decomposition and bacterial production

In most biomass models, decomposition is either ignored or assumed to break down a constant or temperature-dependent fraction of the available organic matter. The second alternative results in very poor estimates of bacterial biomass (since bacterial growth is assumed to be independent of bacterial biomass), which would not be a serious problem if bacteria were not an important food source for some consumers. Accordingly, we have tried to develop a more reasonable picture of bacterial gains and losses by using a set of rules that consider both aerobic and anaerobic processes; we caution that this submodel remains completely untested.

Bacterial consumption of dissolved and particulate organic matter is assumed to depend on bacterial biomass, substrate concentration, oxygen concentration and temperature. The model first estimates organic matter uptake according to a Michaelis–Menten relationship, with the maximum uptake rate dependent on temperature in the same way as basal algal metabolism (see above). The K_s value for this relationship is assumed to be 100 mg dry wt of organic matter per m^3, and the V_m value is assumed to be 3 times bacterial biomass per day at optimum temperature. Next, a basal respiration rate is calculated as a function of temperature. The organic uptake estimate is used to calculate a potential growth rate (50% of uptake) and, by difference, a potential aerobic respiration demand which is added to the basal respiration. This potential total aerobic respiration rate is compared to an estimate of potential oxygen uptake, which in turn comes from another Michaelis–Menten relationship $(K_s = 1000 \text{ mg O}_2 \text{ m}^{-3})$. If the potential oxygen consumption is sufficient to meet aerobic demand based on organic matter intake, then that aerobic demand is taken as the realized respiration rate. However, if potential oxygen consumption is insufficient, the fraction of the organic uptake that cannot be used aerobically is assumed to go into anaerobic respiration. The anaerobic respiration is assumed to result in re-release of most (95%) of the aerobically unusable organic intake as small molecules (e.g. acetate) with consequent reduction in growth efficiency.

Nutrient release due to bacterial respiration is assumed to be no different (per unit organic matter respired) from the respiratory release by other organisms. Thus we have ignored some special bacterial functions such as nitrogen fixation and sulphur cycling.

10.5.1.4. Consumer feeding and metabolism

Major consumer types considered for the IBP tests are shown in fig. 10.2; all of these biomass pools are assumed to be governed by essentially the same dynamic rules, with different parameter values. A key overall assumption is that consumer organisms do not change trophic roles over their life histories; in other words biomass flows are assumed to be due only to feeding, metabolism and mortality. This is obviously a poor assumption, especially for the predacious zooplankters whose early stages may be algal feeders. The consumer calculations must be viewed as an attempt to predict *potential* biomass changes in each pool as a function of food available, metabolic constraints and predation; when checking these potentials against real data, we are assuming that evolution and invasion have led to the presence of enough different species that the potential is sure to be fully used, even if no single species can change rapidly enough due to life history constraints on reproduction and growth.

Rates of food intake by consumers are assumed to depend on their own biomass, the biomass of prey and temperature. In any time step the model first calculates a basic feeding rate based on food availability, using the ration curve shown in fig. 10.3*a*, then modifies this rate according to the temperature

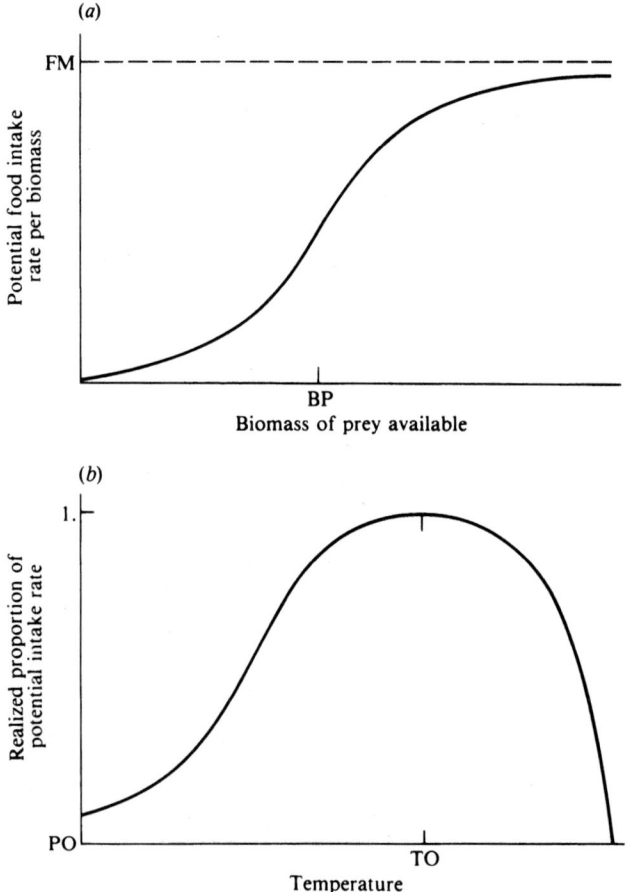

Fig. 10.3. Factors affecting consumer food intake in the IBP comparison model. These functional relationships are computerized as series of connected straight line segments rather than as complex equations. Parameter values (BP, FM, PO, TO) in table 10.2.

Table 10.2 *Food intake and loss parameters for consumer biomass pools; symbols FM, BP, etc. defined in fig. 10.3*

	Max. intake per biomass per day FM	Prey biomass g ving $\frac{1}{2}$FM (mg m^{-3}) BP	Assimilation efficiency	Optimum temperature for feeding TO (°C)	Optimum temperature for metabolism (°C)	Minimum feeding, or metabolic rate PO	Foods electivity coefficients* Prey 1	Prey 2	Prey 3	Max. metabolic rate per biomass per day	Daily death rate
Filter feeding zooplankton	0.6	300†	0.5	25	30	0.1‡	1.0(h)	1.0(a)	0.6(b)	0.07	0.0015
Raptorial zooplankton	0.2	2000	0.8	25	30	0.1	1.0(d)	0.01(c)	—	0.02	0.00015
Microbenthos	0.35	500	0.4	25	30	0.2	0.1(b)	1.0(e)	—	0.1	0.0007
Macrobenthos	0.09	20	0.55	25	30	0.2	1.0(h)	1.0(e)	0.4(b)	0.015	0.00015
Fish	0.07	3000	0.85	25	30	0.1	1.0(g)	0.1(d)	0.1(i)	0.015	0.00007

*Proportion by weight that each prey type would make of total diet if all prey types were present at equal biomass. (a) nannoplankton, (b) bacteria, (c) net phytoplankton, (d) filter feeding zooplankton, (e) epibenthic algae, (f) microbenthos, (g) macrobenthos, (h) particulate organic matter, (i) raptorial zooplankton.

†1000 used for Lake Wingra, and George (Uganda); 150 used for Vorderer Finstertaler See and Char.

‡0.3 used for cold lakes: George (USA), Char, Vorderer Finstertaler See.

relationship in fig. 10.3*b*. The sigmoid ration curve is used rather than an Ivlev (1961) or Michaelis–Menten equation in order to reflect the idea that consumer populations have a higher proportion of adults when food is scarce, and these adults have lower intake rates per unit biomass. Thus the sigmoid form is a compositie of two Ivlev-type curves, one for growing young animals and one for adults. Assimilation efficiencies for consumers are assumed to be independent of feeding rate, and the non-assimilated fraction of the total intake in each time step is transferred to the particulate organic matter pool.

Feeding parameter values for the various consumer pools are given in table 10.2. The most critical of these parameters are BP and PO as defined graphically in fig. 10.3. BP determines how low any consumer type can crop its prey resources before its own biomass growth must cease. PO determines how strongly the consumer will respond to seasonal changes in temperature, and thus the whole seasonal pattern of biomass change. As Rigler (1971) has pointed out, the PO value can be strikingly different in different species; in general we would expect natural selection to favour higher PO values (less response to low temperatures) in colder lakes.

For consumers with multiple prey types, the prey density for fig. 10.3*a* is taken to be a weighted sum of the biomasses of the prey types, where the weightings are electivity coefficients. The possibility of explicit, behavioural switching (Murdoch, 1969) from one prey type to another is not considered. The intake for each prey type is calculated as the total intake multiplied by the weighted relative contribution that the prey type makes to total weighted prey biomass; thus any prey type will be eaten less rapidly when other prey types are more abundant.

Consumer metabolic and excretion rates are assumed to have two components similar to those for algae and bacteria: a basal rate dependent on biomass and temperature, and an active rate correlated with food intake rate. The basal rates (table 10.2) are found from a temperature curve similar to fig. 10.3*b*, but with the optimum temperature higher for metabolism than for feeding. The active rate is assumed to be simply proportional to food intake rate (20% of assimilated ration), so it reflects both specific dynamic action and actual movement or activity. For each pool there is also assumed to be a low physiological or non-predatory mortality rate that takes a constant proportion of the biomass in each time step, independent of temperature or food conditions.

It should be noted that net productivity estimates or *P/B* ratios are never used explicitly in the consumer calculations. Instead only the components of production actually enter the rules for change. The reason for the extra complexity is simple: without knowing the components of change it is impossible to establish a bookkeeping procedure to remove the losses from each pool that are associated with gains (production) in other pools.

10.5.2. Validation criteria

What are reasonable criteria for success of an overall ecosystem biomass model? That is, how do we judge whether it adequately describes several sets of data? There is no simple answer to this question. For one thing we must keep in mind the admonition from basic statistical inference that it is never possible to prove any hypothesis; one can only disprove or make a relative judgement between alternatives. No model, intuitive or mathematical, can capture all the factors that influence aquatic populations, so it would be unreasonable to expect it to reproduce every data point faithfully. On the other hand, the model should at least predict average levels and major seasonal changes accurately. It is essential that the model should not be judged against some ill-defined standard of perfection: instead it should be compared to alternative theoretical frameworks that attempt to predict the same set of observations. Also, one must keep in mind the difference between interpolation and extrapolation: our general model tries to extrapolate to field observations from experimental parameter estimates and relationships, while the regression model of Brylinsky (Chapter 9) simply describes existing field data and could not be meaningfully extrapolated beyond those data. This dichotomy of model types, however, can be quite useful when their use is coordinated.

10.5.3. General patterns predicted by the model

In this section we present a general review of the predictions made by the biomass model and compare them to the predictions in the literature, many of which have come from mathematical analysis not involving computer simulation (Smith, 1969; May, 1973; Phillips, 1973).

Fig. 10.4 shows simulated seasonal changes in a hypothetical eutrophic lake. If some random sampling error were added to distort the curves in fig. 10.4, it would be easy to accept the resulting 'data' as having come from some real lake. However, as examples in the next section will show, this apparent realism is deceptive.

The relatively simple set of rules for change as described above can interact to produce a complex pattern of seasonal change, especially in the plankton. The typical simulated year for a eutrophic lake begins with low plankton populations limited in their growth rates by temperature and, in the case of algae, by light. As light levels increase in early spring, simulated phytoplankton growth begins but zooplankton response is not immediate due to low temperature. Soon the combination of high phytoplankton levels and increasing temperature results in rapid zooplankton increase, which helps to prevent phosphate exhaustion by the phytoplankton. However, considerable

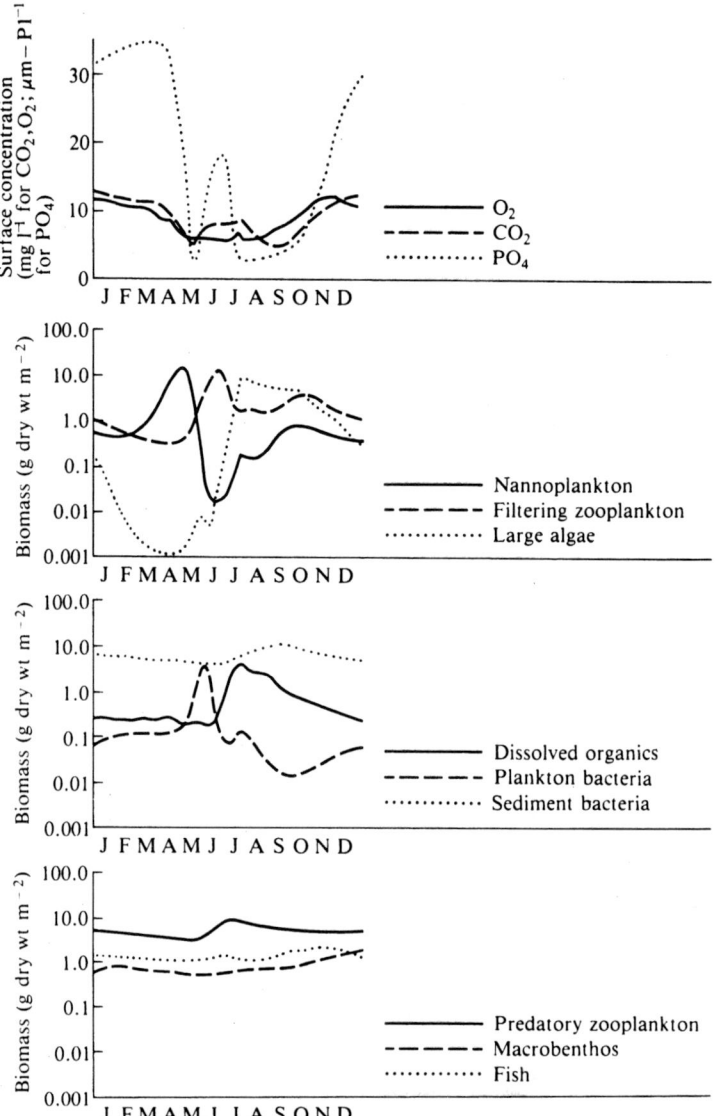

Fig. 10.4. Simulated seasonal changes in a hypothetical eutrophic lake similar to Lake Wingra, by the IBP comparison model.

phosphate soon accumulates in the zooplankton biomass, so low nutrient concentrations, self-shading and heavy grazing combine to produce a sudden sharp decline in the nannophytoplankton biomass. Phosphate levels then begin a short increase, but nannophytoplankton show no response because the grazing rate is still high. At this point the net phytoplankton bloom is initiated, and the biomass of large algae is soon high enough to reduce light penetration sufficiently to prevent any resurgence of nannophytoplankton. Production of dissolved and particulate organic matter by the net phytoplankton contributes to increased bacterial productivity, which helps maintain high zooplankton populations. This pattern persists until fall, when decreasing light levels result in a decline in net phytoplankton and the possibility of a short autumn bloom of nannophytoplankton. Decreasing grazing rates due to decreasing temperature also tend to favour this autumn bloom.

Overall productivity relationships predicted by biomass models are in general agreement with the empirical findings by Brylinsky (Chapter 9). The following predictions relate to summer average biomasses and production rates integrated over the growing season:

(1) *Primary production rates should not be clearly related to free nutrient levels,* simply because any excess of nutrient should be quickly taken in and passed up the food chain; this prediction is clearest in those biomass models that consider the phytoplankton population as a series of pools with different Michaelis constants for nutrient uptake (in our case, nannoplankton and large algae).

(2) *Primary production rates per surface area should be similar in a wide variety of lakes.* This prediction follows from the assumed effect of standing crop on light penetration along with the assumed effect of water depth on diffusion transport rates of nutrients from bottom to surface water.

(3) *Nannoplankton standing crops should not be well correlated with total primary production or nutrient levels,* because grazing should be able to control the abundance of this algal component. In fact, summer nannoplankton standing crops should be even lower in eutrophic than oligotrophic lakes, due to a combination of shading effects by large algae plus high grazing rates. The high grazing rates result in turn from higher zooplankton production during the spring bloom, supplemented during the summer by bacterial production based on organic materials released by the large ungrazeable algae.

(4) *Planktonic bacteria should be very scarce except in highly eutrophic lakes,* due to zooplankton grazing. In lakes where most of the primary production is due to grazeable algae, bacteria and zooplankton must compete indirectly for the same energy source, while the bacteria must also contend with grazing by this competitor.

(5) *The biomass of grazers should be well correlated with primary production and total phosphorus present* (organic plus inorganic). This prediction follows from (1) above, combined with the relatively lower seasonal response rates of planktonic predators to changing prey density.

(6) *Benthos standing crops should not be clearly correlated with primary production.* Instead this pool should be controlled by predation, but no general prediction is possible because predator feeding patterns (e.g. mud digging) and access limitations (e.g. anoxic bottom waters in deep lakes) are so variable. It

does not appear that general biomass models can be expected to give reasonable predictions about higher trophic levels; the variety of behavioural responses and life history strategies is too great.

The most interesting predictions made by the biomass model concern relationships between overall productivity and the persistence and seasonal stability of various community components. In general, increased productivity

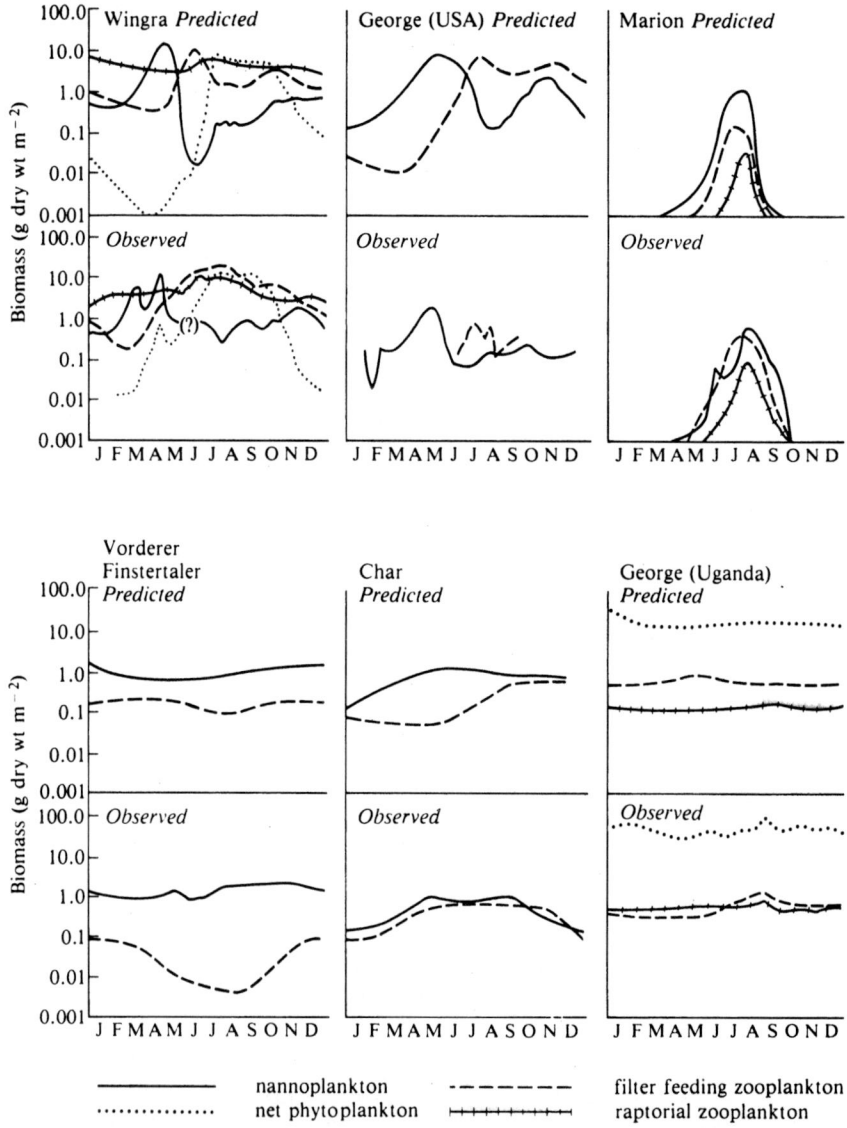

Fig. 10.5. Simulated and observed seasonal plankton dynamics for six IBP lakes.

should lead to increased seasonal variation (decreased stability) in standing crops, especially of plankton organisms; this prediction is compared to IBP data in fig. 10.5 (see description of lakes in following section). However, as shown with Lake George (Uganda) in fig. 10.5, it appears that seasonal variation in environmental conditions is necessary for this potential instability to be expressed even in very eutrophic lakes. Good discussions of this prediction have been presented by Rosenzweig & MacArthur (1963) and McAllister, Le Brasseur & Parsons (1972). As to persistence of components, the parameter values in tables 10.1–10.3 lead to the following predictions for temperate lakes:

(1) Large ungrazeable algal forms should show summer blooms only when total phosphorus levels exceed about 25 μg/l.
(2) Predacious zooplankters should persist only when total phosphorus levels exceed about 10 μg/l.

No persistence predictions would be valid for higher trophic levels since terrestrial food sources can be extremely important. The functional relationship that is most critical in stability-persistence predictions is the grazing or ration curve (fig. 10.3), since this curve determines how efficient the consumer in question will be at depleting food standing crops. Greater efficiency relative to food productivity leads to lower stability. The efficiency–stability question is considered again below in relation to tests of the model against Lake Wingra data.

None of the general predictions outlined above is unique or new; all have appeared in the literature as hypotheses devised by looking at individual parts of the ecosystem. The model simply asserts that these hypotheses remain reasonable when viewed in an overall system context. Further, there is no reason for choosing a biomass dynamics model as opposed to simpler regression models for making general productivity forecasts about lakes that have not yet been well studied, except that the dynamic model represents some fundamental processes that may limit the applicability of linear extrapolations. It remains in the following section to see whether the biomass model can go further, and make reasonable predictions about more specific patterns of seasonal change.

10.5.4. Prediction of seasonal patterns in specific lakes

Six lakes were chosen in initial attempts to apply the biomass model for seasonal predictions, and results from these attempts made it clear that additional tests would not be necessary. The six lakes vary from the highly eutrophic lakes, Lake Wingra (Wisconsin, USA) and Lake George Uganda, to the very cold and unproductive Char Lake (Canadian Arctic). Physical characteristics of these lakes that constitute the basic model input are summarized in table 10.3. An attempt was made to apply the model to one

Table 10.3 *Physical characteristics of lakes used for IBP model tests*

	Latitude (for isolation variation)[a]	Background light extinction coefficient[b] (m^{-1})	Simulated water column depth (m)	Water exchange rate per year min.	peak(s)	Surface temperature (°C) min.	max.	Available phosphate[d] (mg m^{-3})
Wingra	45°N	0.5	3	1.0	1.0	3.0	25.0	30.0
George (USA)	43°N	0.25	10	0.5	0.5	3.0	24.0	8.0
Vorderer Finstertaler	47°N	0.1	30	0.5	0.5	3.0	8.0	1.0
Char	75°N	0.1	30	0	0	1.0	4.5	2.0
Marion	49°N	0.2	2	12.0	3650.0	2.0	23.0	15.0
George (Uganda)	0°N	0.2	2.25	0.0	5.6	25.0	25.0	2.0

[a]Relative cloudiness of different areas not considered.
[b]From equation $I_z = I_0 e^{-kz}$ in absence of phytoplankton and particulate organic materials.
[c]Empirical curves over time were used; only peak rates or values given here.
[d]Peak winter values.

even more extreme situation, Pond B at Pt Barrow, Alaska (10 cm deep), but this attempt failed completely. Lake George (New York, USA) and Vorderer Finstertaler See (Austria) are deep and oligotrophic and are intermediate between Wingra and Char. Marion Lake (British Columbia, Canada) is shallow and might be highly productive if it were not flushed so rapidly at some times of year.

We will not attempt to describe results for all biomass pools for every lake; instead we will concentrate on a few representative components and the problems that arose in dealing with these, as examples of what might be learned from dynamic modelling in a more comprehensive study. In the following discussion no emphasis is placed on deviations in absolute magnitude between predicted and observed biomass levels; only deviations in seasonal patterns of change are considered. The reason for this weak comparison is that we could not be sure of units of measurement or conversion errors on the original data; in several cases where order of magnitude deviations between model and data were observed, we found similar inconsistencies within the original data.

10.5.4.1. Lake Wingra

Initial simulations gave a very distorted picture of phytoplankton and zooplankton changes in this lake (fig. 10.5). We predicted that grazing should deplete nannoplankton during the summer, unless effective grazing rates are much lower than have been reported in most experimental studies and certainly much lower than would be necessary to sustain grazers in the other test lakes. However, W. Lawascz (personal communication) has found much higher concentrations of suspended organic detritus in Lake Wingra than were used in the model, though in a year very different from the one to which model comparisons were made. We hypothesize that in reality this suspended material is filtered out along with nannoplankton, thus competing with the phytoplankton for passage through the filter feeder's gut. The detritus must also be of much lower food value than the algae; otherwise higher zooplankton stocks would be achieved (fig. 10.5). It appears that detritus acts as a mechanism to stabilize phytoplankton–zooplankton interactions in this lake.

The model produces a realistic bloom of net phytoplankton during the summer, even though no special assumptions were made about nutrient requirements of the large algae. The bloom occurs in the model when phosphate levels are very low, as has been reported for blue-green blooms, but the model does not even consider other factors (pH, alkalinity) that have been considered important in producing such blooms.

The model gives good predictions of macrobenthos (fig. 10.6) and fish (*Lepomis macrochirus*) biomass. The macrobenthos biomass in Lake Wingra is very low (1 g m^{-2}), while the fish biomass is very high (1 g m^{-2}) relative to other lakes, and the macrobenthos biomass peak occurs in winter. The model argues that this combination of observations is entirely due to fish predation.

10.5.4.2. Lake George, USA

Only phytoplankton and zooplankton data were available for this lake; predicted and observed values are compared in fig. 10.5. The pattern of phytoplankton seasonal response is well predicted by the model, though the average standing crop is not. Since Lake George has a total phosphorus (per square metre), temperature and incident light conditions not too different from Lake Wingra, the differences in predicted patterns between these lakes are largely due to the effects of water depth. The very poor fit to Lake George zooplankton data in fig. 10.5 may be due simply to our failing to include all biomass components in the data points.

The model predicts that predacious zooplankters should disappear from Lake George, when in fact cyclopoid copepods are quite abundant. We have fair confidence in our estimates of the ration curve parameters for predacious zooplankton, since reported values for these parameters are similar for a wide variety of organisms in different habitats (McQueen, 1969; Monakov, 1972; Fedorenko, 1975); thus we are not likely to have simply underestimated predator feeding efficiency. Simulations were attempted with much higher predator efficiency values than have been reported, but no great improvement was found; the herbivorous zooplankters were simply not productive enough to support observed cyclopoid populations. It appears that this failure of the model is due instead to a more fundamental omission: organisms may move from one trophic role to another. The cyclopoid copepods may require an algal feeding stage in order to persist in Lake George.

10.5.4.3. Vorderer Finstertaler See and Char Lake

The model predicted that phytoplankton and zooplankton biomasses should be relatively stable and similar over time in these cold, oligotrophic lakes (fig. 10.5). Char Lake has relatively higher total phosphorus levels, but receives less incident radiation. For both lakes we predicted that large algal forms and predacious zooplankters should not be present, and that even slight decreases in productivity would lead to disappearance of grazing zooplankters. In fact, the major grazer in Vorderer Finstertaler See is a cyclopoid copepod whose larger individuals are predacious and spend a considerable proportion of the year on the lake bottom where food resources are probably concentrated. It

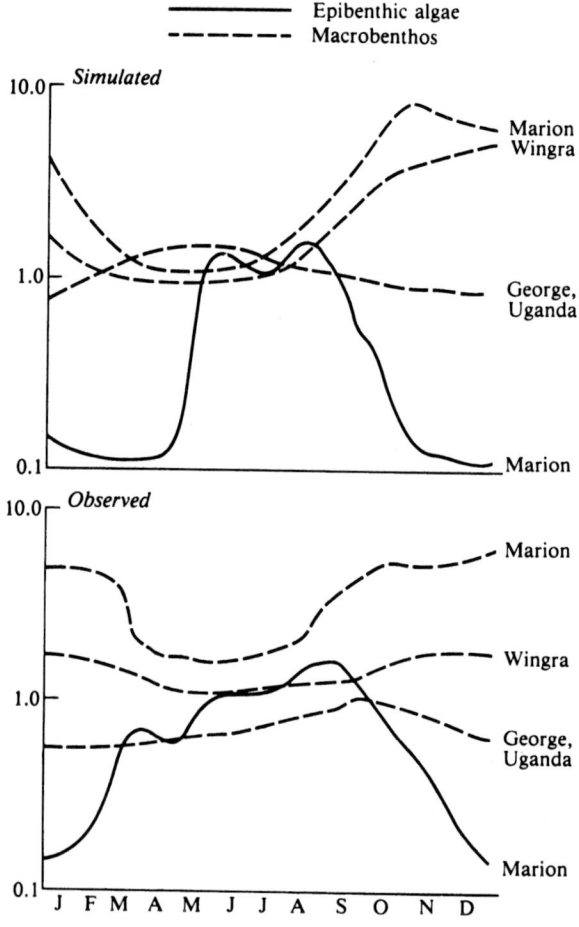

Fig. 10.6. Simulated and observed macrobenthos biomass for three IBP lakes, and simulated versus observed epibenthic algae biomass for Marion Lake; the general comparison model predicted very low epibenthic algal biomass for the other lakes compared in fig. 10.5.

would be interesting to see if this lake could support an obligate grazer who must feed throughout the year, for example a large *Diaptomus* species.

10.5.4.4. Marion Lake

The model gives good phytoplankton and zooplankton predictions for Marion Lake, but this is not surprising since plankton change is dominated by the simple physical process of water turnover. The model also predicted that Marion Lake, in contrast to the other shallow system (Lake Wingra), should have high epibenthic algae and benthos biomasses (fig. 10.6).

In one respect, results for this lake are especially encouraging. Allochthonous organic material and macrophytes are important in overall energy flow (Efford, 1967). The macrophytes are not even represented in the model, yet reasonable predictions were obtained for other components.

10.5.4.5. Lake George Uganda

This lake was included in the analysis to see whether the model would still predict dynamic instability under highly eutrophic conditions even in the absence of seasonal environmental change. As shown in fig. 10.5, the model meets this test very well, though it predicts lower biomasses for blue-green algae than are actually observed. Also, it was necessary to assume that the particulate organic materials produced by dying blue-green algae are not available to filter feeders; otherwise the model would predict extremely high population sizes of filtering zooplankton, which would in turn lead to the disappearance of small algae and planktonic bacteria.

10.6. Discussion and recommendations

Ecosystem modelling capabilities have increased greatly in recent years, partly through the impetus provided by IBP. In this chapter we have touched on some of these capabilities and have gone on to apply a composite model representing some of the features of two of the more advanced IBP models. By utilizing data sets collected from diverse lakes under the auspices of IBP it is possible to evaluate the generality of the model and, consequently, to indicate areas where further research is appropriate.

After testing the general biomass model with data from several of the IBP lakes, we might have concluded that the general framework and approach was sound but that new parameter values would be needed for each lake. However, the failures discussed in the previous section suggest that this conclusion is not warranted; the biomass approach has some more fundamental limitations that we might have overlooked if we had simply tried to force the curves through the data. By examining these limitations we can suggest some topics that may be fruitful for future research, and that are likely to be important in the development of truly predictive ecosystem models.

In the first place, parameter estimates were not completely independent of the data from the IBP lakes; initial simulation trials for several lakes gave very poor results that in hindsight led to obvious parameter adjustments to account for ecological adaptation to very cold or very productive situations. As shown in table 10.2 it was necessary to use higher values for relative respiration and feeding rates at low temperatures in the colder lakes, and it was necessary to assume that filtering efficiencies of zooplankters are lower in more productive lakes (higher BP values in table 10.2). These adjustments would not have been necessary if we had had the foresight to include as part of the model a set of 'site-parameter' relationships giving different parameter values as functions of physical conditions and general trophic status. The development of such relationships from basic evolutionary arguments and empirical observation would seem to be a fruitful area for future research.

A major limitation of most biomass models is in the representation of response to physical conditions, especially as related to seasonal patterns of water movement and stratification. As mentioned above, the usual approach to this problem involves dividing the water mass into more homogeneous layers or cells. Physical mixing models are adequate for representing transfers among these cells, although they greatly increase the computational load. However, some special biological problems arise, because some organisms can actively select which layers or cells to occupy. There has been little work on *a priori* models of habitat selection; such models would have to incorporate an understanding of the trade-offs that organisms must face in choosing among different environments.

A third major limitation is that biomass pools, or even species populations, are not internally homogeneous. Average productivities or efficiencies may be reasonable to use for predicting seasonal average biomass levels, but dramatic short-term changes in biomass can occur as specialist organisms encounter proper conditions for development. Thus the spring phytoplankton peak in Lake Wingra actually consists of a series of shorter blooms, with biomass declines between them; the overall model does not adequately represent any of these peaks and troughs. An obvious solution to this problem would be to use several more specific biomass pools, but again it would be necessary to have some *a priori* model for deciding what parameter combinations to expect as a function of physical conditions.

Finally, it appears necessary to face the problem of changing trophic roles in consumer organisms that have complex life histories. This problem cannot be solved simply by moving biomass from pool to pool over time according to observed changes in feeding patterns; to do so would be to bias the simulation by introducing a time-dependency without a functional basis. One alternative would be to use more narrowly defined pools, with feeding rules for each pool that reflect life history changes of the constituent organisms. This approach is facilitated by using a numbers–biomass algorithm. Unfortunately, the variety of possible life history strategies is almost infinite, and attempts to make general statements that could provide the basis for an *a priori* model have been sadly lacking. This problem is resolved by modelling the dominant species. However, as Walters & Efford (1972) have pointed out, species-level predictive modelling is not likely to be fruitful in many types of ecosystem studies: species compositions change in response to disturbance, and it is not usually possible to forecast the dominants that will appear under the new set of conditions.

Even in the face of these problems, it would be unwise to conclude that ecosystem modelling is premature and that much more descriptive work is needed before we can even think about predictive ecological theories. We find it difficult to see how even a very large number of descriptive studies would help much in answering the questions posed in this section. Thus we argue that model building should continue to be an important tool in ecosystem studies, not as an end in itself but as a means to help anticipate the questions that will have to be answered when the time for general theories does come. From a practical standpoint, the general biomass model described above appears to have enough predictive ability for wide-ranging situations to justify its use for anticipating eutrophication and thermal pollution effects in lakes that have not yet been studied.

The research of R.A. Park and J.F. Koonce was supported by the Eastern Deciduous Forest Biome, US IBP, funded by the National Science Foundation under Interagency Agreement AG-199, 40-193-69 with the Atomic Energy Commission, Oak Ridge National Laboratory.

11. General characteristics of freshwater ecosystems based on Soviet IBP studies

G.G. WINBERG

11.1. Introduction

It was evident from the beginning that the 'Productivity of freshwater communities (PF)' section had to concentrate on studies of all stages of the process of production from primary to final production, i.e. the whole ecosystem. This was realized in the Soviet Union to the utmost extent, stimulated by the organization and traditions of hydrobiological studies in this country and by the pattern of research and data already obtained at Drivyaty Lake in 1964. This pattern presented different levels of the production process as a flow of biologically transformable energy (Winberg, 1969, 1970a, b).

The first results of freshwater production studies were discussed at the International PF Symposium at Kazimierz Dolny (Poland) in 1970, with fourteen papers by Soviet hydrobiologists published in the Proceedings (Kajak & Hillbricht-Ilkowska, 1972). Soviet IBP/PF results were also presented briefly at the Symposium held at the Biological Station of the Byelorussian University at Naroch Lake in May 1972 and published as 'Productivity Studies of Freshwater Ecosystems' (Winberg et al., 1973). Material from both these symposia is used for this chapter, which is thus based on earlier generalizations from Soviet studies (Winberg et al., 1971; Winberg, 1972b). The general characteristics of freshwater ecosystems and some data on primary production and bacterioplankton are reviewed here; data on zooplankton and on benthos were given by Ivanova (1975) and Alimov (1975) respectively.

IBP activities included testing, evaluation and improvement of techniques and methods, for example methods of calculating values of animal production. In this way the quantitative indices cited below were obtained; many of them are liable to further verification and are at present not quite comparable. This is one reason for difficulties in comparing data obtained in different water bodies by different teams. The general distinctions of the production process emerge in spite of deficiencies in methods and data; many important but more specialized items of each particular study cannot be considered. Full results have not yet been published for most IBP projects. Data concerning the productivity of L. Baykal are not included here both because of the unique character of this water body and the lack of necessary information in the published data.

A quantitative comparison of the stages of the production process has become possible only due to the trophic levels concept of solar radiation utilization in ecosystems. In practice the application of this general idea meets many difficulties. So the main result of the IBP studies lies not only in the immediate results, but also in the experience gained for the evaluation of methods used and in the underlining of fundamental problems.

The data used here were obtained at lakes and reservoirs in the Soviet Union and characterized in table 11.1

The lakes Krivoye (oligotrophic) and Krugloye (dystrophic) are situated in North Karelia on the shore of the Kandalaksha Bay of the White Sea, 30 km south of the Arctic Circle. These were studied in 1968 and 1969 by the freshwater and experimental biology laboratory of the Zoological Institute of the USSR Academy of Sciences (Alimov et al., 1972; Winberg et al., 1973). In 1970 and 1971 the same laboratory studied L. Zelenetskoye, one of many lakes on the stony tundra shore of the Barents Sea (Winberg et al., 1973). Lake Kharbey in the Bolshezemelskaya tundra, another northern lake (67.5°N), was studied in 1969 by the Komi Branch of the USSR Academy of Sciences (Vlasova et al., 1973). The Northern Institute for Lake and River Fisheries studied the Karelian L. Chedenjarvi in the Shuya basin (Gorbunova et al., 1973). Lake Krasnoye (Punnus-Yarvi) of the Karelian Isthmus was studied for many years by the Institute of Limnology of the USSR Academy of Sciences (Andronikova et al., 1972, 1973; Kalesnik, 1971).

The Byelorussian lakes Naroch (mesotrophic), Myastro (slightly eutrophic) and Batorin (eutrophic) have been studied by the Byelorussian University since 1947 (Winberg et al., 1971, 1972; Ostapenya et al., 1973). Earlier, the same team had studied the eutropic lake Drivyaty (Biological Productivity of an Eutrophic Lake, 1970). The Rybinsk Reservoir was the main subject of studies by the Institute of Biology of Inland Waters of the USSR Academy of Sciences (Kuzin, 1972; Sorokin, 1972; Romanenko, 1973), and the Kiev Reservoir the main subject for the Institute of Hydrobiology of the Academy of Sciences of the Ukrainian Republic (Kievskoye Vodokhranilishche 1972, Gak et al., 1972, Tseeb et al., 1973).

Karakul Lake, 120 km from the city of Alma-Ata in the Ili basin, was studied in 1967 and 1968 by the Department of Hydrobiology and Ichthyology of Kazakh University (Khusainova et al., 1973).

11.2. Lake characteristics and primary production

Table 11.1 lists the lakes in order of increasing primary productivity, measured by oxygen modification of dark and light bottle techniques, except in the least productive Lake Zelenetskoye and the Rybinsk Reservoir where carbon-14

Table 11.1 *Some characteristics of the lakes studied and primary production for the growing season* (P_1)

Lake	Latitude	Area (km²)	Mean depth (m)	Water transparency (m)	Approximate number of ice-free days	P_1 (kJ m⁻² g⁻¹)
Zelenetskoye	69°N	0.24	8.5	4.0–4.7	100	88
Krugloye	65°N	0.24	1.5	1.5	130	150
Krivoye	66°N	0.5	11	4.5	130	565
Gr. Kharbey	67°N	24.3	4.6	2.5–3.8	90	627
Naroch	54°N	79.6	9.0	5–6	170	2 037
Rybinsk Res.	59°N	4450	5.6	0.7–1.5	180	2 092
Karakul	39°N	1.4	1.4	0.95–1.6	240	2 929
Krasnoye	66°N	9.1	6.6	1.5–2.7	170	4 414
Drivyaty	55°N	32.6	5.2	1.5–2.0	170	5 021
Myastro	54°N	13.1	5.4	1–2	170	6 585
Kiev Res.	50°N	929	4.0	—	210	6 987
Batorin	54°N	6.3	3.0	0.3–0.6	170	7 355

was used. P_1 values listed in the table designate conventional (net) phytoplankton production as 80% of gross production, i.e. $P_1 = 0.8\ A_1$ (except for the lakes Zelenetskoye and Krugloye where $P_1 = 0.9\ A_1$).

The great range of P_1 is noticeable, namely eighty-four times from the least productive to the most productive water body. This difference is stressed by the fact that in the least productive water bodies (L. Zelenetskoye, L. Krivoye) there are practically no other nutrient sources for primary production, whereas in some other productive water bodies (Kiev Reservoir) there are important other sources.

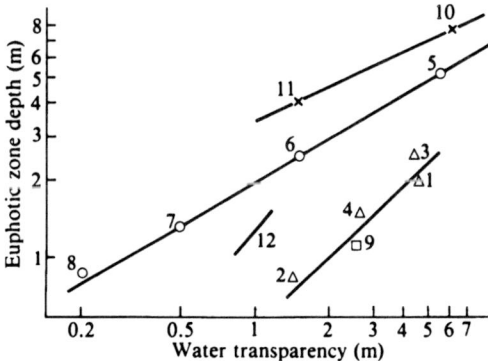

Fig. 11.1. Relationship between water transparency and the relative depth of the euphotic zone in July–August in waters at various latitudes. 1, 2, 3, 4 and 9 – northern lakes (1, Krivoye; 2, Krugloye; 3, Zelenetskoye; 4, Akulkino; 9, Kharbey); 5, 6, 7 and 8–Byelorussian lakes (5, Naroch; 6, Myastro; 7 and 8, Batorin); 10 and 11 – Mingechaur Reservoir on the Kura (Salmanov, 1961); 12 – Rybinsk Reservoir according to Romanenko's equation (1973) $(F\ m^{-2}) = (F_m\ m^{-3})\ 2.1\ S$, where S is the water transparency by Secchi disk. Horizontal axis – transparency of water by Secchi disk; vertical axis – the ratio $(F\ m^{-2})/(F_m\ m^{-3})$.

A comparison of the data on primary production and on the water transparency shows that, if the transparency is the same, the depth of the euphotic layer decreases rapidly to the north. Accordingly the ratio of the maximum rate of photosynthesis per unit of water ($F_m\,m^{-3}$) to photosynthesis under a unit of surface ($F\,m^{-2}$) also decreases, as demonstrated in fig. 11.1. Among other details this demonstrates that the relationship determined by Romanenko between F and S is true only for water bodies situated at latitudes of the Rybinsk Reservoir.

11.3. *P/B* ratios at different trophic levels

Table 11.2 demonstrates that P/B ratios for the growing season vary considerably but are generally within the values determined by possible rates of cell division under natural conditions. These differences in P/B values cannot yet be discussed as results are contradictory and further studies are needed. Some hydrobiological publications contain very high P/B values for phytoplankton which are not easily explained by physiological reasons. Thus Kozhova (1973) considered the high P/B ratios for phytoplankton in Baykal Lake while its biomass is very low to be a typological peculiarity of Baykal Lake. The clarification of conditions determining the P/B values of phytoplankton is an actual and urgent problem in hydrobiology.

This problem may be solved only by special studies with adequate techniques and with parallel observations on cell multiplication, measurements of photosynthetic intensity and phytoplankton biomass by chlorophyll content or other objective methods.

The bacterial biomass for the water bodies considered was calculated from

Table 11.2 *P/B ratios for growing season*

Lake	Phyto-plankton	Bacterio-plankton	Zooplankton		Benthos	
			non-carn-ivorous	carnivorous	non-carn-ivorous	carnivorous
Zelenetskoye	10.5	11	3.5	7.9	2.8	1.8
Krugloye	100	40	12.4	11.2	2.1	3.0
Krivoye	85	30	13	5	1.3	2.6
Gr. Kharbey	110	—	4.7	1.8	1.1	1.6
Naroch	39	63	16.3	29.5	2.4	2.3
Rybinsk Res.	35	64	(50)	14.6	2.4	3.0
Karakul	—	—	25.4	30.6	12.8	25.0
Krasnoye	112	95	21.1	10.8	4.2	5.0
Drivyaty	22	61	20	9	0.7	3.7
Myastro	46	70	13.9	26.4	4.7	3.5
Kiev Res.	42	54	27.4	13.7	3.6	3.9
Batorin	38	38	18.4	21.2	4.8	4.7

the general quantity of bacteria determined by direct count and by the average volume (weight) of one bacterial cell in each of the waters. The production of bacterioplankton is determined by the observed rate of increase in numbers of bacteria in water freed from zooplankton by filtration. Only for the Rybinsk Reservoir was the production of bacterioplankton determined by the rate of heterotrophic carbon-14 assimilation (Romanenko, 1973).

The bacterioplanktonic production (P_b) increases, in general, parallel to the increase in primary production of plankton (P_1). It is characteristic and quite natural that the P_b/P_1 ratio (demonstrated in table 11.3) is highest (0.78) in the dystrophic Lake Krugloye with its swampy drainage basin, and in the Rybinsk (0.70) and Kiev (0.44) reservoirs, with a considerable or most part of bacterioplanktonic production due to allochthonous organic matter. In all the other lakes this ratio is within strikingly narrow limits, 0.20–0.46, i.e. bacterioplanktonic production is from 20 to 46% (mean 31%) of phytoplankton production, which in these lakes increases from 21 to 1758 kcal (88–7355 kJ) $m^{-2} a^{-1}$. This tendency points to the decrease in P_b/P_1 from the least to the most productive waters.

It is more informative to compare with the primary production not the bacterioplankton production, but its food requirements expressed as the food energy assimilated during the same period (A). This value may be estimated if the growth efficiency is known for the evaluation of the production; the growth efficiency may be expressed by the quotient $K_2 = P/A$. Where A is the assimilation: $A = P + R$, where R is metabolic expense. It is obvious that $A = (1/K_2)P$.

The meaning of K_2 for bacterioplankton has been repeatedly discussed. For all water bodies considered except the Kiev Reservoir the value $K_2 = 0.25$ is assumed for bacterioplankton. The same value is used by Sorokin in the IBP Handbook on aquatic bacterial production (Sorokin & Kadota, 1972). For the Kiev Reservoir Gak assumed $K_2 = 0.4$. There are many reasons to believe that at least for eutrophic waters the average K_2 for bacterioplankton is closer to the latter figure. At $K_2 = 0.25$, $A_b = 4P_b$, if $K_2 = 0.4$, $A_b = 1.5P_b$. It is more reasonable to assume that A_b is 2–3 times more than P_b. Then if P_b constitutes 20–46% of P_1, A_b would be from 40–92% to 60–138% of P_1. In principle it is quite possible that $P_b > P_1$ due to external sources of primary production not included in P_1; and in waters with a relatively high input of allochthonous organic matter, as in Krivoye Lake, Rybinsk and Kiev reservoirs, it is a general rule that $P_b > P_1$.

The ratio of the production for the growing season to the mean biomass of the bacterioplankton (P_b/B) is within the limits 38–95 for all the waters considered except Zelenetskoye lake, where it has a very low value of 11. In all the waters considered P/B for the bacterioplankton is expressed by the same range of figures as for the phytoplankton P/B, the averages (56 and 54) coinciding (Table 11.2).

These quoted values for bacterioplanktonic P/B ratios were determined by experienced authors: Yu. I. Sorokin and V.I. Romanenko (Rybinsk Reservoir), V.G. Drabkova (Krasnoye Lake), D.Z. Gak (Kiev Reservoir), Yu. S. Balyatskaya (Byelorussian lakes), N.K. Kuzmitskaya, T.V. Zharova, M.V. Fursenko (Krivoye, Krugloye, Zelenetskoye lakes). The unexpected absence of notable and regular differences in P/B of the bacterioplankton in northern waters with low productivity and the southern waters with higher water temperatures and a longer growing season, deserves special attention and comparative studies.

IBP studies greatly stimulated the improvement of techniques for determining the production of aquatic animals, as well as the biological peculiarities of the abundant species of freshwater animals and of general ecological–physiological relationships. The latter have been discussed at the USSR IBP/PF symposia (Winberg *et al.*, 1973; Gutelmacher *et al.*, 1974). The production of the abundant species may now be determined with the same accuracy as for their numbers or biomass, and specific production (daily P/B etc.) may probably be determined more exactly.

The P/B ratios of zooplankton for the growing season (close to the annual P/B (see table 11.2)). are definitely lower in the four northern lakes (1.8–12.4) with an average of 7.5, compared with values of 9–30.6, average 19.9 for other waters.* For zoobenthos the summed P/B values are generally more conventional. As zoobenthos is unevenly distributed the evaluation of a reliable average biomass is complicated. The biological heterogeneity of the zoobenthos and big differences in the environment produce high fluctuations of P/B, their determination also being difficult due to insufficient knowledge of the ecological–physiological peculiarities of various species. Nevertheless, table 11.2 gives P/B values proposed by different authors. High P/B values exist for the benthos of Karakul Lake where *Chironomus plumosus* predominate in the non-carnivorous benthos, for these have three generations annually in this southern water body.

For small chironomid larvae considered as carnivores, the P/B value taken is double and obviously overestimated. The lowest P/B for Drivyaty Lake is connected with a very high biomass of large bivalve molluscs. In other cases P/B values for non-carnivorous benthos are within 1.14–4.8 (with an average of 2.74), for carnivorous benthos P/B is 1.6–5.0 (the average being 4.10). The P/B for zooplankton is considered in more detail by Ivanova (1975) and for zoobenthos by Alimov (1975).

---

* P/B for non-carnivorous zooplankton of the Rybinsk Reservoir is not included as the calculations for the production of separate species have not been made and the plankton production was determined by the assumed P/B which seems to be too high.

11.4. The relative values of production at different trophic levels compared with primary production

In table 11.3 the values for zooplankton production (P_z) and zoobenthos production (P_d) are expressed as percentages of P_1. For the calculation of the ecological efficiency ratio, i.e. the ratio of the food requirements (C) of the animals of the given trophic level to the production of the previous level (C_n/P_{n-1}), it is necessary to know the C values determined with similar but somewhat different K_2 values. It has recently been found that species of both planktonic (M.B. Ivanova) and benthic (A.F. Alimov) animals having a longer life span use the assimilated food for growth less efficiently, i.e. they are characterized by lower K_2 values than species having a short life span. This conclusion of great general ecological interest has also been mentioned by other authors (McNeil & Lawton, 1970).

Taking the above into consideration, the values of C included in table 11.3 are calculated as $C = (1/aK_2)P$ and with the average K_2 value assumed to be 0.3. The assimilated food (A) for non-carnivores is taken as 0.7, for carnivores as 0.8.

The dystrophic L. Krugloye and both Rybinsk and Kiev reservoirs occupy a special position because of their great inflow of allochthonous organic matter. It is characteristic that for these three waters the relative values of P and C reach maximum values (table 11.3). For the other nine water bodies the production of non-carnivorous zooplankton fluctuates irregularly within

Table 11.3 *Relative values of bacterioplanktonic production (P_b), non-carnivorous zooplankton (P_{2z}), carnivorous zooplankton (P_{3z}), carnivorous benthos (P_{3d}), and of nutritive requirements $(\Sigma C_2$ and $\Sigma C_3)$ as percentages of primary production $P_1)$*

| Lake | P_b | Non-carnivorous | | | | Carnivorous | | | | $\Sigma C_3/\Sigma P_2$ | $\Sigma P_2/\Sigma P_3$ |
		P_{2z}	P_{2d}	ΣP_2	ΣC_2	P_{3z}	P_{3d}	ΣP_3	ΣC_3		
Zelenetskoye	29	1.7	11.3	13.0	62	0.5	0.5	1.0	4.2	0.32	13
Krugloye	78	30.5	12.5	43.0	204	5.6	1.9	7.5	31.3	0.73	6
Krivoye	46	12.6	1.4	13.0	62	1.9	0.3	2.2	9.2	0.71	6
Gr. Kharbey	—	2.2	2.1	4.3	20	0.2	0.2	0.4	1.7	0.40	11
Naroch	35	13.7	2.5	16.2	77	0.4	0.1	4.1	17.1	1.05	4
Rybinsk Res.	70	15.6	1.6	17.2	82	1.3	0.1	1.4	5.8	0.34	12
Karakul	—	5.8	3.4	9.2	44	0.4	0.4	0.8	3.3	0.36	11.5
Krasnoye	23	8.7	1.1	9.8	47	1.0	0.1	1.1	4.6	0.47	9
Drivyaty	37	10.0	0.9	10.9	52	2.6	0.2	2.8	11.7	1.07	4
Myastro	26	7.6	0.2	7.8	37	2.8	0.04	2.8	11.7	1.50	2.5
Kiev Res.	44	20.7	9.2	29.9	142	0.7	0.2	0.9	3.7	0.12	33
Batorin	20	7.8	0.7	8.5	40	3.1	0.1	3.2	13.3	1.57	2.5

limits of 1.7–13.7%, the average being 7.8%.* The total production of the non-carnivorous plankton and zoobenthos in the same nine lakes ranges within limits of 4.3–16.2% (the average being 10.3% of P_1), and their total nutritive requirements for the growing season are 20–77% (averaging 49% of P_1).

The values P_3 and C_3, i.e. the production and nutritive requirements of carnivores, are less accurate as there are always uncertainties as to the carnivorous nature of a species or a stage and the food composition in definite circumstances. It is quite probable that the high ratio of food requirements of carnivores to the production of non-carnivores ($\Sigma C_3/\Sigma P_2$) in four Byelorussian lakes (1.05, 1.07, 1.50, 1.57) may be explained by the inclusion of only partly carnivorous animals. The lowest ratio $\Sigma C_3/\Sigma P_2$ 0.12 is for Kiev Reservoir. In the other seven cases it is 0.32–0.73, with a mean of 0.47.

The ratio of productions of animals having carnivorous and non-carnivorous feeding ($\Sigma P_2/\Sigma P_3$) is uncommonly high for Kiev Reservoir and equals 33. For the other eleven waters it fluctuates irregularly within limits of 2.5–13, (the average being 7.4% of P_1, or if Kiev Reservoir is included 9.6% of P_1).

Thus in spite of the great differences in the productivity and in the other properties of the lakes studied, no regular differences in the relative values of P and C were found. Otherwise, the production of animals of the plankton and benthos and their food requirements have the same ratio to the primary production of the phytoplankton.

11.5. Metabolic costs for phytoplankton and bacterioplankton

The same studies also enable us to form an idea of the relative participation of phyto-, bacterio- and zooplankton in the breakdown or mineralization of organic matter in water. As indicated above, it was assumed that $P_1 = 0.8A_1$. It follows that the phytoplankton respiration $R = 0.2\ A_1 = 0.25P_1$ (except for Zelenetskoye Lake and Krugloye Lake where it was assumed $R = 0.1A$).

The cost of the breakdown of bacterio- and zooplankton may be calculated from the ratio $R = (1 - K_2)/K_2$. For zooplankton it is assumed, as above, that on average $K_2 = 0.3$. Consequently $R = 7/3\ P_3 = 2.33P_3$. For bacterioplankton the mean value is taken as between the two supposed values 0.25 and 0.4 (see above), i.e. $K_2 = 1/3 = 0.333$. In this case $R_b = 2P_b$. The sum of three values calculated by this method $\Sigma R = R_1 + R_b + R_z$ is given in table 11.4.

It is noticeable that there is a good agreement between ΣR calculated by the production and A_1 measured independently (fig. 11.2). The calculated values of ΣR are generally very close to A_1. It is especially remarkable that as one

---

* The lowest value 1.7 was determined for Zelenetskoye Lake which cannot be considered to be characteristic for this lake with the lowest primary production as these data were obtained for 1971, whereas in 1970 the zooplankton production in the same lake was several times higher (Winberg *et al.*, 1973).

Table 11.4 *Gross primary production* (A_1), *the total calculated metabolic expenses* (ΣR) *for phytoplankton* (R_1), *bacterioplankton* (R_b) *and zooplankton* (R_z), *and their relative percentage values of the total*

Lakes	Primary production (kJ m^{-2} g^{-1})	Total metabolic expenses (ΣR) (kJ m^{-2} g^{-1})	Percentage of ΣR		
			R_1	R_b	R_z
Zelenetskoye	113	79	32	63	5
Krugloye	167	376	4	62	34
Krivoye	628	753	8	67	25
Naroch	2 548	2 657	19	53	28
Rybinsk Res.	2 594	4 021	12	73	15
Krasnoye	5 518	4 067	27	50	23
Drivyaty	6 276	6 418	19	58	23
Myastro	8 234	6 594	25	51	24
Kiev Res.	8 979	11 573	17	53	30
Batorin	9 192	6 581	28	44	28

could expect (see above) just for Lake Krugloye, Rybinsk and Kiev Reservoirs $\Sigma R \gg A_1$.

The values for the relative participation of the phyto-, bacterio- and zooplankton in the breakdown are equally interesting (table 11.4). As percentages of ΣR the phytoplankton part is 4–32% (average 19.1%), the bacterioplankton part 44–73% (average 57.4%) and the zooplankton part 5–34% (average 23.5%). It is remarkable that these figures are very close to the

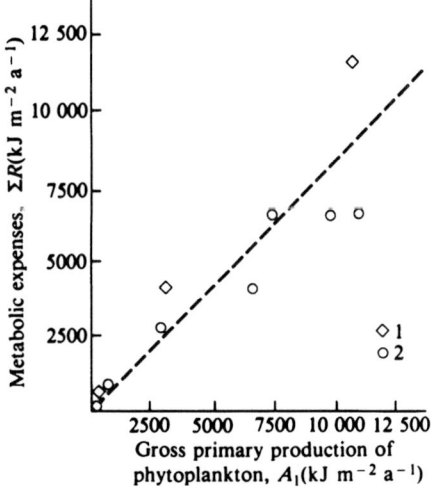

Fig. 11.2. Comparison of gross primary phytoplankton production (A_1 kJ m^{-2} a^{-1}) and the calculated value of metabolic expenses of phyto-, bacterio- and zooplankton (ΣR kJ m^{-2}a^{-1}). 1, reservoirs; 2, lakes.

directly measured relative values of bacterio- and zooplankton in the oxygen consumption. Thus according to Romanenko (1973) the mean rate of oxygen consumption of the water from the Rybinsk Reservoir, freed from plankton by filtration ('bacterial respiration') is 70.5% of its consumption by unfiltered water. In table 11.4 R_b is 73% for the Rybinsk Reservoir. According to Shcherbakov (1967) and Shcherbakov & Ivanova (1969), the oxygen consumption by zooplankton in the mesotrophic Glubokoye Lake is 5–21% of the total oxygen consumption in various months, the average being 15%; in the highly eutrophic Beloye Lake it is 9–23%, the average being 17%.

These and other data may produce the impression that the figures in table 11.4 are very realistic, but the bacterial share might be somewhat higher, and that of the zooplankton lower. Such a result would be produced if $K_2 = 0.25$ were taken to calculate R_b. Then $R_1 = 15.5$, $R_b = 66$ and $R_z = 18.5\%$ of ΣR, however much absolute values of R_b and ΣR increased, and the ΣR would be considerably higher than A_1 for most lakes. These questions may be solved only by further studies.

Above all the values were related to the primary production of phytoplankton (P_1). In some waters phytoplankton constitutes the main source of the primary production, while in others a considerable part of the primary production is contributed by macrophytes, periphyton, microphytobenthos. The latter is true, for example in Naroch Lake and Kiev Reservoir, where about 50% of the total primary production is by the phytoplankton. The extreme case is Karakul Lake, where the annual phytoplankton production is evaluated as 560 kcal (2352 kJ) m^{-2} and macrophyte production as 8000 kcal (33 600 kJ) m^{-2}. It would seem that if the significance of other sources of primary production is so great in some waters, then the phytoplankton primary production cannot be used as the main basis for comparison. Actually, and as the above testifies, P_1 turned out to be unexpectedly very well connected with the values of the other levels of the production process. This gives reason to assume that the essentially different ways in which primary production of phytoplankton and of macrophytes enter into the biological turnover, and the low efficiency of utilization of the macrophyte production, result in the pelagic ecosystem being probably to some extent autonomous.

11.6. Comments on fish production

In some USSR IBP studies data are presented for the final utilizable production (fish catch). This material is of considerable interest in some cases, but is not considered here. Reliable values of fish stocks are obtainable only with great difficulty. For some waters (Karakul Lake, Chedenjarvi Lake, four lakes in Byelorussia) the catch of non-predatory fish was calculated as 0.32–0.72% of P_1, which agrees with earlier information. It is usually supposed that the catch forms a considerable or even the largest part of the annual fish production in lakes commercially fished. Contrary to this, data from the ichthyological laboratory of the Institute of Biology of Inland Waters of the USSR Academy of Sciences (Sorokin, 1972) suggest that fish production in the Rybinsk Reservoir is 281.5 kg ha^{-1} (including 61.2 kg ha^{-1} of carnivorous fish), while the catch is only 8.3 kg ha^{-1}. According to these data the production of only non-carnivorous fish constitutes 4.14% of P_1. This value is quite uncommon for lakes, and twice as high as the similar index for extensively exploited carp ponds, where the efficiency of primary production utilization is several times higher than in lakes.

Although fish have a low immediate participation in the energy flow of a water body, the present author has stressed many times that this does not indicate their low importance in the ecosystem. The biomass of, for example, zooplankton is determined by the food resources available to it, but its size and species composition depends on its consumption by fishes. The fish population is an important constituent of the ecosystem with a strong influence on the other elements. Unfortunately the significance of fish in the ecosystem has been inadequately studied, being limited by the 'ichthyocentric' ideas of the 'fish environment', which in practice is taken to be independent of the influence of the fish.

Finally it should be remembered that studies of the general rules of biotic transformations of matter and energy in aquatic ecosystems are the necessary but not exclusive basis for their study; they may 'stimulate the biological analysis of the ecosystem components' (Winberg, 1962). Any biological fact under natural conditions is in some way limited by the general properties of the ecosystem. So the above comparison of the final data from many authors has revealed some general qualities of freshwater ecosystems and given orientation for more differentiated studies which may discover the characteristic distinctions of various types of aquatic ecosystem.

Appendix I. Sites of IBP/PF investigations

To facilitate comparative analyses of the results from widely-distributed PF sites participant scientists were required to tabulate the vital data, as known in 1973, on forms called Data Reports (DR). In many cases the preparation of these preceded the publication of results in scientific papers. The 92 Data Reports (DR 1–93, as DR 36/37 were combined) summarized data from 120 sites: 91 from *lakes*, 11 from *reservoirs*, 8 from *ponds*, with only 10 from *running water*.

The *global distribution* of these Data Reports (see endpaper map) was: 32 from Europe plus 14 from USSR west of the Urals, 5 from USSR east of the Urals; 12 from Asia south of the USSR (including 6 from India, 5 from Japan); 16 from North America (12 from USA, 4 from Canada); 5 from Africa; 4 from Australia; 3 from South America; 1 from the Middle East (Israel) and 1 from Antarctica. The majority of them came from below 500 m *altitude*, but at least 11 sites (DR 1, 2, 5, 8, 19, 26, 30, 32, 34, 48, 49) were from higher altitudes. Their *latitudinal distribution* was as follows: 62 from the north temperate zone; 12 from north of 60 °N (DR 2, 39, 59, 60, 61, 62, 64, 70, 71, 72, 79, 82); 9 from the intertropical zone, 23 °N to 23 °S, (DR 4, 5, 9, 20, 31, 41, 76, 77, 90); 6 from south of 24 °S (DR 14, 26, 51, 52, 54, 92); 1 from south of 60 °S (DR 83). Thus work in the north temperate zone dominated the total IBP/PF effort, and lake studies predominated, with very few in running water.

List of data reports

Original Data Reports are deposited with the Freshwater Biological Association, Ambleside, England. The list here indicates latitude, longitude, altitude, area and depth (maximum depth, z, unless mean depth, $\bar{z}$, is indicated) where information is readily available, together with the name of the Chief Investigator and address while the work was in progress.

DR 1. *Vorderer Finstertaler See*: Austria. 47° 12′ N, 11° 2′ E; 2237 m alt.; 15.7 ha; z 28.5 m.
R. Pechlaner, Zoological Institute, University of Innsbruck.

DR 2. *Øvre Heimdalsvatn*: Norway. 61° 25′ N, 8° 43′ E; 1090 m alt.; 0.77 km²; z 13 m.
P. Larsson, Zoological Museum, Sarsgaten, Oslo.

DR 3. *Loch Leven*: Scotland. 56° 10′ N, 3° 30′ W; 107 m alt.; 13.31 km²; z 25.5 m.
N. Morgan, Nature Conservancy, Edinburgh.

DR 4. *L. Chad*: Republic of Chad, Africa. 12°–14° N, 13°–15° E; 282 m alt.; 8000–25000 km²; z 3.8 m.
C. Lévêque, ORSTOM, Fort Lamy, Chad.

DR 5. *L. Chilwa*: Malawi, Africa. 15° 30′ S, 36° E; 623 m alt.; 700 km²; z 3 m.
C. Howard-Williams, c/o Mrs M. Kalk, University of Malawi, Limbe.

DR 6. *Abbot's Pond & Priddy Pool*: Bristol, UK 51° 28′ N, 2° 40′ W.
F. Round, Botany Department, University of Bristol.

DR 7. *Reservoir Saidenbach*: DDR, E. Germany. 50° 44′ N, 13° 14′ E; 438 m alt.; 146 ha; z 45 m.
F. Uhlmann, Sektion Wasserwesen, Technische Universität, Dresden.

DR 8. *L. Yunoko*: Japan. 36° 48′ N, 139° 30′ E; 1478 m alt.; 0.33 km²; z 12 m.
S. Schiniura, Y. Yamaguchi, Department of Botany, Kyoiku University, Otsuka, Tokyo.

DR 9. *Temple 'tank'* (pond): India. 20° 15′ N, 73° E; 423 m alt.; 4884 m²; z 6.3 m.
S.V. Ganapati, Department of Biochemistry, N.S, University of Baroda, Baroda.

DR 10. *Jezárko Pond*: Czechoslovakia. 4.0 ha; z 1.3 m.
J. Fott, Department of Hydrobiology, Charles University, Prague, 2.

DR 11. *4 lakes, Washington*: USA. ca 47° N, ca 122° W.
Sammamish, 12 m alt., 19.8 km²; z 32 m; *Chester Morse*, 473 m alt., 6.54 km², z 35 m; *Findley, Washington*; and 4 *Cedar River drainage lakes*.
E. Welch, Department of Civil Engineering, University of Washington, Seattle.

DR 12. *L. Wingra & L. George*: New York, USA, ca 43° 31′ N, 73° 50′ W; 97 m alt.; (primary production; macrophytes).
M.S. Adams, Department of Botany, University of Wisconsin, Madison.

DR 13. *L. Wingra & L. George*: NY, USA. (See DR 12); (phytoplankton and chemistry).
J.F. Koonce, Institute for Environmental Studies, University of Wisconsin, Madison.

DR 14. *4 saline lakes in Victoria*: Australia. ca 38° S, 143° E. *Coragulac*, 105 m alt.; 22.4 ha, z 2 m; *Corangamite*, 115 m alt., 23 370 ha, z 5 m; *Red Rock Tarn*, 164 m alt., ca 1 ha, z 2 m; *Pink Lake*, 115 m alt., 13.4 ha, z 70 cm.
U.T. Hammer, Department of Biology, University of Saskatchewan, Saskatoon, Canada.

DR 15. 4 *lakes in Saskatchewan*: Canada, *ca.* 50° N, 104° W. *Buffalo Pound*, 509 m alt., 29.5 km², *z* 5.6 m; *Echo Lake*, 479 m alt., 12.5 km², *z* 22 m; *Katepwa*, 478 m alt., 16 km², *z* 23 m; *Pasqua*, 479 m alt., 19.9 km², *z* 15 m.

U.T. Hammer (as DR 14).

DR 16. 3 *lakes near Saskatoon*: Canada, *ca.* 52° N, 105° W. *Blackstrap*, 535 m alt., 7.6 km², *z* 8.8 m; *Burton*, 530 m alt., 0.4 km³, *z* 7 m; *Humboldt*, 530 m alt., 4.4 km², *z* 7 m.

D.T. Waite, U.T. Hammer (as DR 14).

DR 17. *L. Wingra*: USA. (See DR 12). Predator–prey model, fishes.

J.F. Kitchell, Laboratory of Limnology, University of Wisconsin, Madison.

DR 18. *Trout streams* (2): Moravia, Czechoslovakia. Brodská & Lušová Brook, tributaries of Bečva River (Danube system), Beskydy Mountains.

J. Libosvárský, Ichthyological Section, Institute for Vertebrate Zoology, Czech. Academy of Sciences, Kvetna, 8, Brno.

DR 19. *Chigonosawa Brook*: Japan. *ca* 35° N, *ca* 137° 30′ E; 700 m alt.; 1500 m².

T. Miura, Otsu Hydrobiological Station, Shimosakamoto, Otsu, Shigakeu.

DR 20. *Upper Paraná River*: Brazil. 22° S, 47° W.

P. de Godoy, Experimental Station of Biology & Fish Culture, 13630 Pirassunungu, São Paulo.

DR 21. *Tjeukemeer*: Netherlands. 52° 50′ N, 5° 45′ E; −1 m alt.; 21 km²; $\bar{z}$ 1.5 m.

H. Golterman, Limnologisch Instituut, Nieuwersluis.

DR 22. *L. Balaton*: Hungary. Fish production (see DR 27, 44). 46° 50′ N, 17° 45′ E.

P. Biró, Biological Institute, Hungarian Academy of Sciences, Tihany.

DR 23. *L. Wingra & L. George*: USA. Zooplankton (see DR 12, 13).

D.C. McNaught, Department of Biological Sciences, State University of New York, Albany, NY.

DR 24. *Spring Rold Kilde*: Denmark. (Leaf decomposition.)

T.M. Iversen, Freshwater Biological Laboratory, Hillerød.

DR 25. *Lawrence Lake, Michigan*, USA. 42° 27′ N, 85° 22′ W; 4.9 ha; *z* 12 m.

R.G. Wetzel, Kellogg Biological Station, Michigan State University, Hickory Corners.

DR 26. *Dam basin, Armidale*: Australia. (Crayfish production.) 30° 30′ N, 151° 36′ E; 975 m alt.; 950 m².

D.J. Woodland, University of New England, Armidale, New South Wales.

DR 27. *L. Balaton*: Hungary. (Plankton) (see DR 22).
J.E. Ponyi (see DR 22).

DR 28. *L. Trummen*: Sweden. 56° 52′ N, 14° 50′ E; 161 m alt.; 0.8 km²; z 2.5 m.
S. Björk, Institute of Limnology, University of Lund.

DR 29. *Locomotive Spring, Utah*: USA.
J. Holment, Department of Wildlife Science, Utah State University, Logan, Utah.

DR 30. *Deep Creek, Idaho*: USA. 42° N, 112° W; 1387–1517 m alt.; 4250–10920 m².
G.W. Minshall, Department of Biology, Idaho State University, Pocatello, Idaho.

DR 31. *Varzea lakes, Amazonas*: Brazil. *ca* 3° S, 60° W; *ca* 0 m alt.; z 15 m HW.
W. Junk, Department of Tropical Ecology, Max-Planck Institute for Limnology, Plön, W. Germany.

DR 32. *Lac de Port-Bielh, Pyrénées centrales*: France. 42° 50′ N, 0° 10′ E; 2285 m alt.; 16.5 ha; z 9 m.
J. Capblancq, Laboratoire d'Hydrobiologie, Université P-Sabatier, Toulouse.

DR 33. *L. Esrom* (35 km N of Copenhagen): Denmark. 56° N, 12° 24′ E; 17.3 km²; z 22 m.
P.U. Jónasson, Freshwater Biological Laboratory, University of Copenhagen, Hillerød.

DR 34. *Srinagar (Kashmir) lakes*: India. *ca* 34° N, 74° E; *ca* 1587 m alt.; *Anchar*, 6.6 *km²*, z 3 m; *L. Dal*, 11.5 km², z 6 m; *Manas-bal lake*, 2.8 km², z 12 m.
V. Kaul, Department of Botany, University of Kashmir, Srinagar.

DR 35. *Fishponds*: Hungary. (General remarks only.)
E. Donassy, Toresvar u.29, 1112 Budapest XI.

DR 36/
37. *Biwa lake*: Japan. 35° N, 136° E; 85 m alt.; 691 km²; z 110 m.
T. Miura (see DR 19)

DR 38. *5 lakes, northern Poland* (*Masurian Lake District*): Poland. 53° N, 22° E. *Flosek*, 4 ha, z̄ 3 m, *Mikołajskie*, 460 ha, z̄ 11 m; *Śniardwy*, 10970 ha, z̄ 5.9 m; *Tałtowisko*, 327 ha, z̄ 14 m; *Warniak* 38 ha, z̄ 1.2 m.
A. Hillbricht-Ilkowska, Department of Hydrobiology, Institute of Ecology, Dziekanów Leśny post Lomianki nr. Warsaw.

DR 39. *L. Hakojärvi*: Finland. 61° 15′ N, 25° 12′ E; 15 ha; z 16 m.
P.O. Lehmusluoto, Department of Limnology, University of Helsinki, Vicki, SF 00710, Helsinki 71.

DR 40. *Bere stream*: England. 50° 44′ N, 2° 12′ W.
D. Westlake, FBA River Laboratory, East Stoke, Wareham, Dorset.

DR 41. *3 types of Amazonian waters*: Brazil. 3° S, 60° W; *ca* 30 m alt.
Lago do Castanho, ca 3–150 ha, *z* 1.12 m.
G.W. Schmidt, Landesanstalt für Fischerei, D 5942, Kirchhunden Albaum, West Germany.

DR 42. *Lake ecosystem model* (*USA East. Decid. Forest Biome*): USA.
R.A. Park, Department of Geology, Rensselaer Polytechnic Institute, Troy, New York.

DR 43. *Stechlinsee*: DDR, E. Germany. 53° 9' N, 13° 1' E; 60 m alt.; 425 km²; *z* 68.5 m.
Nehmitzsee, 53° 6' N, 12° 59' E; 60 m alt.; 171 km²; *z* 18 m.
S.J. Casper, Abt. Limnologie, Zentralinstitut f. Mikrobiologie und Exper. Therapie, 69 Jena, Beuthenbergstrasse 11.

DR 44. *L. Balaton*: Hungary. (Benthos and fish) (see DR 22, 27).
S. Herodek (as in DR 22).

DR 45. *Lake decomposition studies* (*Lakes Washington, Sammamish, Chester Morse, Findley*): USA. (See DR 11.)
Frieda B. Taub, College of Fisheries, University of Washington, Seattle.

DR 46. *L. Biwa*: Japan. (Additional data to DR 36/37.)
S. Mori (as DR 19).

DR 47. *Peatbog, Glenamoy*: Eire. 54° 12' N, 9° 45' W; 30 m alt.; 4.2 ha; *z* 1 m.
G. McCall, Department of Zoology, University of Dublin, Trinity College, Dublin.

DR 48. *Kashmir lakes*: India. *L. Dal*, 34° 0' N, 74° 53' E; 1587 m alt.; 15 km²; *z* 6 m; *Kounsarnag*, 33° 45' N, 75° 0' E; 4666 m alt.; 1.125 km²; *z* 20 m; also *Anchar* (see DR 34), *Alpather, Nagin, Neelnag, Wular*.
M. Das, Department of Zoology, Kashmir University, Srinagar.

DR 49. *Lipno Reservoir*: Czechoslovakia. 48° 48' N, *ca* 14° E; 726 m alt.; 46.5 km²; *z* 6 m.
Z. Brandl, Hydrobiological Laboratory, Czech Academy of Sciences, Lipno Field Station, 38278 Lipno nad Vetavou.

DR 50. *Neusiedlersee*: Austria. *ca* 47° 50' N, 16° 50' E; 115 m alt.; 175 km²; *z* 2 m.
M. Dokulil, Limnologische Lehrkamzel, Universität Wien, Berggasse 18/19, A-1090 Vienna.

DR 51. *Tasmanian lakes*: Tasmania (Australia). *ca* 42° S.
Leake, 550 ha, *z* 5 m; *Tooms*, 600 ha, *z* 5 m; *Crescent*; *Sorell*; *Macquarie Island & Iles Kerguelen*, 54° 30' S, 158° 57' E.
P. Tyler, Botany Department, University of Tasmania, Hobart.

DR 52. *Lake Rotoiti*: New Zealand. 38° S, 176° E; 271 m alt.; 33.89 × 10⁶ m; *z* 93 m.
B.T. Coffey, Botany Department, University of Auckland, NZ.

DR 53. *Lake Tatsu-Numa* (*Urabandai Lake Group*): Japan. 37° 39' N, 140° E.

(Primary production, zoobenthos, detritus.)

E. Takahashi, Department of Biology, Faculty of Science, University of Kobe, Kobe.

DR 54. *Lake Nhlange*: South Africa, N. Zululand. 27° S, 33° E; 30.7 km²; z 7 m.

R.E. Boltt, Institute of Freshwater Studies, Rhodes University, Grahamstown.

DR 55. *Fish-ponds, Mirwart*: Belgium. 50° N, 5° 16′ E.

G. Marlier, Laboratoire d'Hydrobiologie, Institut Royal des Sciences Naturelles, Brussels.

DR 56. *Naini Lake, Delhi*: India. 28° 50′ N, 77° 10′ E; 4000 m²; z 5 m.

J.L. Bhat, Swami Shraddhavard College, University of Delhi, Alipur, Delhi 36.

DR 57. *Ramgarh Lake, Gorakhpur, UP*: India. 25° 40′ N, 83° 25′ E; 95 m alt.; 22 km²; z 2.5 m. (Phytoplankton)

A.B. Sinha.

DR 58. *Lake Chedenjarvi (Shuyu R. basin)*: USSR (see DR 64).

L. Ryshkov, Northern Scientific Research Institute on Fish Economy in Lakes and Rivers, Petrozavodsk.

DR 59. *Tundra pond, Alaska*: USA.

J.E. Hobbie, Department of Zoology, North Carolina State University, Raleigh, NC 27607.

DR 60. *Lake Aleknagik, Alaska*: USA. 59° 17′ N, 158° 37′ W; 83 km²; z 110 m.

R.L. Burgner, Fisheries Research Institute, University of Washington, Seattle.

DR 61. *Lake Zelenetskoye*: USSR (Murmansk Sea coast). 24 ha; z 25 m.

L. Akulkino, 4 ha; z 5 m.

G.G. Winberg, Zoological Institute, Academy of Sciences, University Embankment, Leningrad B 164, USSR.

DR 62. *L. Krasnoye*: USSR. 60° 16′ N, 29° 08′ E; 9 km²; z 26.6 m.

I.A. Andronikova, c/o G.G. Winberg (as DR 61).

DR 63. *L. Karakul*: USSR. 43° 46′ N, 78° 05′ E; 480 m alt.; 1.4 km²; z 3.8 m.

V.P. Mitrofanov, c/o G.G. Winberg (as DR 61).

DR 64. *L. Chedenjarvi*: USSR. *ca* 62° N. 150.6 m alt.; z 3.4 m.

Z.A. Gorbunova, c/o G.G. Winberg (as DR 61). (See DR 58.)

DR 65. *Meteliai lakes, Lithuanian SSR*: USSR. *ca* 54° 18′ N, 23° 45′ E.

Dusia; Obelija; Galstas; Slavantas. (As DR 61.)

DR 66. *Kiev Reservoir (Dnieper R.)*: USSR. 50° 30′ N, 30° 30′ E; − 103 m alt.; 922 km²; z̄ 4 m.

Y.Y. Tseeb, c/o G.G. Winberg (as DR 61).

DR 67. (a) *L. Baykal*: USSR: 51°–55° N, 104°–109° E; 31 500 km²; z 1620 m.

(b) *Bratsk Reservoir*: USSR: *ca* 56° N, 101° E. (As DR 61.)

DR 68. *Lakes Naroch, Myastro, Batorin*: USSR. *ca* 54° 88′ N, 47°–49° E.
Naroch, 80 km², $\bar{z}$ 9.0 m; *Myastro*, 13 km², z 5.4 m; *Batorin*, 6 km²,
z 3.0 m.
J.U. Mikheyeva, c/o G.G. Winberg (as DR 61).

DR 69. *Rybinsk Reservoir (Volga R.)*: USSR. *ca* 59° 14′ N, 39° 02′ E; 102 m
alt.; 4550 km²; $\bar{z}$ 5.6 m.
N. Romanenko, c/o G.G. Winberg (as DR 61).

DR 70. *Lake Krivoye*: USSR. *ca* 65° N, 35° E; 0.5 km²; $\bar{z}$ 11.75 m.
G.G Winberg (as DR 61).

DR 71. *Lake Krugloye*: USSR. *ca* 65° N, 35° E; 0.1 km²; $\bar{z}$ 2.15 m.
G.G. Winberg (as DR 61).

DR 72. *Lake Bolshoy Kharbey*: USSR. 67° 30′ N, 62° 50′ E; 200 m alt.;
2134 km²; z 18.5 m.
T.A. Vlasova, c/o G.G. Winberg (as DR 61).

DR 73. *Sturgeon fish-pond, Lower Don River*: USSR.
Anon, c/o G.G. Winberg (as DR 61).

DR 74. (a) *Lago Maggiore*: Italy. 45° 57′ N, 8° 39′ E; 194 m alt.; 212 km².
(b) *Lakes Vico, Bracciano, Bolsena*: Italy.
L. Tonolli, Istituto Italiano di Idrobiologia, Verbania/Pallanza.

DR 75. *Fish-pond near Warsaw*: Poland. *ca* 52° 15′ N, 21° 0′ E.
T. Backiel, Inland Fisheries Institute, Zabieniec/Warsaw, Post
Piaseczuo 05-500.

DR 76. *Tasek Bera*: Malaya.
J.I. Furtado, School of Biological Sciences, University of Malaya,
Kuala Lumpur.

DR 77. *Lake George, Uganda*: Africa. 0° N/S, 30° 12′ E; 250 km²; z 3 m.
G.G. Ganf, IBP Royal Society African Freshwater Biological Team,
Uganda.

DR 78. *Lake Marion*: Canada. 49° 19′ N, 122° 33′ W; 300 m alt.; 13.3 ha;
z 6 m.
I. Efford, Institute of Resource Ecology, University of British
Columbia, Vancouver.

DR 79. *Lake Char (Cornwallis Island)*: Canada. 74° 42′ N, 94° 53′ W; 30 m
alt.;52.6 ha; z 27.5 m.
F. Rigler, Department of Zoology, University of Toronto, Ontario.

DR 80. *River Thames, Reading*: England. 51° 30′ N, 0° 55′ W.
A. Berrie, Department of Zoology, The University, Reading.

DR 81. *Metropolitan Water Board reservoirs*: England. 51° 25′ N, 0° 30′ W.
A. Duncan (A. Steel), Zoology Department, Royal Holloway College
(University of London), Surrey.

DR 82. *Lake Pääjärvi*: Finland. 61° 1′ N, 25° 1′ E; 13.5 km².
V. Ilmavirta, Lammi Biological Station, SF 16900-Lammi.

DR 83. *Two small Antarctic lakes (Algal & Skua lakes, McMurdo Sound)*:

Antarctica. *ca* 76° S, 160° E.

E.C. Goldman, Division of Environmental Studies, University of California, Davis, California, USA.

DR 84. *Tsimlyansk Reservoir (Don Valley)*: USSR. *ca* 48° N, 43° E; 2702 km^2; z 32 m.

N. Tsiba, (as DR 61).

DR 85. *Lake Dalnee (Kamchatka, Paratunka R.)*: USSR. *ca* 56° N, *ca* 160° E; 1.36 km^2; z 60 m.

F.V. Krogius (as DR 61).

DR 86. *R. Daugava, Latvia*: USSR. 57° N, 24° E.

I. Kumsare (as DR 61).

DR 87. *Lake Onega & L. Ladoga*: USSR. (Macrophytes.)

M. Raspopoo (as DR 61).

DR 88. *Lake Kinneret (Tiberias)*: Israel. 32° 48′ N, 35° 35′ E; 174 km^2; z 24 m.

T. Berman, Kinneret Limnological Laboratory, Tiberias.

DR 89. *Lough Neagh*: N. Ireland. 54° 36′ N, 6° 30′ W.

D.H. Jewson, New University of Ulster, Coleraine, Co., Londonerry.

DR 90. *Indian waters*: India.

A. Sreenivasan, Hydrobiological Research Station, Madras-10.

DR 91. *Lake Mikołajskie*: Poland.

R. Kowalczewski (see DR 38).

DR 92. *Lake Sibaya*: S. Africa. 27° 25′ S, 32° 40′ E.

B.R. Allanson, Rhodes University, Grahamstown.

DR 93. *L. Ivano-Arakhley (Siberia)*: USSR. *ca* 52° N, *ca* 112° 58′ E.

E.N. Shiskin (as DR 61).

Appendix II. List of IBP/PF publications

A. *Handbooks of Methods*. Published for SCIBP by Blackwell Scientific Publications, Oxford. (The number is that in the total IBP series of handbooks.)

No. 3: *Methods for Assessment of Fish Production in Fresh Waters,*
ed. W.E. Ricker (1968, reprinted 1970; 2nd edition 1971; 3rd edition, ed. T. Bagenal, 1978).

No. 8: *Methods for Chemical Analysis of Fresh Waters,*
ed. H.L. Golterman asst. R.S. Clymo (1969, reprinted 1970; 3rd revised printing 1971).
Methods for Physical and Chemical Analysis of Fresh Waters,
H.L. Golterman, R.S. Clymo & M.A.M. Ohnstad (1978).

No. 12: *A Manual on Methods for Measuring Primary Production in Aquatic Environments,*
ed. R. Vollenweider with collaboration of J. Talling, D.F. Westlake (1969, reprinted 1971, 2nd edition 1974).

No. 17: *Methods for Estimation of Secondary Productivity in Fresh Waters,*
ed. W.T. Edmondson, with collaboration of G.G. Winberg (1971).

No. 21: *Project Aqua: A Source Book of Inland Waters Proposed for Conservation,*
eds. H. Luther & J. Rzoska (1969 preliminary issue by IBP Central Office, revised edition 1971).

No. 23: *Techniques for the Assessment of Microbial Production and Decomposition in Fresh Waters,*
eds. Y. Sorokin & H. Kadota (1972).

Symbols, Units and Conversion Factors in Studies of Fresh Water Productivity,
ed. G.G. Winberg & collaborators (1971, IBP Central Office).

Quantities, Units and Symbols. Recommendations for use in IBP Synthesis,
Report of a SCIBP Working Group (1974, SCIBP publication, ICSU).

B. *Proceedings of PF Scientific Meetings*

Primary productivity in aquatic environments. Proceedings of an IBP/PF symposium at Pallanza, Italy. Ed. C.R. Goldman. *Memorie dell'Istituto italiano di Idrobiologia*, 18, suppl. (1965; reprinted 1966 by University of California Press).

Chemical Environment in the Aquatic Habitat. Proceedings of an IBP symposium held in Amsterdam and Nieuwersluis, 10–16 October 1966. Ed. H.L. Golterman & R.S. Clymo. Konninklijke Nederlandse Akademie van Wetenschappen, Amsterdam (1967).

501

The Biological Basis of Freshwater Fish Production. Proceedings of an IBP/PF symposium at Reading, UK, 1966. Ed. S.D. Gerking. Blackwell Scientific Publications, Oxford (1967; revised edition entitled *Ecology of Freshwater Fish Production* 1978).

Productivity Problems of Freshwaters. Proceedings of the IBP/UNESCO symposium held in Kazimierz, Dolny, Poland, 6–12 May 1970. Ed. Z Kajak & A. Hillbricht-Ilkowska. PWN Polish Scientific Publishers, Warsaw-Cracow (1972).

Detritus and its Role in Aquatic Ecosystems. Proceedings of an IBP/UNESCO symposium at Pallanza, Italy, May 1972. Ed. U. Melchiorri-Santolini & J.W. Hopton. *Memorie dell'Istituto Italiano di Idrobiologia,* 29, suppl. (1972).

Humic Substances: Their Structure and Function in the Biosphere. Proceedings of an International Meeting at Nieuwersluis, Netherlands, 29–31 May 1972, sponsored by IBP. Ed. D. Povoledo & H.L. Golterman. PUDOC, Wageningen (1975).

C. *Publications resulting from IBP/PF activities in intertropical areas*

Primera Reunion Regional de Limnologia Latinoamericana, Santa Fé, Argentina, 14–18 Marzo 1968. Resoluciones y Recomendaciones. Informes (with English summaries). Consejo Nacional (1968).

Report of the Regional Meeting of Hydrobiologists in Tropical Africa. 20–28 May 1968. Makerere, Uganda. UNESCO Regional Centre for Science and Technology, Nairobi. (Roneographed) (1969).

SE Asian Regional Meeting: Biology of Inland Waters. Proceedings of a meeting held at Kuala Lumpur and Malacca, 5–11 May 1969. UNESCO Regional Office, Djakarta (1969) (Roneographed) (1970).

D. *In collaboration with other sections of IBP (PP, PT, CT)*

Proceedings of the international symposium on the biological productivity of aquatic and swamp macrophytes, held at Maliuc, Romania, 1–10 September 1970. *Hidrobiologia,* **12**, 5–408 (1971).

Proceedings of the IBP Wetlands Symposium, 11–18 June 1972. Ed. A. Szczepánski. *Polskie Archiwum Hydrobiologii,* **20**(1) (1973).

E. *Publications for the International Hydrological Decade, in collaboration with IBP/PF*

Ecology of Water Weeds in the Neotropics. D.S. Mitchell & P.A. Thomas. UNESCO Technical Papers in Hydrology, No. 12. A contribution to the International Hydrological Decade. UNESCO, Paris (1972).

Aquatic Vegetation and its Use and Control, ed. D.S. Mitchell. A contribution to the International Hydrological Decade. UNESCO, Paris (1974).

Aquatic Weeds in SE Asia. Proceedings of a regional seminar on noxious aquatic vegetation, December 1973, New Delhi. Ed. C.K. Varsney & J. Rzoska. Sponsored by UNESCO/IHD/IBP and the Indian National Academy (1976).

F. *National syntheses of IBP programmes that include PF sections.*

Canada: *Energy Flow–Its Biological Dimensions. The IBP in Canada*: 1964–74, ed. T.W.M. Cameron & L.W. Billingsley. Royal Society of Canada, Ottawa, for the Canadian Committee for the IBP (1975). (Includes Chapter 10: The Char lake project. An introduction to limnology in the Canadian Arctic, by F.H. Rigler, pp. 171–98. Chapter 11: Marion Lake. An analysis of a lake ecosystem, by I.E. Efford & K.J. Hall, pp. 199–226. Chapter 12: IBP/PF synthesis. Estimating the productivity of lakes and reservoirs by K.H. Mann, pp. 221–6.)

Czechoslovakia: Záverecná Zpráwa o Československé Účasti v Mezinárodním Biologickém Programu. Československý Nārodni Komitét pro Mezinárodni Biologický Program při Československé Akademii Véd. Praha 1975. (In Czech.) (Report on the participation of Czechoslovakia in the IBP. The Czechoslovak National Committee for the IBP in the Czechoslovak Academy of Sciences, Prague, 1975.) Contains the appraisal of the main results of six sections of IBP (excludes marine PM). The PF section, pp. 146–75, enumerates projects tackled and personnel involved. A rich bibliography is attached.

Japan: *Productivity of communities in Japanese Inland Waters*, ed. S. Mori & G. Yamamoto. JIBP Synthesis vol. 10, University of Tokyo Press for Japanese Committee of IBP (1975).

Norway: The lake Øvre Heimdalsvatn a subalpine freshwater ecosystem. The Norwegian contribution to the freshwater section of the International Biological Programme, ed. R. Vik, *Holarctic Ecology*, **1**, 81–320.

Poland: *Productivity Freshwater*. Section 8, pp. 182–269 by Z. Kajak in *Polish Participation in the IBP* 1964–1973, Polish Academy of Sciences, Warsaw, 1975 (includes a list of over 300 published papers).

UK: *A review of the United Kingdom contribution to the IBP*, ed. A.R. Clapham, C.E. Lucas & N.W. Pirie. *Philosophical Transactions of the Royal Society of London, Series B*, **274**, 275–553. (PF represented by papers by E.D. Le Cren on The productivity of freshwater communities, pp. 359–74, and P.H. Greenwood on L. George, Uganda, pp. 375–91.) The three main UK projects produced the following number of scientific contributions: Lake George, Uganda, 52; Loch Leven, Scotland, 60; River Thames, England, 23; another 170 studies were published on various subjects relevant to the IBP. (These figures taken from the *Philosophical Transactions of the Royal Society*, 1976.)

USSR: Results of studies in productivity of freshwater communities at all trophic levels by G.G. Winberg, pp. 145–57, in vol. 2 of *Resources of the Biosphere* (Synthesis of the Soviet Studies for the IBP), ed. O.N. Bauer & N.N. Smirnov, Publishing House 'Nauka', Leningrad Branch. (For translation see Chapter 11 of this volume.)

References*

Aasa, R. (1970). Plankton in Lilla Ullevifjärden. Ph.D. thesis, Uppsala University.

Aass, P. (1969). Crustacea, especially *Lepidurus arcticus* Pallas as trout food in Norwegian mountain reservoirs. *Report. Institute of Freshwater Research, Drottningholm*, **49**, 183–201.

Abeliovich, A. & Shilo, M. (1972). Photo-oxidative death in blue-green algae. *Journal of Bacteriology*, **111**, 682–9.

Aberg, B. & Rodhe, W. (1942). Über die Milieufaktoren in einigen Sudschwedischen Seen. *Symbolae Botanicae Upsalienses*, **3**, 3–256.

Adams, M.S. & McCracken, M.D. (1974). Seasonal production of the *Myriophyllum* component of the littoral of Lake Wingra, Wisconsin. *Journal of Ecology*, **62**, 457–67.

Adams, M.S. & Stone, W. (1973). Field studies on photosynthesis of *Cladophora glomerata* (Chlorophyta) in Green Bay, Lake Michigan. *Ecology*, **54**, 853–62.

Adams, M.S., Titus, J. & McCracken, M. (1974). Depth distribution of photosynthetic activity in a *Myriophyllum spicatum* community in Lake Wingra. *Limnology and Oceanography*, **19**, 377–89.

Ahl, T. (1973). Malarensbelastning och vatenkvalitet. *Scripta Limnologica, Uppsala*, **332**, 76 pp.

Ahlgren, G. (1970). Limnological studies of Lake Norrviken, a eutrophic Swedish lake. II. Phytoplankton and its production. *Schweizerische Zeitschrift für Hydrobiologie*, **32**, 353–96.

Albrecht, M.L. (1966). Untersuchungen über die photosynthese der Algen bei sehr niedrigen Wassertemperaturen während des Winters. *Verhandlungen der Internationalen Vereinigung für Theoretische und Angewandte Limnologie*, **16**, 358–63.

Aleem, A.A. & Samaan, A.A. (1969). Productivity of Lake Mariut, Egypt, Part II. Primary production. *Internationale Revue der Gesamten Hydrobiologie und Hydrographie*, **54**, 491–527.

Alimov, A.F. (1975). Obzor issledouvaniï po biologicheskoï produktivnosti donnykh zhivotnÿkh v presnovodnÿkh vodoemakh sovetskogo soyuza. *Izvestiya Akademii nauk SSSR, Seriya Biologicheskaya* **1**, 94–103.

Alimov, A.F., Boullion, V.V., Finogenova, N.P., Ivanova, M.B., Kuzmitskaya, N.K., Nikulina, V.N., Ozeretskovskaya, N.G. & Zharova, T.V. (1972). Biological productivity of Lakes Krivoe and Krugloe. In *Productivity Problems of Freshwaters*, ed. Z. Kajak & A. Hillbricht-Ilkowska, pp. 39–56. Warsaw & Kraków: Polish Scientific Publishers.

Alimov, A.F. & Winberg, G.G. (1972). Biological productivity of two northern lakes. *Verhandlungen der Internationalen Vereinigung für Theoretische und Angewandte Limnologie*, **18**, 65–70.

Allen, H.L. (1969). Chemo-organotrophic utilization of dissolved organic compounds by planktonic algae and bacteria in a pond. *Internationale Revue der Gesamten Hydrobiologie*, **54**, 1–33.

Allen, H.L. (1971a). Primary productivity, chemo-organotrophy, and nutritional interactions of epiphytic algae and bacteria on macrophytes in the littoral of a lake. *Ecological Monographs*, **41**, 97–127.

Allen, H.L. (1971b). Dissolved organic carbon utilization in size-fractionated algal and

* See p. 570 for references added in proof.

bacterial communities. *Internationale Revue der Gesamten Hydrobiologie und Hydrographie.* **56**, 731–49.

Allen, H.L. (1972). Phytoplankton photosynthesis, micronutrient interactions, and inorganic carbon availability in a soft-water Vermont lake. In *Nutrients and Eutrophication*, ed. G.E. Likens, pp 63–83. Lawrence: Allen Press.

Allen, K.R. (1951). The Horokiwi Stream, a study of a trout population. *Fisheries Bulletin. New Zealand Marine Department*, **10**, 231 pp.

Allen, T.F.H. & Koonce, J.F. (1973). Multivariate approaches to algal strategems and tactics in systems analysis of phytoplankton. *Ecology*, **54**, 1234–46.

Ambach, W. & Habicht, H.L. (1962). Unterschungen der Extinktionseigenschaften des Gletschereises und des Schnees. *Archiv für Meteorologie, Geophysik und Bioklimatologie*, B **11**, 512–32.

Ambach, W. & Mocker, H. (1959). Messungen zur Strahlungsextinktion mittels eines kugelförmigen Empfängers in der oberflachennahen Eisschicht eines Gletschers und im Altschnee, *Archiv für Meterologie, Geophysik und Bioklimatologie*, B **10**, 84–99.

Ambasht, R.S. (1971). Ecosystem study of a tropical pond in relation to primary production of different vegetational zones. *Hydrobiologia*, **12**, 57–61.

Ambühl, H. (1967). Die Temperatur-und Sauerstoffverhältnisse des Bodensee – Untersees in den Jahren 1961 bis 1963. *Bericht der Internationalen Gewasserschutzkommission für den Bodensee*, No. 5, 91–128.

American Chemical Society. (1967). *Equilibrium Concepts in Natural Water Systems. Advances in Chemistry Series*, 67, ed. W. Stumm, 344 pp. Washington. DC: American Chemical Society.

Amrén, H. (1964). Temporal variation of the rotifer *Keratella quadrata* (Müll.) in some ponds on Spitsbergen. *Zoologiska Bidrag från Uppsala*, **36**, 193–208.

Andersen, J.M. (1974). Nitrogen and phosphorus budgets and the role of sediments in six shallow Danish lakes. *Archiv für Hydrobiologie*, **74**, 528–50.

Andersen, J.M. (1977). Importance of the denitrification process for the rate of degradation of organic matter in lake sediments. In *Interactions Between Sediments and Fresh Water*, ed. H.L. Golterman. Proceedings of an International Symposium, Amsterdam, 1976, pp 357–62.

Anderson, E.R. (1952). Energy budget studies. Water loss investigations: Volume 1, Lake Hefner studies technical report. *United States Geological Survey Circular*, **229**, 71–119.

Anderson, G.C. & Zeutschel, R.P. (1970). Release of dissolved organic matter by marine phytoplankton in coastal and offshore areas of the Northeast Pacific Ocean. *Limnology and Oceanography*, **15**, 402–7.

Anderson, R.S. (1970). Predator-prey relationship and predation rates for crustacean zooplankters from some lakes in western Canada. *Canadian Journal of Zoology*, **48**, 1229–40.

Andersson, G., Gronberg, G. & Gelin, C. (1973). Planktonic changes following the restoration of Lake Trummen, Sweden. *Ambio*, **2**, 44–7.

Andrews, W.B. (1947). *The Response of Crops and Soils to Fertilizers and Manures.* Mississippi State College. 459 pp.

Andronikova, I.N., Drabkova, V.G. Kuzmenko, K.N., Michaïlova, N.F. & Stravinskaya, E.A. (1972). Biological productivity of the main communities of the Red Lake. In *Productivity Problems of Freshwaters*, ed. Z. Kajak & A. Hillbricht-Ilkowska, pp. 57–71. Warsaw & Kraków: Polish Scientific Publishers.

Andronikova, I.N., Drabkova, V.G., Kuzmenko, K.N., Mokievskii, K.A., Stravinskaya, E.A. & Trifonova, I.S. (1973). Productivity of the main communities of the Red Lake and its biotic balance. In *Productivity Studies of Freshwater Ecosystems*, ed. G.G. Winberg, pp. 5–19. Minsk: Byelorussian State University. (In Russian.)

Angström, A. (1925). On the albedo of various surfaces of ground. *Geografiska annaler*, **7**, 323–42.

Antia, N.J. & Cheng, J.Y. (1970). The survival of axenic cultures of marine planktonic algae from prolonged exposures to darkness at 20 °C. *Phycologia*, **9**, 179–83.

Anwand, K. (1968). Fischbiomasse, Fischanzahl und Antenzusammensetzung des Fischbestandes einiger Kleiner Seen. *Deutsche Fischerizeitung, Rodebeul*, **15** (10), 275–9.

Arai, J. (1964). Some relations between the thermal property of lake and its fetch size. *Geographical Review of Japan*, **37**, 131–7.

Arnemo, R., Berzins, B., Grönberg, B. & Mellgren, I. (1968). The dispersal in Swedish waters of *Kellicottia bostoniensis* (Rousselet) (Rotatoria). *Oikos*, **19**, 351–8.

Assman, A.V. (1958). Some data on the "grazing down" of benthos by fish. *Doklady Akademii Nauk SSSR*, **122** (5), 932–5. (In Russian.)

Assman, A.V. (1960a). O dostupnosti lichinok Chironomid dla ryb. *Trudy Soveshchanii. Ikhtiologicheskaya Komissiya*, **13**, 361–3.

Assman, A.V. (1960b). Izmenenie dostupnosti lichinok Chironomid pri vyedanii rybami. *Izvestiya Akademii Nauk SSSR, Seriya Biologicheskaya*, **5**, 670–85.

Azam, K.M. & Anderson, N.H. (1969). Life history and habits of *Sialis rotunda* and *S. californica* in Western Oregon. *Annals of the Entomological Society of America*, **62**, 549–58.

Bachmann, R.W. & Jones, J.R. (1974). Phosphorus inputs and algal blooms in lakes. *Iowa State Journal of Research*, **49**, 155–60.

Backiel, T. (1971). Production and food consumption of predatory fish in the Vistula River. *Journal of Fish Biology*, **3**, 369–405.

Backiel, T. & Le Cren, E.D. (1967). Some density relationships for fish population parameters. In *The Biological Basis of Freshwater Fish Production*, ed. S.D. Gerking, pp. 261–93. Oxford: Blackwell.

Bailey-Watts, A.E. (1973). Observations on the Phytoplankton of Loch Leven, Kinross, Scotland. Ph.D. thesis, University of London. 455 pp.

Bailey-Watts, A.E. (1974). The algal plankton of Loch Leven, Kinross. *Proceedings of the Royal Society of Edinburgh*, B **74**, 135–56.

Bailey-Watts, A.E. & Lund, J.W.G. (1973). Observations on a diatom bloom in Loch Leven, Scotland. *Biological Journal of the Linnean Society of London*, **5**, 235–53.

Bajkov, A. (1935). How to estimate the daily food consumption of fish under natural conditions. *Transactions of the American Fisheries Society*, **65**, 288–9.

Baranov, I.V. (1962). *Limnological types of the USSR lakes*. Leningrad: Gidrometeorizdat. 274 pp.

Barica, J. (1974). Some observations on internal recycling, regeneration and oscillation of dissolved nitrogen and phosphorus in shallow self-contained lakes. *Archiv für Hydrobiologie*, **73**, 334–60.

Bärlocher, F. & Kendrick, B. (1973a). Fungi and food preferences of *Gammarus pseudolimnaeus*. *Archiv für Hydrobiologie*, **72**, 501–16.

Bärlocher, F. & Kendrick, B. (1973b). Fungi in the diet of *Gammarus pseudolimnaeus* (Amphipoda). *Oikos*, **24**, 295–300.

Baumann, P.C., Kitchell, J.F., Magnuson, J.J. & Kayes, T.B. (1974). Lake Wingra, 1837–1973: a case history of human impact. *Transactions of the Wisconsin Academy of Sciences, Arts and Letters*, **62**, 57–94.

Bayly, I.A.E. & Williams, W.D. (1972). The major ions of some lakes and waters in

Queensland, Australia. *Australian Journal of Marine and Freshwater Research*, **23**, 121–31.

Beadle, L.C. (1974). *The Inland Waters of Tropical Africa. An Introduction to Tropical Limnology*. London: Longman. 365 pp.

Beattie, M., Bromley, H.J., Chambers, M., Goldspink, R., Vijverberg, J., Van Zalinge, N.P. & Golterman, H.L. (1972). Limnological studies on Tjeukemeer – a typical Dutch "polder reservoir". In *Productivity Problems of Freshwaters*, ed. Z. Kajak & A. Hillbricht-Ilkowska, pp. 421–46. Warsaw & Kraków: Polish Scientific Publishers.

Beeton, A.M. (1969). Changes in the environment and biota of the Great Lakes. In *Eutrophication: Causes, Consequences, Correctives*, An International Symposium on Eutrophication, University of Wisconsin, Madison, 1967, pp. 150–87. Washington, DC: National Academy of Sciences.

Bella, D.A. (1970). Simulating the effect of sinking and vertical mixing on algal population dynamics. *Journal of the Water Pollution Control Federation*, **42** (5), 140–52.

Belyavskaya, L.I. & Konstantinov, A.S. (1956). Pitanie lichinok *Procladius choreus* Meig. (Chironomidae, Diptera) i ushcherb nanosimyi imi kormovoi baze ryb. *Voprosy Ikhtiologii*, **7**, 193–203.

Benndorf, J. & Stelzer, W. (1973). Untersuchungen über das Regulations-verhalten hydrischer Ökosysteme und seine Bedeutung für die biogene Phosphatelimination. *Internationale Revue der Gesamten Hydrobiologie*, **58**, 599–616.

Berg, A. (1961). Rôle écologique des eaux de la cuvette congolaise sur la croissance de la jacinthe d'eau. (*Eichhornia crassipes* (Mart.) Solms). *Mémoires. Académie Royale des Sciences d'Outre-Mer. Classe des Sciences Naturelles et Médicales. Collection-8°*, **12**(3), 120 pp.

Berg, A. (1962). Exposé des méthodes d'analyse chimique et physico-chimique des eaux humiques. *Memorie dell'Istituto Italiano di Idrobiologia*, **15**, 183–206.

Berger, F. (1955). Die Dichte natürlicher Wässer und die Konzentrationsstabilität in Seen. *Archiv für Hydrobiologie*, **22** (supplement), 286–94.

Berglund, T. (1968). The influence of predation by brown trout on *Asellus* in a pond. *Report. Institute of Freshwater Research, Drottningholm*, **48**, 77–101.

Berman, T. (1976). Light penetrance in Lake Kinneret, *Hydrobiologia*, **41**, 41–8.

Berman, T. & Pollinger, U. (1974). Annual and seasonal variations of phytoplankton, chlorophyll and primary production in Lake Kinneret. *Limnology and Oceanography*, **19**, 31–54.

Berman, T. & Rodhe, W. (1971). Distribution and migration of *Peridinium* in Lake Kinneret. *Mitteilungen der Internationalen Vereinigung für Theoretische und Angewandte Limnologie*, **19**, 266–76.

Berrie, A.D. (1972a). Productivity of the River Thames at Reading. In *Conservation and Productivity of Natural Waters*, ed. R.W. Edwards & D.J. Garrod, *Symposia of the Zoological Society of London*, 29, pp. 69–86. New York & London: Academic Press.

Berrie, A.D. (1972b). The occurrence and composition of seston in the River Thames and the role of detritus as an energy source for secondary production in the river. *Memorie dell'Istituto Italiano di Idrobiologia*, **29**. (supplement), 473–85.

Bezler, F.I. (1969). On possible correlation between organic and mineral contents of water. *Informatsionnyi Byulleten. Institut Biologii Vnutrennikh vod*, **4**, 68–72. (In Russian.)

Biffi, F. (1963). Determinazione del fattore tempo come caratterestica del potere di autodepurazione del Lago d'Orta in relazione ad un inquinamento constante. *Atti dell'Istituto Veneto di Scienze, Lettere ed Arti*, **121**, 131–6.

Bindloss, M.E. (1974). Primary productivity of phytoplankton in Loch Leven, Kinross. *Proceedings of the Royal Society of Edinburgh*, B **74**, 157–81.

Bindloss, M.E. (1976). The light climate of Loch Leven, a shallow Scottish lake, in relation to primary production of phytoplankton. *Freshwater Biology*, **6**, 501–18.

Bindloss, M.E., Holden, A.V., Bailey-Watts, A.E. & Smith, I.R. (1972). Phytoplankton production, chemical and physical conditions in Loch Leven. In *Productivity Problems of Freshwaters*, ed. Z. Kajak & A. Hillbricht-Ilkowska, pp. 639–59. Warsaw & Kraków: Polish Scientific Publishers.

Birge, E.A. (1891). List of Crustacea Cladocera from Madison, Wisconsin. *Transactions of the Wisconsin Academy of Sciences, Arts and Letters*, **8**, 379–98.

Biro, P. (1969). The spring and summer nutrition of the 300–500 g pike-perch (*Lucioperca lucioperca* L.) in Lake Balaton in 1968. II. The calculation of the consumption, daily and monthly rations. *Annales Instituti Biologici, Tihany*, **36**, 151–62.

Björk, S. (1972). Ecosystem studies in connection with the restoration of lakes. *Verhandlungen der Internationalen für Theoretische und Angewandte Limnologie*, **18**, 379–87.

Black, J.N. (1956). The distribution of solar radiation over the earth's surface. *Archiv für Meteorologie, Geophysik und Bioklimatologie*, B **7**, 165–80.

Black, M.A. (1973). Exogenous Carbon Sources in Photosynthesis of Submerged Aquatic Plants. M.Sc. thesis, University of St Andrews, Scotland.

Blackman, F.F. & Smith, A.M. (1911). Experimental researches on vegetable assimilation and respiration. IX. On assimilation in submerged water plants and its relation to the concentration of carbon dioxide and other factors. *Proceedings of the Royal Society of London*. B **83**, 389–412.

Blanton, J.O. (1973). Vertical entrainment into the epilimnia of stratified lakes. *Limnology and Oceanography*, **18**, 697–704.

Blažka, P. (1966). The ratio of crude protein, glycogen and fat in the individual steps of one production chain. *Hydrobiological Studies*, **1**, 395–408.

Bliss, E.I. (1958). Periodic regressions in biology and climatology. *Bulletin. Connecticut Agricultural Experiment Station*, **615**, 1–55.

Bloesch, J. (1974). Sedimentation und Phosphorhaushalt im Vierwaldstattersee (Horwer Bucht) und im Rotsee. *Schweizerische Zeitschrift für Hydrologie*, **36**, 71–186.

Bloesch, J., Stadelmann, P. & Bührer, H. (1977). Primary production, mineralization, and sedimentation in the euphotic zone of two Swiss lakes. *Limnology and Oceanography*, **22**, 511–26.

Bloomfield, J.A., Park, R.A., Scavia, D. & Zahorcak, C.S. (1974). Aquatic modeling in the Eastern Deciduous Forest Biome, U.S.—International Biological Program. In *Modeling the Eutrophication Process*, ed. E.J. Middlebrooks, D.H. Falkenborg & T.E. Maloney, pp. 139–69. Ann Arbor: Ann Arbor Science.

Bodin, K. & Nauwerck, A. (1968). Produktionsbiologische Studien über die Moosvegetation eines klaren Gebirgssees. *Schweizerische Zeitschrift für Hydrologie*, **30**, 318–52.

Bogoslovskii, B.B. (1960). *Ozerovedeniie*. Moskva: Izdatel'stvo Moskovskogo Universitet. 335 pp.

Bolas, P.M. & Lund, J.W.G. (1974). Some factors affecting the growth of *Cladophora glomerata* in the Kentish Stour. *Journal of the Society for Water Treatment and Examination*, **23**, 25–51.

Boling, R.H., Goodman, E.D., Zimmer, J.O., Cummins, K.W., Reice, S.R., Petersen, R.C. & Van Sickle, J.A. (1974a). Towards a model of detritus processing in a woodland stream. *Ecology*, **56**, 141–51.

Boling, R.H., Petersen, R.C. & Cummins, K.W. (1974b). Ecosystem modeling for small woodland streams. In *Systems Analysis and Simulation in Ecology*, ed. B.C. Patten, vol. 3, New York & London: Academic Press.

Bombówna, M. (1972). Primary production of a montane river. In *Productivity Problems of Freshwater*, ed. Z. Kajak & A. Hillbricht-Ilkowska, pp. 661–71. Warsaw & Kraków: Polish Scientific Publishers.

Bondareva, E.I., Gorlachev, V.P., Morozova, T.N., Topolov, A.A., Shishkin, B.A. & Shishkina, K.A. (1973). Some regional features of biological turnover rate of matter in Ivano-Arakhlei Lakes (Chita district, Siberia). In *Productivity Studies of Freshwater Ecosystems*, ed. G.G. Winberg, pp. 163–73. Minsk: Byelorussian State University. (In Russian.)

Bonetto, A., Dioni, W. & Pignalberi, C. (1969). Limnological investigations in the Middle Parana River Valley. *Verhandlungen der Internationalen Vereinigung für Theoretische und Angewandte Limnologie*, **17**, 1035–50.

Borodičova, N.D. (1962). Počet generacii hlanych predstavitelov bentickych organizmov rybnika Mala Podvinica (Južne Čechy). *Práce Laboratoria Rybactva*, **1**, 29–54.

Borutskii, E.V. (1950). Materialy po dinamike biomassy macrofitov ozer. *Trudy Vsesoyuznogo Gidrobiologicheskogo Obshchestva*, **2**, 43–68.

Borutskii, E.V. (1963). Emergence of Chironomidae (Diptera) imagines from continental waterbodies of different climatic belts as a factor of food supply of fishes. *Zoologicheskii Zhurnal*, **42**, 233–47. (In Russian.)

Böttger, K. (1970). Die Ernährungsweise der Wassermilben (Hydrachnellae, Acari). *International Revue der Gesamten Hydrobiologie*, **55**, 895–912.

Bottrell, H.H. (1975). The relationship between temperature and duration of egg development in some epiphytic Cladocera and Copepoda from the River Thames, Reading, with a discussion of temperature functions. *Oecologia*, **18**, 63–84.

Braginskii, L.P. (1961). O sootnoshenii mezhdu sostavom prudovogo fitoplanktona i proyavleniem ego "potrebnosti" v biogennykh elementakh. In *Pervichnaya Produktsiya Morei i Vnutrennikh Vod*, ed. G.G. Winberg, pp. 139–47. Minsk.

Brandl, Z. (1973). Relation between the amount of net zooplankton and the depth of station in the shallow Lipno Reservoir. *Hydrobiological Studies*, **3**, 7–51.

Bray, J.R. & Curtis, J.T. (1957). An ordination of the upland forest communities of southern Wisconsin. *Ecological Monographs*, **27**, 325–49.

Bregman, Yu. E. (1968). Growth rate and production of *Asplanchna priodonta* in eutrophic Lake Drivyaty. In *Methods of Assessment of Production of Aquatic Animals*, ed. G.G. Winberg, pp. 184–93. Minsk: Vysheishaya Shkova. (In Russian.)

Bretschko, G. (1974). The Chironomid fauna of a high-mountain lake (Vorderer Finstertaler See, Tyrol, Austria, 2237 m. L.D.). *Entomologisk Tidskrift*, **95** (Supplement). 22–33.

Bretschko, G. (1975). Annual benthic biomass distribution in a high-mountain lake (Vorderer Finstertaler See, Tyrol, Austria). *Verhandlungen der Internationalen Vereinigung für Theoretische und Angewandte Limnologie*, **19**, 1274–85.

Bretthauer, R. (1971). Quantitative estimation of low-molecular ninhydrin-positive matter in waters rich in autumn shed leaves. *Internationale Revue der Gesamten Hydrobiologie*, **56**, 123–8.

Brettum, P. (1971). Fordeling og biomasse av *Isoëtes lacustris* og moen *Scorpidium scorpioides* i Øvre Heimsdalsvatn, et høyfjellsvann i Sør-Norge. *Blyttia*, **29**, 1–11. (In Norwegian with English summary.)

Brezonik, P.L. & Lee, G.F. (1968). Denitrification as a nitrogen sink in Lake Mendota, Wisconsin. *Environmental Science and Technology*, **2**, 120–5.

Brezonik, P.L. & Shannon, E.A. (1971). Trophic state of lakes in North Central Florida. *Publications. Water Resources Research Centre, University of Florida*, **13**, 102 pp.

Bridgman, H.A. (1969). The radiation balance of the Southern Hemisphere. *Archiv für Meteorologie, Geophysik und Bioklimatologie*, B **17**, 325–44.

Bristow, J.M. (1975). The structure and function of roots in aquatic vascular plants. In *The Development and Function of Roots*, ed. J.G. Torrey & D.T. Clarkson, pp. 221–36. New York & London: Academic Press.

Broecker, W.S. (1971). A kinetic model for the chemical composition of sea water. *Quaternary Research*, **1** (2), 188–207.

Brooks, J.L. (1969). Eutrophication and changes in the composition of the zooplankton. In *Eutrophication: Causes, Consequences, Correctives*. International Symposium on Eutrophication, University of Wisconsin, Madison, 1967, pp. 536–55. Washington, DC: National Academy of Sciences.

Brooks, J.L. & Deevey, E.S. (1966). New England. In *Limnology in North America*, ed. D.G. Frey, pp. 117–62. Madison: University of Wisconsin Press.

Brooks, J.L. & Dodson, S.I. (1965). Predation, body size and composition of plankton. *Science, New York*, **150**, 28–35.

Brown, A.H. & Webster, G.C. (1953). The influence of light on the rate of respiration of blue-green alga *Anabaena. American Journal of Botany*, **40**, 753–7.

Brown, D.L. & Tregunna, E.B. (1967). Inhibition of respiration during photosynthesis by some algae. *Canadian Journal of Botany*, **45**, 1135–43.

Brown, M.E. (1946). The growth of brown trout (*Salmo trutta* L.). III. The effect of temperature on the growth of two-year-old trout. *Journal of Experimental Biology*, **22**, 145–55.

Brylinsky, M. & Mann, K.H. (1973). An analysis of factors governing productivity in lakes and reservoirs. *Limnology and Oceanography*, **18**, 1–14.

Brylinsky, M. & Mann, K.H. (1975). The influence of morphometry and of nutrient dynamics on the productivity of lakes. *Limnology and Oceanography*, **20**, 666–7.

Buckney, R.T. & Tyler, P.A. (1973). Chemistry of Tasmanian inland waters. *Internationale Revue der Gesamten Hydrobiologie*, **50**, 61–78.

Budyko, M.I. (1974). *Climate and Life*. English translation of *Klimat i Zhizn*, Leningrad, Gidrometeorizdat (1971) (in Russian). New York & London: Academic Press. 507 pp.

Bunt, J.S. & Heeb, M.A. (1971). Consumption of O_2 in the light by *Chlorella pyrenoidosa* and *Chlamydomonas reinhardtii. Biochimica et Biophysica Acta*, **226**, 354–9.

Burgis, M.J. (1971). The ecology and production of copepods, particularly *Thermocyclops hyalinus*, in the tropical Lake George, Uganda. *Freshwater Biology*, **1**, 169–92.

Burgis, M.J. (1973). Observations on the Cladocera of Lake George, Uganda. *Journal of Zoology, London*, **170**, 339–49.

Burgis, M.J., Darlington, J.P.E., Dunn, I.G., Ganf, G.G., Gwahaba, J.J. & McGowan, L.M. (1973). The biomass and distribution of organisms in Lake George, Uganda. *Proceedings of the Royal Society, London*, B **184**, 271–98.

Burlatskaya, V.M. & Samoilenko, V.S. (1962). On conditions of air-mass formation in the Northern Pacific Ocean. *Trudy Instituta Okeanologii*, **57**, 93–116. (In Russian.)

Burns, C.W. (1968). The relationship between body size of filter-feeding Cladocera and the maximum size of particle ingested. *Limnology and Oceanography*, **13**, 675–8.

Burns, N.M., Williams, J.D.H., Jaquet J.M., Kemp, A.L.W. & Lam, D.C.L. (1975). A phosphorus budget for Lake Erie. *Journal of the Fisheries Research Board of Canada*, **33**, 564–73.

Button, D.K., Dunker, S.S. & Morse, M.L. (1973). Continuous culture of *Rhodotorula rubra*, kinetics of phosphato-arsenate uptake, inhibition and phosphate-limited growth. *Journal of Bacteriology*. **113**, 599–611.

Calow, P. (1973). The food of *Ancylus fluviatilis* (Mull.) a littoral stone-dwelling herbivore. *Oecologia*, **13**, 113–33.

Campbell, A.D. (1974). The parasites of fish in Loch Leven, Kinross. *Proceedings of the Royal Society of Edinburgh*, B **74**, 347–64.

Campbell, R.M. & Spence, D.H.N. (1976). Preliminary studies on the primary productivity of macrophytes in Scottish freshwater lochs. In *Underwater Research*, ed. E.A. Drew, J.N. Lythgoe & J.D. Woods, pp. 347–55. New York & London: Academic Press.

Canale, R.P. (1969). Predator–prey relationship in a model for the activated process. *Biotechnology and Bioengineering*, **11**, 887–907.

Canale, R.P. & Vogel, A.H. (1974). Effects of temperature on phytoplankton growth. *Journal of the Environmental Engineering Division, American Society of Civil Engineers*, **100** (1), 231–41.

Canter, H.M. & Lund, J.W.G. (1948). Studies on plankton parasites, I. Fluctuations in numbers of *Asterionella formosa* Hass. in relation to fungal epidemics. *New Phytologist*, **47**, 238–61.

Canter, H.M. & Lund, J.W.G. (1966). The periodicity of planktonic desmids in Windermere, England. *Verhandlungen der Internationalen Vereinigung für Theoretische und Angewandte Limnologie*, **16**, 163–72.

Capblancq, J. (1972). Phytoplancton et productivité primarie de quelques lacs d'altitude dans les Pyrénées. *Annales de Limnologie*, **8**, 231–321.

Capblancq, J. & Laville, H. (1972). Etude de la productivité du lac de Port-Bielh, Pyrénées centrales. In *Productivity Problems of Freshwaters*, ed. Z. Kajak & A. Hillbricht-Ilkowska, pp. 73–88. Warsaw & Kraków: Polish Scientific Publishers.

Caperon, J. & Meyer, J. (1972). Nitrogen limited growth of marine phytoplankton. *Deep-Sea Research*, **19**, 619–32.

Carlander, K.D. (1955). The standing crop of fish in lakes. *Journal of the Fisheries Research Board of Canada*, **12**, 543–70.

Carr, J.L. (1969). The primary production and physiology of *Ceratophyllum demersum*. II. Micro primary productivity, pH and the P/R ratio. *Australian Journal of Marine and Freshwater Research*, **20**, 127–42.

Casey, H. (1975). Variation in chemical composition of the River Frome, England, from 1965–1972. *Freshwater Biology*, **5**, 507–14.

Casey, H. & Downing, A. (1976). Levels of inorganic nutrients in *Ranunculus penicillatus* var. *calcareus* in relation to water chemistry. *Aquatic Botany*, **2**, 75–9.

Casey, H. & Westlake, D.F. (1974). Growth and nutrient relationships of macrophytes in Sydling Water, a small unpolluted chalk stream. *Proceedings of the European Weed Research Council 4th International Symposium on Aquatic Weeds*. 1974, 69–76.

Caspers, H. (1964). Characteristics of hypertrophic lakes and canals in cities. *Verhandlungen der Internationalen Vereinigung für Theoretische und Angewandte Limnologie*, **15**, 631–8.

Caspers, H. & Karbe, L. (1967). Vorschlage für eine saprobiologische Typisierung der Gewässer. *Internationale Revue der Gesamten Hydrobiologie*, **52**, 144–62.

Chapman, D.W. (1967). Production in fish populations. In *The Biological Basis of Freshwater Fish Production*, ed. S.D. Gerking, pp. 3–12. Oxford: Blackwell.

Charles, W.N., East, K., Brown, D., Gray, M.C. & Murray, T.D. (1974). The production of larval Chironomidae in the mud at Loch Leven, Kinross. *Proceedings of the Royal Society of Edinburgh*, B **74**, 241–58.

Chebotarev, A.I. (1960). *Obstchaya Gidrologiia*. Leningrad: Gidrometeorizdat. 538 pp.

Chen, C.W. & Orlob, G.T. (1972). *Ecologic Simulations of Aquatic Environments*. Walnut Creek, California: Water Resources Engineers Inc. 156 pp.

Chislenko, L.L. (1968). *Nomograms to Determine the Weight of Water Organisms on the Basis of their Shape and Size*. Moscow & Leningrad. 105 pp.

Chua, K.E. & Brinkhurst, R.O. (1973). Evidence of interspecific interactions in the respiration of Tubificid Oligochaeta. *Journal of the Fisheries Research Board of Canada*. **30**, 617–22.

Cichon-Lukanina, E.A. & Soldatova, I.N. (1973). Usvoenie pišči vodnymi bespozvonočnymi. In *Trofologia vodnych zivotnych*, eds. G.V. Nikolskij, P.L. Pirožnkov. Moskva, Nauka, pp. 108–121.

Clarke, F.W. (1911). *The Data of Geochemistry*, 2nd ed. Washington: Government Printing Office.

Clarke, F.W. (1924). The data of geochemistry, 5th ed. *Bulletin of the United States Geological Survey*, **770**, 1–841.

Clarke, G.L. (1967). *Elements of Ecology*. New York: Wiley. 560 pp.

Coffman, W.P. (1973). Energy flow in a woodland stream ecosystem. II. The taxonomic composition and phenology of the Chironomidae as determined by the collection of pupal exuviae. *Archiv für Hydrobiologie*, **71**, 281–322.

Coffman, W.P., Cummins, K.W. & Wuycheck, J.C. (1971). Energy flow in a woodland stream ecosystem: I. Tissue support trophic structure of the autumnal community. *Archiv für Hydrobiologie*, **68**, 232–76.

Colby, F.J., Spangler, G.R., Hurley, D.A. & McCombie, A.M. (1972). Effects of eutrophication on salmonid communities in oligotrophic lakes. *Journal of the Fisheries Research Board of Canada*, **29**, 975–83.

Cole, G.R. (1967). A look at simulation through a study on plankton population dynamics. *Laboratory Report. Batelle Memorial Institute, Pacific Northwest*, **485**, 1–19.

Cole, G.A. (1975). *Textbook of Limnology*. St Louis: Mosby. 283 pp.

Collier, B.D., Cox, G.W., Johnson, A.W. & Miller, P.C. (1973). *Dynamic Ecology*. New York: Prentice Hall. 563 pp.

Conway, E.J. (1942). Mean geochemical data in relation to oceanic evolution. *Proceedings of the Royal Irish Academy*, B **48**, 119–59.

Corbett, J.R. (1974). *The Biochemical Mode of Action of Pesticides*. New York & London: Academic Press. 330 pp.

Crawford, N.H. & Linsley, R.K. (1966). Digital simulation in hydrology: Stanford Watershed Model. IV. *Technical Report, Stanford University*. **39**.

Cremer, G.A. & Duncan, A. (1969). A seasonal study of zooplankton respiration under field conditions. *Verhandlungen der Internationalen Vereinigung für Theoretische und Angewandte Limnologie*, **17**, 181–90.

Crisp, D.T. (1966). Input and output of minerals for an area of Pennine moorland: The importance of precipitation, drainage, peat erosion and animals. *Journal of Applied Ecology*, **3**, 327–48.

Cuinat, R. (1971). Diagnoses écologiques dans quatre rivières à truite de Normandie. *Annales d'Hydrobiologie*, **2** (1), 69–134.

Culver, D.A. & Brunskill, G.J. (1969). Fayetteville Green Lake, New York. V. Studies of primary production and zooplankton in a meromictic marl lake. *Limnology and Oceanography*, **14**, 862–73.

Cummins, K.W. (1972). Predicting variations in energy flow through a semi-controlled lotic ecosystem. *Technical Report. Institute Water Research, Michigan State University*, **19**, 21 pp.

Cummins, K.W. (1973). Trophic relations of aquatic insects. *Annual Review of Entomology*, **18**, 183–206.

Cummins, K.W. (1974). Structure and function of stream ecosystems. *Bioscience*, **24**, 631–41.

Cummins, K.W., Coffman, W.P. & Roff, P.A. (1966). Trophic relationships in a small woodland stream. *Verhandlungen der Internationalen Vereinigung für Theoretische und Angewandte Limnologie*, **16**, 627–38.

Cummins, K.W., Klug, J.J., Wetzel, R.G., Petersen, R.C., Suberkropp, K.F., Manny B.A., Wuycheck, J.C. & Howard, F.O. (1972). Organic enrichment with leaf leachate in experimental lotic ecosystems. *Bioscience*, **22**, 719–22.

Cummins, K.W., Petersen, R.C., Howard, F.O., Wuycheck, J.C. & Holt, V.I. (1973). The utilization of leaf litter by stream detritivores. *Ecology*, **54**, 336–45.

Curds, C.R. (1971). A computer-simulation study of predator–prey relationships in a single-stage continuous-culture system. *Water Research*, **5**, 793–812.

Curl, H.C. & Sandberg, J. (1961). The measurement of dehydrogenase activity in marine organisms. *Journal of Marine Research*, **19**, 123–38.

Cushing, D.H. (1973). Production in the Indian Ocean and the transfer from primary to the secondary level. *The Biology of the Indian Ocean*, ed. B. Zeitschel, *Ecological Studies* 3, pp. 475–86. Berlin: Springer Verlag.

Dake, J.M.K. & Harlemann, D.R.F. (1969). Thermal stratification in lakes: analytical and laboratory studies. *Water Resources Research*, **5**, 484–95.

Darbyshire, J. & Colclough, M. (1972). Measurement of thermal conductivities in a freshwater lake. *Geofisica Pura i Applicata*, **93**, 151–8.

Davies, G.S. (1970). Productivity of macrophytes in Marion Lake, British Columbia. *Journal of the Fisheries Research Board of Canada*, **27**, 71–81.

Davies, R.W. & Reynoldson, T.B. (1971). The incidence and intensity of predation on lake-dwelling triclads in the field. *Journal of Animal Ecology*, **40**, 191–214.

Davis, G.E. & Warren, C.E. (1968). Estimation of food consumption rates. In *Methods for Assessment of Fish Production in Fresh Waters*, ed. W.E. Ricker, IBP Handbook No. 3, pp. 204–25. Oxford: Blackwell.

Dawson, F.H. (1973). The Production Ecology of *Ranunculus penicillatus* var. *calcareus* in Relation to the Organic Input into a Chalk Stream. Ph.D. thesis, University of Aston, Birmingham, 350 pp.

Dawson, F.H. (1976). The annual production of the aquatic macrophyte, *Ranunculus penicillatus* var. *calcareus* (R.W. Butcher) C.D.H. Cook. *Aquatic Botany*, **2**, 51–73.

Deevey, E.S. (1940). Limnological studies in Connecticut. V. A contribution to regional limnology. *American Journal of Science*, **238**, 717–41.

Deevey, E.S. (1941). Limnological studies in Connecticut. VI. The quantity and composition of the bottom fauna of thirty six Connecticut and New York Lakes. *Ecological Monographs*, **11**, 413–55.

Deevey, E.S. (1957). Limnological studies in Middle America. *Transactions of the Connecticut Academy of Arts and Sciences*, **39**, 213–328.

Deevey, E.S. & Stuiver, M. (1964). Distribution of natural isotopes of carbon in Linsley Pond and other New England lakes. *Limnology and Oceanography*, **9**, 1–11.

De Haan, H. (1972). Some structural and ecological studies on soluble humic compounds from Tjeukemeer. *Verhandlungen der Internationale Vereinigung für Theoretische und Angewandte Limnologie*, **18**, 685–95.

De Haan, H. (1974). Effect of a fulvic acid fraction on the growth of a *Pseudomonas* from Tjeukemeer. *Freshwater Biology*, **4**, 301–310.

De Haan, H. (1975). Limnologische aspecten van Humusverbindungen in het Tjeukemeer. Ph.D. thesis, Groningen: Rijksuniversiteit. 93 pp.

Delmendo, M.N. (1966). *An evaluation of the fishery resources of Laguna de Bay*. Indo-Pacific Fisheries Council, 12th Session, Honolulu, Technical Paper (C 66) 400, IPFC/FAO Occasional paper 67/5.

Denny, P. (1972). Sites of nutrient absorption in aquatic macrophytes. *Journal of Ecology*, **60**, 819–29.

Desortová, B. (1976). Productivity of individual algal species in natural phytoplankton assemblage determined by means of autoradiography. *Archiv für Hydrobiologie*, **49** (supplement), 415–49.

Dillon, P.J. (1974). A critical review of Vollenweider's nutrient budget model and other related models. *Water Resources Bulletin*, **10**, 969–89.

Dillon, P.J. (1975). The phosphorus budget of Cameron Lake, Ontario: The importance of flushing rate to the degree of eutrophy of lakes. *Limnology and Oceanography*, **20**, 28–39.

Dillon, P.J. & Kirchner, W.B. (1975). The effects of geology and land use on the export of phosphorus from watersheds. *Water Resources*, **9**, 135–48.

Dillon, P.J. & Rigler, F.H. (1974a). A test of a simple nutrient budget model predicting the phosphorus concentration in lake water. *Journal of the Fisheries Research Board of Canada*, **31**, 1771–8.

Dillon, P.J. & Rigler, F.H. (1974b). The phosphorus–chlorophyll relationship in lakes. *Limnology and Oceanography*, **19**, 767–73.

Dillon, P.J. & Rigler, F.H. (1975). A simple method for predicting the capacity of a lake for development based on lake trophic status. *Journal of the Fisheries Research Board of Canada*, **32**, 1519–31.

Dingmann, S.L. & Johnson, A.H. (1971). Pollution potential of some New Hampshire Lakes. *Water Resources Research*, **7**, 1208–15.

Di Toro D.M., O'Connor, D.J. & Thomann, R.V. (1971). A dynamic model of phytoplankton population in the Sacramento-San Joaquin Delta. *Advances in Chemistry Series*, **106**, 131–80.

Di Toro D.M., O'Connor, D.J., Thomann, R.V. & Mancini, J.L. (1973). Preliminary phytoplankton, zooplankton nutrient model of Western Lake Erie. In *Systems Analysis and Simulation in Ecology*, ed. B.C. Patten, vol. 3. New York & London: Academic Press.

Dobrowolski, K.A. (1973). Role of birds in Polish wetland ecosystems. *Polskie Archiwum Hydrobiologii*, **20**, 217–21.

Dodson, S.I. (1970). Complementary feeding niches sustained by size-selective predation. *Limnology and Oceanography*, **15**, 131–7.

Dokulil, M. (1971). Atmung und Anaerobioseresistenz von Süsswasseralgen. *Internationale Revue der Gesamten Hydrobiologie und Hydrographie*, **56**, 751–68.

Dokulil, M. (1975). Horizontal- und Vertikal-gradienten in einem Flachsee (Neusiedlersee, Österreich) *Verhandlungen der Gesellschaft für Ökologie, Wien*, 1975, 177–87.

Doohan, M. (1973). Energetics of Planktonic Rotifers Applied to Populations in Reservoirs. Ph.D. thesis, London University. 226 pp.

Doty, M.S. & Oguri, M. (1957). Evidence for a photosynthetic daily periodicity *Limnology and Oceanography* **2**, 37–40.

Drabkova, V.G. (1965). The dynamics of bacterial numbers, generation time and bacterial production in the water of Red Lake. *Mikrobiologiya*, **34**, 1063–9. (In Russian.)

Drabkova, V.G. (1971). The microflora and its activity in the water and bottom deposits of Red Lake during the annual cycle. In *Ozera Karel'skogo Peresheika*, ed. S.V. Kalesnik, pp. 258–325. Leningrad, Nauka. (In Russian.)

Drischel, H. (1940). Chlorid- Sulfat- und Nitratgehalt der atmosphärischen Niederschläge in Bad Reiners und Oberschreiberhau im Vergleich zu bisher bekannten Wertën anderer Orte. *Balneologe.* 7, 321–34.

Droop, M.R. (1968). Vitamin B_{12} and marine ecology: IV. The kinetics of uptake, growth and inhibition in *Monochrysis lutheri*. *Journal of the Marine Biological Association of the United Kingdom*, 48, 687–733.

Drummond, A.J., Hickey, J.R., Scholes, W.J. & Laue, E.G. (1968). New values for the solar constant of radiation. *Nature, London*, 218, 259–61.

Dugdale, R.C. (1967). Nutrient limitation in the sea: dynamics, identification and significance. *Limnology and Oceanography*, 12, 685–95.

Duncan, A., Cremer, G.A. & Andrew, T. (1970). The measurement of respiratory rates under field conditions and laboratory during an ecological study on zooplankton. *Polskie Archiwum Hydrobiologii*, 17 (30), 149–60.

Duncan, A., Schiemer, F. & Klekowski, R.Z. (1974). A preliminary study of feeding rates on bacterial food by adult females of a benthic nematode, *Plectus palustris* de Man 1880. *Polskie Archiwum Hydrobiologii*, 21, 249–58.

Dvořáková, M. (1976). Analytical Model of an Aquatic Ecosystem. Thesis, Faculty of Science, Charles University, Praha. 63 pp.

Dykyjová, D. (1973). Specific differences in vertical structure and radiation profiles in the helophyte stands (a survey of comparative measurements). In *Ecosystem Study of Wetland Biome in Czechoslovakia*, ed. S. Hejný, Czechoslovakian IBP/PT-PP Report No. 3, pp. 121–31. Třeboň: Czechoslovakian Academy of Science.

Eckstein, H. (1964). Untersuchungen uber den Einfluss des Rheinwassers auf die Limnologie des Schluchsees. *Archiv für Hydrobiologie*, 28, (supplement), 47–118.

Edmondson, W.T. (1960). Reproductive rates of rotifers in natural populations. *Memorie dell'Istituto Italiano di Idrobiologia*, 12, 21–77.

Edmondson, W.T. (1965). Reproductive rate of planktonic rotifers as related to food and temperature in nature. *Ecological Monographs*, 35, 61–111.

Edmondson, W.T. (1968). Water-quality management and lake eutrophication: the Lake Washington case. In *Water Resources Management and Public Policy*, vol. 11, pp. 139–78. Seattle: University of Washington Press.

Edmondson, W.T. (1970). Phosphorus, nitrogen and algae in Lake Washington after diversion of sewage. *Science, New York*, 169, 690–1.

Edmondson, W.T. (1972a). The present conditions of Lake Washington. *Verhandlungen der Internationalen Vereinigung für Theoretische und Angewandte Limnologie*, 18, 184–291.

Edmondson, W.T. (1972b). Nutrients and phytoplankton in Lake Washington. In *Nutrients and Eutrophication*, ed. G.E. Likens, pp. 172–93. Lawrence: Allen Press.

Edmondson, W.T. & Winberg, G.G. (1971). *A Manual on Methods for the Assessment of Secondary Productivity in Fresh Waters*. IBP Handbook No. 17. Oxford: Blackwell. 313 pp.

Edwards, R.W., Egan, H., Learner, M.A. & Maris, P.J. (1964). The control of chironomid larvae in ponds using TDE (DDD). *Journal of Applied Ecology*. 1 (1), 97–117.

Edwards, R.W. & Owens, M. (1962). The effects of plants on river conditions: IV. The oxygen balance of a chalk stream. *Journal of Ecology*, 50, 207–20.

Edwards, R.W. & Owens, M. (1965). The oxygen balance of streams. In *Ecology and the Industrial Society*, ed. G.T. Goodman, R.W. Edwards & J.M. Lambert, pp. 149–72. Oxford: Blackwell.

Efford, I.E. (1967). Temporal and spatial differences in phytoplankton productivity in Marion Lake, British Columbia. *Journal of the Fisheries Research Board of Canada,* **24,** 2283–307.

Efford, I.E. (1972). An interim review of the Marion Lake Project. In *Productivity Problems of Freshwaters,* ed. Z. Kajak & A. Hillbricht-Ilkowska, pp. 89–109. Warsaw & Kraków: Polish Scientific Publishers.

Egglishaw, H.J. (1964). The distributional relationship between the bottom fauna and plant detritus in streams. *Journal of Animal Ecology,* **33,** 463–76.

Egglishaw, H.J. (1968). The quantitative relationship between bottom fauna and plant detritus in streams of different calcium concentrations. *Journal of Applied Ecology,* **5,** 731–40.

Egglishaw, H.J. (1970). Production of salmon and trout in a stream in Scotland. *Journal of Fish Biology,* **2,** 117–36.

Egglishaw, H.J. (1972). An experimental study on the breakdown of cellulose in fast-flowing streams. *Memorie dell'Istituto Italiano di Idrobiologia,* **29** (supplement), 405–28.

Einsele, W. (1936). Ueber die Beziehungen des Eisenkreislaufs zum Phosphatkreislauf im eutrophen See. *Archiv für Hydrobiologie,* **29,** 664–86.

Einsele, W. (1937). Physikalisch-chemische Betrachtung einiger Probleme des limnischen Mangan- und Eisenkreislaufs. *Verhandlungen der Internationalen Vereinigung für Theoretische und Angewandte Limnologie,* **5,** 69–84.

Einsele, W. (1938). Ueber chemische und Kolloidchemische Vorgänge in Eisen-Phosphat-Systemen unter limnochemischen und limnogeologischen Gesicht-spunkten. *Archiv für Hydrobiologie,* **33,** 361–87.

Eisenberg, R.M. (1966). The regulation of density in a natural population of the pond snail, *Limnaea elodes. Ecology,* **47** (6), 889–905.

Ekman, S. (1964). Die jährliche Populationsentwicklung des planktischen Kopepoden *Diaptomus graciloides* im subarktischen Nordschweden. *Zoologiska Bidrag från Uppsala,* **36,** 277–93.

Elizarova, V.A. (1974). Content of the photosynthetical pigments in a unit of phytoplankton biomass of Rybinsk Reservoir. *Trudy Instituta Biologii Vnutrennikh Vod,* **28** (31), 46–66. (In Russian.)

Elliott, J.M. (1971). Some methods for the statistical analysis of benthic invertebrates. *Scientific Publications. Freshwater Biological Association,* No. 25, 144 p.

Elliott, J.M. (1973). The diel activity pattern, drifting and food of the leech *Erpobdella octoculata* (L.) (Hirudinea, Erpobdellidae) in a Lake District stream. *Journal of Animal Ecology,* **42,** 449–59.

El Shamy, F. (1973). A comparison of Feeding and Growth of Bluegill (*Lepomis macrochirus*) in Lake Wingra and Lake Mendota, Wisconsin. Ph.D. thesis, University of Wisconsin, Madison. 154 pp.

Elster, H.H. (1963). Die Stoffwechseldynamik der Binnengewässer. *Verhandlungen der Deutschen Zoologischen Gesellschaft,* **57,** 335–87.

Elster, H.J., Einsle, U. & Muckle, R. (1968). In *Bodensee-Projekt der Deutschen Forschungsgemeimschaft. Zweiten Bericht* pp. 74–7. Wiesbaden: Franz Steiner Verlag.

Eppacher, T. (1968). Physiographie und Zooplankton des Gossenköllesees. *Bericht des Naturwissenschaftlich-medizinischen Vereins in Innsbruck,* **56,** 31–123.

Eppley, R.W. (1972). Temperature and phytoplankton in the sea. *Fisheries Bulletin No. 8, National Marine Fisheries Series,* NOAA 70 (4), 1065–85.

Eppley, R.W., Rogers, J.N. & McCarthy, J.J. (1969). Halfsaturation constants for uptake of nitrate and ammonium by marine phytoplankton. *Limnology and Oceanography,* **14,** 911–20.

Epstein, E. (1965). Mineral metabolism. In *Plant Biochemistry*, ed. J. Bonner & J.E. Verner, pp. 438–66. New York & London: Academic Press.

EUTROSYM '76. (1976). Proceedings of an International Symposium on Eutrophication and Rehabilitation of Surface Waters, Karl-Marx-Stadt Berlin, DDR, 20–25 September 1976. Institut für Wasserwirtschaft der Ministeriums für Unweltschutz und Wasserwirtschaft, p. 300.

Fabris, G.L. & Hammer, U.T. (1975). Primary production in four small lakes in the Canadian Rocky Mountains. *Verhandlungen der Internationalen Vereinigung für Theoretische und Angewandte Limnologie*, **19**, 530–41.

FAO/UN. (1966). Report to the Government of the Philippines on freshwater fisheries investigations. Based on the work of J.W. Parsons. *Report. Food and Agricultural Organization, Expanded Program of Technical Assistance*. No. 1565, 11 pp. + 3 app.

Farnworth, E.G. & Golley, F.B. (1974). *Fragile Ecosystems*. Berlin: Springer Verlag. 258 pp.

Fast, A.W., Moss, B. & Wetzel, R.G. (1973). Effects of artificial aeration on the chemistry and algae of two Michigan lakes. *Water Resources Research*, **9**, 624–47.

Fedorenko, A.Y. (1975). Feeding characteristics and predation impact of *Chaoborus* (Diptera, Chaoboridae) larvae in a small lake. *Limnology and Oceanography*, **20**, 250–8.

Fee, E.J. (1973a). A numerical model for determining integral primary production and its application to Lake Michigan. *Journal of the Fisheries Research Board of Canada*, **30**, 1447–68.

Fee, E.J. (1973b). Modelling primary production in water bodies: a numerical approach that allows vertical inhomogeneities. *Journal of the Fisheries Research Board of Canada*, **30**, 1469–73.

Feth, J.H. (1971). Mechanisms controlling world water chemistry: evaporation–crystallization process. *Science, New York*, **172**, 870–2.

Findenegg, I. (1943). Untersuchungen, uber die Ökologie und die Produktions-verhältnisse des Planktons im Kärtner Seengebiete. *Internationale Revue der Gessamten Hydrobiologie und Hydrographie*, **43**, 368–429.

Findenegg, I. (1953). Kärtner Seen naturkundlich betrachtet. *Carinthia II*, **15**, 5–101.

Findenegg, I. (1964). Produktionsbiologische Plankton-untersuchungen an Ostalpenseen. *Internationale Revue der Gesamten Hydrobiologie und Hydrographe*, **49**, 381–416.

Findenegg, I. (1965). Relationship between standing crop and primary productivity. *Memorie dell'Istituto Italiano di Idrobiologia*, **18** (supplement), 271–89.

Findenegg, I. (1966). Die Bedeutung kurzwelliger Strahlung für die planktishe Primärproduktion in Seen. *Verhandlungen der Internationalen Vereinigung für Theoretische und Angewandte Limnologie*, **16**, 314–20.

Findenegg, I. (1967). Die Bedeutung des Austausches für die Entwicklung des Phytoplanktons in den Ostalpenseen. *Schweizerische Zeitschrift für Hydrologie*, **29** (1), 125–44.

Findenegg, I. (1971). Die Produktionsleistung einiger planktischer Algenarten im ihrem naturlichen Milieu. *Archiv für Hydrobiologie*, **69**, 273–93.

Fish, G.R. (1963). Limnological conditions and the growth of trout in three lakes near Rotorua. *Proceedings of the New Zealand Ecological Society*, **10**, 1–7.

Fish, G.R. (1968). An examination of the trout population of five lakes near Roturua, New Zealand. *New Zealand Journal of Marine and Freshwater Research*, **2**, 333–62.

Fisher, S.G. (1971). Annual Energy Budget for a Small Forest Stream Ecosystem, Bear Brook, West Thornton, New Hampshire. Ph.D. thesis, Dartmouth College.

Fisher, S.G. & Likens, G.E. (1972). Stream ecosystem: organic energy budget. *Bioscience*, **22**, 33–5.

Fisher, S.G. & Likens, G.E. (1973). Energy flow in Bear Brook, New Hampshire; an integrative approach to stream ecosystem metabolism. *Ecological Monographs*, **43**, 421–39.

Fitzgerald, G.P. (1968). Detection of limiting or surplus nitrogen in algae and aquatic weeds. *Journal of Phycology*, **4**, 121–6.

Fitzgerald, G.P. (1969). Some factors in the competition or antagonism among bacteria, algae and aquatic weeds. *Journal of Phycology*, **5**, 351–9.

Flemer, D.A. (1970). Primary productivity of the north branch of the Raritan River, New Jersey. *Hydrobiologia*, **35**, 273–96.

Floodgate, G.D. (1972). The mechanism of bacterial attachment to detritus in aquatic systems. *Memorie dell'Istituto Italiano di Idrobiologia*, **29** (supplement), 309–23.

Fogg, G.E. (1971). Extracellular products of algae in freshwater. *Archiv für Hydrobiologie, Beihefte, Ergebnisse den Limnologie*, **5**, 1–25.

Fogg, G.E. (1974). Oxygen – versus ^{14}C – methodology. In *A manual of Methods for Measuring Primary Production in Aquatic Environments*, ed. R.A. Vollenweider, IBP Handbook No. 12. 2nd ed., pp. 95–7. Oxford: Blackwell.

Fogg, G.E. (1975). Biochemical pathways in unicellular plants. In *Photosynthesis and Productivity in Different Environments*, ed. J.P. Cooper, *International Biological Programme* 3, pp. 437–57. Cambridge University Press.

Fogg, G.E. & Walsby, A.E. (1971). Buoyancy regulation and the growth of planktonic blue-green algae. *Mitteilungen der Internationalen Vereinigung für Theoretische und Angewandte Limnologie*, **19**, 182–8.

Fogg, G.E. & Westlake, D.F. (1955). The importance of extracellular products of algae in freshwater. *Verhandlungen der Internationalen Vereinigung für Theoretische und Angewandte Limnologie*, **13**, 219–32.

Forsberg, C. (1960). Sub-aquatic vegetation in Ösbysjön, Djursholm. *Oikos*, **11**, 183–99.

Foster, J.M. & Idso, S.B. (1975). Light and assimilation number in a small desert, recharged groundwater pond. *Oecologia*, **18**, 155–64.

Fott, J. (1972). Observations on primary production of phytoplankton in two fishponds. In *Productivity Problems of Freshwaters*, ed. Z. Kajak & A. Hillbricht-Ilkowska pp. 673–83. Warsaw & Kraków: Polish Scientific Publishers.

Fott, J., Korinek, V., Prazakova, M., Vondrus, B. & Forjt, K. (1974). Seasonal development of phytoplankton in fish ponds. *Internationale Revue der Gesamten Hydrobiologie und Hydrographie*, **59**, 629–41.

Frantz, T.C. & Cordone, A.J. (1967). Observations on deepwater plants in Lake Tahoe, California and Nevada. *Ecology*, **48**, 709–14.

Frost, W.E. & Smyly, W.J.P. (1952). The brown trout of a moorland fishpond. *Journal of Animal Ecology*, **21**, 62–86.

Fryer, G. (1968). Evolution and adaptative radiation in the Chydoridae (Crustacea: Cladocera). A study in comparative functional morphology and ecology. *Philosophical Transactions of the Royal Society*, B **254**, 221–385.

Gächter, R. (1968). Phosporhaushalt und planktische Primärproduction im Vierwaldstättersee (Hower Bucht). *Schweizerische Zeitschrift für Hydrobiologie*, **30**, 1–66.

Gächter, R. (1973). Phosphorus losses from the soil and the implications for water pollution control. *EAWAG News*, **1**, 4–5.

Gächter, R. & Furrer, J.O. (1972). Der Beitrag der Landwirtschaft zur Eutrophierung der Gewasser im der Schweiz. *Schweizerische Zeitschrift für Hydrologie*, **34**, 41–70.

Gaevskaya, N.S. (1966). *Rol Vysshikh Rastenii v Pitanii Zhivotnykh Presnykh Vodoemov*.

(The Role of Higher Aquatic Plants in the Nutrition of the Animals of Freshwater Basins) Moskva: Nauka. 327 pp. (In Russian.)

Gak, D.Z. (1967). K paschetu bakterial'noi produktsii vodoema. (On the calculation of bacterial production in water bodies). *Gidrobiologicheskii Zhurnal*, **3**, 93–96. (In Russian.)

Gak, D.Z., Gurvich, V.V., Korelyakova, I.L., Kostikova, L.E., Konstantinova, N.A., Olivari, G.A., Priimachenko, A.D., Tseeb, Y. Ya., Vladimirova, K.S. & Zimbalevskaya, L.N. (1972). Productivity of aquatic organism communities of different trophic levels in Kiev Reservoir. In *Productivity Problems of Freshwaters*, ed. Z. Kajak & A. Hillbricht-Ilkowska, pp. 447–55. Warsaw & Kraków: Polish Scientific Publishers.

Galkovskaya, G.A. (1968). Production of the plankton rotifers. In *Methods of Assessment of Production of Aquatic Animals* ed. G.G. Winberg, pp. 135–41. Minsk: Vysheishaya Shkova. (In Russian.)

Gambaryan, M.E. (1968). *Mikrobiologischeskii Issledovaniya Ozera Sevan*. (Microbiological Studies of Lake Sevan.) Eravan: Akademia Nauk Armenian SSR. 166 pp. (In Russian.)

Ganapati, S.V. & Sreenivasan, A. (1972). Energy flow in aquatic ecosystems in India. In *Productivity Problems of Freshwaters*, ed. Z. Kajak & A. Hillbricht-Ilkowska, pp. 457–75. Warsaw & Kraków: Polish Scientific Publishers.

Ganf, G.G. (1972). The regulation of net primary production in Lake George, Uganda, East Africa. In *Productivity Problems of Freshwaters*, ed. Z. Kajak & A. Hillbricht-Ilkowska, pp. 693–708. Warsaw & Kraków: Polish Scientific Publishers.

Ganf, G.G. (1974a). Incident solar irradiance and underwater light penetration as factors controlling the chlorophyll *a* content of a shallow equatorial lake (Lake George, Uganda). *Journal of Ecology*, **62**, 593–610.

Ganf, G.G. (1974b). Diurnal mixing and vertical distribution of phytoplankton in a shallow equatorial lake (Lake George, Uganda). *Journal of Ecology*, **62**, 611–29.

Ganf, G.G. (1974c). Rates of oxygen uptake by the planktonic community of a shallow equatorial lake (Lake George, Uganda). *Oecologia*, **15**, 17–32.

Ganf, G.G. (1974d). Phytoplankton biomass and distribution in a shallow eutrophic lake, (Lake George, Uganda). *Oecologia*, **16**, 9–29.

Ganf, G.G. (1975). Photosynthetic production and irradiance–photosynthesis relationships of the phytoplankton from a shallow equatorial lake (Lake George, Uganda). *Oecologia*, **18**, 165–83.

Ganf, G.G. & Blažka, P. (1974). Oxygen uptake, ammonia and phosphate excretion by zooplankton of a shallow equatorial lake (Lake George, Uganda). *Limnology and Oceanography*, **19**, 313–25.

Ganf, G.G. & Horne, A.J. (1975). Diurnal stratification, photosynthesis and nitrogen-fixation in a shallow, equatorial lake (Lake George, Uganda) *Freshwater Biology*, **5**, 13–39.

Ganf, G.G. & Viner, A.B. (1973). Ecological stability in a shallow equatorial lake (Lake George, Uganda). *Proceedings of the Royal Society, London*, B **184**, 321–46.

Gasith, A., Koonce, J.F. & Hasler, A.D. (1972). Allochthonous organic carbon in Lake Wingra – preliminary report. *Eastern Deciduous Forest Biome Memo-Report*, 72–120, 12 pp.

Gavis, J. (1976). Munk and Riley revisited: nutrient diffusion transport and rates of phytoplankton growth. *Journal of Marine Research*, **34**, 161–79.

Gelin, C. (1971). Primary production and chlorophyll *a* content of nannoplankton in a eutrophic lake. *Oikos*, **22**, 230–34.

Gerking, S.D. (1966). Annual growth cycle, growth potential, and growth compensation in the bluegill sunfish in northern Indiana lakes. *Journal of the Fisheries Research Board of Canada*, **23**, 1923–56.

Gerking, S.D. (ed.) (1967). *The Biological Basis of Freshwater Fish Production*, Symposium on Productivity of Freshwater Communities of the IBP, Reading, 1966. Oxford: Blackwell. 495 pp. (See p. 570.)

Gerletti, M. (1974). Indagini limnologische sui laghi di Bolsena, Bracciano, Vico e Trasimeno: 7. Caratteristiche chimiche delle acque pelagiche dei laghi di Bolsena, Bracciano e Vico. *Quaderni dell'Istituto di Ricerca sulle Asque*, **17**, 55–87.

Gerloff, G.C. (1963). Comparative mineral nutritions of plants. *Annual Review of Plant Physiology*, **14**, 107–24.

Gerloff, G.C. (1969). Evaluating nutrient supplies for the growth of aquatic plants in natural waters In *Eutrophication: Causes, Consequences, Correctives*, International Symposium on Eutrophication, University of Wisconsin, Madison, 1967, pp. 536–55. Washington, DC: National Academy of Sciences.

Gerloff, G.C. (1975). Nutritional ecology of nuisance aquatic plants. *Ecological Research Series, United States Environmental Protection Agency*, EPA – 660/3-75-027, 78 pp.

Gerloff, G.C. & Fishbeck, K.A. (1973). Plant content of elements as a bioassay of nutrient availability in lakes and streams. In *Bioassay Techniques and Environmental Chemistry*, ed. G. Glass, pp. 159–76. Ann Arbor: Ann Arbor Scientific Publications.

Gerloff, G.C. & Krombholz, P.H. (1966). Tissue analysis as a measure of nutrient availability for the growth of angiosperm aquatic plants. *Limnology and Oceanography*, **11**, 529–37.

Gerloff, G.C. & Skoog, F. (1954). Cell contents of nitrogen and phosphorus as a measure of their availability for growth of *Microcystis aeruginosa* in Southern Wisconsin lakes. *Ecology*, **35**, 348–53.

Gerloff, G.C. & Skoog, F. (1957). Nitrogen as a limiting factor for the growth of *Microcystis aeruginosa* in Southern Wisconsin lakes. *Ecology*, **38**, 556–61.

Gessner, F. (1938). Die Beziehungen zwischen Lichintensität und Assimilation bei submersen Wasserpflanzen. *Jahrbuch für Wissenschaftlche Botanik*, **86**, 491–526.

Gessner, F. (1959). *Hydrobotanik Die physiologischen Grundlagen der Pflanzenverbreitung im Wasser II. Stoffhaushalt*. Berlin: VEB Deutscher Verlag der Wissenschaften. 701 pp.

Geus-Kruyt, M. de & Segal, S. (1973). Notes on the productivity of *Stratiotes aloides* in two lakes in the Netherlands. *Polskie Archiwum Hydrobiologii*, **20**, 195–205.

Gibbs, M. (1962). Respiration. In *Physiology and Biochemistry of Algae*, ed. R.A. Lewin, pp. 61–90. New York & London: Academic Press.

Gibbs, R.J. (1970). Mechanisms controlling world water chemistry. *Science, New York*, **170**, 1088–90.

Gibson, C.E. (1975). A field and laboratory study of oxygen uptake by planktonic blue-green algae. *Journal of Ecology*, **63**, 867–80.

Gibson, C.E., Wood, R.B., Dickson, E.L. & Jewson, D.H. (1971). The succession of phytoplankton in L. Neagh. 1968–70. *Mitteilungen der Internationalen Vereinigung für Theoretische und Angewandte Limnologie*, **19**, 146–60.

Gliwicz, Z.M. (1967). The contribution of nannoplankton in pelagial primary production in some lakes with varying trophy. *Bulletin de l'Academie Polonaise des Sciences, Classe II. Série des Sciences Biologiques*, **15**, (6), 343–7.

Gliwicz, Z.M. (1969a). The share of algae, bacteria and trypton in the food of the pelagic zooplankton in lakes with various trophic characteristics. *Bulletin de l'Academie Polonaise des Sciences, Classe II, Serie des Sciences Biologiques*, **17**, 159–65.

Gliwicz, Z.M. (1969b). Studies on the feeding of pelagic zooplankton in lakes with varying trophy. *Ekologia Polska*, A **17**, 663–708.

Gliwicz, Z.M. (1974). Trophic statuses of freshwater zooplankton species *Wiadomosci Ekologiczne*, **20**, 197–206.

Gliwicz, Z.M. (1975). Effect of zooplankton grazing on photosynthetic activity and composition of phytoplankton. *Verhandlungen der Internationalen Vereinigung für Theoretische und Angewandte Limnologie*, **19**, 1490–7.

Gliwicz, Z.M. (1976). Plankton photosynthetic activity and its regulation in two neotropical man-made lakes. *Polskie Archiwum Hydrobiologii*, **23**, 61–93.

Goldman, C.R. (1963). The measurement of primary productivity and limiting factors in freshwater with carbon-14. In *Proceedings of the Conference on Primary Productivity Measurement, Marine and Freshwater*, University of Hawaii, 1961, ed M.S. Doty, pp. 103–13. US Atomic Energy Commission TID-7633.

Goldman, C.R. (1964). Primary productivity and micronutrient limiting factors in some North American and New Zealand lakes. *Verhandlungen der Internationalen Vereinigung für Theoretische und Angewandte Limnologie*, **15**, 365–74.

Goldman, C.R. (1972). The role of minor nutrients in limiting the productivity of aquatic ecosystems. In *Nutrients and Eutrophication*, ed. G.E. Likens, pp. 21–33. Lawrence: Allen Press.

Goldman, C.R. (1974). Eutrophication of Lake Tahoe emphasizing water quality. *Ecological Research Series, US Environmental Protection Agency*, EPA-66013-74-034, 408 pp.

Goldman, C.R., Mason, D.T. & Hobbie, J.R. (1967). Two Antarctic desert lakes. *Limnology and Oceanography*, **12**, 295–310.

Goldman, C.R., Mason, D.T. & Wood, B.J.B. (1963). Light injury and inhibition in Antarctic freshwater phytoplankton. *Limnology and Oceanography*, **8**, 313–22.

Goldman, C.R., Mason, D.T. & Wood, B.J.B. (1972). Comparative study of the limnology of two small lakes on Ross Island, Antarctica. In *Antarctica Terrestrial Biology*, ed. G.A. Llano, *Antarctic Research Series* 20, pp. 1–50. Washington, DC: American Geophysical Union.

Goldman, C.R. & Wetzel, R.G. (1963). A study of the primary production of Clear Lake, Lake County, California. *Ecology*, **44**, 283–94.

Goldman, J.C. & Carpenter, E.J. (1974). A kinetic approach to the effect of temperature on algal growth. *Limnology and Oceanography*, **19**, 756–66.

Goldsworthy, A. (1970). Photorespiration. *Botanical Review*, **36**, 321–40.

Golterman, H.L. (1964). Mineralization of algae under sterile conditions or by bacterial breakdown. *Verhandlungen der Internationalen Vereinigung für Theoretische und Angewandte Limnologie*, **15**, 593–8.

Golterman, H.L. (ed.) (1968). *Methods for Chemical Analysis of Fresh Waters*. IBP Handbook No. 8, 1st edn. Oxford: Blackwell. 172 pp.

Golterman, H.L. (1971). The determination of mineralization losses in correlation with the estimations of net primary production with the oxygen method and chemical inhibitors. *Freshwater Biology*, **1**, 249–56.

Golterman, H.L. (1972). The role of phytoplankton in detritus formation. *Memorie dell'Istituto Italiano di Idrobiologia*, **29** (supplement), 89–103.

Golterman, H.L. (1973a). Deposition of river silts in the Rhine and Meuse Delta. *Freshwater Biology*, **3**, 267–81.

Golterman, H.L. (1973b). Natural phosphate sources in relation to phosphate budgets: a contribution to the understanding of eutrophication. *Water Research*, **7**, 3–17.

Golterman, H.L. (1975a) *Physiological Limnology. An approach to the Physiology of Lake Ecosystems*. Amsterdam: Elsevier. 489 pp.

Golterman, H.L. (1975b). Chemistry of running waters. In *River Ecology*, ed. B.A. Whitton, pp. 39–80. Oxford: Blackwell.

Golterman, H.L. (1976). Zonation of mineralization in stratifying lakes. In *The Role of Terrestrial and Aquatic Organisms in Decomposition Processes*, ed. J.M. Anderson & A. Macfadyen, pp. 3–22. Oxford: Blackwell.

Golterman, H.L. (1977a). Sediments as a source of phosphate for algal growth. In *Interactions Between Sediments and Fresh Water*, ed. H.L. Golterman, Proceedings of an International Symposium, Amsterdam, 1976 pp. 286–93. The Hague: W. Junk.

Golterman, H.L. (ed.) (1977b). *Interactions Between Sediments and Fresh Water*, Proceedings of an International Symposium, Amsterdam, 1976. The Hague and Wageningen: W. Junk and Centre for Agricultural Publications and Documentation, 474 pp.

Golterman, H.L., Viner, A.B. & Lee, G.F. (1977). Preface. In *Interactions Between Sediments and Fresh Water*, ed. H.L. Golterman, Proceedings of an International Symposium, Amsterdam, 1976, pp. 1–9. The Hague: W. Junk.

Goodall, D.W. (1972). Potential applications of biome modelling. *Terre et la Vie*, **1**, 118–38.

Goodall, D.W. (1973). Ecosystem modelling on the Desert Biome. In *Systems Analysis and Simulation in Ecology*, ed. B.C. Patten, vol. 3. New York & London: Academic Press.

Goodling, J.S. & Arnold, T.G. (1972). Deep reservoir thermal stratification model. *Water Resources Bulletin*, **8**, 745–9.

Gorbunova, Z.A., Gordeeva, L.N., Gritsevskaya, G.L., Dmitrenko, Yu.S., Zabolotskii, A.A. & Ryzhkov, L.P. (1973). (Biological productivity of Lake Cheden'yarvi.) In *Productivity Studies of Freshwater Ecosystems*, ed. G.G. Winberg, pp. 44–54. Minsk: Byelorussian State University. (In Russian).

Gorham, E. (1958). The physical limnology of northern Britain: an epitome of the bathymetric survey of the Scottish Freshwater Lochs, 1897–1909. *Limnology and Oceanography*, **3**, 40–50.

Gorham, E. (1961). Factors influencing supply of major ions to inland waters, with special reference to the atmosphere. *Bulletin of the Geological Society of America*, **72**, 795–840.

Gorham, E. (1964). Morphometric control of annual heat budgets in temperate lakes. *Limnology and Oceanography*, **9**, 525–9.

Goulder, R. (1970). Day-time variations in the rates of production by two natural communities of submerged freshwater macrophytes. *Journal of Ecology*, **58**, 521–8.

Goulder, R. & Boatman, D.J. (1971). Evidence that nitrogen influences the distribution of a freshwater macrophyte, *Ceratophyllum demersum. Journal of Ecology*, **59**, 783–92.

Green, K.L. (1974). Experiments and observations on the feeding behaviour of the freshwater leech *Erpobdella octoculata* (L.) (Hirudinea, Erpobdellidae). *Archiv für Hydrobiologie*, **74**(1), 87–99.

Grim, J. (1950). Versuche zur Ermittlung der Produktionskoeffizienten einiger Phytoplankter im einem flachen See. *Biologisches Zentralblatt*, **69**, 147–74.

Grimaldi, E., Peduzzi, R., Cavicchiolie, G., Giussani, G. & Spreafico, E. (1973). Diffusa infezione branchiale da funghii attribuiti al genere Branchiomyces Plehn (*Phycomycetes saprolegniales*) a carico dell'ittiofauna di laghi situati a nord e a sud delle Alpi. *Memorie dell'Istituto Italiano di Idrobiologia*, **30**, 61–96.

Grimas, U. (1961). The bottom fauna of natural impounded lakes in Northern Sweden

(Ankarvattnet and Blasjon). *Report. Institute of Freshwater Research, Drottning-holm*, **42**, 183–237.

Grönberg, B. (1973). Djurplanktonundersökningar i Ekoln (Mälaren) 1967–1969. *Meddelanden från Naturvårdsverkets Limnologiska Undersökning*. No. 54, 35 pp.

Grygierek, E. (1973). The influence of phytophagous fish on pond zooplankton. *Aquaculture*, **2**, 197–208.

Grygierek, E. & Wolny, P. (1962). The influence of carp fry on the quality and abundance of snails in small ponds. *Rocznik Nauk Rolniczych*, B **81**, 211–30. (In Polish, English summary.)

Grygierek, E. & Wolny, P. (1972). Estimation of production of natural fish food in fish ponds. In *Productivity Problems of Freshwaters*, ed. Z. Kajak & A. Hillbricht-Ilkowska, pp. 775–9. Warsaw & Kraków: Polish Scientific Publishers.

Gurzeda, A. (1965). The effect of increasing fish stock density in the feeding habits of young carp. *Rocznik Nauk Rolniczych*, B **86**, 241–63. (In Polish, English summary.)

Gutelmacher, B.L. (1975). Relative significance of some species of algae in plankton primary production. *Archiv für Hydrobiologie*, **75**, 318–28.

Gutelmacher, B.L., Finogenova, N.P. & Stepanova, L.A. (1974). Seminar po voprosam ekologicheskoi fiziologii presnovodnykh bespozvonochnykh. *Gidro-biologicheskii Zhurnal*, **10** (4), 127–30.

Hall, D.J. (1964). The dynamics of a natural population of *Daphnia. Verhandlungen der Internationalen Vereinigung für Theoretische und Angewandte Limnologie*, **15**, 660–4.

Hall, D.J. (1971). The use of instananeous production models. In *A Manual on Methods for the Assessment of Secondary Productivity in Freshwaters*, ed. W.T. Edmondson & G.G. Winberg, IBP Handbook No. 17 pp. 316–17. Oxford: Blackwell.

Hall, D.J., Cooper, W.E. & Werner, E.E. (1970). An experimental approach to the production, dynamics and structure of freshwater animal communities. *Limnology and Oceanography*, **15**, 839–928.

Hall, K.J., Kleiber, P.M. & Yesaki, I. (1972). Heterotrophic uptake of organic solutes by microorganisms in the sediment. *Memorie dell'Istituto Italiano di Idrobiologia*, **29** (supplement), 441–71.

Halldal, P. (1964). Ultraviolet action spectra of photosynthesis and photosynthetic inhibition in a green and red alga. *Physiologia Plantarum*, **17**, 414–21.

Halldal, P. (1966). Induction phenomena and action spectra analyses of photosynthesis in ultraviolet and visible light studied in green and blue-green algae, and in isolated chloroplast fragments. *Zeitschrift für Pflanzenphysiologie*, **54**, 28–44.

Halldal, P. & Taube, Ö. (1972). Ultraviolet action and photoreactivation in algae. In *Photophysiology*, ed. A.C. Giese, vol. 7, pp. 163–88. New York & London: Academic Press.

Hammann, von, A. (1957). Assimilationzahlen submerser Phanerogamen und ihre Beziehung zur Kohlensäureversorgung. *Schweizerische Zeitschrift für Hydrologie*, **19**, 579–612.

Hammer, U.T., Walker, K.F. & Williams, W.D. (1973). Derivation of daily phytoplank-ton production estimates from short-term experiments in some shallow eutrophic Australian saline lakes. *Australian Journal of Marine and Freshwater Research*, **24**, 259–66.

Haney, J.F. (1973). An *in situ* examination of the grazing activities of natural zooplankton communities. *Archiv für Hydrobiologie*, **72**, 87–132.

Hansmann, E.W., Lane, C.B. & Hall, J.D. (1971). A direct method of measuring benthic primary production in streams. *Limnology and Oceanography*, **16**, 822–6.

Happey, C.M. & Moss, B. (1967). Some aspects of the biology of *Chrysococcus diaphanus* in Abbot's Pond, Somerset. *British Phycological Bulletin*, 3, 269–79.

Hargrave, B.T. (1969). Epibenthic algal production and community respiration in the sediments of Marion Lake. *Journal of the Fisheries Research Board of Canada*, 26, 2003–26.

Hargrave, B.T. (1970). The utilization of benthic microflora by *Hyalella azteca* (Amphipoda). *Journal of Animal Ecology*, 39, 427–37.

Hargrave, B.T. (1971). An energy budget for a deposit-feeding Amphipod. *Limnology and Oceanography*, 16, 99–103.

Hargrave, B.T. (1972a). Aerobic decomposition of sediment detritus as a function of particle surface area and organic content. *Limnology and Oceanography*, 17, 583–96.

Hargrave, B.T. (1972b). Prediction of egestion by the deposit-feeding Amphipod *Hyalella azteca*. *Oikos*, 23, 116–24.

Hargrave, B.T. (1973). Coupling carbon flow through some pelagic and benthic communities. *Journal of the Fisheries Research Board of Canada*, 30, 1317–26.

Harlin, M.M. (1973). Transfer of products between epiphytic marine algae and host plants. *Journal of Phycology*, 9, 243–8.

Harrison, M.J., Wright, R.T. & Morita R.Y. (1971). Method for measuring mineralization in lake sediments. *Applied Microbiology*, 21, 698–702.

Hata, Y. (1975). Biotic production and eutrophication in Lake Kojima. Mechanism of eutrophication. In *Productivity of Communities in Japanese Inland Waters*, ed. S. Mori & G. Yamamoto, J.I.B.P. Synthesis. Vol. 10, pp. 219–22. Tokyo: University of Tokyo Press.

Hatch, R.W. & Webster, D.A. (1961). Trout production in four central Adirondack mountain lakes. *Memoirs. Cornell University Agricultural Experiment Station*, 373, 81 pp.

Hayes, F.R. (1957). On the variation in bottom fauna and fish yield in relation to trophic level and lake dimensions. *Journal of the Fisheries Research Board of Canada*, 14, 1–32.

Hayes, F.R. & Anthony, E.H. (1964). Productive capacity of North American lakes as related to the quantity and the trophic level of fish, the lake dimensions, and the water chemistry. *Transactions of the American Fisheries Society*, 93, 53–7.

Hayne, W. & Ball, R.C. (1956). Benthic productivity as influenced by fish predation. *Limnology and Oceanography*, 1, 162–75.

Healey, F.P. (1975). Physiological indicators of nutrient deficiency in algae. *Technical Report. Department of the Environment, Fisheries and Marine Service, Research and Development Directorate*, 585, 30 pp.

Healy, M. (1972). Bioenergetics of a sand goby (*Gobius minutus*) population. *Journal of the Fisheries Research Board of Canada*, 29, 187–94.

Heitmann, M.L. (1972). Zur Modellierung physikalischer Bedingungen von Seen. *Verhandlungen der Internationalen Vereinigung für Theoretische und Angewandte Limnologie*, 18, 419–25.

Heitmann, M.L. Richter, D. & Schumann, P. (1969). Der Wärme- und Wasserhaushalt des Stechlin und Nehmitzsees. *Abhandlungen des Deutschen Meteorologischen Diensts in der Sowjetisch Besetzten Zone Deutschlands*, 96, 98 pp.

Hellström, B.G. & Nauwerck, A. (1971). Zur Biologie and Populationsdynamik von *Polyartemia forcipata* (Fischer). *Report Institute of Freshwater Research, Drottningholm*, 51, 47–66.

Henderson, H.F., Ryder, R.A. & Kudhongania, A.W. (1973). Assessing fishery potentials of lakes and reservoirs. *Journal of the Fisheries Research Board of Canada*, 30, 2000–9.

Henson, E.B., Bradshaw, A.S. & Chandler, D.C. (1961). The physical limnology of Cayuga Lake, New York. *Memoirs. Cornell University Agricultural Experiment Station*, **378**, 3–63.

Henson, E.B., Lind, A.O. & Potash, M. (1975). Limnological utilization of ERTS-1 satellite imagery. *Verhandlungen der Internationalen Vereinigung für Theoretische und Angewandte Limnologie*, **19**, 179–88.

Hepher, B. (1962a). Primary production in fish ponds and its application to fertilization experiments. *Limnology and Oceanography*, **7**, 131–6.

Hepher, B. (1962b). Ten years of research in fishponds fertilization in Israel. I. The effect of fertilization on fish yields. *Bamidgeh*, **14**, 29–38.

Herbert, D. (1964). Multi-stage continuous culture. In *Continuous Cultivation of Microorganisms*, ed. J. Malek, K. Beran & J. Hospodka, Proceedings of the 2nd Symposium, Prague, 1962, pp. 23–44. Prague: Academia.

Hetling, L.J. & Sykes, R.M. (1973). Sources of nutrients in Canadarago Lake. *Journal of the Water Pollution Control Federation*, **45**, 145–56.

Hickman, M. (1971a). The standing crop and primary productivity of the epiphyton attached to *Equisetum fluviatile* L. in Priddy Pool, North Somerset. *British Phycological Journal*, **6**, 51–9.

Hickman, M. (1971b). Standing crops and primary productivity of the epipelon of two small ponds in North Somerset, UK. *Oecologia*, **6**, 238–53.

Hickman, M. (1973). The standing crop and primary productivity of the phytoplankton of Abbot Pond, North Somerset. *Journal of Ecology*, **61**, 269–87.

Hickman, M. & Round, F.E. (1970). Primary production and standing crop of epipsammic and epipelic algae. *British Phycological Journal*, **5**, 247–55.

Hillbricht-Ilkowska, A. (1972). Interlevel energy transfer efficiencies in plankton food chain. Manuscript on speech given at IBP Symposium, Reading, 1972.

Hillbricht-Ilkowska, A. & Pourriot, G. (1970). Production of experimental populations of *Brachionus calyciflorus* Pallas (Rotatoria) exposed to artificial predation of different rates. *Polskie archiwum hydrobiologii*, **17**, 241–8.

Hillbricht-Ilkowska, A., Prejs, A. & Węgleńska, T. (1973). Experimentally increased fish stock in the pond type lake Warniak. VIII. Approximate assessment of the utilization by fish of the biomass and production of zooplankton. *Ekologia Polska*, **21**, 553–62.

Hillbricht-Ilkowska, A. & Spodniewska, I. (1969). Comparison of the primary production of phytoplankton in three lakes of different trophic type. *Ekologia Polska*, A **17**, 241–61.

Hillbricht-Ilkowska, A., Spodniewska, I., Węgleńska, T. & Karabin, A. (1972). The seasonal variation of some ecological efficiencies and production rates in the plankton community of several Polish lakes of different trophy. In *Productivity Problems of Freshwaters*, ed. Z. Kajak & A. Hillbricht-Ilkowska, pp. 111–27. Warsaw & Kraków: Polish Scientific Publishers.

Hillbricht-Ilkowska, A. & Węgleńska, T. (1970). The effect of sampling frequency and the method of assessment on the production values obtained for several zooplankton species. *Ekologia Polska*, **18**, 539–57.

Hillbricht-Ilkowska, A. & Węgleńska, T. (1973). Experimentally increased fish stock in the pond type lake Warniak. VII. Numbers, biomass and production of zooplankton. *Ekologia Polska*, **21**, 533–52.

Hobbie, J.E. (1964). Carbon-14 measurements of primary production in two arctic Alaskan lakes. *Verhandlungen der Internationalen Vereinigung für Theoretische und Angewandte Limnologie*, **15**, 360–4.

Hobbie, J.E., Barsdate, R.J., Alexander, V., Stanley, D.W., McRoy, C.P., Stross, R.G., Bierle, D.A., Dillon, R.D., Miller, M.C., Coyne, P.I. & Kelley, J.H. (1972). Carbon flux through a tundra pond ecosystem at Barrow, Alaska. *United States Tundra Biome Report*, **72** (1), 1–29.

Hobbie, J.E. & Crawford, C.C. (1969). Respiration corrections for bacterial uptake of dissolved organic compounds in natural waters. *Limnology and Oceanography*, **14**, 528–32.

Hobbie, J.E. & Likens, G.E. (1973). Output of phosphorus, dissolved organic carbon, and fine particulate carbon from Hubbard Brook Watershed. *Limnology and Oceanography*, **18**, 734–42.

Hobbie, J.E. & Wright, R.T. (1965). Competition between planktonic bacteria and algae for organic solutes. In *Primary Productivity in Aquatic Environments*, ed. C.R. Goldman, pp. 175–85. Berkeley: University of California Press.

Hoch, G., Owens, O.H. & Kok, B. (1963). Photosynthesis and respiration. *Archives of Biochemistry and Biophysics*, **101**, 171–80.

Hodkinson, I.D. (1975). Dry weight loss and chemical changes in vascular plant litter of terrestrial origin, occurring in a beaver pond system. *Journal of Ecology*, **63**, 131–42.

Holčik, J. (1972). Abundance, ichthyomass and production of fish populations in three types of water-bodies in Czechoslovakia (man-made lake, trout lake, arm of the Danube river). In *Productivity Problems of Freshwaters*, ed. Z. Kajak & A. Hillbricht-Ilkowska, pp. 843–55. Warsaw & Kraków: Polish Scientific Publishers.

Holčik, J. & Pivnička, K. (1972). The density and production of fish populations in the Kličava reservoir (Czechoslovakia) and their changes during period 1957–1970. *Internationale Revue der Gesamten Hydrobiologie*, **57**, 883–94.

Holland, R. (1968). Correlation of *Melosira* species with trophic conditions in Lake Michigan. *Limnology and Oceanography*, **13**, 555–6.

Holliday, F.G.T., Tytler, P. & Young, A.H. (1974). Activity levels of trout (*Salmo trutta*) in Airthrey Loch, Stirling and Loch Leven, Kinross. *Proceedings of the Royal Society of Edinburgh*, B **74**, 315–31.

Holmgren, S. (1968). Phytoplankton Production in a Lake North of the Arctic Circle. Fil. Lic. thesis, Uppsala, Institute of Limnology, 43 pp.

Holmgren, S. & Lundgren, A. (1972–74). Fytoplankton och primarproduction i Stugsjon och Hymenjaure (1971–73). In *Experiment Med Gödsling av Sjöar i Kuokkelområdet*, Reports 1–3 from the Kuokkel project. Uppsala: Institute of Limnology.

Hoogers, B.J. & Van der Weij, H.G. (1971). The development cycle of some aquatic plants in the Netherlands. *Proceedings of the European Weed Research Council 3rd International Symposium on Aquatic Weeds*, 1971, 3–18.

Hooper, F.F., Ball, R.C. & Tanner, H.A. (1953). An experiment in the artificial circulation of a small Michigan lake. *Transactions of the American Fisheries Society*, **82**, 222–41.

Hora, S.L. & Pillay, T.V.R. (1962). Handbook on fish culture in the Indo-Pacific region. *FAO Fisheries Biology Technical Papers*, No. 14, 203 pp.

Horie, S. (1962). Morphometric features and the classification of all the lakes in Japan. *Memoirs of the College of Science, Kyoto University*, B **29** (3), 191–262.

Horne, A.J., Newbold, J.D. & Tilzer, M.M. (1975). The productivity mixing modes and management of the world's lakes. *Limnology and Oceanography*, **20**, 663–5.

Horne, A.J. & Viner, A.B. (1971). Nitrogen fixation and its significance in tropical Lake George, Uganda. *Nature, London*, **232**, 417–18.

Horton, P.A. (1961). The bionomics of brown trout in a Dartmoor stream. *Journal of Animal Ecology*, **30**, 311–38.

Horvath, R.S. (1972). Microbial co-metabolism and the degradation of organic compounds in nature. *Bacteriological Reviews*, **36**, 146–55.

Hought, R.A. (1974). Photorespiration and productivity in submerged aquatic vascular plants. *Limnology and Oceanography*, **19**, 912–27.

Hough, R.A. & Wetzel, R.G. (1972). A ^{14}C assay for photorespiration in aquatic plants. *Plant Physiology*, **49**, 987–90.

Houghton, H.G. (1954). On the annual heat balance of the Northern Hemisphere. *Journal of Meteorology*, **11**, 1–9.

Howard, H.H. & Prescott, G.W. (1971). Primary production in Alaskan tundra lakes. *American Midland Naturalist*, **85**, 108–23.

Hrbáček, J. (1962). Species composition and the amount of zooplankton in relation to the fishstock. *Rozpravy Československé Akademie Věd. Řada Matematických a Prírodnich Ved*, **72** (10), 115 pp.

Hrbáček, J. (1966). A morphometrical study of some backwaters and fish ponds in relation to the representative plankton samples, with an appendix by C.O. Junge on depth distribution for quadric surfaces and other configurations. *Hydrobiological Studies*, **1**, 221–65.

Hrbáček, J. (1969). Relations between some environmental parameters and the yield of fish as a basis for a predictive model. *Verhandlungen der Internationalen Vereinigung für Theoretische und Angewandte Limnologie*, **17**, 1069–81.

Hrbáček, J. (1972). Unpublished lecture. IBP/PF/UNESCO Symposium. Results of PF projects, September 1972, Reading, UK.

Hrbáček, J., Dvořáková, M., Kořinek, V. & Procházková, L. (1961). Demonstration of the effect of the fish stock on the species composition of zooplankton and the intensity of metabolism of the whole plankton association. *Verhandlungen der Internationalen Vereinigung für Theoretische und Angewandte Limnologie*, **14**, 192–5.

Hrbáčková, M. & Hrbáček, J. (1978). The growth rate of *Daphnia pulex* and *Daphnia pulicaria* (Crustacea: Cladocera) at different food levels. *Věstnik Československe Spolecnosti Zoologické*, **42**, 115–27.

Hruška, V. (1961). An attempt at a direct investigation of the influence of the carp stock on the bottom fauna of two ponds. *Verhandlungen der Internationalen Vereinigung für Theoretische und Angewandte Limnologie*, **14**, 732–6.

Huff, D.D. (1968). Simulation of the Hydrologic Transport of Radioactive Aerosols. Ph.D. dissertation, Stanford University.

Huff, D.D., Koonce, J.F., Ivarson, W.R.F., Weiler, P.R., Dettman, E.H. & Harris, R.F. (1973). Simulation of urban runoff, nutrient loading, and biotic response of a shallow eutrophic lake. In *Modeling the Eutrophication Process* ed. E.J. Middlebrooks, D.H. Falkenborg & T.E. Maloney, pp. 33–55. Ann Arbor: Ann Arbor Science.

Hunt, R.L. (1966). Production and angler harvest of wild brook trout in Lawrence Creek, Wisconsin. *Technical Bulletin. Wisconsin Conservation Department*, **35**, 1–52.

Hurlbert, S., Zedler, J. & Fairbanks, D. (1971). Ecosystem alteration by mosquitofish (*Gambusia affinis*) predation. *Science, New York*, **175**, 639–41.

Hutchinson, G.E. (1957). *A Treatise on Limnology, Volume 1. Geography, Physics and Chemistry*. New York: Wiley. 1015 pp.

Hutchinson, G.E. (1967). *A Treatise on Limnology, Volume II. Introduction to Lake Biology and the Limnoplankton*. New York: Wiley. 1115 pp.

Hutchinson, G.E. (1975). *A Treatise on Limnology, Volume III. Limnological Botany*. New York: Wiley. 660 pp.

Hutchinson, G.E. & Löffler, H. (1956). The thermal classification of lakes. *Proceedings of the National Academy of Sciences of the United States of America*, **42**, 84–6.

Hynes, H.B.N. (1970). *The Ecology of Running Waters.* Toronto: University of Toronto Press. 555 pp.

Idso, S.B. (1973). On the concept of lake stability. *Limnology and Oceanography*, **18**, 681–2.

Idso, S.B. & Cole, G.A. (1973). Studies on a Kentucky Knobs Lake: V. Some aspects of the vertical transport of heat in the hypolimnion. *Journal of Ecology*, **61**, 413–20.

Idso, S.B. & Foster, J.M. (1975). An analytical study of three characteristic forms of light-forced primary production in aquatic ecosystems. *Oecologia*, **18**, 145–54.

Ikusima, I. (1965). Ecological studies on the productivity of aquatic plant communities: I. Measurement of photosynthetic activity. *Botanical Magazine, Tokyo*, **78**, 202–11.

Ikusima, I. (1966). Ecological studies on the productivity of aquatic plant communities: II. Seasonal changes in standing crop and productivity of a natural submerged community of *Vallisneria denseserrulata. Botanical Magazine, Tokyo*, **79**, 7–19.

Ikusima, I. (1967). Ecological studies on the productivity of aquatic plant communities: III. Effect of depth on daily photosynthesis in submerged macrophytes. *Botanical Magazine, Tokyo*, **80**, 57–67.

Ikusima, I. (1970). Ecological studies on the productivity of aquatic plant communities: IV. Light conditions, and community photosynthetic production. *Botanical Magazine, Tokyo*, **83**, 330–41.

Ikusima, I. (1975). Productivity of freshwater communities in Lake Biwa. Primary production of aquatic macrophytes. In *Productivity of Communities in Japanese Inland Waters*, ed. S. Mori & G. Yamamoto. JIBP Synthesis, vol. 10, pp. 28–31. Tokyo: University of Tokyo Press.

Ilmavirta, K., Ilmavirta, V. & Kotimaa, A.-L. (1974). Pääjärven kasviplankton. *Luonnon Tutkija*, **78**, 133–48.

Imboden, D.M. (1973). Limnologische Transport- und Nährstoffmodelle. *Schweizerische Zeitschrift für Hydrologie*, **35**, 29–68.

Imboden, D.M. (1974). Phosphorus model of lake eutrophication. *Limnology and Oceanography*, **19**, 297–304.

Infante, A. (1973). Untersuchungen über die Ausnutzbarkeit verscheidener Algen durch das Zooplankton. *Archiv für Hydrobiologie*, **42** (supplement), 340–405.

Ingham, L., & Arme, C. (1973). Intestinal helminths in rainbow trout, *Salmo gairdneri* (Richardson). *Journal of Fish Biology*, **5**, 309–14.

International Council of Scientific Unions Special Committee for the International Biological Programme (1974). *Quantities, Units and Symbols. Recommendations for Use in IBP Synthesis.* London: ICSU. 64 pp.

Ioffe, T.J. (1972). The improvement of reservoir productivity through acclimatization of invertebrates. *Verhandlungen der Internationalen Vereinigung für Theoretische und Angewandte Limnologie*, **18**, 818–21.

Iovino, A.J. & Bradley, W.H. (1969). The role of larval Chironomidae in the production of lacustrine copropel in Mud Lake, Marion County, Florida. *Limnology and Oceanography*, **14**, 898–905.

Irwin, J. (1972). New Zealand lakes bathymetric surveys 1965–70. *New Zealand Oceanographic Institute Records*, **1** (6), 107–26.

Isirimah, N.O., Keenery, D.R. & Dettmann E.H. (1976). Nitrogen cycling in Lake Wingra. *Journal of Environmental Quality*, **5**, 182–8.

Ivanova, M.B. (1973). Estimation of accuracy for the calculation of production and elimination of planktonic crustaceans (*Eudiaptomus gracilis*) in the Lake Krasavitza taken as an example. *Zoologischeskii Zhurnal*, **52**, 111–20. (In Russian with English summary.)

Ivanova, M.B. (1975). Biologo-produktsionnye issledouvaniya zooplanktona v ozerakh i vodokhranilishchakh SSSR. *Izvestiya Akademii nauk SSSR, Seriya Biologicheskaya*, **1**, 104–13.

Iverson, T.M. (1973). Decomposition of autumn-shed beech leaves in a spring brook and its significance for the fauna. *Archiv für Hydrobiologie*, **72**, 305–12.

Ivlev, V.S. (1945). Biologicheskaya produktivnost' vodoemov. *Uspekhi Sovremenn oi Biologii*, **19**(1), 98–120. (Translation by W.E. Ricker published in 1966. The biological productivity of waters. *Journal Fisheries Research Board of Canada*, **23**, 1727–59.)

Ivlev, V.S. (1961). *Experimental Ecology of the Feeding of Fishes*. Translated from the Russian by D. Scott. New Haven: Yale University Press.

Izvekova, E.I. (1967). Pitanie lichinok nekotorykh khironomid Uchinskogo vodokhranilischcha. *Informatsionnyi Buylleten. Institut Biologii Vnutrennikh Vod*, **1**, 42–4.

Izvekova, E.I. (1971a). On the feeding habits of Chironomid larvae. *Limnologica, Berlin*, **8**(1), 201–2.

Izvekova, E.I. (1971b). K vaprosu o roli khironomid-fitratorov v protsessakh samoochishcheniya vodoemov. In *Kompleksnye Issledovaniya Vodokhranilishch*, ed. V.D. Bykov, N.I. Sokolova & K.K. Edelstein, pp. 204–7. Moskva: Izdatel'stvo Moskovskogo Universitet.

Izvekova, E.I. & Lvova-Kachanova, A.A. (1972). Sedimentation of suspended matter by *Dreissena polymorpha* Pallas and its subsequent utilisation by Chironomidae larvae. *Polskie Archiwum Hydrobiologii*, **2**, 203–10.

Jackson, D.D. (1905). The normal distribution of chlorine in the natural waters of New York and New England. *Water-Supply and Irrigation Paper, Washington*, **144**, 31 pp.

Jackson, W.S. & Volk, R.J. (1970). Photorespiration. *Annual Review of Plant Physiology*, **21**, 385–432.

James, H.R. & Birge, E.A. (1938). A laboratory study of the absorption of light by lake water. *Transactions of the Wisconsin Academy of Sciences, Arts and Letters*, **31**, 1–154.

Jannasch, H.W. & Pritchard, P.H. (1972). The role of inert particulate matter in the activity of aquatic microorganisms. *Memorie dell'Istituto Italiano di Idrobiologia*, **29** (supplement), 289–308.

Javornický, P. (1979). The changes of plankton photosynthetic activity: a system regulated by chlorophyll content. *Internationale Revue der Gesamten Hydrobiologie und Hydrographie*, (in press).

Javornický, P. & Komárková, J. (1973). The changes in several parameters of plankton primary productivity in Slapy Reservoir 1960–1967, their mutual correlations and correlations with the main ecological factors. *Hydrobiological Studies*, **2**, 155–211.

Jenkins, R.M. (1968). The influence of some environmental factors on standing crop and harvest of fishes in US reservoirs. In *Reservoir Fishery Resources Symposium, Athens: Georgia, April 1967*. University of Georgia Press. pp. 298–321.

Jenkins, R.M. & Morais, D.I. (1971). Reservoir sport fishing effort and harvest in relation to environmental variables. In *Reservoir Fisheries and Limnology*, ed. G.E. Hall, *Special Publication, American Fisheries Society, No. 8*, pp. 371–84.

Jewson, D.H. (1975). The relation of incident-radiation to diurnal rates of photosynthesis in Lough Neagh. *Internationale Revue der Gesamten Hydrobiologie und Hydrographie*, **60**, 759–67.

Jewson, D.H. (1976). The interaction of components controlling net phytoplankton photosynthesis in a well-mixed lake (Lough Neagh, Northern Ireland) *Freshwater Biology*, **6**, 551–76.

Jewson, D.H. (1977a). Light penetration in relation to phytoplankton content of the euphotic zone of Lough Neagh, N. Ireland. *Oikos*, **28**, 74–83.

Jewson, D.H. (1977b). A comparison between *in situ* photosynthetic rates determined using ^{14}C uptake and oxygen evolution methods in Lough Neagh, Northern Ireland. *Proceedings of the Royal Irish Academy*, B **77** (3), 87–99.

Jewson, D.H. & Wood, R.B. (1975). Some effects on integral photosynthesis of artificial circulation of phytoplankton through light gradients. *Verhandlungen der Internationalen Vereinigung für Theoretische und Angewandte Limnologie*, **19**, 1037–44.

Jónasson, P.M. (1972). Ecology and production of the profundal benthos in relation to phytoplankton in Lake Esrom. *Oikos*, supplementum 14, 1–148.

Jónasson, P.M. & Kristiansen, J. (1967). Primary and secondary production in Lake Esrom. Growth of *Chironomus anthracinus* in relation to seasonal cycles of phytoplankton, and dissolved oxygen. *Internationale Revue der Gesamten Hydrobiologie und Hydrographie*, **52**, 163–317.

Jónasson, P.M. & Thorhauge, F. (1972). Life cycle of *Potamothrix hammoniensis* (Tubificidae) in the profundal of a eutrophic lake. *Oikos*, **23**, 151–8.

Jones, J.R. & Bachmann, R.W. (1976). Prediction of phosphorus and chlorophyll levels in lakes. *Journal of the Water Pollution Control Federation*, **48**, 2176–82.

Jørgensen, E.G. & Steemann Nielsen, E. (1965). Adaption in plankton algae. *Memorie dell'Istituto Italiano di Idrobiologia*, **18** (supplement), 39–46.

Jørgensen, S.E. (1977). The influence of phosphate and nitrogen in sediment on restoration of lakes. In *Interactions Between Sediments and Fresh Water*, ed. H.L. Golterman, Proceedings of an International Symposium, Amsterdam, 1976, pp. 387–9. The Hague: W. Junk.

Jørgensen, S.E., Kamp Nielsen, L. & Jacobsen, O.S. (1975). A submodel for anaerboic mud–water exchange of phosphate. *Ecological Modelling*, **1**, 133–46.

Jupp, B.P., Spence, D.H.N. & Britton, R.H. (1974). The distribution and production of submerged macrophytes in Loch Leven, Kinross. *Proceedings of the Royal Society of Edinburgh*, B **74**, 195–208.

Kajak, Z. (1958). On the investigation on the living conditions of benthos. *Ekologia Polska*, B **4**, 190–201. (In Polish with English summary.)

Kajak, Z. (1962). Review of literature on the benthos of reservoirs in connection with the building of the Debe reservoir on the Bug and Narew Rivers. *Ekologia Polska*, B **8**, 4–27.

Kajak, Z. (1964). Remarks on conditions influencing the appearance of new generations of Tendipedidae larvae. *Ekologia Polska*, A **12**, 173–83.

Kajak, Z. (1966). Abundance and production of benthos and the factors governing them. *Zeszyty Problemowe "Kosmosu"*, **13**, 69–92. (In Polish.)

Kajak, Z. (1968). *Experimental analysis of factors decisive for benthos abundance (in particular Chironomidae)*. Warszawa. 94 pp. (In Polish with English summary.)

Kajak, Z. (1972). Analysis of the influence by fish on benthos by the method of enclosures. In *Productivity Problems of Freshwaters*, ed. Z. Kajak & A. Hillbricht-Ilkowska, pp. 781–93. Warsaw & Kraków: Polish Scientific Publishers.

Kajak, Z. & Dusoge, K. (1970). Production efficiency of *Procladius choreus* Mg. (Chironomidae, Diptera) and its dependence on the trophic conditions. *Polskie Archiwum Hydrobiologii*, **17**, 217–24.

Kajak, Z. & Dusoge, K. (1971). The regularities of vertical distribution of benthos in bottom sediments of three Masurian Lakes. *Ekologia Polska*, A **19**, 485–99.

Kajak, Z. & Hillbricht-Ilkowska, A. (eds.) (1972). *Productivity Problems of Freshwaters*. IBP-UNESCO Symposium, Kazimierz Dolny, Poland, 1970. Warsaw & Kraków: Polish Scientific Publishers. 918 pp.

Kajak, Z., Hillbricht-Ilkowska, A. & Pieczyńska, E. (1972). The production processes in several Polish lakes. In *Productivity Problems of Freshwaters*, ed. Z. Kajak & A. Hillbricht-Ilkowska, pp. 129–47. Warsaw & Kraków: Polish Scientific Publishers.

Kajak, Z. & Kajak, A. (1975). Some trophic relations in the benthos of shallow parts of Marion Lake. *Ekologia Polska*, **23**, 573–86.

Kajak, Z. & Pieczyński, E. (1966). The influence of invertebrate predators on the abundance of benthic organisms (chiefly Chironomidae). *Ekologia Polska*, B **12**, 175–9.

Kajak, Z. & Rybak, J.I. (1966). Production and some trophic dependencies in benthos against primary production and zooplankton production of several Masurian Lakes. *Verhandlungen der Internationalen Vereinigung für Theoretische und Angewandte Limnologie*, **16**, 441–51.

Kajak, Z. & Rybak, J. (1970). Food conditions for larvae of Chironomidae in various layers of bottom sediments. *Bulletin de l'Académie Polonaise des Sciences. Classe II. Série des Sciences Biologiques*, **18** (4), 193–6.

Kajak, Z. & Warda, J. (1968). Feeding of benthic non-predatory Chironomidae in lakes. *Annales Zoologici Fennici*, **5**, 57–64.

Kajak, Z. & Wiśniewski, R. (1966). An attempt at estimating the intensity of consumption of Tubificidae by predators. *Ekologia Polska*, B **12**, 181–4.

Kajak, Z. & Zawisza, J. (1973). Experimentally increased fish stock in the pond type lake Warniak XIV. The relations between the fish and other biocoenotic components (Summing up the studies) *Ekologia Polska*, **21**, 631–43.

Kalff, J. (1967). Phytoplankton abundance and primary production rates in two Arctic ponds. *Ecology*, **48**, 558–65.

Kalff, J. (1969). A diel periodicity in the optimum light intensity for maximum photosynthesis in natural phytoplankton populations. *Journal of the Fisheries Research Board of Canada*, **26**, 463–8.

Kalff, J. (1972). Net plankton and nannoplankton production and biomass in a north temperate zone lake. *Limnology and Oceanography*, **17**, 712–20.

Kalff, J. & Welch, H.E. (1974). Phytoplankton production in Char Lake, a natural polar lake and in Meretta Lake, a polluted polar lake, Cornwallis Island, Northwest Territories. *Journal of the Fisheries Research Board of Canada*, **31**, 621–36.

Kalff, J., Welch, H.E. & Holmgren, S. (1972). Pigment cycles in two high arctic Canadian lakes. *Verhandlungen der Internationalen Vereinigung für Theoretische und Angewandte Limnologie*, **18**, 250–6.

Kalff, J. & Wetzel, R.G. (1979). Rates of primary production and chemoorganotrophy by phytoplankton and mosses in Char Lake, Cornwallis Island, Canadian arctic (in preparation).

Kalleberg, H. (1958). Observations in a stream tank of territoriality and competition in juvenile salmon and trout (*Salmo salar* L. and *S. trutta* L.). *Report. Institute of Freshwater Research, Drottningholm*, **39**, 55–98.

Kamp Nielsen, L. (1977). Modelling the temporal variation in sedimentary phosphorus fractions. In *Interactions Between Sediments and Fresh Water*, ed. H.L. Golterman, Proceedings of an International Symposium, Amsterdam, 1976, pp. 277–85. The Hague: W. Junk.

Kansanen, A., Niemi, R. & Överlund, K. (1974). Pääjärven makrofyytit. *Luonnon Tutkija*, **78**, 111–8.

Karabin, A. (1974). Studies on the predatory role of the cladoceran *Leptodora kindtii* (Focke) in secondary production of two lakes with different trophy. *Ekologia Polska*, **22**, 295–310.

Karbaum, H. (1966). Über die Wassertemperaturflacher Binnenseen. *Beiträge zur Geophysik*, **75** (196), 66–78.

Karpevich, A.F. & Mordukhai-Boltovskoi, F.D. (1966). Obzor sumpoziuma "Rekonstruktsya fauny i flory vodoemov SSSR". In *Biologicheskii Resursy Vodoemov Puti'ikh Rekonstruktsii i Ispol'zovaniya*, pp. 133–45. Moskva: Nauka.

Karzinkin, G.S. (1967). Razvitie problemy biologicheskoi produktivnosti vodoemov za pyat'desyat let Sovetskoi vlasti. (Development of problem of biological productivity in water bodies in the USSR for 50 years.) *Voprosy Ikhtiologii*, **7**, 879–905. (In Russian.)

Kaul, V. & Vass, K.K. (1972). Production studies of some macrophytes of Srinagar lakes. In *Productivity Problems of Freshwaters*, ed. Z. Kajak & A. Hillbricht-Ilkowska, pp. 725–31. Warsaw & Kraków: Polish Scientific Publishers.

Kaushik, N.K. & Hynes, H.B.N. (1971). The fate of the dead leaves that fall into streams. *Archiv für Hydrobiologie*, **68**, 465–515.

Keast, A. & Welsh, L. (1968). Daily feeding periodicities, food uptake rates – and dietary changes with hour of day in some lake fishes. *Journal of the Fisheries Research Board of Canada*, **25**, 1133–44.

Keeney, D.R. (1973). The nitrogen cycle in sediment-water systems. *Journal of Environmental Quality*, **2**, 15–29.

Kelly, M.G., Hornberger, G.M. & Cosby, B.J. (1974). Continuous automated measurements of rates of photosynthesis and respiration in an undisturbed river community. *Limnology and Oceanography*, **19**, 305–12.

Kerekes, J.J. (1974). Limnological conditions in five small oligotrophic lakes in Newfoundland. *Journal of the Fisheries Research Board of Canada*, **31**, 555–83.

Kerekes, J.J. (1975). The relationship of primary production to basin morphometry in five small oligotrophic lakes in Terra Nova National Park in Newfoundland. *Symposia Biologica Hungarica*, **15**, 35–48.

Khailov, K.M. (1971). *Ecological Metabolism of the Sea*. Kiev: Izdatel'stvo Naukova. 252 pp. (In Russian.)

Kho.nskis, V. (1969). *Dynamics and Thermics of Small Lakes*. Vilnyus: Mintis 204 pp. (In Russian.)

Khomskis, W.R. & Filatova, T.N. (1972). Principles of typology of stratified lakes in relation to vertical exchange. *Verhandlungen der Internationalen Vereinigung für Theoretische und Angewandte Limnologie*, **18**, 528–36.

Khusainova, N.Z., Mitrofanov, V.P., Mamilova, R.Kh. & Sharapova, L.I. (1973). Biologicheskaya produktivnost oz Karakul. In *Produktsionnobiologicheskie Issledovaniya Ekosistem Presnykh Vod*, ed. G.G. Winberg, pp. 32–43. Minsk: Byelorussian State University.

Kiefer, D.O., Holm-Hansen, O., Richards, R.C., Goldman, C.R. & Berman, T. (1972). Phytoplankton in Lake Tahoe: Deep living populations. *Limnology and Oceanography*, **17**, 418–22.

Kimerle, A. & Anderson, N.H. (1971). Production and bioenergetic role of the midge *Glyptotendipes barbipes* (Staeger) in a waste stabilization lagoon. *Limnology and Oceanography*, **16**, 646–59.

Kimmel, B.L. & Lind, O.T. (1972). Factors affecting phytoplankton production in a eutrophic reservoir. *Archiv für Hydrobiologie*, **71**, 124–41.

Kipling, C. & Frost, W.E. (1970). A study of the mortality, population numbers, year class strength, production and food consumption of pike, *Esox lucius* L. in Windermere. *Journal of Animal Ecology*, **39**, 115–57.

Kirchner, W.B. (1975). An examination of the relationship between drainage basin morphology and the export of phosphorus. *Limnology and Oceanography*, **20**, 267–70.

Kirillova, T.V. (1970). *The Radiation Climate in Lakes and Reservoirs.* Leningrad: Hydrometeorological Publishing House, 253 pp. (In Russian.)

Kirillova, T.V. & Smirnova, N.P. (1972). The radiation balance of lakes on the territory of the USSR. *Verhandlungen der Internationalen Vereinigung für Theoretische und Angewandte Limnologie*, **18**, 554–62.

Kitchell, J.F., Koonce, J.F., O'Neill, R.V., Shugart Jr, H.H., Magnuson, J.J. & Booth, R.S. (1974). Model of fish biomass dynamics. *Transactions of the American Fisheries Society*, **103**, 786–98.

Kleerekoper, H. (1952). The mineralization of plankton. *Journal of the Fisheries Research Board of Canada*, **10**, 284–91.

Klekowski, R.Z. & Shushkina, E.A. (1966). Ernahrung, Atmung, Wachstum und Erargieumformung in *Macrocyclops albidus* (Jurine). *Verhandlungen der Internationalen Vereinigung für Theoretische und Angewandte Limnologie*, **16**, 399–418.

Klingenberg, M. (1968). The respiration chain. In *Biological Oxidations*, ed. T.P. Singer, pp. 3–54. New York: Interscience.

Kondratev, G.P. (1969). Velichina filtratsionnoi raboty u nekotorykh dvustvorok Volgogradskogo Vodokhranilishcha. In *Volgogradskoie Vodokhranilischche*, pp. 32–36. Saratov: Saratovskii Universitet.

Konstantinov, A.S. (1958). Biologiya khirónomid i ikh razvedenie. *Trudy Saratovskogo Otdeleniya VNIRO*, **5**, 363.

Konstantinov, A.S. (1960). On the method of determination of the production of animals which serve as food for fishes. *Nauchnye Doklady Vysshei Shkoly. Biologicheskie Nauki*, **4**, 59–62. (In Russian.)

Konstantinov, A.S. (1961). O pitanii nekotorykh khischchnykh lichinok Chironomidae *Voprosy Ikhtiologii*, **1**, 570–82.

Konstantinov, A.S. (1972). *Obshchaya gidrobiologiya* (General hydrobiology). Moscow Vysshaya Shkola. 472 pp. (In Russian.)

Kořínek, V. (1972). Results of the study of some links of the food chain in a carp pond in Czechoslovakia. In *Productivity Problems of Freshwaters*, ed. Z. Kajak & A. Hillbricht-Ilkowska, pp. 541–53. Warsaw & Kraków: Polish Scientific Publishers.

Kořínkova, J. (1971). Quantitative relations between submerged macrophyte and populations of invertebrates in a carp pond. *Hidrobiologia*, **12**, 377–82.

Koshinsky, G.D. (1970). The morphometry of shield lakes in Saskatchewan. *Limnology and Oceanography*, **15**, 695–701.

Kouwe, F.A. & Golterman, H.L. (1976). Rol van bodemfosfaten in het eutrofierings-process. I. Onderzoek in een proefvijver in het Veluwerandmeer. H_2O, **9**, 84–6.

Kowalczewski, A. (1975). Periphyton primary production in the zone of submerged vegetation of Mikolajskie Lake. *Ekologia Polska*, **23**, 509–43.

Kowalczewski, A. & Lack, T.J. (1971). Primary production of the River Thames and Kennet at Reading. *Freshwater Biology*, **1**, 197–212.

Kowalczewski, A. & Mathews, C.P. (1970). The leaf litter as a food source of aquatic organisms in the River Thames. *Polskie Archiwum Hydrobiologii*, **17**, 133–4.

Kozhova, O.M. (1973). Phytoplankton and production–destruction processes in Bratsk Reservoir. In *Productivity Studies of Freshwater Ecosystems*, ed. G.G. Winberg, pp. 71–82. Minsk: Byelorussian State University. (In Russian.)

Kozlovsky, D.G. (1968). A critical evaluation of the trophic level concept. I. Ecological efficiences. *Ecology*, **49**, 48–60.

Krambeck, H.J. (1974). Energiehaushalt und Stofftransport eines Sees. *Archiv für Hydrobiologie*, **73**, 137–92.

Kramer, J.R. (1964). Theoretical model for the chemical composition of freshwater with application to the Great Lakes. *Publications. Great Lakes Research Division.* 11, 147–60.

Kraus, E.B. & Turner, J.S. (1967). A one-dimensional model of the seasonal thermocline. The general theory and its consequences. *Tellus*, 19, 98–106.

Krause, H.R. (1959). Biochemische Untersuchungen über den postmortalen Abbau von totem Plankton unter aeroben und anaeroben Bedingungen. *Archiv für Hydrobiologie*, 24 (supplement), 297–337.

Krause, H.R. (1962). Investigation of the decomposition of organic matter in natural waters. *FAO Fisheries Biology Report*, 34 (FB/R34), 19 pp.

Krogius, F.V., Krokhin, E.M. & Menshutkin, V.V. (1972). The modelling of the ecosystem of Lake Dalnee on an electronic computer. In *Productivity Problems of Freshwaters*, Z. Kajak & A. Hillbricht-Ilkowska, pp. 149–64. Warsaw & Kraków: Polish Scientific Publishers.

Kubitschek, H.E. (1970). *Introduction to Research with Continuous Cultures.* Englewood Cliffs, New Jersey: Prentice-Hall, 195 pp.

Kuznetsov, S.I. (1959). *Die Rolle der Mikroorganismen im Stoffkreislauf der Seen.* Berlin: VEB Deutschen Verlag der Wissenschaften. 301 pp.

Kuznetsov, S.I. (1970). *Mikroflora Ozer i ee Geokhimicheskaya Dejatel'nost.* (The Microflora of Lakes and its Geochemical Activity). Leningrad: Nauka. 440 pp. (In Russian.)

Květ, J., Szczepanski, A. & Westlake, D.F. (eds.). *Ecology of Wetlands.* International Biological Programme. Cambridge: Cambridge University Press. (In prep.).

Lack, J.T. (1973). Studies on the Macrophytes and Phytoplankton of the River Thames and Kennet at Reading. Ph.D. thesis, University of Reading. 96 pp.

Ladle, M. (1972). Larval Simuliidae as detritus feeders in chalk streams. *Memorie dell'Istituto Italiano di Idrobiologia*, 29 (supplement), 429–39.

Langbein, W.B. & Dawdy, D.R. (1964). Occurrence of dissolved solids in surface waters in the United States. *Professional Papers. United States Geological Survey*, 501-D, 115–17.

Larkin, P.A. (1964). Canadian lakes. *Verhandlungen der Internationalen Vereinigung für Theoretische und Angewandte Limnologie*, 15, 76–90.

Larkin, P.A. & Northcote, T.G. (1958). Factors in lake typology in British Columbia, Canada. *Verhandlungen der Internationalen Vereinigung für Theoretische und Angewandte Limnologie*, 13, 252–63.

Larsen, D.P. & Mercier, H.T. (1976). Phosphorus retention capacity of lakes. *Journal of the Fisheries Research Board of Canada*, 33, 1742–50.

Larson, D.W. (1972). Temperature, transparency, and phytoplankton productivity in Crater Lake, Oregon. *Limnology and Oceanography*, 17, 410–17.

Larson, G.L. (1973). A limnological study of a high mountain lake in Mount Rainier National Park, Washington State, USA. *Archiv für Hydrobiologie*, 72(1), 10–48.

Larsson, P. & Tangen, K. (1975). The input and significance of particulate terrestrial organic carbon in a sub-alpine freshwater ecosystem. In *Fennoscandian Tundra Ecosystems. Part 1 Plants and Microorganisms*, ed. F.E. Wielgolaskie, *Ecological Studies. Analysis and Synthesis*, vol. 16, pp. 351–9. Berlin: Springer-Verlag.

Lasker, R. (1970). Utilization of zooplankton energy by a Pacific sardine population in the California current. In *Marine Food Chains*, ed. J.H. Steele, pp. 265–84. Edinburgh: Oliver and Boyd.

Lauscher, F. (1951). Über die Verteilung der Windgeschwindigkeit auf der Erde. *Archiv für Meteorologie, Geophysik und Bioklimatologie*. B 2, 427–35.

Laws, E.A. (1975). The importance of respiration losses in controlling the size distribution of marine phytoplankton. *Ecology*, 56, 419–26.

536   References

Lean, D.R.S. (1976). Phosphorus kinetics in lake water: Influence of membrane filter pore size and low pressure filtration. *Journal of the Fisheries Research Board of Canada*, **33**, 2800–4.

Lean, D.R.S., Charlton, M.N., Burnison, B.K., Murphy, T.P., Millard, S.E. & Young, K.R. (1975). Phosphorus: changes in ecosystem metabolism from reduced loading. *Verhandlungen der Internationalen Vereinigung für Theoretische und Angewandte Limnologie*. **19**, 249–57.

Lebedev, Yu. M. & Mal'tsman, T.S. (1967). Pervichnaya produktsiya planktona i ee ispol'zovanie v domashkinskom orostitel'nom vodokhranilischche orenburgskoi oblasti. (Planktonic primary production and its utilisation in Domashkinsky irrigating reservoir of Orenburg district.) *Trudy Instituta Biologii Vnutrennikh Vod*, **15** (18), 154–74. (In Russian.)

Le Cren, E.D. (1962). The efficiency of reproduction and recruitment in freshwater fish. In *The Exploitation of Natural Animal Populations*, ed. E.D. Le Cren & M.W. Holdgate, pp. 283–96. Oxford: Blackwell.

Le Cren, E.D. (1969). Estimates of fish populations and production in small streams in England. In *Symposium on Salmon and Trout in Streams*, ed. T.G. Northcote, pp. 269–80. University of British Columbia.

Le Cren, E.D. (1976). The productivity of freshwater communities. *Philosophical Transactions of the Royal Society, London*, B **274**, 359–74.

Le Cren, E.D., Kipling, C. & McCormack, J.C. (1972). Windermere: effects of exploitation and eutrophication on the salmonid community. *Journal of the Fisheries Research Board of Canada*, **29**, 819–32.

Lee, A.H. (1972). Some thermal and chemical characteristics of Lake Ontario in relation to space and time. *Publications. Institute of Environmental Science and Engineering, Great Lakes Institute*, EG-6, 162 pp.

Lefèvre, M., Jakob, H. & Nisbet, M. (1952). Auto- et heteroantagonisme chez les algues d'eau douce. *Annales de la Station Centrale de Hydrobiologie Appliquée*, **4**, 5–197.

Legner, M., Punčochář, P. & Straškrabová, V. (1976). Development of the microbial component of a river community. In *Continuous Culture 6: Applications and New Fields*, ed. A.C.R. Dean, D.C. Ellwood, C.G.T. Evans & J. Helling, pp. 329–44. Chichester: E. Horwood.

Lehman, J.T., Botkin, D.B. & Likens, G.F. (1975). The assumptions and rationales of a computer model of phytoplankton population dynamics. *Limnology and Oceanography*, **20**, 343–64.

Lellak, J. (1966a). Zur Frage der experimentalen Untersuchung der Einflusses der Fresstätigkeit des Fischbestandes auf die Bodenfauna der Teiche. *Verhandlungen der Internationalen Verein für Theoretische und Angewandte Limnologie*, **16**, 1383–91.

Lellak, J. (1966b). Influence of the removal of the fish population on the bottom animals of the five Elbe backwaters. *Hydrobiological Studies*, **1**, 323–80.

Lemoalle, J. (1969). Premières données sur la production primaire dans la région de Bol (avril-octobre 1968) (lac Tchad). *Cahiers ORSTOM, série hydrobiologie*, **3**, 107–19.

Lemoalle, J. (1973). L'énergie humineuse et l'activité photosynthetique du phytoplankton dans le lac Tchad. *Cahiers ORSTOM, série hydrobiologie*, **7**, 95–116.

Lerman, A. (1971). Time to chemical steady-states in lakes and ocean. *Advances in Chemistry Series*, **106**, 30–76.

Lerman, A. & Stiller, M. (1969). Vertical eddy diffusion in Lake Tiberias. *Verhandlungen der Internationalen Vereinigung für Theoretische und Angewandte Limnologie*, **17**, 323–33.

Lettau, H. & Lettau, K. (1972). *Manual to Calculate Climatic Time Series of Global and*

Diffuse-Radiation. Madison: University of Wisconsin Department of Meteorology and Center for Climatic Research. Cyclostyled Report. 46 pp.

Levanidov, V. Ya. (1949). Znachenie allokhtonnogo materiala kak pishchevogo resursa v vodoeme na primere pitaniya vodyanogo oslika (*Asellus aquaticus* L.) *Trudy Vsesoyuznogo Gidrobiologicheskogo Obshchestva* **1**, 100–17.

Levanidov, V. Ya & Kurenkov, I.I. (1973). Znachenie trofologicheskikh issledovanii pri izuchenii biologicheskoi productivnosti. In *Trofologiya Vodnykh Zhivotnykh*, ed. G.V. Nikol'skii & P.L. Pirozhnikov, pp. 95–107. Moskva: Nauka.

Lévêque, C., Carmouze, J.P., Dejoux, C., Durand, J.R., Gras, R., Iltis, A., Lemoalle, J., Loubens, G., Lauzanne, L. & Saint-Jean, L. (1972). Recherches sur les biomasses et la productivité du Lac Tchad. In *Productivity Problems of Freshwaters*, ed. Z. Kajak & A. Hillbricht-Ilkowska, pp. 165–81. Warsaw & Kraków: Polish Scientific Publishers.

Lewis, W.M. (1973). The thermal regime of Lake Lanao (Philippines) and its theoretical implication for tropical lakes. *Limnology and Oceanography*, **18**, 200–17.

Lewis, W.M. (1974). Primary production in the plankton community of a tropical lake. *Ecological Monographs*, **44**, 377–409.

Lewkowicz, M. (1971). Biomass of zooplankton and production of some Rotatoria species and *Daphnia longispina* in ponds for carp fry. *Polskie Archiwum Hydrobiologii*, **18**, 215–23.

Lex, M., Silvester, W.B. & Stewart, W.D.P. (1972). Photorespiration and nitrogenase activity in the blue green alga, *Anabaena cylindrica*. *Proceedings of the Royal Society, London*, B **180**, 87–102.

Libosvarsky, J. (1971). Umrtnost a produkce pstruha v pstruhovych tocich. (*Mortality and production of brown trout in trout streams*). *Vertebratologicke Zpravy*, 1971 (2), 87.

Libosvarsky, J. & Lusk, S. (1970). The bionomics and net production of brown trout (*Salmo trutta* morpha *fario* L.) in the Loucka Creek, Czechoslovakia. *Ekologia Polska*, **18** (16), 361–82.

Lien, L. (1979). The energy budget of the brown trout (*Salmo trutta* L.) population of Øvre Heimdalsvatn. *Holarctic Ecology*, **1**, 279–300.

Lieth, H. (1964). Versuch einer kartographischen Erfassung der Stoffproduktion der Erde. In *Geographischen Taschenbuch*, 1964/1965, pp. 72–80. Wiesbaden: Steiner Verlag.

Lieth, H. (1972). Modeling the primary productivity of the world. *Archivio Botanico Biogeographico Italiano*, 47, ser. 4, vol. 17, (1–2), 11–16.

Lieth, H. (1975). Primary productivity in ecosystems: comparative analysis of global patterns *Perspectives in Ecology*, 67–88.

Lieth, H. & Box, E. (1972). Evapotranspiration and primary productivity: C.W. Thornwaite memorial model. In *Papers on Selected Topics in Climatology*, ed. J.R. Mather, pp. 37–47. New York: Elmer.

Likens, G.E. (ed.) (1972). *Nutrients and Eutrophication*. Lawrence: Allen Press. 328 pp.

Likens, G.E. (1975). Primary production of inland aquatic ecosystems. *Ecological Studies*, **14**, 185–202.

Likens, G.E. & Borman, F.M. (1974). Linkages between terrestrial and aquatic ecosystems. *Bioscience*, **24**, 447–56.

Linacre, E.T. (1969). Empirical relationships involving the global radiation intensity and ambient temperature at various latitudes and altitudes. *Archiv für Meteorologie, Geophysik und Bioklimatologie*, B **17**, 1–20.

Lind, O.T. (1971). The organic matter budget of a Central Texas reservoir. *Special Publications, American Fisheries Society*, **8**, 193–202.

Linskens, H.F. (1963). Beitrag zur Frage der Beziehungen zwischen Epiphyt und Basiphyt bei marinen Algen. *Pubblicazioni della Stazione Zoologica di Napoli*, **33**, 274–93.

List, R.J. (1951). Smithsonian meteorological tables. 6th revised edition. *Smithsonian Miscellaneous Collections*, **114**, 527 pp.

Livingstone, D.A. (1954). On the orientation of lake basins. *American Journal of Science*, **252**, 547–54.

Livingstone, D.A. (1963). Chemical composition of rivers and lakes. In *Data of Geochemistry*, 6th ed., ed. M. Fleischer, Chapter G, 1–64. *Professional Papers. United States Geological Survey*, 440-G, 64 pp.

Löffler, H. (1968). Tropical high-mountain lakes. Their distribution, ecology and zoogeographical importance. *Colloquium Geographicum*, **9**, 57–76.

Lorenzen, C.J. (1963). Diurnal variation in photosynthetic activity of natural phytoplankton populations. *Limnology and Oceanography*, **8**, 56–62.

Lorenzen, M. & Mitchell, R. (1973). Theoretical effects of artificial destratification on algal production in impoundments. *Environmental Science and Technology*, **7**, 939–44.

Luferov, V.P. (1961). O pitanii lichinok Pelopiinae (Diptera, Tendipedidae). *Trudy Instituta Biologii Vodokhranilishch*, **4** (7), 232–45.

Luferov, V.P. (1963). K voprosu o potreblenii rybami epibiontov zatoplennykh lesov. In *Materialy po Biologii i Gidrologii Volzhskikh Vodokhranilishcha*, pp. 66–8. Moskva: Akademiya Nauk SSSR.

Lukanin, V.S. (1957). Produkcja chironomid pribojnoj zony skalistogo poberezia Azovskogo mora. In *Sbornik Rabot Studenčeskogo Naučnogo Obščestva*. Moskovskij Techničeskij Institut Rybnoj Promyšlennosh: Ihoziajztva. Moskva. pp. 23–32.

Lund, J.W.G. (1954). The seasonal cycle of the plankton diatom, *Melosira italica* (Ehr.) Kütz. subsp. *subarctica* O. Müll. *Journal of Ecology*, **42**, 151–79.

Lund, J.W.G. (1965). The ecology of freshwater phytoplankton. *Biological Reviews*, **40**, 231–93.

Lund, J.W.G. (1970). Primary production. *Water Treatment and Examination*, **19**, 332–58.

Lund, J.W.G. (1971a). Eutrophication. In *The Scientific Management of Animal and Plant Communities for Conservation*, ed. E. Duffey & A.S. Watt, pp. 225–40. Oxford: Blackwell.

Lund, J.W.G. (1971b). The seasonal periodicity of three planktonic desmids in Windermere. *Mitteilungen der Internationalen Vereinigung für Theoretische und Angewandte Limnologie*, **19**, 3–25.

Lund, J.W.G. (1972). Preliminary observations on the use of large experimental tubes in lakes. *Verhandlungen der Internationalen Vereinigung für Theoretische und Angewandte Limnologie*, **18**, 71–7.

Lund, J.W.G., Jaworski, G.H.M. & Bucka, H. (1971). A technique for bioassay of freshwater with specific reference to algal ecology. *Acta Hydrobiologica, Kraków*, **13**, 235–49.

Lund, J.W.G., Jaworski, G.H.M. & Butterwick, C. (1975). Algal bioassay of water from Blelham Tarn, English Lake District, and the growth of planktonic diatoms. *Archiv für Hydrobiologie*, **49** (supplement), 49–69.

Lush, D.L. & Hynes, H.B.N. (1973). The formation of particles in freshwater leachates of dead leaves. *Limnology and Oceanography*, **18**, 968–77.

Lusk, S. & Zdrazilek, P. (1969). Contribution to the bionomics and production of the brown trout (*Salmo trutta* m. *fario* L.) in the Lusova Brook. *Zoologicke Listy*, **18**, 381–402.

Lvova-Kachanova, A.A. (1971). O roli dreisseny (*Dreissena polymorpha* Pallas) v protsessakh samoochishcheniya vody Uchinskogo vodokhranilishcha. In *Kompleksnye Issledovaniya Vodokhranilishch*, pp. 196–202. Moskva: Izdatel'stvo Moskovskogo Universitet.

Lvova-Kachanova, A.A. & Izvekova, E.I. (1973). In *Kompleksnye Issledovaniya Vodokhranilishch*, vol. 2, pp. 130–6. Moskva: Izdatel'stvo Moskovskogo Universitet.

Lyakhnovich, V.P. (1965). Estestvennaya kormovaya baza ryb v prudovykh khozyaistvakh BSSR. *Trudy Belorusskogo Nauchno-Issledovatel'skogo Instituta Rybnogo Khozyaistva*, **5**, 3–9.

McAllister, C.D. (1963). Measurements of diurnal variation in productivity at ocean station "P". *Limnology and Oceanography*, **8**, 289–92.

McAllister, C.D., Le Brasseur, R.J. & Parsons, T.R. (1972). Stability of enriched aquatic ecosystems. *Science, New York*, **175**, 562–4.

MacArthur, R. (1955). Fluctuations of animal populations and a measure of community stability. *Ecology*, **36**, 533–6.

MacArthur, R.H. (1972). *Geographical Ecology*. New York: Harper & Row, 269 pp.

McColl, R.H.S., White, E. & Waugh, J.R. (1975). Chemical run-off in catchments converted to agricultural use. *New Zealand Journal of Science*, **18**, 67–84.

McComish, T. (1970). Laboratory Experiments on Growth and Food Conversion by the Bluegill. Ph.D. thesis, University of Missouri.

McConnell, W.J. & Sigler, W.F. (1959). Chlorophyll and productivity in mountain rivers. *Limnology and Oceanography*, **4**, 335–51.

McCormack, A.J.A., Loucks, O.L., Koonce, J.F., Kitchell, J.F. & Weiler, P.R. (1973). An ecosystem model for the pelagic zone of a lake. In *Proceedings of the American Bar Association, National Institute for Environmental Litigation*. Chicago: American Bar Association.

McCracken, M.D., Adams, M.S., Titus, J. & Stone, W. (1975). Diurnal course of photosynthesis in *Myriophyllum spicatum* and *Oedogonium*. *Oikos*, **26**, 355–61.

McCullough, E.C. (1968). Total daily radiant energy available extraterrestrially as a harmonic series in the day of the year. *Archiv für Meteorologie, Geophysik und Bioklimatologie*, B **16**, 129–43.

McCullough, E.C. & Porter, W.P. (1971). Computing clear day solar radiation spectra for the terrestrial ecological environment. *Ecology*, **52**, 1008–15.

McDonnell, A.H. (1971). Variations in oxygen consumption by aquatic macrophytes in a changing environment. *Proceedings of the 14th Conference on Great Lakes Research*, 52–8.

McGahee, C.F. & Davis, G.J. (1971). Photosynthesis and respiration in *Myriophyllum spicatum* L. as related to salinity. *Limnology and Oceanography*, **16**, 826–9.

McIntire, C.D. (1966). Some factors affecting respiration of periphyton communities in lotic environments. *Ecology*, **47**, 918–30.

MacIntosh, D.H. & Thom, A.S. (1969). *Essentials of Meteorology*. London: Wykeham Publications, 239 pp.

McIntire, C.D., Garrison, R.L., Phinney, H.K. & Warren, C.E. (1964). Primary production in laboratory streams. *Limnology and Oceanography*, **9**, 92–102.

MacIntyre, F. (1970). Geochemical fractionation during mass transfer from sea to air by breaking bubbles. *Tellus*, **22**, 451–62.

Mackay, R.J. & Kalff, J. (1973). Ecology of two related species of caddis fly larvae in the organic substrates of a woodland stream. *Ecology*, **54**, 499–511.

McKenzie, D.H. (1975). A simulation model of interactive algal and protozoan continuous culture populations, Ph.D. thesis, University of Washington.

McLaren, I.A. (1964). Zooplankton of Lake Hazen, Ellesmere Island, and nearby ponds, with special reference to the copepod *Cyclops scutifer* Sars. *Canadian Journal of Zoology*, **42**, 613–29.

McLaren, I.A. (1969). Primary production and nutrients in Ogac Lake, a landlocked fjord in Baffin Island. *Journal of the Fisheries Research Board of Canada*, **26**, 1561–76.

McNeil, S. & Lawton, J.H. (1970). Annual production and respiration in animal populations. *Nature, London*, **225**, 472–4.

McQuate, A.G. (1956). Photosynthesis and respiration of the phytoplankton in Sandusky Bay. *Ecology*, **37**, 834–9.

McQueen, D.J. (1969). Reduction of zooplankton standing stocks by predaceous *Cyclops bicuspidatus thomasi* in Marion Lake, BC *Journal of the Fisheries Research Board of Canada*, **26**, 1605–18.

Madsen, B.L. (1972). Detritus on stones in small streams. *Memorie dell'Istituto Italiano di Idrobiologia*, **29** (supplement), 385–403.

Magnuson, J. & Kitchell, J. (1971). Energy-nutrient flux through fishes. *Eastern Deciduous Forest Biome Memo Report*, No. 71–58, 41 pp.

Mahler, H.R. & Cordes, E.H. (1966). *Biological Chemistry*. New York: Harper & Row.

Maier, R. (1973). Produktions-und Pigmentanalysen an *Ultricularia vulgaris* L. In *Okosystemforschung* ed. H. Ellenberg, pp. 87–101. Berlin: Springer-Verlag.

Maksimova, L.P. (1961). Pitanie i stepen' ispol'zovaniya estestvennykh i isskusstvennykh kormov gibridami karpa s amurskim sazanom. *Izvestiya Gosudarstvennogo Nauchno-Issledovatel'skogo Instituta Ozernogo i Rechnogo Rybnogo Khozyaïsta*, **51**, 65–95.

Malaisse, F., Freson, R., Goffinet, G. & Malaisse-Mousset, M. (1975). Litter fall and litter breakdown in Miombo. In *Tropical Ecological Systems. Trends in Terrestrial and Aquatic Research*, ed. F.B. Golley & E. Medina. *Ecological Studies*, **11**, pp. 137–52. New York: Springer-Verlag.

Malone, T.C. (1971). The relative importance of nannoplankton and netplankton as primary producers in tropical oceanic and neritic phytoplankton communities *Limnology and Oceanography*, **16**, 633–9.

Mann, K.H. (1965). Energy transformation by a population of fish in the River Thames. *Journal of Animal Ecology*, **34**, 253–7.

Mann, K.H. (1967). The cropping of the food supply. In *The Biological Basis of Freshwater Fish Production*, ed. S.D. Gerking, pp. 243–57. Oxford: Blackwell.

Mann, K.H. (1969). The dynamics of aquatic ecosystems. *Advances in Ecological Research*, **1**, 1–81.

Mann, K.H., Britton, R.H., Kowalczewski, A., Lack, T.J., Mathews, C.P. & McDonald, I. (1972). Productivity and energy flow at all trophic levels in the River Thames, England. In *Productivity Problems of Freshwaters*. ed. Z. Kajak & A. Hillbricht-Ilkowska, pp. 579–96. Warsaw & Kraków: Polish Scientific Publishers.

Mann, R.H.K. (1971). The populations, growth and production of fish in four small streams in southern England. *Journal of Animal Ecology*, **40**, 155–90.

Manning, W.M. & Juday, R.E. (1941). The chlorophyll content and productivity of some lakes in northeastern Wisconsin. *Transactions of the Wisconsin Academy of Sciences, Arts and Letters*, **33**, 363–93.

Manny, B.A. & Wetzel, R.G. (1973). Diurnal changes in dissolved organic and inorganic carbon and nitrogen in a hardwater stream. *Freshwater Biology*, **3**, 31–43.

Marker, A.F.H. (1976a). The benthic algae of some streams in southern England. I. Biomass of the epilithon in some small streams. *Journal of Ecology*, **64**, 343–58.

Marker, A.F.H. (1976b). The benthic algae of some streams in southern England.

II. The primary production of the epilithon in a small chalk stream. *Journal of Ecology*, **64**, 359–71.

Marker, A.F.H. (1976c). The production of algae growing on gravel in a chalk stream. *Report. Freshwater Biological Association*, **44**, 46–52.

Mason, D.T. (1967). Limnology of Mono Lake, California. *University of California Publications in Zoology*, **83**, 1–102.

Mathews, C.P. (1971). Contribution of young fish to total production of fish in the River Thames near Reading. *Journal of Fish Biology*, **3**, 157–80.

Mathews, C.P. & Kowalczewski, A. (1969). The disappearance of leaf litter and its contribution to production in the River Thames. *Journal of Ecology*, **57**, 543–52.

May, R.M. (1973). On relationships among various types of population models. *American Naturalist*, **107**, 46–57.

Mecom, J.O. (1972). Feeding habits of Trichoptera in a mountain stream. *Oikos*, **23**, 401–7.

Megard, R.O. (1970). Lake Minnetonka: nutrients, nutrient abatement, and the photosynthetic system of the phytoplankton. *Interim Report. Limnological Research Center, University of Minnesota*, No. 7, 210 pp.

Megard, R.O. (1972). Phytoplankton, photosynthesis and phosphorus in Lake Minnetonka, Minnesota. *Limnology and Oceanography*, **17**, 68–87.

Meier-Brock, C. (1969). Substrate relations in some *Pisidium* species (Eulamellibranchiata: Sphaeriidae). *Malacologia*, **9**, 121–5.

Melack, J.M. & Kilham, P. (1974). Photosynthetic rates of phytoplankton in East African alkaline, saline lakes. *Limnology and Oceanography*, **19**, 743–55.

Menshutkin, V.V. (1971). *Mathematical Modelling of Populations and Aquatic Faunal Communities*. Leningrad: Nauka. 194 pp. (In Russian.)

Menshutkin, V.V. & Prikhodko, T.I. (1971). Analogue investigation of phytoplankton vertical distribution and production. *Gidrobiologicheskii Zhurnal*, **7** (2), 5–10. (In Russian.)

Meybeck, M. (1976). Total mineral dissolved transport by world major rivers. *Hydrological Sciences Bulletin*, **21**, 265–84.

Middlebrooks, E.J., Falkenborg, D.H. & Maloney, T.E. (1974). *Modeling the Eutrophication Process*. Ann Arbor: Ann Arbor Scientific Publishers. 228 pp.

Mikheev, V.P. (1967). Filtratsionnoe pitanie dreisseny. *Trudy Vsesoyuznyi Nauchno-Issledovatel'skii Institut Prudovogo Rybnogo Khozyaistva, Moskva*, **15**, 117–29.

Mikheeva, T.M. (1970). Evaluation of potential production of a unit of phytoplankton biomass. In *Biological Productivity of a Eutrophic Lake*, ed. G.G. Winberg, pp. 50–70. Moskva: Nauka. (In Russian.)

Miller, A.G., Cheng, Y.H. & Coleman, B. (1971). The uptake and oxidation of glycolic acid by blue-green algae. *Journal of Phycology*, **7**, 97–100.

Miller, J.D.A. & Fogg, G.E. (1957). Studies on the growth of Xanthophyceae in pure culture. I. The mineral nutrition of *Monodus subterraneus* Petersen. *Archiv für Mikrobiologie*, **28**, 1–17.

Miller, M.C. (1972). *The Carbon Cycle in the Epilimnion of Two Michigan Lakes*. Ph.D. thesis, Michigan State University, 214 pp.

Minshall, G.W. (1967). Role of allochthonous detritus in the trophic structure of a woodland spring-brook community. *Ecology*, **48**, 139–49.

Mitchell, D.S. (1969). The ecology of vascular hydrophytes on Lake Kariba. *Hydrobiologia*, **34**, 448–64.

Mokievskii, K.A. (1961). The technique and some results in the studying of solar radiation penetration into water. In *Primary Production in Seas and Inland Waters*, ed. G.G. Winberg, pp. 273–80. Minsk. (In Russian.)

Monakov, A.V. (1968). Some data on the feeding of *Heterocope saliens* Lill. (Copepoda, Calanoida). *Trudy Instituta Biologii Vnutrennikh Vod*, **17** (20), 27–32. (In Russian.)

Monakov, A.V. (1972). Review of studies on feeding of aquatic invertebrates conducted at the Institute of Biology of Inland Waters, Academy of Sciences, USSR. *Journal of the Fisheries Research Board of Canada*, **29**, 363–83.

Monakov, A.V. & Sorokin, Yu. I. (1972). Some results on investigations on nutrition of water animals. In *Productivity Problems of Freshwaters*, ed. Z. Kajak & A. Hillbricht-Ilkowska, pp. 765–73. Warsaw & Kraków: Polish Scientific Publishers.

Monteith, J.L. (1972). Solar radiation and productivity in tropical ecosystems. *Journal of Applied Ecology*, **9**, 747–66.

Monteith, J.L. (1973). *Principles of Environmental Physics*. London: Edward Arnold. 241 pp.

Moon, P. (1940). Proposed standard radiation curves. *Journal of the Franklin Institute*, **230**, 583–617.

Mordukhai-Boltovskoi, F.D. (1955). K voprosu o formirovanii bentosa v krupnykh vodokhranilishchakh (na primiere Rybinskogo vodokhranilishcha) *Zoologicheskii Zhurnal*, **34**, 975–85.

Mordukhai-Boltovskoi, F.D. & Dzyuban, N.A. (1966). Formirovanie fauny bespozvonochnykh krupnykh vodokhranilishch. In *Ekologiya Vodnykh Organizmov*, pp. 98–103. Moskva: Nauka.

Morgan, N.C. (1956). The biology of *Leptocerus aterrimus* Steph. with reference to its availability as a food for trout. *Journal of Animal Ecology*, **25**, 349–65.

Morgan, N.C. (1966). Fertilization experiments in Scottish freshwater lochs: II. Sutherland, 1954, 2. Effects of bottom fauna. *Freshwater and Salmon Fisheries Research*, No. 36, 19 pp.

Morgan, N.C. (1972). Productivity studies at Loch Leven (a shallow nutrient rich lowland lake) In *Productivity Problems of Freshwaters*, ed. Z. Kajak & A. Hillbricht-Ilkowska, pp. 183–205. Warsaw & Kraków: Polish Scientific Publishers.

Morgan, R.I. (1974). The energy requirements of trout and perch populations in Loch Leven, Kinross. *Proceedings of the Royal Society of Edinburgh*, B **74**, 332–45.

Moriarty, D.J.W. (1973). The physiology of digestion of blue-green algae in the cichlid fish, *Tilapia nilotica*. *Journal of Zoology*, **171**, 25–39.

Moriarty, D.J.W., Darlington, J.P.E.C., Dunn, I.G., Moriarty, C.M. & Tevlin, M.P. (1973). Feeding and grazing in Lake George, Uganda. *Proceedings of the Royal Society, London*, B **184**, 299–319.

Mortimer, C.H. (1941). The exchange of dissolved substances between mud and water in lakes. Introduction, I and II. *Journal of Ecology*, **29**, 280–329.

Mortimer, C.H. (1942). The exchange of dissolved substances between mud and water in lakes. III and IV, Summary and References. *Journal of Ecology*, **30**, 147–201.

Mortimer, C.H. (1974). Lake hydrodynamics. *Mitteilungen der Internationalen Vereinigung für Theoretische und Angewandte Limnologie*, **20**, 124–97.

Morton, S.D. & Lee, G.F. (1968). Calcium carbonate equilibria in lakes. *Journal of Chemical Education*, **45**, 511–13.

Moskalenko, B.K. & Votinsev, K.K. (1972). Biological productivity and balance of organic substance and energy in Lake Baikal. In *Productivity Problems of Freshwaters*, ed. Z. Kajak & A. Hillbricht-Ilkowska, pp. 207–26. Warsaw & Kraków: Polish Scientific Publishers.

Moss, B. (1968). The chlorophyll *a* content of some benthic algal communities. *Archiv für Hydrobiologie*, **65**, 51–62.

Moss, B. (1969a). Algae of two Somersetshire pools: Standing crops of phytoplankton and epipelic algae as measured by cell numbers and chlorophyll *a*. *Journal of Phycology*, **5**, 158–68.

Moss, B. (1969b). Vertical heterogeneity in the water column of Abbot's Pond: II. The influence of physical and chemical conditions on the spatial and temporal distribution of the phytoplankton and of a community of epipelic algae. *Journal of Ecology*, **57**, 397–414.

Moss, B. (1970a). Seston composition in two freshwater pools. *Limnology and Oceanography*, **15**, 504–13.

Moss, B. (1970b). The algal biology of a tropical montane reservoir (Mlungusi Dam, Malawi). *British Phycological Journal*, **5**, 19–28.

Moss, B. (1972). The influence of environmental factors on the distribution of freshwater algae: An experimental study. I. Introduction and the influence of calcium concentration. *Journal of Ecology*, **60**, 917–32.

Moss, B. (1973a). The influence of environmental factors on the distribution of freshwater algae. An experimental study – Part IV. Growth of test species in natural lake water, and conclusion. *Journal of Ecology*, **61**, 193–211.

Moss, B. (1973b). Studies on Gill Lake, Michigan – I. Seasonal and depth distribution of phytoplankton. *Freshwater Biology*, **2**, 289–307.

Moss, B. (1976). The effects of fertilization and fish on community structure and biomass of aquatic macrophytes and epipelic algal populations: An ecosystem experiment. *Journal of Ecology*, **64**, 313–42.

Moss, B. & Moss, J. (1969). Aspects of the limnology of an endorheic African lake (L. Chilwa, Malawi). *Ecology*, **50**, 109–18.

Moss, B. & Round, F.E. (1967). Observations on standing crops of epipelic and epipsammic algal communities in Shear Water, Wilts. *British Phycological Bulletin*, **3**, 241–8.

Moyle, J.B. (1956). Relationships between the chemistry of Minnesota surface waters and wildlife management. *Journal of Wildlife Management*, **20**, 303–20.

Mozley, S.C. (1970). Morphology and ecology of the larvae of *Trissocladius grandis* (Kieffer) Diptera, Chironomidae a common species in the lakes and rivers of Northern Europe. *Archiv für Hydrobiologie*, **67**, 433–51.

Müller, H. (1966). Die planktische Primärproduktion in Karpfenteichen unter Einfluss verschiedener Phosphatdüngung. *Verhandlungen der Internationalen Vereinigung für Theoretische und Angewandte Limnologie*, **16**, 1333–9.

Müller, H. (1972). Wachstum und Phosphatbedarf von *Nitzschia actinastroides* (Lemm.) v. Goor im statischer und homokontinuierlicher Kultur unter Phosphat-limitierung. *Archiv für Hydrobiologie*, **38** (supplement), 399–484.

Munk, W.H. & Riley, G.A. (1952). Absorption of nutrients by aquatic plants. *Journal of Marine Research*, **11**, 215–41.

Murdoch, W.W. (1969). Switching in general predators: experiments on predator specificity and stability of prey populations. *Ecological Monographs*, **39**, 335–54.

Murphy, G.I. (1962). Effect of mixing depth and turbidity on the productivity of fresh-water impoundments. *Transactions of the American Fisheries Society*, **91**, 69–76.

Nakamura, K. (1967). City temperature of Nairobi. *Japanese Progress in Climatology*, 61–5.

Nalewajko, C. (1966). Photosynthesis and excretion in various planktonic algae. *Limnology and Oceanography*, **11**, 1–10.

Nalewajko, C. & Marin, L. (1969). Extracellular production in relation to growth of four planktonic algae and of phytoplankton populations from Lake Ontario. *Canadian Journal of Botany*, **47**, 405–13.

Naumann, E. (1921). Spezielle Untersuchungen über die Ernährungsbiologie des tierischen Limnoplanktons. I. Über die Technik des Nährungerwerks bei den Cladoceran und ihre Bedeutung für die Biologie der Gewässertypen. *Acta Universitatis Lundensis, NF*, **17**, 3–26.

Naumann, E. (1932). Grundzüge der regionalen Limnologie. *Binnengewasser*, **11**, 1–176.

Nauwerck, A. (1963). Die Beziehungen zwischen Zooplankton und Phytoplankton im See Erken. *Symbolae Botanicae Upsalienses*, **17** (5), 163 pp.

Nauwerck, A. (1966). Beobachtungen uber das Phytoplankton Klarer Hochgebirgsseen. *Schweizerische Zeitschrift für Hydrologie*, **28**, 3–28.

Nauwerck, A. (1968). Das Phytoplankton des Latnjajaure 1954–1965. *Schweizerische Zeitschrift für Hydrologie*, **30**, 188–216.

Nauwerck, A. & Persson, G. (1971). Ekmanjaure – en aterfödd sjö. *Fauna och Flora*, **4**, 130–40.

Neil, J.H. & Owen, G.E. (1964). Distribution, environmental requirements and significance of *Cladophora* in the Great Lakes. *Publications. Great Lakes Research Institute*, **11**, 113–21.

Nelson, D.J. & Scott, D.C. (1962). Role of detritus in the production of a rock-outcrop community in a Piedmont stream. *Limnology and Oceanography*, **7**, 395–413.

Neumann, J. (1959). Maximum depth and average depth of lakes. *Journal of the Fisheries Research Board of Canada*, **16**, 923–7.

Nilsson, N.A. (1965). Food segregation between salmonoid species in North Sweden. *Report. Institute of Freshwater Research, Drottningholm*, **46**, 58–78.

Nilsson, N.A. (1967). Interactive segregation between fish species. In *The Biological Basis of Freshwater Fish Production*. ed. S.D. Gerking, pp. 296–313. Oxford: Blackwell.

Nilsson, N.A. (1972). Effects of introductions of salmonids into barren lakes. *Journal of the Fisheries Research Board of Canada*, **29**, 693–7.

Nilsson, N.A. & Pejler, B. (1973). On the relation between fish fauna and zooplankton composition in North Swedish lakes. *Report. Institute of Freshwater Research, Drottningholm*, **53**, 51–77.

Noble, R. (1972). A method of direct estimation of total food consumption with application to young yellow perch. *Progressive Fish Culturist*, **34**, 191–4.

Northcote, T.G. & Larkin, P.A. (1956). Indices of productivity in British Columbia lakes. *Journal of the Fisheries Research Board of Canada*, **13**, 515–40.

Novozhilova, M.I. (1957). Vremya generatsii bakterii i produktsiya bakterialnoi biomassy v vode Rybinskogo vodokhranilishcha. (The generation time of bacteria and the production of bacterial biomass in the water of Rybinsk Reservoir). *Mikrobiologiya*, **26**, 202–9. (In Russian.)

Nümann, W. (1972). The Bodensee: effects of exploitation and eutrophication on the salmonid community. *Journal of the Fisheries Research Board of Canada*, **29**, 833–47.

Nydegger, P. (1957). Vergleichende limnologische Untersuchungen an sieben Schweizer Seen. *Beiträge zur Geologie der Schweiz, Hydrologie*, **9**, 1–80.

Nyggard, G. (1955). On the productivity of five Danish waters. *Verhandlungen der Internationalen Vereinigung für Theoretische und Angewandte Limnologie*, **12**, 123–33.

Nykvist, N. (1963). Leaching and decomposition of water-soluble organic substances from different types of leaf and needle litter. *Studia Forestalia Suecica*, **3**, 1–31.

O'Brien, W.J. (1974). The dynamics of nutrient limitation of phytoplankton algae: a model reconsidered. *Ecology*, **55**, 135–41.

O'Connor, D.J., Thomann, R.V. & Di Toro, D.M. (1973). Dynamic water quality forecasting and management. *Ecological Research Series, US Environmental Protection Agency*, EPA-660/3-73-009, 201 pp.

Odum, E.P. (1971). *Fundamentals of Ecology*. Philadelphia: Saunders, 574 pp.

Odum, E.P. & de la Cruz, A.A. (1967). Particulate organic detritus in a Georgia salt marsh-estuarine ecosystem. In *Estuaries*, ed. G.H. Lauff, pp. 383–8. Washington: American Association for the Advancement of Science.

Odum, H.T. (1956). Primary production in flowing waters. *Limnology and Oceanography*, 1, 102–17.

Odum, H.T. (1957a). Primary production measurements in eleven Florida springs and a marine turtle-grass community. *Limnology and Oceanography*, 2, 85–97.

Odum, H.T. (1957b). Trophic structure and productivity of Silver Springs, Florida. *Ecological Monographs*, 27, 55–112.

Ohle, W. (1956). Bioactivity, production and energy utilization of lakes. *Limnology and Oceanography*, 1, 139–49.

Ohle, W. (1958). Diurnal production and destruction rates of phytoplankton in lakes. *Rapport et Procès-Verbaux des Réunions. Conseil Permanent International pour l'Exploration de la Mer*, 144, 129–31.

Ohle, W. (1961). Tagesrhythmen der Photosynthese von Planktonbiocoenosen. *Verhandlungen der Internationalen Vereinigung für Theoretische und Angewandte Limnologie*, 14, 113–9.

Ohle, W. (1965). Nährstoffanreicherung der Gewässer durch Dungemittel und Meliorationen. *Münchner Beiträge zur Abwasser- Fischerei- und Flussbiologie*, 12, 54–83.

Ohle, W. (1972). Die Sedimente des Grossen Plöner Sees als Dokumente der Zivilisation. *Jahrbuch für Heimatkunde, Plön*, 2, 7–27.

Olah, J. (1972). Leaching, colonization and stabilization during detritus formation. *Memorie dell'Istituto Italiano di Idrobiologia*, 29 (supplement), 105–27.

Oliver, D.R. (1971). Life history of the Chironomidae. *Annual Review of Entomology*, 16, 211–30.

Olszewski, P. (1971). Trofia i saprobia. *Zeszyty Naukowe Wyszzej Szkoly Rolniczej v Olsztynie*, C 3, 3–10.

O'Melia, C.R. (1972). An approach to the modelling of lakes. *Schweizerische Zeitschrift für Hydrologie*, 34, 1–34.

O'Neill, R.V. (1971). Systems approaches to the study of forest floor arthropods. In *Systems Analysis and Simulation in Ecology*, ed. B.C. Patten, vol. 1, pp. 441–77. New York: Academic Press.

O'Neill, R.V., Goldstein, R.A., Shugart, H.H. & Mankin, J.B. (1972). Terrestrial ecosystem energy model. *EDFB Memo Report*, 72–19, 39 pp.

Opuszynski, K. (1969). Production of plant-eating fishes (*Ctenopharyngodon idella* Val. and *Hypophthalmichthys molitrix* Val.) in carp ponds. *Rocznik Nauk Rolniczych*, H 91 (2), 219–309. (In Polish with English summary.)

Opuszynski, K. (1971). Present state and the perspectives in culture of phytophagous fish in Europe. In *Proceedings of a Symposium on New Ways of Freshwater Fishery Intensification*. Fisheries Research Institute (Vodnany). 58 pp.

Orlob, G.T. & Selna, L.G. (1968). Mathematical simulation of thermal stratification in deep impoundments. In *Proceedings of a Special Conference on Current Research into the Effects of Reservoirs on Water Quality. Technical Report, Department of Environmental and Water Resources Engineering*, 17, pp. 121–67.

Orlob, G.T. & Selna, L.G. (1970). Temperature variations in deep reservoirs. *Journal of the Hydraulics Division, American Society of Civil Engineers*, 96, (HY2), 391–410.

Ostapenya, A.P. (1965). Polnota okisleniya organicheskogo veshchestva vodnykh bezpozvonochnykh metodom biochromatnogo okisleniya. *Doklady Akademii Nauk SSSR*, **9**, 273–6. (In Russian.)

Ostapenya, A.P., Petrovich, P.G., Mikheeva, T.M., Kovalevskaya, R.Z., Kryutchkova, N.M., Potaenko, Yu. S. & Gavrilov, S.I. (1973). Peculiarities of biological productivity of the Lakes Naroch, Myastro, Batorin. In *Productivity Studies of Freshwater Ecosystems*, ed. G.G. Winberg, pp. 83–94. Minsk: Byelorussian University. (In Russian.)

Otsuki, A. & Wetzel, R.G. (1972). Coprecipitation of phosphate with carbonates in a marl lake. *Limnology and Oceanography*, **17**, 763–7.

Otsuki, A. & Wetzel, R.G. (1973). Interactions of yellow-organic acids with calcium carbonate in freshwater. *Limnology and Oceanography*, **18**, 490–3.

Otsuki, A. & Wetzel, R.G. (1974). Release of dissolved organic matter by autolysis of a submersed macrophyte; *Scirpus subterminalis*. *Limnology and Oceanography*, **19**, 842–5.

Ott, J. & Schiemer, F. (1973). Respiration and anaerobiosis of free-living nematodes from marine and limnic sediments. *Netherlands Journal of Sea Research*, **7**, 233–43.

Overbeck, J. (1972). Distribution pattern of phytoplankton and bacteria, microbial decomposition of organic matter and bacterial production in eutrophic, stratified lakes. In *Productivity Problems of Freshwater*, ed. Z. Kajak & A. Hillbricht-Ilkowska, pp. 227–37. Warsaw & Kraków: Polish Scientific Publishers.

Owens, M., Knowles, G. & Clark, A. (1969). The prediction of the distribution of dissolved oxygen in rivers. In *Advances in Water Pollution Research*, ed. S.H. Jenkins, Proceedings of the Fourth International Conference on Water Pollution Research, Prague, pp. 125–47. Oxford: Pergamon Press.

Owens, M., Learner, M.A. & Maris, P.J. (1967). Determination of the biomass of aquatic plants using an optical method. *Journal of Ecology*, **55**, 671–6.

Owens, M. & Maris, P.J. (1964). Some factors affecting the respiration of some aquatic plants. *Hydrobiologie*, **23**, 533–43.

Paasche, E. (1973). Silicon and the ecology of marine phytoplankton diatoms. II. Silicate uptake kinetics in five diatom species. *Marine Biology*, **19**, 262–9.

Packard, T.T. (1971). The measurement of respiratory electron-transport activity in marine phytoplankton. *Journal of Marine Research*, **29**, 235–44.

Packard, T.T., Healy, M.L. & Richards, F.A. (1971). Vertical distribution of the activity of the respiratory electron transport system in marine plankton. *Limnology and Oceanography*, **16**, 60–70.

Packard, T.T. & Taylor, P.B. (1968). The relationship between succinate dehydrogenase activity and the oxygen consumption in the brine shrimp, *Artemia salina*. *Limnology and Oceanography*, **13**, 552–5.

Padan, E., Raboy, B. & Shilo, M. (1971). Endogenous dark respiration of the blue-green alga *Plectonema boryanum*. *Journal of Bacteriology*, **106**, 45–50.

Park, R.A., O'Neill, R.V., Bloomfield, J.A., Shugart, H.H. Jr, Booth, R.S., Koonce, J.F. et al. (1974). A generalized model for simulating lake ecosystems. *Simulation*, August 1974, 33–50.

Parsons, T.R. & Strickland, J.D.H. (1962). On the production of particulate organic carbon by heterotrophic processes in sea water. *Deep Sea Research*, **8**, 211–22.

Patalas, K. (1960a). Mieszanie wody jako czynnik określajacy intensywnoše krażenia materii w różnych morfologicznie jeziorach okolic Wegorzewa. *Roczniki Nauk Rolniczych*, B **77**, 223–42.

Patalas, K. (1960b). Punktowa ocena pierwotnej produktywonsci jezior okolic Wegorzewa. *Roczniki Nauk Rolniczych*, B **77**, 299–326.

Patalas, K. (1961). Wind und morphologiebedingte wasserwegungstypen als bestimmender Faktor für die Intersität des Stoffkreislaufes in nordpolnischen Seen. *Verhandlungen der Internationalen Vereinigung für Theoretische und Angewandte Limnologie*, **14**, 59–64.

Patalas, K. (1970). Primary and secondary production in a lake heated by thermal power plant. In *The Environmental Challenge of the 70's*, Proceedings, Institute of Environmental Sciences in 16th Annual Technical Meeting, 1970, pp. 267–71. Boston, Massachusetts.

Patalas, K. (1971). The crustacean plankton communities in 45 lakes of the Experimental Lakes Area (ELA), north-western Ontario. *Journal of the Fisheries Research Board of Canada*, **28**, 231–48.

Patten, B.C. (1959). An introduction to the cybernetics of the ecosystem. *Ecology*, **40**, 221–31.

Patten, B.C. (1966). The biocoenetic process in an estuarine phytoplankton community. *Oak Ridge National Laboratory, United States Atomic Energy Commission*, 3946, UC-48-Biology & Medicine, 97 pp.

Pavoni, M. (1963). Die Bedeutung des Nannoplanktons im Vergleich zum Netzplankton. *Schweizerische Zeitschrift Hydrologie*, **25**, 220–31.

Pavoni, M. (1969). Beziehung zwischen Biomasse und Stickstoffgehalt des Phytoplanktons und die daravs ableitbare Anwendung der Bestimmungsmethoden für die Praxins. *Verhandlungen der Internationalen Vereinigung für Theoretische und Angewandte Limnologie*, **17**, 987–97.

Pchelkina, N.V. (1950). O pitanii nekotorych vodnykh lichinok dvukrylykh. *Trudy Vsesoyuznogo Gidrobiologicheskogo Obschchestva*, **2**, 150–68.

Pearre, S. (1964). Metabolic Activity as an Indicator of Zooplankton Abundance. M.Sc. thesis, Halifax, Dalhousie University. 78 pp.

Pearsall, W.H. (1923). A theory of diatom periodicity. *Journal of Ecology*, **9**, 165–82.

Pechlaner, R. (1964). Plankton production in natural lakes and hydro-electric water-basins in the alpine region of the Austrian Alps. *Verhandlungen der Internationalen Vereinigung für Theoretische und Angewandte Limnologie*, **15**, 375–83.

Pechlaner, R. (1966). Die Finstertaler Seen (Kühtai Österr.) I. Morphometrie, Hydrographie, Limnophysik und Limnochemie. *Archiv für Hydrobiologie*, **62**, 165–230.

Pechlaner, R. (1967). Die Finstertaler Seen (Kühtai Österr.) II. Der Phytoplankton *Archiv für Hydrobiologie*, **63**, 145–93.

Pechlaner, R. (1970). The phytoplankton spring outburst and its conditions in Lake Erken (Sweden). *Limnology and Oceanography*, **15**, 113–30.

Pechlaner, R. (1971). Factors that control production rates and biomass of phytoplankton in high-mountain lakes. *Mitteilungen der Internationalen Vereinigung für Theoretische und Angewandte Limnologie*, **19**, 125–45.

Pechlaner, R., Bretschko, G., Gollman, P., Pfeifer, H., Tilzer, M. & Weissenbach, H.P. (1972a). Ein Hochgebirgssee (Vorderer Finstertaler See, Kühtai, Tirol) als Modell des Energie transportes durch ein limnisches Okosystem. *Verhandlungen der Deutschen Zoologischen Gesellschaft*, **65**, 47–56.

Pechlaner, R., Bretschko, G., Gollmann, P., Pfeifer, H., Tilzer, M. & Weissenbach, H.P. (1972b). The production processes in two high-mountain lakes (Vorderer and Hinterer Finstertaler See, Kühtai, Austria.). In *Productivity Problems of Freshwaters*, ed. Z. Kajak & A. Hillbricht-Ilkowska, pp. 239–69. Warsaw & Kraków: Polish Scientific Publishers.

Penman, H.L. (1948). Natural evaporation from open water, bare soil and grass. *Proceedings of the Royal Society, London*, A **193**, 120–45.

Pennak, R.W. (1957). Species composition of limnetic zooplankton communities. *Limnology and Oceanography*, **2**, 222–32.

Persson, G. (1973). Zooplankton i Stugsjön och Hymenjaure under år 1972. In *Experiment med gödsling av sjöar i Kuokkelområdet. Kuokkelprojektets rapport 2.* 8 pp. Stencil. Uppsala: Limnological Institute.

Peters, R.H. (1975a). Orthophosphate turnover in Central European lakes. *Memorie dell'Istituto Italiano di Idrobiologia*, **32**, 297–311.

Peters, R.H. (1975b). Phosphorus excretion and the measurement of feeding and assimilation by zooplankton. *Limnology and Oceanography*, **18**, 270–9.

Peters, R.H. & Lean, D.R.S. (1973). The characterization of soluble phosphorus released by limnetic zooplankton. *Limnology and Oceanography*, **18**, 270–9.

Peters, R.H. & Rigler, F.H. (1973). Phosphorus release by *Daphnia*. *Limnology and Oceanography*, **18**, 821–39.

Petersen, R.C. & Cummins, K.W. (1974). Leaf processing in a woodland stream. *Freshwater Biology*, **4**, 343–68.

Petrov, V.V. (1972). Bentofauna udobrennych ozer Morozovskoj gruppy Karelśkogo perešejka. *Izdatel'stvo Gos NIORCh.*, **79**, 60–76.

Pfeifer, H. (1974). Das Phytobenthos der Vorderen Finstertaler Sees. Dissertation, Universitat Innsbruck. 119 pp.

Phillips, O.M. (1973). The equilibrium and stability of simple biological systems: I. Primary nutrient consumers. *American Naturalist*, **107**, 73–93.

Phinney, H.K. & McIntire, C.D. (1965). Effects of temperature on the metabolism of periphyton communities developed in laboratory streams. *Limnology and Oceanography*, **10**, 341–4.

Pidgaiko, M.L., Grin, V.G., Kititsina, L.A., Lechina, L.G., Polivannaya, M.F., Sergeeva, O.A. & Vinogradskaya, T.A. (1972). Biological productivity of Kurakhov's Power Station cooling reservoir. In *Productivity Problems of Freshwaters*, ed. Z. Kajak & A. Hillbricht-Ilkowska, pp. 477–91. Warsaw & Kraków: Polish Scientific Publishers.

Pieczyńska, E. (1972a). Ecology of the eulittoral zone of lakes. *Ekologia Polska*, **20**, 637–732.

Pieczyńska, E. (1972b). Production and decomposition in the eulittoral zone of lakes. In *Productivity Problems of Freshwaters*, ed. Z. Kajak & A. Hillbricht-Ilkowska, pp. 271–85. Warsaw & Kraków: Polish Scientific Publishers.

Piontelli, R. & Tonolli, E.V. (1964). Il tempo di residenza dell acque lacustri in relazione ai fenomeni di arricchimento in sostanze immesse, con particolare riguardo al Lago Maggiore. *Memorie dell'Istituto Italiano di Idrobiologia*, **17**, 247–66.

Pivnička, K. (1975). Abundance growth and production of the roach (*Rutilus rutilus* L.) population in the Klíčava Reservoir during the years 1964 and 1967–1972. *International Revue der Gesamten Hydrobiologie*, **60**, 209–20.

Platt, T., Denman, K.L. & Jassby, A.D. (1975). The mathematical representation and prediction of phytoplankton productivity. *Technical Report. Fisheries and Marine Service, Canada*, **523**, 110 pp.

Poddubnaya, T.L. (1961). Materialy po pitaniyu massovykh vidov tubifitsid Rybinskogo Vodokhranilishcha. *Trudy Instituta Biologii Vodokhranilishch*, **4** (7), 219–31.

Pomeroy, L.R. (1960). Residence time of dissolved phosphate in natural waters. *Science*, **131**, 1731–2.

Pourriot, R. & Deluzarches, M. (1971). Recherches sur la biologie des Rotifères. II. Influence de la température sur la durée du développement embryonnaire et post-embryonnaire. *Annales de Limnologie*, **7** (1), 25–52.

Pourriot, R. & Hillbricht-Ilkowska, A. (1969). Recherches sur la biologie de quelques Rotifers planctoniques. I. Résultats préliminaires. *Bulletin de la Société Zoologique de France,* **94** (1), 111–18.

Powers, C.F., Schultz, E.W., Malueg, K.W., Brice, R.M. & Schuldt, M.D. (1972). Algal responses to nutrient additions in natural waters. In *Nutrients and Eutrophication,* ed. G.E. Likens, pp. 141–54. Lawrence: Allen Press.

Procházková, L. (1975a). Long term studies on nitrogen in two reservoirs related to field fertilization. *Proceedings of a Conference on Nitrogen as a Water Pollutant, International Association of Water Pollution Research,* 11. Copenhagen.

Procházková, L. (1975b). Balances in man-made lakes (Bohemia). 2.1 Nitrogen and phosphorus budgets: Slapy Reservoir. In *Coupling of Land and Water Systems,* ed. A.D. Hasler, *Ecological Studies,* **10,** pp. 65–73. Berlin, Heidelberg & New York: Springer-Verlag.

Procházková, L., Blažka, P. & Králová, M. (1970). Chemical changes involving nitrogen metabolism in water and particulate matter during primary production experiments. *Limnology and Oceanography,* **15,** 797–807.

Procházková, L., Straskrabová, V. & Popovský, J. (1973). Changes of some chemical constituents and bacterial numbers in Slapy Reservoir during eight years. *Hydrobiological Studies,* **2,** 83–154.

Prosser, M.V., Wood, R.B. & Baxter, R.M. (1968). The Bishoftu crater lakes: A bathymetric and chemical study. *Archiv für Hydrobiologie,* **65,** 309–24.

Provasoli, L., McLaughlin, J.J.A. & Pinter, I.J. (1954). Relative and limiting concentrations of major mineral constituents for the growth of algal flagellates. *Transactions of the New York Academy of Sciences,* **16,** 412–17.

Prowse, G.A. (1972). Some observations on primary and fish production in experimental fish ponds in Malacca, Malaysia. In *Productivity Problems of Freshwaters,* ed. Z. Kajak & A. Hillbricht-Ilkowska, pp. 555–61. Warsaw & Kraków: Polish Scientific Publishers.

Pyrina, I.L. (1967a). Photosynthesis of freshwater phytoplankton in different light conditions in a water body. In *Krugovorot Veshchestva i Energii v Ozernykh Vodolmakh,* ed. G.I. Galazii & K.K. Votintsev, pp. 202–10. Novosibirsk, Nauka. (In Russian.)

Pyrina, I.L. (1967b). Dependence of photosynthesis of phytoplankton on its biomass and chlorophyll content. *Trudy Instituta Biologii Vnutrennikh Vod,* **15** (18), 94–103. (In Russian.)

Pyrina, I.L. (1974). Spectral measurements of photosynthetically available radiation by the photointegrator with filters. In *Radiation Processes in Atmosphere and on Earth Surface,* ed. N.I. Goica, Proceedings of the 9th All-Soviet Scientific Conference on Actinometry, pp. 423–8. Leningrad. Hydrometeorological Publishing House. (In Russian.)

Pyrina, I.L., Rutkovskaya, V.A. & Ilyinski, A.L. (1972). Influence of phytoplankton on penetrating solar radiation into water of the Volga reservoirs. *Trudy Instituta Biologii Vnuttrennikh Vod,* **23** (26), 93–106. (In Russian.)

Rabinowitch, E.I. (1951). *Photosynthesis and Related Processes,* vol. 2. New York: Interscience, 608 pp.

Rasumov, A.S. (1962). Mikrobial nyiplankton vody (The microbial plankton of water) *Trudy Vsesoyuznogo Gidrobiologicheskogo Obshchestva,* **12,** 60–190. (In Russian.)

Rau, G.M. (1976). Dispersal of terrestrial plant litter into a subalpine lake. *Oikos,* **27,** 153–60.

Rawson, D.S. (1952). Mean depth and fish production in large lakes. *Ecology,* **33,** 513–21.

Rawson, D.S. (1953a). The standing crop of net plankton in lakes. *Journal of the Fisheries Research Board of Canada*, **10**, 224–37.

Rawson, D.S. (1953b). The bottom fauna of Great Slave Lake. *Journal of the Fisheries Research Board of Canada*, **10**, 486–520.

Rawson, D.S. (1955). Morphometry as a dominant factor in the productivity of large lakes. *Verhandlungen der Internationalen Vereinigung für Theoretische und Angewandte Limnologie*, **12**, 164–75.

Rawson, D.S. (1960). A limnological comparison of twelve large lakes in northern Saskatchewan. *Limnology and Oceanography*, **5**, 195–211.

Rawson, D.S. (1961). A critical analysis of the limnological variables used in assessing the productivity of North Saskatchewan lakes. *Verhandlungen der Internationalen Vereinigung für Theoretische und Angewandte Limnologie*, **14**, 160–6.

Reed, E.B. (1970). Annual heat budget and thermal stability in small mountain lakes, Colorado, USA. *Schweizerische Zeitschrift für Hydrologie*, **32**, 397–404.

Reimers, H. (1954). Untersuchungen über die Wirkung des Laubfalles auf die produktionsbiologischen Faktoren in einem fischereilich genutzten Gewasser, Ph.D. thesis, University of Hamburg.

Reynolds, C.S. (1972). Growth, gas vacuolation and buoyancy in a natural population of aplanktonic blue-green alga. *Freshwater Biology*, **2**, 87–106.

Reynolds, C.S. (1973). Growth and buoyancy of *Microcystis aeruginosa* Kütz. emend. Elenkin in a shallow lake. *Proceedings of the Royal Society, London*, B **184**, 29–50.

Reynolds, C.S. & Walsby, A.E. (1975). Water-blooms. *Biological Reviews*, **50**, 437–81.

Reynoldson, T.B. & Davies, R.W. (1970). Food niche and co-existence in lake-dwelling triclads. *Journal of Animal Ecology*, **39**, 599–617.

Rich, P.H., Wetzel, R.G. & Van Thuy, N. (1971). Distribution, production and role of aquatic macrophytes in a southern Michigan marl lake. *Freshwater Biology*, **1**, 3–21.

Richardson, J.L. (1975). Morphometry and lacustrine productivity. *Limnology and Oceanography*, **20**, 661–3.

Ricker, W.E. (ed.) (1968). *Methods for the Assessment of Fish Production in Fresh Waters*. IBP Handbook No. 3. Oxford: Blackwell. 313 pp. (See p. 570.)

Ricker, W.E. & Foerster, R.E. (1948). Computation of fish production. *Bulletin of the Bingham Oceanographic Collection, Yale University*, **11** (4), 173–211.

Ried, A. (1970). Energetic aspects of the interaction between photosynthesis and respiration. In *Prediction and Measurement of Photosynthetic Productivity*, Proceedings IBP/PP Technical Meeting, Třeboň, 1969, pp. 231–46. Wageningen: PUDOC.

Rigler, F.H. (1956). A tracer study of the phosphorus cycle in lake water. *Ecology*, 37, 550–62.

Rigler, F.H. (1964). The phosphorus fractions and the turnover time of inorganic phosphorus in different types of lakes. *Limnology and Oceanography*, **9**, 511–18.

Rigler, F.H. (1966). Radiobiological analysis of inorganic phosphorus in lake water. *Verhandlungen der Internationalen Vereinigung für Theoretische und Angewandte Limnologie*, **16**, 465–70.

Rigler, F.H. (1968). Further observations inconsistent with the hypothesis that molybdenum blue method measures orthophosphate in lake water. *Limnology and Oceanography*, **13**, 7–13.

Rigler, F.H. (1971). Feeding Rates: Zooplankton. In *A Manual of Methods for the Assessment of Secondary Productivity in Fresh Waters*, ed. W.T. Edmondson & G.G. Winberg, IBP Handbook No. 17, pp. 228–55. Oxford: Blackwell.

Rigler, F.H. (1972). *Limnocalanus* feeding. In *Char Lake Project Annual Report 1971–1972*.

Rigler, F.H. (1975). Nutrient kinetics and the new typology. *Verhandlungen der Internationalen Vereinigung für Theoretische und Angewandte Limnologie,* **19**, 197–210.

Rigler, F.H., MacCallum, M.E. & Roff, J.C. (1974). Production of zooplankton in Char Lake. *Journal of the Fisheries Research Board of Canada,* **31**, 637–46.

Riley, G.A. (1970). Particulate organic matter in sea water. *Advances in Marine Biology,* **8**, 1–118.

Riley, G.A. (1973). Particulate and dissolved organic carbon in the oceans. In *Carbon and the Biosphere,* ed. G.M. Woodwell & E.V. Pecan, pp. 204–20. Washington: US Atomic Energy Commission.

Riznyk, R.Z. & Phinney, H.K. (1972). Manometric assessment of the interstitial microalgae production in two estuarine sediments. *Oecologia,* **10**, 193–203.

Roback, S.S. (1969). Notes on the food of Tanypodinae larvae. *Entomological News,* **80**, 13–19.

Robinson, N. (1966). *Solar Radiation.* New York: Elsevier. 347 pp.

Rodhe, W. (1949). The ionic composition of lake waters. *Verhandlungen der Internationalen Vereinigung für Theoretische und Angewandte Limnologie,* **10**, 377–86.

Rodhe, W. (1958a). The primary production in lakes: some results and restrictions on the ^{14}C method. *Rapport et Procès-Verbaux des Réunions. Conseil Permanent International pour l'Exploration de la Mer,* **144**, 122–8.

Rodhe, W. (1958b). Primärproduction und Seetypen. *Verhandlungen der Internationalen Vereinigung für Theoretische und Angewandte Limnologie,* **13**, 121–41.

Rodhe, W. (1962). Sulla produzione di fitoplancton in laghi transparenti di alta montagna. *Memorie dell'Istituto Italiano di Idrobiologia,* **15**, 21–8.

Rodhe, W. (1965). Standard correlations between photosynthesis and light. *Memorie dell'Istituto Italiano di Idrobiologia,* **18** (supplement), 365–81.

Rodhe, W. (1969). Crystallization of eutrophication concepts in Northern Europe. In *Eutrophication: Causes, Consequences, Correctives.* An International Symposium on Eutrophication, University of Wisconsin, Madison, 1967, pp. 50–64. Washington, DC: National Academy of Sciences.

Rodhe, W. (1972). Evaluation of primary parameters of production in Lake Kinneret (Israel). *Verhandlungen der Internationalen für Theoretische und Angewandte Limnologie,* **18**, 93–104.

Rodhe, W., Vollenweider, R.A. & Nauwerck, A. (1958). The primary production and standing crop of phytoplankton. In *Perspectives of Marine Biology,* ed. A.A. Buzzati-Traverso, pp. 299–322. Berkeley: University of California Press.

Rodina, A.G. (1965). *Metody Vodnoi Mikrobiologii* (Methods in Aquatic Microbiology) Moscow: Nauka. 363 pp. (In Russian.)

Rodina, A.G. (1971). *Methods in Aquatic Microbiology.* Translation by R.R. Colwell & M.S. Zambriski of *Metody Vodnoi Mikrobiologii* (1965). Baltimore: University Park Press. 352 pp.

Romanenko, V.I. (1973). Primary production and bacterial decomposition of organic matter in Rybinsk Reservoir. In *Productivity Studies of Freshwater Ecosystems,* ed. G.G. Winberg, pp. 110–25. (In Russian.)

Rosenzweig, M.L. & MacArthur, R.H. (1963). Graphical representation and stability conditions of predator–prey interactions. *American Naturalist,* **97**, 209–23.

Rosinoer, I.M., Saprykina, A.P. & Poryskaya, S.M. (1973). On the organic substances in the atmospheric precipitation in the Voronezh Area. *Gidrobiologicheskii Zhurnal,* **9**, 100–4. (In Russian.)

Ross, P.E. & Kalff, J. (1975). Phytoplankton production in Lake Memphremagog, Quebec (Canada) – Vermont (USA). *Verhandlungen der Internationalen Vereinigung für Theoretische und Angewandte Limnologie,* **19**, 760–9.

Rossolimo, L.L. (1971). Antropogennoe evtrofirovanie vodoemov, ego sushchnost i zadachi issledovaniya. *Gidrobiologicheskii Zhurnal*, **7** (3), 98–108. (In Russian.)

Round, F.E. & Eaton, J.W. (1966). Persistent vertical-migration rhythms in benthic microflora: III. The rhythm of epipelic algae in a freshwater pond. *Journal of Ecology*, **54**, 609–16.

Rounsefell, G.A. (1946). Fish production in lakes as a guide for estimating production in proposed reservoirs. *Copeia*, 1946, No. 1, 29–40.

Rutkovskaya, V.A. (1972). Relation between total and photosynthetically active radiation over oceans. *Meteorologiya i Gidrologiya*, **9**, 53–8. (In Russian.)

Ruttner, F. (1931). Die Schichtung in tropischen Seen. *Verhandlungen der Internationalen Vereinigung für Theoretische und Angewandte Limnologie*, **5**, 44–67.

Ruttner, F. (1952). Planktonstudien der Deutschen Limnologischen Sunda-Expedition. *Archiv für Hydrobiologie*, **21** (supplement), 1–271.

Ruttner, F. & Sauberer, F. (1938). Durchsichtigkeit der Wassers und Planktonschichtung. *Internationale Revue der Gesamten Hydrobiologie und Hydrographie*, **37**, 405–19.

Růžička, J. & Simmer, J. (1970). Measurement of productivity of algal strains by characteristic constants. *Algological Studies*, **1**, 33–40.

Rybak, J.I. (1969). Bottom sediments of the lakes of various trophic type. *Ekologia Polska*, A **17**, 611–62.

Rybak, J.I. (1972). Spatial and time changes of some environmental factors in the pelagial of Mikołajskie Lake. *Ekologia Polska*, **20** (40), 541–60.

Ryder, R.A. (1965). A method for estimating the potential fish production of north-temperate lakes. *Transactions of the American Fisheries Society*, **94**, 214–18.

Ryther, J.H. (1956). Photosynthesis in the sea as a function of light intensity. *Limnology and Oceanography*, **1**, 61–70.

Ryther, J.H. & Guillard, R.R.L. (1962). Studies of marine planktonic diatoms. III. Some effects of temperature on respiration of five species. *Canadian Journal of Microbiology*, **8**, 447–53.

Ryther, J.H. & Yentsch, C.S. (1957). The estimation of phytoplankton production in the ocean from chlorophyll and light data. *Limnology and Oceanography*, **3**, 281–6.

Ryther, J.H. & Yentsch, C.S. (1958). Primary production of continental shelf waters off New York. *Limnology and Oceanography*, **3**, 327–35.

Rzóska, J. (1961). Observations on tropical rainpools and general remarks on temporary waters. *Hydrobiologia*, **17**, 265–85.

Rzóska, J., Brook, A.J. & Prowse, G.A. (1955). Seasonal plankton development in the White and Blue Nile near Khartoum. *Verhandlungen der Internationalen Vereinigung für Theoretische und Angewandte Limnologie*, **16**, 716–18.

Sadler, W.O. (1935). Biology of the midge *Chironomus tentans* Fabricius, and methods for its propagation. *Memoirs. Cornell University Agricultural Experiment Station*, No. 173, 25 pp.

Sakamoto, M. (1966). Primary production by phytoplankton community in some Japanese lakes and its dependence on depth. *Archiv für Hydrobiologie*, **62**, 1–28.

Salmanov, M.A. (1961). Estimations of the phytoplankton primary production and photosynthesis in the Mingechur Reservoir by ^{14}C-method. In *Primary Production in Seas and Inland Waters*, ed. G.G. Winberg, pp. 223–7. Minsk. (In Russian.)

Sarvala, J. (1974). Pääjärven energiatalous. *Luonnon Tutkija*, **78**, 181–90. (In Finnish with English summary.)

Sauberer, F. (1953). Der Strahlungshaushalt eines alpinen Seen. *Archiv für Meteorologie, Geophysik und Bioklimatologie*, **4**, 253–74.

Sauberer, F. & Ruttner, F. (1941). *Die Strahlungsverhältnisse der Binnengewässer*. Leipzig: Akademische Verlagsgesellschaft. 240 pp.

Saunders, G.W. (1957). Interrelations of dissolved organic matter and phytoplankton. *Botanical Review*, **23**, 389–410.

Saunders, G.W. (1963). The biological characteristics of freshwater. *Publications, Great Lakes Research Institute*, **10**, 245–57.

Saunders, G.W. (1969). Some aspects of feeding zooplankton. In *Eutrophication: Causes, Consequences, Correctives*, pp. 556–73. Washington: National Academy of Sciences.

Saunders, G.W. (1971). Carbon flow in the aquatic system. In *The Structure and Function of Fresh Water Microbial Communities*, ed. J. Cairns, *Research Division Monograph, Virginia Polytechnic Institute and State University*, **3**, pp. 31–45. Charlottesville, Va: University Press of Virginia.

Saunders, G.W. (1972a). The transformation of artificial detritus in lake water. *Memorie dell'Istituto Italiano di Idrobiologia*, 29 (supplement), 261–88.

Saunders, G.W. (1972b). Summary of the general conclusions of the symposium. *Memorie dell'Istituto Italiano di Idrobiologia*, **29** (supplement), 533–40.

Saunders, G.W. (1972c). The kinetics of extracellular release of soluble organic matter by plankton. *Verhandlungen der Internationale Vereinigung für Theoretische und Angewandte Limnologie*, **12**, 140–6.

Saunders, G.W. (1972d). Potential heterotrophy in a natural population of *Oscillatoria agardhii* var. *isothrix* Skuja. *Limnology and Oceanography*, **17**, 704–11.

Scavia, D. & Chapra, S.C. (1977). Comparison of an ecological model of Lake Ontario and phosphorus loading models. *Journal of the Fisheries Research Board of Canada*, **34**, 286–90.

Schelske, G.L. & Stoermer, E.S. (1972). Phosphorus, silica and eutrophication of Lake Michigan. In *Nutrients and Eutrophication*, ed. G.E. Likens, pp. 157–70. Lawrence: Allen Press.

Schiemer, F., Duncan, A. & Klekowski, R.Z. (1980). For details of this reference see list of references added in proof on page 570.

Schindler, D.W. (1971a). Light, temperature, and oxygen regimes of selected lakes in the Experimental Lakes Area, Northwestern Ontario. *Journal of the Fisheries Research Board of Canada*, **28**, 157–69.

Schindler, D.W. (1971b). A hypothesis to explain differences and similarities among lakes in the Experimental Lakes Area, Northwestern Ontario. *Journal of the Fisheries Research Board of Canada*, **28**, 295–301.

Schindler, D.W. (1972). Production of phytoplankton and zooplankton in Canadian Shield Lakes. In *Productivity Problems of Freshwaters*, ed. Z. Kajak & A. Hillbricht-Ilkowska, pp. 311–31. Warsaw & Kraków: Polish Scientific Publishers.

Schindler, D.W. (1975). Whole lake eutrophication experiments with phosphorus, nitrogen and carbon. *Verhandlungen der Internationalen Vereinigung für Theoretische und Angewandte Limnologie*, **19**, 3221–31.

Schindler, D.W. (1978). Factors regulating phytoplankton production and standing crop in the world's fresh waters. *Limnology and Oceanography*, **23**, 478–86.

Schindler, D.W. & Fee, E.J. (1973). Diurnal variation of dissolved organic carbon and its use in estimating primary production and CO_2 invasion in Lake 227. *Journal of the Fisheries Research Board of Canada*, **30**, 1501–10.

Schindler, D.W. & Fee, E.J. (1974). Primary production in freshwater. In *Proceedings of the First International Congress of Ecology*, ed. W.H. van Dobben & R.H. Lowe-McConnell, pp. 155–8, Wageningen: PUDOC.

Schindler, D.W. & Fee, E.J. (1975). The roles of nutrient cycling and radiant energy in aquatic communities. In *Photosynthesis and Productivity in Different Environments*, ed. J.P. Cooper, *International Biological Programme* 3, pp. 323–43. Cambridge University Press.

Schindler, D.W. & Holmgren, S.K. (1971). Primary productivity in the Experimental Lakes Area, Northwestern Ontario, and other low carbonate waters, and a scintillation method for determining ^{14}C activity in photosynthesis. *Journal of the Fisheries Research Board of Canada*, **28**, 189–201.

Schindler, D.W. & Nighswander, J.E. (1970). Nutrient supply and primary production in Clear Lake, Eastern Ontario. *Journal of the Fisheries Research Board of Canada*, **27**, 2009–36.

Schindler, D.W., Welch, H.E., Kalff, J., Brunskill, G. & Kritsch, N. (1974). The physical and chemical limnology of Char Lake, Cornwallis Island (Lat. 74° 42′ N; Long. 94° 50′ W). *Journal of the Fisheries Research Board of Canada*, **31**, 558–607.

Schmidt, G.W. (1973). Primary production of phytoplankton in the three types of Amazonian waters: II. The limnology of a tropical flood-plain lake in central Amazonia (Lago do Castanho). *Amazoniana*, **4** (2), 139–204.

Schnitnikoff, A.V. (1973). Some geographical characteristics of Eurasia according to the water balance of lakes and ionic composition of their water. In *Voprosy Sovremennoi Limnologii*, ed. S.V. Kalesnik, pp. 38–56. Leningrad: Nauka. (In Russian.)

Sculthorpe, C.D. (1967). *The Biology of Aquatic Vascular Plants*, London: Arnold. 610 pp.

Seaburg, K. & Moyle, J. (1964). Feeding habits, digestion rates, and growth of some Minnesota warmwater fishes. *Transactions of the American Fisheries Society*, **93**, 269–85.

Seckel, G.R. (1970). The trade wind zone oceanography pilot study. Part VIII. Sea-level meteorological properties and heat exchange processes. *Special Scientific Report. United States Department of the Interior, Fish and Wildlife Service, Fisheries*, **612**, 1–129.

Sedell, J.R., Triska, J.F., Hall, J.D., Anderson, N.H. & Lyford, L.H. (1974). Sources and fates of organic inputs in coniferous forest streams. In *Integrated Research in the Coniferous Forest Biome*, ed. R.H. Waring & R.L. Edmonds, pp. 57–69. Seattle: University of Washington.

Seki, H., Nakai, T. & Otobe, H. (1974). Turnover rate of dissolved materials in Philippine Sea at winter of 1973. *Archiv für Hydrobiologie*, **73**, 238–44.

Sellers, W.D. (1965). *Physical Climatology*. Chicago: University of Chicago Press. 272 pp.

Serruya, C. & Pollingher, U. (1971). An attempt at forecasting the *Peridinium* bloom in Lake Kinneret (Lake Tiberias). *Mitteilungen der Internationalen Vereinigung für Theoretische und Angewandte Limnologie*, **19**, 277–91.

Serruya, C. & Serruya, S. (1972). Oxygen content in Lake Kinneret: physical and biological influences. *Verhandlungen der Internationalen Vereinigung für Theoretische und Angewandte Limnologie*, **18**, 580–7.

Šesták, Z., Čatský, J. & Jarvis, P.G. (eds.) (1971). *Plant Photosynthetic Production. Manual of Methods*. The Hague: W. Junk. 818 pp.

Šetlík, I., Sust, V. & Malek, I. (1970). Dual purpose open circulation units for large scale culture of algae in temperate zones: I. Basic design considerations and scheme of a pilot plant. *Algological Studies*, **1**, 111–66.

Shapiro, J. (1973). Blue-green algae: why they become dominant. *Science*, **174**, 382–4.

Sharp, J.H. (1973). Size classes of organic carbon in seawater. *Limnology and Oceanography*, **18**, 441–7.

Shcherbakov, A.P. (1967). Role of zooplankton in destruction of organic matter in a lake. *Zhurnal Obschchei Biologii*, **28**, 131–8. (In Russian.)

Shcherbakov, A.P. & Ivanova, A.I. (1969). The role of zooplankton in the organic matter destruction in a eutrophic lake. *Zhurnal Obshchei Biologii*, **30**, 140–6. (In Russian.)

Sherstyankin, P.P. (1975). Pronikoveniye solnechnogo sveta v vody Baikala. In *Krugovorot Veshchestra i Energii v Ozernykh Vodolmakh*, ed. G.I. Galazii & K.K. Votintsev, pp. 357–61. Novosibirsk. Nauka.

Shushkina, E.A. (1966). Relationship between production and biomass of lake zooplankton. *Gidrobiologicheskii Zhurnal*, **2** (1), 27–35. (In Russian with English summary.)

Sillen, L.G. (1961). The physical chemistry of sea water. *Publications of the American Association for the Advancement of Science*, **67**, 549–81.

Sioli, H. (1964). General features of the limnology of Amazonia. *Verhandlungen der Internationalen Vereinigung für Theoretische und Angewandte Limnologie*, **15**, 1053–8.

Sioli, H. (1975). Tropical river: The Amazon. In *River Ecology*, ed. B.A. Whitton, pp. 461–88. Oxford: Blackwell.

Sivko, T.H. & Lyakhnovich V.P. (1967). Nablyudeniya na prudakh biologicheskoi ochistki stochnykh vod sakharnogo zavoda. *Gidrobiologicheskii Zhurnal*, **3** (1), 54–61.

Slack, H.D. (1955). Factors affecting the productivity of *Coregonus clupeoides* Lacepede in Loch Lomond. *Verhandlungen der Internationalen Vereinigung für Theoretische und Angewandte Limnologie*, **12**, 183–6.

Slobodkin, L.B. (1959). Energetics in *Daphnia pulex* populations. *Ecology*, **40**, 232–43.

Slobodkin, L.B. (1960). Ecological energy relationships at the population level. *American Naturalist*, **94**, 213–36.

Smirnov, N.N. (1964). On the quantity of allochthonous pollen and spores received by the Rybinsk Reservoir. *Hydrobiologia*, **24**, 421–9.

Smirnov, N.N. (ed.) (1973). *Many-year Zooplankton Data for Lakes*. Academy of Science USSR, Soviet National Committee for International Biological Programme. Moskva: Nauka. 202 pp.

Smith, F.E. (1969). Effects of enrichment in mathematical models. In *Eutrophication: Causes, Consequences, Correctives*. Proceedings of an International Symposium on Eutrophication. University of Wisconsin, Madison, 1967, pp. 631–45. Washington, DC: National Academy of Sciences.

Smith, I.R. (1974). The structure and physical environment of Loch Leven, Scotland. *Proceedings of the Royal Society of Edinburgh*, B 74, 81–100.

Smith, I.R. & Sinclair, I.J. (1972). Deep waves in lakes. *Freshwater Biology*, **2**, 387–99.

Smith, R.C., Tyler, J.E. & Goldman, C.R. (1973). Optical properties and color of Lake Tahoe and Crater Lake. *Limnology and Oceanography*, **18**, 189–99.

Smithsonian Institution (1966). *Smithsonian Meteorological Tables*. Washington: Smithsonian Institution. 527 pp.

Smyly, W.J.P. (1972). The crustacean zooplankton past and present, in Esthwaite Water. *Verhandlungen der Internationalen Vereinigung für Theoretische und Angewandte Limnologie*, **18**, 320–6.

Smyly, W.J.P. (1973). Bionomics of *Cyclops strenuus abyssorum* Sars. (Copepoda: Cyclopoida). *Oecologia*, **11**, 163–86.

Snodgrass, W.J. & O'Melia, C.R. (1975). Predictive model for phosphorus in lakes. *Environmental Science and Technology*, **9**, 937–44.

Sobolev, Yu. A. (1971). Natural forage reserve of ponds during breeding of phytophagous fishes together with carp. *Gidrobiologicheskii Zhurnal*, 7 (5), 59–66. (In Russian.)

Soeder, C.J. (1967). Tagesperiodische Wandemungen bei begeisselten Planktonalgen. *Die Umschau*, 12, 338.

Sokolov, A.A. (1966). The dependence of lake and reservoir water budget on the relation between the lake surface area and the catchment area. *Publications of the International Association of Scientific Hydrology*, 70, 203–9.

Sokolova, N. Ya. (1970). The macrobenthos and the specification of its formation in the small reservoirs of the Upper Volga Basin, Communication 1. *Byulletin Moskovskogo Obshchestva Ispytalelei Prirody, otdel biologicheski*, 75 (4), 54–67. (In Russian with English summary.)

Sokolova, N. Ya. (1971). Sravnitelnaya ochenka sposobov opredeleniya produktsii Lichinok Chironomidae. *Zoologicheskii Zhurnal*, 41, 1618–30. (In Russian.)

Solomon, D. & Brafield, A. (1972). The energetics of feeding, metabolism and growth of perch (*Perca fluviatilis* L.) *Journal of Animal Ecology*, 41, 699–718.

Sorokin, H.M. (1968). Hydrological characteristics of small lakes of some lake areas in the North-West. In *Ozera Razlichnych Landschaftor Severo-Zapada SSSR*, Vol. 1, pp. 91–121. Leningrad: Nauka. (In Russian.)

Sorokin, Yu. I. (1966). O primenenii radioaktivnogo ugleroda dlya izucheniya pitaniya i pishchevykh svyazei vodnykh zhivotnykh. *Trudy Instituta Biologii Vnutrennikh Vod*, 12 (15), 75–119.

Sorokin, Yu. I. (1967). Nekotorye itogi izucheniya troficherskoi roli bakterii v vodoemakh. (Some results of studying the trophic role of bacteria in reservoirs). *Gidrobiologicheskii Zhurnal*, 3 (5), 32–42. (In Russian.)

Sorokin, Yu. I. (1972). Biological productivity of the Rybinsk Reservoir. In *Productivity Problems of Freshwaters*, ed. Z. Kajak & A. Hillbricht-Ilkowska, pp. 493–503. Warsaw & Kraków: Polish Scientific Publishers.

Sorokin, Yu. I. & Kadota, H. (1972). *Techniques for the Assessment of Microbial Production and Decomposition in Freshwaters*, IBP Handbook No. 23. Oxford: Blackwell. 112 pp.

Sorokin, Yu. I. & Meshkov, A.N. (1959). O primenenii radioaktivnogo izotopa ugleroda dlya izucheniya pitaniya vodnykh bespozvonochnykh. *Trudy Instituta Biologii Vodokhranilisch*, 2 (5), 7–14.

Sorokin, Yu. I. & Paveljeva, E.B. (1972). On the quantitative characteristics of the pelagic ecosystem of Lake Dalnee (Kamchatka). In *Biological Oceanography of the Northern North Pacific Ocean*, ed. A.Y. Takenouti, pp. 621–6. Tokyo: Idemitsu Shoten.

Sorokin, Yu. I. & Wyshkwarzew, D.I. (1973). Feeding on dissolved organic matter by some marine animals. *Aquaculture*, 2, 141–8.

Spence, D.H.N. (1976). Light and plant response in freshwater. In *Light as an Ecological Factor II*, ed. G.C. Evans, R. Bainbridge & O. Rackham, pp. 93–133. Oxford: Blackwell.

Spence, D.H.N. & Chrystal, J. (1970a). Photosynthesis and zonation of freshwater macrophytes. I. Depth distribution and shade tolerance. *New Phytologist*, 69, 205–15.

Spence, D.H.N. & Chrystal, J. (1970b). Photosynthesis and zonation of freshwater macrophytes. II. Adaptability of species of deep and shallow water. *New Phytologist*, 69, 217–27.

Spodniewska, I., Pieczynska, E. & Kowalczewski, A. (1975). Ecosystem of the Mikołajskie Lake, primary production. *Polskie Archiwum Hydrobiologie*, 22, 17–37.

Sreenivasan, A. (1972). Energy transformation through primary productivity and fish production in some tropical impoundments and ponds. In *Productivity Problems of Freshwaters*, ed. Z. Kajak & A. Hillbricht-Ilkowska, pp. 505–14. Warsaw & Kraków: Polish Scientific Publishers.

Stańczykowska, A. (1960). Beobachtungen über die Gruppierungen von *Viviparus fasciatus* Mull. in dem Weichselarm "Konfederatka". *Ekologia Polska*, A **2**, 21–48.

Stańczykowska, A. (1964). On the relationship between abundance, aggregations and "condition" of *Dreissena polymorpha* Pall. in 36 Masurian lakes. *Ekologia Polska*, A **12**, 653–90.

Stańczykowska, A. (1968). The filtration of populations of *Dreissena polymorpha* Pall. in different lakes as a factor affecting circulation of matter in the lake. *Ekologia Polska*, B **14**, 265–70.

Stańczykowska, A. (1969). Z zagadnién odzywiania sie *Viviparus fasciatus* Mull. *Ekologia Polska*, B **5**, 271–3.

Stańczykowska, A., Pliński, M. & Magnin, E. (1972). Etude de trois populations de *Viviparus malleetus* (Reeve) (Gastropoda, Prosobranchia) de la région de Montréal. II. Etude qualitative et quantitative de la nourriture. *Canadian Journal of Zoology*, **50**, 1617–24.

Stankovič, S. (1960). The Balkan lake Ohrid and its living world. *Monographiae Biologicae*, **9**, 357 pp.

Stanley, R.A. & Naylor, A.W. (1972). Photosynthesis in Eurasian water milfoil (*Myriophyllum spicatum* L.). *Plant Physiology*, **50**, 149–51.

Steel, J.A.P. (1972). The application of fundamental limnological research in water supply system design and management. In *Conservation and Productivity of Natural Waters*, ed. R.W. Edwards & D.J. Garrod, *Symposia of the Zoological Society of London* 29, pp. 41–67. New York & London: Academic Press.

Steel, J.A.P. (1975). The management of Thames Valley reservoirs. In *Proceedings of a Symposium on the Effect of Storage on Water Quality*, Reading University, March 1975, pp. 371–419. Medmenham Water Research Centre.

Steel, J.A.P. (1978). Reservoir algal productivity. In *Mathematical Models in Water Pollution Control*, ed. A. James. Chichester: J. Wiley & Sons.

Steel, J.A.P., Duncan, A. & Andrew, T.E. (1972). The daily carbon gains and losses in the seston of Queen Mary Reservoir, England, during early and mid 1968. In *Productivity Problems of Freshwaters*, ed. Z. Kajak & A. Hillbricht-Ilkowska, pp. 515–27. Warsaw & Kraków: Polish Scientific Publishers.

Steele, J.H. (1962). Environmental control of photosynthesis in the sea. *Limnology and Oceanography*, **7**, 137–50.

Steele, J.H. (1965). Notes on some theoretical problems in production ecology. *Memorie dell'Istituto Italiano di Idrobiologia*, **18** (supplement), 383–98.

Steele, J.H. (1974). *The Structure of Marine Ecosystems*. Oxford: Blackwell. 128 pp.

Steele, J.H. & Baird, I.E. (1962). Carbon–chlorophyll relations in cultures. *Limnology and Oceanography*, **7**, 101–2.

Steele, J.H. & Menzel, P.W. (1961). Conditions for maximum primary production in the mixed layer. *Deep-Sea Research*, **9**, 39–49.

Steemann Nielsen, E. (1947). Photosynthesis of aquatic plants with special reference to the carbon-sources. *Dansk Botanisk Arkiv*, **12** (8), 1–71.

Steemann Nielsen, (1962a). Inactivation of the photochemical mechanism in photosynthesis as a means to protect the cells against too high light intensity. *Physiologia Plantarum*, **15**, 161–71.

Steemann Nielsen, E. (1962b). On the maximum quantity of plankton chlorophyll per surface unit of a lake or the sea. *Internationale Revue der Gesamten Hydrobiologie*, **47**, 333–98.

Steemann Nielsen, E. (1965). On the determination of the activity of C/14 ampoules for measuring primary production. *Limnology and Oceanography*, **10** (supplement), 247–52.

Steemann Nielsen, E. (1975). *Marine Photosynthesis. With Special Emphasis on the Ecological Aspects.* Amsterdam: Elsevier. 140 pp.

Steemann Nielsen, E. & Hansen, V.K. (1959). Light adaption in marine phytoplankton populations and its interrelation to the temperature. *Physiologia Plantarum*, **12**, 358–70.

Steemann Nielsen, E. & Jørgensen, E.G. (1968). The adaption of plankton algae. I. General part. *Physiologia Plantarum*, **21**, 401–13.

Stepanova, L.A. (1971). Production of mass forms of planktonic crustaceans in Lake Ilmen. *Gidrobiologecheskii Zhurnal*, **7** (6), 19–30. (In Russian with English summary.)

Stewart, K.M. & Markello, S.J. (1974). Seasonal variations in concentrations of nitrate and total phosphorus, and calculated nutrient loading for six lakes in western New York. *Hydrobiologia*, **44**, 61–89.

Stewart, W.D.P. & Alexander, G. (1971). Phosphorus availability and nitrogenase activity in aquatic blue-green algae. *Freshwater Biology*, **1**, 389–404.

Stewart, W.D.P. & Pearson, H.W. (1970). Effects of aerobic and anaerobic conditions on growth and metabolism of blue-green algae. *Proceedings of the Royal Society, London*, B **175**, 293–311.

Stockner, J.G. (1968). Algal growth and primary productivity in a thermal stream. *Journal of the Fisheries Research Board of Canada*, **25**, 2037–58.

Stockner, J.G. (1971). Preliminary characterization of lakes in the Experimental Lakes Area, Northwestern Ontario using diatom occurrences in sediments. *Journal of the Fisheries Research Board of Canada*, **28**, 265–75.

Stockner, J.G. (1972). Paleolimnology as a means of assessing eutrophication. *Verhandlungen der Internationalen Vereinigung für Theoretische und Angewandte Limnologie*, **18**, 1018–30.

Stockner, J.G. & Armstrong, F.A.J. (1971). Periphyton of the Experimental Lakes Area. Northwestern Ontario. *Journal of the Fisheries Research Board of Canada*, **28**, 215–29.

Storch, T.A. (1971). *Production of Extracellular Dissolved Organic Carbon by Phytoplankton and its Utilization by Bacteria.* Ph.D. thesis, University of Michigan. 147 pp.

Straškraba, M. (1968). Der Anteil der höheren Pflanzen an der Produktion der stehenden Gewässer. *Miteilungen der Internationalen Vereinigung für Theoretische und Angewandte Limnologie*, **14**, 212–30.

Straškraba, M. (1972). Physical structure of the environment and the trophic efficiency of aquatic ecosystems. Presented at the "Final Symposium on Synthesis of PF Results". IBP meeting. Reading.

Straškraba, M. (1973). Limnological basis for modelling reservoir ecosystems. In *Man-Made Lakes: Their Problems and Environmental Effects*, ed. W.C. Ackermann, G.F. White & E.B. Worthington, *Geophysical Monographs*, **17**, pp. 517–35. Washington DC: American Geophysical Union.

Straškraba, M. (1976a). Development of an analytical phytoplankton model with parameters empirically related to dominant controlling variables. In *Umweltbiophysik, Arbeitstagung veranstaltet von der Gesellschaft für physikalische und mathematische Biologie der DDR vom 29.10. bis 1.11.1973 in Kuhlungsborn*, 33–65.

Straškraba, M. (1976b). Empirical and analytical models of eutrophication. *Eutro-symposium 1976*, Proceedings of an International Symposium on Eutrophication and Rehabilitation of Surface Waters, *Karl-Marx-Stadt*, **3**, 352–371.

Straškraba, M. & Javornický, P. (1973). Limnology of two re-regulation reservoirs in Czechoslovakia. *Hydrobiological Studies*, **2**, 249–316.

Straškraba, M., Legner, M., Komárková, J., Fott, J., Peréz, M., Holcik, J. & Holcíková (1969). Primera contribución al conocimiento limnologico de las lagunas y embalses de Cuba. *Serie Biologia, Academi Ciencias, Habana, Cuba*, **4**, 1–44.

Straškraba, M. & Pieczyńska, E. (1970). Field experiments on shading effect by emergents on littoral phytoplankton and periphyton production. *Rozpravy Československé Akademie Věd, Rada MPV*, **80**, 7–32.

Straškraba, M. & Straskrabová, V. (1975). Management problems of Slapy Reservoir, Bohemia, Czechoslovakia. In *Proceedings of a Symposium on the Effects of Storage on Water Quality*, Reading University, March 1975, pp. 450–83. Medmenham Water Research Centre.

Straškrabová, V. (1968). Bakteriologische Indikation der Wasservenunreinigung mit abbaubaren Stoffen. *Limnologica*, **6**, 29–36.

Straškrabová, V. (1973). Methods for counting water bacteria: comparison and significance. *Acta Hydrochimica et Hydrobiologica*, **1**, 433–54.

Straškrabová, V. & Legner, M. (1969). The quantitative relation of bacteria and ciliates to water pollution. *Advances in Water Pollution Research*, **4**, 57–67.

Straškrabová-Prokesova, V. (1966). Seasonal changes in the reproduction rate of bacteria in two reservoirs. *Verhandlungen der Internationalen Vereinigung für Theoretische und Angewandte Limnologie*, **16**, 1527–33.

Stross, R.G., Chisholm, S.W. & Downing, T.A. (1973). Cause of daily rhythms in photosynthetic rates of phytoplankton. *Biological Bulletin, Marine Biological Laboratory, Woods Hole*, **145**, 200–9.

Stross, R.G., Neess, J.C. & Hasler, A.D. (1961). Turnover time and production of the planktonic Crustacea in limed and reference portion of a bog lake. *Ecology*, **42**, 237–45.

Stross, R.G. & Pemrick, S.M. (1974). Nutrient uptake kinetics in phytoplankton: a basis for niche separation. *Journal of Phycology*, **10**, 164–9.

Stull, E.A., Amezaga, De, E. & Goldman, C.R. (1973). The contribution of individual species of algae to the seasonal patterns of primary productivity in Castle Lake, California. *Verhandlungen der Internationalen Vereinigung für Theoretische und Angewandte Limnologie*, **18**, 1776–83.

Stumm, W. & Morgan, J.J. (1970). *Aquatic Chemistry. An Introduction Emphasizing Chemical Equilibria in Natural Waters*. New York: Wiley Interscience. 583 pp.

Sumari, O. (1971). Structure of the perch populations of some ponds in Finland. *Annales Zoologici Fennici*, **8**, 406–21.

Sundaram, T.R. & Rehm, R.G. (1973). The seasonal thermal structure of deep temperate lakes. *Tellus*, **25**, 157–68.

Sundaram, T.R., Rehm, R.G., Rudinger, G. & Merritt, G.E. (1970). A study of some problems on the physical aspects of thermal pollution. *Report. Cornell Aeronautical Laboratory*, VT-2790-A-1, 188 pp.

Sushchenya, L.M. (1970). Food rations, metabolism and growth of crustaceans. In *Marine Food Chains*, ed. J.H. Steele, pp. 127–41. Berkeley: University of California Press.

Sverdrup, H.U., Johnson, M.W. & Fleming, R.H. (1942). *The Oceans. Their Physics, Chemistry and General Biology*. New York: Prentice-Hall. 1087 pp.

Swynnerton, G.H., & Worthington, E.B. (1939). Brown Trout growth in the Lake District. A study of the conditions in "acid" lakes and tarns. *Salmon and Trout Magazine*, **97**, 337–55.

Szczepańska, W. (1968). Vertical distribution of periphyton in the Lake Mikołojskie. *Polskie Archiwum Hydrobiologie*, **15**, 177–82.

Szczepanski, A, (1965). Deciduous leaves as a source of organic matter in lakes. *Bulletin de l'Académie Polonaise des Sciences, Classe II.*, **13**, 215–17.

Takahashi, M., Fujii, K. & Parsons, T.R. (1973). Simulation study of phytoplankton photosynthesis and growth in the Frazer River Estuary. *Marine Biology*, **19**, 102–16.

Talling, J.F. (1957a). Photosynthetic characteristics of some freshwater plankton diatoms in relation to underwater radiation. *New Phytologist*, **56**, 29–50.

Talling, J.F. (1957b). The phytoplankton population as a compound photosynthetic system. *New Phytologist*, **56**, 133–49.

Talling, J.F. (1957c). Diurnal changes of stratification and photosynthesis in some tropical African waters. *Proceedings of the Royal Society, London.* B **147**, 57–83.

Talling, J.F. (1960). Self-shading effects in natural populations of a planktonic diatom. *Wetter und Leben*, **12**, 235–42.

Talling, J.F. (1961). Photosynthesis under natural conditions. *Annual Review of Plant Physiology*, **12**, 133–54.

Talling, J.F. (1965). The photosynthetic activity of phytoplankton in East African lakes. *Internationale Revue der Gesamten Hydrobiologie und Hydrographie*, **50**, 1–32.

Talling, J.F. (1966a). The annual cycle of stratification and phytoplankton growth in Lake Victoria (East Africa). *Internationale Revue der Gesamten Hydrobiologie und Hydrographie*, **51**, 545–621.

Talling, J.F. (1966b). Photosynthetic behaviour in stratified and unstratified lake populations of a planktonic diatom. *Journal of Ecology*, **54**, 99–127.

Talling, J.F. (1969). The incidence of vertical mixing, and some biological and chemical consequences in tropical African lakes. *Verhandlungen der Internationalen Vereinigung für Theoretische und Angewandte Limnologie*, **17**, 998–1012.

Talling, J.F. (1970). Generalized and specialized features of phytoplankton as a form of photosynthetic cover. In *Prediction and Measurement of Photosynthetic Productivity*, Proceedings IBP/PP Technical Meeting, Třeboň, 1969, pp. 431–45. Wageningen: PUDOC.

Talling, J.F. (1971). The underwater light climate as a controlling factor in the production ecology of freshwater phytoplankton. *Mitteilungen der Internationalen Vereinigung für Theoretische und Angewandte Limnologie*, **19**, 214–43.

Talling, J.F. (1975). Primary production of freshwater microphytes. In *Photosynthesis and Productivity in Different Environments*, ed. J.P. Cooper, *International Biological Programme* 3, pp. 225–47. Cambridge University Press.

Talling, J.F. (1976). The depletion of carbon dioxide from lake water by phytoplankton. *Journal of Ecology*, **64**, 79–121.

Talling, J.F. & Rzóska, J. (1967). The development of plankton in relation to hydrological regime in the Blue Nile. *Journal of Ecology*, **55**, 637–62.

Talling, J.F. & Talling, I.B. (1965). The chemical composition of African lake waters. *Internationale Revue der Gesamten Hydrobiologie*, **50**, 421–63.

Talling, J.F., Wood, R.B., Prosser, M.V. & Baxter, R.M. (1973). The upper limit of photosynthetic productivity by phytoplankton: evidence from Ethiopian soda lakes. *Freshwater Biology*, **3**, 53–76.

Tamiya, H. (1957). Mass culture of algae. *Annual Review of Plant Physiology*, **8**, 309–34.

Tanaka, S., Mitzutani, T. & Hanya, T. (1975). Productivity of the community in Lake Yunoko. 2.6. The productivity of the entire lake. In *Productivity of Communities in Japanese Inland Waters*, ed. S. Mori & A. Yamamoto, JIBP Synthesis, vol. 10, pp. 192–4. Tokyo: University of Tokyo Press.

Tanimizu, K. & Miura, T. (1976). Studies on the submerged plant community in Lake Biwa: I. Distribution and productivity of *Egeria densa*, a submerged plant invader, in the South basin. *Physiology and Ecology, Japan*, **17**, 283–90.

Tarent, A.A. & Colman, B. (1972). Dark assimilation of acetate- ^{14}C by *Anabaena flos-aquae. Canadian Journal of Botany*, **50**, 2067–71.

Tarwid, M. (1969). Analysis of the alimentary tract of predatory Pelopiinae larvae (Chironomidae). *Ekologia Polska*, A **17**, 125–31.

Taub, F.B. (1969a). A biological model of a freshwater community: A gnotobiotic ecosystem. *Limnology and Oceanography*, **14**, 136–42.

Taub, F.B. (1969b). Gnotobiotic models of freshwater communities. *Verhandlungen der Internationalen Vereinigung für Theoretische und Angewandte Limnologie*, **17**, 485–96.

Taub, F.B. (1969c). A continuous gnotobiotic (species defined) ecosystem. In *The Structure and Function of Fresh-Water Microbial Communities*, ed. J. Cairns, Jr, *Research Division Monograph, Virginia Polytechnic Institute and State University*, **3**, 101–20.

Taub, F.B. (1979). A continuous gnotobiotic (species-defined) ecosystem. In *Aquatic Microbial Communities*, ed. J. Cairns, Garland Reference Library of Science and Technology, vol. 15, pp. 105–38. New York: Garland Publishing Company.

Taub, F.B. & Dollar, A.M. (1968). The nutritional inadequacy of *Chlorella* and *Chlamydomonas* as food for *Daphnia pulex. Limnology and Oceanography*, **13**, 607–17.

Taub, F.B. & McKenzie, D.H. (1973). Continuous cultures of an alga and its grazer. In *Modern Methods in the Study of Microbial Ecology*, ed. T. Rosswall, pp. 371–8. Stockholm: NFR.

Teal, J.M. (1957). Community metabolism in a temperate cold spring. *Ecological Monographs*, **27**, 283–302.

Tessenow, U. (1964). Experimentaluntersuchungen zur Kieselsäurerückführung aus dem Schlamm der Seen durch Chironomidenlarven (*plumosus*-Gruppe). *Archiv für Hydrobiologie*, **60**, 497–504.

Tessenow, U. (1972). Lösungs-, Diffusions- und Sorptionsprozesse in Seesedimenten. *Archiv für Hydrobiologie*, **38** (supplement), 353–98.

Thienemann, A. (1927). Der Ban des Seebeckens in seiner Bedeutung für das Leben in See. *Verhandlungen der Zoologisch-botanischen Gesellschaft in Wien*, **77**, 87–91.

Thienemann, A. (1931). Tropische Seen und Seetypenlehre. *Archiv für Hydrobiologie*, **9** (supplement), 205–31.

Thomann, R.V. (1972). *Systems Analysis and Water Quality Management*. Stanford: ESSC Publications.

Thomann, R.V. (1977). Comparison of lake phytoplankton and loading plots. *Limnology and Oceanography*, **22**, 370–3.

Thomas, C.W. (1963). On the transfer of visible radiation through sea ice and snow. *Journal of Glaciology*, **4** (34), 481–4.

Thomas, E.A. (1950). Beitrag zur Methodik der Produktionsforschung in Seen. *Schweizerische Zeitschrift für Hydrologie*, **12**, 25–37.

Thomas, E.A. (1956/57). Der Zürichsee, sein Wasser und sein Boden. *Jahrbuch des Zurichsee*, **17**, 173–208.

Thomas, E.A. (1969). The process of eutrophication in Central European lakes. In *Eutrophication: Causes, Consequences, Correctives*, ed. G.A. Rohlich, pp. 29–49. Washington, DC: National Academy of Sciences.

Thomas, E.A. (1973). Phosphorus and eutrophication. In *Environmental Phosphorus Handbook*, ed. E.J. Griffith, A. Beeton, J.M. Spenser & D.T. Mitchell, pp. 585–611. New York, NY: John Wiley.

Thomas, N.A. & O'Connell, R.L. (1966). A method for measuring primary production by stream benthos. *Limnology and Oceanography*, **11**, 386–92.

Thomasson, K. (1956). Reflection on Arctic and Alpine lakes. *Oikos*, **7**, 117–43.

Thorpe, J.E.(1974a). Trout and perch populations at Loch Leven, Kinross. *Proceedings of the Royal Society of Edinburgh*, B **74**, 295–313.

Thorpe, J.E. (1974b). Movements of brown trout (*Salmo trutta* L.) in Loch Leven, Kinross. *Journal of Fish Biology*, **6**, 153–80.

Thorpe, J.E. & Roberts, R.J.(1972). An aeromonad epidemic in the brown trout (*Salmo trutta* L.). *Journal of Fish Biology*, **4**, 441–51.

Tilly, L.J.(1968). The structure and dynamics of Cone Spring. *Ecological Monographs*, **38**, 169–97.

Tilzer, M. (1972). Dynamik und Producktivität von Phytoplankton und palagischen Bakterien in einem Hochgebirgsee (Voderer Finstertaler See, Osterreich) *Archiv für Hydrobiologie*, **40** (supplement), 201–73.

Tilzer, M. (1973). Diurnal periodicity in the phytoplankton population of a high-mountain lake. *Limnology and Oceanography*, **18**, 15–30.

Tilzer, M., Goldman, C.R. & Amezaga, De, E. (1975). The efficiency of photosynthetic light energy utilized by lake phytoplankton. *Verhandlungen der Internationalen Vereinigung für Theoretische und Angewandte Limnologie*, **19**, 800–7.

Tilzer, M.M., Goldman, C.R., Richards, R.C. & Wrigley, R.C. (1976). Influence of sediment inflow on phytoplankton primary productivity in Lake Tahoe (California-Nevada). *Internationale Revue der Gesamten Hydrobiologie und Hydrographie*, **61**, 169–82.

Tilzer, M.M., Hillbricht-Ilkowska, A., Kowalczewski, A., Spodniewska, I. & Turczinska, J. (1977). Diel phytoplankton periodicity in Mikołajskie Lake, Poland, as determined by different methods in parallel. *International Revue der Gesamten Hydrobiologie und Hydrographie*, **62**, 279–89.

Tilzer, M. & Schwarz, K. (1976). Seasonal and vertical patterns of phytoplankton light adaption in a high mountain lake. *Archiv für Hydrobiologie*, **77**, 488–504.

Timms, B.V. (1975). Morphometric control of variation in annual heat budgets. *Limnology and Oceanography*, **20**, 110–12.

Tirén, T.(1977). Denitrification in sediment-water systems of various types of lakes. In *Interactions Between Sediments and Fresh Waters*, ed. H.L. Golterman. Proceedings of an International Symposium, Amsterdam, 1976, pp. 363–9. The Hague: W. Junk.

Titus, J., Goldstein, R.A., Adams, M.S., Mankin, J.B., O'Neill, R.V., Weiler, P.R., Shugart, H.H. & Booth, R.S. (1975). A production model for *Myriophyllum spicatum* L. *Ecology*, **56**, 1129–38.

Tominaga, K. & Ichimura, S. (1966). Ecological studies on the organic matter production in a mountain river. *Botanical Magazine, Tokyo*, **79**, 815–29.

Tooming, H. & Nijlisk, H. (1967). Coefficients for the ratio of integral radiation in natural conditions. In *Fitoaktinometricheskie Issledovaniya Rastitelnogo Pokrova*, pp. 140–50. Tallin: Valgus. (In Russian.)

Triska, F.J.(1970). Seasonal Distribution of Aquatic Hyphomycetes in Relation to the Disappearance of Leaf Litter from a Woodland Stream. Ph.D. dissertation, University of Pittsburgh. 189 pp.

Tseeb, Ya. Ya., Denisova, A.I., Priimachenko, A.D., Vladimirova, K.S., Zimbalvskaya, L.N., Mikhailenko, L.E., Enaki, G.A., Zhdanova, G.A. & Sergeev, A.I. (1973). Produktivnost soobshchestv vodnykh organizmov Kievskogo vodokhranilish-cha. (The productivity relations of the aquatic organisms of Kiev Reservoir.) In *Productivity Studies of Freshwater Ecosystems*, ed. G.G. Winberg, pp. 60–71. Minsk: Byelorussian State University. (In Russian.)

Tseeb, Ya. Ya. & Maistrenko. Yu. G. (1972). *Kievskoe Vodokhranilische, Gidrokhimiya, Biologiya, Produktivnost*. (The Kiev Reservoir, Hydrochemistry, Biology, Productivity). Kiev: Naukova Dumka, 460 pp.

Tsuda, M. (1972). Interim results of the Yoshino River productivity survey, especially on benthic animals. In *Productivity Problems of Freshwaters*, ed. Z. Kajak & A. Hillbricht-Ilkowska, pp. 827–41. Warsaw & Kraków: Polish Scientific Publishers.

Tubb, R.A. & Dorris, T.S. (1965). Herbivorous insect populations in oil refinery effluent holding pond series. *Limnology and Oceanography*, 10, 121–34.

Tudorancea, C. (1972). Studies on Unionidae populations from the Crapina–Jijila complex of pools (Danube zone liable to inundation). *Hydrobiologia*, 39 (4), 572–561.

Turner, W.R. (1960). Standing crops of the fishes in Kentucky farm ponds. *Transactions of the American Fisheries Society*, 89, 333–7.

Tusa, I. (1969). On the feeding biology of the brown trout (*Salmo trutta* M. *fario* L.) in the course of day and night. *Zoologicke Listy.*, 18, 275–84.

Tyler, J.E. (1959). Natural light as a monochromator. *Limnology and Oceanography*, 4, 102–5.

Tyurin, P.N. (1962). Faktor estestvennoi smertnosti ryb i ego znachenie pri regulirovanii rybolovstva. *Voprosy Ikhtiologii*, 2, 403–27.

Tzur, Y. (1973). One-dimensional diffusion equations for the vertical transport in an oscillating stratified lake of varying cross-section. *Tellus*, 25, 266–71.

Uhlmann, D. (1958). Untersuchungen über die biologische Selbstreinigung häuslichen Abwassers in Teichen. *Wissenschaftlische Zeitschrift der Karl-Marx-Universität, Leipzig*, 8, 17–66.

Uhlmann, D., Benndorf, J. & Albert, W. (1971). Prognose des Stoffhaushaltes von Staugewässer mit Hilfe kontinuierlicher oder semikontinuierlicher biologischer Modelle. I. Grundlagen. *Internationale Revue der Gesamten Hydrobiologie*, 56, 513–39.

Uhlmann, D., Weisse, G. & Gnauck, A. (1974). Gleichgewicht und Stabilität in Ökosystemen. *Biologische Rundschau*, 12 (6), 365–81.

Umnov, A.A. (1973). Matematicheskaya model ozera Myastro. In *Productivity Studies of Freshwater Ecosystems*, ed. G.G. Winberg, pp. 95–109. Minsk: Byelorussian State University.

United States Department of Agriculture (1941). *Yearbook of Agriculture.* Washington. 1248 pp.

Unni, K.S. (1976). Production of submerged aquatic plant communities of Doodhadhari Lake, Raipur (M.P. India). *Hydrobiologia*, 48, 175–7.

Uspienski, S.M. (1965). *Die Wildgänse Nordeurasiens*. Wittenberg-Lutherstadt: Ziemsen.

Vaccaro, R.F., Hicks, E., Jannasch, N.W. & Casey, F.G. (1968). The occurrence and role of glucose in sea water. *Limnology and Oceanography*, 13, 356–60.

Van Oosten, J. (1960). Temperature of Lake Michigan, 1930–32. *Special Scientific Report. United States Department of the Interior, Fish and Wildlife Service, Fisheries*, 322, 1–34.

Ventz, D. (1973a). Beitrag zur Hydrographie von Seen. *Acta Hydrophysica*, 17, 307–16.

Ventz, D. (1973b). Die Einzugsgebietsgrösse, ein Geofaktor für den Trophiczustand stehender Gewässer. *Fortschnitte der Wasserchemie und ihrer Grenzgebiete*, 14, 105–18.

Viner, A.B. (1975a). The supply of minerals to tropical rivers and lakes. (Uganda). In *Coupling of Land and Water Systems*, ed. A.D. Hasler, pp. 227–61. Berlin: Springer-Verlag.

Viner, A.B. (1975b). Non-biological factors affecting phosphate recycling in the water of a tropical eutrophic lake. *Verhandlungen der Internationalen Vereinigung für Theoretische und Angewandte Limnologie*, 19, 1404–15.

Viner, A.B. (1975c). The sediments of Lake George (Uganda). II. The release of ammonia and phosphate from an undisturbed mud surface. *Archiv für Hydrobiologie*, **76**, 368–78.

Viner, A.B. (1975d). The sediments of Lake George (Uganda). III. The uptake of phosphate. *Archiv für Hydrobiologie*, **76**, 393–410.

Viner, A.B. & Horne, A.J. (1971). Nitrogen fixation and its significance in tropical Lake George, Uganda. *Nature, London*, **232**, 417–18.

Viner, A.B. & Smith, I.R. (1973). Geographical, historical and physical aspects of Lake George. *Proceedings of the Royal Society, London*, B **184**, 235–70.

Vitousek, P.M. & Peiners, W.A. (1975). Ecosystem succession and nutrient retention: a hypothesis. *Bioscience*, **25**, 376–81.

Vlasova, T.A., Baranovskaya, V.K., Gezen, M.V., Popova, E.I. & Sidorov, G.P. (1973). Biological productivity of Harbey Lakes, Vorkuta Tundra. In *Produktsionnobiologicheskie Issledobaniya Ekosistem Presnykh Vod*, ed. G.G. Winberg, pp. 147–63. Minsk: Byelorussian State University. (In Russian.)

Vollenweider, R.A. (1960). Beiträge zur Kenntnis optischer Eigenschaften der Gewässer und Primärproduktion. *Memorie dell'Istituto Italiano di Idrobiologia*, **12**, 201–44.

Vollenweider, R.A. (1964). Ueber oligomiktische Verhältnisse des Lago Maggiore und einiger anderer insubrischer Seen. *Memorie dell'Istituto Italiano di Idrobiologia*, **17**, 191–206.

Vollenweider, R.A. (1965). Calculation models of photosynthesis–depth curves and some implications regarding day rate estimates in primary production measurements. *Memorie dell'Istituto Italiano di Idrobiologia*, **18** (supplement), 427–57.

Vollenweider, R.A. (1968). Scientific fundamentals of the eutrophication of lakes and flowing waters, with particular reference to nitrogen and phosphorus as factors in eutrophication. *OECD Technical Report*, DAS/CSI/68.27. 159 pp.

Vollenweider, R.A. (1969a). Moglichkeiten und Grenzen elementarer Modelle der Stoffbilanz von Seen. *Archiv für Hydrobiologie*, **66**, 1–36.

Vollenweider, R.A. (ed.) (1969b). *A Manual on Methods for Measuring Primary Production in Aquatic Environments.* IBP Handbook No. 12, 1st edition. Oxford: Blackwell.

Vollenweider, R.A. (1970). Models for calculating integral photosynthesis and some implications regarding structure properties of the community metabolism of aquatic systems. In *Prediction and Measurement of Photosynthetic Productivity*, Proceedings IBP/PP Technical Meeting, Třebõn, 1969, pp. 455–72. Wageningen: PUDOC.

Vollenweider, R.A. (1971). *Water Management Research, Scientific Fundamentals of the Eutrophication of Lakes and Flowing Water With Particular Reference to Nitrogen and Phosphorus as Factors in Eutrophication.* Paris: UNESCO. 155 pp.

Vollenweider, R.A. (1974). Environmental factors linked with primary production. In *A Manual on Methods for Measuring Primary Production in Aquatic Environments*, ed. R.A. Vollenweider, IBP Handbook No. 12, 2nd edition, pp. 157–77. Oxford: Blackwell.

Vollenweider, R.A. (1975). Input–output models; with special reference to the phosphate loading concept in limnology. *Schweizerische Zeitschrift für Hydrologie*, **37**, 53–84.

Vollenweider, R.A. (1976). Advances in defining critical loading levels for phosphorus in lake eutrophication. *Memorie dell'Istituto Italiano di Idrobiologia*, **33**, 53–83.

Vollenweider, R.A. & Dillon, P.J. (1974). The application of the phosphorus loading concept to eutrophication research. *Report. Associate Committee on Scientific*

Criteria for Environmental Quality, National Research Council of Canada, **7**, 42 pp.

Vollenweider, R.A. & Nauwerck, A. (1961). Some observations on the C14 method for measuring primary production. *Verhandlungen der Internationalen Vereinigung für Theoretische und Angewandte Limnologie*, **14**, 134–9.

Voronkov, P.P. (1970). *Hydrochemistry of Local Runoff of the European Part of the USSR*. Leningrad: Gidrometerorizdat. (In Russian.)

Wagner, G. (1976). Simulationsmodelle der Seenentrophierung, dargestellt am Beispiel des Bodensees Obersees. Teil II: Simulation der Phosphorhaushaltes der Bodensee-Obersees. *Archiv für Hydrobiologie*, **78**, 1–41.

Walker, K.F. (1973). Studies on a saline lake ecosystem. *Australian Journal of Marine and Freshwater Research*, **24**, 21–71.

Walters, C.J. & Efford, I.E. (1972). Systems analysis in the Marion Lake IBP Project. *Oecologia*, **11**, 33–44.

Ward, J.C. & Karaki, S. (1971). Evaluation of effect of impoundment on water quality in Cheney Reservoir. *Water Research Technical Publications, United States Department of the Interior, Bureau of Reclamation, Research Report*, **25**, 1–69.

Warren, C. (1971). *Biology and Water Pollution Control*. Philadelphia: Saunders. 434 pp.

Warren, C.E. & Davis, G.E. (1967). Laboratory studies on the feeding, bioenergetics and growth of fish. In *The Biological Basis of Freshwater Fish Production*, ed. S.D. Gerking, pp. 175–214. Oxford: Blackwell.

Warren, C.E., Davis, G.E., Phinney, H.K., McIntyre, C.D. & Brocksen, R.W. (1965). Trophic dynamics of simplified communities in laboratory streams. Progress Report, Oregon, 109 pp.

Waters, T.F. (1969). The turnover ratio in production ecology of freshwater invertebrates. *American Naturalist*, **103**, 173–85.

Watt, K.E.F. (1973). *Principles of Environmental Science*. New York: McGraw-Hill. 319 pp.

Watt, W.D. (1966). Release of dissolved organic material from the cells of phytoplankton populations. *Proceedings of the Royal Society, London*, B **164**, 521–5.

Watt, W.D. (1971). Measuring the primary production rates of individual phytoplankton species in natural and mixed populations. *Deep-Sea Research*, **18**, 329–39.

Węglenska, T. (1971). The influence of various concentrations of natural food on the development, fecundity and production of planktonic crustacean filtrators. *Ekologia Polska*, **19**, 427–73.

Welch, E.B. (1969). Factors initiating phytoplankton blooms and resulting effects on dissolved oxygen in Duwanish River Estuary, Seattle, Washington. *Water-Supply and Irrigation Paper*, 1873–A.

Welch, E.B. (1974). The water quality of nearshore areas in Lake Vänern: causes and prospects. *Report. National Swedish Environmental Protection Board, Limnological Survey, Uppsala*, SNV PM 509 NLU, **77**, 1–31.

Welch, E.B. & Ball, R.C. (1966). Food consumption and production of pond fish. *Journal of Wildlife Management*, **30**, 527–36.

Welch, E.B. Henrey, G.R. & Stoll, R.K. (1975). Nutrient supply and the production and biomass of algae in four Washington lakes. *Oikos*, **26**, 47–54.

Westlake, D.F. (1964). Light extinction, standing crop and photosynthesis within weed beds. *Verhandlungen der Internationalen Vereinigung für Theoretische und Angewandte Limnologie*, **15**, 415–25.

Westlake, D.F. (1965). Some basic data for investigations of the productivity of aquatic macrophytes. *Memorie dell'Istituto Italiano di Idrobiologia*, **18** (supplement), 229–48.

Westlake, D.F. (1966a). The light climate for plants in rivers. In *Light as an Ecological Factor*, ed. R. Bainbridge, G.C. Evans & O. Rackham, pp. 99–119. Oxford: Blackwell.

Westlake, D.F. (1966b). A model for quantitative studies of photosynthesis by higher plants in streams. *International Journal of Air and Water Pollution*, **10**, 883–96.

Westlake, D.F. (1966c). Some effects of low velocity currents on the metabolism of macrophytes (Abstract). *Verhandlungen der Internationalen Vereinigung für Theoretische und Angewandte Limnologie*, **16**, 678–9.

Westlake, D.F. (1967). Some effects of low-velocity currents on the metabolism of aquatic macrophytes. *Journal of Experimental Botany*, **18**, 187–205.

Westlake, D.F. (1968). The weight of water-weed in the River Frome. *Yearbook. River Authorities' Association*, 1968, 59–68.

Westlake, D.F. (1975a). Macrophytes. In *River Ecology*, ed. B.A. Whitton, pp. 106–28. Oxford: Blackwell.

Westlake, D.F. (1975b). Primary production of freshwater macrophytes. In *Photosynthesis and Productivity in Different Environments*, ed. J.P. Cooper, *International Biological Programme* 3, pp. 189–206. Cambridge University Press.

Westlake, D.F., Casey, H., Dawson, H., Ladle, M., Mann, R.H.K. & Marker, F.H. (1972). The chalk-stream ecosystem. In *Productivity Problems of Freshwaters*, ed. Z. Kajak & A. Hillbricht-Ilkowska, pp. 615–35. Warsaw & Kraków: Polish Scientific Publishers.

Wetzel, R.G. (1964). A comparative study of the primary productivity of higher aquatic plants, periphyton and phytoplankton in a large shallow lake. *Internationale Revue der Gesamten Hydrobiologie und Hydrographie*, **49**, 1–61.

Wetzel, R.G. (1966a). Productivity and nutrient relationships in marl lakes of northern Indiana. *Verhandlungen der Internationalen Vereinigung für Theoretische und Angewandte Limnologie*, **16**, 321–32.

Wetzel, R.G. (1966b). Variations in productivity of Goose and hypereutrophic Sylvan Lakes, Indiana. *Investigations of Indiana Lakes and Streams*, **7**, 147–84.

Wetzel, R.G. (1967). Dissolved organic compounds and their utilization in two marl lakes. *Hidrologiai Közlöny*, **47**, 298–303.

Wetzel, R.G. (1968). Dissolved organic matter and phytoplankton productivity in marl lakes. *Mitteilungen der Internationalen Vereinigung für Theoretische und Angewandte Limnologie*, **14**, 261–70.

Wetzel, R.G. (1969). Factors influencing photosynthesis and excretion of dissolved organic matter by aquatic macrophytes in hard-water lakes. *Verhandlungen der Internationalen Vereinigung für Theoretische und Angewandte Limnologie*, **17**, 72–85.

Wetzel, R.G. (1974). The enclosure of macrophyte communities. In *A Manual on Methods for Measuring Primary Productivity in Aquatic Environments*, ed. R.A. Vollenweider, IBP Handbook No. 12, 2nd edition, pp. 100–7. Oxford: Blackwell.

Wetzel, R.G. & Allen, H.L. (1972). Functions and interactions of dissolved organic matter and the littoral zone in lake metabolism and eutrophication. In *Productivity Problems of Freshwaters*, ed. Z. Kajak & A. Hillbricht-Ilkowska, pp. 333–47. Warsaw & Kraków: Polish Scientific Publishers.

Wetzel, R.G. & Hough, R.A. (1973). Productivity and role of aquatic macrophytes in lakes. An assessment. *Polskie Archiwum Hydrobiologii*, **20**, 9–19.

Wetzel, R.G. & Manny, B.A. (1972a). Decomposition of dissolved organic carbon and nitrogen compounds from leaves in an experimental hard-water stream. *Limnology and Oceanography*, **17**, 927–31.

Wetzel, R.G. & Manny, B.A. (1972b). Secretion of dissolved organic carbon and nitrogen by aquatic macrophytes. *Verhandlungen der Internationalen Vereinigung für Theoretische und Angewandte Limnologie*, **18**, 162–70.

Wetzel, R.G. & Manny, B.A. (1977). Seasonal changes in particulate and dissolved organic carbon and nitrogen in a hard-water stream. *Archiv für Hydrobiologie*, **80**, 20–39.

Wetzel, R.G. & Otsuki, A. (1973). Allochthonous organic carbon of a marl lake. *Archiv für Hydrobiologie*, **73**, 31–56.

Wetzel, R.G., Rich, P.H., Miller, M.C. & Allen, H.L. (1972). Metabolism of dissolved and particulate detrital carbon in a temperate hard-water lake. *Memorie dell'Istituto Italiano Idrobiologia*, **29** (supplement), 185–243.

Whitcamp, M. (1966). Decomposition of leaf litter in relation to environment, microflora, and microbial respiration. *Ecology*, **47**, 194–201.

Whitcamp, M. & Crossley, D.A. (1966). The role of arthropods and microflora in the breakdown of white oak litter. *Paedobiologia*, **6**, 293–303.

Whitcamp, M. & Olsen, J.S. (1963). Breakdown of confined and non-confined oak litter. *Oikos*, **14**, 138–47.

White, D.E., Hem, J.D. & Waring, G.A. (1963). Chemical composition of subsurface waters, *Professional Papers. United States Geological Survey*, **440-F**, 1–67.

Whiteside, M.C. (1970). Danish chydorid cladocera: modern ecology and core studies. *Ecological Monographs*, **40**, 79–118.

Whitford, L.A. & Schumacher, G.J. (1961). Effect of current on mineral uptake and respiration by freshwater algae. *Limnology and Oceanography*, **6**, 423–5.

Whitford, L.A. & Schumacher, G.J. (1964). Effect of current on respiration and mineral uptake in *Spirogyra* and *Oedogonium*. *Ecology*, **45**, 168–70.

Whitton, B.A. (1970). Biology of *Cladophora* in freshwaters. *Water Research*, **4**, 457–76.

Whitwer, E.E. (1955). Efficiency of finely-divided *vs.* tape-like aquatic plant leaves. *Ecology*, **36**, 511–12.

Wiktor, J. (1969). Biology of *Dreissena polymorpha* (pall) and its ecological role in the Szczecin Firth. *Studia i materialy Morski Instytut Rybacki* A 5. (In Polish with English summary.)

Williams, P.J. (1973). The validity of the application of simple kinetic analysis to heterogeneous microbial populations. *Limnology and Oceanography*, **18**, 159–64.

Williams, W.D. & Hang Fong Wan (1972). Some distinctive features of Australian inland waters. *Water Research*, **6**, 829–36.

Winberg, G.G. (1960). Rate of metabolism and food requirements of fishes. *Fisheries Research Board of Canada, Translation Series*, **194**, 202 pp.

Winberg, G.G. (1962). The energy principle in studying food associations and the productivity of ecological systems. *Zoologicheskii Zhurnal*, **41**, 1618–30. (In Russian.)

Winberg, G.G. (1969). Energy flow in the ecosystem of a eutrophic lake. *Doklady Akademii Nauk SSSR*, **186** (1), 198–201. (In Russian.)

Winberg, G.G. (1970a). Features of the Drivyaty lake ecosystem. In *Biological Productivity of a Eutrophic Lake* ed. G.G. Winberg, pp. 185–196. Moskva: Nauka. (In Russian.)

Winberg, G.G. (1970b). Energy flow in aquatic ecological systems. *Polskie Archiwum Hydrobiologii*, **17** (1/2), 11–19.

Winberg, G.G. (ed.) (1971a). *Symbols, Units and Conversion Factors in Studies of Freshwater Productivity*. London: IBP Central Office. 23 pp.

Winberg, G.G. (ed.) (1971b). *Methods for the Estimation of Production of Aquatic Animals*. Translated from the Russian by A. Duncan. New York & London: Academic Press. 175 pp.

Winberg, G.G. (1971c). A comparison of several widespread methods of calculating the production of aquatic bacteria. *Gidrobiologicheskii Zhurnal*, **7** (4), 86–96. (In Russian.)

Winberg, G.G. (1972a). Edgardo Baldi Memorial Lecture–Etudes sur la bilan biologiques énergetiques et la productivité des lacs en Union Soviétique. *Verhandlungen der Internationalen Vereinigung für Theoretische und Angewandte Limnologie*, **18**, 39–64.

Winberg, G.G. (1972b). Some interim results of Soviet IBP investigations on lakes. In *Productivity Problems of Freshwaters*, ed. Z. Kajak & A. Hillbricht-Ilkowska, pp. 363–81. Warsaw & Kraków: Polish Scientific Publishers.

Winberg, G.G. (ed.) (1973). *Productivity Studies of Freshwater Ecosystems*. Minsk: Byelorussian State University. 207 pp. (In Russian.)

Winberg, G.G. (1976). Results of studies in productivity of freshwater communities of all trophic levels. In *Resources of the Biosphere*, Synthesis of the Soviet Studies for the IBP, vol. 2, pp. 145–57. Leningrad: Nauka.

Winberg, G.G., Alimov, A.F., Boullion, V.V., Ivanova, M.B., Korobtsova, E.V., Kuzmitskaya, N.K., Nikulina, V.N., Finogenova, N.P. & Fursenko, M.V. (1973). Biological productivity of two subarctic lakes. In *Productivity Studies of Freshwater Ecosystems*, ed. G.G. Winberg, pp. 125–47. Minsk: Byelorussian State University. (In Russian.)

Winberg, G.G., Alimov, A.F., Galkovskaya, G.A., Ivanova, M.B., Kititsyna, L.A., Kryuchkova, N.M., Monakova, A.V., Ostapenya, A.P., Pechen-Finenko, G.A., Sokolova, N. Yu. & Khlebovich, T.V. (1973). The progress and state of research on the metabolism, growth, nutrition and production of freshwater invertebrate animals. *Gidrobiologicheskii Zhurnal*, **9** (3), 123–29. (In Russian.)

Winberg, G.G., Babitsky, V.A., Gavrilov, S.I., Gladky, G.V., Zakharenkov, I.S., Kovalevskaya, R.Z., Mikheeva, T.M., Nevyadomskaya, P.S., Ostapenya, A.P., Petrovich, P.G., Potaenko, J.S. & Yakushko, O.F. (1971). In *Bioproductivity of Lakes in Byelorussia*, ed. P.G. Petrovich, pp. 5–33.

Winberg, G.G., Babitsky, V.A., Gavrilov, S.I., Gladky, G.V., Zakharenkov, I.S., Kovalevskaya, R.Z., Mikheeva, T.M., Nevyadomskaya, P.S., Ostapenya, A.P., Petrovich, P.G., Potaenko, J.S. & Yakushko, O.F. (1972). Biological productivity of different types of lakes. In *Productivity Problems of Freshwaters*, ed. Z. Kajak & A. Hillbricht-Ilkowska, pp. 383–404. Warsaw & Kraków: Polish Scientific Publishers.

Winberg, G.G. & Lomonosova, M.S. (1953). Obchee chislo bakterii i skorost potreblenia kisloroda v vodakh raznoi stepeni zagriaznenia (Total counts of bacteria and the rate of oxygen consumption in waters of different degrees of contamination). *Mikrobiologiya*, **22**, 294–303. (In Russian.)

Winberg, G.G. & Lyakhnovich, V.P. (1965). *Udobrenie prudov* (Fertilization of ponds). Moskva: "Pishchevaya Promyshlennost". 269 pp. (In Russian.)

Winberg, G.G., Pechen, G.A. & Shushkina, E.A. (1965). Production of planktonic crustaceans in three lakes of different type. *Zoologicheskii Zhurnal*, **44**, 676–87. (In Russian.)

Wium-Anderson, S. (1971). Photosynthetic uptake of free CO_2 by the roots of *Lobelia dortmanna*. *Physiologia Plantarum*, **25**, 245–8.

Wójcik-Migaɫa, I. (1965). The influence of the carp fry stock on the dynamics of benthic fauna. *Rocznik Nauk Rolniczyck*, B **86**, 195–214. (In Polish with English summary.)

Wolny, P. (1962). The use of purified town sewage for fish rearing. *Rocznik Nauk Rolniczych*, B **81**, 231–49. (In Polish with English summary.)

Wolny, P. (1970). Results of six-year research on effectiveness of nursery pond fertilization. *Rocznik Nauk Rolniczych*, H **91** (4), 565–88. (In Polish with English summary.)

Wolny, P. & Grygierek, E. (1972). Intensification of fish ponds production. In *Productivity Problems of Freshwaters*, ed. Z. Kajak & A. Hillbricht-Ilkowska, pp. 563–71. Warsaw & Kraków: Polish Scientific Publishers.

Wright, J.C. (1959). Limnology of Canyon Ferry Reservoir. II. Phytoplankton standing crop and primary production. *Limnology and Oceanography*, 4, 235–45.

Wright, J.C. (1960). The limnology of Canyon Ferry Reservoir. III. Some observations on the density dependence of photosynthesis and its cause. *Limnology and Oceanography*, 5, 356–61.

Wright, J.C. (1961). The limnology of Canyon Ferry Reservoir. IV. The estimation of primary production from physical limnological data. *Limnology and Oceanography*, 6, 330–7.

Wright, J.C. (1965). The population dynamics and production of *Daphnia* in Canyon Ferry Reservoir, Montana. *Limnology and Oceanography*, 10, 583–90.

Wright, R.T. & Hobbie, J.E. (1966). Use of glucose and acetate by bacteria and algae in aquatic ecosystems. *Ecology*, 47, 447–64.

Wróbel, S. (1970). Primary production of phytoplankton and production of carp in ponds. *Polskie Archiwum Hydrobiologii*, 17, 103–7.

Wróbel, S. (1972). Some remarks to the production of basic communities in ponds with inorganic fertilization. *Verhandlungen Internationalen Vereinigung für Theoretische und Angewandte Limnologie*, 18, 221–7.

Yamamoto, G. (1975). Productivity of the communities of acid lakes in Urabandai, with special reference to Lake Tatsu-numa, one of the disharmonic lakes 4.7. Consideration. In *Productivity of Communities in Japanese Inland Waters*, ed. S. Mori & A. Yamamoto, JIBP Synthesis, vol. 10, pp. 192–4. Tokyo: University of Tokyo Press.

Yashouv, A. (1969). The fish pond as an experimental model for study of interaction within and among fish population. *Verhandlungen der Internationalen Vereinigung für Theoretische und Angewandte Limnologie*, 17, 582–93.

Yashouv, A. (1972). The carrying capacity and ecological niche as management concepts of fish production in ponds. In *Productivity Problems of Freshwaters*, ed. Z. Kajak & A. Hillbricht-Ilkowska, pp. 573–8. Warsaw & Kraków: Polish Scientific Publishers.

Yentsch, C.S. & Lee, R.W. (1966). A study of photosynthetic light reactions and a new interpretation of sun and shade phytoplankton. *Journal of Marine Research*, 24, 319–47.

Yentsch, C.S. & Ryther, J.H. (1957). Short term variations in phytoplankton chlorophyll and their significance. *Limnology and Oceanography*, 2, 140–2.

Yentsch, C.S. & Scagel, R.F. (1958). Diurnal study of phytoplankton pigments. *Journal of Marine Research*, 17, 567–84.

Yoshimura, S. (1936). A contribution to the knowledge of deep water temperature of Japanese lakes. *Japanese Journal of Astronomy and Geophysics*, 13 (1/2), 61–120.

Yount, J.L. (1966). A method of rearing large numbers of pond midge larvae, with estimates of productivity and standing crop. *American Midland Naturalist*, 76, 230–8.

Zahner, R. (1964). Beziehungen zwischen dem Auftreten von Tubificiden und der Zufur organischer Stoffe in Bodensee. *Internationalen Revue der Gesamten Hydrobiologie*, 49, 417–54.

Zaika, V.E. (1972). *Udel'naya Produktsiya Vodnykh Berpozvonochnykh* (The Specific Production of Aquatic Invertebrates). Kiev: Naukova Dumka. 145 pp. (In Russian with English summary.)

Zawisza, J. & Backiel, T. (1970). Gonad development, fecundity and egg survival in

Coregonus albula. In *Biology of Coregonid Fishes*, ed. C.C. Lindsey & C.S. Woods, pp. 363–397. Winnipeg: University of Manitoba Press.

Zelitch, I. (1971). *Photosynthesis, Photorespiration and Plant Productivity*. New York & London: Academic Press. 347 pp.

Zeuthen, E. (1953). Oxygen uptake as related to body size in organisms. *Quarterly Review of Biology*, **28**, 1–12.

Zhadin, V.I. (ed.) (1950). *Zhizn Presnykh Vod SSSR. III*. Moskva: Akademii Nauk SSSR. 910 pp.

Zieba, J. (1971). Production of macrobenthos in fingerling ponds. *Polskie Archiwum Hydrobiologii*, **18**, 235–46.

Zieba, J. (1973). Macrobenthos of ponds with sugar factory wastes. *Acta Hydrobiologica, Kraków*, **15**, 113–29.

Zimbalevskaya, L.N. (1966). Ekologicheskie gruppirovki fauny zaroslei Dnepra. *Gidrobiologicheskii Zhurnal*, **2** (5), 34–41.

Zobell, C.E. (1946). *Marine Microbiology*. Waltham: Chronica Botanica Company. 240 pp.

Zuromska, H. (1967). Mortality estimation of roach (*Rutilus rutilus* L.) eggs and larvae in lacustrine spawning grounds. *Rocznik Nauk Rolniczych*, H. **90** (3), 539–56. (In Polish with English summary.)

References added in proof

Andronikova, I.M. & Drabkova, V.G. (1972). Specificity of annual limnological cycles in Lake Krasnoe versus climatic factors. *Verhandlungen der Internationalen Vereinigung für Theoretische und Angewandte Limnologie*, **18**, 522–7.

Bagenal, T. (ed.) (1978). *Methods for the Assessment of fish Production in Fresh Waters*. 3rd edn. IBP Handbook No. 3. Oxford: Blackwell. 365 pp.

Bottrell, H.H., Duncan, A., Gliwicz, Z.M., Grygierek, E., Herzig, A. Hillbricht-Ilkowska, A., Kurasawa, H., Larsson, P. & Weglenska, T. (1976). A review of some problems in zooplankton production studies. Contribution from the Plankton Ecology Group (IBP). *Norwegian Journal of Zoology*, **24**, 419–56.

Gerking, S.D. (ed.) (1978). *Ecology of Freshwater Fish Production*. Oxford: Blackwell Scientific Publications. 520 pp.

Kalesnik, S.V. (ed.) (1971). Ozera Karel'skogo peresheika. (The lakes of the Karelian isthmus.) Leningrad: Nauka. 531 pp.

Kuzin, B.S. (ed.) (1972). Rybinskoe vodokhranilishche i ego zhizn'. (Rybinsk reservoir and its life.) Leningrad: Nauka. 364 pp.

Macan, T.T. (1949). Survey of a moorland fish pond. *Journal of Animal Ecology*, **18**, 160–86.

Pyrina, I.L., Devyatkin, V.G. & Elizarova, V.A. (1975). Experimental investigation of heating influence on phytoplankton development and photosynthesis. In *Anthropogenous Factors in the Fresh Basin Life*. Transactions Institute Biology Inland Waters Academy of Sciences, USSR, vol. 30(33), ed. N.V. Batorin, pp. 67–84. Leningrad: Publishing House 'Nauka'. (In Russian.)

Schiemer, F., Duncan, A. & Klekowsi, R.Z. (1980). A bioenergetic study of a benthic nematode, *Plectus palustris* de Man 1880, throughout its life cycle. II. Growth and fecundity at different densities of bacterial food and general ecoenergetical considerations. *Oecologia*, (in press).

Index

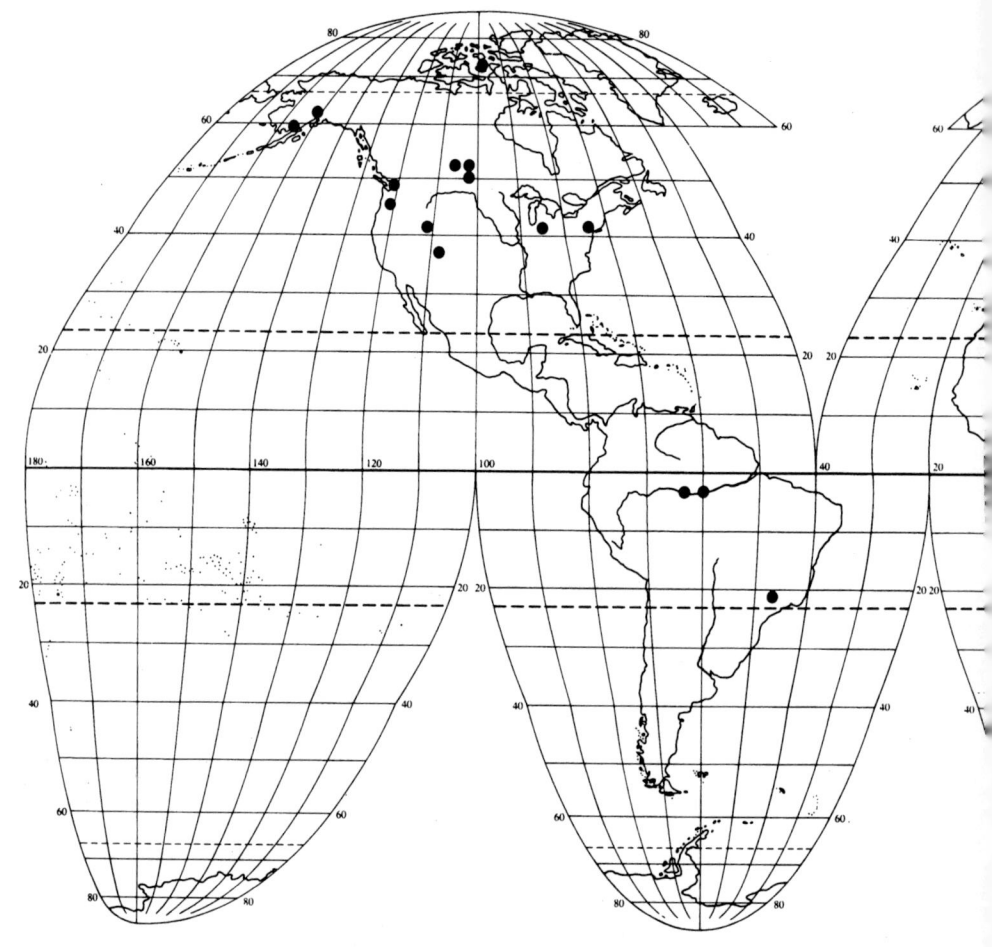

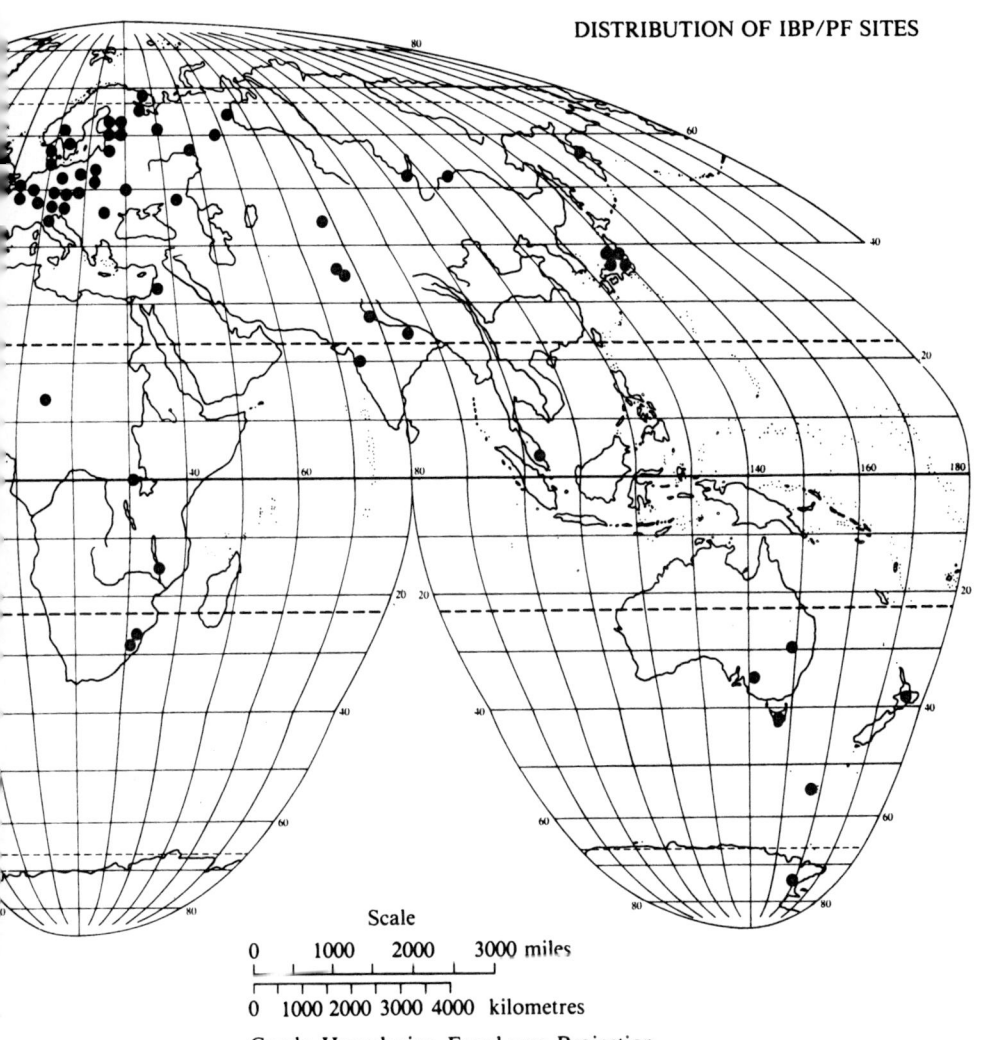

DISTRIBUTION OF IBP/PF SITES

Scale

| 0 | 1000 | 2000 | 3000 miles |

| 0 | 1000 | 2000 | 3000 | 4000 | kilometres |

Goode Homolosine Equal-area Projection

Prepared by Henry M. Leppard
© 1961 by The University of Chicago